D0554455

ELECTRONICS AND ELECTRON DEVICES

THE MACMILLAN COMPANY
NEW YORK • CHICAGO
DALLAS • ATLANTA • SAN FRANCISCO
LONDON • MANILA
IN CANADA
BRETT-MACMILLAN LTD.
GALT, ONTARIO

ELECTRONICS AND ELECTRON DEVICES

ARTHUR LEMUEL ALBERT, M.S.
Professor of Communication Engineering
Oregon State College

Third Edition of
FUNDAMENTAL ELECTRONICS AND VACUUM TUBES

THE MACMILLAN COMPANY *New York*

Published simultaneously in Canada

Printed in the United States of America

Fourth Printing, 1959

Previous editions published under the title, *Fundamental Electronics and Vacuum Tubes*, copyright 1938 and 1947 by The Macmillan Company

PREFACE

The introduction of transistors in 1948 has resulted in important changes in electronics. Semiconductor theory and the principles of semiconductor devices must be presented to undergraduate students in electrical engineering. In *Electronics and Electron Devices,* semiconductor theory is incorporated in the chapter on Basic Electronic Theory, and two chapters on semiconductor devices and circuits are included.

This book is the third edition of *Fundamental Electronics and Vacuum Tubes,* the title having been modified in accordance with present usages. This edition has been rewritten almost entirely, and much of the descriptive material of the earlier editions has been eliminated to provide space for new material. The objectives remain as for preceding editions: To provide a textbook for junior and senior college and university courses on basic electronics and electron devices; a book that is well balanced between theory and illustrative applications; and a book that is suited to the needs of students interested in the power field, as well as those who are looking forward to careers in communication and control.

An important feature of this book is the chapter on Magnetic Amplifiers. Although these amplifiers are not electronic in the usual sense, magnetic devices are important in control and often supplement, and sometimes supplant, electron devices. Because basic training in magnetic amplifiers so often is overlooked, the chapter was included.

In addition to the changes mentioned, this book differs from its predecessor by including a chapter on Wave-Shaping and Control Circuits, covering differentiating, integrating, limiting, clipping, multivibrator, and similar circuits.

As in previous editions, the Standards of the American Institute of Electrical Engineers and of The Institute of Radio Engineers have been followed closely. It is, of course, important to follow revised editions of the various Standards as these new editions are published. Also, references, review questions, and problems are included at the end of each chapter.

The contributions of the many authorities whose writings were consulted are acknowledged with gratitude. The American Institute of Electrical Engineers, The Institute of Radio Engineers, Electronics, Radio Corporation of America, General Electric Company, Western Electric Company, Bell Telephone Laboratories, and other organizations have been helpful in releasing material and information, and in other ways. The chapter on Magnetic Amplifiers was written at my request by my good friend and former colleague, J. J. Wittkopf, an outstanding teacher and an engineer of experience in control applications. I am indeed grateful to him for his excellent contribution. As for previous editions, my wife has been helpful with good suggestions and with her intelligent and careful typing of the manuscript.

Arthur L. Albert

CONTENTS

ELECTRONICS AND ELECTRON DEVICES

CHAPTER 1

BASIC ELECTRONIC THEORY

This book is devoted to a study of the fundamentals of electronics, to the application of electronic principles to electron devices, and to the use of these devices in electric circuits.

Electronics is defined [1] as "that field of science and engineering which deals with electron devices and their utilization." An **electron device** is defined [1] as "a device in which conduction by electrons takes place through a vacuum, gas, or semiconductor."

Electron devices include the **electron tube,** defined [1] as "an electron device in which conduction by electrons takes place through a vacuum or gaseous medium within a gas-tight envelope." An important type is the **vacuum tube,** defined [1] as "an electron tube evacuated to such a degree that its electrical characteristics are essentially unaffected by the presence of residual gas or vapor." Another important tube is the **gas tube,** defined [1] as "an electron tube in which the pressure of the contained gas or vapor is such as to affect substantially the electrical characteristics of the tube." Semiconductor devices include the crystal rectifier, copper oxide and selenium rectifiers, and the transistor. The three basic types of electron devices will be considered in this book.

The explanations used in electronics are founded on theories of the electrical nature of matter; these theories will be discussed in this chapter. Simplified explanations and illustrations, justifiable in an engineering textbook, will be used. It should be noted, however, that the studies of physics have gone far beyond the material presented.

1-1. *THE ELEMENTARY PARTICLES OF MATTER*

Prior to about 1930 it was thought that the smallest particles of matter were the negative electron and the positive proton. Later it was shown that

[1] Superior numbers refer to references at end of chapter.

there were other particles,[2] and additional particles are being found as studies continue. These elementary particles are "the building blocks of the universe," and their detailed study is in the field of nuclear physics, not electronics. Although certain of these particles will be mentioned in the pages that follow, the discussions largely will be confined to the electron and to electronics.

1-2. *THE ELECTRON*

This word was used by the ancient Greeks and by early writers. It was Stoney, however, who in 1891 first used the word in the modern sense.[3] The **electron** is defined [4] as "the natural, elementary quantity of negative electricity." Note that the electron is defined merely as a "quantity" of negative electricity. Engineering usage is to regard an electron as a very small negatively charged *particle* of matter.

The quantity of negative electricity on an electron is defined [4] as 4.774×10^{-10} statcoulomb, or 1.592×10^{-19} coulomb (or 1.602×10^{-19} coulomb [7]), and the mass *at rest* (or below one-tenth the velocity of light) as 9.00×10^{-28} gram, or 9.00×10^{-31} kilogram. The *apparent* radius of an electron is about 1.9×10^{-13} centimeter.

The ratio of the electronic charge e to the mass m first was measured by J. J. Thomson in about 1900. This ratio is about 5.3×10^{17} statcoulombs per gram, or 1.77×10^{11} coulombs per kilogram. Millikan, working between 1908 and 1917, first measured the charge on an electron.

An electron would possess mass by virtue of its charge alone [5] for the following reasons. Electrons in motion constitute a current of electricity. The magnitude of the current is determined by the *number* of electrons passing a given point *in unit time.* If the electrons in an electron beam are accelerated, the rate of electron flow and the current magnitude are increased. When the magnitude of a current is increased, additional energy is transferred to the accompanying magnetic field, and hence when electrons are accelerated more energy is transferred to the magnetic field. A *force* is required to produce electron acceleration, and the opposing inertia effect and the mass are the result of the electric charges on the electrons. Since mass is a fundamental property of matter, it may be stated that matter is electrical in nature.

The mass of an electron is reasonably constant below one-tenth the velocity of light (electromagnetic radiation) in free space, which is about 3×10^{8} meters per second. The mass m at any velocity v is given, in terms of the mass m_o at rest, by the Lorentz formula as

$$m = \frac{m_o}{\sqrt{1 - \left(\frac{v}{c}\right)^2}} \tag{1-1}$$

where c is the velocity of light, measured in the same units as v. A possible source of error in this and similar equations has been discussed.[6]

1-3. *ELECTRONS IN ELECTRIC FIELDS*

Suppose that the parallel-plate electrodes of Fig. 1-1 are in free space, and that a voltage of E volts is impressed as shown. The voltage will establish lines (or tubes[4]) of electric force having the direction indicated by the arrows; that is, from positive to negative. The electric field intensity equals numerically

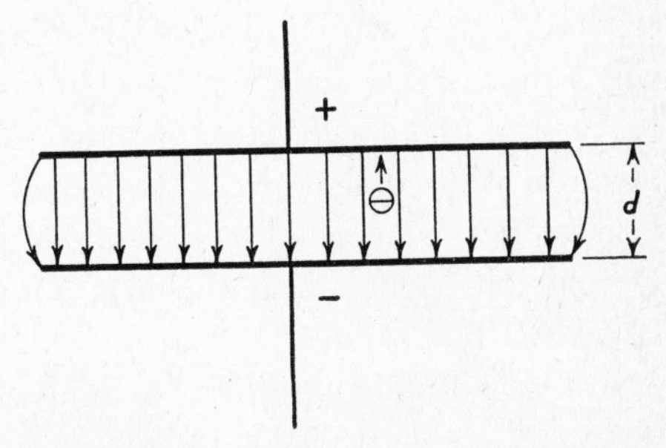

FIG. 1-1. An electron in an electric field produced by a potential difference between two electrodes will be drawn toward the positive electrode.

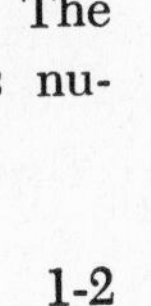

$$\text{E} = \frac{E}{d} \tag{1-2}$$

where E is the field intensity in volts per meter when E is the impressed voltage in volts, and d is the distance between the electrodes in meters.* This equation also gives the numerical magnitude of the voltage gradient, but the positive directions of electric field intensity and of voltage gradient are opposite.[4]

A negative electron at a given point in an electric field will be acted on by the electric field at that point and will experience a force tending to move it in a direction *against* the electric field and *toward* the more positive electrode (Fig. 1-1). The magnitude of the force at the point will be

$$F = \text{E}e \tag{1-3}$$

where F is the force in **newtons** when the electric field intensity E is in volts per meter, and e is the electron charge in coulombs.

* The Standards of The Institute of Radio Engineers and of the American Institute of Electrical Engineers are followed for letter symbols,[4,7] graphic symbols, and similar purposes, such as numerical values for constants. An exception is the use of E instead of *E* for the magnitude only of electric field intensity.

If the electron of Fig. 1-1 is free to move, it will experience an acceleration

$$a = \frac{F}{m} = \frac{Ee}{m} \qquad 1\text{-}4$$

where a is in meters per second per second when E is in volts per meter, e is in coulombs, and m is the mass at rest in kilograms.*

When the electron of Fig. 1-1 moves a distance d in the direction of the arrow near the electron, the work done is

$$W = Fd = \text{E}ed = \frac{Eed}{d} = Ee \qquad 1\text{-}5$$

where W is the work in **joules** when E is in volts, and e is in coulombs. Since this value is the same for all electrons, the work done when an electron moves through a difference of potential of E volts is commonly measured in **electron volts,** defined [4] as "the amount of energy gained by an electron in passing from one point to another when the potential of the second point is one volt higher than the first. One electron volt $= 1.592 \times 10^{-12}$ erg." (1.592×10^{-19} joule.)

In Fig. 1-1 an electron is shown moving toward the *positive* electrode and *against* the electric field. In this instance work is done *on* the electron by the electric circuit, because the positive charges on the positive electrode attract the electron, and the negative charges on the negative electrode repel the electron. In this way the *kinetic energy* of the electron is *increased* and its *potential energy* is *decreased.* When the electron strikes the positive metal electrode, the positive charge on the electrode is reduced. Some of the accumulated kinetic energy of the electron will be transferred to the electric circuit, some will probably be radiated as light (or other electromagnetic radiation), and the rest of the kinetic energy will, perhaps, be used in liberating electrons from the metal. The exact nature of the phenomena will depend on circumstances such as the velocity of the electron and the nature of the metal surface.

If the electron of Fig. 1-1 *is caused* to move toward the *negative* electrode and *with* the electric field (for example, if an electron is shot by some means toward the negative electrode), work is done *by* the electron on the electric circuit in this way: As the speeding negative electron approaches the nega-

* The CGS electrostatic system is extensively used in electronics and may be used readily with equations such as those just given. For instance, the force in dynes equals the product of the electric field intensity in statvolts per centimeter and the charge on the electron in statcoulombs. Also, the acceleration in centimeters per second per second equals the product of field intensity in statvolts per centimeter and electron charge in statcoulombs divided by the mass of the electron in grams.

tive electrode, the negative charges on the electrode repel the electron and decrease its velocity; also, the approaching electron repels negative charges and drives them from the electrode. Hence, the speeding electron will *lose kinetic energy* and will *gain potential energy.*

When the electron is freely falling (moving unimpeded) through a difference of potential, the work done on it must equal the kinetic energy gained by it. If the electron starts from rest, the kinetic energy is given by the usual equation for a freely falling body,

$$W = \frac{mv^2}{2} \tag{1-6}$$

The velocity of an electron starting from rest and having freely fallen, or passed, through a difference of potential of E volts is found from equations 1-5 and 1-6 to be

$$Ee = \frac{mv^2}{2} \quad \text{and} \quad v = \sqrt{\frac{2Ee}{m}} = 5.95 \times 10^5\sqrt{E} \tag{1-7}$$

where v will be the final velocity in meters per second when an electron, initially at rest, falls unimpeded through a difference of potential of E volts.

1-4. *ELECTRONS IN MAGNETIC FIELDS*

An electron or other electrically charged particle *in motion* constitutes an electric current. When electricity flows at the rate of one coulomb per second, the current is one ampere. The reciprocal of the charge per electron, or $1/(1.592 \times 10^{-19})$, is 6.28×10^{18}, and hence current flow at the rate of 6.28×10^{18} electrons per second is a current flow of one ampere.

Electrons *at rest* in a magnetic field have no magnetic forces exerted on them. Electrons in motion, as in an electronic stream or beam, constitute an electric current and establish a magnetic field just as does any electric current. This magnetic field will react with another magnetic field and will produce a force which acts on the moving electrons or other charged particles. The magnitude of the force exerted on a straight wire carrying a current in a magnetic field in free space, or air, is

$$F = IlB \tag{1-8}$$

where F is the force in newtons, when I is the current magnitude in amperes, B is magnitude of the magnetic induction (flux density in webers per square meter) at *right angles to the wire,* and l is the length in meters of wire immersed in the magnetic field. The force is exerted at right angles to the axis of the wire.

Since the current carried by the wire is a flow of electrons, it might be expected that a force would be exerted by a magnetic field on a beam, or ray, of electrons. This is true. Demonstrating the magnetic deflection of a "cathode ray" was one of the earliest experiments in electronics. It is found that the force exerted on a beam, or ray, of electrons is at right angles to the direction of the electron beam, and that at each instant the force exerted on an individual electron in the beam is at right angles to the direction of motion of the electron at that instant.

From the fundamental relations in the opening paragraph of this section, the current of electricity resulting from the motion of an individual electron is $I = e/t$, in which e is the electron charge in coulombs, and t is in seconds. Also, for an electron moving with velocity v, the length of path covered in unit time is $l = vt$. Thus the magnitude of the force on an individual electron is, from equation 1-8,

$$F = IlB = \frac{e}{t} \times vtB = evB = 1.592 \times 10^{-19} Bv \qquad 1\text{-}9$$

where F is in newtons, B is in webers per square meter, and v is in meters per second. If the direction of motion is not at right angles to the magnetic field, then v is the component that is at right angles; the same reasoning applies to length of wire l in equation 1-8. In determining the direction of the force exerted on an electron beam by the conventional methods so extensively used in electrical studies, it must be remembered that current flow and *electron* current flow are opposite in direction.

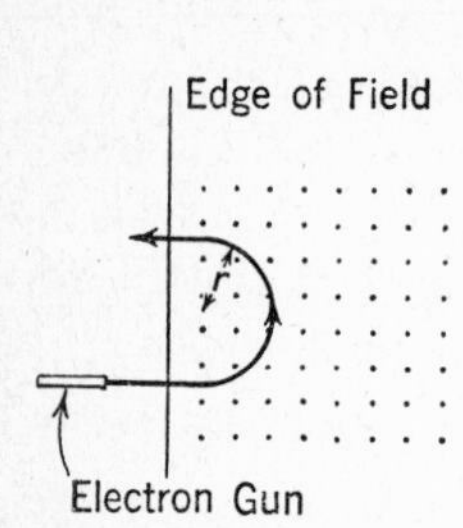

FIG. 1-2. An electron shot *into* a magnetic field will be ejected from the field. The dots represent magnetic lines of force up out of the page. For description of an electron gun see section 17-3.

The principle that the force at each instant on a moving electron is at right angles to the direction of travel at that instant is used to determine the path of an electron in a magnetic field. Two conditions will be considered: first, an electron being shot *from without* into a magnetic field, and second, a moving electron being released *within* a magnetic field.

The first condition is depicted in Fig. 1-2. An electron having a velocity less than one-tenth the velocity of light is shot at right angles into a constant uniform magnetic field limitless in extent except at the left edge. Because at each instant the force on the electron is at right angles to the direction of motion at that instant, the path will be circular as shown. The force given by equation 1-9, which is driving the electron inward, is equal in magnitude to the opposing centrifugal force mv^2/r tending to drive it outward. Hence

$$1.592 \times 10^{-19}Bv = \frac{9 \times 10^{-31}v^2}{r} \quad \text{and} \quad r = \frac{5.65 \times 10^{-12}v}{B} \tag{1-10}$$

where r is the radius of the electron path in meters, v is the velocity of the electron in meters per second, and B is the magnetic induction, or flux density, in webers per square meter. From Fig. 1-2 it is seen that a moving electron that enters a magnetic field from outside the field will be forced out of the field.

The second condition is shown in Fig. 1-3. An electron gun in a constant, uniform, and limitless magnetic field is shooting an electron into the field. The direction of motion is at right angles to the field. The electron will take a circular path given by equation 1-10.

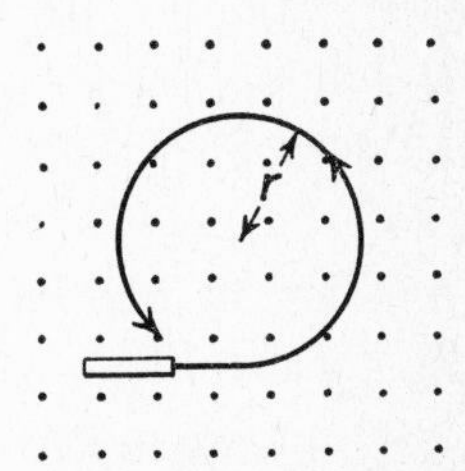

FIG. 1-3. Path of an electron "shot" from an electron gun located in a limitless magnetic field.

As has been stressed, the force exerted by a magnetic field on a moving electron is *always* at right angles to the direction of motion. Thus a *magnetic field can do no work on an electron* or other charged particle. A magnetic field can, and does, alter the direction of motion, but the magnitude of the kinetic energy is not affected. This is in contrast with the action of an electric field which can, and does, increase and decrease the kinetic energy of a charged particle.

When a moving electron, or other charged particle, is in a combined electric and magnetic field, each field acts independently on the electron. The direction of motion at each instant is determined by the resultant of the forces acting.

1-5. *THE QUANTUM OF ENERGY*

Energy is transmitted through space by variations in electric and magnetic fields constituting electromagnetic wave action. Such forms of radiant energy include radio waves, heat waves (infrared), visible light, and x-rays. These electromagnetic waves are emitted and absorbed only in discrete amounts, or "bundles" of energy of definite magnitudes. These bundles of energy are called **quanta.** If the frequency is such that the energy lies in the frequency band of visible light, the quanta are called **photons.** The terms sometimes are used interchangeably.[2]

A quantum of energy, such as a photon, has the properties of wave motion, and also of a charged particle. For example, both wave theories and corpuscular theories of light have been proposed. The situation was once summarized by A. H. Compton who stated: [8] "We continue to think of light as propagated as electromagnetic waves; yet the energy of the light is concentrated in particles associated with the waves, and whenever the light does something, it does it as particles."

The amount of energy in joules contained in a quantum (or photon) is equal to hf, where h is Planck's constant 6.624×10^{-34} joule-second,[7] and f is the frequency of the electromagnetic wave in cycles per second. Electromagnetic radiations, therefore, are quanta of energy, and each quantum may be pictured as consisting of "spurts," or bundles, of waves, each bundle containing its characteristic energy hf. Ultraviolet light is of shorter wave length and higher frequency than red light, for example. Hence a quantum of ultraviolet light contains more energy than one of red light. Similarly, x-rays are of yet higher frequency and contain more energy. The various bundles, or quanta, arriving from a source of radiation "blend" together, giving a continuous energy flow. A feeble beam of light contains few quanta; a strong beam consists of many quanta arriving per unit time.

1-6. *EARLY ATOM MODELS*

An **atom** is the smallest particle of an element that can exist. Rutherford is credited with the idea that an atom might consist of a central nucleus surrounded by rotating electrons. The electrostatic forces of attraction between the positive nucleus and the negative electrons were considered to be equal and opposite to the centrifugal forces of the electrons. It early was recognized that this atom model was inadequate, one reason being that the rotating electrons would radiate energy and the atom would disintegrate.

In the model proposed by Bohr about 1913, the disintegration difficulties were avoided by assuming that the electrons traveled about the nucleus in fixed orbits at definite, different, energy levels above zero level at the nucleus. Also, it was considered that an electron did not radiate energy while in a particular orbit, but did radiate a quantum of energy if it fell from an orbit at one energy level to an orbit at a lower level. It was thought that if an electron were to absorb a quantum of energy of the correct amount, it could be raised to an orbit at a higher energy level. The distance between the nucleus and the revolving electrons was considered small in the usual sense, but great compared to the dimensions of these particles. The "volume" of an atom consists largely of space, and an atom and "solid" matter are not actually *solid* but are largely space.

1-7. *LATER ATOM MODELS*

Certain of the ideas of the Bohr atom model are used today. The atomic nucleus is thought to be composed of positive particles called **protons** and uncharged particles called **neutrons.** These particles have large mass compared to an electron, and the nucleus largely determines the mass of an atom.

A normal **hydrogen atom** is considered to have a nucleus consisting of a single proton, and to have one revolving electron. It is the simplest atom.

A normal **helium atom,** which is the next in complexity, is pictured to have a nucleus of two protons and two neutrons, and to have two revolving electrons.

A normal **lithium atom** may be one of two types, called **isotopes** of lithium. The simplest of these two atoms is thought to have a nucleus of three protons and four neutrons.

The *atomic weights* of the three elements just discussed are: hydrogen, 1.008; helium, 4.00; and lithium, 6.940. The *atomic numbers* are: hydrogen, 1; helium, 2; and lithium, 3. Hence, in accordance with the atom models just considered, *the atomic number indicates the number of electrons revolving at one or more energy levels about the central nucleus.* These electrons have negligible mass compared to the particles in the nucleus. *In the nucleus there are as many positive protons as there are revolving negative electrons,* making the normal atom as a whole electrically neutral. *The number of neutrons in the nucleus equals the atomic weight minus the number of protons.* Thus a **neon atom** has the structure shown in Fig. 1-4.

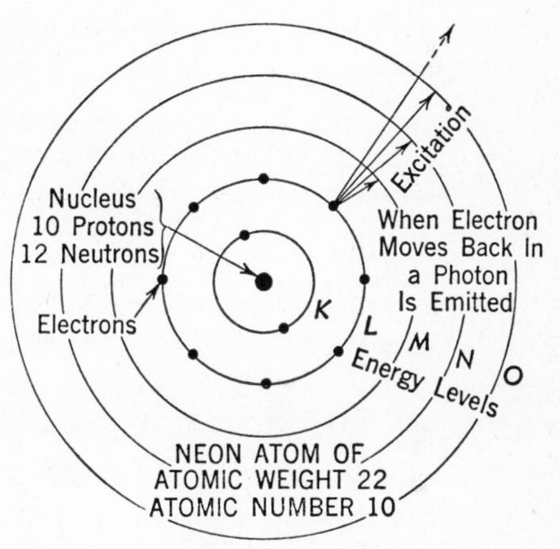

FIG. 1-4. Showing the nucleus and the electrons in energy-level orbits for a neon atom. Excitation occurs when electrons move to higher energy levels. Ionization occurs when an electron is removed completely from the atom. (Adapted from reference 2.)

Several **energy levels,** or **energy shells,** are shown. These levels, or shells, accommodate different maximum numbers of electrons, the number being $2n^2$, where n is the shell number. For example, the ***K* level,** or shell, is the first, and can have a maximum of two electrons. The ***L* level** is the second, and can contain a maximum of $2 \times 2^2 = 8$ electrons, etc. The *K* shell contains electrons having the lowest kinetic energy, the *L* shell electrons have energies higher, etc. Usually the low-level shells are filled before higher shells contain electrons. The outermost shell is occupied by the **valence electrons,** and the number of electrons in this shell determines the chemical nature of an element. The maximum number of electrons in the outer shell is eight. These shells also are called **quantum energy levels,** or **quantum levels.**

Certain elements important in electronics are included in Table 1. The elements from helium to radon have the full number of valence electrons and are the stable **inert gases** which form no chemical compounds with other atoms. Certain of these gases are used extensively in gas tubes. The elements from lithium to cesium have only one outer, or valence, electron. In general, having but few valence electrons results in an unstable valence structure, and these elements are good electron emitters, and sometimes good conductors. The elements selenium, silicon, and germanium have four valence electrons. These are the semiconductors widely used in rectifiers and transistors.

Table 1

ARRANGEMENT OF ELECTRONS IN CERTAIN ATOMS

ELEMENT	ATOMIC NUMBER	NUMBER OF ELECTRONS AT EACH QUANTUM LEVEL					
		1	2	3	4	5	6
Helium	2	2					
Neon	10	2	8				
Argon	18	2	8	8			
Krypton	36	2	8	18	8		
Xenon	54	2	8	18	18	8	
Radon	86	2	8	18	32	18	8
Lithium	3	2	1				
Sodium	11	2	8	1			
Potassium	19	2	8	8	1		
Rubidium	37	2	8	18	8	1	
Cesium	55	2	8	18	18	8	1
Copper	29	2	8	18	1		
Silver	47	2	8	18	18	1	
Gold	79	2	8	18	32	18	1
Boron	5	2	3				
Aluminum	13	2	8	3			
Gallium	31	2	8	18	3		
Arsenic	33	2	8	18	5		
Antimony	51	2	8	18	18	5	
Carbon	6	2	4				
Silicon	14	2	8	4			
Germanium	32	2	8	18	4		
Selenium	34	2	8	18	6		

Atom models are useful in presenting a possible picture of atomic structure, but often the picture is too exact. The studies of deBroglie, Schrödinger, Heisenberg, and others have shown mathematically that electrons have the properties, not only of particles, but also of waves. Davisson and Germer in 1927 proved this experimentally.[5] Thus the electrons, and other particles as well, that make up the atoms from which all matter is composed, behave as particles and as waves, and actually are both.[5]

1-8. *THE ISOLATED ATOM*

In both gases and solids the atoms are closely packed together and in general tend to influence each other. Before studying atoms under these conditions, however, it is advisable to consider an isolated atom. The normal hydrogen atom, consisting of a positively charged nucleus and one revolving negative electron, is the simplest atom and will be considered first.

Force on an Electron. Assume that the electron is located at infinity. Now suppose that the negative electron is permitted to approach the positive nucleus by moving through free space. As this occurs, the electron will experience at each position a force (given by Coulomb's law) having a magnitude given by the equation

$$F = \frac{Q_n Q_e}{\epsilon_v r^2} = \frac{Q_e^2}{\epsilon_v r^2} = \frac{e^2}{\epsilon_v r^2} \qquad \text{1-11}$$

the Q_e (and e) being used because the positive charge on the nucleus and the negative charge Q_e (or e) on the electron are numerically the same. The force will be in newtons when Q_e (or e) is 1.592×10^{-19} coulomb, ϵ_v (permittivity of free space or vacuum) is 1.113×10^{-10}, and r is the distance in meters between the nucleus and electron. Because one particle is positive and the other is negative the force will be one of attraction.

Potential Energy of an Electron. If the electron is assumed to have zero potential energy at infinity, the potential energy of the electron at any point in space a distance r from the isolated nucleus is the work required to move the electron from infinity to that point, or

$$W = \int_{\infty}^{r} \frac{Q_e}{\epsilon_v r^2}\, dr = -\frac{Q_e^2}{\epsilon_v r} \quad \text{or} \quad W = \frac{e^2}{\epsilon_v r} \qquad \text{1-12}$$

Using the values of the preceding paragraph, the work will be in joules, or in electron volts (section 1-3). As indicated, the potential energy of the electron at any point r will be negative because a force of attraction exists between the nucleus and the electron. These relations are shown in a general way in Fig. 1-5.

Electron Charge Density. The relations just given are not intended to hold when distance r approaches a small value. When r is small, the formerly

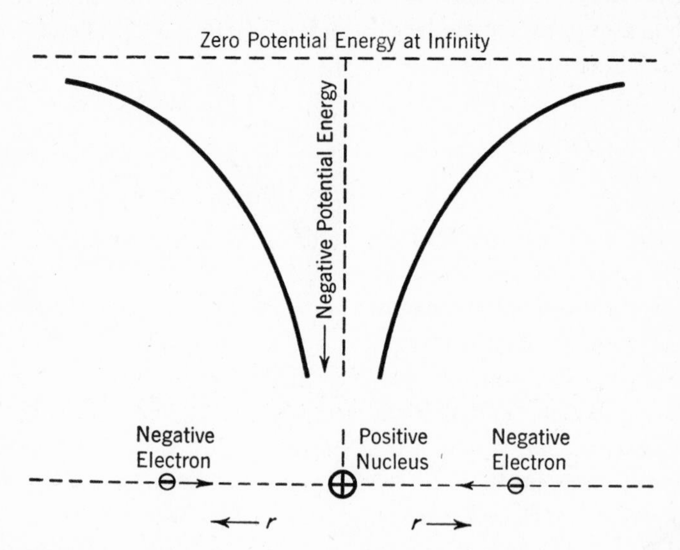

FIG. 1-5. Potential energy curves for an electron as it approaches a positive nucleus.

isolated particles in a sense "join forces," forming a normal hydrogen atom consisting of a nucleus and a revolving "satellite" electron. The attractive force of the atomic nucleus tends to cause the revolving electron to fall into the nucleus; this is prevented by the centrifugal force resulting from the orbital path around the nucleus. This orbital path should not be regarded as, for instance, a definite circle of given radius traversed each revolution. Rather, if visible, the "trail" left by the satellite electron would enclose the nucleus. The electron in its motion in the space about a nucleus spends different amounts of time in each region. This leads to the concept of an **electron charge density,** about the nucleus, somewhat as shown in Fig. 1-6. The density of the dots is proportional to the amount of time the electron spends in a small volume of space about the nucleus. Hence, regions where the dots are dense may be considered as more negative than regions where the dot density is less.

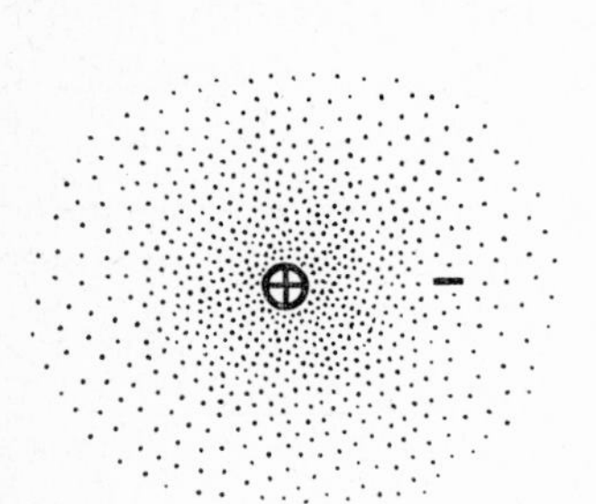

FIG. 1-6. Electron charge density of an isolated atom.

The paths of the electrons about the nucleus of a carbon atom are illustrated in Fig. 1-7. This shows the nucleus, the two electrons at the first quantum energy level, or the *K* shell, and four electrons at the next level, or *L* shell (Table 1). These electrons in their motion around the nucleus would produce a negative electron charge density for an isolated carbon atom somewhat like that of Fig. 1-6.

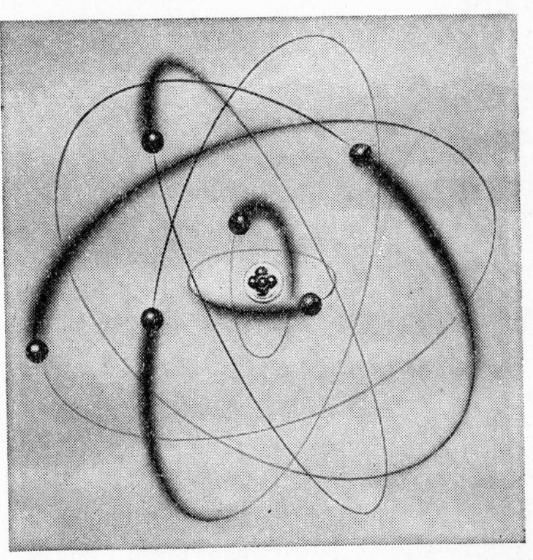

FIG. 1-7. A carbon atom consists of a dense central nucleus and electrons in motion about this nucleus. (Courtesy Western Electric Co.)

1.9. *ATOMS IN GASES*

A **gas molecule** may be a single atom, or a cluster of two or more atoms. The molecules of the chemically inert, **monatomic gases** commonly used in electronic devices are single atoms. The *diameters* of the molecules of four widely used gases are considered to be: helium, 2.64×10^{-8} centimeter; argon, 2.87×10^{-8} centimeter; neon, 2.34×10^{-8} centimeter; and mercury (vapor) 2.37×10^{-8} centimeter. The number of molecules per cubic centimeter of *any* gas at atmospheric pressure and zero degrees Centigrade is called the **Loschmidt number** and is [9] approximately 2.7056×10^{19}.

The molecules of a gas are in rapid motion. They bombard the walls of an enclosing tube, exerting a force on the walls called **gas pressure.** Normal air is a mixture of several gases, including about 76 per cent nitrogen. If molecules are pumped *out* of a tube, the molecules per unit volume, and hence the pressure, are reduced. The pressure in a so-called "vacuum" tube is about 10^{-6} millimeter of mercury. At this pressure and at zero degrees Centigrade, there remain about 3.56×10^{10} gas molecules in a cubic centimeter of nitrogen, truly a large number. Gas pressure can be increased by pumping more gas molecules *in*, or by heating the gas so that the molecular motion is increased.

Because of their random motions, gas molecules strike each other. The **mean free path** *of a molecule* is the *average* distance a molecule moves without collision with another molecule. The mean free path is [9]

$$\lambda_m = \frac{1}{17.8\, N\, \sigma^2} \qquad \text{1-13}$$

where λ_m is the mean free path in centimeters for the particular gas and conditions of study, N is the number of molecules per cubic centimeter, and σ is the molecular *radius* in centimeters.

Of much importance in electron gas tubes is the mean free path *of an electron.* The apparent radius of an electron is less than for a molecule, and hence the electronic mean free path is [9]

$$\lambda_e = 5.66\, \lambda_m \qquad 1\text{-}14$$

where λ_m is as given by equation 1-13.

The factor N of equation 1-13 is the number of gas molecules per cubic centimeter, and is

$$N = \frac{P}{KT} \qquad 1\text{-}15$$

where P is the gas pressure in dynes per square centimeter, K is the Boltzmann constant (1.37×10^{-16} erg, or 1.37×10^{-23} joule, per degree Kelvin), and T is the absolute temperature in degrees Kelvin (degrees Centigrade plus 273).

Of interest is the application of equations 1-13 and 1-14 to a vacuum tube. As previously stated, in a vacuum tube there may remain some 3.56×10^{10} molecules per cubic centimeter at zero degrees Centigrade. If these are nitrogen molecules having a diameter of 3.18×10^{-8} centimeter, the mean free path is 6370 centimeters, or 209 feet, and the mean free path of an electron in the tube is 36,000 centimeters, or 1182 feet. It is evident why a *vacuum* tube is "essentially unaffected" by the presence of the gas or vapor, even though a large number of gas molecules remain.

1-10. *IONIZATION IN GASES*

A *normal* gas atom contains equal amounts of positive and negative electricity and is electrically neutral. An atom may, however, be broken into charged particles called **ions,** defined [4] as an "electrified portion of matter of subatomic, atomic, or molecular dimensions."

The smallest ion is an electron which has been torn away from an atom. This electron is a **negative ion.** Such an ion has little mass and readily is accelerated by an electric field. Sometimes electrons attach themselves to otherwise neutral atoms, forming massive negative ions. This seldom occurs in the gases used in electron tubes.

If an atom is ionized by the removal of an electron, the remaining part of the atom is a **positive ion.** Such ions are massive compared to negative ions (electrons), and are not accelerated so readily. However, because they are massive, much kinetic energy is stored in even a low-speed positive ion. This fact will be found of importance in subsequent chapters.

Ions are created from normal atoms by the process of **ionization.** To disrupt an atom and produce ionization requires energy. When gases and vapors * are ionized, the required energy may come from several sources.

1-11. *IONIZATION BY COLLISION*

High-speed particles, such as electrons, may produce ionization. When a rapidly moving electron "collides" with a normal atom, one of four things may happen.[10]

First, nothing at all may result from the so-called collision, which is not an impact in the usual sense, but rather is an electromagnetic field reaction. At kinetic energies of less than about one electron volt, an electron may pass through an atom (which largely is space) with no loss of energy and but little chance of deflection.[10]

Second, at higher electron velocities and energies, collisions become more frequent. If the kinetic energy is still below critical values, the electron is slightly deflected but loses little or no energy.[10]

Third, if an electron has yet higher velocity, its kinetic energy may be sufficient to place the atom in an **excited state.** In this state one or more of the outer-shell electrons have absorbed quanta of energy from the colliding electron, and as a result these outer-shell electrons have been raised to higher than normal energy, or quantum, levels. These electrons have not left the atom, however.

Fourth, at still higher velocities and kinetic energies, the colliding electron may transfer sufficient energy to the atom to separate completely one (or more) of the outer-shell electrons from the parent atom. This process is **ionization by collision.**

The length of time an atom remains in an excited state is of importance in ionization by collision because an excited atom is unstable and hence susceptible to ionization. Most atoms remain in excited states only for short periods, of the order of 10^{-7} or 10^{-8} second. Then, for most atoms the excited electrons fall back to their normal energy levels. In so doing, these electrons must lose their excess energies, which they do by radiating quanta of electromagnetic energy, perhaps as photons of visible light.

* Gas tubes first are evacuated, and then the desired amount of an inert gas, such as argon, is added. Vapor tubes are evacuated and contain an element such as mercury in liquid form. Because of the low pressure, some of the mercury vaporizes and exists within the tube as a vapor, which is in every sense a gas. Temperature has little effect on gas tubes, but affects the performance of vapor tubes by varying the amount of vaporization and hence the vapor pressure.

Some few of the excited atoms do not immediately give up their excess energies, but persist in a **metastable condition** [10] for periods as long as perhaps 0.1 second. Then it appears that the excess energy is not radiated, but is passed on to other atoms, or is given up to the walls of the enclosing tube. Atoms in a metastable condition are, for a relatively long time, highly susceptible to complete ionization by successive collisions with high-speed electrons. Or, the required additional energy may be obtained by absorbing quanta of radiant energy of the proper frequency. Also, an excited atom may obtain the additional energy for complete ionization by absorption from another excited atom. Ionization in successive steps is **multistage ionization,** or **cumulative ionization.**

Positive and negative ions produced by collision, or by other means, will themselves experience an acceleration, and if the electric field intensity is sufficient, they, in turn, may achieve kinetic energies sufficient to cause ionization by collision, and then the process may be repeated. Thus, ionization by collision may increase as an avalanche if not controlled.

Data for the first excitation potentials, and for the final ionization potentials, of several inert gases and for mercury vapor are listed in Table 2. The voltages given are the potential differences, through which an electron, starting from rest, must fall unimpeded to have the required kinetic energy to produce excitation and ionization.

Table 2

IONIZATION CHARACTERISTICS OF GASES USED IN ELECTRONIC DEVICES

(From the work of P. T. Smith, and compiled from reference 10)

GAS	FIRST CRITICAL EXCITATION POTENTIAL VOLTS	IONIZING POTENTIAL VOLTS	MEAN FREE PATH, IN CENTIMETERS, OF ELECTRON, 1MM. MERCURY, 0° CENTIGRADE
Helium	19.73	24.48	0.1203
Neon	16.60	21.47	0.0753
Argon	11.57	15.69	0.0431
Mercury vapor	4.66	10.39	0.0142

Note: Because mercury vapor rather than a gas is considered here, the value 0.0142 centimeter for mercury should be used only at low pressures (Reference 10).

1-12. *NATURAL IONIZATION OF GASES*

A few ions exist in gas under normal conditions.[11] The energies to produce these ions may come from several possible *natural* sources. One cause of such ionization is high-velocity electrons and other particles from disintegrating radioactive substances, traces of which are widely distributed. Another cause of natural ionization is radiation such as cosmic rays, x-rays from radioactive substances, and ultraviolet rays from the sun. Natural ionization in *air* under normal conditions produces about 2 ions per second per cubic centimeter. At high altitudes, cosmic rays may cause ionization at 100 times this rate.[10]

The effects of natural ionization are shown in Fig. 1-8 and are studied as follows: Two small flat electrodes are spaced one centimeter apart in air under normal conditions. A direct voltage is impressed between the electrodes, and as the voltage is increased slowly the resulting electric current that flows through the air between the electrodes is measured. If there were no ionization by natural (or artificial) causes, the gas would be a perfect insulator. However, between zero and about 30 volts the current will increase as in region *A*. This current is carried through the air by the free ions resulting from natural ionization.

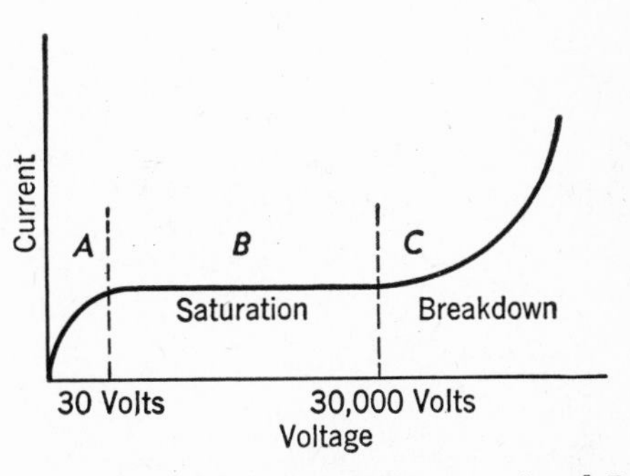

FIG. 1-8. The current in regions *A* and *B* consists of ions produced by natural ionizing agents. In region *C* ionization by collision occurs. (Figure not to scale.)

A constant **saturation current** is reached at voltages above 30 volts as shown in region *B* of Fig. 1-8. The voltage now is sufficient to remove the ions as fast as they are produced by natural ionization. At sea level the saturation current for air can be calculated in the following manner.[10] If 2 ions per second are created between the electrodes, then $2 \times 1.592 \times 10^{-19}$ coulomb of electricity can be carried between the electrodes per second. One coulomb per second is one ampere, hence the current is about 3×10^{-19} ampere.[10]

When the voltage between the flat electrodes one centimeter apart in air reaches about 30,000 volts, the current increases rapidly, and if the voltage is further increased, breakdown of the air occurs, a spark will pass, and an arc may form. A voltage of about 30,000 volts per centimeter is required before ionization by collision can occur between electrodes in closely parallel planes to provide additional carriers and thus permit the current to exceed the saturation current of region *B* in Fig. 1-8. This high voltage is necessary for the following reasons. In air at normal temperature and pressure, the molecular density is great and the mean free path accordingly is low. Thus, a high voltage must be applied before the naturally produced ions can achieve ionizing energies in the short free runs that are possible.

1-13. *RECOMBINATION AND DEIONIZATION*

An ionized gas contains positive and negative ions, and because of the attractive force between the ions, there will be a tendency for **recombination** to take place, and for the gas to **deionize.** The **deionization time** determines the upper frequency limit beyond which some gas tubes may not be used.

During a given interval of time the probability of a positive ion recombining with a negative ion is proportional to the number of *negative* ions present, and similarly for a negative ion meeting a positive ion. Hence the rate of recombination is αn^2, where α is the **coefficient of recombination,** and n is the *total* number of positive and negative ions present per unit volume of gas.

The ions of a gas may travel, or diffuse, to the walls of a tube. In many devices recombination appears to occur largely at the walls. The arriving ions impart energy to the walls, causing local heating. When ions accumulate on the glass (insulating) walls of a tube, they form localized charged regions, or "islands." In certain devices charged surface regions may affect operation; for example, by deflecting electron beams. Conducting coatings and collecting rings are used to drain off such charges.

1-14. *ION SHEATHS IN GAS TUBES*

Certain types of electron tubes contain inert gas, or vapor, which is ionized during operation. The distribution of ions is usually not uniform in gas tubes, particularly when operating voltages are impressed between the electrodes.

An important factor in causing a nonuniform distribution of ions is the difference in the mobilities of the ions. The mass of an electron is small compared with that of a positive ion; hence, the mobility of an electron is large compared with that of a positive ion. The **mobility** of an ion often is expressed as the velocity in centimeters per second that a particle has in an electric-field strength of one volt per centimeter.

The velocity attained by a charged particle can be computed by equation 1-7, which indicates that the velocity is inversely proportional to the square root of the mass of the particle. The mass of a mercury atom is about 3.3×10^{-25} kilogram, and that of an electron is 9.0×10^{-31} kilogram. The ratio of the *square roots* of these numbers is about 1 to 607. Hence, when an electron and a massive positive mercury ion are in the same electric field, the electron attains a velocity about 607 times that of the massive ion.

Thus suppose that two electrodes are in a gas **plasma,** a region in which the number of positive and negative ions is equal and their distribution is uniform. If a voltage is impressed between the electrodes, the *negative* electrode will repel electrons and attract positive ions. Because of their low mass and high mobility, the electrons quickly leave the vicinity of the negative electrode, but the massive positive ions tend to remain. The result is that a *positive space charge,* or *positive ion sheath,* forms in a thin layer about the negative electrode where formerly a uniform plasma existed.

This positive ion sheath will act as a shield around the negative electrode because the electric lines of force will extend between positive ions and the charges on the negative electrode, and will not extend into the main body of the gas. The negative potential on the electrode cannot, therefore, exert an influence beyond the ion-sheath region. Increasing or decreasing the voltage on this negative electrode merely varies the thickness of the positive ion sheath. A positive ion sheath may also exist at the positive electrode if the electrons are drawn away from the positive electrode by the external electric circuit at a sufficiently high rate.

1-15. *ELECTRIC DISCHARGES IN GASES*

Electric discharges in gas that conduct under the influence of the applied voltage alone are **self-maintained discharges.**[12,13] Although external effects may assist in *starting* the discharge, once it is started the discharge itself must produce the required ions for maintaining the discharge current. For instance, it might be necessary initially to heat a filament in a tube to *start* a discharge in the tube, after which the heating current could be removed and a self-maintained discharge would continue. **Non-self-maintained discharges** require an external agent to supply ions for the discharge current. Thus, operation of the common gas rectifier tube requires a hot filament. Self-maintained discharges will be discussed briefly in the following pages. The non-self-maintained type will be considered extensively throughout the book.

The self-maintained discharge depends for existence on ionization by collision by *both* positive and negative ions.[12] For instance, suppose that the voltage has just been applied between the two electrodes of Fig. 1-9. Also, assume that conditions are such that the *positive* ions produce no ionization. Some few ions exist in the gas because of natural ionization (section 1-12). If an electron starting from point 1 achieves sufficient velocity, it will cause ionization of a gas atom at point 2. Assume that the newly produced electron and the original electron again cause ionization at point 3. In this way an appreciable current will reach the positive electrode. However, the positive ion proceeding to the negative electrode is assumed to produce no ionization. Hence, after a short time the gas is "cleaned up," and no current can flow (except as external agencies produce ions). But if the positive ion on its way to the negative electrode causes ionization, then the phenomenon can continue as a self-maintained discharge. Also, such a discharge may occur if the positive ions knock, or release, electrons from the negative electrode.

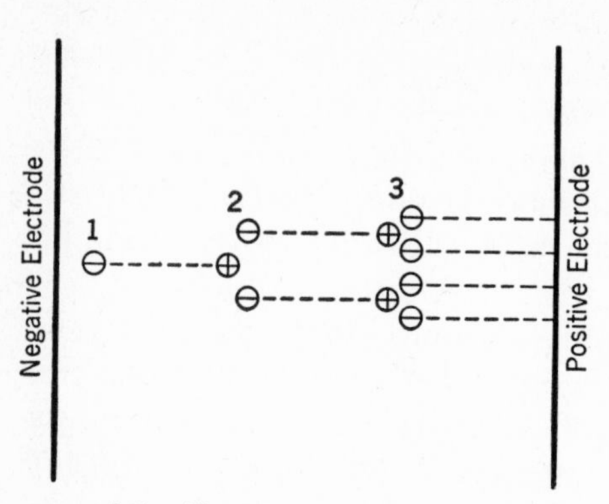

FIG. 1-9. If positive ions do not produce ionization by collision, a gas soon will be cleared of ions.

Electric Spark. A common form of self-maintained discharge is the electric spark. Suppose that a voltage is impressed between two flat parallel metal electrodes one centimeter apart in normal air and that the voltage is increased slowly. Because of the few free ions in the air (produced by natural ionization), a small current of ions will flow to each electrode. When the voltage has reached about 30,000 volts (per centimeter), ionization by collision occurs, the air breaks down, and an **electric spark** passes between the electrodes.

Electric Arc. If the voltage is maintained at a sufficiently high value, the spark will develop into an **electric arc,** which is a self-maintained discharge. If the voltage is not maintained sufficiently high, the spark will be transient in nature. In the arc discharge, electrons are drawn from the negative electrode, or **cathode,** and flow in large numbers to the positive electrode, or **anode.** For the arrangement of two flat parallel metal electrodes, the energy required to release the electrons from the (initially) cold cathode probably results from (1) the kinetic energy of the massive positive ions striking the negative cathode, and (2) the high positive ion sheath formed near the cathode surface. The arc has a negative resistance: The greater the arc current, the less the arc resistance, and hence less voltage is required. An arc is unstable, and unless the current is limited by resistance in series with the electrodes, or by other means, an arc may reach a destructive value. The preceding discussion is for an arc in air, but the phenomenon is similar for self-maintained arcs in tubes; this will be considered in more detail in subsequent chapters.

Corona. In air or gas in the normal state, the magnitude of the voltage gradient is directly proportional to the electric field intensity. For this reason, if electrodes are shaped so that the electric field is concentrated at sharp edges or points (for example as in a needle spark gap), the strong electric field may cause the voltage gradient to reach a value such that the air or gas will ionize and break down. However, the discharge may be limited to the vicinity of the electrode, because elsewhere between the electrodes the electric field may not be concentrated, and the critical voltage gradient may not be exceeded. If the discharge consists largely of small "streamers" extending into the ionized region, the term **brush discharge** sometimes is used. However, if the discharge is more of a uniform glow, the terms **glow discharge,** or **corona,** are employed.

Glow Discharge.[11,12] This is a self-maintained discharge, and for the case under consideration the electric discharge is between two *cold* flat electrodes in a glass tube at least several inches long and about an inch in diameter, the tube containing air at low pressure. Phenomena accompanying the glow discharge are utilized in many electronic devices. The discharge and its characteristics are depicted in Fig. 1-10.

Under normal operating conditions a **cathode glow** exists at the (cold) negative electrode. The glow may be of two types: the **normal glow** in which the discharge does *not* completely cover the cathode, and the **abnormal glow** in which the cathode is completely covered. For the *normal* glow, the cathode area covered by the glow increases and decreases directly with the magnitude of the current flowing through the tube.

As the curve of Fig. 1-10 shows, most of the voltage drop occurs near the cathode in the **Crookes dark space.** This drop is called the **cathode fall** (see Table 3). For the normal glow, the cathode fall is relatively independent of the current magnitude. This dark space is the distance the electrons (either drawn from the cathode or released by ionization near the cathode) must travel before they have sufficient energy to produce ionization by collision.

The negative glow starts at the edge of the Crookes dark space and termi-

Table 3

NORMAL CATHODE FALLS FOR CERTAIN GASES AND ELECTRODES

(Values, in volts, from reference 12)

TYPE OF CATHODE	AIR	HELIUM	NEON	ARGON
Platinum	277	160	152	131
Aluminum	229	141	120	100
Iron	269	161		131
Copper	252	177		131

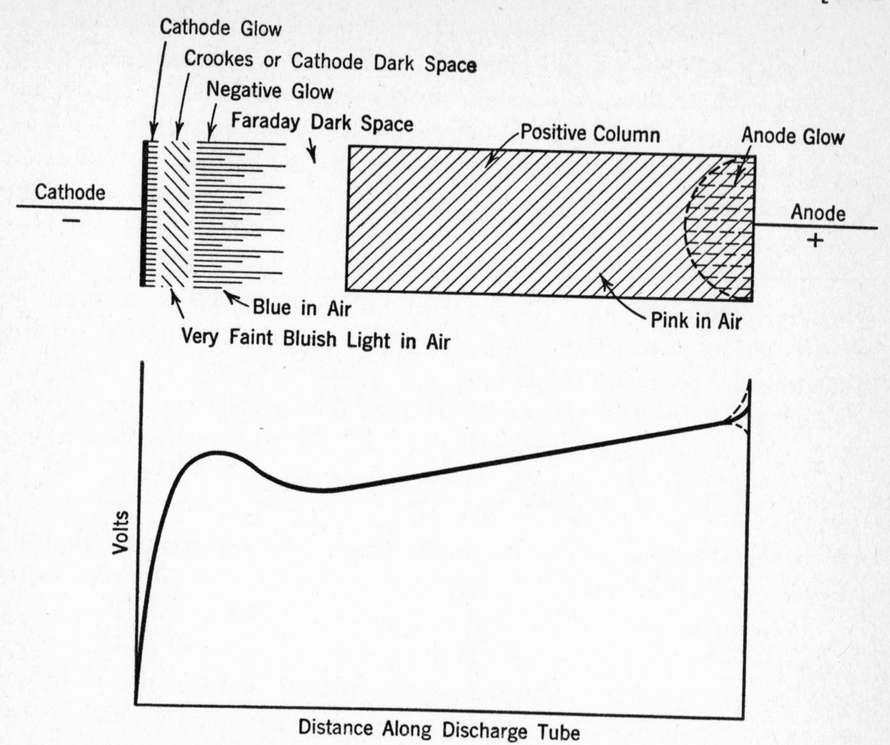

FIG. 1-10. Appearance, nomenclature, and potential distribution for the glow discharge. (Adapted from reference 12.)

nates in the **Faraday dark space.** In the region of the negative glow, gas atoms are excited and some are ionized by electrons from the Crookes dark space. Light is emitted when excited atoms revert to normal state.

The Faraday dark space starts where the electrons, because of collisions with gas atoms, have lost their ionizing abilities. The current in the Faraday dark space is largely electrons that are again accelerated to the extent that they cause ionization by collision in the **positive column.**

The positive column extends through most of the length of the gas tube. Ionization and excitation of gas atoms occur in the positive column. The accompanying phenomena of recombination and loss of excitation energies provide the uniform glowing (positive) column of light for such devices as neon tubes.

1-16. *ELECTRONIC PHENOMENA IN SOLIDS* [14,15,16,17]

The development of electronic devices such as crystal detectors and copper-oxide and selenium rectifiers, and in particular the invention, about

1948, of the amplifying transistor, have focused attention on the field of solid-state electronics.

In the inert, or monatomic, gases previously considered, the gas atoms are free to move within the enclosing tube. By contrast, in a solid the atoms are held in fixed positions although they may, and do, vibrate. A solid, such as the semiconductor germanium used in transistors, consists of crystals, which are composed of row upon row of germanium atoms arranged in a three-dimensional lattice structure.

The following pages will consider the electronic principles involved in solids such as germanium. Relatively speaking, this is a new study, and the development of appropriate theories is in progress, rather than accomplished. Many of the statements and illustrations used should be regarded in this light. As progress continues, new theories and explanations will result.

1-17. *ATOMS IN SOLIDS*

In solid matter, as in the metal copper or the semiconductor germanium, the atoms are packed so closely together that there are about 10^{22} atoms per

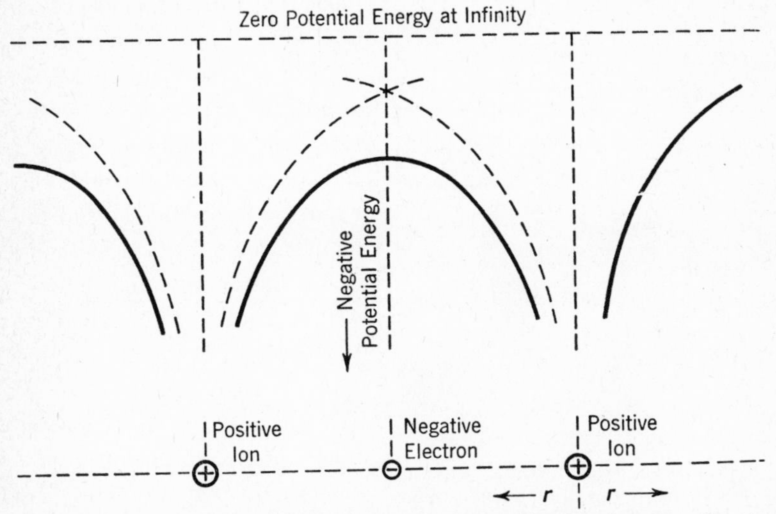

FIG. 1-11. When a negative electron approaches a *single isolated* positive ion (such as a metal atom minus an electron) the work done on the electron is as shown by the broken curves (Fig. 1-5). When an electron approaches two such ions, the work involved is greater. (Imagine an electron approaching from the bottom of the page.) This is indicated by the solid curves. For an electron to leave an *isolated* atom, energies equal to the broken curves are required. For an electron to leave an atom of a metal and travel within the metal, energies equal to the solid curves are necessary. If energies are measured with respect to zero at the nucleus, then if an electron has kinetic energy greater than the "humps" of potential energy shown by the solid curve, the electron will be free to wander about the solid, and is called a free electron. These curves indicate that a free electron will not, necessarily, have energies sufficient to escape from the surface of a metal.

cubic centimeter in a metal. Because of the close atomic spacing, the diagram for the isolated atom, Fig. 1-5, does not hold for atoms in a solid.

When atoms are close together, the situation (simplified) is somewhat like Fig. 1-11. Here are shown two **metal ions,** which are metal atoms minus one electron. The ion at the right (inadequately) represents the edge of the metal. The ion at the left represents an ion formed (for purposes of illustration) from one of the many rows of atoms in the crystal lattice structure.

If the ions are isolated, the curve for the *negative* work done in moving the electron from infinity to the immediate vicinity of one of ions is as shown by the broken lines of Fig. 1-11. The work and potential energies are negative because a force of attraction exists between the electron and the nucleus. The work done in separating the electron from the nucleus would be *positive.* Now, if the positive ions are not isolated, an electron in the electric fields of these ions experiences two forces acting on it, and hence its potential energy with respect to infinity is greater, as the solid curves indicate. If the potential energy arbitrarily is measured *with respect to zero at the ion nucleus,* instead of with respect to zero at infinity, then an electron at a distance r between two ions inside a metal would have *less* potential energy than if it were a distance r outside the metal. This means that an electron with a certain amount of kinetic energy may move about within a solid, yet may not be able to escape from the surface into the space beyond.

FIG. 1-12. In a solid material such as carbon, the paths of the outer electrons are linked, and the electrons are "shared" by adjacent atoms. (Courtesy Western Electric Co.)

Because of the small spacings between atoms, the electron paths, or orbits, or wave functions, interlink as illustrated for *carbon atoms* by Fig. 1-12. This interlinking means that electrons are shared by adjacent atoms. This causes the electron charge density to be different from that of Fig. 1-6 and to resemble the charge density of Fig. 1-13.

A force of repulsion exists between the two positive atomic nuclei. Because of the electron sharing, through path interlinking, and the resulting increased negative electron charge density between the nuclei, the force of repulsion is *less* than if the paths were not interlinked. This lessened force is, in fact, an **electric bond** between the adjacent atoms. This bond and similar bonds with other atoms hold the various atoms in their positions in the solid state.

The illustration given here is for carbon atoms. From Table 1 it is seen that silicon and germanium atoms also have four outer-shell, or level, valence electrons, like the carbon atom. For this reason, silicon atoms form bonds as just explained; so do germanium atoms. These are referred to as **electron-pair bonds,** or as **covalent bonds.**[17]

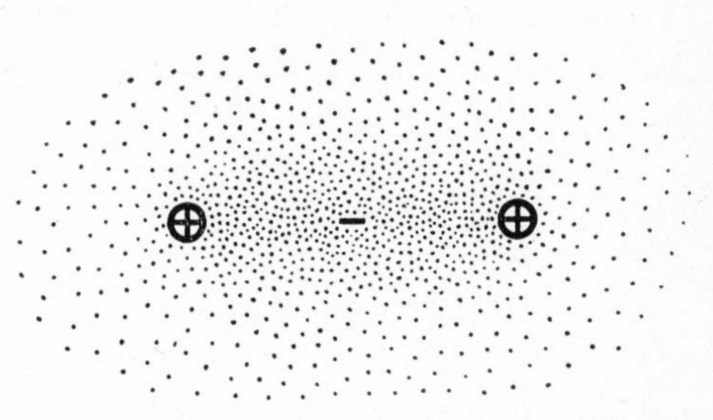

FIG. 1-13. Electron charge density of two atoms of a solid material.

Thus the atoms of carbon, silicon, and germanium arrange themselves, in the solid state, as shown in Fig. 1-14. A *single* **crystal** of one of these elements may be physically large, perhaps 6 inches long and an inch in diameter. A crystal contains "innumerable" atoms, all arranged in the uniform crystal **lattice structure.**[17,18] The interlinked paths of the shared electrons are *illustrated* by the two rods between each pair of atoms. These rods represent the electric forces binding the atoms in their fixed positions in the solid. If each atom has four outer, or valence, electrons, and if each atom shares one electron with each neighbor atom, then the crystal lattice structure must be as in Fig. 1-14.

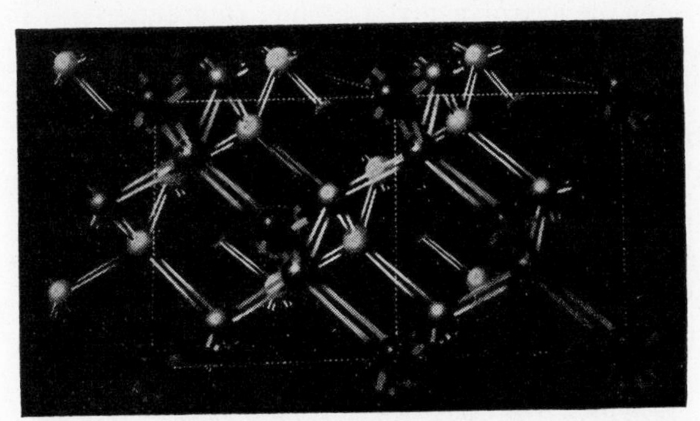

FIG. 1-14. Showing the arrangement of the atoms in a semiconductor such as silicon, or germanium. All atoms are identical. The rods represent the shared electrons and the resulting bonds. (Courtesy General Electric Co.)

1-18. *ENERGY BANDS*

In addition to having a certain potential energy because of its position in space, an electron in motion about a nucleus has velocity and hence kinetic energy. It is very important to note, however, that in accordance with the theories of **quantum mechanics** [17,19] a given electron in a particular atom may have only certain energies and modes of motion about the nucleus. It is, in fact, for reasons such as this that the structure of atoms is as shown in Fig. 1-4 and as given in Table 1. In quantum mechanics, these modes of motion, and energies and velocities, are termed **quantum states.**[17,19]

Pauli Exclusion Principle. If an isolated normal atom could be examined in detail, it would be found that no two satellite electrons were simultaneously in exactly the same quantum state; that is, no two electrons of a *given* atom would simultaneously have identical energy levels or be in exactly the same positions. This is in accordance with the theories of quantum mechanics: The quantum state of an electron determines its mode of motion, and from the **Pauli exclusion principle,** no two electrons of a *given* atom may be simultaneously in the same quantum state. This has been compared [17] to the fact that no two objects can occupy the same position at the same time. Of course, electrons could (and would) have the same quantum states in all normal atoms of the same element, if the atoms were sufficiently far apart for them not to influence each other appreciably.

Formation of Energy Bands. When atoms are very close together, as they are in solids such as the metal copper and the semiconductor germanium, the electron paths interlink (Fig. 1-12), and the quantum states overlap.[17,19] Nevertheless, the Pauli exclusion principle still applies. Hence, when atoms are closely packed, as in a germanium crystal, the quantum states no longer are those of a single isolated atom. Instead, the relatively narrow permitted energy levels of a given atomic quantum state broaden into a band of energies, or into an **energy band,** when atoms are packed into a crystal.[14,15,17,18] Thus, the energy levels or shells (Fig. 1-4 and Table 1) at which electrons are permitted are narrow in isolated atoms, but the permitted, or allowed, energy bands are relatively wide when atoms are closely spaced, as in a crystal of germanium.

An analogy [17,20] exists between electron energy bands and electric circuit theory. Thus, a given electron in a given quantum state in an isolated atom may "oscillate," or travel, only within narrow limits, but in closely packed crystals, the permitted limits are much wider. Similarly, an isolated resonant electric circuit will oscillate at only one frequency, but if it is inductively coupled with other resonant circuits, the combination oscillates at several frequencies, or over a band of frequencies.

1-19. *ENERGY-LEVEL DIAGRAMS*

When a diagram such as Fig. 1-11 giving (with respect to zero at the nucleus) potential energies of electrons in a solid is combined with a diagram showing the kinetic energy bands that electrons in a solid may occupy, a diagram such as Fig. 1-15 results. This figure is for the semiconductor silicon, which has (Table 1) two electrons in the first band, eight in the second, and four in the third, or valence, band. This diagram is for *pure* silicon and is descriptive only, and the widths of the bands and the spacing between

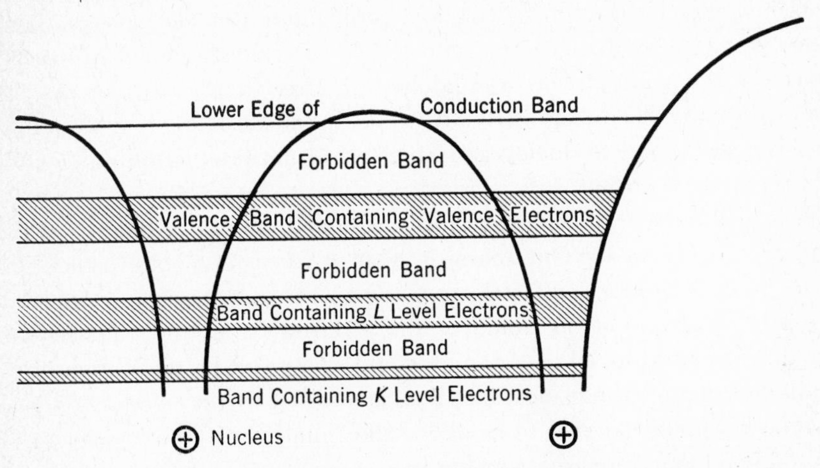

FIG. 1-15. Kinetic energy bands occupied by electrons in a perfect crystal.

bands are not to scale. Electrons of the proper kinetic energies are in the **allowed bands,** but no electrons are in the **forbidden bands** because these bands are arranged in accordance with the structures of the atoms composing the crystal (Table 1). The energy-level diagram for pure germanium would contain another band.

In Fig. 1-15 the upper edge of the valance band has been placed below the "humped" curves representing the potential energy distribution between atoms of the semiconductor silicon. The upper edge of the valence band represents the maximum kinetic energy that an electron may have in pure silicon at absolute zero. Since this energy is below the "potential humps," the electrons in the valence band in silicon at absolute zero cannot wander about freely. Instead, these valence electrons would be bound to the nucleus and would be utilized in forming bonds. Of course, this is an idealized "picture" and would, in fact, result in silicon being an insulator at absolute zero.

As previously stated, Fig. 1-15 is for silicon atoms. A similar figure for more complex atoms would have additional energy bands. Since bands below the valence band are not involved in describing many electronic phenomena, such as conduction, it simplifies energy-level diagrams to omit bands below the valence band, and this will be done in subsequent figures.

1-20. *INSULATORS*

As explained in the preceding sections, the four outer, or valence, electrons of elements such as carbon, silicon, and germanium form electron-pair bonds which serve to bind the atoms into a solid crystal structure. Also as explained, if the material were absolutely pure (free from foreign atoms) and were at absolute zero, the material would be an insulator because all the electrons are bound to nuclei, and in this theoretical situation no electrons are free to carry current.[17,18]

As is well known, the elements just mentioned are not ordinarily used as insulators in practice.[21] One reason is that at room temperature the atoms would be in vibration, and at least a few of the electron bonds would be disrupted. By this action, some of the electrons from the valence band would receive additional energy and would as a result be raised to a higher **conduction band.** Those electrons that are raised to the conduction band have energies about equal to or above the "humps" of potential energy of Fig. 1-15 and therefore are no longer bound to nuclei. They are, accordingly, free to move when a difference of potential applied to the crystal structure of the solid establishes an electric field within the solid. These moving electrons constitute a current, and the material that was an insulator at absolute zero would not be an insulator at room temperature. However, if it is diffi-

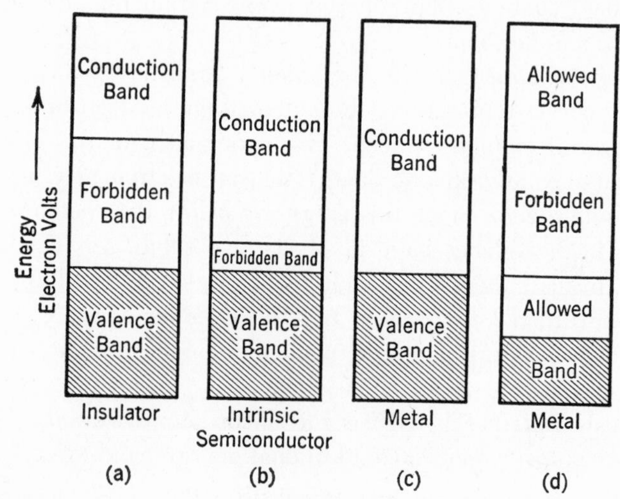

FIG. 1-16. Energy-level diagrams for solids. Shaded areas represent bands filled with electrons. (Reference 18.)

cult, at all except very high temperatures, to move electrons to the conduction band, then at all except these very high temperatures the material remains an insulator.

As is well known, the insulators extensively used in practice are mixtures, such as porcelain, and are not pure elements. It is possible to present only a general explanation of the reason for insulating properties. In an insulator the valence band is full of electrons, and conduction cannot occur by electrons in this band. The situation has been likened [17] to a garage floor *completely* filled with automobiles; obviously none can move. Furthermore, in an insulator there is a large energy gap of several electron volts between the filled valence band and the next higher permitted band, which is the normally unoccupied conduction band. Because of the strong atomic bonds, or in other words, because of the wide energy gap (Fig. 1-16a) between the valence band and the conduction band, an insulator (as used in electrical practice) must be raised to high temperature before bonds can be broken and electrons raised to the conduction band. The direct-current volume resistance of a typical insulator (polystyrene) is 10^{18} ohms per centimeter cube at 25°C.

1-21. *INTRINSIC SEMICONDUCTORS*

Pure silicon and germanium are elements that are inherent or **intrinsic semiconductors.** At absolute zero they are theoretically perfect insulators. Energies required to break the bonds and transfer electrons from the valence band to the conduction band Fig. 1-16b are given [19] as follows: for diamond (which is an intrinsic semiconductor of carbon atoms), about 6 to 7 electron volts; for silicon, about 1.1 electron volts; and for germanium, about 0.7 electron volts. Because of the high energy required to break the bonds and permit conduction, diamond "crystal counters" are used in studies of high-energy radiation in nuclear, and other, experimentation. On the other hand, silicon and germanium become conducting because of thermal energies received at room temperature. The resistance of "pure," or intrinsic germanium is about 60 ohms per centimeter cube at room temperature.[22] At higher temperatures more valence bonds are broken, and the resistance decreases; at lower temperatures the reverse is true; hence, intrinsic germanium has a negative temperature coefficient of resistance.

When, in an intrinsic semiconductor, an electron is raised from the valence band across the relatively narrow forbidden band, and to the conduction band, a positive **hole** is left in the valence band. This hole is represented by a vacancy in the crystal bond structure. It is positive because there is now a deficiency of negative electricity in the crystal structure since an electron is missing from its normal position. A nearby valence electron may readily change atoms, and fall into this hole. But, when this occurs, a hole exists where the second electron left. In this manner, a positive hole may migrate within the valence band from atom to atom. If a germanium crystal contains less than one part in 10^{12} of impurity atoms it is *intrinsic*.[22]

Although conduction of electricity by holes is possible, this involves an intermittent, and accordingly a slower, current conduction than that of electrons in the conduction band.[17,22,23] The negative electron current carriers, are *above* the potential energy "humps," are free to move readily within the crystal, and transport electricity through the semiconductor when a difference of potential is applied.

1-22. *METALS*[15]

It sometimes is difficult to distinguish between a semiconductor and a metal. However, several possibilities exist.[14,17,18] First, if *no* forbidden band exists between the last *filled* band and the conduction band above (Fig. 1-16c), then electrons may move readily to the conduction band, and the element is a good conductor. Magnesium is an example.[18] Second, if a forbidden band *does* exist between the last band containing electrons and the empty conduction band above it, but if the last occupied band is only *partly filled* (Fig. 1-16d), then conduction is possible by electrons in this band.[18] The garage floor mentioned earlier now is *not* completely filled with automobiles, and motion among them is possible.[17] The alkali metals are examples.[18] It is also of interest to note that each of three of the best electrical conductors—copper, silver, and gold—has only one outer, or valence, electron.

In a metal atom, such as of copper, the single valence electron has kinetic energy at room temperature well above the potential energy "humps" of Fig. 1-15 and is at liberty to move with ease from atom to atom. Accordingly it is called a **free electron,** and acts to carry current through a metal conductor when a difference of potential is applied to the conductor. The remainder of the (copper) atom is positively charged. There are about 10^{22} atoms in every cubic centimeter of copper, and if each atom contributes an electron, there are about 10^{22} free electrons per cubic centimeter. Thus, there are many free carriers to conduct electricity through a metal such as copper. These electrons are free at all times, and the number per unit volume is approximately constant, at least over the usual range of temperatures. It is, therefore, possible to picture a metal as consisting of positive metal ions "floating" in a **free-electron gas.**

1-23. *IMPURITY SEMICONDUCTORS*

Intrinsic semiconductors, such as chemically *pure* germanium, have been discussed. Theoretically, such semiconductors are insulators at absolute zero. In many semiconductor applications pure material is not used; in fact, the material is intentionally made impure by the controlled addition of foreign atoms. Important semiconductor devices, such as the **transistor** (Chapter 10), are made from impurity semiconductors.

A single crystal of pure germanium has been represented as composed of germanium atoms that have formed electron-pair, or covalent, bonds (Fig. 1-14). A single germanium crystal with perfect uniformity of the bond structure would be in a **perfect,** or **reference, state.**[17] Single crystals of germanium are made, or grown, by immersing the end of a single crystal seed in molten germanium, and slowly withdrawing the seed crystal (and the attached growing crystal) from the molten germanium in a controlled manner.[22] Atoms of elements other than germanium may be introduced as desired to the molten germanium. When this is done the final grown crystal will *not* be pure germanium. It will (for the usual crystal) be a crystal composed largely of germanium atoms, but with an occasional foreign, or impurity, atom in a position otherwise occupied by a germanium atom in pure, or intrinsic, germanium. A crystal composed in this manner contains flaws, or imperfections, and is in an **imperfect state.**[23]

Donor Impurities. Suppose that a small amount of the element arsenic is introduced into the germanium crystal as explained in the preceding paragraph. The difference between these elements is that the arsenic atom contains *five* valence electrons and the germanium atom contains *four* valence electrons. The arsenic atom will enter into the crystal structure, and will form bonds with four germanium atoms. There is, however, no place in the crystal structure for the fifth electron of the arsenic atom. This results in an imperfection in the crystal structure because of the excess negative electron. (See reference 22 for an illustration.) Furthermore, this electron is but loosely attached to the arsenic atom, and what binding force does exist is readily overcome, even at low temperatures. For germanium the energy required is about 0.04 electron volt.[17] This excess electron is essentially free to move through the space between the atoms of the crystal structure, and will conduct current if a difference of potential is applied. An atom such as arsenic (or antimony) that provides an excess of negative electrons is called a **donor**, or a **donor impurity**. A semiconductor containing a donor impurity and hence carrying current by virtue of excess negative electrons is known as a ***n*-type semiconductor**. The energy-level diagram for an *n*-type germanium semiconductor is shown in Fig. 1-17. The donor atoms, such as arsenic, have valence electrons at levels higher than the valence band for germanium because of the feeble bonds of the extra electrons in the donor atoms. Thus, as shown in Fig. 1-17, electrons have left the valence band of the donor, and are in the conduction band; holes accordingly are created in the valence band of the donor.

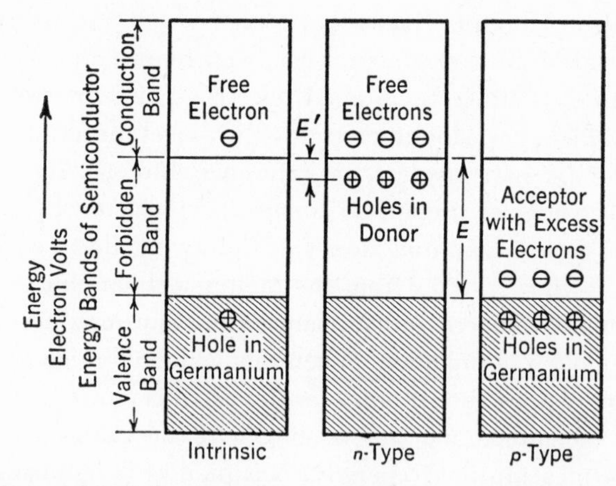

FIG. 1-17. Energy-level diagrams for germanium at temperatures above absolute zero, but below room temperature. In the intrinsic, or pure germanium, one electron has been driven by thermal excitation from the valence band to the conduction band. The arsenic of the *n*-type germanium has valence electrons near the conduction band, illustrated by comparing E' with E. Electrons have been driven by thermal excitation to the conduction band, leaving holes in the valence band of the donor arsenic. The gallium of the *p*-type germanium has a valence band with holes near the germanium valence band. Electrons have been taken from the germanium valence band, leaving holes in this band.

Acceptor Impurities. If a small amount of the element gallium is introduced into the germanium crystal, gallium atoms will take the place of some germanium atoms. Because the gallium atom has only three valence electrons, one bond, occasionally, will be incomplete (Fig. 1-14). This results in an imperfection in the crystal structure. (See reference 22.) The incomplete bond is termed a **hole**; it is in the acceptor valence band, which is at a slightly higher energy level than the germanium valence band. An electron from a neighboring germanium atom may enter the acceptor-atom hole, leaving a hole in the germanium atom the electron just left. This action may continue, and the hole may migrate about in the germanium. The migration is a step-by-step process from atom to atom. In a sense a hole behaves like a positive electron, and current can be carried by a semiconductor because of the presence of the holes. The acceptor atoms are pictured (Fig. 1-17) as having taken an electron from the germanium, leaving a hole in the otherwise filled valence band. At all except very low temperatures, electrons from germanium atoms will have sufficient energies to "fall up into" the holes in the gallium atoms, and hence transfer the holes, as shown, to the valence band of the germanium. An atom such as gallium (or aluminum) has one valence electron less than germanium. Accordingly gallium or aluminum provide holes for electrons, and such an element is called an **acceptor**, or an **acceptor impurity.**[17,18] A semiconductor containing an acceptor impurity, and accordingly conducting electricity because of the *positive* holes, is called a ***p*-type semiconductor.**

The conduction phenomena in the three types of semiconductors can be summarized [17,18,23] as follows. *Intrinsic* semiconductors will be insulators at a temperature of absolute zero; at room temperature there will be an equal number of electrons and holes, and conduction will be by both electrons and holes. In *n-type semiconductors* conduction will be by electrons at low temperatures, and by electrons and a few holes at room temperature; at very high temperatures conduction will be by approximately equal numbers of electrons and holes, and the impurity semiconductor will behave as an intrinsic semiconductor. In *p-type semiconductors* conduction will be by holes at low temperature, and by holes and a few electrons at room temperature; at very high temperatures conduction will be by approximately equal numbers of holes and electrons, and the impurity semiconductor will act like an intrinsic semiconductor.

1-24. *THE HALL EFFECT* [17,18]

If a thin rectangular strip of metal carrying a current in the direction of its length is subjected to a magnetic field normal to the surface of the strip, a difference of potential will appear between the two sides of the metal strip, as Fig. 1-18 indicates. Assuming that the strip is *n*-type germanium at room temperature, conduction will be by negative electrons, and the magnetic field will cause the electrons to deviate from a straight path through the metal strip. The electrons will be forced to one side of the strip, and this excess charge will establish a difference of potential between the edges of the flat strip. This will produce an electric field of magnitude and direction such as to cancel the effect of the magnetic field, and hence subsequent electrons will flow through the strip in "straight" lines.

Much information regarding semiconductors is obtained by a study of the Hall effect.[17,18] For instance, if conduction is largely by electrons, as in *n*-type material, one edge of the strip will be positive and the other will be negative. If, however, the conduction is largely by positive holes, the polarity will be reversed. The magnitude of this potential difference is [18]

$$V = \frac{10^{-8}RIH}{t} \qquad 1\text{-}16$$

where V is in volts, when I is the current in amperes, H the magnetic field in gausses, t the thickness of the sample in centimeters, and R the Hall coefficient in centimeters cubed per coulomb.

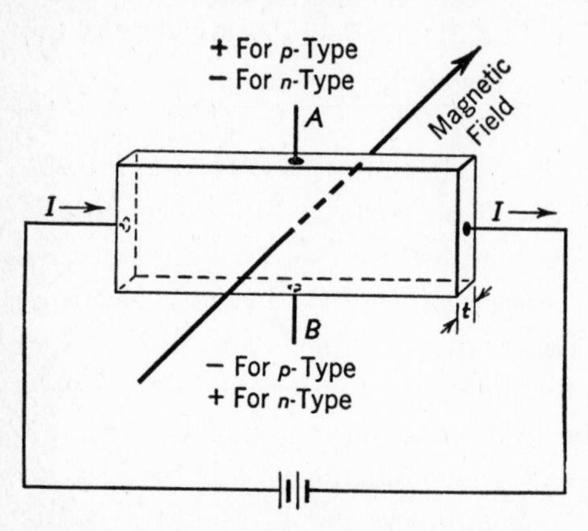

FIG. 1-18. For semiconductors, Hall-effect differences of potential of the polarities indicated will exist between terminals A and B connected to the edges of a strip of the material of thickness t. (From reference 18.)

The number of free negative electrons or of positive holes in a semiconductor can be found from measurements of the Hall effect from the relation [18]

$$n = \pm \frac{3\pi}{8Re} \qquad 1\text{-}17$$

where n is the number of electrons or holes per cubic centimeter, R is the Hall coefficient in centimeters cubed per coulomb, and e is the electron charge in coulombs and is negative for electrons and positive for holes. As an example of the use of this equation, the following data [18] are given: A sample of silicon contained 3 boron atoms per 10,000 silicon atoms giving (Table 1) a *p-type* impurity, or acceptor, semiconductor. From measurements, the Hall coefficient in equation 1-16 was +0.5 centimeter cubed per coulomb. From equation 1-17, this gives $n = 1.5 \times 10^{19}$ holes per cubic centimeter. There are 5.2×10^{22} silicon atoms per cubic centimeter.[18] There should, therefore, be $3 \times 5.2 \times 10^{22}/10^4$ or 1.56×10^{19} holes, assuming one hole per boron atom. As previously stated, experimentation gave 1.5×10^{19} holes.

The conductivity of an electrical conductor is [18]

$$\sigma = ne\mu \qquad 1\text{-}18$$

where n is the number of current carriers per cubic centimeter, e is the carrier charge in coulombs, and μ is their mobility in centimeters per second per volt per centimeter. Hence,

$$\mu = \frac{\sigma}{en} = \frac{8R\sigma}{3\pi} = \frac{0.85}{\rho} \qquad 1\text{-}19$$

where ρ is the resistivity in ohms per centimeter cube. The mobility of the holes in this sample of silicon is calculated to be 33 centimeters per second per volt per centimeter. The mean free path can be shown to be 2.3×10^{-7} centimeters.[18]

1-25. *CURRENT CONDUCTION BY ELECTRONS AND BY HOLES*

As has been mentioned, holes behave in many ways like positive electrons. From this, there is a tendency to assume that current-flow phenomena are similar. The phenomena are, in fact, different, as has been shown experimentally as follows. A filament of *n-type* germanium was arranged[17] as shown in Fig. 1-19, with point contacts x and y firmly touching the material.

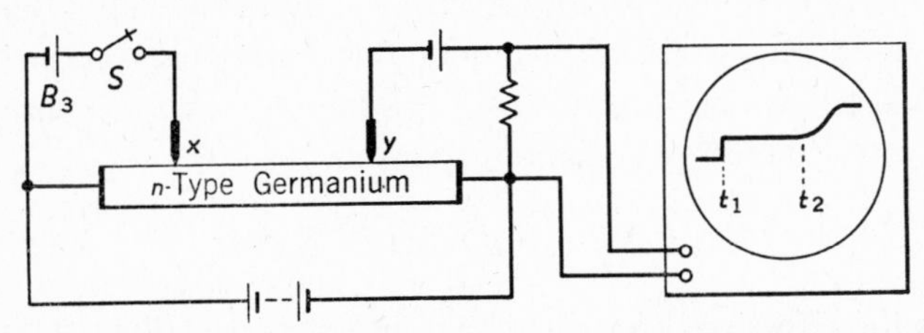

FIG. 1-19. If the switch is closed at time t_1, the voltage "immediately" rises at point y as indicated by the trace on the oscillograph at the right. This voltage rise is propagated as an electromagnetic-wave impulse with a speed approximating that of light. A second voltage rise occurs at a later time t_2 when the low-mobility holes, "injected" at point x, arrive at point y.

The end contacts merely carried current to and from the germanium and otherwise had no effect. Prior to closing switch S at time t_1, the magnitude of the voltage indicated by the oscillograph was as shown to the left of t_1. At the instant switch S was closed, the voltage at the oscillograph jumped up to the value shown between t_1 and t_2. This is what is expected from conventional electrical theory: When the switch is closed, electrons are withdrawn at point x from the germanium by the battery B_3 in series with this contact. This withdrawal of electrons causes a localized positive charge at point contact x. This establishes a positive electric field which is propagated toward contact y as an electromagnetic wave and travels with essentially the speed of light from point contact x to y. This does *not* mean that the electrons have traveled at the speed of light, however. At time t_2 after closing switch S, the voltage at the oscillograph again increased, and for the following reason: When electrons were withdrawn at point contact x, some positive holes were "injected" by removing some electrons from valence bonds. These holes are deep within the germanium, and migrate in a step-by-step manner with relatively low mobilities. The time interval between t_1 and t_2 is the time required for the holes to migrate from the point contact x to point contact y. Typical values of mobility in germanium are: for electrons 3600, and for holes 1700, centimeters per second per volt per centimeter.[24] Attention is again directed to the fact that in Fig. 1-19 electrons do *not* have to migrate to point y to "deliver" a voltage change; however, holes *must* migrate from x to y to cause a voltage change at point y. These phenomena will be treated in more detail in Chapter 10.

1-26. *THERMAL AGITATION IN CONDUCTORS*

The random velocities of the free electrons in a metallic conductor depend on temperature, increasing as the temperature is raised. Because the motion is random, at a given instant more electrons may be moving in one direction than in another. As a result of this motion, random noise-producing voltage variations exist between the ends of a conductor such as a piece of copper wire, or a wire-wound resistor, the magnitude being [25]

$$E = \sqrt{4KTR(f_2 - f_1)} \qquad \text{1-20}$$

where E is the effective value of voltage in volts, K is Boltzmann's constant in joules per degree T (section 1-9), T is the temperature in degrees Kelvin, R is the effective resistance of the conductor in ohms, and f_2 and f_1 are the frequency limits in cycles of the band over which the noise voltage is measured.

Such thermal-agitation noise voltages, if amplified, may reach bothersome levels in communication equipment. The phenomenon was early investigated [26] by Johnson, and is sometimes called the **Johnson effect**. The name **thermal noise** (and sometimes resistance noise) commonly is employed.

In general, carbon and composition-type resistors produce more random noise voltage than given by equation 1-20. Furthermore, the noise usually increases with increase in current flow, probably because of variations in the microphonic contacts comprising the current path through the carbon or other material.

REFERENCES

1. Institute of Radio Engineers. *Standards on Electron Tubes: Definitions of Terms,* 1950. Proc. I.R.E., April 1950, Vol. 38, No. 4.
2. Stranathan, J. D. *Elementary particles of physics.* Electronics, Aug. 1943, Vol. 16, No. 8.
3. *Origin of the word electron.* Jl. of the A.I.E.E., Oct. 1930, Vol. 49, No. 10.
4. American Institute of Electrical Engineers. *American Standard Definitions of Electrical Terms.* 1941.
5. Waterman, A. T. *Fundamental properties of the electron.* Electrical Engineering, Jan. 1934, Vol. 53, No. 1.

6. Boddie, C. A. *The motional mass of the electron.* Electrical Engineering, Jan. 1947, Vol. 66, No. 1.
7. Institute of Radio Engineers. *Standards on Abbreviations, Graphical Symbols, Letter Symbols, and Mathematical Signs,* 1948. *Standards on Abbreviations,* Proc. I.R.E., April 1951, Vol. 39, No. 4; *Standards on Graphical Symbols,* Proc. I.R.E., June 1954, Vol. 42, No. 6.
   American Institute of Electrical Engineers. *American Standard Graphical Symbols for Telephone, Telegraph, and Radio Use,* 1944.
   ASA Z32.5–1944; also, *American Standard Graphical Symbols for Electronic Devices,* 1944, ASA Z32.10–1944.
8. Compton, A. H. *The corpuscular properties of light.* Physical Review Supplement, Reviews of Modern Physics, July and Oct. 1929, Vol. 1.
9. Loeb, L. B. *The Nature of a Gas.* John Wiley & Sons.
10. Hull, A. W. *Fundamental electrical properties of mercury vapor and monatomic gases.* Electrical Engineering, Nov. 1934, Vol. 53, No. 11.
11. Crowther, J. A. *Ions, Electrons, and Ionizing Radiations.* Longmans, Green & Co.
12. Slepian, J., and Mason, R. C. *Electric discharges in gases.* Electrical Engineering, April 1934, Vol. 53, No. 4.
13. Darrow, K. K. *Electric discharges in gases.* Electrical Engineering, March 1934, Vol. 53, No. 3.
14. Wilson, A. H. *Semi-conductors and Metals.* Cambridge University Press.
15. Seitz, F. *Modern Theory of Solids.* McGraw-Hill Book Co.
16. Seitz, F., and Johnson, R. P. *Modern theory of solids.* Jl. of Applied Physics, Feb. and March 1937, Vol. 8, Nos. 2 and 3.
17. Shockley, W. *Electrons and Holes in Semiconductors.* D. Van Nostrand Co.
18. Pearson, G. L. *The physics of electronic semiconductors.* Trans. A.I.E.E., Vol. 66, 1947.
19. Shockley, W. *The quantum physics of solids.* Bell System Technical Journal, Vol. 18, Oct. 1939.
20. Herzfeld, K. F. *The present theory of electric conduction.* Electrical Engineering, April 1934, Vol. 53, No. 4.
21. Federal Telephone and Radio Corporation. *Reference Data for Radio Engineers.*
22. Jordan, J. P. *The ABC's of germanium.* Electrical Engineering, July 1952, Vol. 71, No. 7.
23. Shockley, W. *Transistor electronics, imperfections, unipolar and analog transistors.* Proc. I.R.E., Nov. 1952, Vol. 40, No. 11.
24. Shockley, W. *Solid state physics in electronics and metallurgy.* Trans. American Institute of Mining and Metallurgical Engineers. Vol. 194, pp. 829–843, 1952.
25. Merchant, C. J. *Thermal noise in a parallel R C circuit.* Electronics, July 1944, Vol. 17, No. 7.
26. Johnson, J. B. *Thermal agitation of electric charge in conductors.* Physical Review, July 1928, Vol. 32, No. 7.

QUESTIONS *

1. Why does the apparent mass of an electron increase with velocity?
2. Why does an electric field, but not a magnetic field, accelerate an electron?
3. How may an electron lose and gain kinetic energy? Potential energy?
4. Is it possible for an electron to travel in a straight line in a magnetic field? Explain.
5. Is it possible for a magnetic field to affect the velocity of an electron?
6. What may be the path of the electron of Fig. 1-3 if the field is not limitless?
7. An electron is caused to approach a negative electrode. What phenomena occur? What is meant by "negative electrode"?
8. What are the meanings of the terms quantum, photon, phonon?
9. What type of negative ion is most important in electronics?
10. What is meant by electron charge density? Is it caused by other than valence electrons?
11. What effect does gas pressure have on mean free path?
12. What relation exists between gas pressure and ionizing voltage?
13. What are monatomic gases, and why are they used in electron tubes?
14. What is the difference between excited state and metastable condition?
15. Why is ionization by natural causes important in electronics?
16. How are positive and negative ion sheaths formed?
17. What is the difference between self-maintained and non-self-maintained discharges?
18. What do the "humps" in the curves of Fig. 1-11 indicate?
19. If a specimen of germanium were made more and more pure, would its resistance at room temperature increase or decrease? Why?
20. Why are germanium and silicon used for electronic devices? What is the meaning of the rods between atoms in Fig. 1-14?
21. At absolute zero would intrinsic and impurity semiconductors have the same resistance? What would be the magnitude?
22. Is it possible to have germanium ions in a solid? Give an example if it is possible.
23. What are important uses of the Hall effect in electronics?
24. Why does the oscillograph trace of Fig. 1-19 flatten beyond t_2? What would be the shape of the curve if the switch were opened at the time corresponding to the end of the curve?
25. What is the fundamental differences between conduction by electrons and by holes?

PROBLEMS

1. Plot a curve for the mass of an electron at 0.1, 0.2, etc., the velocity of light.
2. The voltage in Fig. 1-1 is 950 volts, and the spacing is 1.25 centimeters. Calculate the force on the electron in both dynes and newtons. Calculate the acceleration in centimeters and meters per second per second. What effects have been neglected? Are they important?

* To encourage "outside reading," certain questions at the ends of the chapters may make advisable the consultation of references listed and other books on electronics.

3. The electron of problem 2 starts from rest and falls unimpeded through the entire 950 volts. What will be its final velocity in meters per second per second? What will be the kinetic energy when the electron strikes the positive electrode? What will become of this energy?
4. Repeat problem 3 for an electron shot upward through a hole in the lower plate with the velocity gained by freely falling through a difference of potential of 250 volts. Express the final velocity in miles per second.
5. If the spacing between two parallel-plane electrodes is 0.12 inch, and if the potential difference is 350 volts, calculate the time in seconds and microseconds that will be required for an electron, starting from rest, to cross from one electrode to the other.
6. An electron, having the velocity obtained by falling unimpeded through 775 volts, is shot, as in Fig. 1-2, at right angles into a magnetic field in vacuum of 2.0 webers per square meter. Calculate the radius of curvature. Calculate if the electron gun of Fig. 1-3 is "below" the page and pointing upward, making an angle of 60° with the plane of the page.
7. Mercury vapor is at 200°C. and the pressure is 1.3×10^{-3} millimeter of mercury. Calculate the mean free path of an electron in the vapor.
8. Calculate the thermal-agitation voltage between the terminals of a 100,000-ohm wire-wound resistor at 20°C. and for a band of frequencies of 5000 cycles.

CHAPTER 2

EMISSION OF ELECTRONS

Many electron devices operate on the principle of the control of electrons by electric and magnetic fields. The ratio of the charge on an electron to the mass of an electron is large. For this reason a field of force readily influences the motion of an electron, and the remarkable control features of electron tubes are the result.

Before such control is possible, however, electrons must be made available in the regions adjacent to the control elements producing the fields. Thus, in thermionic vacuum, and gas tubes, the electrons are emitted by a hot cathode; in phototubes they are released from a cathode by the energy of impinging electromagnetic waves such as visible light. This chapter will be devoted to a study of these and other methods of *releasing* electrons from solids and from mercury pools.

2-1. *FREE ELECTRONS IN METALS*

There are many free electrons in a metal at room temperature. They are free because at ordinary temperatures they have kinetic energies above the potential energy "humps" between the atoms of the metal (Figs. 1-11 and 1-15). These electrons are able to break free from their parent atom and are "jostled about" in the metal because of thermal agitation. The free electrons have random velocities covering a wide range of values. With no applied external field forces, the electron drift in one direction is offset by that in the opposite direction, so the net drift during a considerable time interval is zero.

A free electron well beneath the surface of a metal is surrounded by a symmetrical grouping of electric charges. Just below the surface of a metal the situation is different. From the atomic viewpoint, the outermost portion of a metal surface is a region having a negative electron charge density (Fig. 1-6). For an electron to escape from the metal into the space just beyond, theoretically it must have, at right angles to the surface, a component of velocity, and kinetic energy, sufficient to carry it through the negative **surface potential barrier**. This effect may be small, and often is neglected.

If an electron succeeds in passing through the surface, it will immediately experience a force attracting it back to the metal. The reason is as follows: The electron carries away from the metal a negative charge, leaving the metal equally positively charged, and a force of attraction results in accordance with equation 1-11. The situation is depicted in Fig. 2-1, where it is assumed that an electron from the metal at the left has had sufficient kinetic energy to escape through the surface barrier and to travel a distance x into a vacuum at the right. The electric field between the electron of charge $-Q_e$ and the equally positively charged metal is shown by the solid lines. The force of attraction is exactly as if the metal were removed and an **image charge** of the electron (that is, a charge of $+Q_e$) existed in evacuated space at a position $-x$ behind the position at which the surface was located. This image charge may also be thought of as a potential barrier. As the electron travels away from the metal its kinetic energy of motion decreases, and its potential energy of position increases in accordance with the curve at the right of Fig. 1-11. Hence, if the electron attempting to escape does not have sufficient energy, and is not influenced by other forces, it must come to rest, and then fall back into the metal.

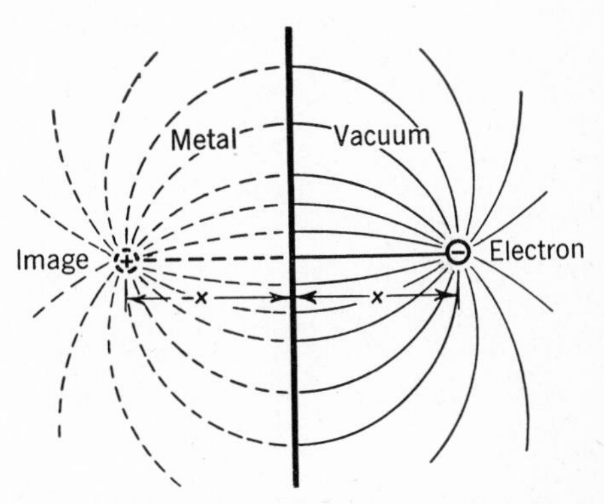

FIG. 2-1. The force on an electron which has just escaped from a metal surface into evacuated space is the same as if an "image charge" of equal magnitude and opposite sign were situated in evacuated space an equal distance behind the location of the surface.

2-2. *ELECTRON ESCAPE*

For an electron *by its own action* to escape completely from beneath a clean surface, the electron must have kinetic energy sufficient to overcome the negative potential energy barrier and to carry it through the surface; also the electron must possess sufficient energy to overcome the image force and to carry it to a region beyond the action of this force. (As will be dis-

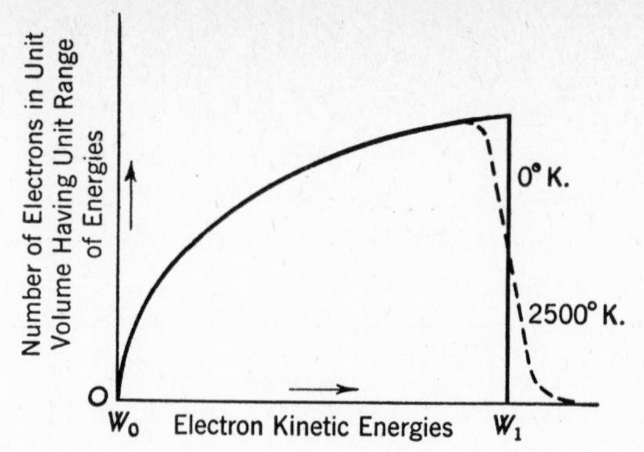

FIG. 2-2. At absolute zero (0°K.), no electrons have energies exceeding W_1, but at some temperature such as 2500°K., thermal agitation causes some electrons to have random velocities and kinetic energies exceeding W_1.

cussed later, this region need not be at infinity.) The free electrons in a metal are often considered to constitute an **electron gas** within the metal. The motions of the electrons of this "gas" are studied statistically by the methods [1,2,3] of Fermi and Dirac. The distribution of kinetic energy among the free electrons in a metal is considered to be as shown in Fig. 2-2.

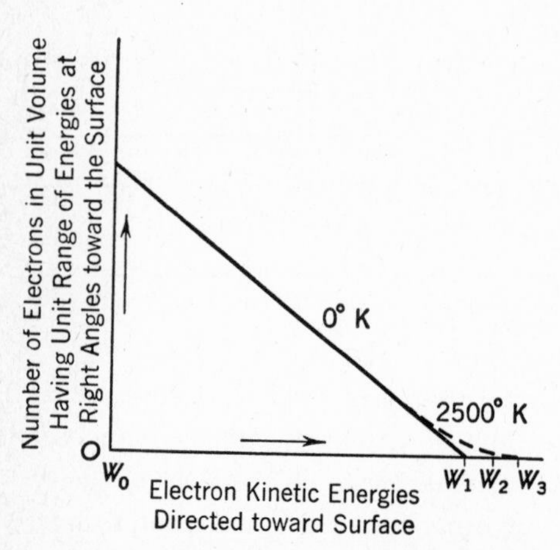

FIG. 2-3. At 0°K., many electrons have *small* kinetic energies (near W_o) directed toward the metal surface, but no electrons have surface-directed energies greater than W_1. However, at 2500°K., a few electrons will have greater surface-directed energies as indicated by W_2 and W_3.

Electron kinetic energies are plotted on the X axis, and the number of electrons in unit volume having unit range of energies are plotted on the Y axis. The total area enclosed in Fig. 2-2 represents the total number of electrons per unit volume. At 0°K., no electrons have kinetic energies above a value often designated W_1, but at some temperature such as 2500°K. (for tungsten) some electrons would have initial kinetic energies greater than the value W_1.

From the standpoint of electron escape, however, the important consideration is the distribution of kinetic energy *directed at right angles toward the boundary surface.* Only those electrons with sufficient velocity, and resulting kinetic energy, at right angles toward the surface, can pass through the barrier potentials. From statistical reasoning, Fig. 2-3 is developed.[1,2,3] This figure differs from Fig. 2-2 in that it considers the free electrons within the metal only from the standpoint of their kinetic energies directed toward the surface. Fig. 2-3 indicates that *many* electrons have small amounts of energy (near W_0) directed toward the surface, but that *no electrons* at 0°K. have energies greater than W_1. This figure shows also that only relatively *few electrons* at 2500°K. (for tungsten) have kinetic energies exceeding the value W_1, such as those with energies W_2.

The situation may now be considered with the aid of Figs. 1-11 and 1-15, and as shown in Fig. 2-4. The *exact* location of energy level W_0 with respect to the potential energy humps is uncertain (section 1-19). Figure 2-4 shows

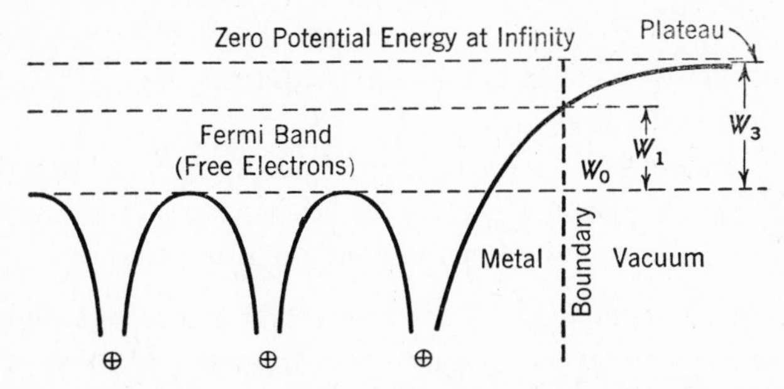

FIG. 2-4. For an electron to escape, unaided, from a metal surface into a vacuum, it must have kinetic energy directed toward the surface to carry it through the surface and to the "plateau" indicated by W_3.

that at 0°K. no electrons can escape from the metal because none has kinetic energy exceeding the value W_1 necessary just to penetrate the surface forces. Thus, an insulated metal body at 0°K. would remain uncharged. If, however, the metal body is at some temperature such as 2500°K. (for tungsten), some electrons will have energies directed perpendicular to the surface as great as W_3, and will leave the surface, and will not return. Thus, as has been known for many years, a sufficiently hot insulated metal body will lose negative electricity. Those electrons of a metal that have energies between W_0 and W_1 of Fig. 2-2 are said to have energies in the **Fermi band.** These are the free electrons, and these are in the Fermi band of Fig. 2-3.

2-3. *WORK FUNCTION*

It will be noted in Fig. 2-4 that the energy W_3 need not carry the electron to infinity to cause escape. Because the phenomenon is so involved and so many factors need consideration, the application of equation 1-12 is approximate, and it is assumed that energy W_3 need carry the escaping electron only to the potential energy "plateau" where the attraction of the metal body for the escaping electron is negligible. The **surface work** that must be expended is

$$W_e = W_3 - W_1 \qquad \text{or} \qquad W_e = Ee = \phi e \tag{2-1}$$

where W_e is the amount of energy that must be added to cause electron escape. For an electron, this energy would be Ee, or ϕe as it is commonly written, and the unit of measure is the electron volt. Since the electric charge e is the same for all electrons, ϕ is often used to designate the energy of escape. The unit is, of course, the electron volt, although it often is stated in volts, electron volt being implied. Thus, the **surface work** function, or merely **work function**, is the smallest amount of energy that must be added to permit the ejection, or escape, of an electron at 0°K. This indicates that the work function of a metal should vary with temperature, but the variation found experimentally is small.[2]

In practical electron devices the energy necessary for escape through the potential barrier and image force may be imparted to the electrons, and they may be liberated, in any one of four ways, as follows:

1. *By photoelectric emission.* If a beam of light is directed onto the surface of certain metals, energy is absorbed from the photons of light, and electrons are emitted from the surface.
2. *By thermionic emission.* If a suitable metal is sufficiently heated, energy will be absorbed by the electrons and they will escape.
3. *By secondary emission.* If a beam of high-velocity electrons is directed against a metal, energy from the electrons constituting the beam may be absorbed by electrons within the metal, and so-called secondary electrons may escape through the metallic surface.
4. *By field emission* (cold cathode or autoelectronic emission). If a metallic surface is subjected to a very intense electric field, electrons may escape through the metallic surface, even at room temperature or below.

2-4. *PHOTOELECTRIC EMISSION*

The photoelectric emission of electrons occurs when radiant energy, in the form of electromagnetic waves, strikes the surface of a metal. These electromagnetic waves may be, and in fact often are, the visible radiations of ordinary light. Although the effect is most pronounced with metals, there is some emission from insulators.[4]

The discovery of the photoelectric effect is attributed to Hertz and Hallwachs. While studying spark discharges to check Maxwell's theoretical prediction of the existence of electromagnetic waves, Hertz observed that a spark gap broke down more readily when it was subjected to ultraviolet radiations. He showed that the radiations must fall on the metal gap terminals (and not merely on the space between), that the metal should be clean and smooth, and that directing the radiant energy onto the negative spark-gap terminal produced the largest effect.[5]

This discovery by Hertz was studied further by Hallwachs. His experiments were performed in air. Elster and Geitel are credited with developing the predecessor of the modern phototube consisting of the photoactive (or light-sensitive) material in an *evacuated* glass bulb. They found that the current output of a phototube is proportional to the intensity of the illumination (or radiation for invisible light).

Theories of Photoelectric Emission. The exact nature of the mechanism by which electrons are ejected from a metal when it is exposed to radiation of the correct wave length is not definitely known. For some time it was thought that the free electrons of a metal merely absorbed and accumulated energy from the oncoming radiation until these electrons had energies sufficiently high for them to escape from the metal.

It was next thought that the photoelectric current came from the more stable of the orbital electrons instead of from the so-called free electrons. It was assumed that the quanta of photons of energy of the radiation "reached" into the atoms and by some triggerlike action imparted energy to, and released, certain of the more stable orbital electrons.

At present, it is believed that the photoelectric effect is largely a surface phenomenon, and that the radiant energy, if of the proper wave length, is able to impart energy to the free electrons near and at the surface, and to permit their escape. The electrons constituting the photoelectric current seem to be of the same fundamental type that are released thermionically in ordinary electron tubes.

Laws of Photoelectric Emission. Although the exact nature of photoemission is obscure (as are the *exact* natures of most fundamental phenomena), the laws are definitely known. These laws will now be discussed.

First, photoelectric emission is almost entirely independent of the temperature of the metallic surface, at least over wide limits about room temperature.[6] Thus it does *not* appear probable that in photoemission the free electrons merely absorb and gradually accumulate energies until they are finally able to escape through the restraining forces at the metallic surface.

Second, the number of **photoelectrons** (electrons released by radiant energy) emitted varies directly as the intensity of the light (or radiation) striking the photosensitive surface. This statement, as for all others regarding photoelectric emission, is strictly true only when spurious effects are eliminated and when only *pure* photoemission is being studied. Such statements hold closely for commercial phototubes.

Third, the initial velocity of the photoelectrons released from a metal is independent of the intensity of the light falling on the surface. A feeble beam of light from a distant star will release electrons with the same velocity as will a powerful light beam of the same wave length. (But of course the light from the distant star will release electrons at a lower rate.) As one author stated,[6] "the ejection of a photoelectron is, so to speak, a private affair between the electron and the quantum concerned in ejecting it."

Fourth, the velocity of the ejected photoelectrons depends on the frequency (or wave length) of the light or other radiation releasing it. If the frequency is sufficiently high to release electrons, then increasing the frequency higher will cause the electrons to be ejected with higher initial velocities.

2-5. *EINSTEIN'S EQUATION FOR PHOTOELECTRIC EMISSION*

In 1905 Einstein proposed that the photoemissive process could be represented mathematically by the equation

$$\frac{1}{2}mv^2 = hf - \phi e \qquad \text{or} \qquad v = \sqrt{\frac{2(hf - \phi e)}{m}} \tag{2-2}$$

According to this principle, an electron near the surface of the metal can escape only if the radiant energy hf that the electron absorbs, equals or exceeds the **photoelectric surface work** ϕe, which represents the energy that an electron must have to pass through the restraining negative potential at the metal surface. If the radiant energy hf that an electron absorbs just equals ϕe, the electron just escapes and with zero velocity; ϕ is the **photoelectric surface work function.** If, however, an electron absorbs more energy than it needs to escape, it has excess energy and will escape with a velocity v as indicated by equation 2-2.

From this explanation, and from equation 2-2, it might be inferred that for a radiation of given wave length and frequency all photoelectrons are ejected from a metal with the *same* velocity, since hf is constant for that particular radiation, and ϕe is a constant for a given metal. This is not true, however, because all electrons do not have the same *initial* energy when they absorb the quantum hf. At ordinary temperatures some of the free electrons have initial velocities with a resultant component directing them toward the surface, and tending to overcome in part the surface forces. Such electrons, when they absorb quanta of radiant energy and escape, will have excess energies and greater velocities than electrons that do not have initial components toward the surface. This effect probably is small.

Another, and probably more important, factor that causes photoelectrons to be ejected with various velocities is that collisions are encountered in different ways as the various electrons escape. Thus, even if they had no initial energies, and if they each absorbed from the radiation the same amount of energy, they would have different energies and velocities after leaving the metal surface.

Equation 2-2 gives, therefore, the relations for the *maximum* velocity with which a photoelectron is ejected. The remainder of the electrons have velocities between this value and zero. The value of hf that will just enable an electron to escape with zero velocity gives the lowest frequency and longest wave length that can possibly cause photoemission from a given surface. Note again that the *total number* of electrons emitted depends on the light intensity.

These critical values of frequency and wave length that are just sufficient to cause photoemission are known as the **threshold frequency** and **threshold wave length** of a surface. From equation 2-2 the relations are

$$f_0 = \frac{\phi e}{h} \qquad \text{2-3}$$

where f_0 is the threshold frequency in cycles per second, and the other units are as listed under equation 2-2. The threshold wave length λ_0 in centimeters equals the velocity of light, 3×10^{10} centimeters per second, divided by the frequency in cycles per second. Thus, $f_0 = c/\lambda_0$, and equation 2-3 becomes

$$\lambda_0 = \frac{hc}{\phi e} \quad \text{or} \quad \phi = \frac{hc}{\lambda_0 e} \qquad \text{2-4}$$

the units being as previously listed under equation 2-2.

If the constants are inserted in equation 2-4, then the photoelectric surface work function becomes approximately

$$\phi = \frac{1235}{\lambda_0} \qquad \text{2-5}$$

where ϕ is in volts, and λ_0 is expressed in **millimicrons**. One millimicron is 10^{-9} meter. Wave lengths of light or other radiation are also measured in **angstrom units**, one angstrom equaling 10^{-8} centimeter, or 10^{-10} meter. The millimicron is the larger unit, equaling 10 angstrom units. Values of the threshold frequency and surface work function for a number of metals are given in Table 4.

Table 4

VALUES OF THE LONGEST WAVE LENGTH λ_0 AT WHICH PHOTOELECTRIC EMISSION OCCURS FROM CERTAIN ELEMENTS, AND CORRESPONDING VALUES OF THE PHOTOELECTRIC WORK FUNCTION ϕ

(*Data from reference 7*)

ELEMENT	LONGEST WAVE LENGTH λ_0 IN MILLIMICRONS ($m\mu$)	SURFACE WORK FUNCTION ϕ IN VOLTS
Platinum	196	6.30
Nickel	246	5.01
Silver	261	4.73
Tungsten	265	4.58
Sodium	500 *	1.90 to 2.46
Lithium	540 *	2.1 to 2.9
Potassium	550 *	1.76 to 2.25
Rubidium	570 *	1.8 to 2.2
Cesium	660 *	1.9

* Thin films.

2-6. *CHARACTERISTICS OF PHOTOELECTRIC EMITTERS*

The surface work ϕe that an electron must do in escaping from a photosensitive surface is a very important characteristic of that surface. Since the charge e is the same for all electrons, ϕ is in reality the factor that specifies the sensitivity characteristic of a surface. The value of ϕ can be determined by the **stopping-potential method** [7] illustrated by Fig. 2-5. The photosensitive surface of the *positive* plate P is illuminated by light passing through the *negative* grid G. Of the photoelectrons ejected from plate P, some will have maximum velocities, some will have minimum velocities, and the bulk of the electrons will have velocities in between these extremes. An electron emitted from the plate is immediately repelled by the negative grid. The difference of potential between the plate and grid is represented by E. If the kinetic energy $mv^2/2$ with which it is ejected is greater than the energy Ee that it would lose in passing to the negative grid, then it is possible for the ejected photoelectron to strike the grid; otherwise, it is not. The negative grid voltage that is *just sufficient* to prevent all electrons from reaching the grid is known as the **stopping poten-**

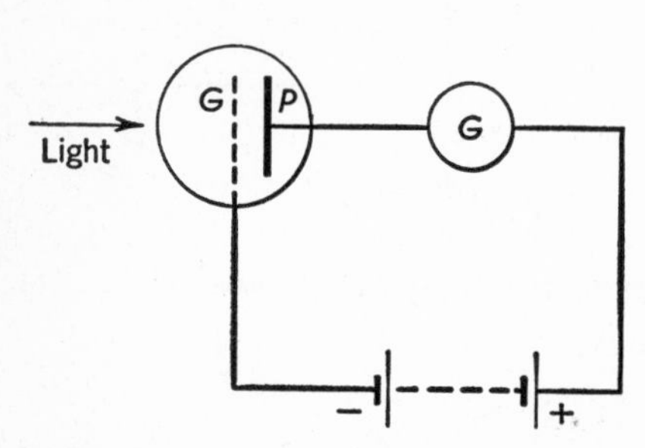

FIG. 2-5. Circuit for studying photoelectron stopping potential.

tial. The velocity of the *fastest* electrons ejected by a given frequency and wave length can be found from the equation

$$v = \sqrt{\frac{2Ee}{m}} \qquad 2\text{-}6$$

The velocity v will be in centimeters per second when the voltage E is in statvolts, the electronic charge e is in statcoulombs, and the mass m of the electron is in grams.

The work function ϕ for a photosensitive surface can be calculated from equation 2-3 or 2-4 if either the threshold frequency or threshold wave length is known. The accurate measurement of either of these requires certain refinements; in particular, the condition of the surface is important. The stopping potentials of the ejected electrons are determined for a given metal at several different wave lengths of monochromatic light (that is, light of one wave length). When these values are corrected for errors and plotted as ordinates with the values of frequency as abscissas, a straight line intercepting the X axis at the lowest frequency at which emission occurs is obtained.[7] Knowing this value, the work function ϕ can be calculated as mentioned early in this paragraph. It is usually expressed in volts as in Table 4.

Color Sensitivity. A second and very important factor to consider is that a photosensitive surface is responsive to light of a definite frequency and wave length as indicated by equations 2-3 and 2-4. The relative response of a surface (or of a phototube) to radiation of various wave lengths is called the **color sensitivity** or **spectral sensitivity**, because, of course, it is the wave length (or wave lengths) present that determine the color characteristics of light. The color sensitivity curves for the so-called alkali metals that are often used for photosensitive surfaces are shown in Fig. 2-6. These curves bring out two important points: *First,* the wave length of maximum sensitivity increases in the same order as the atomic number of the elements increases (Table 1); and *second,* the relative number of emitted photoelectrons, and hence the magnitude of the phototube current, decreases as the atomic number increases.

Selective Photoelectric Effect. A third characteristic usually of minor importance is as follows: If curves for the metals of Fig. 2-6 are taken with polarized light instead of ordinary unresolved light, it is found that the spectral sensitivity curves are changed in part. For certain types of polarized light, and for certain angles of incidence, the photoelectric emission is more pronounced. Apparently, a selective effect, determined by the atomic structure, exists.[6,7]

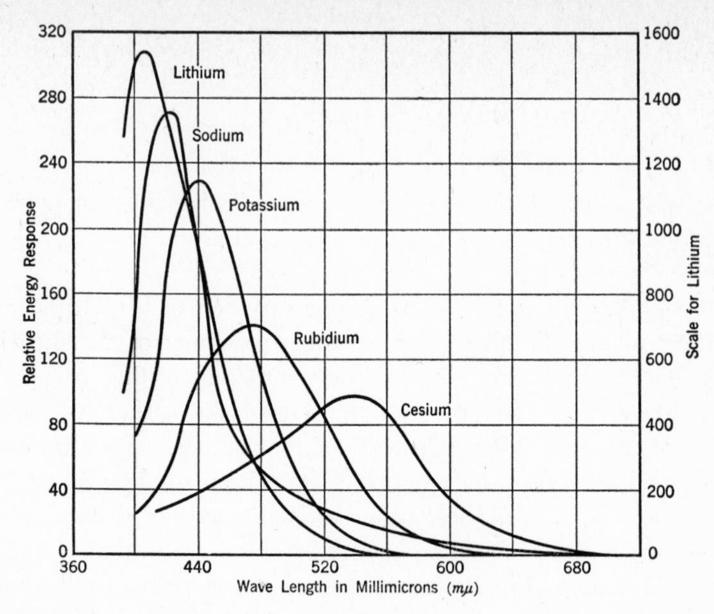

FIG. 2-6. Color sensitivity curves for certain metals. (Data by Miss Seiler, *Astrophysical Journal,* Vol. 52, 1920.)

Photoemission from Composite Films. Thus far, the discussions have been limited to photoemission from the surface of a solid metallic electrode, rather than from thin films of a given metal deposited on some other material. The emission curves of the various elements are greatly modified [4,7] when they are studied in layers only a few atoms thick. Composite films made up of several elements have been developed and are far superior for many purposes either to electrodes of a single metal, or to thin films of a single element. Thus, the cesium-oxygen-silver surface early replaced the pure-metal type of cathode in phototubes for general use with ordinary light.

2-7. *THERMIONIC EMISSION* [8,9,10,11]

In emission of this type the energy required for liberating the electrons from the metal comes from the energy source used to heat the metal; that is, the emission is thermal in nature. Thermionic effects date back at least to 1725 when DuFay discovered that the *space* in the vicinity of a red-hot body is a conductor of electricity. Little additional theoretical information on thermionics was obtained until 1887 when Elster and Geitel found that an electric charge could be made to pass from a hot body through a vacuum to another body in its vicinity. The phenomenon of thermionic emission was discovered by Edison in 1883.

In discussing the theory of emission it is generally assumed that the free electrons in a metal move in random directions and with various velocities. The higher the temperature of the metal, the larger will be the percentage of rapidly moving electrons. If these have sufficient velocity directed at right angles toward the surface, they may escape through the retarding forces at the surface of the metal. For an electron to escape from the metal and into the region (for instance, a vacuum) beyond, the kinetic energy of the electron because of its motion must equal (or exceed) the work required to overcome the retarding surface forces. That is,

$$\frac{1}{2}mv^2 = \phi e \qquad 2\text{-}7$$

where m is the mass of the electron, v is the velocity, ϕe is the **thermionic surface work,** and ϕ is the **thermionic work function** of the metal.

Just as heat energy is required to evaporate a liquid, heat energy is required to "evaporate" electrons, and the emission of electrons causes cooling of the metal. Davisson and Germer measured the energy input to a filament when the plate was *negative* (no current left the filament) and when it was *positive,* and determined the heat required for electron evaporation and the work function.[4] The numerical values for the photoelectric work function and the thermionic work function should be the same for a given metal. Of course, variations in samples, test methods, and measurement errors prevent these data from being always in agreement. A comparison between the work functions listed in Tables 4 and 6 is of interest.

Richardson's Theoretical Emission Equation. In 1903 Richardson developed theoretically an equation for thermionic emission. This equation is

$$I_s = aT^{1/2}\epsilon^{-b/T} \qquad 2\text{-}8$$

In this I_s is the **saturation current,** the maximum electron current that an emitting surface can supply per unit area at temperature T. It is called the saturation current for the following reason: Assume that the tube contains only the emitting cathode and an electron-collecting anode. If the anode is sufficiently positive, it will take *all* the electrons that are emitted from the cathode. Under these conditions the tube is said to be saturated.

Values of a and b and typical operating temperatures are given in Table 5. Using these values in equation 2-8, and with the operating temperature T in degrees Kelvin, I_s will be the maximum current in amperes per square centimeter of filament area that can be obtained under saturation conditions as previously explained. The value of ϵ is 2.718.

Table 5

VALUES OF CONSTANTS FOR EQUATION 2-8

(*Data from references 11 and 12*)

ELEMENT	a	b	T, DEGREES K.
Calcium	1.74×10^4	36,500	
Carbon	2.37×10^6	48,700	2000
Molybdenum	2.1×10^7	50,000	2000
Nickel	4.61×10^6	34,000	
Platinum	1.195×10^7	49,300	1600
Tantalum	4.3×10^2	44,200	2000
Tungsten	1.05×10^7	53,000	2000

It has been pointed out [8] that since the early electron theory has been modified, as a *theoretical* relation equation 2-8 is not correct. However, as an *empirical* formula, equation 2-8 is satisfactory. Richardson's early equation probably represents the first attempt to express electronic phenomena *quantitatively*.

Accepted Emission Equation. Richardson showed that the equation

$$I_s = AT^2\epsilon^{-b/T} \qquad 2\text{-}9$$

would be in accordance with observed emission data. M. v. Laue, in 1918, and Dushman in 1922, independently derived this equation.[9] Equation 2-9 sometimes is referred to as Dushman's equation, but more generally as Richardson's equation, and is usually employed. Although equation 2-9 may be derived by theoretical means,[1,9] assumptions must be made. Based on elec-

Table 6

VALUES OF THE CONSTANTS FOR EQUATION 2-9

(*Data from references 9, 11, and 12*)

ELEMENT	A	b	ϕ VOLTS	I_s AMP. PER SQ. CM.	T DEGREES K.
Calcium	60.2	26,000	2.24	4×10^{-3}	1100
Carbon	5.93	45,700	3.93	2.84×10^{-3}	2000
Cesium	162	21,000	1.81	2.5×10^{-11}	500
Molybdenum	60.2	50,900	4.38	2.34×10^{-3}	2000
Nickel	26.8	32,100	2.77		
Platinum	60.2	59,000	6.27	9.2×10^{-10}	1600
Tantalum	60.2	47,200	4.07	1.38×10^{-2}	2000
Thorium	60.2	38,900	3.35	4.3×10^{-3}	1600
Tungsten	60.2	52,400	4.52	1×10^{-3}	2000

tron theory, $A = 120.4$, yet experimentally $A = 60.2$ for most pure metals. From an engineering viewpoint equation 2-9 is regarded as an empirical equation, with the values for A and b supplied by experiment. Values for A and b for the temperatures indicated are given in Table 6. Note that a and b of Table 5 and A and b of Table 6 are different, and are used in different equations.

The value of I_s of equation 2-9 is in amperes per square centimeter of filament or cathode emitting area when A and b are as listed in Table 6, T is the operating temperature in degrees Kelvin, and ϵ equals 2.718. Typical values of saturation current I_s at operating temperatures T are listed. Values of the thermionic surface work function ϕ also are given. The values of ϕ were computed from the theoretical relation [9]

$$\phi = \frac{Kb}{e} \tag{2-10}$$

where ϕ will be in volts, when K is the Boltzmann constant (section 1-9), and e is the charge on the electron.

2-8. *DETERMINATION OF EMISSION CONSTANTS*

Early investigators believed that electron emission could not occur in a perfect vacuum. Langmuir is credited [9] with disproving this, and with establishing important laws of thermionic emission. In 1913, he found [9] that the electron current passing from a heated electron-emitting filament or **cathode** to a positive electron-collecting plate or **anode** in a vacuum depended, *first,* on the temperature of the cathode, and *second,* on the voltage between cathode and anode. Only the first of these phenomena will be considered now; the second will be treated in section 3-5. Langmuir found that if the voltage on the plate or anode were sufficiently high, so that all the emitted electrons were pulled over to the positive plate, then the *saturation current,* or maximum current of electrons from the cathode to the anode, and back to the cathode through the external wire circuit, was determined for a given cathode material by the *area* of the cathode-emitting surface and the *temperature* of the cathode.

Considerable care must be employed in determining emission characteristics. Adsorbed and absorbed gases must be removed from the emitter, and a very high vacuum must be maintained. Also, the temperature of the surface must be accurately known, and a correction must be made for supporting leads. These conduct away heat, causing a lower temperature near the ends of a filament.

For determining A and b of equation 2-9, the *saturation* currents are determined for a given specimen at various operating temperatures. These data are plotted with $\log_{10}(I_s/T^2)$ on the Y axis and $(1/T)$ on the X axis as in Fig. 2-7. This is done because the logarithm of equation 2-9 is

$$\log_{10}\frac{I_s}{T^2} = \log_{10} A - 0.434b\frac{1}{T}$$

and when the experimentally determined emission data are plotted in this way, the result (Fig. 2-7) is a straight line.[4,9] Furthermore, the slope of the

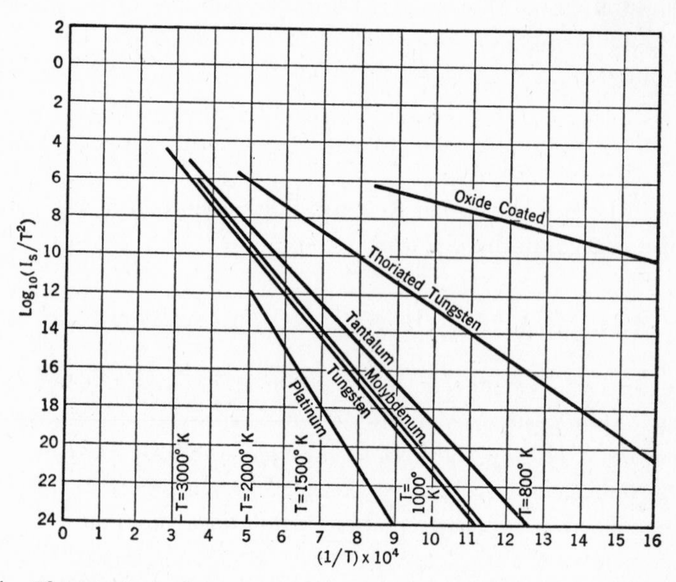

FIG. 2-7. Thermionic emission curves for oxide-coated filament (barium and strontium oxides on a base of platinum with 5 per cent nickel) and for several pure metals. These curves were determined experimentally. The fact that data give straight lines when plotted to these co-ordinates indicates that equation 2-9 accurately represents thermionic emission. The fact that the curves for pure metals intercept the Y axis at about the same point indicates that for pure metals the experimental value of A is approximately the same. (Data from reference 9.)

curve will be $-0.434b$, and the intercept on the Y axis will be $\log A$. From these values, b and A for equation 2-9 can be found. Typical values are listed in Table 6. As an example of the use of Fig. 2-7, the slope of the curve for tantalum is numerically $[(24.0 - 6.7)/(12.7 - 4.0)] \times 10^4 = 19{,}900$. Since this value equals numerically $0.434b$, the value of b will be $19{,}900/0.434 = 45{,}800$. The intercept on the Y axis is 1.8, and this equals $\log A$; then $A = 63$ approximately. Checking with Table 6 gives $b = 47{,}200$, and $A = 60.2$.

2-9. *THE SCHOTTKY EFFECT*

In determining saturation current the electron-collecting anode is made sufficiently positive to collect *all* the emitted electrons. This potential establishes an electric field between the cathode and anode; it is this field of force that causes the emitted electrons to travel to the anode. This electric field usually is spoken of as terminating *on* the surface of the cathode. Considering atomic dimensions, the surface of an emitting filament is difficult to define and would be irregular. Also some of the lines of force of the externally applied field might terminate on electrons that were just emerging from the heated cathode.

The applied electric field appears to overcome, in part, the electric image forces (section 2-1). When an electron has the required energy and leaves a metal, it takes a definite amount of negative electricity away and leaves the metal positive. This immediately produces an attractive force at the surface of the metal tending to cause the electron to return. The effect is as though an equal positive charge or electric image were located an equal distance behind the metallic surface. The voltage applied between the heated emitting cathode and the collecting anode, and the field it produces, act in a direction opposite to that of the field of the image force, and reduce the energy that an electron must have to leave the immediate vicinity of the surface and become completely divorced from the cathode. This increases the *apparent* saturation current for a given temperature. Thus the effect of the electron-collecting voltage applied between cathode and anode is the same as a small reduction in the surface work function. The phenomenon is illustrated in Fig. 2-8. For the voltages commonly applied in small vacuum tubes this effect is small.

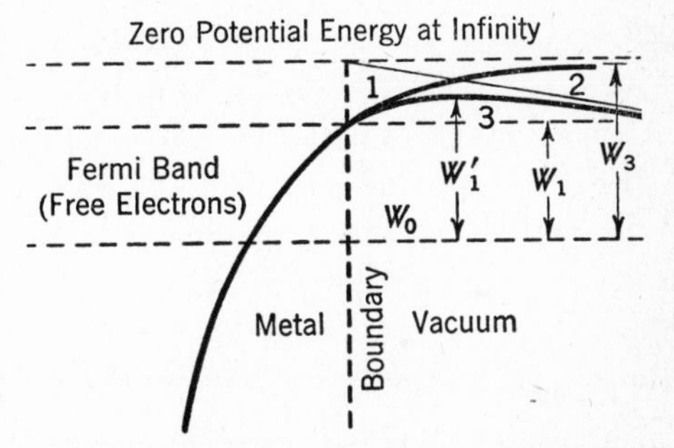

FIG. 2-8. When an electron escapes into evacuated space from a metal, such as a cathode, curve 1 and W_3 show the energy, directed toward the surface, that is required. However, if an external electric field exists between the cathode and anode, and if the field is directed (curve 2) so as to assist the electron, then the energy required for escape is shown by curve 3 and W_1'. Since the external field counteracts in part the image force, it in effect reduces the surface work function.

The magnitude of the effect of the externally applied electric field on the thermionic emission from a cathode was determined by Schottky and is given by the relation [9]

$$I = I_s \epsilon^{4.39\sqrt{E}/T} \qquad \text{(approx.)} \tag{2-11}$$

where I represents the apparent saturation current at any field strength E (in volts per centimeter at the surface of the filament), I_s is the true saturation current (the value given by the emission equation 2-9), and T is the temperature in degrees Kelvin. Thus, when emission data are being taken it is necessary to correct these data to obtain the true emission current.[4,9]

2-10. *TEMPERATURE DETERMINATION*

In studying thermionic emission it is necessary to know accurately the cathode temperature for use in equation 2-9. An indirect method usually is employed to determine cathode temperature, as follows: If a cathode having a resistance of R ohms carries a current of I amperes, the power input is I^2R. This will cause the temperature of the cathode to increase until equilibrium is reached and the heat generated equals the heat lost. Because the cathode often is a slender filament, little heat is conducted away by the leads, and because of the vacuum, negligible heat convection is possible. Most of the energy loss is by radiation, largely in the infrared.

The energy radiated in unit time from a filament depends on two factors: *first,* the type and nature of the radiating surface, and *second,* the temperature. These relations are expressed by the Stefan-Boltzmann fourth-power law

$$P = \sigma e_t T^4 \tag{2-12}$$

where P is the power (energy per second) radiated in ergs per square centimeter per second, T is the temperature in degrees Kelvin, and σ is the Stefan-Boltzmann constant of about 5.7×10^{-8} watt per square meter per second per degree. The value of e_t is the total **radiation emissivity,** and for pure tungsten is 0.260 at 2000°K.

Referring to equation 2-12, if the radiation emissivity e_t, the Stefan-Boltzmann constant σ, and the temperature T are known, then the power radiated per square meter can be determined, and if the area is known, the total power input can be calculated. Conversely, if the power input is measured, and all factors but the temperature are known, then the operating temperature T can be computed.[4] The entire cathode in a tube may not operate at the same temperature but may be slightly cooler where heat is conducted away by the lead wires. The resistance of a filament is less near the end portions, and the voltage drop along a filament is not uniform. Hence, the brightness along a filament and the emission of electrons are not uniform.

2-11. *CATHODE MATERIALS*

The filaments of the tubes used by the early investigators were made of metals such as platinum, and of carbon. Wehnelt, in 1905, investigated emission from metallic oxides, such as those of barium, calcium, and strontium coated on a platinum wire, and found these to be excellent emitters. In 1913 Langmuir and Rogers discovered [9] that if a tungsten wire containing 1 or 2 per cent of thorium oxide were properly heat treated, the emission would greatly exceed that of pure tungsten. Cathodes used as a source of electrons in thermionic tubes are of three types: *first,* pure metals, usually tungsten; *second,* oxide-coated metals; and *third,* thoriated tungsten. They will be treated in the order of historical development.

Tungsten Filaments.[13] These are used extensively for the source of electrons in large high-voltage high-vacuum tubes. This includes the types used for large radio amplifiers, for high-voltage rectification, and for x-ray tubes. One reason is that tungsten has no active surface layer (such as an oxide coating) that would be damaged by positive ion bombardment. Also, the filament will withstand high overloads and other misuses. However, tungsten is not so efficient an electron emitter as are the other cathodes and is being supplanted for many purposes by thoriated-tungsten filament.

The constants of tungsten are given in Table 6. Although the work function of tungsten is comparatively high (4.52), the melting point is also high (about 3600°K.); therefore it can be operated at the high temperature of about 2500°K., thus providing copious electron emission. During operation the tungsten slowly evaporates, reducing the size of the filament and eventually lowering the emission.

Oxide-Coated Cathodes. The oxide-coated cathode widely used in thermionic electron tubes consists essentially of a layer of barium and strontium oxides on the surface of a metal wire, or on the surface of a metal cylindrical sleeve, or on a metal structure of other shape. Oxide-coated filaments first were used commercially about 1913 in amplifiers for long telephone lines. Credit is due Arnold and his associates for developing these filaments for practical application. Platinum, tungsten, molybdenum, and nickel are among the materials that have been used for the wire core of filament-type oxide-coated cathodes. Nickel and nickel alloys commonly are used.[14,15] Some [16] have claimed that the type of core material had little or no effect on electron emission; it now is believed that the core material has important effects on emission. [4,15,16,17,18] Also, an interface layer that may form [16] between the oxide layer and the core material may affect emission.

Several processes have been developed for depositing material for the emitting layer on the core.[4,9] One method has been to draw a wire through a suitable paste, and then bake the paste at low temperature. In another method, the base metal is dipped repeatedly into the proper solutions. A common method is to spray the metal cathode structure with suitable mixtures,[14] such as $Ba \cdot Sr \cdot CO_3$, to which a small percentage of $CaCO_3$ is sometimes added.

The cathode must be activated, or formed, after the tube is assembled. The details of this process vary, one method being to heat the cathode in the evacuated tube for several minutes at a temperature well above the operating range. Then, the cathode is operated at a temperature near normal for a longer period with a voltage applied between anode, or plate, and cathode. A typical operating temperature is 900°K. when the tube is in use. After activation, the structure within the tube may be pictured as in Fig. 2-9. A very thin, perhaps monatomic, layer of barium and strontium atoms sometimes is considered to have been formed during activation on the oxide surface.[1,14,15] The emission properties have been attributed to the barium, the strontium being added to improve the mechanical characteristics.

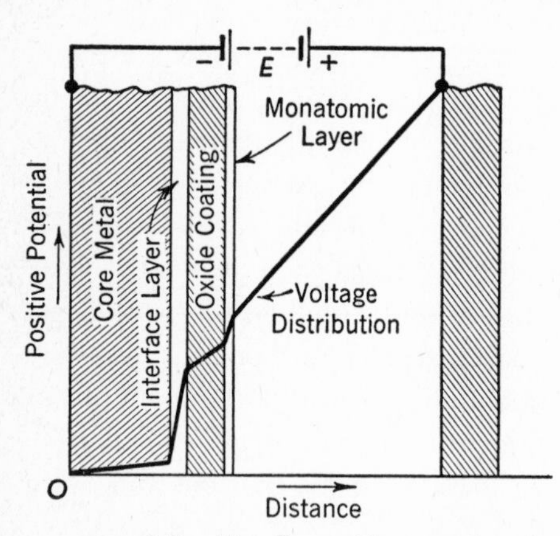

FIG. 2-9. For the oxide-coated cathode, it is probable that an interface layer and a monatomic surface layer exist as indicated. When a voltage E is applied between the cathode at the left and anode at the right, the voltage distribution will be somewhat as shown. (Adapted from reference 14.)

Many theories have been advanced to explain electron emission from oxide-coated cathodes, but because of the complexities no entirely satisfactory explanation exists.[1,14,15] A theory accepted for many years is that the barium atoms become ionized, or otherwise electrically distorted, to form a positive surface layer that lowers the thermionic work function for the emission of electrons from the oxide coating beneath. Other theories have been proposed.[15] It has been stated [15] (1953) that the functioning of the oxide-coated cathode "may be the least understood phenomena in the entire science of electronics." Emission curves for oxide-coated filaments are shown in Fig. 2-10. For such cathodes typical values corresponding to those in Table 6 are $A = 0.01$, $b = 11{,}600$, $\phi = 1.0$ volt, and $I_s = 0.092$ ampere per square centimeter at 1000°K.

In addition to having a high electron-emitting efficiency, the oxide-coated cathode will stand more positive ion bombardment than the thoriated-tungsten type to be discussed in the following paragraphs. Because these positive ions come from ionization by collision with the residual gas atoms, the oxide-coated cathode need not be operated in as high a degree of vacuum as the thoriated-tungsten cathode. In fact, it is difficult to produce and maintain an extremely high degree of vacuum if an oxide-coated cathode is used, because some gas is evolved from the oxide coating during its operating life. The oxide-coated cathode works well in gas-filled tubes, and is well adapted to heat-shielded cathode construction (section 2-15).

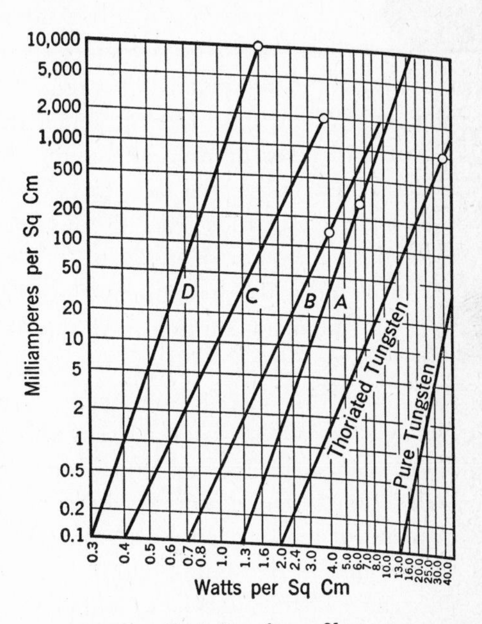

FIG. 2-10. Emission from filaments as a function of power input. The lettered curves are for oxide-coated filaments. Curve *A* is for an early specimen; curves *B*, *C*, and *D* indicate improvements as later coatings were developed. (From reference 9.)

Thoriated-Tungsten Filaments. In the development of ductile tungsten wire for lamp filaments, the addition of a small amount of thorium oxide or thoria (ThO_2) was found to prevent the "offsetting" of tungsten crystals at right angles to the filament when such filaments were heated by alternating currents.[11] As a tube is exhausted, the glass and all metal parts are heated to free occluded gases by baking the tube in an oven and by heating the electrodes with a high-frequency induction furnace. This heating process "flashes" a getter (section 2-13) that absorbs residual gas and further *produces* and *maintains* a high vacuum.

In the activation process, two steps are included.[1,4,9,14] *First,* the filament is burned for a minute or two at a temperature higher than 2700°K. This high temperature cleans the surface of the tungsten and reduces some of the thoria inside the filament to metallic thorium. At this high temperature any thorium atoms that diffuse to the surface are immediately evaporated, and the emission is substantially the same as for pure tungsten. *Second,* the temperature is decreased to from 2000–2200°K. The rate of evaporation is lowered so that a layer of thorium atoms accumulates on the surface. The emission gradually builds up until it is about 1000 times that of pure tungsten at the same temperature.

The metallic thorium diffuses to the surface along the grain boundaries of the tungsten particles and then spreads out over the surface. The grain size is therefore important; a small grain structure should exist. If the grains are too large, the number of paths available for the metallic thorium to pass to the surface would be so reduced that the surface of the filament would not be covered adequately, thus reducing the emission. The thorium spreads along the filament surface in a monatomic layer. The work function of a monatomic layer of thorium on tungsten is *lower* than for pure thorium alone. The action of the monatomic layer is to lower the surface work function so that electrons from the interior of the filament more readily pass through.

During normal operation the layer of thorium atoms slowly evaporates from the filament surface, and this layer is continuously replenished by thorium atoms from within. If the tube is mistreated (for example by raising the filament to an excessive temperature), the thorium layer may be completely evaporated. The emission will then be greatly decreased, and the filament must be reactivated. In one method the filament is operated at a high temperature by impressing about 3½ times normal voltage across the filament for ten to twenty seconds, and then at about 1½ times normal voltage for one to two hours. During both operations *no voltage* is applied to the plate. If this does not return the filament to normal condition, the thorium content of the filament has probably been exhausted, and the characteristics permanently altered.

So many variables enter that it is not possible to include one set of data for A and b of equation 2-9 for all thoriated-tungsten filaments. The values $A = 3.0$, $b = 30{,}500$, $\phi = 2.63$ volts, and $I_s = 0.04$ ampere per square centimeter for a thoriated-tungsten filament at 1600°K. are typical.

A process of **carburization** has been developed that greatly improves thoriated-tungsten filaments.[4,19] The filament is heated in napthalene (or other suitable) vapor. Carbon from this vapor diffuses into the tungsten, forming a surface layer of tungsten carbide. With a properly carburized filament the rate of evaporation of the thorium layer is about one-sixth that of the noncarburized type, and its characteristics are in general improved. Carburized thoriated-tungsten filaments have supplanted pure-tungsten filaments to some extent [19] even in large power tubes.

Although the thoriated-tungsten filament is a more efficient emitter than pure tungsten, it is not so good as an oxide-coated filament. Furthermore, the thoriated tungsten is sensitive to small amounts of residual gas, and must, therefore, be operated in a high vacuum and cannot be used in gas tubes as can the oxide-coated type. It can be operated at temperatures much below that required for tungsten, and requires less heating power. Tubes using thoriated tungsten can be evacuated to a very high degree. This is necessary in tubes employing high voltages.

2-12. *THE SCHROTEFFEKT*

Translated, this means *small-shot effect*, and is due to the fact that the electron flow from the cathode to the plate consists of discrete charges. Furthermore, the emission of electrons at a given instant may be greater or less than the average emission over a given interval.[20] Hence, the emission and the electron flow to the plate resemble a "rain of small shot." Such an effect causes noise in a loud-speaker connected to a vacuum-tube amplifier. This generated noise is one of the factors limiting the amount of amplification that can be used satisfactorily. For example, if the speech or signal strength impressed on an amplifier is so low that it has about the same magnitude as the variations due to the shot effect, then both the useful signal and the noise would have about the same strength in the output of the amplifier. To overcome this trouble, it would be necessary to use a higher signal strength so that the signal-to-noise ratio would be high. For maximum shot effect, a tube must be operated so that the space current to the plate is limited by temperature. If the space current is limited by the space charge (section 3-4), then this accumulation of electrons "damps out" the variations due to shot-effect irregularities in emission.

2-13. *RESIDUAL GAS*

Although tubes are evacuated to a high degree, many residual gas atoms remain (section 1-9). Three important effects [9,12] are caused by gases: *First*, they may form monomolecular or monatomic films on the surface of the emitting cathode; this may greatly lower the emission. *Second*, if residual gas is present, it may be ionized by collision, resulting in the formation of massive *positive* ions. These will travel toward the negative cathode, and if the voltage gradient is sufficient, they will attain a velocity sufficient to knock atoms off the cathode surface. *Third*, the accumulation of positive ions may neutralize, in part, the negative space charge, thus interfering with the normal operation of a tube. Positive ion bombardment of a tungsten filament has little effect on the electron emission; it does, however, slowly disintegrate the filament and reduce its life. With thoriated-tungsten and oxide-coated filaments, the bombardment tends to disintegrate the active surface layer, lowering the emission.

In the manufacture of glass tubes, gases are driven out of the glass walls by heating the glass in a flame to a temperature of about 400°C., depending on the type of glass. Metal parts are treated both before assembly and during exhaust to remove gases. The materials most widely used for the electrode structures include tungsten, molybdenum, nickel, iron, and graphite; special treatments have been developed for preparing these for use in tubes. After assembly and during exhaust, the metal parts of tubes with glass envelopes are heated to high temperatures by high-frequency induction coils. Tubes with metal envelopes are heated with a gas flame. The filament is burned at a high temperature to remove occluded gases. Even with such care, residual gas remains, and this is "cleaned up" by a **getter,** and also by electrical means. Substances such as phosphorus, calcium, magnesium, barium, strontium, aluminum, zirconium, and alloys and mixtures of these are used for getters. An alloy of barium and aluminum is used extensively as a **flash getter** in the small low-voltage receiving-type vacuum tubes.[21] In these the getter material commonly is in the form of a pellet if the tube has a glass envelope. The getter is flashed, or vaporized, by the induction coil at the end of the pumping period. If the tube has a metal envelope, the getter is flashed by passing an electric current through a trough containing the getter. The vaporized metal atoms "sweep through" the tube, and to the coldest portions of the tube, such as the glass or metal envelope, where the vaporized metal condenses. The getter material adsorbs, absorbs, and occludes [21] residual gas atoms. This shortens the exhaust period required.

During the life of the tube, the getter remains active and continues to "clean up" gas as it may be liberated from the walls and electrodes of the tube, thus maintaining the vacuum. Certain types of getter materials also are used in gas tubes to maintain the purity of the gas. These materials must adsorb and absorb unwanted impurity gases as they are liberated, but must not react with the gas intentionally placed in the tube.

For large power-handling tubes, and for other tubes that operate at high temperature, **bulk getters** and **coating getters** are employed.[21] For the first type, sheets or wires of gas-absorbing metal are attached to hot electrodes in the tube. For the second type, suitable nonvolatile metal powders are sintered onto the electrode surfaces.

An electrical method of assisting with the clean-up process consists of drawing an electron current from the cathode to the anode during evacuation. The residual gas is ionized by these currents, and these ions acquire velocities sufficient to *drive* them into the walls of the tube and the electrodes. There are at least two sources of ions in thermionic vacuum tubes. The *first* of these is (as previously discussed) the formation of ions through ionization by collision of the electrons (flowing to the plate) and residual gas atoms. The *second* is the emission of both positive and negative ions by the filament. These ions are massive charged particles of the filament material.[22]

2-14. ELECTRON-EMISSION EFFICIENCY

Many factors enter into the selection of the best material for a cathode. Included among these are adaptability to manufacture, ability to withstand positive-ion bombardment, electron-emitting characteristics, and heat-radiating characteristics. The last-named is important because if a cathode radiates heat readily, then more electric energy will be required to maintain it at the electron-emitting temperature (section 2-15).

The rate at which heat is radiated (largely in the infrared) depends on the temperature and the heat radiation emissivity of the surface. This characteristic is determined by the chemical nature of the surface and by its physical condition. The heat radiation emissivity for tungsten was given in section 2-10. For an oxide coating a typical value at usual operating temperatures is 0.70. The radiation emissivity of thoriated tungsten is about the same as for pure tungsten, but for carburized thoriated tungsten it is about 1.2 times larger.[4]

Because the heat radiation emissivity is different for various materials, the electron emission per unit area for cathodes at the same temperature cannot be taken as a basis of comparison. Thermionic emitters should be compared on the basis of electron emission per watt of input power at the same value of watts input per unit area.[4] The total electron emission from a cathode is given by an equation of the general form

$$I = CP^n \qquad \text{2-13}$$

where C is a constant depending on the surface, P is the electric power supplied to heat the filament, and n is a constant depending on the type of emitter. Figure 2-10 gives the emission per square centimeter as a function of the power input in watts per square centimeter. It is, therefore, a comparison of the relative efficiencies of various emitters. When plotted on power-emission charts such as Fig. 2-10, emission curves should be almost straight lines.

The emission characteristics of vacuum tubes of the small-signal type can be determined with a circuit arranged as in Fig. 2-11. The filament (or heater) is energized in the usual manner, but the grid and plate (and the other grids if it is a multielectrode tube) are connected together and made about 45 volts positive with respect to the filament. The values of filament current *must be low* or the tube may be permanently damaged. The filament current and thus the power input should be varied, and the resulting values of the emission current (not exceeding normal plate current) should be determined by reading the anode milliammeter. These curves can be extended to give the emission at normal operating conditions. The power input in watts (or watts per square centimeter if the dimensions are known) is then plotted on the X axis, and the emission current (or current per square centimeter) is plotted on the Y axis. If the curves cannot be plotted in terms of power and emission *per square centimeter,* the results may be misinterpreted.

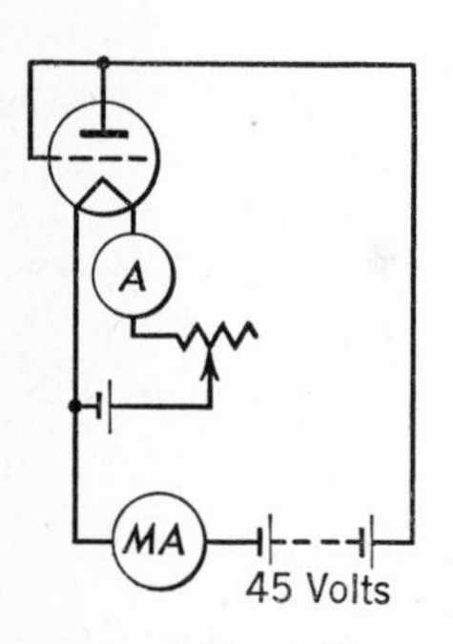

FIG. 2-11. Circuit for obtaining the emission efficiency.

If an emission curve is not substantially a straight line when plotted on a power-emission chart, but bends *downward,* the reasons may be as follows: *First,* the rate of cooling may not be according to the Stefan-Boltzmann law, equation 2-12; *second,* the anode voltage may be too low to draw off all the emitted electrons; *third,* there may be considerable cooling due to the evaporation of the electrons. If the curve bends *upward,* the reasons may be, *first,* residual gas (poor vacuum) or *second,* heating of the electrodes by the electron current.

2-15. *MECHANICAL STRUCTURE OF CATHODES*

The source of electrons in a thermionic tube is either a directly heated filament or an indirectly heated cathode structure.

Directly heated filaments are of tungsten or thoriated-tungsten wire, often in the form of a ribbon; or they consist of a nickel wire or ribbon on which has been deposited an oxide coating. Usually, filaments are used only in tubes for battery operation, or in tubes where the signal level will be high. With weak signals they tend to cause hum. One source of hum is variations in temperature caused by the periodic fluctuations if alternating heating current is used. When the current is maximum in either direction, the power supplied, and the resulting temperature, is greatest; hence, the hum produced has *twice* the fundamental frequency. This effect is minimized, *first,* by making the filament massive so that the temperature variations are small, and *second,* by using a material that will supply *large* quantities of electrons even if the temperature is lowered somewhat. Filament-type cathodes used in amplifier tubes designed for output stages must be capable of emitting large electron currents. The series-parallel combination of the third filament of Fig. 2-12 provides a large emitting surface for power-output tubes. Filament-type cathodes are also used in many rectifier tubes.

Indirectly heated cathodes were developed so that tubes heated with alternating current could be used for amplification without causing excessive hum in circuits having low signal strength. For such cathodes a heater wire is enclosed in a metal cylinder, often of nickel, which is covered with an oxide coating. Because of its thermal storage, the temperature of the oxide-coated cylinder does not change appreciably during the alternating-current variations. Also, the cylindrical cathode is all at the same electrical potential, a condition necessary for low noise. Furthermore, the cylindrical construction makes possible close spacing of electrodes, and offers other advantages. The heater wire is usually tungsten, and must be insulated electrically from the cylinder walls. One method has been to separate the heater wire from the metal cylinder by an insulating refractory material; another was to hold the wire so that it was spaced from the cylinder, but without insulating material. In the first type, the cathode heats slowly because of the mass of insulating material that must be brought up to temperature; in the second type, the wire would sometimes contact the cylinder. A "hairpin" of wire wound into a double helical coil has been used effectively. With this construction adjacent wires carry the heating current in opposite directions, and the magnetic effects cancel. The helical coil is suspended under tension to hold it in position, and often is covered with a thin layer of insulating material. Sometimes a "hairpin" of the insulated heater wire [23] is folded back and forth within the cylinder.

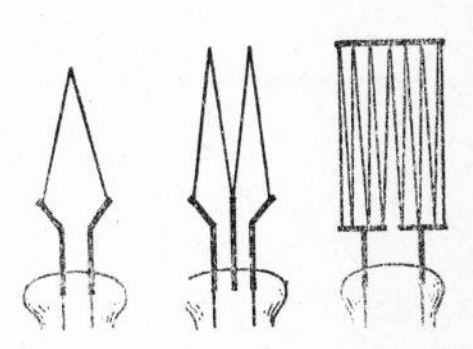

FIG. 2-12. Typical filament-type cathodes.

Heat-shielded cathodes were developed because most of the heat lost from a cathode is by radiation. An oxide-coated cathode with a heat radiation emissivity of about 0.7 will lose heat at a high rate. Nevertheless, the need for high electron emission, and the other desirable characteristics of the oxide-coated cathode (such as its ability to operate in a gas), necessitate its use in tubes for handling large power currents such as in gas and vapor rectifiers.

Koller explains as follows how oxide-coated cathodes are designed so that both the electron emission and the thermal efficiency are high.[4] Assume that to obtain the needed electron emission an oxide coating is applied to a nickel cylinder (emissivity of nickel is 0.15) whose total end area is 0.1 of the total area, and whose side area is 0.9 of the total. Assume that the ends have an emissivity of 1.0 and the coated side, 0.75. Then, the heat radiated would be $(0.1 \times 1) + (0.9 \times 0.75) = 0.775$ unit. Now suppose that the oxide coating is applied to the *inside* of the cylinder instead of the outside. The heat radiation will now be $(0.1 \times 1) + (0.9 \times 0.15) = 0.235$ unit, or about one-third the former value.

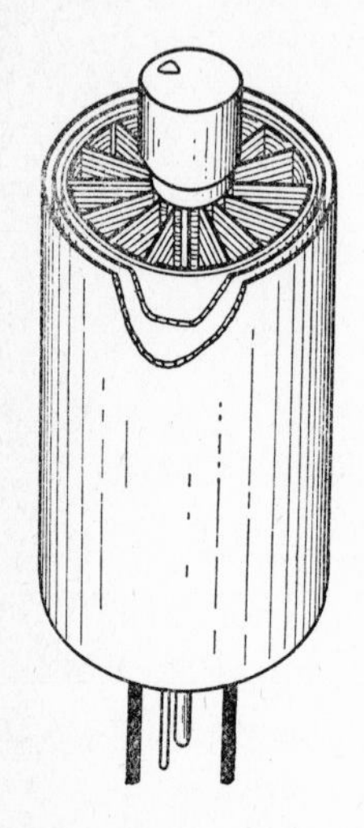

FIG. 2-13. Special heat-shielded cathode. Electrons emerge from open end. The heat required for adequate thermal emission is about 1/24 as much as for a filament of same electron-emitting area. (Courtesy General Electric Co.)

Oxide coatings may also be applied to the vanes extending out to the cylinder. Also, additional non-coated cylinders, as shown in Fig. 2-13, reflect heat back to the emitting elements. Holes are often placed in these cylinders to facilitate the passage of the electrons. Koller states [4] that three such cylinders reduce the radial heat flow to 4 per cent that of a coated surface. This saving is of *great* importance where tubes are designed to rectify large amounts of power.

There are other methods of decreasing radiation. Thus if oxide-coated ribbons are wound on edge to form a spiral as in Fig. 2-14, or if they are folded "accordion fashion," adjacent turns will act as heat shields, and the radiation will be greatly reduced.[15] These structures, and the cathode discussed in the preceding paragraph, are suited only to gas-filled tubes in which positive ions are produced by collision. If these positive ions are not present to neutralize the electrons that accumulate in the deep slots and holes, these accumulations of electrons will form strong negative space charges, and will prevent electrons from flowing out the slots and holes to the positive anode, or plate.

2-16. *SECONDARY EMISSION*

If an electron moving with high velocity strikes a metal plate, the energy of impact may be sufficient to knock out or release **secondary electrons** from the surface of the metal. The number of secondary electrons thus produced depends on (1) the number of the bombarding or **primary electrons**; (2) the velocities of the primary electrons; (3) the type of material used for the bombarded surface; and (4) the physical condition of the surface. Secondary emission is nearly independent of temperature, except as a high temperature may change the nature of the surface by changing the structure or by releasing gas.

Secondary emission can be studied with the circuit of Fig. 2-15. The grid is made *more positive* than the plate. When the plate is at zero potential with respect to the cathode, the grid will collect all the electron current flowing from the cathode. As the plate is made positive by moving the contact to the left, the *plate* begins to collect electrons which pass through the wires of the grid. As the plate is made more positive, some of the electrons attracted by it are accelerated to such a degree that they have sufficient kinetic energy to cause *secondary emission* from the metal plate by knocking electrons out of it. This occurs at *A* of Fig. 2-16. As the plate voltage is made more positive, the plate current *decreases* to *B*. The reason is as follows: Although the plate potential is sufficiently positive to accelerate further the electrons passing through the grid until they are able to cause secondary emission when they strike the plate, the plate is unable to capture these secondary electrons because they are pulled away from the region of the plate by the grid, which is more positive. Thus at *B* the net current in the plate circuit is zero. Just as many *secondary* electrons are produced and lost to the grid as there are primary electrons striking the plate.

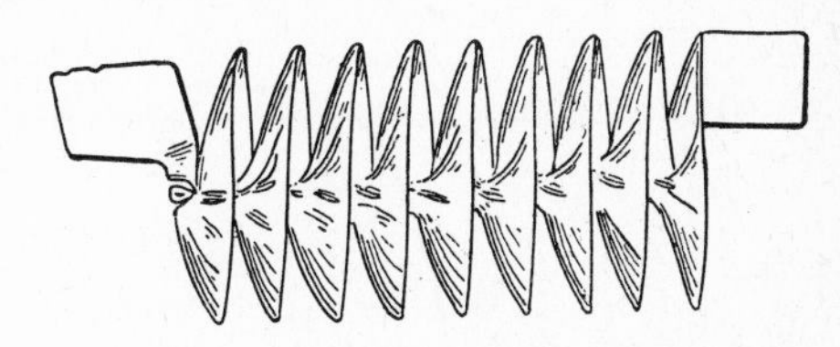

FIG. 2-14. A heat-shielded cathode consisting of a spiral of oxide-coated metal. Adjacent turns act as heat shields. Such cathodes can be used only in gas tubes.

If the plate is made more positive, the net electron current to it reverses in direction and goes to *C*, because *more secondary electrons are produced than there are primary electrons striking the plate.* Hence, some of the highspeed primary electrons are producing *more than one* secondary electron. But, as the plate is made *more* positive approaching the potential of the grid, the plate is able to attract the secondary electrons back to it and the plate current then becomes positive and flattens off between *D* and *E* because of saturation (section 3-3). It is interesting to note that secondary emission occurs in both two-electrode and three-electrode thermionic vacuum tubes under usual operating conditions. Since the plate under these conditions is the only *positive* electrode within the structure, however, the effect is not apparent, at least in small tubes.

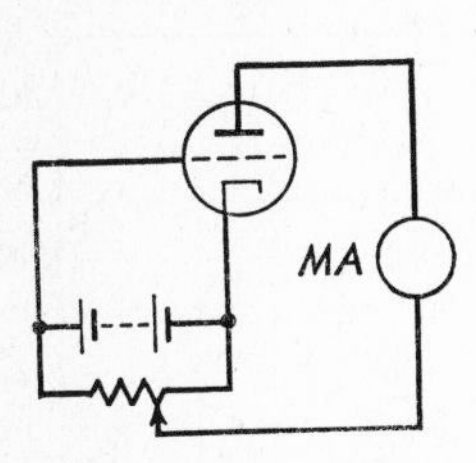

FIG. 2-15. Circuit for studying secondary emission in a thermionic vacuum triode.

Little is known regarding the exact nature of secondary emission. It appears that the primary or bombarding electrons impart energy to the electrons of the metal, thus permitting them to escape through the surface forces as secondary electrons. It is not known if the secondary electrons come from the free electrons in a metal or from the electrons more closely bound to the atoms.[24] The secondary emission from certain composite surfaces is higher than from pure-metal surfaces. The cesium–cesium-oxide–silver surface [25] is a good source of secondary electrons as Fig. 2-17 indicates.

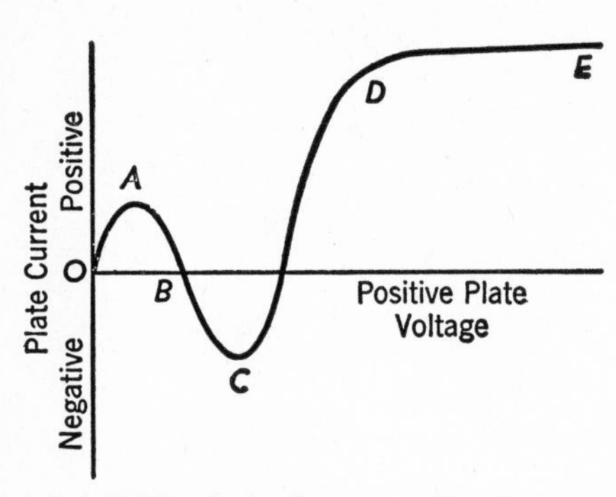

FIG. 2-16. Secondary emission from the plate of the tube of Fig. 2-15 will cause the plate current to vary as shown.

Important facts [24,25] regarding secondary emission are as follows: Secondary emission depends on the condition of the surface. Emission increases from a low value, rises to a maximum for primary electrons having energies equivalent to several hundred volts, and then decreases slowly. The *maximum* ratio of secondary to primary electrons is about 1.5 for well-degassed ordinary metals, about 4 for metals not specially treated, and as high as 10 for films of alkali metal on oxidized metal surfaces. The velocities of the secondary electrons are small, even for primary electrons of a thousand volts. Electrons may be released by positive-ion bombardment. The phenomenon of secondary emission is utilized in electron multipliers used in photoelectric devices, Chapter 16.

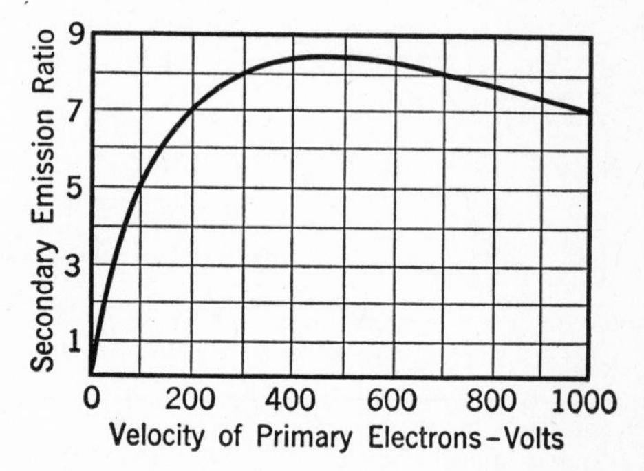

FIG. 2-17. Ratio of secondary to primary electrons for cesium–cesium-oxide–silver. (From reference 25.)

2-17. *FIELD EMISSION*

It has been found that very strong electric fields will cause emission of electrons from a metal. One author states that very intense fields having a voltage gradient of 10^6 or 10^7 volts per centimeter at the surface of a metal will draw electrons out of the metal at room temperature.[9] This is usually called **field emission,** or **cold-cathode emission,** and is sometimes called an **autoelectronic emission.** It occurs in a circuit such as Fig. 2-18. A fine point is used as a *cathode*, and emission occurs at moderate applied voltages of 1000 to 10,000 volts. It has been found that the emission varies exponentially with field strength.[9] Because of the concentration of the electric lines of force at the small surface area of the fine point, the high voltage gradient previously mentioned is produced. In this discussion it is assumed that a perfect vacuum exists. The currents extracted from metals by strong electric fields are independent of temperature (until a point is reached where thermionic emission becomes appreciable).

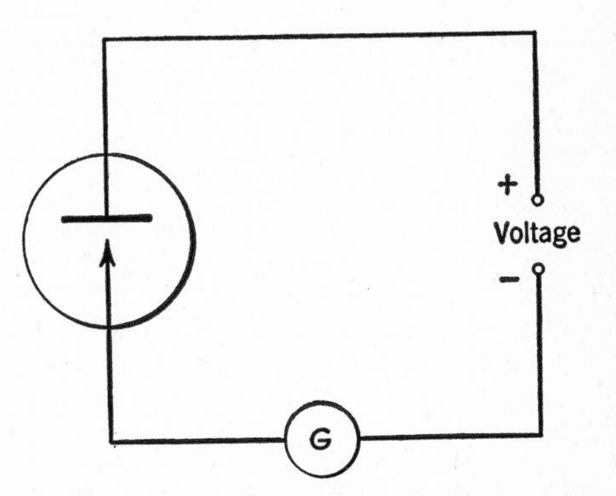

FIG. 2-18. A moderately high voltage will cause an intense electric field at a point. In vacuum, 1000 to 10,000 volts may be sufficient to pull electrons out of a point.

The electrons are able to "tunnel" through the potential barrier without "passing over" the entire potential difference.[4,26] The strong electric field is thought to *decrease the width* of the potential barrier, making it possible for even those electrons having low kinetic energies to tunnel the barrier. The probability of the occurrence of this phenomenon increases rapidly with increase in electric field strength. Field emission should not be confused with the Schottky effect (section 2-9).

2-18. *CONTACT DIFFERENCE OF POTENTIAL*

When two surfaces of dissimilar metals are in *intimate* contact, a **contact difference of potential** is found to exist between portions of the metal objects *not* in contact. The magnitude of the difference of potential is equal to the difference between the surface work functions of the two metals. The metal of low work function ϕ will be positive with respect to the metal of high work function ϕ'. Contact difference of potential can be explained by the use of energy-level diagrams based on Figs. 1-15 and 2-4.

When the two metal bodies are *not* in intimate contact, a condition of equilibrium exists for each metal between the kinetic energies of the free electrons and the potential barriers at the surfaces; hence, no electrons escape from either metal. For the metal of low work function, energy $W_3 - W_1$ is required for escape. For the metal of high work function, energy $W_3' - W_1'$ is required for escape. As indicated in Fig. 2-19, $W_3 - W_1$ is less than $W_3' - W_1'$, which means that it is easier for electrons to leave the metal having the low work function ϕ than for them to leave the metal of high work function ϕ'.

Immediately after the two metals are brought into *intimate* contact, there is an exchange of electrons because of their kinetic energies. More electrons will flow from the metal of low work function to the metal of high work function than will flow in the opposite direction. This unequal flow will continue until the metal of low work function is made positive by a net loss of electrons, and a state of equilibrium is reached in which there is as much electron migration in one direction as the other. When this equilibrium condition is reached, as in Fig. 2-19b, the tops of the Fermi bands are aligned, and a contact difference of potential ϕ_c given by the relation

$$(W_3' - W_1') - (W_3 - W_1) = \phi_c e \qquad 2\text{-}14$$

is established, with the metal of low work function positive with respect to the metal of high work function. Usually this contact difference of potential is less than a volt; it is, however, sometimes important in electron devices.[27]

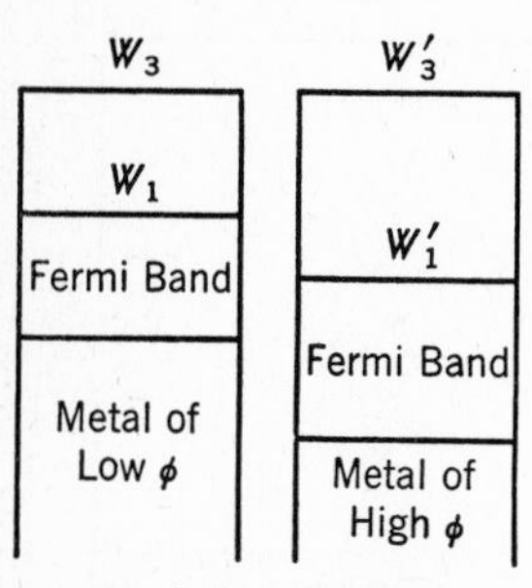

(a) Before Contact

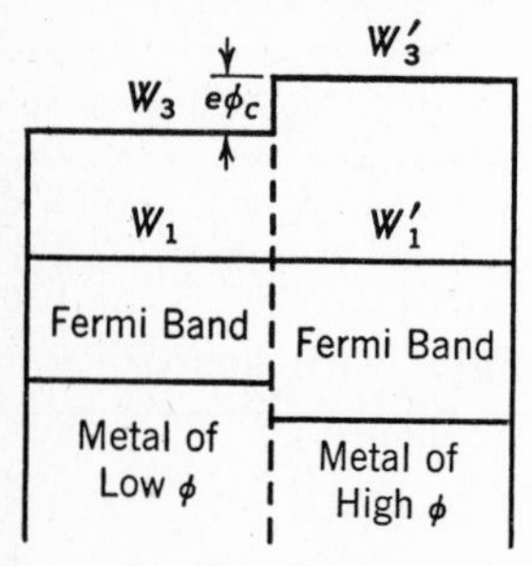

(b) After Contact

FIG. 2-19. When two dissimilar metals are brought into intimate contact, the potential energy bands become aligned, with the tops of the Fermi bands coinciding. This causes the metal of low work function to become positive, and the metal of high work function to become negative, the difference of potential being ϕ_c. This assumes that no other contact potentials exist.

REFERENCES

1. Danforth, W. E. *Elements of thermionics.* Proc. I.R.E., May 1951, Vol. 39, No. 5.
2. M.I.T. Staff. *Applied Electronics.* John Wiley & Sons.
3. Dow, W. G. *Fundamentals of Engineering Electronics.* John Wiley & Sons.
4. Koller, L. R. *The Physics of Electron Tubes.* McGraw-Hill Book Co.
5. Richtmyer, F. K. *Introduction to Modern Physics.* McGraw-Hill Book Co.
6. Hughes, A. L. *Fundamental laws of photoelectricity.* Electrical Engineering, Aug. 1934, Vol. 53, No. 8.
7. Hughes, A. L., and DuBridge, L. A. *Photoelectric Phenomena.* McGraw-Hill Book Co.
8. Reimann, A. L. *Thermionic Emission.* John Wiley & Sons.
9. Dushman, Saul. *Electron emission.* Electrical Engineering, July 1934, Vol. 53, No. 7.
10. King, R. W. *Thermionic vacuum tubes and their applications.* Bell System Technical Journal, Oct. 1923, Vol. 2, No. 4.

11. Dushman, Saul. *Thermionic emission.* Reviews of Modern Physics, Oct. 1930, Vol. 2, No. 4.
12. *International Critical Tables.* McGraw-Hill Book Co.
13. Jones, H. A., and Langmuir, I. *The characteristics of tungsten filaments as functions of temperature.* General Electric Review, 1927, Vol. 30.
14. Bounds, A. M., and Briggs, T. H. *Nickel alloys for oxide-coated cathodes.* Proc. I.R.E., July 1951, Vol. 39, No. 7.
15. Bounds, A. M., and Hambleton, P. N. *The nickel base indirectly heated oxide cathode.* Communication and Electronics, A.I.E.E., May 1953, No. 6; also, Electrical Engineering, A.I.E.E., June 1953, Vol. 72, No. 6.
16. Michaelson, H. B. *The metal-oxide interface in oxide-coated cathodes.* The Sylvania Technologist, Jan. 1949, Vol. 2, No. 1.
17. Poehler, H. A. *The influence of the core material on the thermionic emission of oxide cathodes.* Proc. I.R.E., Feb. 1952, Vol. 40, No. 2.
18. Becker, J. A. *Thermionic electron emission.* Bell System Technical Journal, July 1935, Vol. 14, No. 3.
19. Ayer, R. B. *Use of thoriated-tungsten filaments in high-power transmitting tubes.* Proc. I.R.E., May 1952, Vol. 40, No. 5.
20. Pearson, G. L. *Fluctuation noise in vacuum tubes.* Bell System Technical Journal, Oct. 1934, Vol. 13, No. 4.
21. Espe, W., Knoll, M., and Wilder, M. *Getter materials for electron tubes.* Electronics, Oct. 1950, Vol. 23, No. 10.
22. Michaelson, H. B. *Metallic-ion emission in vacuum tubes.* The Sylvania Technologist, July 1948, Vol. 1, No. 3.
23. Klemperer, H. *Heater-cathode insulation performance.* Electrical Engineering, Sept. 1936, Vol. 55, No. 9.
24. Pomerantz, M. A., and Marshall, J. F. *Fundamentals of secondary electron emission.* Proc. I.R.E., Nov. 1951, Vol. 39, No. 11.
25. Zworykin, V. K., Morton, G. A., and Malter, L. *The secondary emission multiplier—A new electronic device.* Proc. I.R.E., March 1936, Vol. 24, No. 3.
26. Herzfeld, K. F. *The present theory of electric conduction.* Electrical Engineering, Apr. 1934, Vol. 53, No. 4.
27. O'Neill, G. D. *The effect of grid contact potential and initial electron velocity on electron tube characteristics.* The Sylvania Technologist, Oct. 1948, Vol. 1, No. 4.

QUESTIONS

1. What evidence indicates the presence of a negative shell or barrier at the surface of a metal?
2. In what four ways are electrons emitted from a metal?
3. How is the energy required imparted to the electrons in each method of emission?
4. Do the electrons emitted photoelectrically and thermionically come from the same group of electrons within the metal? Cite experimental evidence in support of your view.
5. Why do emitted photoelectrons have velocities between maximum and minimum values?

6. Referring to the stopping-potential method of section 2-6, could you use a source of radiation having continuously variable wave length, and vary the source until a wave length is found such that with the grid G positive no electrons are collected? This would give λ_0 of equation 2-4, and ϕ could be found. What source would you use?
7. It is stated that the initial energies of the electrons have but little effect on their photoemission. What experimental evidence supports this statement?
8. The statement sometimes is made that electrons have velocities of a few volts. What is the meaning of this?
9. Referring to Fig. 2-6, which of these metals would be best for operation with incandescent lamps as a source of radiation? Which would operate best if a mercury-vapor tube were the source?
10. Name one composite surface used in phototubes. Is this extensively used?
11. Thermionic emission has been likened to evaporation. Is there experimental evidence supporting this view?
12. What is the difference between adsorbed and absorbed gases?
13. How does an electron escape from a surface in thermionic emission?
14. What is meant by the term thermionic work function, and how does it differ from photoelectric work function?
15. What is your understanding of the term saturation current?
16. What factors must be known to determine the saturation current experimentally?
17. Electric energy is used to heat the cathode in a vacuum tube containing negligible gas. What becomes of this energy? What becomes of it if the tube contains gas?
18. What are the important characteristics of tungsten when used for filaments in vacuum tubes? Is this the only pure metal used as filaments in vacuum tubes?
19. Discuss the thoriated-tungsten filament. What are its outstanding characteristics?
20. Why are oxide-coated filaments used? Are indirectly heated cathodes made of tungsten or thoriated tungsten?
21. How is the oxide placed on wires for filaments? On indirectly heated cathodes?
22. What is meant by carburization, and what are its advantages?
23. Discuss the *Schroteffekt*. Is it of practical importance?
24. On what basis should thermionic emitters be compared?
25. What are the advantages of the heat-shielded cathode?
26. How can a voltage of, say, 5000 volts between a point and a plane produce a very high voltage gradient causing field emission?
27. Discuss the Schottky effect. Is it of considerable practical importance?
28. What are similarities and differences between field emission and Schottky effect?
29. Can heat-shielded cathodes be used in gas tubes? Why?
30. In evacuating tubes to operate at, say, 5000 volts and to handle several hundred watts, great care is used to obtain a high degree of vacuum. Why is this of special importance? What type or types of filaments would be used in such tubes?

PROBLEMS

1. Plot a curve showing the variations in the velocity of photoelectrons emitted from cesium for different wave lengths of radiation less than the long-wave limit. What assumption is made? What effect would a more intense beam have on the velocity? Would the velocity be the same at a higher temperature?
2. Assume that the values of the long-wave limits λ_0 of Table 4 had been determined experimentally, and calculate the photoelectric surface work function ϕ.
3. Referring to Fig. 2-5, assume that the distance between the plate and grid is 0.25 centimeter, and that they are large plane surfaces compared to this spacing. What voltage in volts must be impressed between these two electrodes to prevent current flow to the grid at the longest wave length λ_0?
4. In an electron multiplier tube there are ten plates, each being a cesium–cesium-oxide–silver surface. Each plate is 100 volts more positive than the preceding one, and the plates are arranged so that an electron starting from a photoemissive surface "bounces" from the first positive plate to the second, etc., down the tube. Assuming ideal operation, what will be the electron multiplying ratio of the assembly? Derive a general equation for a tube of this nature.
5. What must be the length of a tungsten filament that is 0.0025 centimeter in diameter if it is to deliver 12.5 milliamperes at 2450° K.?
6. Make the calculations required to check the saturation current for tungsten given in Table 6.
7. Calculate the saturation current for tungsten at 1000, 1500, 2000, and 2500° K., and plot saturation current against temperature. What assumptions have been made? Are they valid?
8. Use the curves of Fig. 2-7 and calculate A and b for oxide-coated and thoriated-tungsten filaments. Do these values agree with those given in the text? Suggest reasons for differences that may be found.
9. Calculate the saturation currents for oxide-coated and thoriated-tungsten filaments and compare these with the values given in the text.
10. Calculate the contact differences of potential, and indicate the polarities, for a platinum-nickel contact, and for a tungsten-nickel contact.

CHAPTER 3

TWO-ELECTRODE ELECTRON TUBES

Four methods of liberating electrons from metals were considered in the previous chapter. This chapter will be devoted to two-electrode electron tubes that depend on thermal emission and on field emission for liberating electrons. Photoelectric devices and secondary-emission multipliers will be considered in Chapter 16. Electron tubes having two electrodes are called **diodes,** defined [1] as a "two electrode electron tube containing an anode and a cathode." The term diode includes both vacuum and gas, or vapor, types. The **cathode** is defined [1] as "an electrode through which a primary stream of electrons enters the interelectrode space," and the **anode** as "an electrode through which a principal stream of electrons leaves the interelectrode space."

Although Edison discovered the rectifying properties of the thermionic vacuum tube in 1883 (section 2-7), little practical use was made of the principle until 1904, when Fleming adapted the diode for use as a demodulator, or detector, of radiotelegraph signals (Chapter 13).

The diode was used for this purpose for some years, but was supplanted (for a time) by the three-electrode tube or triode. The diode now is almost universally used as a detector in modern radio-receiving sets. The diode is suitable for alternating-current power rectification, and has achieved wide use for this purpose. It is used for rectifying large amounts of power in industrial equipment, and also for supplying direct current to small devices such as radio receiving sets. The theory of vacuum and gas diodes will be considered in this chapter; applications will be covered in subsequent chapters, particularly Chapter 7.

3-1. *RATINGS OF RECTIFIER TUBES*

When a diode is used as a rectifier, the voltage impressed between the cathode and anode, or plate, is *positive* during one half of the cycle, and *negative* during the other half. For the part of the cycle that the anode is positive, current flows. For the other half, no current flows. The rectifier tube, therefore, is alternately a *conductor* and an *insulator*. When the tube conducts, voltage drops occur across the secondary of the transformer, the load resistor,* and the tube. During the negative half-cycle, no drops occur; the tube now acts as an insulator, and the *full secondary voltage is impressed between the electrodes.* It is very important that the tube does not "flash back" or arc across during the part of the cycle that the tube acts as an insulator and has the entire voltage impressed between the electrodes.

Definitions applying to diodes (and to other tubes as well) will now be given. The **electrode voltage** is defined [1] as "the voltage between an electrode and the cathode or a specified point of a filamentary cathode." Usually this reference point is the *negative* cathode terminal if direct current is used for heating a filamentary cathode, or a center tap on the filament or on the filament-supply transformer if alternating current is used.

The **peak forward anode voltage** is [1] "the maximum instantaneous anode voltage in the direction in which the tube is designed to pass current." The **peak inverse anode voltage** is [1] "the maximum instantaneous anode voltage in the direction opposite to that in which the tube is designed to pass current." If the peak inverse anode voltage is exceeded, the tube may flash back or arc across as previously explained. This might damage or ruin the tube, and also might injure other equipment. Usually the peak inverse anode voltage is the open-circuit secondary voltage, but under abnormal and switching conditions, higher transient voltages may occur.

Under normal conditions, electrons from the heated cathode are drawn over to the anode during the positive half-cycle. This is called the **anode current,** or **space current.** The **peak anode current** (maximum instantaneous value [1]) is an important factor to consider, particularly with *gas* tubes. These may be damaged, for reasons that will be explained later, if this current value is exceeded. Vacuum diodes are not so easily damaged in this way, but damage may result to them as will be explained.

* Load is defined [2] as the power that a device delivers. In electronics the word load commonly is used to designate that which takes power from the device. To avoid confusion, the term load resistor, or load impedance, will usually be employed to designate that which is taking power from a tube or other device. There will be a few exceptions, but not where confusion will result.

During normal operation, heat is evolved at the anode or plate. This is because the electrons drawn from the cathode give up their kinetic energy to the anode when they strike it. The amount of energy and the heating of the plate is influenced by the **average anode current**, which is, of course, the direct-current value obtained by integrating the instantaneous anode current over an interval of time, and dividing by this time interval.[1] Another and important factor enters into determining the heating of the plate; this is the anode or plate voltage. This is because the acceleration of the electrons, and their final velocities and energies, are determined by the anode voltage. Thus, a tube might stand a high average current at a low voltage, but would not stand the same average current at a higher voltage. If an anode overheats, gases may be driven from it, and a vacuum tube will then contain gas, which may limit its use, if not render it useless. Also, an overheated anode may buckle or warp, or may have holes melted in it.

3-2. *THE THERMIONIC VACUUM DIODE* *

This consists of a thermionic electron-emitting cathode and an electron-collecting anode, or plate. The plate is made positive with respect to the cathode so that electrons will flow to it. The electrode structure is housed in a glass or metal container that is evacuated.

Suppose that the circuit connections are as in Fig. 3-1. The "A" battery supplies the current I_f for heating the filament. The desired positive potential E_b impressed on the plate is obtained by selecting with the **voltage divider** [2] the portion of the "B" battery voltage desired. The electron current I_b flowing to the plate is measured by a milliammeter placed as shown. The negative filament terminal (or the cathode) is taken as the point of reference.[1] This terminal may be grounded if desired.

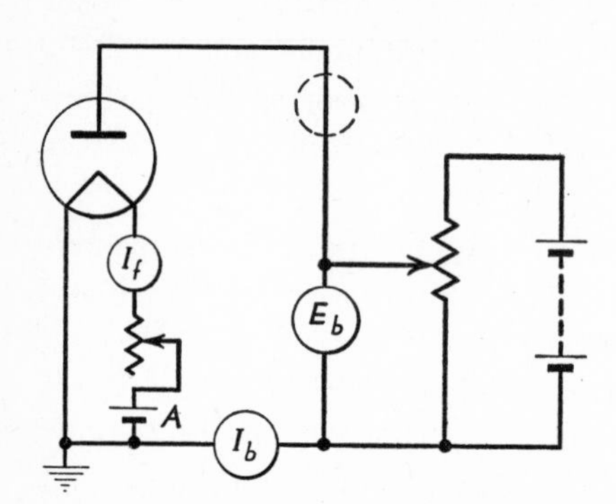

FIG. 3-1. Circuit for studying a diode. The "A" battery heats the filament; the other battery is the "B" battery for supplying plate voltage.

* Various trade names have been applied by different manufacturers to electron tubes. In a book on fundamentals it seems particularly important that only accepted nomenclature and descriptive names rather than trade names should be used. In general this will be done, but certain of the more widely used trade names will be introduced so that industrial literature will be more understandable.

It will be noted that the plate-circuit milliammeter for measuring the current I_b is connected for safety near the common (grounded) terminal. The dotted position sometimes used is at a higher potential than the one shown.

3-3. *THERMIONIC VACUUM DIODE CHARACTERISTICS*

Suppose that the voltage between plate and filament is held constant at E_b volts and that the filament current I_f is varied. The values of the direct-current plate current I_b when plotted give a curve such as E_b of Fig. 3-2. Now suppose that the plate voltage is held at a new value E_b' and the filament current is varied as before. A new curve E_b' will result. No electrons are emitted by the filament and hence no current can flow to the plate until a certain temperature corresponding to the current at point A is reached. After this, the plate current increases rapidly until region B, where the curve gradually flattens. This leveling in region B is called **temperature saturation.** Here, the electrons are being taken by the positive plate at a constant rate independent of filament temperature (the limiting factor being **space charge,** section 3-4). If the plate voltage is increased to E_b', then the plate is more positive, more electrons can be taken, and the leveling off occurs at a higher value of plate current.

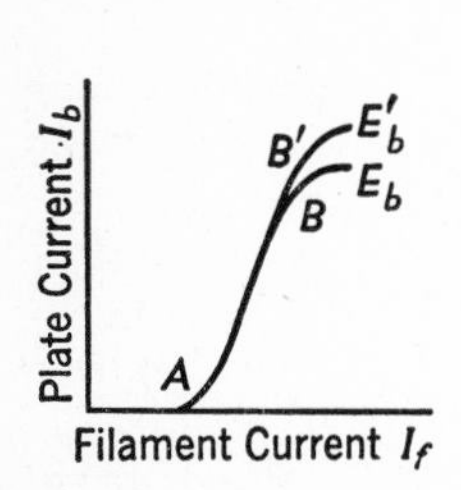

FIG. 3-2. Variations in diode plate current with filament current for two different plate voltages E_b and E_b'. The flattening is termed temperature saturation.

Now suppose that the filament current I_f is held constant at some value providing electron emission, and that the plate voltage E_b is reduced to zero. It will be found that even with zero potential some current will flow to the plate, and that the plate voltage must be made negative to stop this current. The reason for this current flow with zero plate voltage is that some of the emitted electrons have sufficient energy to carry them to the plate and on around the circuit, even with zero plate voltage or with a slightly negative plate voltage. This is illustrated in Fig. 3-3. With the filament current I_f constant, the plate current I_b will vary as shown with increasing values of plate voltage E_b. The plate current reaches a constant value as in Fig. 3-3 because of **voltage saturation.** If the filament is operated at a higher temperature by increasing the filament current to I_f', the saturation plate current will be higher as indicated. When the region of voltage saturation is reached, the plate is at a positive potential sufficient to attract substantially all the electrons to it as they are emitted. Thus, increasing the temperature provides more emission, and a larger plate current results. Although the plate-current curve is almost flat after voltage saturation is reached, it usually rises slightly. One reason for this is the Schottky effect (section 2-9). The operating plate voltage chosen must be below the voltage-saturation value if the operation of the tube is to be independent of filament current and temperature. This is very important in tubes used as amplifiers and for similar purposes.

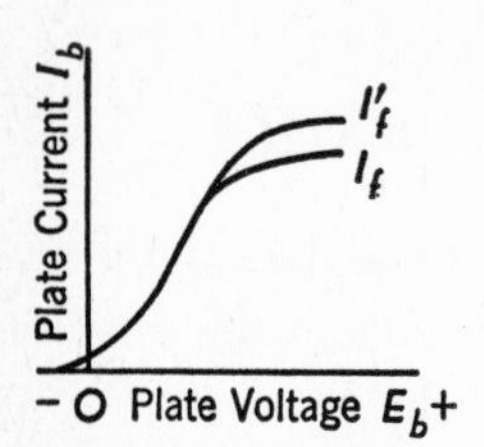

FIG. 3-3. Variation as in diode plate current with plate voltage at two filament temperatures. The flattening is termed voltage saturation.

In performing tests as just explained the anode may be damaged if the power dissipated at the anode or plate exceeds the rating of the tube. As was mentioned (section 3-1) this will depend on both the plate current and the plate voltage. Because tubes usually are designed to be operated at temperature saturation, it is advisable to use a value of filament heating current considerably below normal when obtaining data for voltage-saturation curves such as Fig. 3-3. Otherwise, it will be necessary to use rather high values of plate voltage before saturation is reached, and the tube may be damaged.

The electrons are drawn to the plate by the positive potential E_b, supplied in Fig. 3-1 by the "B" battery. This potential imparts energy to the electrons, the energy being supplied by the battery. When they suddenly strike the plate, they give it most of their energy, causing the plate to heat. Any such loss may be imagined to be an $I_b^2R_p$ loss, where R_p is the **direct-current plate resistance.**

3-4. *SPACE CHARGE IN THERMIONIC VACUUM DIODES*

The presence of the electrons in the space between the filament and plate has an important effect on the operation of a vacuum tube, and this effect will now be considered. Although explained for diodes, the principles apply to other vacuum tubes. First, consider conditions in a diode when the filament is at operating temperature, and the *plate voltage is zero.* Neglecting the few high-velocity electrons that cross to the plate, there are few electrons in the vicinity of the plate. Most of the emitted electrons hover for a time about the filament in a dense cloud. As many electrons return to the filament during a given time as are emitted by it. Now consider conditions in the same tube when the filament is at operating temperature, and the *plate-to-filament voltage is highly positive* so that voltage saturation has been reached. Under these conditions the electrons are almost all taken by the plate *as soon as they are emitted,* and there is no longer as dense a cloud of electrons around the filament. Above voltage saturation, the electron current is limited by the cathode temperature as explained in section 3-3.

These negative electrons are charges in space and so constitute a **negative space charge** in the region between the filament and plate. In the first instance, with *zero plate voltage,* the space charge was concentrated about the filament. In the second, *after voltage saturation* had been reached, the negative space charge was *more* uniformly distributed throughout the space between the electrodes. This statement needs further explanation, because it might appear that it would be uniformly distributed, an assumption that is incorrect. Assuming a uniform current flow from filament to plate, more electrons must be moving near the cathode, where they have low velocities, than near the plate, where their velocities are higher.[3] There is, accordingly, a higher electron density near the cathode, even after voltage saturation.

The negative space charge actually extends throughout the space between the filament and plate. For ordinary small-signal tubes under normal operating conditions, the density of the charges is relatively so very much greater near the filament (or cathode) surface that it may be assumed that the space charge is confined to a thin sheath about the filament varying in thickness from a few thousandths to a few hundredths of a centimeter.[4] This sheath of negative charges produces a field near the surface of the filament that tends to force emitted electrons back into the filament. The positive plate potential tends to draw electrons out of this negative sheath to the plate. The plate current (from a very low value to the region of voltage saturation) of Fig. 3-3 is determined largely by the extent to which the positive electric field produced by the plate neutralizes the negative electric field produced near the filament by the negative space charge.

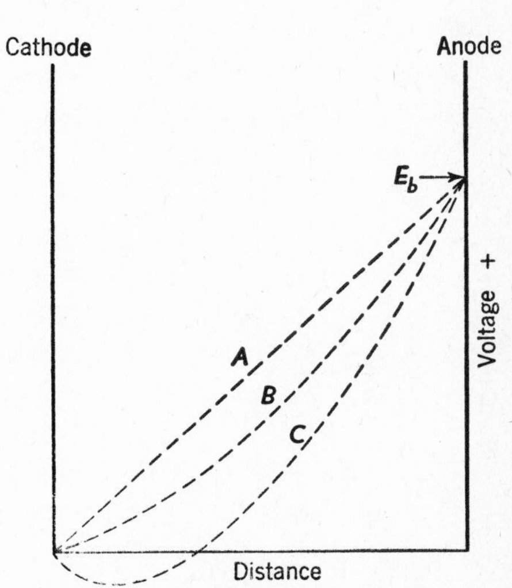

FIG. 3-4. Showing the approximate voltage distribution between the cathode and anode, or plate, in a vacuum diode. The electrodes are considered to be plane surfaces close together.

Potential Distribution between Electrodes. The influence of the negative space charge alters greatly the distribution of the voltage applied between the filament and the plate. Figure 3-4 represents a parallel-plane cathode and a positive anode, respectively. Distance and voltage are plotted as indicated. Several conditions now will be investigated. *First,* with a voltage E_b applied between the two (plane) electrodes, and with the cathode *cold,* the distribution of the voltage between the electrodes will be as shown by curve *A.* Since there is no emission, there are no electrons and hence *no space charge* between the cathode and anode. *Second,* with the cathode *heated* and with a positive voltage E_b applied to the anode, but *assuming* all the electrons are emitted with zero velocity, there will be a space charge, and the voltage distribution will be as shown by *B.* But electrons are *not* all emitted with zero velocity; many of them have velocities sufficient to carry them far out into the space between the electrodes. The *third* condition, curve *C,* represents the *actual* voltage distribution in a thermionic vacuum tube operated *well below voltage saturation* as is normally done for small-signal tubes.

Curve *C* is exaggerated because in reality the negative space charge in a thermionic vacuum tube usually is a thin sheath *close* to the filament. Nevertheless, the negative space-charge region is at a potential *lower* (more negative) than the filament, and hence the space charge tends to repel electrons back to the filament. The positive plate draws the plate current from the negative space-charge region.

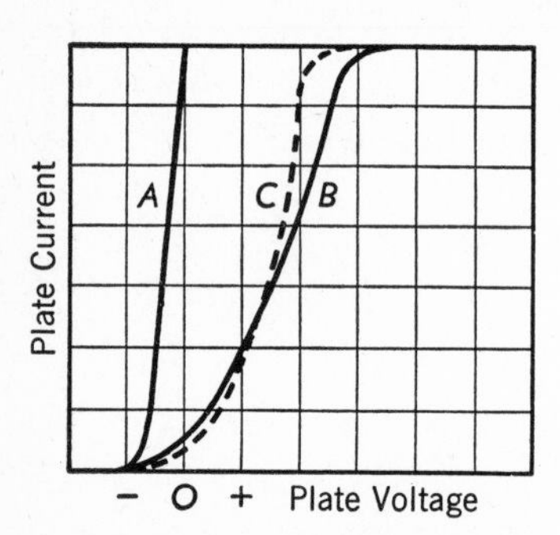

FIG. 3-5. Illustrating the effect of space charge on plate current. Curve *A,* no space charge; curve *B,* space charge in good vacuum; and curve *C,* space charge and residual gas.

Effect of Space Charge on Plate Current. Typical plate-current–plate-voltage curves for a diode were shown in Fig. 3-3. Note the effect of the space charge on this plate current as represented in Fig. 3-5. If some electrons *had* small initial velocities, but if *no space-charge effect existed,* some plate current would flow as shown by *A,* even for *negative* plate potentials. As soon as the plate became just slightly positive it would take *all* the electrons emitted. But with space charge, the plate is not able to do this until it is sufficiently positive to overcome the negative space-charge effects as shown in *B.* In the practical case there are three parts to curve *B.* The *first* part is determined by the initial velocity of the electrons; the *second* part by the space charge; and the *third* part by filament emission.[5]

The knee of the curve, where saturation is reached and where the current is no longer determined by space charge, would be very abrupt but for several reasons.[6] *First,* the electrons have different initial velocities, and thus the applied plate voltage does not affect each the same. *Second,* there is an *IR* drop along a filament; thus when saturation is reached for the negative end, it has not been reached for the positive end. After saturation is reached, increasing the plate voltage does not increase the plate current. Such a tube is therefore a *constant-current* device, and can be inserted in *series* in a circuit to hold the current essentially constant for wide voltage fluctuations. The space charge offers opposition to the flow of electron current from the filament or cathode to the plate. There is, therefore, a relatively high voltage drop between these electrodes in *vacuum* tubes.

3-5. *EQUATIONS FOR PLATE CURRENT*

Langmuir derived the equation for the theoretical relation between the voltage applied between a thermionic cathode and an anode and the resulting electron current between them when in a high vacuum. Child had previously derived the same formula for currents carried by positive ions in low-pressure arcs.[6,7]

Excluding the first and the last parts of curve *B* of Fig. 3-5, and considering *only the central portion,* where the plate current is controlled by the *space charge,* and with the filament at a constant temperature, the current over this central portion is given by the so-called theoretical **three-halves power law** (also called Child's law, Langmuir-Child law, and Child-Langmuir-Schottky law,[1] or equation)

$$I_b = GE_b^{3/2} \qquad \text{3-1}$$

In this expression I_b is the direct-current plate, or anode, current flowing, G is a constant determined by the units used, the spacing, and geometrical dimensions of the electrodes,[7] and E_b is the voltage between the anode or plate and the cathode. The constant G is called the **perveance,** defined[1] as "the quotient of the space-charge-limited cathode current by the three-halves power of the anode voltage in a diode."

As explained, equation 3-1 applies to the portion of the curve where the electron current from the cathode to anode is limited by the negative space charge. Over this region the current to the anode or plate varies theoretically as plate voltage raised to the three-halves power. For ordinary vacuum tubes the actual value of this power may differ from the theoretical three-halves value. The actual value of this power, and the value of the constant G can be obtained as follows: Taking the logarithm of each side of equation 3-1 gives

$$\log_{10} I_b = \log_{10} G + m \log_{10} E_b \qquad 3\text{-}2$$

where m, the actual power to be found, has been substituted for the theoretical 3/2 value.

Now if values of $\log_{10} I_b$ are plotted on the Y axis, and values of $\log_{10} E_b$ are plotted on the X axis, and if I_b and E_b have been determined experimentally for a particular vacuum tube, then the intercept on the Y axis is $\log_{10} G$, from which G can be found, and the slope of the line gives m, the actual value of the power to which E_b should be raised for this particular tube. Another method is available for finding G; also, certain precautions and corrections should be used.[1] In particular, initial velocities and contact differences of potential cause deviations from the theoretical three-halves power law. Also, with a filament, a drop of potential occurs along its length, and hence all parts are not at the same potential with respect to the plate. As a result, the voltage E_b of equation 3-1 will be different for each part of the filament.

For two infinite parallel-plane electrodes, the equation applying is [7,8]

$$I_b = \frac{2.33 \times 10^{-6} E_b^{3/2}}{x^2} \qquad 3\text{-}3$$

where I_b is the plate current in amperes per square centimeter of either anode or cathode area, E_b is in volts, and x is the spacing between the filament and plate in centimeters.

For a straight, single-wire cathode of radius a, placed at the axis of a cylindrical anode of radius r, the equation for the anode current in amperes per centimeter length of cathode is (where r/a is greater than 10)

$$I_b = \frac{14.65 \times 10^{-6} E_b^{3/2}}{r} \qquad 3\text{-}4$$

where E_b is in volts and r is in centimeters.

3-6. *EFFECT OF RESIDUAL GAS*

Although in a vacuum tube the residual gas is removed to such a degree that it has no appreciable effect on the performance, it is of interest to investigate the influence of residual gas on the plate current. This is illustrated by curve C of Fig. 3-5.

If the velocity of the emitted electrons in traveling to the plate is sufficient, they may collide with and ionize atoms of the residual gas. The resulting positive ions have a pronounced effect on the space charge and hence on the plate current. Thus, the effect of residual gas on the plate current (aside from possible effects on emission) is largely confined to anode voltages above the ionization potential of the residual gas. For this reason, curve *C* of Fig. 3-5 follows curve *B* closely until ionization by collision begins.

In studying the effect of the positive ions, it will be recalled that they are massive and move very slowly compared to the electrons. This was discussed in section 1-14, where it was shown that for mercury vapor the negative ions (electrons) would attain velocities 607 times the velocities of the positive particles when both types of particles were acted on by the same electric field. Because of this, positive mercury ions will remain in an electric field between a cathode and an anode 607 times as long as electrons. Although a positive ion can neutralize the electric field of only one electron *at a given time*, the *time interval* that the positive ion spends in the field is relatively so much greater that in effect the positive ion neutralizes the negative field (space charge) produced between the electrodes by many electrons. This neutralizing effect greatly increases the flow of electrons to the plate; that is, the plate current that flows for a given voltage. Or, if the plate current is the same in identical tubes, except that one contains residual gas, the voltage drop across the tube will be less for the gaseous tube. Hence, neutralization of space charge by positive ions is important, but the additional current flow caused by the motion of positive ions themselves is not important in increasing plate current. For this reason the magnitudes of the saturation currents as given by the maximum value of curves *B* and *C*, Fig. 3-5, are about the same for a well-evacuated tube as for one containing a small amount of residual gas.

3-7. *TYPICAL THERMIONIC VACUUM DIODES*

Such tubes are available in many sizes, from those that will rectify but a few microamperes at low voltages, to those that will rectify currents measured in amperes at thousands or hundreds of thousands of volts. One such tube has a peak inverse voltage rating of 150,000 volts and a peak anode current of 1.0 ampere. In the small tubes the cathodes may be oxide-coated filaments or indirectly heated cathodes. In the high-voltage types thoriated-tungsten or tungsten filaments are used.

The plates of such tubes tend to become very hot, and this is one of the important factors limiting the amount of current they can pass at given voltages. This heating is due largely to the energy imparted by the rapidly moving electrons to the positive plate and to the energy received by radiation from the filament. The power dissipated by the electrons is equal to E_bI_b, where E_b is the voltage applied between the *plate* and *filament.* Two general methods are used for cooling the plates. *First,* the tube is cooled by air. In some installations forced air is used. *Second,* some of the large tubes are cooled by circulating water. Water-cooling of the plate was first made possible by a copper-glass seal developed [9] by Housekeeper.

3-8. *LOW-PRESSURE THERMIONIC GAS DIODES*

There are two types of thermionic gas diodes, those in which the gas is at low pressure, and those in which the gas is at high pressure. Most gas diodes are the low-pressure type, and these will be considered first.

The low-pressure gas diodes are of two types: *first,* those using one of the monatomic inert gases such as argon, and *second,* those using mercury vapor. The theory of operation of these is essentially the same. In the tubes containing gas, ordinary temperature changes do not appreciably affect the gas pressure and hence do not vary the characteristics of these tubes. In the type containing mercury vapor, an excess of condensed mercury exists, and ordinary temperature changes cause more or less of the excess mercury to vaporize, thus appreciably affecting the mercury-vapor pressure and the operating characteristics.

The discharge occurring in low-pressure gas diodes is usually classified as an arc, because of the low value of voltage drop at the cathode, called the cathode fall.[10] As mentioned in section 1-15, this arc is a non-self-maintained discharge since the electrons required are furnished by a heated cathode. This cathode is usually of the oxide-coated type, may be either directly or indirectly heated, and is often heat-shielded.

The effect of residual gas in a tube was discussed in a preceding section of this chapter. It was shown that the positive ions, resulting from collisions of the electrons with the gas atoms, neutralized the negative space charge and thereby permitted a large plate current to flow. The phenomena occurring in a gas tube are similar.

The gas diodes used as rectifiers consist essentially of a hot cathode which emits electrons and a positive anode to which these electrons are attracted. These electrodes are enclosed in a glass or metal tube. The containers are evacuated, and then a small amount of gas is introduced. The gases used are neon, helium, argon, and mercury vapor, the last two being most common. In the mercury-vapor tube a small amount of condensed mercury is introduced. At the low pressure, and at the operating temperatures, the tube becomes filled with mercury vapor.

Assume that the cathode of a mercury-vapor tube is at operating temperature, but that the positive anode voltage is *not* applied. The tube is now filled with *neutral* mercury atoms. When the anode voltage is applied, electrons are drawn to the positive anode and are accelerated to high velocities. These electrons collide with and ionize the neutral mercury atoms. The electrons produced by this action are mobile and are *quickly* drawn to the anode; they add slightly to the anode (or plate) current, but this is unimportant. The *massive* positive ions are (relatively) *slowly* drawn to the negative cathode, and it is these ions which largely determine the operating characteristics of the tube. A positive mercury ion remains in the interelectrode space 607 times as long as an electron (section 3-6). These ions form a **sheath** about the cathode, and the *positive* electric field of this sheath neutralizes, to a large extent, the *negative* electric field of the space charge.

3-9. *VOLTAGE DISTRIBUTION IN LOW-PRESSURE THERMIONIC GAS DIODES*

The effect of the space charge on the voltage distribution in a vacuum thermionic tube is shown in Fig. 3-4. Corresponding curves for a mercury-vapor tube are shown in Fig. 3-6. Curves for a similar tube containing a gas such as argon instead of mercury vapor would have the same general shape. Note that there is no negative space-charge region in a gas or vapor diode.

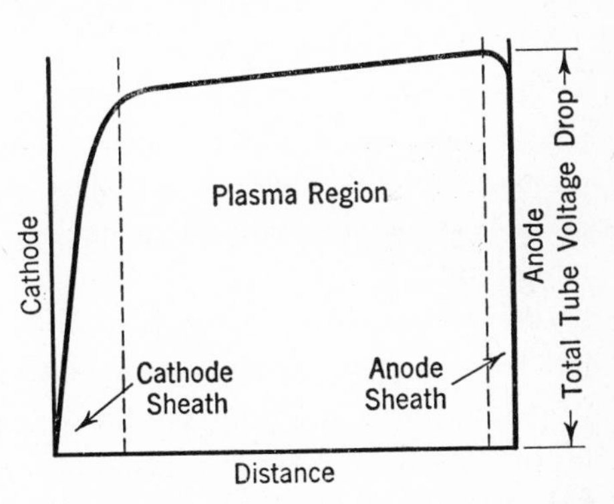

FIG. 3-6. In gas and vapor tubes most of the voltage drop occurs in a thin sheath near the cathode. Distances not to scale. Compare with Fig. 3-4.

The space between the cathode and anode of a gas or vapor thermionic tube is divided into three regions, of which the cathode sheath and the plasma are the most important. The cathode sheath of Fig. 3-6 also is called a positive ion sheath, or positive space charge, and is established as explained in section 1-14 and in the preceding section. In the plasma there are electrons, positive ions, and gas atoms. The electrons and positive ions have random velocities, and some excitation and ionization occurs. When electrons in excited atoms fall back to their normal positions, radiation results, and the plasma glows. Recombination and deionization occur in this region and particularly at the walls of the envelope. Very little voltage can exist across the plasma; that shown in Fig. 3-6 is the voltage required to cause a "drift" of ions and a net current flow. There is a slight change in voltage near the anode; this is usually a slight drop in potential, but there may be no change, and even a slight rise. This is determined by the shape of the anode, by the anode-current magnitude, and by such factors as the nature and pressure of the gas. The potential change in the region near the anode is of little importance.

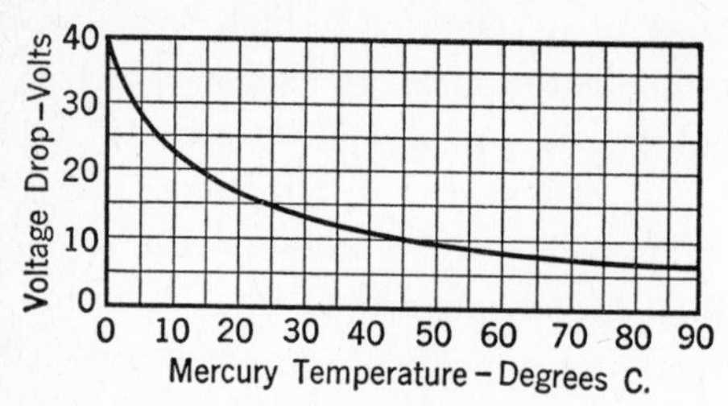

FIG. 3-7. Effect of temperature on the voltage drop in a mercury-vapor diode with anode current constant. (Data from reference 3.)

A characteristic of a gas tube of great practical importance is this: The gas (or mercury-vapor) tube will pass *large* currents with a *low internal voltage drop* between the cathode and anode. Theoretically, because of multistage ionization of metastable atoms (section 1-11), it is necessary for the voltage *between anode and cathode* to equal only the first excitation potential (Table 2) of the gas.[10] Actually, the voltage drop is higher, and is *about* equal to the ionizing potential of 10.4 volts for mercury vapor. This low voltage drop is substantially the same for both small and large tubes. For mercury-vapor tubes it varies with the temperature of the condensed mercury determined by the bulb temperature, as shown in Fig. 3-7. This low drop results in high efficiency. Furthermore, the small voltage drop is almost constant over the entire operating range. This makes possible a rectifier with good regulation; that is, the output voltage varies little with the magnitude of the load current. These characteristics for high-vacuum and gas tubes of comparative sizes are shown in Fig. 3-8.

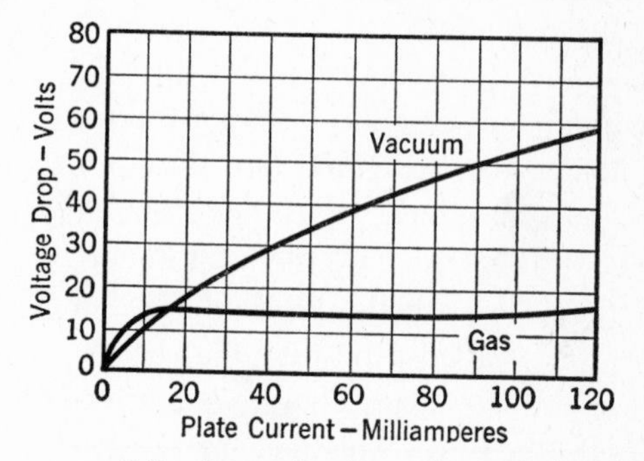

FIG. 3-8. Internal voltage drops between cathode and anode for comparable gas and vacuum rectifier tubes.

A circuit for studying the operation of a gas diode can be arranged as in Fig. 3-9. The curve of Fig. 3-8 can be obtained by varying the anode current I_b by changing the voltage divider, and by measuring the corresponding voltage drop E_b. For determining the curve of Fig. 3-7, the temperature of the rectifier bulb must be varied. This can be accomplished by mounting the tube in a temperature-controlled oil bath.

3-10. *OPERATION OF LOW-PRESSURE THERMIONIC GAS DIODES*

Greater care must be used in the operation of gas tubes than is needed for high-vacuum tubes.[11] There are at least two important rules to follow. These will be explained for mercury-vapor tubes, but they also apply to similar tubes using argon and other gases.

First, the cathode should reach operating temperature before the anode voltage is applied. This is *very important* in the larger tubes. These tubes have massive oxide-coated cathodes (which may be heat-shielded) and come up to operating temperature slowly, often requiring several minutes or more. Suppose that the anode voltage is applied *before* the electrons are emitted. Then there will be no drop in voltage across the load outside the tube, and the entire voltage to be rectified will be impressed across the tube. As a few electrons are emitted, ionization by collision occurs and positive ions are produced. Owing to the high voltage existing across the tube, these massive positive ions are drawn to the cathode with high velocities, and when they strike the cathode, they will knock off the active material and permanently injure the cathode.

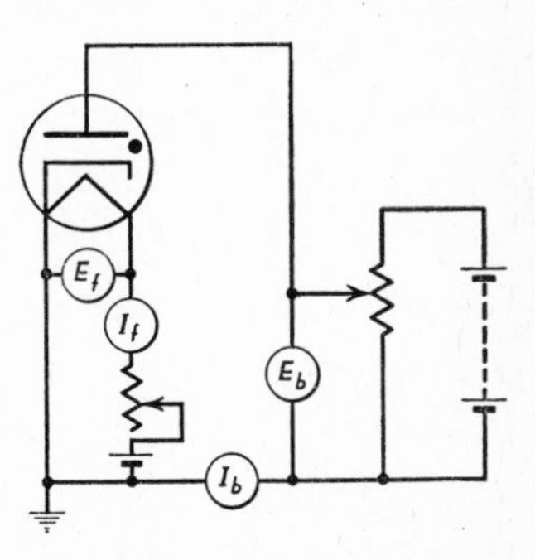

FIG. 3-9. A circuit for testing a gas diode. The dot in the tube symbol indicates that the tube contains gas, or mercury vapor. When testing or operating a gas tube, resistance should be placed in series with each electrode unless care otherwise is used to ensure that currents do not become excessive.

The ionizing potential for mercury vapor is only 10.4 volts. The total drop in the tube (Fig. 3-8) is about 14 volts. The potential drop across the tube that will accelerate the massive positive ions sufficiently to cause disintegration is approximately 22 volts. Thus, in operation, the voltage across a mercury-vapor tube should never be allowed to approach closely this value. Slow-operating relays commonly are used to delay the application of the anode voltage until cathode operating temperature is reached.

Second, the rated peak anode current should not be exceeded. This current is determined by the maximum emission from the cathode and by the heat-dissipating characteristics of the tube. If this current is exceeded, the voltage drop within the tube may be sufficient to cause cathode disintegration by positive-ion bombardment; also, the tube may be overheated, causing damage to the electrodes. The rated peak inverse anode voltage should not be exceeded, or the tube may arc back, permanently damaging it. The peak inverse voltage that a mercury-vapor tube will stand depends on the temperature as shown by Fig. 3-10. This is because the vapor pressure depends on the temperature of the tube, and the vapor pressure determines the vapor density and hence the operating characteristics.

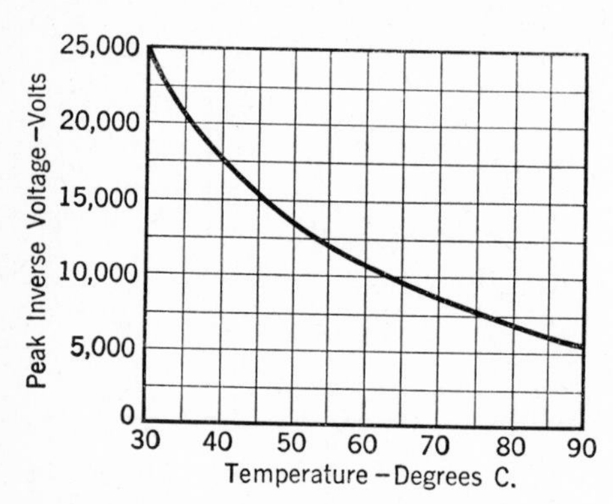

FIG. 3-10. Indicating the way in which the peak inverse anode voltage that a mercury-vapor tube will stand varies with temperature. (Data from reference 3.)

3-11. *TYPICAL LOW-PRESSURE THERMIONIC GAS DIODES*

As for the vacuum tubes, gas diodes are made in many sizes ranging from those that pass peak anode currents of a few hundred milliamperes and capable of withstanding peak inverse of only about 1000 volts, to those that will pass peak anode currents of many amperes and that are capable of withstanding many thousands of volts. As an illustration of this latter type, data are included for a type of thermionic mercury-vapor rectifier tube used in large radio broadcasting stations.[12]

Cathode heater voltage	5 volts
Cathode heater current	30 amperes
Peak anode current	40 amperes
Average anode current	10 amperes
Peak inverse anode voltage	22,000 volts

The data given for anode currents and voltages are maximum values. At 35°C. and peak anode current of 40 amperes the voltage drop across the tube is 14 volts. A heat-shielded oxide-coated cathode is used, and a heating period of at least one minute is needed for the cathode to reach operating temperature. The tube is air-cooled.

It should be recalled (section 2-15) that the heat-shielded cathode can be used only in gas tubes. In these, the positive ions neutralize the space charge that would otherwise prevent the electrons from flowing out from the shielding structure. In a vacuum tube the heat shields would act as electrostatic shields and would prevent the positive electric field of the plate from reaching the emitted electrons within the heat-shielding structure.

3-12. *VACUUM DIODES AND LOW-PRESSURE GAS DIODES AS RECTIFIERS*

Briefly, the operating characteristics of two comparative types of diodes are as follows: *First*, more power is required to heat the filament of the vacuum tube than is required by the heat-shielded cathode of the gas tube. *Second*, the voltage drop across the vacuum tube is greater than that across the gas diode; also, the drop across the former increases with increase in current, but that across the latter is substantially constant. *Third*, the rectifying efficiency of the vacuum diode is lower than for the gas tube. This is important for tubes handling large amounts of power, but of little consequence for such applications as rectifiers for radio receiving sets.

The vacuum diode has certain characteristics that make its use desirable. *First*, it is rugged and not easily damaged by misuse. A momentary short circuit, which would ruin a gas tube, would probably merely overheat the plate of a vacuum tube. *Second*, small variations in the voltage of the filament supply for a vacuum tube are not objectionable. The gas tube is very critical as to filament supply. Thus, for a typical mercury-vapor tube rated at 2.5 volts on the filament, the voltage should be maintained within 5 per cent of this value. If the filament is operated above 2.5 volts, the life of the tube will be greatly reduced, and if operated at voltages below the lower limit, the tube may fail immediately. *Third*, the vacuum tube does not usually produce radio interference, but the gas diode does. *Fourth*, the vacuum diode has been developed to withstand much higher peak inverse anode voltages, and will, therefore, rectify much higher alternating voltages. In a sense, the low-pressure gas diode should be regarded as a high-current, low-voltage rectifier tube, and the vacuum diode should be thought of as a low-current, high-voltage rectifier tube, although, of course, their fields of application overlap.

Low-pressure gas diodes are sometimes called **Phanotrons,** and high-vacuum diodes are sometimes referred to as **Kenotrons.**

3-13. *HIGH-PRESSURE THERMIONIC GAS DIODES*

The essential elements of these tubes are a short, heavy thoriated-tungsten filament close to a heavy graphite anode. Argon gas at the relatively high pressure of about 5 centimeters of mercury is introduced into the tube. This argon gas serves two purposes. *First*, through ionization by collision, the argon gas provides positive ions for reducing the space charge, thereby permitting large currents to flow. *Second,* the argon gas at this high pressure reduces evaporation of the filament. This is important because the cathode is operated at high temperature to provide the required electrons from a simple cathode structure. As an example of the protective action of the gas, one author states [13] that in similar tubes using a barium-coated nickel cylinder for a cathode, the life of the filament was 4000 hours when in mercury vapor at 1 to 3 millimeters pressure, but was only 20 hours when the vapor pressure was 0.01 millimeter. In each instance the tube was passing 5 amperes.

These high-pressure gas diodes can be used *only* at low voltages, less than about 200 volts. If higher voltages are used, they will arc across on the negative half-cycle. This limits the use of these tubes to such applications as storage-battery chargers. These tubes are made in various sizes, from several amperes to 20 amperes continuous capacities. They have trade names **Tungar** and **Rectigon.**

3-14. *X-RAY TUBES*

A form of electromagnetic radiation commonly known as **x-rays** is emitted when a rapidly moving stream of electrons strikes a metal electrode. They are also known, particularly in the medical profession, as **roentgen rays,** named after the scientist who discovered the phenomena in 1895. The first x-ray tubes consisted essentially of a cold metal cathode and a metal anode in a glass tube containing traces of residual gas. No heated filament was used, the initial current starting from the free ions present. This tube was not entirely satisfactory, largely because of erratic performance and lack of control.

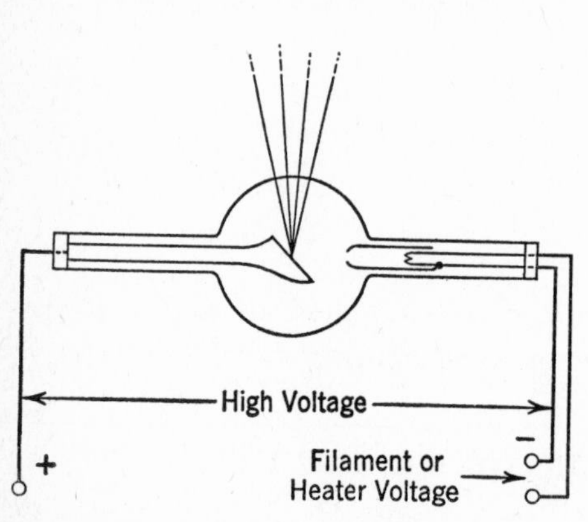

FIG. 3-11. Basic circuit of an x-ray tube.

A high-vacuum, hot-cathode, x-ray tube with a heated filament as the source of electrons was invented by Coolidge in 1913. This tube is stable in operation, and the quality of the rays is readily controllable. A simplified circuit is shown in Fig. 3-11. The construction depends on the use to which the tube is to be put. The cathode usually is a tungsten filament surrounded by a small metal cylinder connected electrically to the filament. The metal cylinder assists in focusing the electron beam on the **target** (also called an anode, and sometimes an anticathode). The target is the electrode on which cathode rays are focused and from which roentgen rays are emitted. The target often contains a tungsten insert in a massive copper electrode.

Tungsten is used because it will withstand the electron bombardment, and copper is employed because of its high heat-conducting ability. In this connection it should be remembered that the target is in a high vacuum; therefore, the heat generated by the impinging electrons must be conducted away largely through the body of the metal electrode. Cooling fins sometimes are placed on the end of the positive electrode exposed to the air. Also, this electrode is sometimes water-cooled. X-ray tubes have been designed with a target that is rotated rapidly to minimize pitting. Glass that readily passes x-rays is used for the envelope.

In the medical profession x-rays are used in several ways. *First*, because of their penetrating power, x-rays can be used with fluorescent screens (which emit *visible* light when struck by *invisible* x-rays) to study the body. *Second*, since x-rays affect photographic films, they can be used to penetrate the body and photograph conditions therein. *Third*, x-rays not only penetrate, but produce important effects on living tissue. They are used for treatment of cancer and diseases of the skin.

Also, x-rays are used for examining welds and castings. In general, each special use requires rays of a different **hardness** or quality. The hardness of x-rays is defined [2] as "the attribute which determines the penetrating ability." Long x-rays do not penetrate readily, but the shorter the wave length, the *harder* the rays, and the greater their penetrating ability. Medical science ordinarily is interested in x-rays of from 0.1 to 1 angstrom unit.[14] (One angstrom unit equals 10^{-8} centimeter.)

The control of the x-ray radiation from a Coolidge tube is relatively simple. The *hardness* (that is, the wave length or frequency) of the radiation depends on the voltage applied between the cathode and target, since this voltage determines the velocity and the energy of the impinging electrons. The *quantity* of the radiation of any degree of hardness, that is, the "amount" of radiation of a given type, is determined by the filament current, because the tube is a diode and is operated at high voltage, and at these high voltages, voltage saturation has been reached. The current is not limited by the space charge, but by the number of electrons emitted; that is, by the temperature of the filament. Increasing the temperature increases the emission and thus the electron current to the target. An increased current to the target gives greater x-ray radiation.

3-15. *GLOW LAMPS AND VOLTAGE-REGULATOR TUBES*

These consist of two initially cold (room temperature) electrodes mounted close together in a small bulb filled with an inert gas, usually neon, argon, or helium gas, or combinations of these. The type of gas, the gas pressure, and the shapes of the electrodes are determined by the purposes for which the device is to be used. These are defined [2] as **electric discharge lamps,** and are forms of the glow tube. Over the usual operating range the phenomenon occurring is that of the self-maintained discharge, and the cathode is either partially or wholly covered with the normal glow, giving the lamp its illuminating properties. The dimensions and electrode arrangements are usually such that the positive column is almost nonexistent, at least from the standpoint of illumination. Most of the voltage applied to the tube is consumed as a drop in voltage (or cathode fall) near the cathode in accordance with Fig. 3-6.

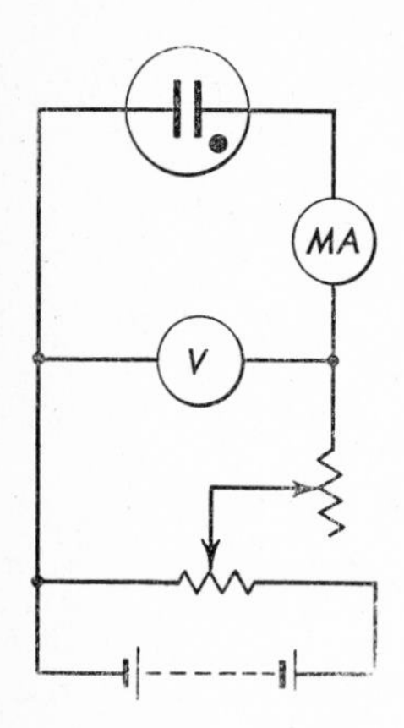

FIG. 3-12. Circuit for determining the characteristics of a glow lamp.

The characteristics of a glow lamp can be obtained with a simplified circuit such as Fig. 3-12, and are as indicated in Fig. 3-13. The series rheostat is set at a value sufficiently high to protect the tube from overload, and the value of the applied voltage is gradually increased from zero by varying the voltage divider. The current values obtained will differ greatly in magnitude, and the data should be plotted on semilog paper to extend the current values obtained at the initial voltage settings. From *A* to *B* the discharge is of the same nature as that described in section 1-12, considering the fact that the glow lamp contains gas at low pressure. At *B* the gas becomes ionized and is said to **break down.** The voltage immediately falls to point *C*. From *C* to *D* the current through the tube is controlled by varying the series rheostat. Over this region the voltage drop across the tube is essentially constant, rising but several volts for a wide range of current values. This phenomenon is used in voltage regulators (Chapter 5). When the tube first ionizes at point *C*, a small portion of the cathode is covered with the normal glow. As the current increases from *C* to *D*, the glow covers more and more of the cathode. Thus, a source of illumination is available, the intensity of which varies directly as the current through the tube. Such lamps were used in an early system of television [15] as the receiving element. This glow lamp, when viewed through a rotating scanning disk operating in synchronism with a similar disk at the sending end, reproduced the television images. By using three separate lamps, glows of different colors were reproduced, and television in color was demonstrated some years ago.[16]

When used as sources of low-intensity constant illumination, the glow lamps are operated with a current-limiting series resistor that is commonly in the base. The operating point for such purposes is usually at about point D. Such glow tubes have been used in oscillators.[17] If the current is permitted to exceed the value at point D, the glow changes from the normal to the abnormal glow, the voltage across the tube rises rapidly, and in the vicinity of E an arc is formed between the electrodes, the voltage drops to F, and this arc will damage the tube, probably rendering it unfit for further use.

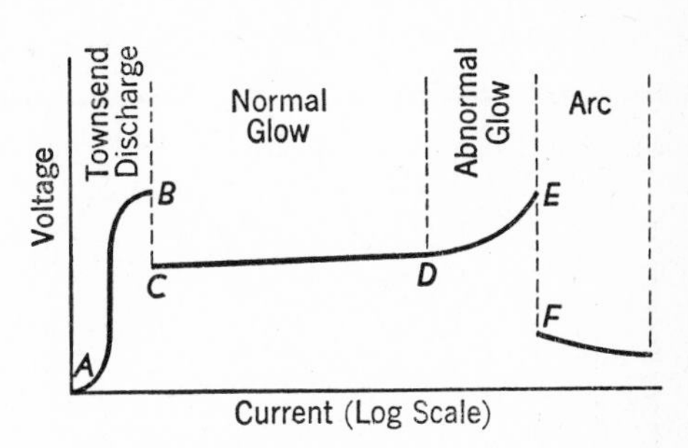

FIG. 3-13. Phenomena occurring in a glow tube at various values of current. The transitions from one condition to another are indefinite and should be regarded as regions rather than definite points.

3-16. *NEON TUBES* [18,19]

This is the name popularly given to that class of electric discharge lamps so widely used for advertising signs and decorative purposes. The basic principle of operation of these tubes is the same as that of the glow lamp just described. The difference is that in the neon tube almost the entire tube is filled with the positive column (section 1-15). The light radiated comes from this column. These lamps were developed by Claude.

A simple neon tube consists essentially of an initially cold electrode in each end of a *long* glass tube filled with an inert gas, such as neon. Before ionization the gas is a good insulator and no current flows. When a high voltage is impressed, however, the gas becomes ionized, current flows, and a self-maintained glow discharge is established. This was discussed in section 1-15, and a drawing of the glow discharge and of a voltage distribution curve was given.

A high-voltage drop exists at the cathode and this drop cannot be neutralized as in hot-cathode thermionic gas tubes, because a high voltage gradient must exist near the cathode to produce ions in this region if the discharge is to be self-maintained. This results in "sputtering." This is caused by the massive positive ions being drawn to the cathode with high velocity. They knock metal atoms off the cathode, causing its gradual disintegration. This can be minimized by proper design so that the current per unit area is low.[19] Disintegration of the cathode by sputtering largely determines the life of the tube. The sputtering forms a thin metallic film on the inside of the tube near the cathode. As this film is deposited, the useful gas of the tube is occluded, *diminishing* the gas pressure and *increasing* the mean free path of the positive ions. The voltage required to operate the tube gradually increases, and this further increases the sputtering. The voltage required to operate the tube approaches in magnitude that supplied by the transformer, and the tube begins to flicker and eventually ceases to operate. The potential gradient in a neon tube is shown in Fig. 3-14. This curve shows that the voltage drop is high at the anode and cathode, and that it is constant along the positive column.

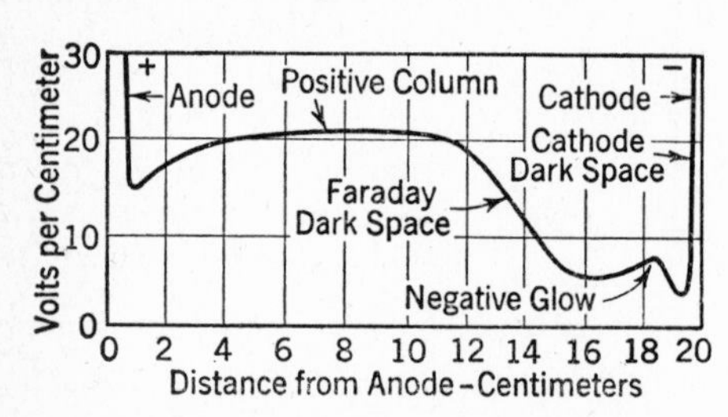

FIG. 3-14. Potential gradient for a typical neon tube. (From reference 19.)

Neon tubes require a high voltage (about 15,000 volts) to produce the initial discharge, but need only a relatively low voltage (depending on the length) to maintain the discharge. Thus, the transformers used should have good *current* regulation, but poor *voltage* regulation, just the opposite from power-system transformers. Glow discharge lamps, or tubes, are used for illuminating purposes in addition to sign advertising. The sodium-vapor lamp is an example.[20]

3-17. *MERCURY-ARC TUBES*

The mercury-arc lamp dates back to 1860 when Way discovered that if an electric circuit was opened by a mercury contact, a brilliant arc was formed.[21] In 1890, Hewitt began developing mercury-arc lamps with the arc enclosed in an evacuated glass tube.

The Hewitt mercury lamp was commercialized about 1901. It consists of a long glass tube with a metallic electrode at one end and a mercury pool as an electrode at the other end. The lamp usually is mounted at an angle with the horizontal, with the mercury pool below the other electrode. The arc is started by tipping the tube until a thin filament of mercury connects with the metal electrode, and then suddenly breaking the mercury path by restoring the tube to the operating position. Ions are released by the arc formed when the current is interrupted, and a self-maintained arc discharge is established within the tube. Mercury-arc lamps sometimes are used in industrial shops. The greenish-blue color of the mercury arc is excellent for high visibility.

3-18. *FLUORESCENT LAMPS* [22,23]

The widely used fluorescent lamp is a form of electric discharge lamp.[2] Several types have been developed, and several circuit arrangements are possible. The discussion will be confined to the common *preheated-cathode fluorescent lamp* and the *glow-switch starter.* A typical lamp of this type consists (Fig. 3-15) of a glass tube 4 feet long and 1.5 inches in diameter containing an electrode at each end. Each electrode is a coiled oxide-coated tungsten filament. The tube contains an inert gas such as argon at low pressure, and a drop of mercury, which at room temperature and low pressure fills the tube with mercury vapor. The inside wall of the glass tube is coated with a material called a *phosphor.*[22,23] Assume that the lamp is "off"; then, the contacts of the glow switch will be open. Now assume that the lamp is "turned on" by closing the switch mounted in the wall of the room. This impresses substantially the entire 115-volt (rms), 60-cycle voltage across the glow switch. This switch consists of a fixed electrode, and a movable bimetallic-strip electrode in a glass bulb containing argon at low pressure. The applied voltage breaks down the gap between the electrodes and establishes a self-maintained glow discharge. The heat from the glow warms the bimetallic strip, which closes the contact. When this contact closes, the glow switch is short-circuited, and the two lamp filaments are connected in series to the supply. These filaments are heated to electron-emitting temperatures. There is a drop of about 10 volts across each cathode. When the short-circuited glow switch has cooled, the bimetallic strip opens the contact, thus interrupting the filament heating current. The collapsing magnetic field in the ballast inductor produces a transient voltage of 1000 volts or more across the tube. This voltage will be between the two end electrodes having cathodes at electron-emitting temperatures. A self-maintained discharge will then be established between the two hot cathodes, and the arc current will maintain at least part of the cathodes at electron-emitting temperatures. The arc has a negative volt-ampere characteristic, and the current magnitude is limited to safe value by the ballast inductor. Because the tube contains mercury vapor, the discharge includes much invisible ultraviolet radiation.

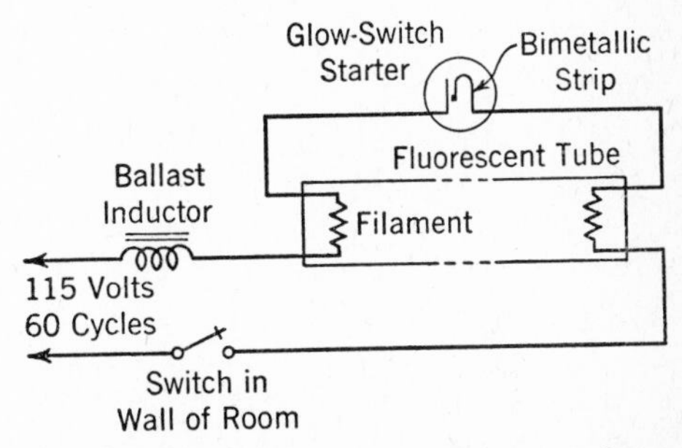

FIG. 3-15. Circuit for a fluorescent lamp using a starter switch and a ballast inductor.

The phosphors used are crystalline inorganic materials such as calcium tungstate and magnesium tungstate, to name but two.[22,23] These materials absorb photons of invisible ultraviolet and radiate energy of visible wave length (Fig. 3-16). An important advantage of the fluorescent lamp is its high efficiency in converting from electric energy to visible radiation.

3-19. *MERCURY-ARC RECTIFIERS* [21,24,25]

Hewitt is also credited with inventing the mercury-arc rectifier. For many years it was used almost exclusively for the rectification of alternating power by means other than rotating machines.

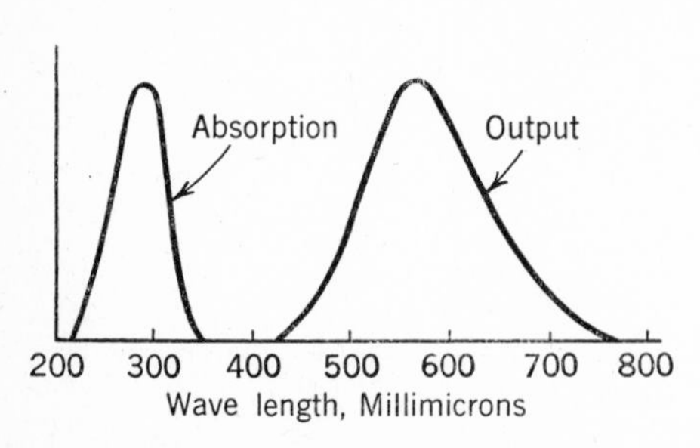

FIG. 3-16. The electric discharge in a fluorescent lamp contains much radiation of wave lengths shorter than the eye can detect. Energy (marked absorption) at these wave lengths is absorbed by the coating of phosphors on the inside of the tube, and is radiated (marked output) by this coating as an output of longer visible wave lengths. The magnitudes indicated are not to scale. (Courtesy Illuminating Engineering Society.)

A circuit of a single-phase mercury-arc rectifier is shown in Fig. 3-17. The arc is started by closing the starting switch, tipping the bulb so that the mercury pool makes contact with the starting electrode, and then suddenly restoring the bulb to the upright position, thus breaking the mercury connection and establishing an arc. This arc forms the **cathode spot** on the mercury pool and produces initial ionization so that current can then flow to the anodes.

These anodes usually are graphite. They are so connected to the transformer that they are alternately positive and negative. When an anode is positive it draws electrons from the ionized mercury vapor. A rectified current will therefore flow through the load, represented in Fig. 3-17 by a battery to be charged. If the mercury arc stops for any reason, recombination occurs, the vapor in the tube deionizes, and the arc will not start again until the tube is tipped as previously explained. The reactor in series with the load is to prevent the arc from stopping when the applied alternating-voltage wave passes through zero. The inductive reactance prevents the current to one anode from dying out until the current to the other has been established within the tube. In rectifiers of larger sizes, two energized auxiliary excitation or "keep-alive" electrodes are provided. These are similar to the main anodes, but are connected to an auxiliary circuit and arranged to conduct a current of at least 5 amperes at all times, thus assuring continuous operation of the arc if the main anode current falls to a low value. The vaporized mercury condenses on walls of the tube and flows back into the mercury pool. The temperature at which the bulb operates determines the point at which condensation occurs. This in turn determines the vapor pressure in the tube, and thus temperature influences the operating conditions.

Theory of the Mercury Arc. It was early thought that the temperature of the cathode spot was high, and that thermal emission of electrons occurred from this spot. It is now considered that the temperature of the spot is too low for much emission. The bombardment of the mercury pool by positive ions was suggested as the cause of the emission, but this is considered inadequate to account for the high emission that occurs during rectification. An acceptable theory is as follows.

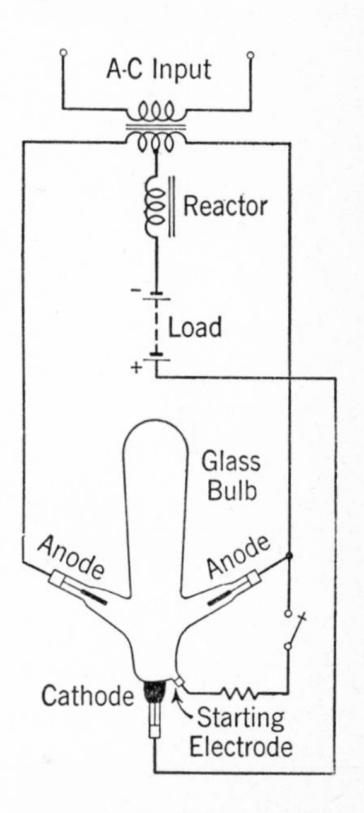

FIG. 3-17. Circuit of a small single-phase mercury-arc rectifier. The load is the battery being charged.

Ionization by collision occurs within the mercury-arc rectifier tube. The mobile *electrons* so released are withdrawn quickly to the anodes, adding but little to the total current flow. The massive positive ions travel relatively slowly to the negative mercury-pool cathode, where they form a very intense *positive* space charge over the surface of the mercury, and thus a strong electric field exists between the positive ions and the surface of the mercury pool. The pressure of the mercury vapor immediately over the cathode spot is high, and it is assumed that the electrons need to travel but a short distance before they encounter and ionize a mercury atom. This action provides positive ions for pulling the electrons from the surface of the mercury.

It was suggested [10] by Langmuir that this high-intensity field extracts electrons from the relatively cold mercury in accordance with field-emission theory caused by positive ions as mentioned in section 2-17. Experiments show that fields of about 10^6 volts per centimeter are required to produce electron emission from metals at low temperature. Since the cathode drop is only about 10 volts, the positive space-charge film would need to be of the order of 10^{-5} centimeter thick.[21] The area of a typical cathode spot is about 2.5×10^{-4} square centimeter. The cathode voltage drop is given by investigators as being from 9 volts to 9.9 volts, which is less than 10.4 volts, the ionizing potential of mercury, but this may be justified on the basis of cumulative ionizing effects.[21,24] The work function of mercury is 4.5 volts. Assuming the arc drop to be 9.9 volts, $9.9 - 4.5 = 5.4$ volts. Referring to Table 2, this is *not* sufficient voltage to produce ionization in one step, but is greater than the first critical excitation potential, thus making cumulative ionization possible.

Arc-Back. Referring to Fig. 3-17, electron current flows from the mercury pool to the anode on the right when this anode is *positive*, and to the one on the left during the part of the cycle that the left one is *positive*. Under normal operating conditions, very little reverse (electron) current flows in the opposite direction. If the tube is overloaded, however, the temperature of the mercury vapor will be increased, and the vapor pressure will also increase. This lowers the insulating properties of the mercury vapor, and decreases the peak inverse anode voltage that a tube will stand between the electrode, negative at a given instant, and the mercury pool. The tube then has a tendency to **arc back** or **backfire,** which may permanently damage the tube.[21,25,26]

3-20. *MULTIANODE MERCURY-ARC RECTIFIERS*

A single-phase glass-enclosed rectifier having two anodes was described in the preceding section. Polyphase rectifiers having more than two electrodes in a glass envelope have been constructed. Multianode rectifiers for polyphase service ordinarily use steel tanks, however. In a typical six-phase rectifier, six main anodes and auxiliary electrodes are placed in the same steel tank and use the same mercury-pool cathode. Six-phase rectifiers have less "ripple" (alternating components) in the rectified output, and have other advantages.[21] Multianode rectifiers have been used extensively for providing direct-current power for electric railways and electrochemical processes, to name but two important uses. However, the use of multianode rectifiers is decreasing, for reasons to be explained.

When a steel tank is used, at least three important problems exist. *First,* the anodes must be insulated from the steel tank, and from each other; *second,* the mercury pool must be insulated from the steel tank; and *third,* some method other than tipping must be used to start the rectifier.

It is apparent why the anodes must be insulated from the steel tank, but it is not evident why the mercury pool must be insulated. The reason is this: In the mercury-arc rectifiers that have been considered thus far, and in the type now under consideration, the cathode spot must be maintained at all times because if this is not done, the arc goes out, and the rectifier must be started again. There is a tendency for the cathode spot to "wander" and for the arc to leave the mercury pool and transfer to the steel tank. This would be disastrous, and would probably melt a hole in the tank. If both the mercury pool and the anodes are insulated, the arc cannot transfer to the tank.

Regarding the third point, starting sometimes is accomplished with an insulated starting anode that is lowered down into the mercury pool, and is then quickly withdrawn, causing an arc, ionization, and the establishment of a cathode spot. Energized excitation anodes then maintain the arc as long as the rectifier is in operation. Large fluctuations of the load current are therefore possible without danger of the arc becoming extinguished. Various types of baffles, shields, and grids are used around the main anodes of multi-anode mercury-arc rectifiers. In general, these serve two main purposes: *First,* they reduce the tendency to arc back, and *second,* in some rectifiers they give control over the starting of anode-current flow. Baffles, shields, and grids, singly or in combination, reduce the tendency to arc back in the following ways: *First,* they protect the anodes from particles and drops of condensed mercury, and from blasts of mercury vapor. Drops of mercury on an anode might have "cathode spots" formed on their surfaces, and this would make possible reverse-current flow and arc-back. *Second,* they decrease the deionizing time, tending to keep the ion density lower and the insulating properties of the vapor near the anode higher. *Third,* they "even out" the voltage distribution near the anodes, reducing strong fields that might pull electrons from the anode.

As previously mentioned, grids in some tank-type rectifiers give control over the anode-current flow. In such devices the grids that surround the anodes are insulated from them, and a connection to each separate grid is brought out of the tank. It is then possible to connect externally a control voltage between the grid and other electrodes and control the anode current.

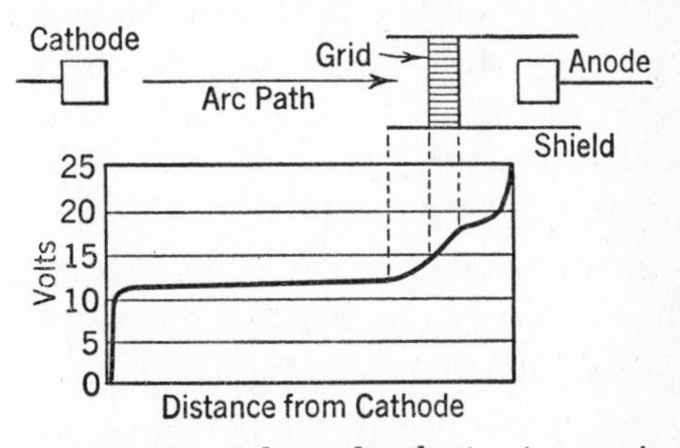

FIG. 3-18. Voltage distribution in a typical mercury-vapor rectifier.

Although these baffles, shields, and grids reduce the tendency to arc back, and also provide control, they reduce the efficiency of the rectifier from a power standpoint. As was previously mentioned, the cathode potential or fall is a little less than 10 volts. This is the first drop in potential across the rectifier. There is then a small drop in voltage along the arc path. A considerable drop in voltage exists near the anode. This may be very low if the anode structure is simple, or fairly high for special structures as Fig. 3-18 indicates. The drop in potential across a moderately sized glass-tube mercury-arc rectifier is about 15 volts, and for large tank-types may be from 20 to 25 volts, depending on the constructional details.

3-21. *SINGLE-ANODE MERCURY-ARC RECTIFIERS*

The tendency to arc back, and other limitations of the multianode mercury-arc rectifier, led to the development of the single-anode type, which largely has superseded the multianode device. Several of the single-anode rectifiers are used for rectification in polyphase circuits where formerly the multianode type with several anodes in one container was employed. The **Ignitron,** developed about 1933, and the **Excitron,** developed a few years later, are the single-anode mercury-arc rectifiers used.

The Ignitron. This consists essentially of an anode, a mercury pool, and an **Ignitor** in a glass or metal container. The tube is "fired" at the part of the cycle desired by passing a current through the Ignitor, which is a semiconductor such as silicon carbide that permanently dips into the mercury pool. The mercury does not make a good contact with this semiconducting tip. If an electric current from some outside source is passed between the Ignitor and the pool, small sparks result at the point of contact, a cathode spot is formed on the mercury surface, and if the anode is positive, an arc is established to the anode.[27,28,29]

As a simple illustration of the operation of the Ignitron, Fig. 3-19 is included. When the alternating voltage is directed so that the anode is positive, the rectifier element passes current through the Ignitor, thus firing the tube. When the direction of the alternating voltage is such that the anode is negative, the tube is deionized, and also, because of the rectifier, the igniting current is negligible. The Ignitron cannot (ordinarily) pass current in the reverse direction because electrons are not emitted from the relatively cold graphite anode. With suitable circuit arrangements, the tube may be fired at any part of the positive half-cycle. This flexibility is one of the important features of the Ignitron, and will be discussed in Chapter 6.

Glass tubes or steel tanks are used for Ignitrons, depending on the size, use, and other factors. Small Ignitrons are of the "sealed off" type, but large Ignitrons are connected to an evacuating system.[30,31] An Ignitron can be designed with an internal voltage drop much below that of the multianode type.

Advantages of the Ignitron are as follows: *First,* arc-back is reduced because the inverse voltage is on the tube when the vapor largely is deionized. *Second,* the anode may be close to the mercury pool, giving a low internal voltage drop and high efficiency. *Third,* the Ignitron can be fired at any point on the cycle, providing control. (A similar control feature exists with certain types of multianode rectifiers, but the Ignitron principle has advantages.) *Fourth,* no excitation or keep-alive electrodes are needed. *Fifth,* the Ignitron has advantages over the thermionic gas tubes in that the mercury-pool cathode instead of a thermionic cathode is used. The mercury-pool cathode has excellent characteristics; its life is unlimited, and it will stand high overloads.

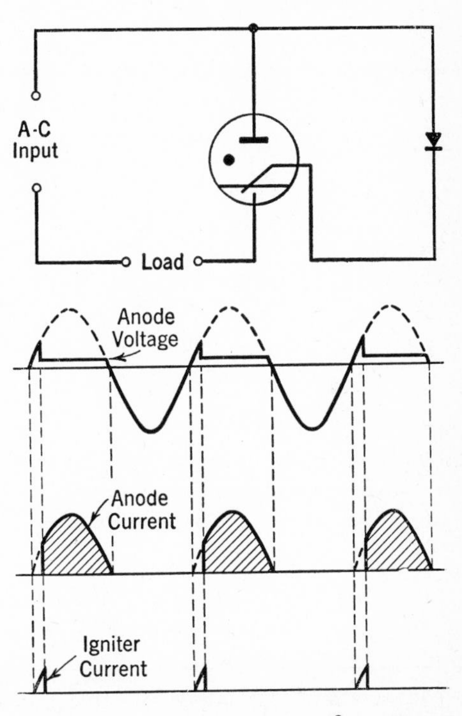

FIG. 3-19. The Ignitron rectifier starts to conduct at the beginning of each positive half-cycle. A resistance load is assumed.

The Excitron. This rectifier consists of a container enclosing a mercury-pool cathode, an anode with a surrounding control grid, and an excitation anode a short distance above the center of the pool. A device "shoots" a jet of mercury against the excitation anode to start an excitation arc. The excitation anode is at all times about +35 volts with respect to the pool, and once the arc is started, this anode continuously maintains it. Accordingly, ionization occurs in the Excitron at all times. The rectified current flow to the main anode is controlled by the potential of the grid.[32,33]

An important question arises: If the excitation anode maintains an arc and ionization occurs at all times, it may appear that arc-back would occur readily in an Excitron. Furthermore, it might be thought that arc-back would not occur in an Ignitron in which no excitation arc is maintained. This opinion early was held, but tests and experience showed that in common types of rectifiers most arc-backs occurred at the conclusion of the conducting period of an anode, and that arc-backs were caused by residual ionization near the anode.[32] A small excitation arc *near the cathode* was found to have negligible effect on the ion current to a anode during the nonconducting period, and the ion current *to the anode* at the end of the rectifying period is considered to be the major cause of arc-back.[32] In reality Ignitrons and Excitrons are three-electrode devices and hence should be in the next chapter; however, it seems advisable to consider them with mercury-arc rectifiers of the older types.

REFERENCES

1. Institute of Radio Engineers. *Standards on Electron Tubes, Definitions of Terms,* Proc. I.R.E., April 1950, Vol. 38, No. 4. *Methods of Testing,* Proc. I.R.E., Aug., Sept., 1950, Vol. 38, Nos. 8 and 9.
2. American Institute of Electrical Engineers. *American Standard Definitions of Electrical Terms.* 1941.
3. McArthur, E. D. *Electronics and Electron Tubes.* John Wiley & Sons. See also articles in General Electric Review, March–Sept., Nov., Dec. 1933.
4. Pidgeon, H. A. *Simple theory of the three-electrode vacuum tube.* Journal of the Society of Motion Picture Engineers, Feb. 1935, Vol. 24.
5. Langmuir, I., and Compton, K. T. *Electrical discharges in gases.* Reviews of Modern Physics, Apr. 1931, Vol. 3, No. 2.
6. Koller, L. R. *The Physics of Electron Tubes.* McGraw-Hill Book Co.
7. Dushman, Saul. *Electron emission.* Electrical Engineering, July 1934, Vol. 53, No. 7.
8. Thompson, B. J. *Space-current flow in vacuum-tube structures.* Proc. I.R.E., Sept. 1943, Vol. 31, No. 9.
9. Wilson, W. *A new type of high power vacuum tube.* Bell System Technical Journal, July 1922, Vol. 1, No. 1.
10. Slepian, J., and Mason, R. C. *Electric discharges in gases.* Electrical Engineering, Apr. 1934, Vol. 53, No. 4.
11. Westinghouse Electric Corporation. *Industrial Electronics Reference Book.* John Wiley & Sons.
12. Radio Corporation of America. *Broadcast Equipment. RCA Tube Handbook—Transmitting Equipment.*
13. Hull, A. W. *Gas-filled thermionic tubes.* Trans. A.I.E.E., July 1928, Vol. 47.
14. Clark, G. L. *X-rays—What should we know about them?* Electrical Engineering, Jan. 1935, Vol. 54, No. 1.
15. Gray, F., Horton, J. W., and Mathes, R. C. *The production and utilization of television signals.* Trans. A.I.E.E., Vol. 46, 1927. Bell System Technical Journal, Oct. 1927, Vol. 6.
16. Ives, H. E., and Johnsrud, A. L. *Television in colors by a beam scanning method.* Journal of the Optical Society of America, Jan. 1930, Vol. 20. Also, Bell System Monograph B-448.
17. Kock, W. E. *Generating sine waves with gas discharge tubes.* Electronics, March 1935, Vol. 8, No. 3.
18. Moore, D. F. *Development of gaseous-tube lighting.* Electrical Engineering, Aug. 1931, Vol. 50, No. 8.
19. Kober, P. A. *Neon tube sign lighting.* Electrical Engineering, Aug. 1931, Vol. 50, No. 8.

20. Dushman, Saul. *Low pressure gaseous discharge lamps.* Electrical Engineering, Aug. and Sept. 1934, Vol. 53, Nos. 8 and 9.
21. Prince, D. C., and Vogdes, F. B. *Principles of Mercury Arc Rectifiers and Their Circuits.* McGraw-Hill Book Co.
22. Amick, C. L. *Fluorescent Lighting Manual.* McGraw-Hill Book Co.
23. Illuminating Engineering Society. *IES Lighting Handbook.*
24. Guntherschulze, A. *Electric Rectifiers and Valves.* John Wiley & Sons.
25. Jolley, L. B. W. *Alternating Current Rectification.* John Wiley & Sons.
26. Slepian, J., and Ludwig, L. R. *Backfires in mercury arc rectifiers.* Trans. A.I.E.E., March 1932, Vol. 51.
27. Slepian, J., and Ludwig, L. R. *A new method of starting an arc.* Electrical Engineering, Sept. 1933, Vol. 52, No. 9.
28. Ludwig, L. R., Maxfield, F. A., and Toepfer, A. H. *An experimental Ignitron rectifier.* Electrical Engineering, Jan. 1934, Vol. 53, No. 1.
29. Marshall, D. E. *Ignitor characteristics and circuit calculations.* Trans. A.I.E.E., Vol. 66, 1947.
30. Herskind, C. C., and Remscheid, E. J. *Performance of pumped Ignitron rectifiers.* Trans. A.I.E.E., Vol. 67, 1948.
31. Herskind, C. C., and Remscheid, E. J. *Development of pumpless Ignitron.* Trans. A.I.E.E., Vol. 70, 1951.
32. Winograd, H. *Development of excitation-type rectifier.* Trans. A.I.E.E., Vol, 63, 1944.
33. Housley, J. E., and Winograd, H. *Alcoa rectifier installation.* Trans. A.I.E.E., Vol. 60, 1941.

QUESTIONS

1. What is meant by diode, cathode, and anode?
2. What was the first practical use of the diode?
3. How is the term load used in electronics?
4. Why is the peak inverse anode voltage important? Is there any relation between this and the direct-voltage output of a complete rectifier unit?
5. What is meant by the term saturation?
6. What causes a vacuum diode and a gas diode to exhibit the property of resistance?
7. What effect does space charge have on plate current?
8. Under what conditions does a vacuum diode operate as a constant-current device?
9. How do you account for the statement in section 3-4 that there is a higher electron density near the cathode, even after voltage saturation?
10. During manufacture a diode is evacuated to a high degree. During normal use will the gas pressure probably increase or decrease? How will this effect the operation of the tube?
11. Referring to Fig. 3-8, why do not the two curves coincide near zero values?
12. Again referring to Fig. 3-8, what causes the decrease near the center of the curve for the gas tube?
13. Is the discharge in a thermionic gas diode rectifier an arc or a glow? Is it maintained or non-self-maintained?

14. Why are oxide-coated thermionic cathodes used in gas diodes? Are they ever of the heat-shielded type?
15. It is stated in section 3-9 that the cathode fall is equal to the lowest excitation potential. What does this mean and why is it true ?
16. In Fig. 3-6 a small drop in potential is indicated at the anode. How is this possible?
17. Why does temperature affect a mercury-vapor tube to a marked degree?
18. How do vacuum and gas diodes compare as power rectifiers?
19. Why should the cathode of a gas diode be brought up to operating temperature before the anode voltage is impressed?
20. How can a high-pressure gas diode operate without filament heating current once it is started?
21. How are the quality and quantity of x-rays controlled in a modern tube?
22. When a glow lamp is operated from a constant-voltage source, what precautions should be taken?
23. What are the outstanding advantages of the mercury pool as a cathode?
24. What causes arc-back and what developments have been made to reduce its occurrence?
25. Compare the efficiencies of the multianode and the single-anode mercury-arc rectifier. Which type probably will supplant the other? Why?

PROBLEMS

1. For a certain vacuum diode I_b is 10 milliamperes and E_b is 60 volts. Assuming the three-halves power law holds, what is the value of the perveance G?
2. Obtain data for a triode experimentally, from a tube manual, or from some other source, plot these data, and find the value of m and G discussed in section 3-5.
3. A plane thermionic cathode and a plane anode are each 10 square centimeters in area and are 0.1 centimeter apart. What current will flow at 10, 50, and 100 volts, assuming that the current is space-charge limited?
4. A straight tungsten filament is 0.0035 centimeter in diameter, is 3.5 centimeters long, and is at the center of a cylinder that is 0.5 centimeter in diameter and 3.5 centimeters long. What current will flow between the cathode and anode when the difference of potential between them is 100 volts? When it is 1000 volts? Assume that the current is space-charge limited.
5. A heated tungsten filament is at the center of a cylinder 1.0 centimeter in diameter. What is the maximum current per square centimeter that can flow if the impressed voltage is 225 volts? What is the minimum diameter of the filament for this condition?
6. Using the data given in section 3-11, calculate the cathode heating power, the peak power loss in the tube, the average power loss at full load, the maximum direct-voltage output, the total power handling capacity, the anode-circuit efficiency, and the total efficiency considering the heating power. Assume the drop to be constant across the tube at a normal value.
7. Assume that when delivering 600 amperes direct current at 600 volts the drop from pool to anode in a mercury-arc rectifier is 15 volts. Calculate the power

loss in the rectifier, and the efficiency of the rectifier, exclusive of auxiliary equipment.

8. Assume that baffles, shields, and grids placed in the arc stream of the rectifier of problem 7 increase the voltage drop to 22 volts, and calculate the power loss and efficiency.

CHAPTER 4

THREE-ELECTRODE ELECTRON TUBES

Although the underlying principle of the thermionic vacuum tube was discovered by Edison [1] in 1883, and the diode was used by Fleming in 1904 to detect high-frequency radio signals, the vacuum tube was relatively unimportant until De Forest, in 1906, introduced a third electrode, the **control grid.** Until 1912 the **Audion,** as this early three-electrode tube was called, was used only as a detector,[2] although it was more sensitive than diode detectors because of the control and addition of energy (amplification) from a battery connected to the anode.[2] By 1912 the Audion and associated circuits for use as an amplifier had been developed.[2] The earliest extensive commercial use of the triode was as an amplifier in wire telephone systems.[3,4]

The early Audion marks the beginning of the field of electronics. A three-electrode tube now is classed as a **triode,** defined [5] as "a three-electrode electron tube containing an anode, a cathode, and a control electrode." Both vacuum and gas triodes will be discussed in this chapter. The discussion will be confined almost entirely to basic electronic principles, leaving applications and special tubes to later chapters.

4-1. *THE CONTROL GRID*

This is defined [5] as "a grid, ordinarily placed between the cathode and an anode, for use as a control electrode." The grid is often in the form of a spiral of fine wires surrounding a cylindrical cathode. The current of electrons flowing from the cathode to the plate must pass through the grid. The *control* grid is, therefore, in a strategic position, and largely can control this plate-current flow.[6] Because there is no other grid in a triode, it is common practice to drop the word control. In the triode (as in the diode), electrons are emitted by a hot cathode. The emission of these negative electrons leaves the cathode positively charged, and hence there are forces tending to draw the electrons back. If the grid and the plate of a triode are connected directly to the hot cathode, a few of the emitted electrons will have initial energies sufficiently high to carry them across the space within the tube, to one of these electrodes, and on around the circuit and back to the cathode. Most of the electrons will remain in a relatively thin region or sheath around the cathode, charging this region negatively and thus forming a negative space charge. They will then repel other electrons as they are emitted from the cathode, and a condition soon will be established where the electrons are repelled to the cathode at the same rate they are emitted (neglecting any high-velocity electrons that may reach the grid or plate).

If the grid is left directly connected to the cathode, and the plate is made positive with respect to the cathode, the positive electric field produced by the plate partly will neutralize the negative space charge, and will draw electrons from this region. The voltage distribution within the tube and between the grid wires will now be approximately as shown in Fig. 3-4. If the plate is made highly positive, the negative electrons will be taken as fast as they are emitted, and *voltage saturation* will be reached as discussed in section 3-3. The space-charge sheath will now be altered, and the electrons will be more uniformly distributed throughout the region between the electrodes. For the conditions discussed, Figs. 3-2 and 3-3 of the preceding chapter apply to triodes. It is important to note that in most instances *triodes are operated well below voltage saturation.* (They are, therefore, temperature saturated, section 3-3.) The space charge can be considered as a thin negative sheath close to the cathode, and the magnitude of this negative charge limits the flow of electrons for a given plate voltage.

The grid is placed between the cathode and the plate, and is closer to this space-charge sheath than is the plate. If the grid is made slightly positive with respect to the cathode, the positive electric field of the grid partly neutralizes the negative space-charge field, thus permitting the positive electric field of the plate to draw over a *larger* electron current. If the grid is made slightly negative, then the negative electric field near the cathode is made stronger and the plate current becomes smaller. The grid commonly is made negative with respect to the cathode so that negligible current flows in the grid circuit, and for other reasons (sections 4-11 and 4-20).

4-2. *STATIC CHARACTERISTIC GRAPHS—GRID NEGATIVE*

The theoretical relations between grid voltage, plate voltage, and plate current considered in the preceding section can be shown by means of static characteristic graphs, or curves, obtained with a circuit arranged as in Fig. 4-1. In discussing such circuits and graphs, certain standardized terms are used. Thus, **electrode voltage** is defined [5] as "the voltage between an electrode and the cathode or a specified point of a filamentary cathode. (Note— The terms grid voltage, anode voltage, plate voltages, etc., are used to designate the voltage between these specific electrodes and the cathode. Unless otherwise stated, electrode voltages are understood to be measured at the available terminals.)"

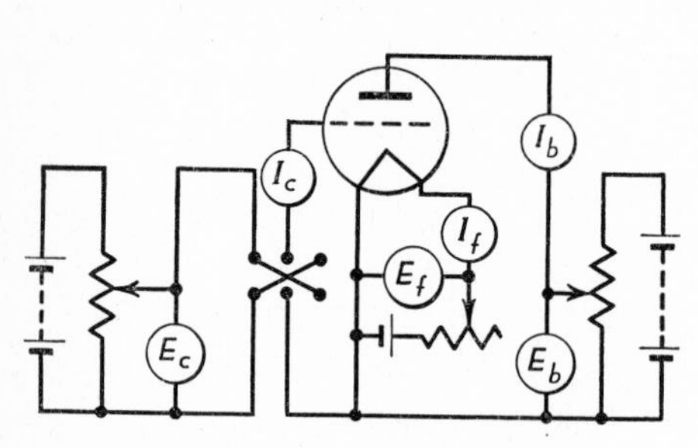

FIG. 4-1. Circuit for determining the static characteristics of a triode.

The magnitude of the direct grid voltage, or **grid bias**,[5] as it commonly is called, is varied by the voltage divider, and is indicated in Fig. 4-1 as E_c. By operating the reversing switch, the grid bias may be made either positive or negative. When the grid is slightly negative, or zero, or positive with respect to the cathode, grid current I_c flows. The magnitude of the plate voltage E_b between plate and cathode may be varied with the voltage divider and the resulting plate current I_b determined.

For obtaining the characteristic curves of Fig. 4-2, the cathode current is held constant at the normal value, the grid voltage is held constant at the values indicated on the curves, and the plate voltage is varied, giving the plate-current values. All the curves of this "family" have approximately the same shape and are about equally spaced.

The curves of Fig. 4-3 may be plotted from Fig. 4-2, or in the following manner: Cathode current is held constant at the normal value, the plate voltage is held constant at the values indicated, and the grid voltage is varied, giving the plate currents, which are plotted as curves. It will be noted that in this family each curve has about the same shape, and for equal plate-voltage increments, the curves are almost equally spaced. For each different plate voltage shown, the plate current can be reduced to zero by making the grid sufficiently negative. The grid will attract electrons and conduct a current when it becomes positive. Grid-current curves have not been plotted, but will be considered in section 4-11. Triodes operated with positive grids will be treated in Chapter 8.

Curves taken in the above manner with direct voltages on the electrodes are known as **static characteristic curves** as distinguished from load, or dynamic, characteristic curves, section 4-16. In the curves of Fig. 4-2, the grid voltage is constant, and in those of Fig. 4-3, the plate voltage is con-

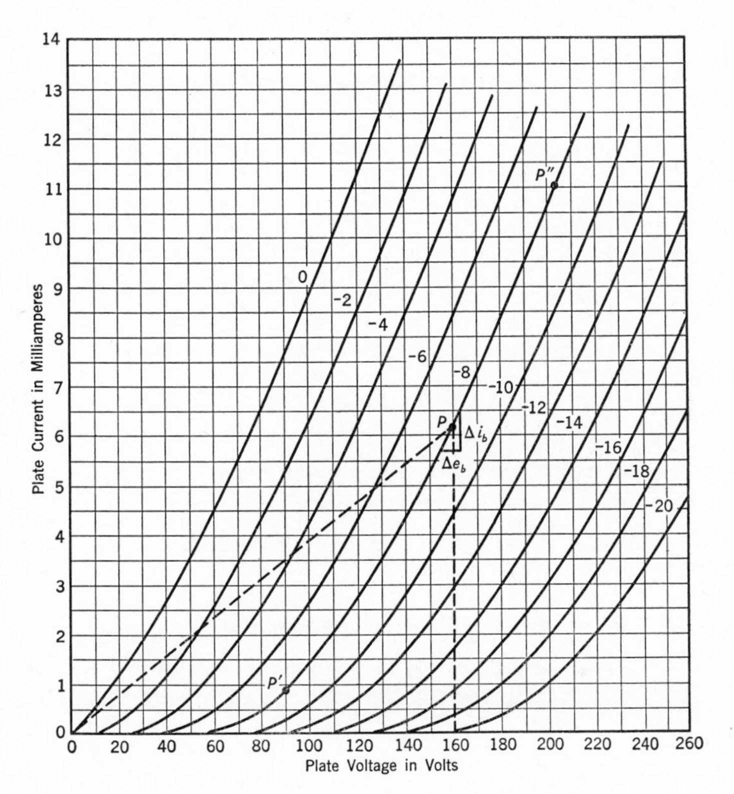

FIG. 4-2. Plate characteristics of a small-signal triode. Numbers on curves are grid voltages in volts.

stant; these are, therefore, **constant grid-voltage curves** and **constant plate-voltage curves.** With proper equipment, characteristic graphs, or curves, can be displayed on an oscillograph[7,8] and point-by-point plotting thereby avoided.

4-3. *VACUUM-TUBE COEFFICIENTS*

In the design of vacuum-tube equipment certain tube coefficients must be known. These coefficients, as they are designated in the standards,[5] have also been termed constants, and parameters. Tube coefficients may be evaluated from characteristic graphs, or curves, or may be measured with approved circuits.[5] For triodes, reliable coefficients are obtained from characteristic curves, but for tetrodes and pentodes, this graphical method is unsatisfactory.

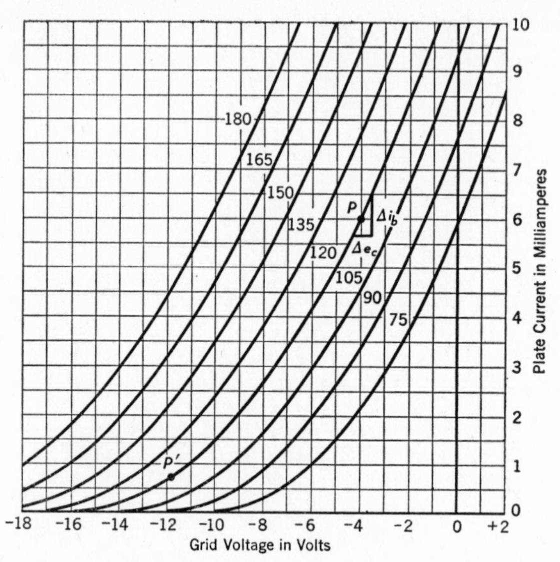

FIG. 4-3. Grid-plate characteristics of a small-signal triode. Numbers on curves are plate voltages in volts.

The definitions of the coefficients are generalized to apply to tubes of many types and over a wide frequency range. The abbreviations to be used as subscripts in specifying the electrodes in a tube are specified [5] as follows:

f, filament
h, heater
k, cathode
g and *c*, grid
p and *b*, plate or anode
s, metal shell or shielding envelope
d, deflecting, reflecting, or repelling electrostatic electrode
j, *l*, *m*, etc., general abbreviation for any electrode
1, 2, 3, etc., grid subscripts for multigrid tubes

The distinctions between *g* and *c*, *p* and *b*, and other details will be explained in section 4-18.

The generalized definitions on which the common tube coefficients are based now will be given essentially as in the standards.[5] Small, sinusoidal alternating voltages and currents, and linear relations are assumed.

Short-Circuit Driving-Point Admittance (of the *j*th Terminal of an *n*-

Terminal Network): The driving-point admittance between that terminal and the reference terminal when all other terminals have zero alternating components of voltage with respect to the reference point. Thus in Fig. 4-4a the short-circuit driving-point admittance is the driving-point, or input, admittance measured between terminals j and m when l and n are short-

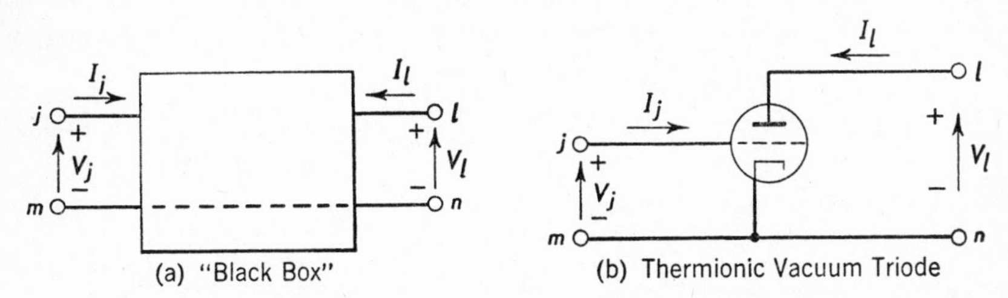

FIG. 4-4. The "black box" serves only to hide from view the enclosed unknown active network, and has no effect on electrical measurements. Conventional directions [5] of voltages and currents are shown.

circuited so that no alternating components of voltage can exist between them.

Electrode Admittance (of the jth Electrode of an n-Electrode Electron Tube): The short-circuit driving-point admittance between the jth electrode and the reference point measured directly at the jth electrode. (To be able to determine the intrinsic electronic merit of an electron tube, the driving-point and transfer admittances must be defined as if measured directly at the electrodes inside the tube.)

Thus in Fig. 4-4b the electrode admittance is the driving-point, or input admittance, measured between electrodes j and m with l and n short-circuited. (If not measured directly at the electrodes, the effect of the electrode leads, base, and tube socket would be included.)

Electrode Impedance: The reciprocal of the electrode admittance.

Electrode Conductance: The real part of the electrode admittance.

Electrode Resistance: The reciprocal of the electrode conductance. (This is the effective parallel resistance and is not the real component of the electrode impedance.)

μ-Factor (of an n-Terminal Electron Tube): The ratio of the magnitude of infinitesmal change in the voltage of the jth electrode to the magnitude of an infinitesmal change in the voltage of the lth electrode under the conditions that the current in the mth electrode remains unchanged, and the voltages of all other electrodes be maintained constant.

For the vacuum triode of Fig. 4-4b, which is an active (as distinguished from passive) network, the output voltage will exceed the input voltage for common circuit conditions, and the μ-factor of interest is the ratio of $\Delta V_l/\Delta V_j$ under the condition that I_l is maintained constant.

Transfer Admittance (from the *j*th Terminal to the *l*th Terminal of an *n*-Terminal Network): The quotient of the complex alternating component I_l of the current flowing to the *l*th terminal from the *l*th external termination by the complex alternating component of voltage V_j applied to the *j*th terminal with respect to the reference point when all other terminals have arbitrary external terminations.

The transfer admittance from input to output terminals of Fig. 4-4a is I_l/V_j for an arbitrary termination, in this instance a short circuit.

Interelectrode Transadmittance (*j-l* Interelectrode Transadmittance of an *n*-Electrode Electron Tube): The short-circuit transfer admittance from the *j*th electrode to the *l*th electrode.

For the triode of Fig. 4-4b the interelectrode transadmittance is I_l/V_j with terminals *l* and *n* short-circuited.

Interelectrode Transconductance (*j-l* Interelectrode Transconductance): The real part of the *j-l* interelectrode transadmittance.

4-4. *AMPLIFICATION FACTOR*

Referring to Fig. 4-3, it is seen that with the plate 135 volts positive, no plate current flows if the grid is made 17 volts negative. Thus the grid, due to its greater influence on the space charge near the cathode, is 135/17 or about 7.95 times more effective than the plate. The **amplification factor** is defined[5] as "the μ-factor for the plate and control-grid electrodes of an electron tube under the condition that the plate current is held constant." As applied to the thermionic vacuum triode, the amplification factor is the ratio of an infinitesmal change in plate voltage to an infinitesmal change in grid voltage under the condition that the plate current remains constant. The symbol is the Greek letter μ.

In the first paragraph of this section was given a numerical example of the greater effect of grid voltage than plate voltage on plate current. These relations hold only for the portions of the curves at very low plate-current values, and when used as an amplifier to increase signal strength, a tube is not operated at these low values. Instead, a tube usually is operated as an amplifier at some point such as *P* of Figs. 4-2 or 4-3. Referring to the latter figure, a graphical determination of the amplification factor, over the region of normal operation, is made as follows: With +135 volts on the plate, and −6 volts bias on the grid, the plate current is 5.6 milliamperes. With +120 volts on the plate, and −4.4 volts on the grid, the plate current remains at 5.6 milliamperes. The approximate amplification factor, represented by μ, for this particular tube when operated at the voltage values being considered is

$$\mu = \frac{135 - 120}{6.0 - 4.4} = \frac{15}{1.6} = 9.4$$

The amplification factor also can be determined from Fig. 4-2. With a grid bias of −6 volts and a plate voltage of +132 volts, the plate current is 5.5 milliamperes. With a grid bias of −4 volts and a plate voltage of +113 volts, the plate current is 5.5 milliamperes. The approximate value of the amplification factor is $\mu = (132 - 113)/(6 - 4) = 9.5$

The amplification factor can be determined from **constant plate-current curves** plotted as in Fig. 4-5. Selecting the 5.0-milliampere curve, −6 volts on the grid and +127 volts on the plate determines the operating point *P*. If, now, the small triangle Δe_b, Δe_c is constructed at this point *P*, the amplification factor is approximately

$$\mu = \frac{\Delta e_b}{\Delta e_c} = \frac{19}{2} = 9.5 \qquad 4\text{-}1$$

The amplification factor for different triodes varies from about 3 to as high as 100, depending on the purposes for which they are designed. General-purpose triodes have amplification factors of about 20.

It will be noted that the amplification factor of equation 4-1 is the slope of the 5.0-milliampere constant-current curve; also, this slope is seen to be negative because grid-voltage values are negative. The amplification factor can be expressed mathematically as $\partial e_b/\partial e_c$, partials being used because there is more than one variable. This expression will be negative, if positive and negative senses are considered, because the slope of the curve is negative. If the amplification factor is to be positive, and if it is to be

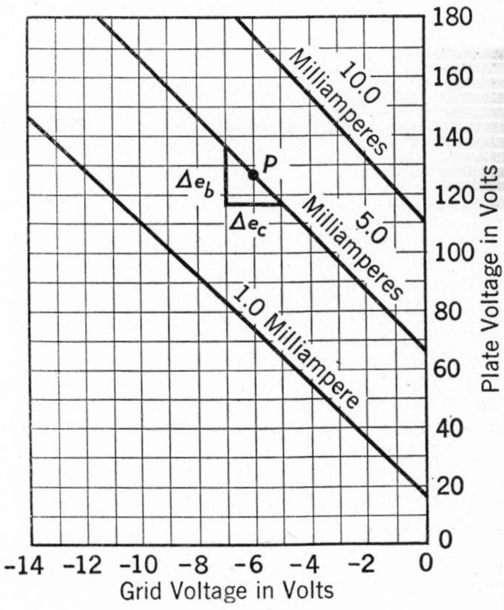

FIG. 4-5. Constant plate-current curves plotted from data of Fig. 4-2.

defined by an equation, then the $\partial e_b/\partial e_c$ must be preceded by a negative sign, giving

$$\mu = -\left[\frac{\partial e_b}{\partial e_c}\right]_{i_b \text{ constant}} \tag{4-2}$$

In fact, the 1933 I.R.E. standards used this equation in defining amplification factor. This caused confusion because some asserted μ defined thus was negative. (See page 385, Aug. 1940, Proc. I.R.E., Vol. 28, No. 8.) To avoid this uncertainty, the 1943 standards did not use equation 4-2, but defined μ as "the ratio of the change in plate voltage to a change in control-electrode voltage, under the conditions that the plate current remains unchanged and that all other electrode voltages are maintained constant. It is a measure of the effectiveness of the control-electrode voltage relative to that of the plate voltage upon the plate current. The sense is usually taken as positive when the voltages are changed in opposite directions. As most precisely used, the term refers to infinitesimal changes." The 1950 standards [5] quoted in the preceding section state clearly that the μ-factor, on which the amplification factor is based, is a *ratio of magnitudes.* Thus, positive and negative sense does not enter the definition, and the amplification factor μ is merely a numerical ratio.

4-5. *PLATE CONDUCTANCE AND RESISTANCE*

The general definitions for electrode admittance, electrode conductance, and electrode resistance were given in section 4-3. The terms plate admittance, etc., are defined if the word *plate* is substituted for *electrode* in these general definitions. More specifically, the plate conductance is [9] the quotient of the in-phase component of plate alternating current by the plate alternating voltage, all other electrode voltages being maintained constant. It should be noted that this is the **alternating-current plate conductance** and *not* the direct-current conductance. Referring to Fig. 4-2, the plate conductance is the slope of a plate-voltage-plate-current curve at *a point of operation* [5] such as point P, infinitesimal values being assumed. For Fig. 4-2, the value of the plate conductance represented by g_p is approximately

$$g_p = \frac{\Delta i_b}{\Delta e_b} = \frac{6.5 - 5.7}{163 - 155} = \frac{0.0008}{8} = 0.0001 \text{ mho} \qquad \text{(or 100 micromhos)} \tag{4-3}$$

Mathematically,

$$g_p = \left[\frac{\partial i_b}{\partial e_b}\right]_{e_c \text{ constant}} \tag{4-4}$$

The **alternating-current plate resistance** is the reciprocal of the alternating-current plate conductance.[5] Numerically, it equals the reciprocal of the slope of the curve; for example, the value of the alternating-current plate resistance, represented by r_p, when the tube is operated at point P of Fig. 4-2 is

$$r_p = \frac{\Delta e_b}{\Delta i_b} = \frac{8}{0.0008} = 10{,}000 \text{ ohms} \tag{4-5}$$

If the plate resistance is calculated at point P' instead of point P, the value will be different. Mathematically,

$$r_p = \left[\frac{\partial e_b}{\partial i_b}\right]_{e_c \text{ constant}} \tag{4-6}$$

For operation as an amplifier, if the total plate current composed of the in-phase and out-of-phase components is divided by the plate alternating voltage, the **plate admittance** will be obtained. The reciprocal of this will be the **plate impedance.** These last two terms take into consideration the alternating current that will flow through internal tube capacitances existing between the various electrodes. Sinusoidal currents and voltages are assumed.

The equivalent *direct-current resistance* of the tube of Fig. 4-2 at point P can be determined from the triangle formed by the broken lines intersecting at this point. This direct-current plate resistance is $R_p = 160/0.0061 = 26{,}300$ ohms, a value over 2.6 times that of the alternating-current plate resistance as previously calculated. The alternating-current plate resistance r_p is the term used for most calculations and is the term listed in giving tube data.

4-6. *CONTROL-GRID–PLATE TRANSCONDUCTANCE*

Returning for a moment to the preceding section, *conductance* is a measure of the *ability* of a circuit to conduct alternating current. The plate conductance is the in-phase alternating-plate current divided by the alternating-plate voltage. Note that the current and voltage are for the *same* electrode.

In tubes several electrodes are acting on the same electron current, and mutual effects exist among these electrodes. Thus a voltage impressed between grid and cathode will affect the current in the plate circuit as has been shown by curves such as Fig. 4-3. This is in the nature of a mutual, or transfer, effect, or a carry-over from one circuit to another. Thus, it is possible to define a transconductance just as it is possible to define a conductance.

The **control-grid–plate transconductance** g_{pg} is the name applied to the interelectrode transconductance (section 4-3) between the grid and the plate of a triode. This is an important coefficient in the design of electronic equipment. The terms **grid–plate transconductance** and **transconductance** g_{pg}, or **mutual conductance** g_m, are generally used in practice.

The transconductance from one electrode to another is the quotient of the in-phase component of the alternating current of the second electrode divided by the alternating voltage of the first electrode, all other electrode voltages being maintained constant.[9] This term refers to infinitesimal sinusoidal values. The numerical value of the transconductance can be determined graphically from curves such as Fig. 4-3. Assuming that P is the operating point, the slope of the curve at this point gives the numerical value of the transconductance, or mutual conductance; then,

$$g_{pg} = g_m = \frac{\Delta i_b}{\Delta e_c} = \frac{6.5 - 5.6}{3.6 - 4.4} = \frac{0.0009}{0.8} = 0.001125 \text{ mho, or } 1125 \text{ micromhos} \qquad 4\text{-}7$$

The expression for transconductance, or mutual conductance, can be written

$$g_{pg} = g_m = \left[\frac{\partial i_b}{\partial e_c}\right]_{e_b \text{ constant}} \qquad 4\text{-}8$$

The reciprocal of the transconductance g_{pg} is the **transresistance** r_{pg}, or, the reciprocal of the mutual conductance g_m is the **mutual resistance** r_m. Thus from equation 4-8,

$$r_{pg} = r_m = \frac{1}{g_{pg}} = \frac{1}{g_m} = \frac{1}{0.001125} = 890 \text{ ohms} \qquad 4\text{-}9$$

4-7. *RELATIONS AMONG THE COEFFICIENTS*

The values of μ, g_{pg} or g_m, and r_p vary with operating conditions. Calculations made at some point P' instead of point P of Figs. 4-2 and 4-3 represent different operating conditions and will give different numerical values, as illustrated in Fig. 4-6.

For a given set of operating conditions the following relations exist among the coefficients.

$$\mu = r_p g_{pg} = \frac{r_p}{r_{pg}}, \qquad r_p = \frac{\mu}{g_{pg}} = \mu r_{pg}, \qquad \text{and } g_{pg} = \frac{\mu}{r_p} \qquad 4\text{-}10$$

That these relations are true can be determined by studying the numerical values given by the curves of Fig. 4-6. Or, from equations 4-5 and 4-7,

$$r_p \times g_{pg} = \frac{\Delta e_b}{\Delta i_b} \times \frac{\Delta i_b}{\Delta e_c} = \frac{\Delta e_b}{\Delta e_c} = \mu \qquad 4\text{-}11$$

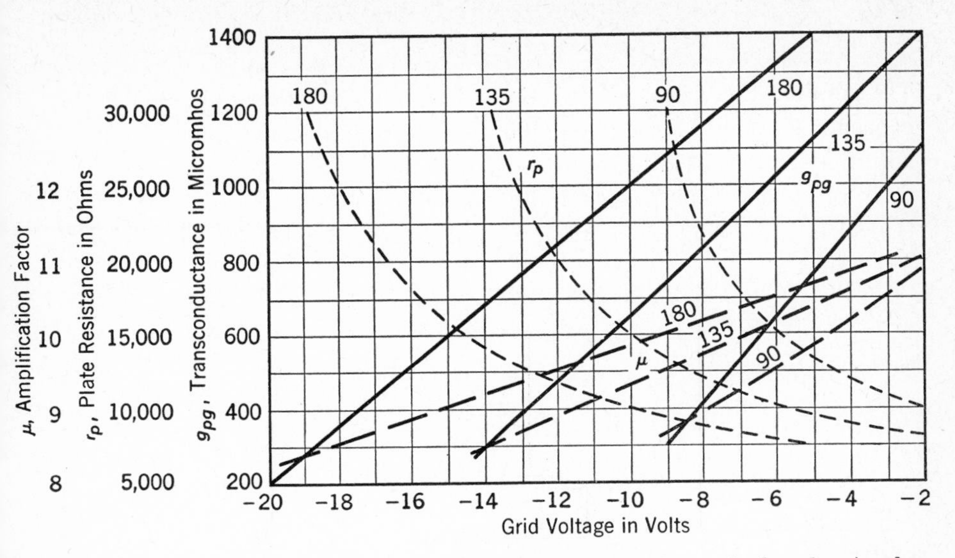

FIG. 4-6. Illustrating variations in amplification factor μ (heavy broken lines), plate resistance r_p (light broken lines), and grid-plate transconductance g_{pg} (mutual conductance g_m) for the small-signal triode having the characteristics of Figs. 4-2 and 4-3. Numbers on curves are plate voltages in volts.

The relationships just given apply correctly only at low frequencies, say up to a few thousand cycles, depending on the tube, circuit, and degree of accuracy desired. For approximate results they may be used at frequencies up to hundreds of thousands of cycles, depending on the situation.

4-8. *MEASUREMENTS OF TRIODE COEFFICIENTS*

Triode coefficients μ, g_{pg}, and r_p may be determined from static graphs as explained in sections 4-4, 4-5, and 4-6. This procedure is important from an instructional standpoint, and the results are reasonably accurate and reliable for low-frequency use. More satisfactory methods use bridge and other special circuits; also, alternating driving voltages. Detailed explanations of these circuits and their use are given in the Standards.[5] The following discussion and included circuits are based on these standards.

Amplification Factor. A circuit for measuring the amplification factor μ of a triode is shown in Fig. 4-7a. When the circuit is balanced by varying C_1 and R_1 or R_2 until a minimum tone is heard in the telephone receivers, the amplification factor is[5]

$$\mu = \frac{R_2}{R_1} \tag{4-12}$$

The amplification factor μ is a ratio with no units of measurement. Variable capacitor C_1 should be of about 100 micromicrofarads, and capable of small adjustments. It is used to compensate for small reactive components so that the null point will be more definite. Either R_1 or R_2 must be variable. In a typical measurement on a triode of $\mu = 20$, R_1 may be 100 ohms and R_2 2000 ohms.

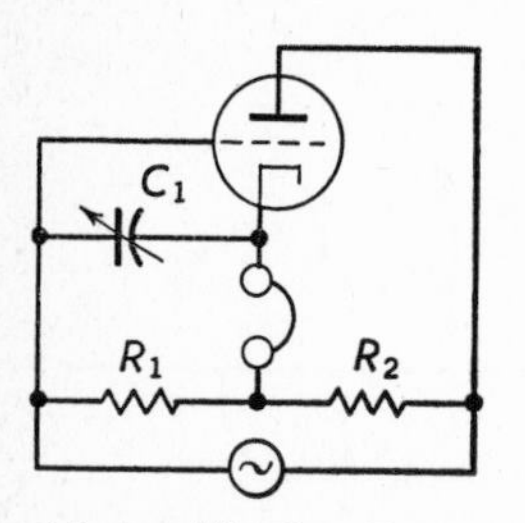

(a) Amplification Factor

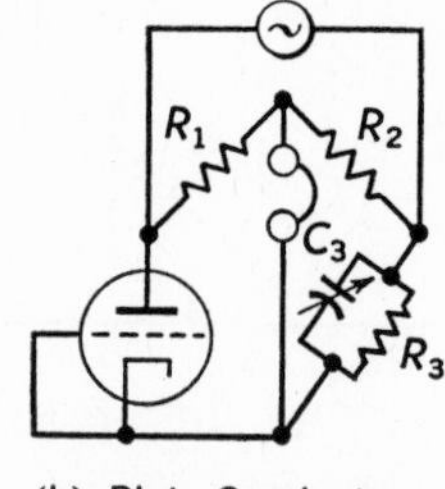

(b) Plate Conductance and Resistance

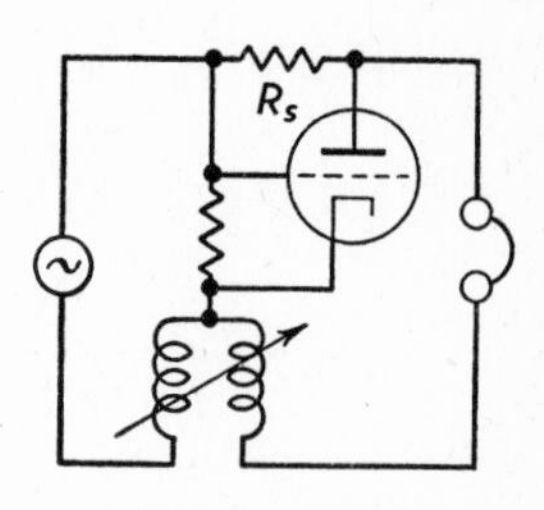

(c) Transconductance

FIG. 4-7. Circuits for measuring triode coefficients. Direct electrode voltages and other details omitted. See *I.R.E. Standards*,[5] from which these diagrams were adapted, for details.

Plate Conductance and Resistance. A circuit for measuring the plate conductance and resistance of a triode is shown in Fig. 4-7b. When the bridge is balanced by varying capacitor C_3 and the resistors until a minimum tone is heard in the telephone receivers, the alternating plate conductance and resistance are [5]

$$g_p = \frac{R_2}{R_1 R_3} \quad \text{and} \quad r_p = \frac{R_1 R_3}{R_2} \tag{4-13}$$

The common units of measure are the mho and micromho for plate conductance, and ohm for plate resistance. The purpose of capacitor C_3 is to "sharpen" the null point as discussed in the preceding paragraph. Resistors R_1, R_2, and R_3 need not all be varied. Resistors R_1 and R_2 are the bridge "ratio arms." As an example, for a triode of $r_p = 20{,}000$ ohms, R_1 may well be 1000 ohms, R_2 100 ohms, and R_3 a variable adjusted to 2000 ohms.

Grid–Plate Transconductance (Mutual Conductance). The transconductance of a triode may be measured using the circuit of Fig. 4-7c. At balance,[5]

$$g_{pg} = g_m = \frac{1}{R_s} \tag{4-14}$$

The units of measure are the mho and the micromho. The variable inductor is to compensate for small reactive components and thus assist in obtaining sharp balance.

Precautions in Measuring Coefficients.[5] The magnitude of the alternating voltage should be so low that varying the magnitude slightly will not affect the measured values. Stray capacitive and inductive couplings that may cause errors or make balancing difficult should be avoided. Grounding and shielding of the circuit may be required.

In general, direct current will flow through the resistors forming the bridge. When this is true, allowances must be made for the direct voltage drops in such resistors, or other devices, so that the correct voltages are actually impressed on the electrodes of the tube under test. In some circuits it is only necessary to place a direct-current milliammeter in the plate circuit, and adjust the applied plate voltage until the desired plate current flows. It is well to use batteries with short leads and tap along the batteries to vary the voltages, rather than to use power supplies, voltage dividers, or other means. Batteries, power supplies, and in particular voltage dividers, should in general have a capacitor connected across them of such value that the capacitive reactance of the test frequency is negligible. However, circuits should be simplified as much as is in keeping with reliable performance.

The types of variable resistors to use in these circuits depend on the types of tubes being tested. The resistors must have sufficient current-carrying capacity. For small triodes the usual laboratory decade resistance box of 10,000 ohms is satisfactory.

The telephone receivers used to indicate balance should be shunted by a coil of high inductance and low resistance, or should be connected through a low-resistance audio-frequency transformer. This greatly reduces or eliminates the direct-current flow through the telephone receivers. An input transformer, often of unity ratio, usually is placed between the bridge and the oscillator to isolate the circuits. Sometimes the secondary of the transformer is shunted with a voltage divider of perhaps a thousand ohms resistance, with the slider grounded. Varying the position of the slider sometimes makes balancing easier. Sometimes the use of shielded and grounded input and output transformers is advisable.

In some circuits one of the bridge elements is shunted with a small variable capacitor to offset reactances and facilitate balancing the bridge; often such capacitors may be omitted. Also, in some circuits mutual inductance is included for ease in balancing, and there are instances where this may be omitted. The extent to which a circuit may be simplified depends largely on experimentation.

A test frequency of 1000 cycles per second is satisfactory. Telephone receivers are sensitive to this frequency. Also, this frequency is sufficiently low for stray capacitive and inductive couplings usually not to be serious. The coefficients measured at 1000 cycles are satisfactory at higher frequencies, the upper limit of reliability depending on circumstances. Additional precautions and other important information will be found in reference 5.

4-9. *DESIGN OF THERMIONIC VACUUM TRIODES*

The design of thermionic vacuum tubes is a highly specialized subject based on theoretical and extensive experimental investigations. This section is limited to citing references on the subject, and to considering in a general way the factors affecting tube coefficients. Early investigators of design equations, in particular for the amplification factor, include King,[10] Miller,[11] Vodges and Elder,[12] and Kusunose [13] who summarized and extended earlier work. More recent writers on the subject include Salzberg,[14] Thompson,[15] Liebmann,[16] Herold,[17] and others.[18]

Amplification Factor. The amplification factor is in reality the ratio of the *effectiveness* of the grid and plate voltages in producing electric fields near the surface of the cathode where the space charge is concentrated. It is the *resultant electric field* acting on the space charge that determines the plate current. This factor depends primarily on the type of grid structure and its location, and will be increased by any modification causing the grid to *shield the cathode* more effectively from the effect of the *plate voltage.* Thus, for a given type of tube, larger grid wires, closer spacing, or greater distance between grid and plate, all *increase* the amplification factor, because all these reduce the effectiveness of the plate.

Plate Resistance. The plate resistance is the property of a triode that causes an alternating voltage drop between the cathode and plate when an alternating current flows through the tube. Looking back into the tube from the cathode and plate terminals, the plate resistance is the internal impedance of the tube at low frequencies where reactive and transit time effects (section 8-23) are negligible. The plate resistance becomes *less* if the area of the plate is increased. It is also made *less* by *increasing* the area of the cathode. *Decreasing* the distance between the anode and cathode also *decreases* the plate resistance, but this may not be advisable since it also lowers the amplification factor. Of course, the grid could be moved nearer the cathode to compensate for such a change in plate-cathode spacing, but because of the heating of the grid by the cathode, there is a practical limit to this.

Transconductance. Since from equation 4-10 the transconductance $g_{pg} = \mu/r_p$, this factor will, of course, vary as either μ or r_p is changed in this equation.

4-10. *PLATE CURRENT IN A THERMIONIC VACUUM TRIODE*

As was shown in section 3-5, the plate current in a *diode* varies (over the central portion of the curve where the plate current is controlled by the space charge) according to a theoretical three-halves power law. In a *triode,* the plate current over the central portion of the curves of Fig. 4-2 is also determined by the potential of the space charge; but, instead of being controlled (as in a diode) only by the plate voltage, the grid voltage must be considered as well. If the negative space-charge potential is increased or decreased by the *fields* from the plate and grid, corresponding *decreases* or *increases* of the plate current will occur.

In a triode the externally applied electric field acting in the negative space-charge region close to the cathode consists of two parts: *first,* the field due to the grid; and *second,* the "stray" field due to the plate which is able to *reach through* the meshes of the grid to the space-charge region.[19] The more effective this shielding, the less will be the influence of the plate with respect to the grid, and the higher will be the amplification factor. The resultant externally applied electric field in the space-charge region near the cathode is the *algebraic sum* of the field produced by the grid-bias voltage E_c and the "stray" field due to the *effective* plate voltage E_b/μ. Hence, the total field is proportional to $E_c + (E_b/\mu)$. This is sometimes called the **equivalent grid voltage.** Or, the total field at the cathode would be proportional to $E_b + \mu E_c$, called the **equivalent plate voltage.** Hence, the plate current is given by the theoretical three-halves power equations

$$I_b = G'\left(E_c + \frac{E_b}{\mu}\right)^{3/2} \qquad 4\text{-}15$$

$$I_b = G(\mu E_c + E_b)^{3/2} \qquad 4\text{-}16$$

The terms G' and G are perveances,[5] and their magnitudes, which are different, are determined by the geometrical construction of the tube. In so far as the plate current is concerned, an **equivalent diode** can be designed for any triode.

As in the case of the diode, various factors may cause the plate current to deviate from the three-halves power law. That is, the exponent may be other than 3/2 or 1.5. This exponent usually lies between the theoretical 1.5 and about 2.5, depending on the tube. Since the value 2.0 is somewhat of an average value, the plate-current curve is sometimes said to follow a square law. The value of this exponent can be determined as for the exponent of equation 3-1. On log-log paper, values of $(E_b/\mu) + E_c$ are plotted on the X axis, and values of the plate current are plotted on the Y axis. The slope of the curve is the value of the exponent applying to that particular tube. At any value of I_b, all terms of the equations except G or G' are then known, and hence these can be determined.

4-11. *GRID CURRENT IN A THERMIONIC VACUUM TRIODE*

Grid Negative. In the preceding discussions, no mention was made of the grid currents that flow in a triode. It may be thought that when the grid is negative, no current will flow since it will *repel* the negative electrons. But, a *negative* grid will attract *positive* ions, and there are other reasons why a current will flow.[5]

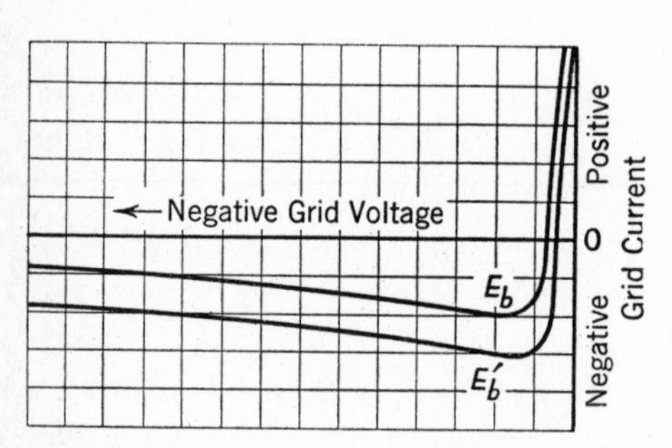

FIG. 4-8. Negative grid-current curves for a triode at plate voltage E_b and a larger plate voltage E_b'. Positive grid current occurs when an excess of electrons over positive ions is flowing to the grid inside the tube.

First, there is always some residual gas, and the plate current produces *positive* ions by collision with gas atoms; these positive ions flow to the negative grid. *Second,* the hot cathode gives off a few positive ions, and these also flow to the negative grid. This effect is usually negligible. However, because of high initial velocities and contact differences of potentials, some of the electrons from the cathode may flow to the grid. *Third,* there is some leakage between the grid and cathode, and this allows some current to flow. *Fourth,* the grid may become so hot that thermal emission may occur from it. With the grid negative, these currents are *very low,* giving a high equivalent resistance. Also, these currents vary widely for different tubes, and with the age of the tube. The manner in which the grid current varies for a *negative* grid is shown in Fig. 4-8. Since curves for different tubes of even the same type vary so widely, numerical values are not shown.

Thus if a circuit such as Fig. 4-1 is studied, it will be found that with the plate at E_b volts and the grid at zero volts, an *electron* current flows to the grid because of the initial velocity of the electrons. As the grid is made negative, *positive ions* will be attracted to the grid and the *electrons* will be repelled. At some grid voltage the number of positive ions and electrons will be equal and the net grid current will be zero, as Fig. 4-8 indicates. As the grid is made more negative, the grid current reverses and increases to a maximum value and then *decreases* very slowly. This peculiar shape of the grid-current-grid-voltage curve may be explained as follows: As the grid is made *more negative,* more positive ions are attracted to it. This cannot continue for long, however, because a more negative grid means *less plate-current flow,* and the plate current determines the amount of ionization of the residual gas. This causes the grid current to decrease slowly after passing through the maximum value. *Increasing* the plate voltage to a new value E_b' *increases* ionization by collision, thus providing more positive ions and a greater current flow, as Fig. 4-8 indicates. In actual operation as an amplifier, the grid is usually maintained negative.

Grid Positive. The variations in the grid current for a *positive* grid (instead of negative as just considered) are shown in Fig. 4-9. The currents that flow to a positive grid are so *very much larger* than those flowing to a highly negative grid that Fig. 4-9 is plotted to a *much smaller scale* than is Fig. 4-8. As the grid is made more positive, the current to it increases quite rapidly. At some point, the electrons have velocities sufficient to produce secondary emission from the grid. As the grid voltage is made more positive, these secondary electrons are released in large numbers and *are attracted to the plate.* A study of Fig. 4-1 will show that this effect will cause the *net* grid current to decrease as in Fig. 4-9.

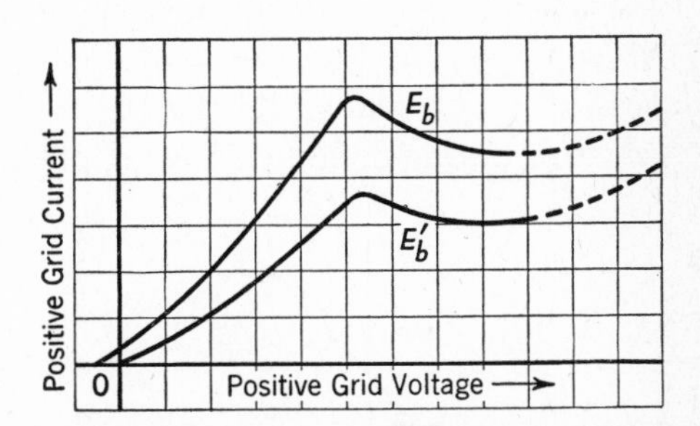

FIG. 4-9. Positive grid-current curves for a triode at plate voltage E_b and a larger plate voltage E_b'.

If the plate is made more positive to some potential E_b', the current flow to the grid will be less, as indicated, because the more positive plate takes a greater percentage of the electrons withdrawn from the space-charge region. Similarly, the secondary emission effects will be less pronounced. Theoretically, as the grid is made more positive it should capture the secondary electrons, and the grid current should again increase as shown by the dotted lines of Fig. 4-9. Actually, however, the tube might be ruined before this action was apparent. Care should be taken in operating a tube with a positive grid as it may be permanently damaged. The resistance of the grid circuit varies from a very high value when the grid is negative to a low value of a few thousand ohms or less when it is positive.

4-12. *STATIC CHARACTERISTIC GRAPHS—GRID POSITIVE*

The characteristics of a small thermionic vacuum triode were given in Figs. 4-2 and 4-3. The tubes that have been discussed are used primarily in circuits where little power is involved, and are designated [5] as **small-signal tubes** (indicated by SS). These are thermionic vacuum tubes "designed primarily to function as devices in which power output is not ordinarily an important consideration." [5] This is in contrast with tubes, usually physically large, designated [5] as **power-output tubes** (indicated by PO), which are thermionic vacuum tubes "designed primarily to function as devices in which power output is an important consideration." [5]

The electronic theory of small-signal and power-output tubes is fundamentally the same, although the physical sizes and mechanical details may differ considerably. Small-signal types have been used in most of the past discussions of basic theory, and this will be largely true for subsequent discussions. Past statements that the grid of a triode is usually maintained negative during operation have applied to small-signal tubes.

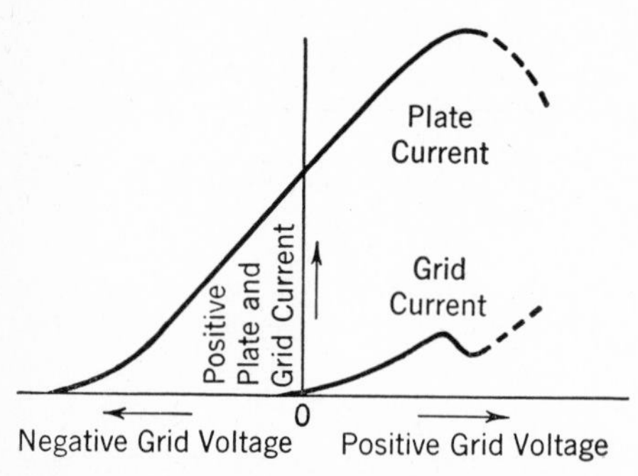

FIG. 4-10. Approximate shapes of grid and plate characteristic curves for a thermionic high-vacuum triode. Negative grid currents omitted. These curves are for one constant value of applied plate-to-cathode voltage.

The grids of power-output triodes commonly are driven positive by the positive peaks of the signal voltage impressed between the grid and cathode. Hence, in the design of circuits for power-output triodes, it is necessary to have available static characteristic graphs showing how the tube functions when the grid is positive.

The characteristics of a thermionic vacuum triode with the grid negative, and with the grid positive, are shown in Fig. 4-10. The part of the curve with the grid negative may be taken with the circuit of Fig. 4-1. This circuit cannot be used for the positive-grid part, because the continuous currents would probably damage, if not destroy, the tube electrodes. In the positive-grid region of Fig. 4-10, the grid current varies as explained in the preceding section. The plate current falls for the following reason: If the grid is sufficiently positive, its action, combined with that of the positive plate, will cause voltage saturation (section 3-3), and the space current will be essentially constant. As the grid is made more and more positive, it will take an increasing percentage of the space current, and hence the plate current will fall.

Static characteristics* with the grid positive are obtained by pulse methods, of which there are two general types.[5] These are point-by-point methods, and curve-tracer methods. The details of these methods will be found in the Standards.[5] Briefly, they are as follows:

Point-by-Point Methods. The cathode of the tube is properly energized. The grid is biased to some reference value which may be cut off. The plate is made positive to the value desired. A positive pulse of voltage is then impressed between the grid and cathode by suitable switching means, commonly electronic. This positive pulse drives the grid positive for an instant to the value desired. Peak magnitudes of the grid and plate currents are observed, conveniently on a cathode-ray oscillograph. If a repetitive pulse is used, observations are simple. The instantaneous grid and plate currents may be large, yet the electrodes will not overheat if the pulse duration is short. The process is repeated point-by-point at different plate and grid voltages.[5,20,21]

Curve-Tracer Methods. With these methods, the characteristic curves are displayed on the tube of a cathode-ray oscillograph, where they may be photographed or directly observed. Such methods are elaborate, and references 5, 22, and 23 are recommended for details.

4-13. *PLATE CURRENT WITH GRID-SIGNAL VOLTAGE—WITHOUT PLATE LOAD*

The effect of a grid-signal voltage on the plate current in a triode can be investigated, using Fig. 4-3. The 135-volt curve has been selected and the required portion reproduced in Fig. 4-11a. The grid of the triode is biased −6 volts, and the curve indicates that the plate current when no signal is impressed is about 5.6 milliamperes. When the source of signal voltage impresses the positive half-cycle of +2 volts, the net voltage between the grid and cathode is −4 volts, and the resulting plate current at that instant will be about 7.7 milliamperes. When the negative half-cycle of −2 volts is impressed, the net voltage between grid and cathode is −8 volts, and the resulting plate current will be 3.8 milliamperes.

From the preceding paragraph it is seen that a pulsating current consist-

* The I.R.E. Standards[5] define the static characteristic as a "relation usually represented by a graph, between a pair of variables such as electrode voltage and electrode current with all other voltages maintained constant." Hence, the term may be applied to characteristics other than those obtained by a circuit using direct voltages such as Fig. 4-1.

ing of a direct-current component of 5.6 milliamperes and an alternating-current component that rises to 7.7 and falls to 3.8 milliamperes will flow in the plate of the tube when operated as has been explained. The alternating-current component is not exactly sinusoidal because the positive half-cycle will be $7.7 - 5.6 = 2.1$ milliamperes, and the negative half-cycle will be $5.6 - 3.8 = 1.8$ milliamperes. This is because the curve is not exactly a straight line, and represents signal distortion. Had the curve been straight, then the maximum and minimum values each would have been about 2.0 milliamperes.

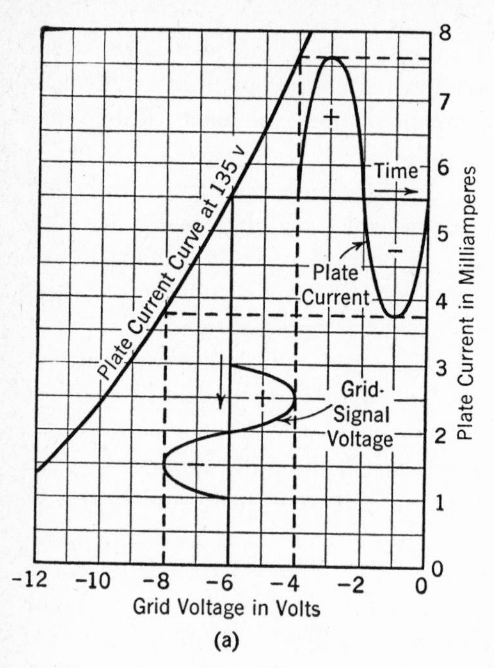

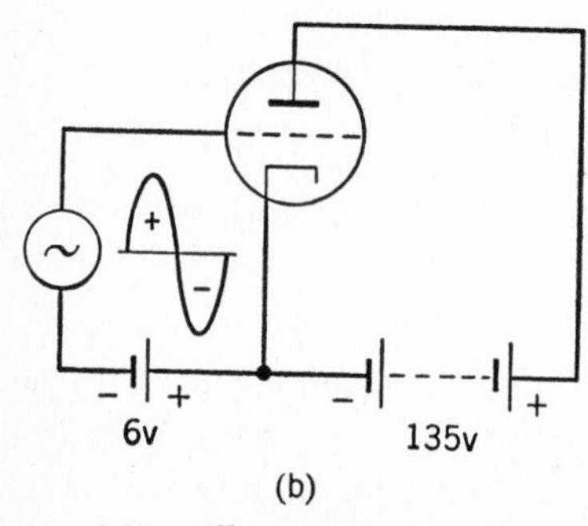

FIG. 4-11. Illustrating the plate-current variations in a thermionic vacuum triode when an alternating signal voltage is impressed in series with the grid-bias voltage.

Attention is called to the fact that the plate-current variations and the grid-voltage variations are in phase. Note that the positive half-cycle of applied grid-signal voltage causes the plate current to rise, and that the negative half-cycle of applied grid-signal voltage causes the plate current to fall.

4-14. *EQUIVALENT CIRCUIT FOR A THERMIONIC VACUUM TRIODE*

Assume that the triode of Fig. 4-11b and all energizing batteries are in a "black box" as in Fig. 4-12a. This situation corresponds to Fig. 4-4a, and the available terminals have been numbered the same; also, the same positive directions of voltages and currents have been indicated. The voltages have been represented by E instead of V. It is desired to find a circuit that is equivalent to the triode.

If the two terminals l and n are connected through an alternating-current measuring device A of zero internal impedance, if an alternating signal voltage of +2 and −2 volts maximum is impressed between terminals j and m, and if the curve of Fig. 4-11a is *assumed straight,* then the current-measuring device will indicate an alternating current of +2.0 and −2.0 milliamperes maximum. Because the grid is always negative, the input impedance between terminals j and m will be assumed infinite, and the current flow assumed zero.

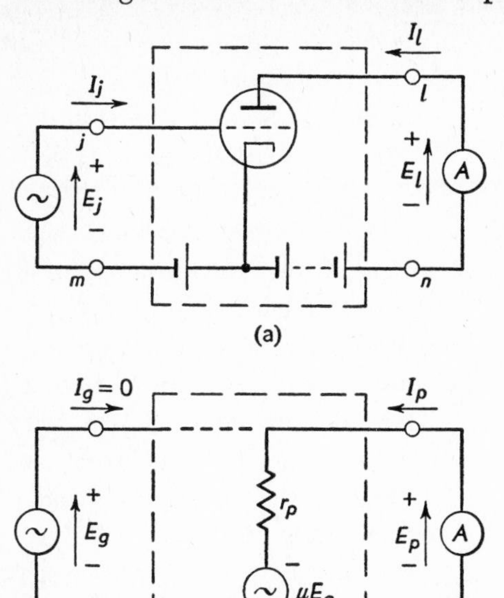

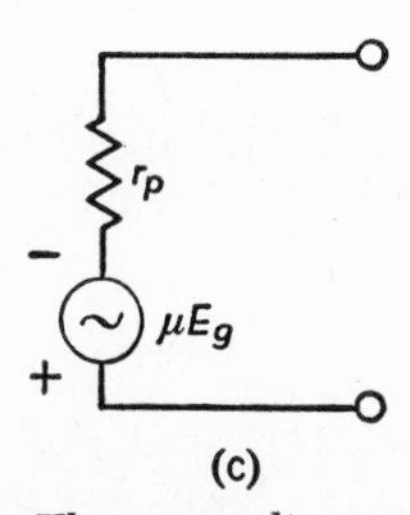

FIG. 4-12. The upper diagram shows the thermionic vacuum tube of Fig. 11b and energizing batteries "hidden from view in a black box," with only input and output terminals available. The triode performs exactly as if it were a generator of voltage μE_g (having the instantaneous polarity shown in the center diagram) and internal resistance r_p. The equivalent circuit is simplified further in the lower diagram.

The slope of the curve of Fig. 4-11a at the operating bias point of −6 volts is (section 4-6) $g_{pg} = 0.001$ mho. That is, $\Delta i_b/\Delta e_c = 0.001$ mho. Thus, $\Delta i_b = \Delta e_c g_{pg} = 2.0 \times 0.001 = 0.002$ ampere, or 2.0 milliamperes. From equation 4-10, $g_{pg} = \mu/r_p$ and it can be written that $\Delta i_b = \Delta e_c \mu/r_p$. Considering alternating quantities,

$$I_p = \frac{\mu E_g}{r_p} \qquad 4\text{-}17$$

From this it follows that the circuits of Fig. 4-12a and 4-12b are equivalent as far as terminals $j-m$ and $l-n$ are concerned. Furthermore, because the grid-input impedance with negative bias may be assumed infinite, in most instances it is satisfactory to use Fig. 4-12c as the equivalent circuit. The statements that have been made apply at low frequencies and for small-signal magnitudes. Otherwise, reactive circuit effects and nonlinear coefficients are involved.

4-15. *PLATE CURRENT WITH GRID-SIGNAL VOLTAGE—WITH RESISTANCE LOAD*

As explained in the preceding section, a triode without plate load and biased to operate on the part of the plate-current curve that is essentially straight can be represented by Fig. 4-12c. If a load resistor is placed in the plate circuit as in Fig. 4-13a, the equivalent circuit will be as shown in Fig. 4-13b. This principle sometimes is stated [24] as an "equivalent plate-circuit theorem." This is, of course, but an application of Thévenin's theorem (section 7-11).

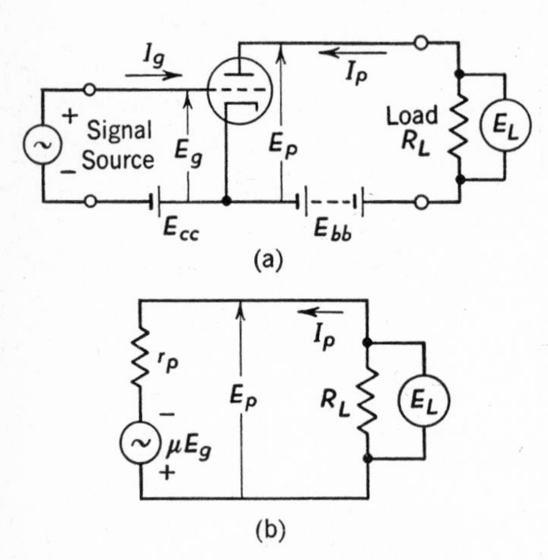

FIG. 4-13. A simple triode amplifier (*a*) and equivalent alternating-current circuit (*b*).

Neglecting all direct currents and voltages, the alternating-voltage equation for Fig. 4-13b is

$$-\mu E_g + I_p r_p + I_p R_L = 0 \quad \text{4-18}$$

The effective, or rms, value of the alternating current I_p, or I_L, flowing in the circuit is

$$I_p = \frac{\mu E_g}{r_p + R_L} \quad \text{4-19}$$

and the voltage drop across the load resistor as indicated by the alternating-current voltmeter is

$$E_L = I_p R_L = \frac{\mu E_g R_L}{r_p + R_L} \quad \text{4-20}$$

This alternating voltage drop will vary as the impressed grid voltage E_g and with correct adjustments will be almost an exact replica of it. The weak alternating voltage impressed on the grid has been amplified, and the amplified voltage E_L may be impressed on the grid of a second tube and further amplified if desired.

If the **voltage amplification factor** A_v for the circuit of Fig. 4-13 is the ratio of the output alternating voltage to the input alternating voltage, then,

$$A_v = \frac{E_L}{E_g} = \frac{\mu E_g R_L}{E_g(r_p + R_L)} = \frac{\mu R_L}{r_p + R_L} \quad \text{4-21}$$

Furthermore, the power output of the tube is

$$P_o = I_p^2 R_L = \frac{\mu^2 E_g^2 R_L}{(r_p + R_L)^2} \quad \text{4-22}$$

If it is desired, the resistor R_L can be replaced by an impedance Z_L, and the equations can be generalized. These equations apply only at low frequencies and for small magnitudes of voltage and current so that distortion is not appreciable.

4-16. *LOAD (DYNAMIC) CHARACTERISTIC GRAPHS*

The *static curves* of Fig. 4-3 were taken *without* a resistor in the plate circuit. But such a load resistor affects the operation of a triode and must be considered. Curves taken (or calculated) with load resistance *in the plate circuit* are known [5] as **load, or dynamic, characteristic curves.** Such curves can be obtained experimentally by slightly modifying the circuit of Fig. 4-1. For obtaining the *dynamic* curves, a resistor of the value at which the load curve is desired, say of 10,000 ohms, should be connected in the plate circuit. The supply voltage should then be *increased* until the plate current at some grid bias (say −6 volts) is the same as with *no* resistance in the plate cir-

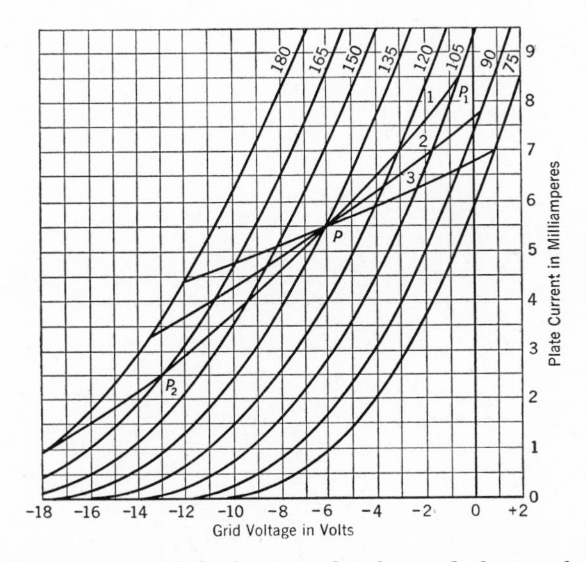

FIG. 4-14. Static curves, and load curves, for the triode having the characteristics shown in Figs. 4-2 and 4-3. Curves 1, 2, and 3 are for resistance loads in the plate circuit of 10,000, 20,000, and 30,000 ohms, respectively. Plate voltages in volts shown on curves.

cuit. The voltage *on the plate* is now the same as with no resistance present. After this adjustment has been made, and with the supply voltage *held constant,* the values of plate current for various grid-bias voltages will give the load characteristic curves of Fig. 4-14. When an impedance load instead of the usual resistance load is used in the plate circuit, the simple method just outlined will not be satisfactory. The cathode-ray oscillograph provides a satisfactory means of obtaining the load characteristic for such conditions.[22,23]

For *calculating* the load characteristic curves, assume that the tube of Fig. 4-13 having the static characteristics of Fig. 4-14 is being used and that the load resistance is 10,000 ohms. With no resistor in the plate circuit, the entire battery voltage E_{bb} is impressed *on the plate*. With a resistor R_L in series, the direct voltage E_b actually *on the plate* is

$$E_b = E_{bb} - I_b R_L \qquad 4\text{-}23$$

since there is a drop in direct voltage across the load resistor caused by I_b, the direct component of the plate current.

If the circuit is adjusted so that the tube operates at point P of Fig. 4-14,

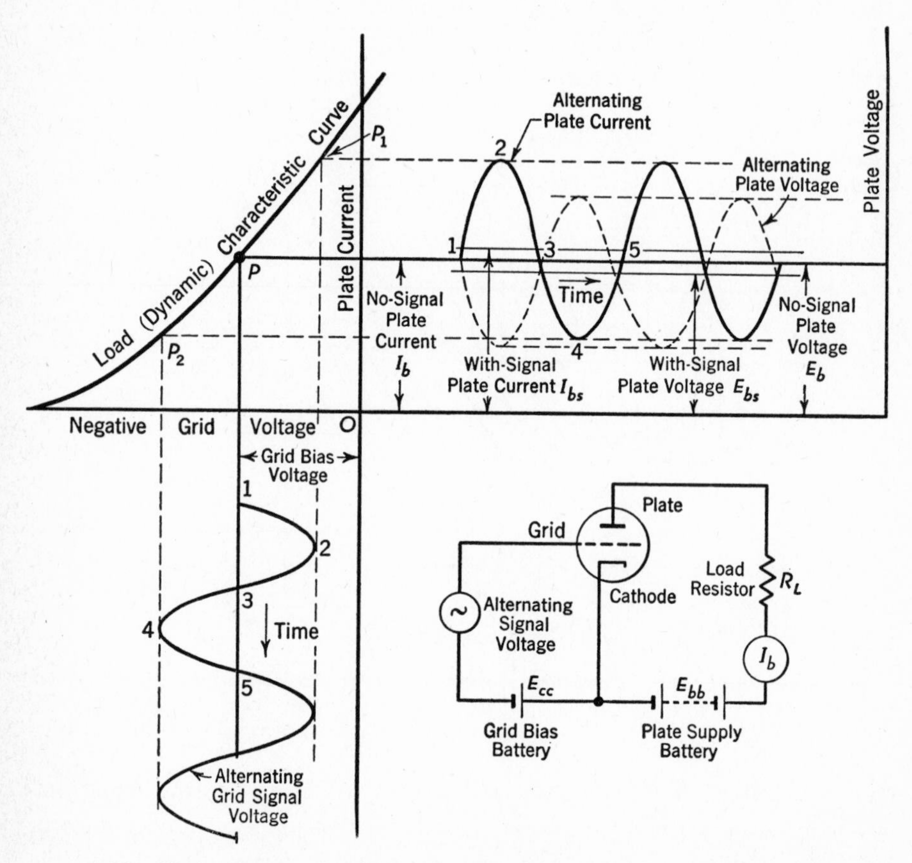

FIG. 4-15. Plate current and voltage variations that occur when an alternating signal voltage is impressed on a triode with negative grid bias and with a plate load resistor. Cathode and plate at normal operating conditions.

it must have +135 volts on the plate and −6 volts on the grid, the plate current will be 5.5. milliamperes, and the total *constant* applied direct voltage will be $E_{bb} = E_b + I_b R_L = 135 + (5.5 \times 10{,}000) = 190$ volts. It is desired to find the plate current that will flow when the negative grid bias is reduced toward zero and the direct voltage actually reaching the plate is 120 volts. For this condition, $I_b \times 10{,}000 = 190 - 120$, and $I_b = 7.0$ milliamperes. This gives a second point on the load curve (the first being point P). It is now desired to find what the current will be when the negative grid bias is further reduced toward zero, and the direct voltage on the plate is 105 volts. For this condition, $I_b \times 10{,}000 = 190 - 105$, and $I_b = 8.5$ milliamperes. This gives a third point P_1 on the load curve, and this curve is then sufficiently well established for grid voltages less than the −6 volt bias value.

Now suppose that the grid voltage is made more negative than −6 volts. This will cause the plate current to decrease, but *not* along the 135-volt plate-voltage curve. The load curve will cross the 150-volt curve at the value $I_b \times 10{,}000 = 190 - 150$, from which $I_b = 4.0$ milliamperes. Similarly, the load curve will cross the 165 plate-voltage curve at point P_2 at the value $I_b = 2.5$ milliamperes. Thus, the load curve for 10,000 ohms in the plate circuit has been established. If the point of operation is at P with −6 volts on the grid and 135 volts on the plate, and if the grid voltage is varied above and below the bias value by an alternating grid-signal voltage as in Fig. 4-15, then the corresponding plate-current variations *will follow this load curve*, and *not* a static curve.

4-17. *PHASE RELATIONS FOR TRIODE AMPLIFIER WITH RESISTANCE LOAD*

A triode operated as an amplifier is shown in Fig. 4-15. Because there is a load resistor R_L in the plate circuit, operation will be in accordance with a load, or dynamic, curve for the resistance R_L. The grid-bias battery fixes the point of operation at point P of the curve. The point of operation is the point on the curve about which the signal varies. With no alternating signal voltage on the grid, the tube is in a **no-signal,** or **quiescent state,** and the plate current indicated by the milliammeter will be constant, having the no-signal value I_b. If the grid-bias battery were increased in voltage, the point of operation would be shifted toward point P_2. This would reduce the plate-current flow. If the grid-bias battery were decreased in voltage, the point of operation would be shifted toward P_1. This would increase the plate-current flow.

Now assume that an alternating signal voltage is impressed in series with the grid-bias battery. During the positive half-cycle of alternating grid voltage, indicated by points 1, 2, 3, the net negative voltage between grid and cathode is less than the negative battery bias voltage. During the negative half-cycle of the alternating voltage, indicated by points 3, 4, 5, the net negative grid voltage between grid and cathode is greater than the bias voltage. The reduction in negative grid voltage causes the plate current to increase as shown by points 1, 2, 3, and the increase in negative grid voltage causes the plate current to decrease as indicated by points 3, 4, 5. Because the load characteristic curve is not exactly a straight line, a sinusoidal grid-signal voltage will not produce a perfectly sinusoidal plate current. The result is that at the instant the alternating signal voltage is applied to the grid, the plate current, as indicated by the direct-current milliammeter, will increase slightly. It no longer indicates I_b, the no-signal value of current flow, but now indicates I_{bs} the average value of the with-signal pulsating current. The no-signal voltage E_b from plate to cathode is shown in Fig. 4-15. When the signal is impressed and the direct plate current rises to I_{bs}, there will be a greater direct voltage drop in the load resistor R_L, and the with-signal direct plate voltage as measured with a direct-current voltmeter between plate and cathode will fall to E_{bs}.

With *no* load resistor in the plate circuit the plate-to-cathode voltage remains constant although the plate current may be varying in accordance with an alternating signal voltage injected in series with the grid bias. This does not hold true, however, if there is a load resistor in the plate circuit.

Referring to Fig. 4-15, when the grid voltage is made less negative by the positive half-cycle 1, 2, 3 of the impressed alternating signal voltage, the plate current increases. This increase in plate current causes an increase in the voltage drop in the load resistor, and the plate voltage decreases below the no-signal or quiescent value.

When the grid voltage is made more negative by the negative half-cycle 3, 4, 5 of the alternating grid-signal voltage, then the plate current decreases, there will be less voltage drop across the load resistor, and the plate voltage will rise above the no-signal value. Thus, with an alternating signal voltage impressed between grid and cathode, a corresponding amplified alternating signal voltage is produced between the plate and cathode. Furthermore, as explained in this section, the positive half-cycle of grid-signal voltage causes the plate current to increase, producing the positive half-cycle of plate current; however, this increases the drop in the load resistor, and the plate-to-cathode voltage falls, resulting in the negative half-cycle of plate voltage. For these reasons, the grid-voltage variations and plate-current changes are *in phase,* but the plate voltage is *180° out of phase* with the grid voltage and plate current. Phasor diagrams will be found in Chapter 7.

4-18. *LETTER SYMBOLS FOR ELECTRON TUBES*

The letter symbols to be used in electronics are listed in the I.R.E. Standards [5] from which Fig. 4-16 has been obtained, and from which the following material is summarized.

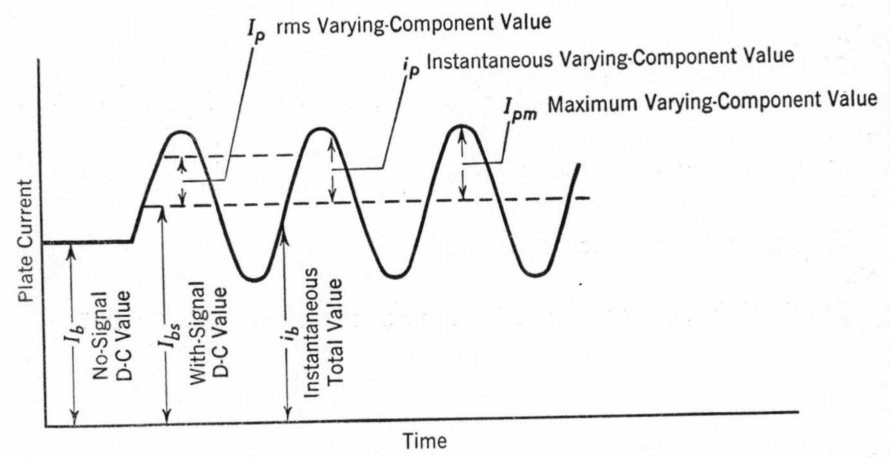

FIG. 4-16. Illustrating letter symbols (for currents) discussed in section 4-18. (Reference 5.)

Instantaneous values of current, voltage, and power that vary with time are represented by lower-case letters; examples, i, e, p, etc. Subscript g is used for grid, and p for plate.

Instantaneous total values of current and voltage (no-signal direct value plus varying-component value, Fig. 4-16) are represented by lower-case letters and subscript c for grid and b for plate; examples, e_c, i_c, e_b, i_b, etc.

Average, maximum, effective (rms) values are represented by upper-case letters; examples, E, I, etc.; if necessary, E_{av} and I_{av} are used for average, and E_m and I_m for effective (rms) values; average values use subscripts b for plate and c for grid; examples, E_b, E_c, etc.; maximum and effective varying components use subscripts p for plate and g for grid; examples, E_p, E_g, etc.

No-signal values are represented by upper-case letters and subscripts b for plate and c for grid; examples, E_b, I_b, E_c, etc.

With-signal average values are represented by adding subscript s to no-signal symbols; examples, E_{bs}, etc.

Supply voltages are represented by upper-case letters and double subscripts; examples, plate supply, E_{bb}, grid supply, E_{cc}, filament supply, E_{ff}, etc.

External values of resistance, impedance, admittance, etc., in the external circuit of an electrode are represented by upper-case letters; examples, R_p, Z_p, Y_p, etc.

Inherent values of resistance, impedance, and admittance, etc., inherent within the tube, are represented by lower-case letters; examples, r_p, z_p, y_p, etc.

Electrode abbreviations to be used as subscripts are given in section 4-3.

Scalar quantities are represented by single letters in italics; examples, E, I, etc.

Phasor quantities (expressed by complex numbers) are represented by bold italic letters; examples, $\boldsymbol{E}$, $\boldsymbol{I}$, etc.

Vector quantities are represented by bold roman letters; examples, **E**, **I**, etc.

Certain of the most common letter symbols are assembled in Table 7.

Table 7

LETTER SYMBOLS FOR ELECTRON TUBES

(*See also section* 5-8)

CONTROL GRID

e_c	Instantaneous total voltage
i_c	Instantaneous total current
E_c	Average voltage
I_c	Average current
e_g	Instantaneous value of alternating-voltage component
i_g	Instantaneous value of alternating-current component
E_g	Effective or maximum value of alternating-voltage component
I_g	Effective or maximum value of alternating-current component
E_{cc}	Supply voltage

CATHODE

E_f	Voltage impressed
I_f	Current
E_{ff}	Supply voltage
I_s	Saturation current

PLATE

e_b	Instantaneous total voltage
i_b	Instantaneous total current
E_b	Average voltage
I_b	Average current
e_p	Instantaneous value of alternating-voltage component
i_p	Instantaneous value of alternating-current component
E_p	Effective or maximum value of alternating-voltage component
I_p	Effective or maximum value of alternating-current component
E_{bb}	Supply voltage

MUTUAL

C_{gp}	Grid-plate capacitance
C_{gk}	Cathode-grid capacitance
C_{pk}	Plate-cathode capacitance

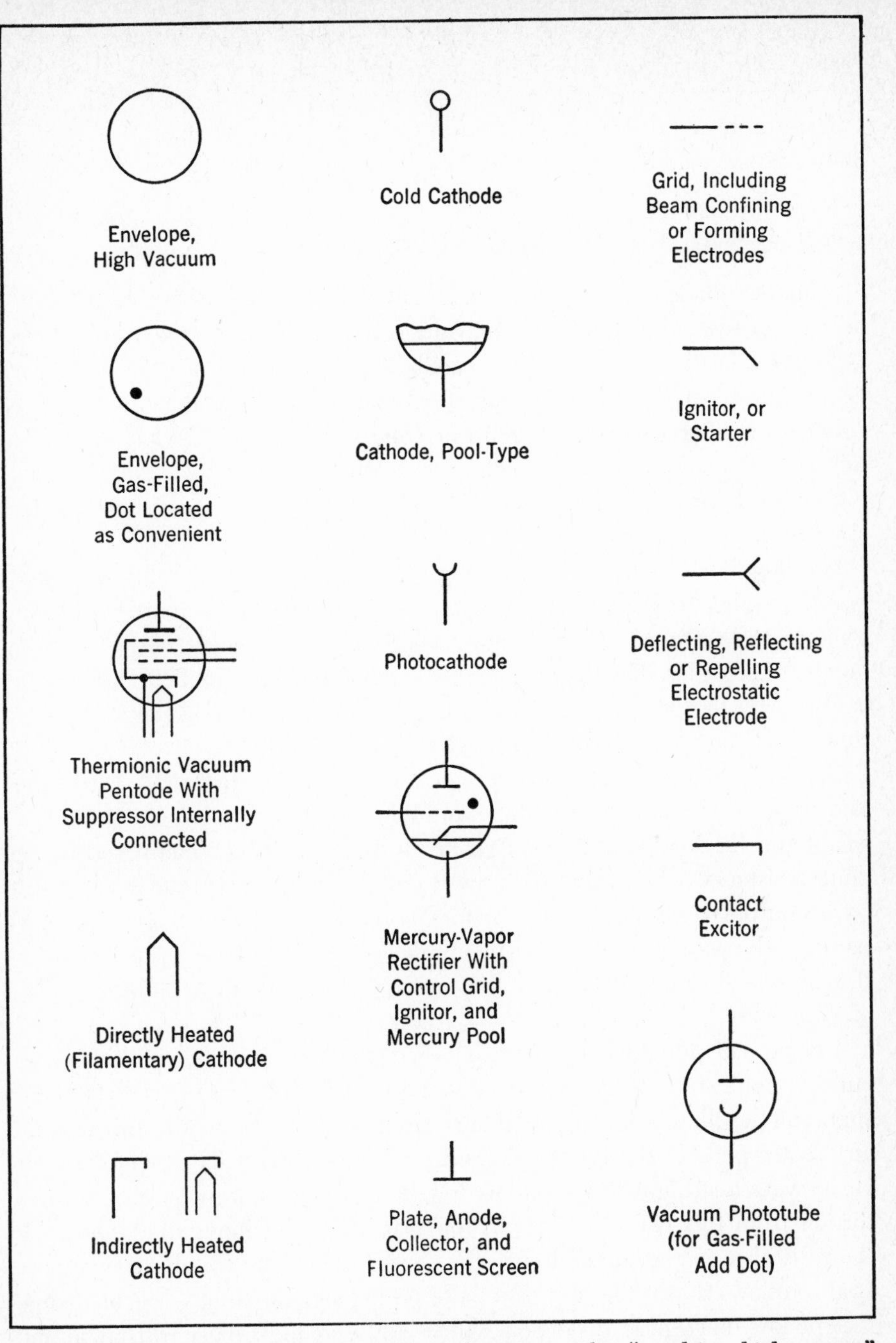

FIG. 4-17. Symbols commonly used in electronics. The "envelope, high-vacuum" symbol sometimes is used to indicate any enclosure, evacuated, or otherwise, a procedure common in drawing transistor diagrams. (Reference 5.)

4-19. *GRAPHICAL SYMBOLS FOR ELECTRON TUBES*

The graphical symbols to be used in electronics are specified in the I.R.E. Standards.[5] Certain of those commonly used are included in Fig. 4-17.

4-20. *VACUUM-TUBE AMPLIFIERS*

Amplification is an important function of thermionic vacuum tubes. Amplifiers are of two fundamental types, **voltage amplifiers,** and **power amplifiers.** These will be discussed in detail in Chapters 7 and 8, but it is important to distinguish at this time between the fundamental principles involved. A voltage amplifier increases the output voltage from some device, such as a microphone or a phonograph pickup, until this voltage is of sufficient magnitude to drive a power amplifier. Small-signal tubes are used in voltage amplifiers.

It follows that in the design of amplifiers both voltage amplification and power amplification must be considered. Or, perhaps it is more correct in some instances to say that voltage amplification and *power output* are the factors to consider. This is true because in some power amplifiers (now to be considered) the grid of the power-output tube is maintained negative at *all* times, and the power input to this tube is negligible.

The basic theory of amplifier design was given in section 4-14. Little was said about distortion. In a voltage amplifier the signal voltages usually are small, and the load resistor is of relatively large resistance compared to the plate resistance when triodes are used. However, the opposite is true with power-output tubes; the load resistor commonly is of low resistance compared to the value used in voltage amplifiers, and the tube often is driven by the impressed signal voltages to the point beyond which distortion becomes objectionable.

A change of signal wave shape, and distortion, occur because the dynamic curve is not a straight line. From Figs. 4-15 and 4-18 it is seen that the wave shape of the plate current, and hence the signal output of the tube, is not an *exact* replica of the input voltage. Also, it is evident that if the dynamic curve were a straight line, then, as Fig. 4-18 shows, there would be no distortion from this cause, at least. From Fig. 4-14 it follows that the larger the resistance in the plate circuit, the straighter will be the dynamic curve. It also follows that if the impressed alternating signal voltage, or **grid swing,** is kept low, negligible distortion will occur. But, high plate resistance and small grid swing limit the power output to low values and thus defeat the purpose of using a power-output tube.

One important reason for maintaining the grid of a power-output tube negative is that if a grid is permitted to go positive with respect to the cathode, the positive grid will draw electrons from the cathode, and the input resistance to the grid circuit will fall to a value of perhaps a few

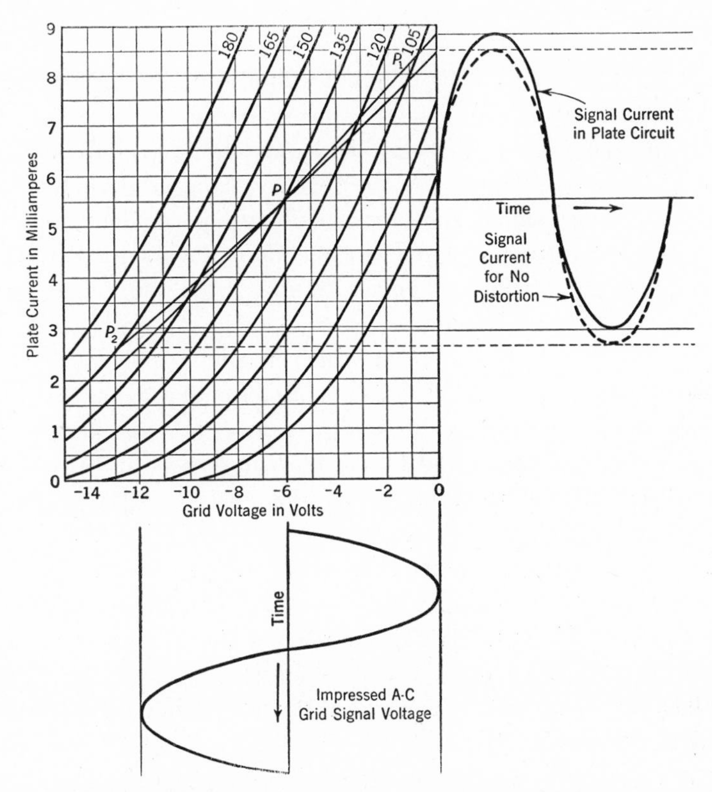

FIG. 4-18. The curvature of line P_2, P, P_1 of Figs. 4-14 and 4-15 distorts the signal output wave so that it is not a replica of the impressed signal. If P_2, P, P_1 were a straight line, the output would follow the dotted curve, resulting in negligible distortion.

thousand ohms. If this happens, the grid will draw power from the driving source, such as a microphone. Unless the signal source can provide this power without producing a distorted voltage, then signal distortion will result. A second important reason for maintaining a negative grid is that such a tube draws less direct-current power from the plate-supply device.

4-21. *MAXIMUM POWER TRANSFER*

It is advisable briefly to discuss at this time the transfer of power from a source to a load, because this is important in amplifier design. Thus in Fig.

4-19 is shown a generator, developing a *constant* open-circuit voltage E and having an *internal* resistance R_G, delivering power to a load resistor R_L. The current flowing through the circuit is $I = E/(R_G + R_L)$, and the power delivered to the load is

$$P = I^2 R_L = \frac{E^2 R_L}{(R_G + R_L)^2} \qquad 4\text{-}24$$

Now let the effect of choosing different values of R_L be studied. If $R_L = 0$, then the current will rise to a high value limited only by the internal resistance of the generator R_G, but since R_L is zero, no power will be delivered. Now let R_L be very large, approaching an infinite value. The current and the power delivered now approach zero. At some value between these two limits **maximum power transfer** from the generator to the load occurs.

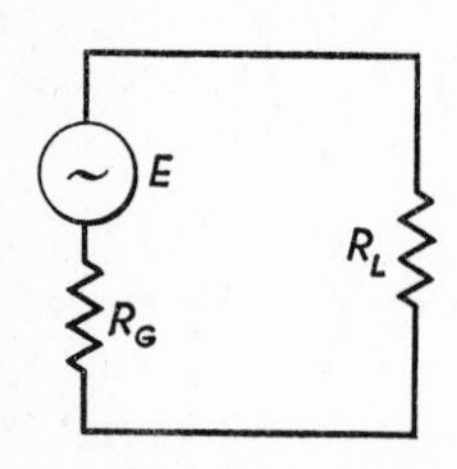

FIG. 4-19. Maximum power flows from a generator to a load when the resistance of the load equals the internal resistance of the generator.

The point at which maximum power is transferred can be found by differentiating equation 4-24 and equating the result to zero. It also may be obtained by assuming values of R_L with respect to R_G (for example $R_L = 0.5R_G$, $R_L = 1.0R_G$, and $R_L = 1.5R_G$) and substituting these values in equation 4-24. It will be found that the maximum power is transferred from a source to a load *when the resistance of the load equals the internal resistance of the source.* This applies to direct as well as alternating power.

The efficiency of transfer is equal to the *power delivered* divided by the *total power generated.*

$$\text{Efficiency} = \frac{I^2 R_L}{I^2 R_G + I^2 R_L} = \frac{R_L}{R_G + R_L} \qquad 4\text{-}25$$

For the condition of maximum power transfer, when $R_G = R_L$, the efficiency becomes $R_L/(2R_L) = 0.5$ or *50 per cent.*

These relations are plotted in Fig. 4-20 and are the basis of much communication design. Communication networks are not constant-potential circuits as are power circuits. Furthermore, in communication *maximum power output* rather than maximum efficiency usually is desired. It should be noted that the curve for power transfer is *quite flat,* indicating that from the standpoint of *power transfer alone,* some deviation from the relation $R_G = R_L$ can be tolerated.

The relations were derived for maximum power transfer between circuits of pure resistance, but can be extended to cover impedances. Thus, it can be proved that the maximum power will be delivered from one circuit to another when the circuits have *equal* resistances, but equal and opposite reactances (that is, **conjugate impedances**).

Further, it can be shown [25] that when the *magnitude* of the load impedance, but not the angle, can be varied, the maximum power will be taken by the load when the *magnitude* of the load impedance equals the magnitude of the impedance of the source. Transformers can be used for changing impedances, and in this manner transformers are useful in **matching** loads to generators for maximum power output.

The application of these relations to the triode is as follows: As previously shown, the triode may be considered as a generator having a voltage of magnitude μE_g and internal resistance of r_p. It will, therefore, deliver *maximum power* when connected to a load having a resistance equal to the plate resistance r_p. However, *maximum undistorted power* (with distortion less than 5 per cent) is delivered when the relations are such that R_L equals approximately $2r_p$ for the triode (section 8-4).

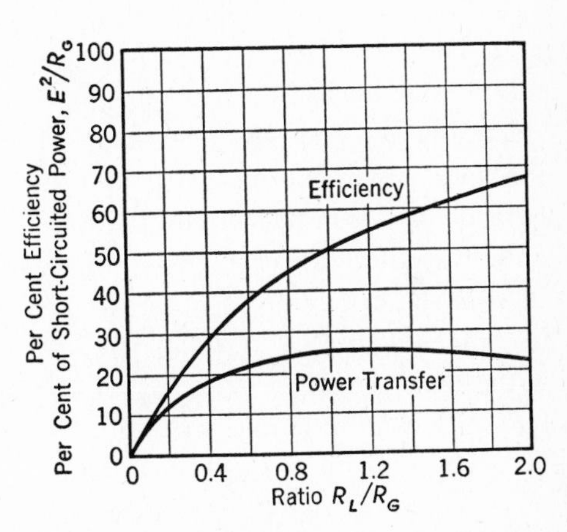

FIG. 4-20. Power output and efficiency for different values of R_L/R_G.

4-22. *VACUUM TRIODES FOR ULTRAHIGH FREQUENCIES*

The triodes that have been described are, in general, useful up to several hundred megacycles. At these frequencies, and at higher frequencies, their performance becomes unsatisfactory for reasons to be considered in Chapters 8 and 9. It is of interest, however, to note that by using a special construction, triodes are made available that will perform satisfactorily at frequencies up to thousands of megacycles.[26,27,28]

4-23. *THREE-ELECTRODE THERMIONIC GAS TUBES*

Vacuum thermionic triodes were discussed in the preceding pages; gas (including mercury-vapor) thermionic triodes now will be considered. These are commonly called by the trade name **Thyratron;** in the past they were sometimes called **grid-glow tubes.**[29]

Vacuum tubes are pumped and treated (section 2-13) to the extent that any remaining residual gas has no appreciable effect on the operation of the tube. Furthermore, in the vacuum triode the grid has complete control over the magnitude of the plate current at all times for normal operation. This is in contrast with the gas triode (Thyratron), in which the grid has control only *before* conduction starts.

In the Thyratron, electrons are also given off by a hot cathode. This cathode is often of the heat-shielded type (section 2-15). With *no* positive anode voltage applied, the electrons form a negative space charge about the cathode. When a positive anode voltage is applied, some of these electrons *may* be drawn over to the anode, and if the voltage is sufficient, they will cause ionization by collision with the gas atoms and thus produce positive ions. The word *may* was used because if the third electrode or *grid* in a gas tube is sufficiently negative (or not sufficiently positive in a type to be considered later), then the negative grid *prevents* the electrons from passing to the anode and thus prevents ionization by collision. If, however, the grid is made *less* negative or the plate is made *more* positive, an anode current will flow, and ionization by collision will result.

As in the gas diode (section 3-8), these positive ions travel to the cathode, and there neutralize the negative space charge. This reduction in opposing space charge makes it easier for the electrons to flow to the anode. This means that *large* currents can flow from the cathode to the anode, and that *the voltage drop is low,* being only about 15 volts for the mercury-vapor tube. A low internal voltage drop means an efficient tube, because the internal heat losses are low.

There is an important distinction between the thermionic vacuum triode and the thermionic gas triode. In the vacuum triode the grid controls the plate current *at all times.* In the gas triode the grid can control the time at which current *starts* to flow from cathode to anode, but as soon as the current starts, the *grid loses control,* and within operating limits, the grid potential *has no further control* effect on the current to the anode.

In using the vacuum triode, control by the grid makes possible such functions as amplification, oscillation, etc. Because the grid in the gas triode loses control, such functions are not possible, at least in the usual sense of the word. However, the gas triode is extensively used as a controlled rectifier, as a relay or switch, and for similar control purposes. The gas triode, or Thyratron, is often referred to as a grid-controlled rectifier. It should be kept in mind, however, that *the grid loses control* as soon as current flows.

The reason the grid loses control when current flows is as follows: After ionization the negatively charged grid is in a plasma or region of approximately equal numbers of positive and negative ions (section 1-14). The negative grid attracts to itself slowly moving, massive, positive ions and repels the highly mobile negative electrons. By this action, the electrons are largely removed from the vicinity of the grid. A *positive-ion sheath* is therefore formed around this negative grid. The positive electric field from this sheath neutralizes the negative field of the grid, thus preventing the grid from further influencing the operation of the tube.[19,30,31]

The grid continues to be ineffective until the gas deionizes. This occurs when the anode current ceases to flow, and this can be caused either by opening the anode-circuit switch, or by reducing the anode voltage to a low value. When the positive and negative ions recombine, the tube is deionized. After this has occurred, the grid is again able to determine the value of the anode voltage at which the gas breaks down and conducts.

The gas triode that has been discussed is of the **negative-grid type.** There is also a **positive-grid type,** in which the grid must be *positive* before discharge starts. These differences are obtained largely by electrode construction. In the negative-grid tube the shielding of the cathode by the grid is not complete, and the positive electric field from the anode is able to *reach through* the grid and draw electrons from the cathode region against the negative potential of the grid. In the positive-grid type, the cathode is completely shielded by the grid from the effects of the anode voltage. The positive anode potential is unable, therefore, to influence, to any appreciable extent, the electrons emitted from the cathode and hence the operation of the tube. A positive grid is necessary before the tube conducts.

These tubes may be filled with any of the inert gases, neon or argon, and more recently, hydrogen,[32] being used. Or, they may be filled with mercury vapor. The gas or vapor employed depends on the use for which the tube is designed. Mercury vapor is slow to deionize, and hence such tubes cannot be used in alternating-current circuits of more than several hundred cycles per second. The inert gases, and hydrogen, may, however, be used at higher frequencies. Because the vapor pressure in mercury Thyratrons depends on the operating temperature, and since the operating characteristic depends to some extent on the vapor pressure, these tubes are less stable under certain conditions than the inert gas tubes. References 33 and 34 are recommended for detailed information on deionization and recovery time of Thyratrons.

4-24. *CHARACTERISTICS OF THREE-ELECTRODE THERMIONIC GAS TUBES*

Characteristic curves can be obtained by the circuit arrangement of Fig. 4-21. The voltages at which ionization and conduction occur can be determined by the first deflection of the milliammeter in the anode circuit. For the negative-grid tube the voltage on the grid should be negative, and for the positive-grid tube, it should be positive.

Curves for a negative-grid mercury-vapor Thyratron are shown in Fig. 4-22. These curves show the positive anode voltage necessary (at the various temperatures) to cause the tube to conduct for different negative-grid voltages. The grid voltages are surprisingly small and the breakdown (or conducting) points quite definite for a mercury-vapor tube. For a gas tube the breakdown points at low grid and plate voltages are indefinite and erratic, and apparently occur at anode voltages less than the ionization potential, at least in some tubes. The relations between grid current and grid voltage after breakdown are shown in Fig. 4-23.

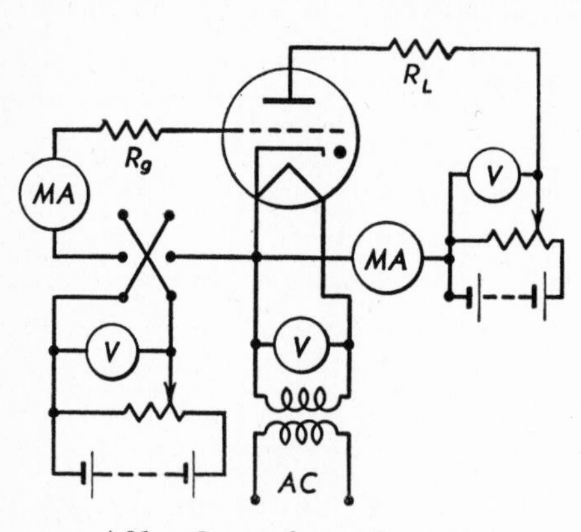

FIG. 4-21. Circuit for studying both positive and negative grid-controlled hot-cathode gas tubes. Mercury-vapor tubes should be immersed in a temperature-controlled oil bath. The resistors R_g and R_L are to limit the current.

Curves for a positive-grid mercury-vapor Thyratron are shown in Fig. 4-24. As these curves indicate, the voltage on the anode has very little effect on the operation, the breakdown at the various temperatures being controlled almost entirely by the positive potential of the grid. As previously mentioned, this is because the grid almost completely shields the cathode from the anode field. The relations between grid current and grid voltage after breakdown for the positive-grid Thyratron are much the same as shown in Fig. 4-23 for the negative-grid type.

4-25. *THREE-ELECTRODE COLD-CATHODE GAS TUBES*

The grid-controlled gas-filled triode, or Thyratron, considered in the preceding pages has a heated cathode to which power must be supplied during operating periods. For many applications a gas triode is needed that takes negligible stand-by power. This requirement is met by the cold-cathode tube. Cold-cathode tubes were considered in the preceding chapter, but these contained only two electrodes and were without the firing-control feature of the Thyratron. The cold-cathode tube now to be considered has a third control electrode. This control feature is not like that of the control grid in a vacuum tube; the control electrode in the cold-cathode tube controls the breakdown point, "firing" the tube. In this respect it resembles the Thyratron.

There are several types of cold-cathode three-electrode gas tubes, or **cold-cathode triodes.** In one type the **main anode** is a wire at the center of the structure and enclosed in a glass tube to within a short distance of the top end. The **starter anode** and the **cathode** are identical circular segments and are coated with barium and strontium.[35] The two segments are interchangeable; either may be used as the cathode or as the starter anode to initiate the discharge and fire the tube. In another form of the cold-cathode tube,[36] the cathode is similar in shape, but is one piece. The starter anode is a wire loop just above the cathode.

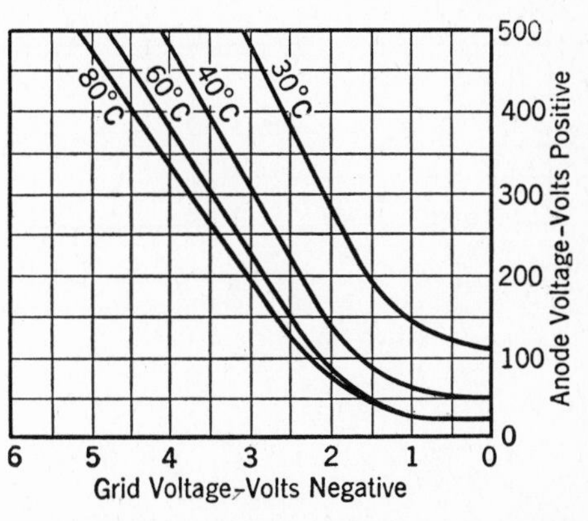

FIG. 4-22. Breakdown characteristics of a hot-cathode *negative* grid-controlled mercury vapor rectifier. (General Electric Review.)

There are two gaps to be broken down and made conducting. One is between the cathode and starter anode, and the other between the cathode and the main anode. The tube is so designed and constructed that the starter gap breaks down at a lower potential than the main gap. For a typical tube the starter gap alone breaks down at about +70 volts, and the main gap alone breaks down at about +175 volts. After the tube fires, a self-maintained discharge exists (section 1-15), and the cathode is partly covered with a glow.

If the current from the main anode to cathode is allowed to increase, the area covered by the glow increases, and the voltage drop between the cathode and main anode remains essentially constant as is typical for the self-maintained discharge. This voltage drop is about 70 volts. If the resistance in the main anode circuit is reduced and the current is permitted to increase until the cathode is fully covered with the glow, then the voltage drop increases. If the current is permitted to increase further, the gap will arc across, damaging, if not ruining, the tube. The cold-cathode triode can be operated with various combinations of signs and magnitudes of voltages on the main and the starter anodes. The tubes are designed primarily for operation with these two electrodes positive, and only this method of operation will be considered.

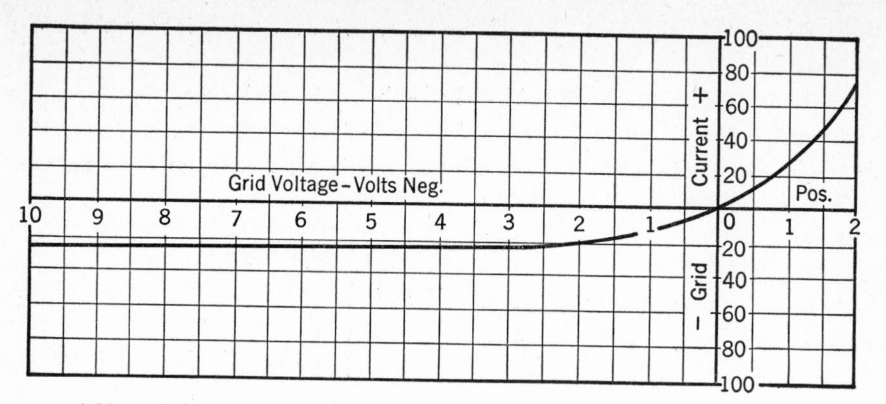

FIG. 4-23. Grid current in milliamperes in the mercury-vapor rectifier of Fig. 4-22 after breakdown. Positive values represent electron current to the grid. Negative values represent positive ion current to the grid. (General Electric Review.)

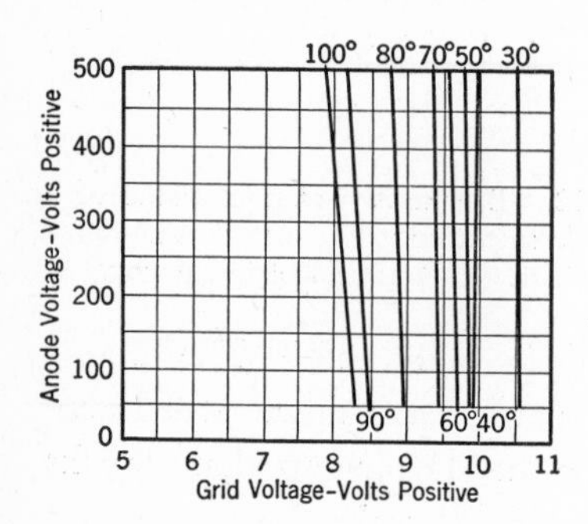

FIG. 4-24. Breakdown characteristics of a hot-cathode *positive* grid-controlled mercury vapor rectifier. (General Electric Review.)

4-26. CHARACTERISTICS OF THREE-ELECTRODE COLD-CATHODE GAS TUBES

The characteristics of interest can be determined by the circuit of Fig. 4-25. The voltage on the main anode should be adjusted to a value somewhat *less* than the voltage that will break down the main anode gap, and rheostat R should be adjusted to a value that will limit current flow to within rated value. Then, the voltage on the starter anode should be increased until the tube fires. This will occur, for a typical tube, at about +70 volts on the starter anode. When the starter gap breaks down the current across it increases, ionization by collision occurs, and a self-maintained glow discharge is established. At this instant the tube is filled with ions, and the main gap breaks down. Only a small current flows through the starter gap after breakdown, because of the limiting protective resistance, but a relatively large current can flow to the main gap. After breakdown the relation between current through the main gap and voltage across the main gap can be determined by varying R of Fig. 4-25. In a typical tube the starter-anode drop after breakdown is about 60 volts, and the current about a milliampere or less, depending on the value of the protective resistor and the tube. The main anode drop is about +70 volts, and varies several volts as the current is changed from about 5 to 30 milliamperes, the upper limit for the tube.

Before breakdown, small, but significant, currents flow across the starter gap and across the main gap. Current flow is possible because of natural ionization of the gas (sections 1-12 and 3-15) and photoelectric emission from the electrodes that are oxide-coated. In fact, cold-cathode triodes sometimes are coated so that light is excluded from the interior, and erratic performance caused by varying light intensity is prevented. In a typical tube, ionization, and the establishment of a self-maintained discharge, may require from a few microseconds to several milliseconds. The addition of a small amount of radioactive material tends to shorten and stabilize the initial ionization period. By this means typical cold-cathode triodes are made to have an ionization time that is about 50 microseconds.[37]

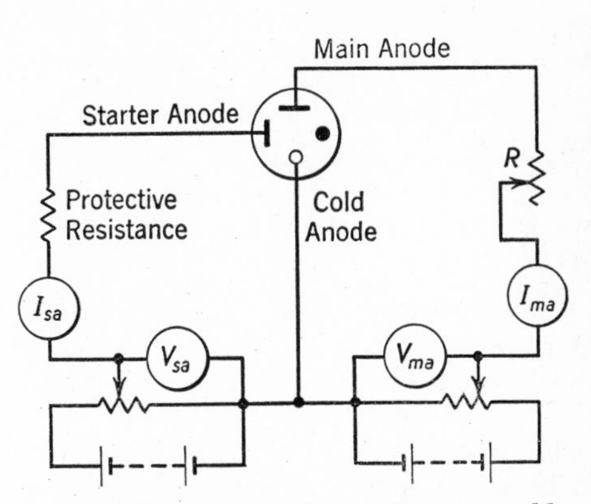

FIG. 4-25. Circuit for studying a cold-cathode three-electrode gas tube. After breakdown the voltage on the main anode will equal the voltage V_{ma} minus the $I_{ma}R$ drop in resistor R.

The cold-cathode tube has many applications, including use as a relay, a rectifier, and a voltage regulator. Typical operation as a relay is as follows: The voltages applied across the main gap and the starter gap are below the breakdown value. However, the voltage across the starter gap is only *slightly* below breakdown, and a feeble incoming signal-voltage pulse, added in series with the starter-gap voltage, will fire the tube. Then, the main gap will break down, and a large current will flow to the device to be operated. Other uses are considered in Chapter 6.

REFERENCES

1. Houston, E. J. *Notes on phenomena in incandescent lamps.* Trans. A.I.E.E., Vol. 1.
2. De Forest, L. *The Audion.* Electrical Engineering, Mar. 1947, Vol. 66, No. 3.
3. Gherardi, B., and Jewett, F. B. *Telephone repeaters.* Trans. A.I.E.E., Nov. 1919. Vol. 38.

4. Espenschied, L. *"Historic first" electronic lecture.* Proc. I.R.E., Feb. 1946, Vol. 34, No. 2.
5. Institute of Radio Engineers. *Standards on Electron Tubes: Definitions of Terms,* 1950, Proc. I.R.E., Vol. 38, No. 4, April 1950; *Methods of Testing,* 1950, Vol. 38, Nos. 8 and 9, Aug. and Sept., 1950. *Standards on Abbreviations, Graphical Symbols, Letter Symbols, and Mathematical Signs,* 1948; *Standards on Abbreviations,* Proc. I.R.E., April 1951, Vol. 39, No. 4; *Standards on Graphical Symbols,* Proc. I.R.E., June 1954, Vol. 42, No. 6.
6. Pidgeon, H. A. *Simple theory of the three-electrode vacuum tube.* Journal of the Society of Motion Picture Engineers, Feb. 1935, Vol. 24.
7. Webking, H. E. *Producing tube curves on an oscilloscope.* Electronics, Nov. 1947, Vol. 20, No. 11.
8. Kuder, M. L. *Electron tube curve generator.* Electronics, Mar. 1952, Vol. 25, No. 3.
9. American Institute of Electrical Engineers. *American Standard Definitions of Electrical Terms.* 1941.
10. King, R. W. *Calculation of the constants of the three-electrode vacuum tube.* Physical Review, Apr. 1920, Vol. 15, No. 4.
11. Miller, J. M. *The dependence of the amplification constant and the internal plate circuit resistance of three-electrode vacuum tubes upon the structural dimensions.* Proc. I.R.E., Feb. 1920, Vol. 8.
12. Vogdes, F. B., and Elder, F. R. *Formulas for the amplification constant for three-electrode tubes in which the diameter of the grid wires is large compared to the spacing.* Physical Review, Dec. 1924, Vol. 24.
13. Kusunose, V. *Calculation of characteristics and the design of triodes.* Proc. I.R.E., Oct. 1929, Vol. 17.
14. Salzberg, B. *Formulas for the amplication factor for triodes.* Proc. I.R.E., Mar. 1942, Vol. 30, No. 3.
15. Thompson, B. J. *Space-current flow in vacuum-tube structures.* Proc. I.R.E., Sept. 1943, Vol. 31, No. 9.
16. Liebmann, G. *The calculation of amplifier valve characteristics.* Jl. I.E.E., May 1946, Vol. 43, Part 3.
17. Herold, E. W. *Empirical formula for amplification factor.* Correspondence, Proc. I.R.E., May 1947, Vol. 35, No. 5.
18. Fremlin, J. H., Hall, R. N., and Shatford, P. A. *Triode amplification factors.* Electrical Communication, Dec. 1946, Vol. 23, No. 4.
19. Koller, L. R. *Physics of Electron Tubes.* McGraw-Hill Book Co.
20. Easton, E. C., and Chaffee, E. L. *Pulse-type tester for high-power tubes.* Electronics, Feb. 1947, Vol. 20, No. 2.
21. Leferson, J. *The application of direct-current resonant-line type pulsers to the measurement of vacuum-tube static characteristics.* Proc. I.R.E., June 1950, Vol. 38, No. 6.
22. Wagner, H. M. *Tube characteristic tracer using pulse techniques.* Electronics, Apr. 1951, Vol. 24, No. 4.
23. Terman, F. E., and Pettit, J. M. *Electronic Measurements.* McGraw-Hill Book Co.
24. Chaffee, E. L. *Theory of Thermionic Vacuum Tubes.* McGraw-Hill Book Co.

25. Everitt, W. L. *Communication Engineering*. McGraw-Hill Book Co.
26. McArthur, E. D. *Disk seal tubes*. Electronics, Feb. 1945, Vol. 18, No. 2.
27. Rose, G. M., Power, D. W., and Harris, W. A. *Pencil-type UHF triodes*. R.C.A. Review, Sept. 1949, Vol. 10, No. 3.
28. Morton, J. A., and Ryder, R. M. *Design factors of the Bell Telephone Laboratories 1553 triode*. Bell System Technical Journal, Oct. 1950, Vol. 29, No. 4.
29. Knowles, D. D., and Sashoff, S. P. *Grid-controlled glow and arc discharge tubes*. Electronics, July 1930, Vol. 1, No. 4.
30. *Industrial Electronics Reference Book*. John Wiley & Sons.
31. McArthur, E. D. *Electronics and Electron Tubes*. John Wiley & Sons.
32. Heins, H. *Hydrogen Thyratrons*. Electronics, July 1946, Vol. 19, No. 7.
33. Malter, L., and Johnson, E. O. *Studies of Thyratron behavior*. R.C.A. Review, June 1950, Vol. 11, No. 2.
34. Malter, L., and Boyd, M. R. *Grid current and grid emission studies in Thyratrons*. Proc. I.R.E., June 1951, Vol. 39, No. 6.
35. Ingram, S. B. *Cold-cathode gas-filled tubes as circuit elements*. Electrical Engineering, July 1939, Vol. 58, No. 7.
36. Bahls, W. E., and Thomas, C. H. A *new gas-filled triode*. Electronics, May 1938, Vol. 11, No. 5.
37. Peck, D. S. *The cold cathode glow discharge tube*. Bell Laboratories Record, Dec. 1951, Vol. 29, No. 12.

QUESTIONS

1. Why are triodes usually operated below voltage saturation as stated in section 4-1?
2. What is meant by the term grid bias?
3. Why are the grids of triode amplifiers usually operated negative with respect to the cathode?
4. What will be the effect on the μ, r_p, and g_{pg} of a triode if the control-grid wires are spaced closer together?
5. What will be the effect on the μ, r_p, and g_{pg} of a triode if the control grid is moved closer to the plate?
6. What is the difference between the direct and alternating plate resistance?
7. What is the meaning of the statement in parentheses in the definition of Electrode Resistance, section 4-3?
8. What are the differences between the two basic methods of determining tube coefficients?
9. Referring to section 4-8, it is stated that low values of alternating test voltage should be used. Why?
10. In section 4-8 it is stated that balancing may be easier if the position of the grounded voltage divider on the input transformer is varied. Why?
11. What is the meaning of amplification factor in terms of the electric fields from the grid and plate?
12. If the amplification factor of a tube is changed by altering the electrode spacing, will the mutual conductance be changed? Why?

13. Referring to section 4-10, what is meant by the terms equivalent grid voltage and equivalent plate voltage?
14. Does current flow to a negative grid? Why?
15. At low frequencies where interelectrode capacitances are negligible, what will be the order of magnitude of the input impedance of a triode with the grid negative, and with it positive?
16. How does an alternating signal voltage impressed on the grid cause an alternating plate-current component?
17. In a voltage amplifier why must impedance be placed in the plate circuit? What is the simplest impedance to use?
18. Referring to Fig. 4-19, why can a vacuum-tube amplifier be represented by the equivalent circuit shown?
19. Explain the difference between static and load characteristic curves. What advantages do the latter have?
20. What is the meaning of the voltage amplification factor?
21. Referring to Fig. 4-15, why is the relationship between the alternating plate current and plate voltage as shown?
22. Again referring to Fig. 4-15, what is the phase relation between the impressed grid-signal voltage and the signal voltage across the load? The cathode is at zero potential.
23. What is meant by instantaneous total current? By instantaneous total voltage?
24. How does the three-electrode thermionic gas tube operate?
25. Referring to the cold-cathode gas tube, is it necessary during operation for electrons to be emitted from the cathode? If so, how are they emitted?

PROBLEMS

1. Derive equations 4-12, 4-13, and 4-14.
2. Referring to section 4-10, determine the equations for the curve of Fig. 4-2 at a grid-bias voltage of −8 volts.
3. Referring to Fig. 4-2, calculate the plate conductance and resistance at points P' and P on the −8-volt curve. Also calculate these at similarly located points on the 0-volt curve and the −20-volt curve.
4. Referring to Fig. 4-3, calculate the mutual conductance at points P' and P on the 120-volt curve, the 180-volt curve, and the 75-volt curve.
5. A signal voltage of 0.5-volt effective value is impressed on the grid of a triode operated with 35,000 ohms in the plate circuit. The plate resistance is 11,000 ohms, and the amplification factor is 9.3. Calculate the alternating signal voltage that will appear across the plate load resistor, and calculate the voltage amplification.
6. Referring to Fig. 4-3, calculate the dynamic curve passing through point P for a resistance in the plate circuit of 25,000 ohms. Transfer to the curves of Fig. 4-2 the dynamic curve just calculated. Note that point P of Fig. 4-3 is at −4 volts.
7. A small-signal triode has an amplification factor of 14 and a plate resistance of 8000 ohms, and an effective signal voltage of 2.5 volts is impressed on the grid. Calculate the voltage output across the following load resistors in the

plate circuit: 4000 ohms, 8000 ohms, 16,000 ohms, 24,000 ohms, 32,000 ohms, and 40,000 ohms. Also calculate the voltage amplification. Plot these two values on the Y axis, with load resistance on the X axis. What conclusions can you draw from these curves?

8. The values of μ and r_p in problem 7 were for 250 volts *on the plate*, and an 8-milliampere plate current. Calculate the direct voltage of the plate supply source so that these same operating conditions will apply with each plate load resistor. Plot these voltage values on the same set of axes as in problem 7.
9. The triode in a power-output stage has $\mu = 3.5$ and $r_p = 1650$ ohms. These figures are for a grid bias of -31.5 volts, a plate voltage of 180 volts, and a plate current of 31 milliamperes. Suppose that the applied signal voltage is such that the grid is driven to zero volts on the positive half-cycle. Neglect distortion, and calculate the power that will be delivered to loading devices having the following equivalent resistances: 500 ohms, 1000 ohms, 1650 ohms, 3300 ohms, 4950 ohms, and 8250 ohms. Plot power output on the Y axis and resistance on the X axis. What conclusions do you draw? At about what value of load resistance would you operate for maximum undistorted power output?
10. Referring to problems 8 and 9, calculate the direct-current plate power dissipation for each tube.

CHAPTER 5

MULTIELECTRODE ELECTRON TUBES

The diodes and the triodes discussed in the preceding chapters were used almost exclusively for many years. About 1928 a four-electrode tube, or tetrode, was introduced, and this was followed perhaps a year later by the five-electrode tube, or pentode. These are classified as **multielectrode tubes,** and defined [1] as "an electron tube containing more than three electrodes associated with a single electron stream." The introduction of tetrodes and pentodes marked the beginning of a new era in the communication art. Until their appearance, the triode was the only tube available for amplification, and it had limitations, particularly in radio-frequency amplifiers.

Since the introduction of the tetrodes and the pentodes, many varieties of tubes have become available. It will be recognized that it is inadvisable to attempt to discuss each of these tubes individually; even if it were possible to do so, it is unnecessary. The many types of tubes have electronic principles that are common. The important matter is to understand these principles thoroughly. Then, when an unfamiliar tube is encountered, it can be classified quickly as to its basic principles of operation, and its general operating characteristics then will be known.

5-1. *MULTIPLE-UNIT TUBES*

It will be noted that in the definition of a multielectrode tube it was specifically stated that the electrodes were associated with a *single* electron stream. This is true for the tetrode, pentode, and certain other types, and these are, accordingly, multielectrode tubes. These tubes will either perform certain functions (such as amplification) better than will a triode, or will perform functions not possible with a triode. There is another type of tube that has more than three electrodes in a single envelope. This is the **multiple-unit tube,** defined [1] as "an electron tube containing within one envelope two or more groups of electrodes associated with independent electron streams." This means that some tubes are in fact combinations of two basic types and perform, simultaneously, more than one function.

For this reason, certain tubes may appear, on first examination, to be multielectrode tubes, but in reality they are multiple-unit tubes, or *two "tubes" in one envelope.* For example, there are duodiode-triodes, consisting of two diode elements and one triode element in one envelope. This multiple-unit tube can simultaneously operate as a diode rectifier (demodulator or detector) and as a triode amplifier. Although the various electrodes may be grouped about the same cathode structure, one set of electrodes is above the other, and they use independent electron streams. If the operating principles of the diode, triode, tetrode, and pentode are understood, then these principles can be applied to the separate units in a multiple-unit tube.

5-2. *SPACE CHARGE IN MULTIELECTRODE THERMIONIC VACUUM TUBES*

In the three-electrode vacuum tube or triode discussed in the preceding chapter, it was shown that a cloud of electrons constituting a negative space charge existed (largely) in a thin sheath around the hot cathode. The current to the plate, and hence the performance of the tube, was seen to depend on the manner in which the electric fields from the positive plate and the (usually) negative grid acted upon this space charge.

In explaining the operation of a multielectrode vacuum tube, it is advisable to quote from the article [2] by Pidgeon.

> In multielectrode tubes, as well as in triodes, the total space current drawn from the cathode is determined by the extent to which the resultant field, due to the electrodes, overcomes the opposing field produced by space charge. While space charge extends throughout the interelectrode space, it is relatively so much more dense in regions of very low electron velocity that, as a first approximation, its effect usually may be neglected in other regions. Except in space-charge-grid tubes and a few other special tubes, the only important space-charge region is confined to a relatively thin sheath near the cathode surface. Consequently, in such structures the total space current is determined largely by the extent to which the resultant positive field due to the electrodes neutralizes the negative field near the cathode surface produced by space charge. An appreciation of this fact is essential to a clear understanding of the characteristics of multielectrode tubes.

Thus, it is seen that the theoretical considerations developed for the vacuum triode apply to multielectrode vacuum tubes. It is necessary only to consider the influence of the additional electrodes on the electron current to the plate.

5-3. *FOUR-ELECTRODE TUBES OR TETRODES*

These tubes were studied [3] by Schottky as early as 1919. A **tetrode** is defined [1] as "a four-electrode vacuum tube containing an anode, a cathode, a control electrode, and one additional electrode that is ordinarily a grid." Thus in the usual tetrode the electrons going from the cathode to the plate pass through *two* grids. There are several types of tetrodes. In one type the grid next to the cathode is made *positive* with respect to the cathode, and the grid next to the plate is the control electrode. In another type the grid next to the cathode is the control grid, and the grid next to the plate is made positive. The first is called the **space-charge-grid tetrode;** the second is the **screen-grid tetrode.**

The **space-charge grid** is [1] "a grid, usually positive, that controls the position, area, and magnitude of a potential minimum or of a virtual cathode in a region adjacent to the grid." The **screen grid** is [1] "a grid placed between a control grid and an anode, and usually maintained at a fixed positive potential, for the purpose of reducing the electrostatic influence of the anode in the space between the screen grid and the cathode." A third type of thermionic vacuum tetrode is the **beam-power tube** discussed in section 5-14.

5-4. *SPACE-CHARGE-GRID THERMIONIC VACUUM TETRODE*

This tube will be but briefly discussed because it is relatively unimportant. A typical tube consists of an indirectly heated cathode, a space-charge grid, a control grid, and a plate, or anode. The space-charge grid commonly is made about 10 volts positive with respect to the cathode. The electric field caused by the positive grid partially neutralizes the negative space charge.[3] The effect of this is to produce a **virtual cathode,** or apparent source of electrons (as far as the control grid and the plate are concerned), which is between the space-charge grid and the control grid. The control grid and the plate then function much the same as in a three-electrode tube. The space-charge-grid vacuum tetrode is characterized by low amplification factor, low plate resistance, and high plate current. It is, therefore, a power-output tube. The positive space-charge grid collects a large fraction of the total emitted current. This is one of the factors limiting the efficiency of the tube.[3]

5-5. *SCREEN-GRID THERMIONIC VACUUM TETRODE*

The electrodes in a screen-grid tetrode are arranged as shown in Fig. 5-1. The tube is essentially a triode with a fourth electrode, the positive screen grid, inserted between the plate and control grid (in some types enclosing the plate) so that the plate is *electrically shielded* from the control grid. This is of importance when amplifying radio-frequency signals because in triodes a signal voltage is fed back to the grid through the grid-plate capacitance, and oscillations may occur. In the early days of triode amplification, special neutralizing circuits (Chapter 8) were necessary to prevent this radio-frequency feedback. The development of the screen-grid tube with its effective shielding simplified circuit design. This tube was studied [4] extensively by Hull and Williams, and introduced commercially in this country about 1928. Originally it was called a **screen-grid tube.**

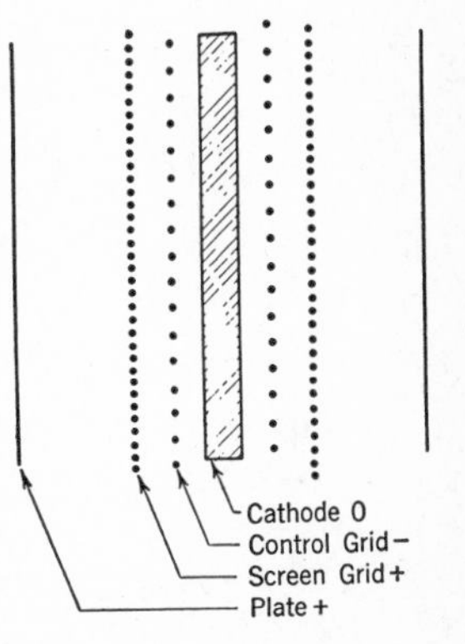

FIG. 5-1. Schematic arrangement of electrodes in a screen-grid vacuum tetrode.

The shielding effect of the screen grid almost completely prevents the electric field from the positive plate from reaching into the cathode region where the space charge exists. The resultant electric field in this region largely controls the plate-current flow. Hence, the shielding of the screen grid not only reduces feedback, but also greatly affects the characteristics of the tube. The screen grid is usually held at a positive direct voltage, and this accelerates the electrons.

Under usual operating conditions the potentials of the electrodes in a multielectrode tube other than those of the control grid and plate should be maintained constant. It will be noted that this condition is specified for the curves shown on the following pages. Thus, when a multielectrode tube, such as a screen-grid tetrode, is amplifying alternating voltages, the impedances to the cathode of these auxiliary electrode circuits must be kept low, otherwise these electrodes would vary in potential. To accomplish this, these electrodes usually are connected to ground (or to the cathode if appreciable impedance exists between ground and cathode) through large capacitors, thus preventing the existence of alternating voltage components between these electrodes and the cathode.

5-6. *STATIC CHARACTERISTICS OF SCREEN-GRID TETRODES*

These may be studied with a circuit arranged as in Fig. 5-2. For obtaining the curves of Fig. 5-3, the filament current, screen-grid voltage, and control-grid voltage are held constant at the values indicated, and current values are observed at different plate voltages. Since the plate is so effec-

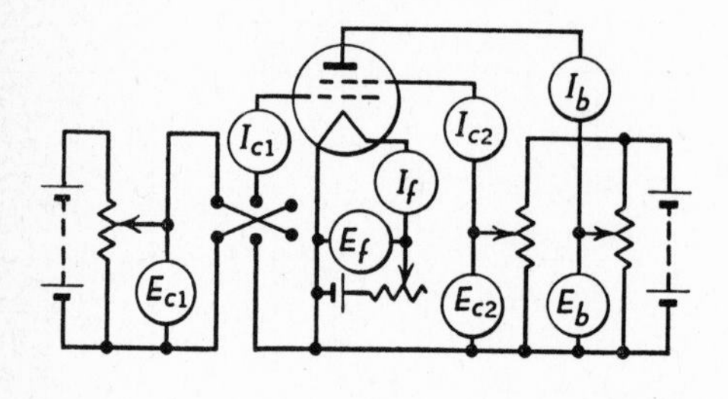

FIG. 5-2. Circuit for determining the static characteristics of a screen-grid vacuum tetrode.

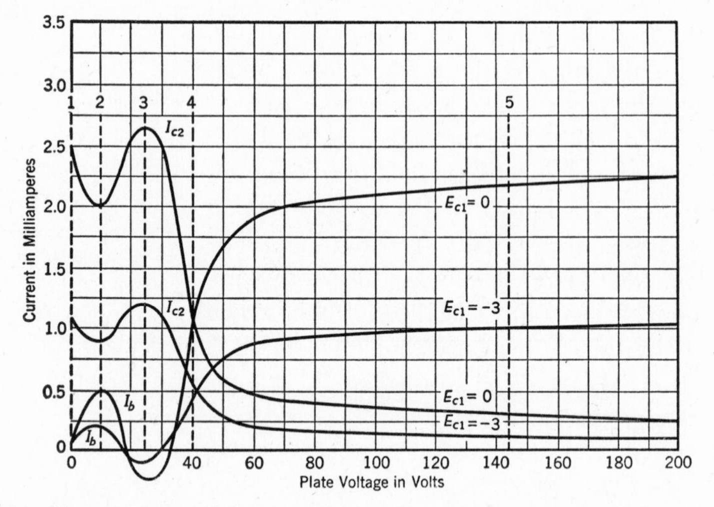

FIG. 5-3. Characteristics of a screen-grid vacuum tetrode as determined by the circuit of Fig. 5-2. Screen-grid voltage E_{c2} held at +45 volts. Control-grid voltage E_{c1} at voltages (in volts) shown on curves. Plate voltage E_b varied. These curves are for an early tetrode; the plates of later tetrodes were treated to minimize secondary emission. Plate-current curves are marked I_b, and screen-grid current curves are marked I_{c2}.

tively shielded, the influence of its positive field in the space-charge region is negligible. Thus at all times the electron current *leaving* this region is determined largely *by the voltages on the control and screen grids and not* by the potential of the plate. Thus, at region 1, where the plate voltage E_b is *zero*, substantially the same total electron current flows from the cathode as when the plate is highly positive. At region 1, the positive screen grid collects the entire electron flow.

At region 2, the plate has been made slightly positive. Some of the rapidly moving electrons pass through the screen grid, and many of these are pulled to the now-positive plate. The resultant current to the screen grid therefore drops, so that the sum of the plate current I_b and the screen-grid current I_{c2} equals (approximately) the current previously collected by the screen grid.

But, at region 2 a new phenomenon occurs. Some of the electrons attracted to the plate have reached sufficient velocity to cause *secondary emission* from the plate. Now, the potential of the plate is much lower than that of the screen grid, so many of these secondary electrons are pulled to the grid. If the number of secondary electrons exceeds the primary electrons, the plate current may even become negative as indicated. Thus at region 3 there is a new source of electrons (namely, secondary emission from the plate) in addition to the electrons from the cathode space-charge region. Thus, the screen-grid current exceeds the (almost constant) value of electrons taken from the space-charge sheath.

As the plate voltage is increased beyond region 3, the plate potential approaches that of the screen grid. The plate is, accordingly, able to capture the electrons released through secondary emission, and the plate current increases. Corresponding decreases in the screen-grid current occur. When region 4 has been reached and the plate is highly positive, no further effect of secondary emission is observed (although it is taking place).

The screen-grid tetrode is usually operated (as an amplifier) over the region of the curves marked 5. As is evident in this region, the plate voltage has but little influence on the plate current. Also, the screen-grid current is very low and may be negative under certain conditions because of secondary emission *from the screen grid.* If the plate voltage is not *at all times* maintained at a value somewhat higher than that of the screen grid (region 5, Fig. 5-3), then the distortion of the signal being amplified will be excessive because of secondary emission from the plate. This limits the operation of the screen-grid tetrode. For the tube of Fig. 5-3, when operated as an essentially distortionless amplifier, the plate potential should not fall below about +80 volts.

Curves showing the effect of control-grid voltage (E_{c1} of Fig. 5-2) on the plate and screen currents are shown in Fig. 5-4. As is evident, the voltage on the control grid has a very *great* effect on the plate current. Thus, in the screen-grid tube the plate is so effectively shielded by the screen grid that the plate potential *does not* greatly affect the plate current, but the tube is so constructed that the control-grid voltage *does* affect the plate current, therefore the tube has a very high amplification factor.

The screen-grid vacuum tetrode is used as a voltage-amplifying tube. It is not necessary to consider dynamic curves or load lines when a tube is used for this purpose.

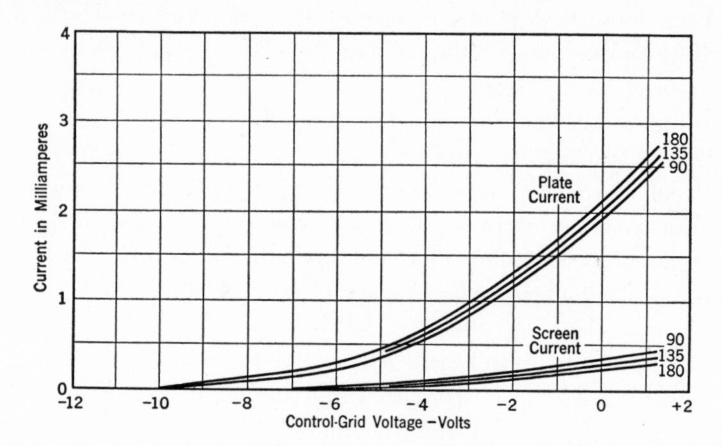

FIG. 5-4. Plate and screen-grid currents for a screen-grid vacuum tetrode. Plate voltages in volts shown on curves. Screen grid at +45 volts.

5-7. *SCREEN-GRID VACUUM TETRODE COEFFICIENTS*

The coefficients defined in section 4-3 apply to the screen-grid tetrode. These coefficients may be evaluated graphically from the static characteristic curves of Figs. 5-3 and 5-4. This procedure now will be explained, largely for purposes of illustration. The errors in using the graphical method with tetrodes are such that the results are approximate.

Referring to Fig. 5-3, if the plate voltage for the grid bias of -3 volts is changed from 100 to 200 volts, the plate current increases only 0.07 milliampere. From Fig. 5-4, using the 135-volt curve, a change of grid bias from -3 to -2.8 volts produces a change in plate current of 0.07 milliampere. Thus, the amplification factor is approximately $(200 - 100)/(3 - 2.8) = 500$. Or, from Fig. 5-4, at a plate voltage of 180 volts and a grid bias of -3 volts, the plate current is the same as at a plate voltage of 90 volts and a grid bias of -2.6 volts. The amplification factor is (see section 4-4): $(180 - 90)/(3 - 2.6) = 225$. As is apparent, these graphical methods do not check, largely because of the errors involved in obtaining data from the curves, and in selecting the same operating points. The accepted methods of finding the coefficients will be discussed in the following pages.

The alternating-current plate resistance can be approximated from Fig. 5-3. Assuming the -3-volt bias curve to be a straight line over the region from 140 to 160 volts, the corresponding current change is about

0.01 milliampere. This gives, from section 4-5, a plate resistance of 20/0.00001 = 2,000,000 ohms. As is apparent, plate-resistance values taken in this manner will vary widely.

Grid–plate transconductance or mutual conductance values can be found satisfactorily from Fig. 5-4. Thus, on the 135-volt curve, at −3 volts bias the plate current is 0.9 milliampere, and at −2 volts it is 1.25 milliamperes. The mutual conductance is, therefore, 0.00035/1 = 0.00035 mho or 350 micromhos.

Other coefficients are specified [1] for multielectrode tubes, because they have more than three electrodes. However, in ordinary use the screen and suppressor grids are held at fixed potentials, and these additional coefficients are not needed in ordinary design.

5-8. *SYMBOLS FOR MULTIELECTRODE TUBES*

The abbreviations used with electron tubes were listed in section 4-3. Attention is called to the fact that the numbers 1, 2, 3, etc., are used to designate the grids. From the standards,[1] "Grids are numbered according to position out from the cathode. When no numerical subscript appears, reference to the control grid is assumed." An example is Fig. 5-3, where I_{c2} is the direct current flowing in the screen-grid circuit, and E_{c1} is the direct grid voltage. The letter and graphical symbols to be used with multielectrode tubes are as specified in sections 4-18 and 4-19.

5-9. *MEASUREMENTS OF MULTIELECTRODE TUBE COEFFICIENTS*

For a thermionic vacuum triode used at low frequencies, the coefficients μ, r_p, and g_{pg} may be determined graphically with sufficient accuracy from the static characteristic curves (sections 4-4, 4-5, and 4-6). As indicated in section 5-7, this method is not reliable for evaluating the coefficients of multigrid tubes.

The standards [1] specify two basic methods for determining coefficients. The first uses a **resistance-ratio method,** and the second a **voltage-ratio method.** A comparison of the two types of circuits used indicates that the resistance-ratio method is better adapted for instruction; hence, this method was presented in Chapter 4. Also, resistance-ratio methods will be included in the present chapter. However, because μ and r_p are much larger for multielectrode tubes than for triodes, the circuits used are different from those for triodes in Chapter 4. The standards [1] should be consulted for details not given here.

Amplification Factor. A resistance-balance circuit for measuring the amplification factor (or μ factor) of a screen-grid thermionic vacuum tetrode is shown in Fig. 5-5a. At balance,

$$\mu = \frac{R_2}{R_1} \tag{5-1}$$

Basically, this circuit is the same as Fig. 4-7a of the preceding chapter, with the addition of direct voltages supplied by batteries to the electrodes. If the voltages are obtained from voltage dividers, these should be shunted with capacitors of low reactance.

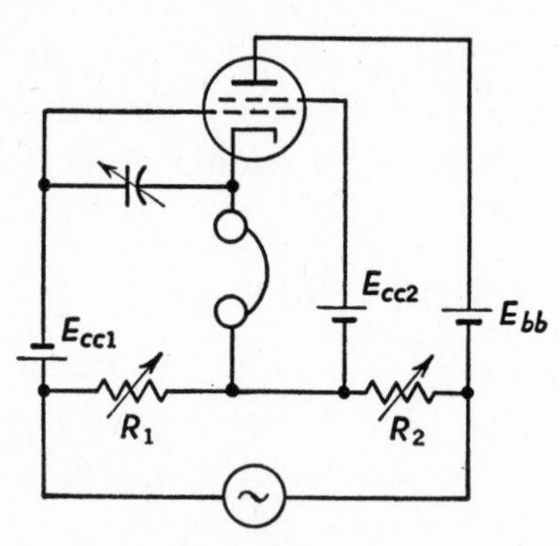

(a) Amplification Factor

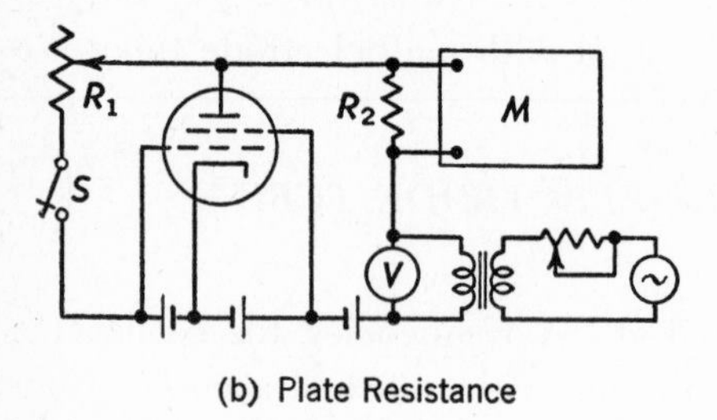

(b) Plate Resistance

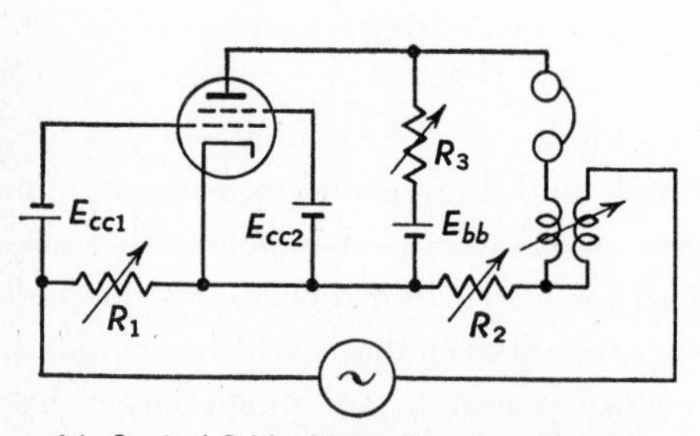

(c) Control-Grid – Plate Transconductance

FIG. 5-5. Simplified circuits for determining μ, r_p, and g_{pg} for a screen-grid tetrode. For precautions and details, see section 4-8 and reference 1.

Plate Resistance. The circuit of Fig. 5-5b may be used for measuring the plate resistance of multielectrode vacuum tubes such as the screen-grid tetrodes. The operation of this circuit is as follows: [1] The tube is operated at normal voltages with switch S open. The value of R_2 is small compared with the plate resistance of the tube. The alternating voltage applied is adjusted until some convenient deflection is obtained on the high-impedance alternating voltmeter *M*. (This may be an amplifier and a vacuum-tube voltmeter.) The tube is then removed from the circuit, and the switch S closed. This action substitutes the rheostat R_1 for the plate resistance of the tube. The value of R_1 is varied until with the same applied alternating voltage the same deflection is obtained on the voltmeter. The value of R_1 is then the same as the plate resistance of the tube. The circuit can be made direct reading if desired.[1] In a screen-grid tetrode the plate is so effectively shielded that changes in plate voltage do not appreciably influence the plate current. Thus the plate resistance is high.

Control-Grid–Plate Transconductance (Mutual Conductance). A circuit for measuring the control-grid–plate transconductance of a multielectrode vacuum tube is shown in Fig. 5-5c. (See also Fig. 4-7c.) At balance,

$$g_{pg} = \frac{R_2}{R_1 R_3} \tag{5-2}$$

This equation holds only when the values of R_1 and R_3 are negligible compared to the plate and grid resistances,[1] a condition met with screen-grid tetrodes.

The precautions and suggestions of section 4-8 should be followed when making measurements on multielectrode tubes. Typical coefficients for a thermionic screen-grid vacuum tetrode are shown in Fig. 5-6. It is of interest to compare these with Fig. 4-6 for a triode.

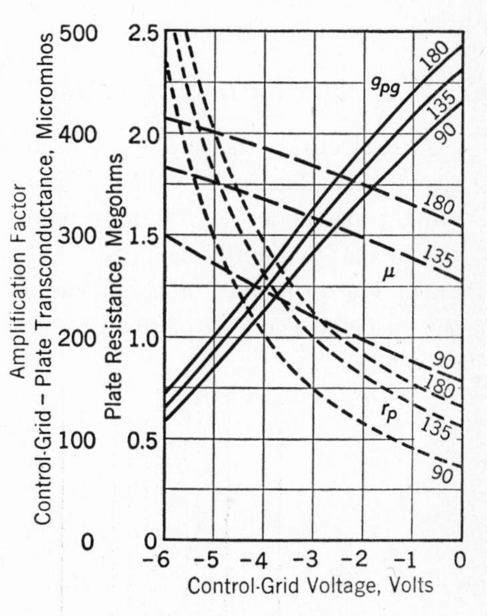

FIG. 5-6. Variations in amplification factor μ, plate resistance r_p, and control grid-plate transconductance g_{pg} for the vacuum tetrode of Figs. 5-3 and 5-4. Numbers on curves are plate voltages in volts.

5-10. *EQUIVALENT CIRCUIT OF A SCREEN-GRID TETRODE*

In the preceding chapter it was shown that for amplification a triode could be represented as a generator producing an open-circuit voltage of μE_g and having an internal resistance of r_p. This same reasoning applies to multielectrode tubes. It is assumed that the tube is being operated as an amplifier over the straight portion of the load curve. The fundamental component only is considered and distortion is neglected. Thus Fig. 4-12c also represents the equivalent circuit of a screen-grid tetrode, and equations 4-19 to 4-22 inclusive apply.

Screen-grid tetrodes are usually operated as voltage amplifiers, and equation 4-21 indicates the amplification obtainable. Both the amplification factor and the plate resistance of screen-grid tetrodes are very high. The high plate resistance limits the amount of voltage amplification obtainable. To secure a large voltage across the load resistor, a large value of load resistance (compared to the plate resistance of the tube) must be used, or most of the generated voltage, μE_g, will be lost *inside* the tube. There are, however, practical limits to the value of the load resistance. The direct component of the plate current must flow through this resistance, and hence an excessive direct voltage drop may occur. This subject will be treated more in detail in Chapter 7.

The thermionic screen-grid vacuum tetrode discussed in the preceding pages is essentially a small-signal audio or radio voltage amplifier. Screen-grid tetrodes are also used as radio power-output tubes,[5] particularly at ultrahigh frequencies.[6] However, from the standpoint of numbers of tubes used, screen-grid tetrodes have limited application compared to other types.

5-11. *SPACE-CHARGE–GRID THERMIONIC VACUUM PENTODE*

Five-electrode thermionic vacuum tubes were introduced in this country about 1929. These are called pentodes, defined[1] as an "electron tube containing an anode, a cathode, a control electrode, and two additional electrodes that ordinarily are grids." Two types of pentodes have been developed. The principles of operation are different.[7] This section will consider the space-charge-grid pentode.

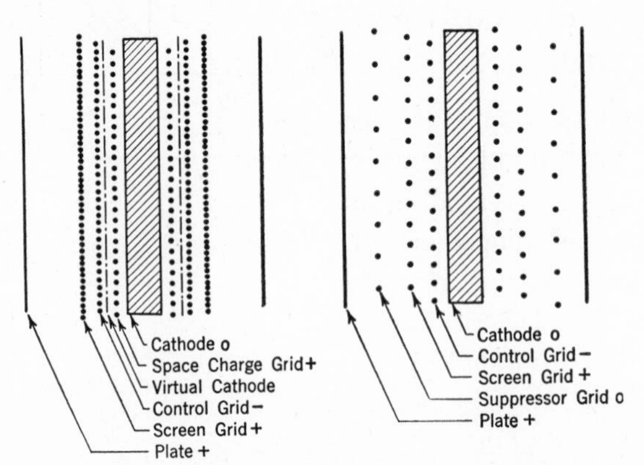

FIG. 5-7. Schematic arrangements of electrodes in a space-charge-grid pentode (left), and suppressor-grid pentode (right). The latter illustration is of the power-output type, with large spacings between the grid wires. The suppressor grid is connected (externally or internally) to the cathode. (From reference 2.)

The arrangement of electrodes in this tube is shown in Fig. 5-7. The space-charge grid is at about +10 volts with respect to the cathode. As explained in section 5-4, the space-charge grid forms a virtual cathode between the space-charge grid and the control grid. With the other electrodes held at the direct potentials indicated in Fig. 5-7, this tube functions as a screen-grid tube. The space-charge-grid pentode exhibits secondary-emission effects like those of the conventional screen-grid tube. Also, this pentode has a high amplification factor, high plate resistance, and high transconductance. This is caused by the additional shielding of the space-charge grid in preventing the plate voltage from affecting the region near the cathode.

The space-charge grid draws a rather large constant current from the cathode, a decided disadvantage. Furthermore, the tube offers little, if any, improvement over the screen-grid tetrode. Space-charge-grid pentodes are seldom used. A pentode in which all grid leads are brought out can be connected as a space-charge-grid pentode for experimentation.

5-12. *SUPPRESSOR-GRID THERMIONIC VACUUM PENTODE*

The arrangement of the electrodes in this tube is shown in Fig. 5-7. In some of these tubes the suppressor grid *internally* is connected directly to the cathode. In others, all three grids are brought out to external terminals, and may be connected as desired outside the tube. Such tubes sometimes are called **triple-grid tubes.** Pentodes usually are connected as suppressor-grid tubes for operation.

The purpose of the **suppressor grid** is to suppress *the effects* of secondary

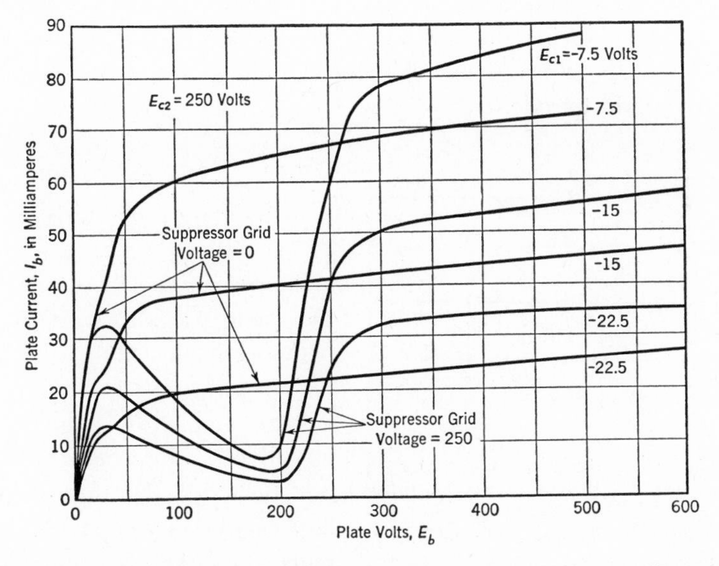

FIG. 5-8. These curves show the effect of a suppressor grid in a small power-output pentode. Curves "suppressor-grid voltage = 0" are for operation with the suppressor grid connected to the cathode. Curves "suppressor-grid voltage = 250" are for operation as a screen-grid tetrode with the suppressor grid connected to the screen grid, and with these grids at +250 volts with respect to the cathode. (From reference 2.)

emission from the plate. Note that it does *not* suppress secondary emission. The suppressor grid is at the potential of the cathode. The impinging electrons release secondary electrons from the plate, but the positive electric field from the *screen grid* is largely unable to "reach through" the suppressor grid and pull these secondary electrons away, and they return to the plate. Thus, even though the plate voltage falls *below* that of the screen grid, the harmful *effects* of secondary emission are largely prevented. No "folds" occur in the plate-voltage–plate-current curves of Fig. 5-8 such as are caused in the curves of the screen-grid tetrode (Fig. 5-3). Also, for amplification the plate may be operated at a direct voltage of the same value as that of the screen grid. This is in contrast with the screen-grid tetrode in which the plate voltage must be much higher than that of the screen.

The action of the suppressor grid is shown by Fig. 5-8, the curves being taken with a circuit similar to Fig. 5-2. These curves are for a triple-grid tube. The curves are for a small power-output pentode and not for a small-signal pentode; this is indicated by the large current values. Curves having similar shapes would be obtained for a small-signal pentode. The first set of curves (suppressor-grid voltage = 0) was taken with the third grid connected to the cathode and therefore acting to suppress secondary emission. The second set of curves (suppressor-grid voltage = 250) was taken with the suppressor grid and the screen grid connected together and at the same potential, thus allowing the secondary electrons from the plate to flow to the grids, and therefore giving the tube the characteristics of the conventional screen-grid tetrode.

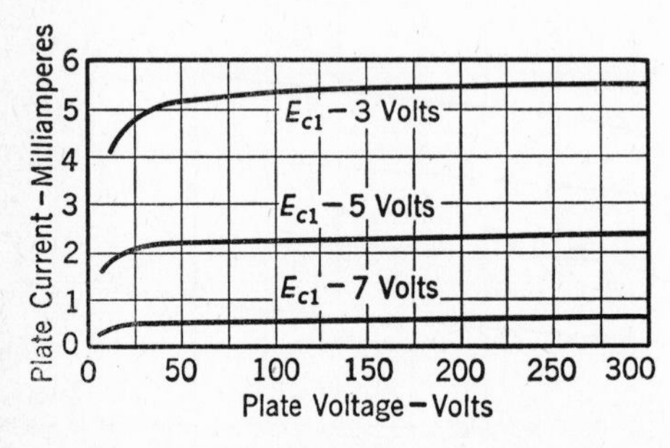

FIG. 5-9. Characteristic curves for a small-signal suppressor-grid pentode with the screen-grid voltage at +135 volts and the control-grid voltage shown on the curves.

5-13. *USES OF SUPPRESSOR-GRID PENTODES*

Suppressor-grid pentodes have two main uses: *First,* small-signal pentodes having high amplification factors are used in circuits where voltage amplification is the main objective; and *second,* power-output pentodes that are physically larger and that will pass larger currents are used as power-output tubes in amplifiers to provide power to drive, for example, a loud-speaker in a radio receiving set.

Small-Signal Pentodes. Typical static characteristic curves for a small-signal suppressor-grid pentode are shown in Fig. 5-9. This tube passes small currents as compared with the power-output pentode of Fig. 5-8. Referring to Fig. 5-9, it is seen that the curves are almost flat. This means that the plate resistance is very high. No attempt will be made to evaluate this resistance from the curves because, as for the screen-grid tetrode, graphical methods for finding this coefficient are not satisfactory. Using circuits discussed early in this chapter gives plate-resistance curves similar to those of Fig. 5-10.

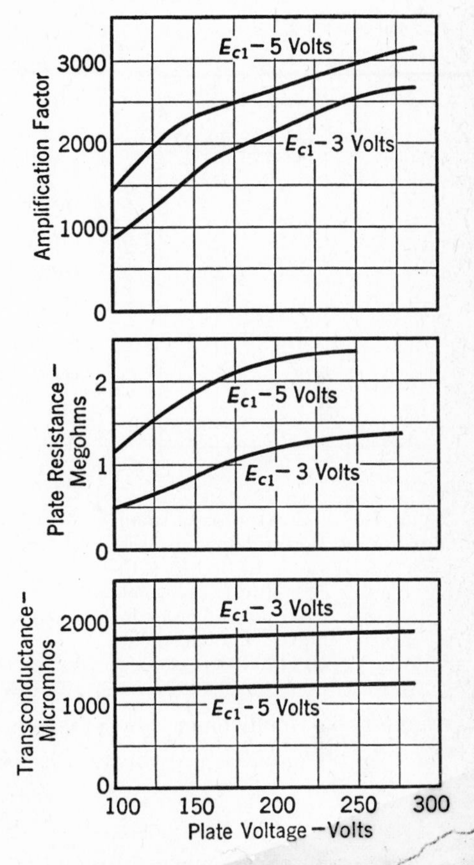

FIG. 5-10. Variations in tube coefficients for the small-signal suppressor-grid pentode of Figs. 5-9 and 5-11. Suppressor grid connected to cathode, and screen-grid voltage at +135 volts.

Grid-voltage–plate-current curves for the same voltage-amplifying pentode are shown in Fig. 5-11. These curves (and those of Fig. 5-9 as well) indicate that the plate voltage has but little effect in determining the plate current (because curves for different plate voltages coincide), but that the grid voltage has a very great effect. This means that a suppressor-grid voltage-amplifying pentode has a high amplification factor. Again, graphical methods are unsatisfactory, but, by means discussed early in this chapter, amplification factors for this tube are as shown in Fig. 5-10.

Power-Output Pentodes. The static characteristics of a small power-output pentode can be determined with a circuit similar to Fig. 5-2. These curves were shown in Fig. 5-8, and are reproduced in Fig. 5-12. By comparison with Fig. 5-9, it is seen that the power pentode passes relatively large currents.

As is evident from Figs. 5-9 and 5-12, the *effects* of secondary emission are largely eliminated. The "knees" of the plate-current curves are rounded and irregular. This is owing mainly to the effect of the zero-potential suppressor-grid wires.[8] Because of these wires the electric field in the region of the suppressor grid is not uniform, but is distorted. There is no definite value of plate voltage where the curves flatten off. Other factors influencing the roundness of the knees of the curves are as follows: Shape and uniformity of the cathode; shapes and pitches of the control and screen grids; relative distances between electrodes and grid side-rods; and the voltages applied to the electrodes.[8]

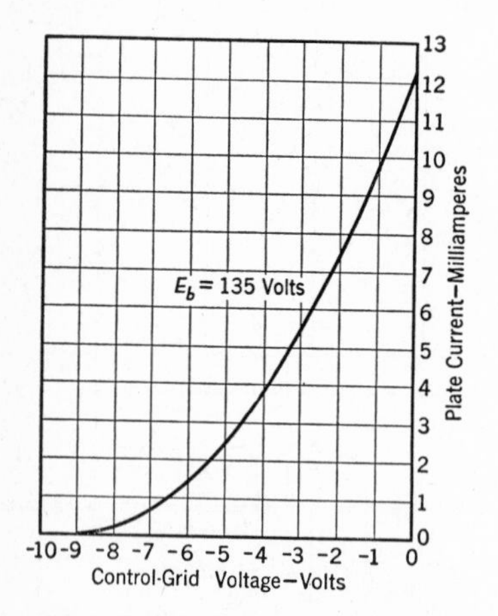

FIG. 5-11. Curves for a small-signal suppressor-grid pentode (Fig. 5-9). Suppressor grid tied to cathode, and screen-grid to cathode voltage E_{c2} +135 volts. Curves for other plate voltages closely coincide with +135-volt curve shown.

The plate-current curves for the power-output pentode of Fig. 5-12 are not so flat as those for the small-signal pentode of Fig. 5-9. This is because the electrodes of the power pentode are designed to pass large currents, but those of the small-signal pentode are designed for high amplification. In the power pentode the wires of the grids are farther apart, giving a more open grid structure and less shielding so that larger currents can flow to the plate.

The way in which the plate current varies with control-grid voltage variations is shown in Fig. 5-13. For the higher plate-voltage values, the *static* curves are parallel, much as for triodes. For the lower plate-voltage values ($E_b = 50$ volts, $E_b = 25$ volts) the curves are no longer parallel. This fact is of importance and will be discussed in Chapter 8. Coefficients for a power-output pentode can be found by the use of circuits previously discussed. Typical values are shown in Fig. 5-14.

Load characteristics (also called dynamic characteristics) of the small power-output suppressor-grid pentode are shown in Fig. 5-13 for three different resistance values. These load curves may be calculated as explained in section 4-15. It is of interest to compare Fig. 5-13 with Fig. 4-14 for a triode. For the triode, the greater the load resistance, the straighter the load, or dynamic, curve will be, and the less will be the distortion. The opposite is true for the power pentode, however. The greater the load resistance, the more curved the load, or dynamic, curves become. Thus, with the power pentode the greater the load resistance, the *greater the distortion* for a given magnitude of alternating input signal voltage applied to the grid (usually called grid "swing").

These relations are further illustrated by the upper curves of Fig. 5-15.

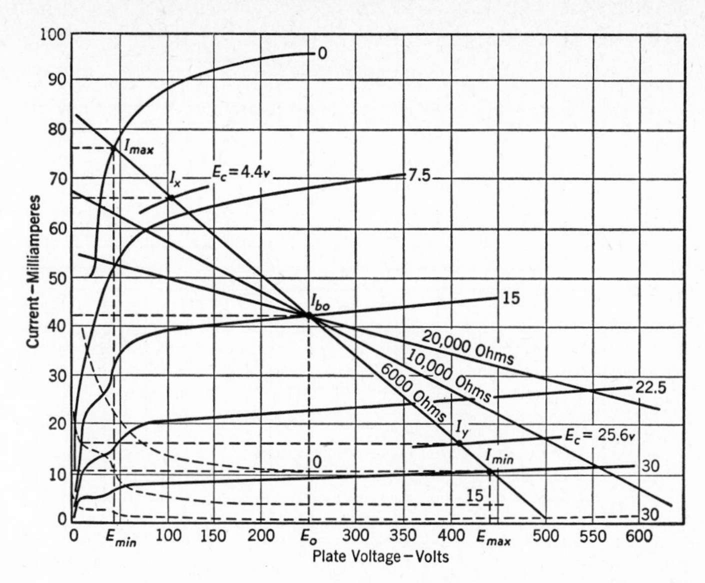

FIG. 5-12. Characteristics of a small power-output pentode. Full lines are plate-current curves; broken curves are for screen-grid current. Negative control-grid voltages shown at ends of curves. The three straight lines will be considered in Chapter 8.

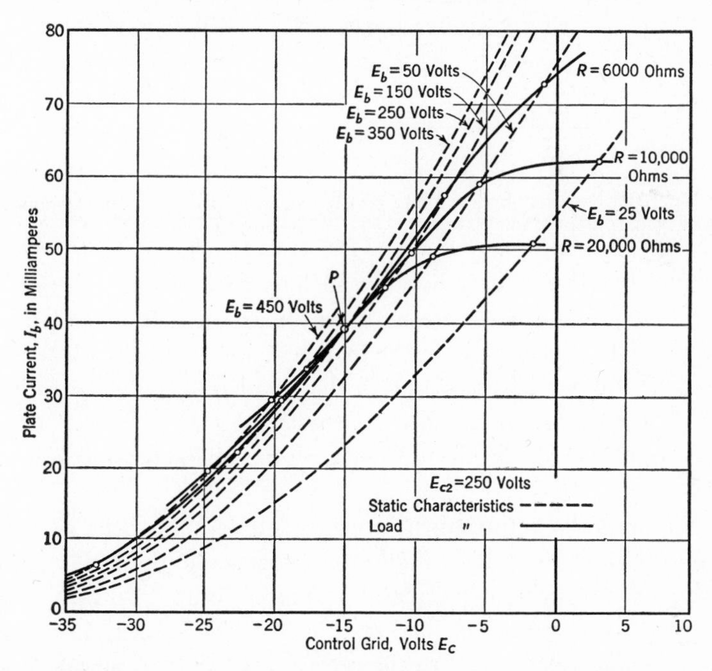

FIG. 5-13. Plate-current–grid voltage characteristics of a small power-output pentode, with load, or dynamic, characteristics for resistance loads of 6,000, 10,000, and 20,000 ohms. (From reference 2.)

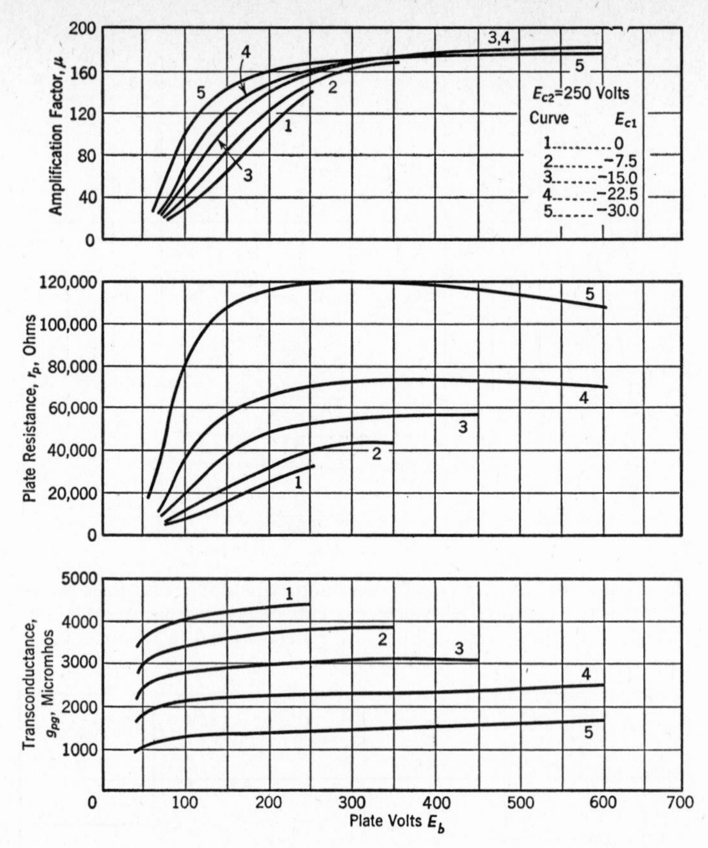

FIG. 5-14. Amplification factor, plate resistance, and grid-plate transconductance for a small power-output pentode. Screen grid at +250 volts. (From reference 2.)

For low input voltages, the *power output increases* as the load resistance is increased. This is in accordance with the theory of section 4-21 that maximum power output occurs when the load resistance and the plate resistance are equal. Of course, this assumes that the characteristics of the tube remain constant. The decrease in power output with increasing load resistance and at higher inputs is due to the decrease of the characteristics as shown in Fig. 5-13.

Thus, although the output of the power pentode follows the simple theory of matched resistances at low values of resistance and input, this theory cannot be applied throughout the operating range. As Fig. 5-13 indicates, it is necessary to use low values of load resistance (compared to the plate resistance) to minimize distortion. For the power pentode here considered, a load resistance of about 6000 ohms would probably be used, sacrificing output to some extent, but obtaining higher quality as the lower sets of curves show.

Further information on distortion in power pentodes will be found in Chapter 8, where their use in amplifiers is considered.

5-14. *BEAM-POWER THERMIONIC VACUUM TUBES*

From the standpoint of the number of electrodes, this tube is a tetrode. However, the operating characteristics are so similar to those of the power-output pentode that it is considered at this point, instead of earlier in the chapter, following tetrodes.

The **beam-power tube** is [1] an "electron-beam tube in which use is made of directed electron beams to contribute substantially to its power-handling capability, and in which the control grid and the screen grid are essentially aligned." The characteristics of an **electron-beam tube** depend [1] "upon the formation and control of one or more electron beams."

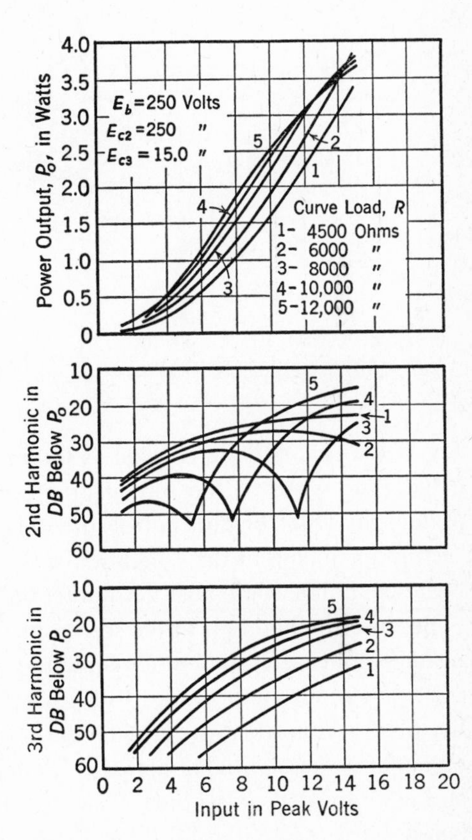

FIG. 5-15. Characteristics of a small power-output pentode. Sinusoidal voltage applied to control grid; peak values as indicated. For definition of *DB*, see Chapter 7. (From reference 2.)

The name beam-power tube is somewhat misleading because a beam of electrons often is considered to be a "pencil-like" ray, such as in the cathode-ray tube. In the beam-power tube, the electron stream is not a ray, but is shaped as in Fig. 5-16. The beam-forming plate, or electrode, in this tube is at cathode potential, usually zero. If this beam-forming electrode is considered to be part of the cathode (as it often is) then the tube is a tetrode; if the beam-forming electrode is considered to be a separate electrode, the tube is a pentode. In any event, the beam-power tube has characteristics similar to the power-output pentode, and the effects of secondary emission from the plate are eliminated without a suppressor grid.

Theory. The special construction of a beam-power tube is shown by

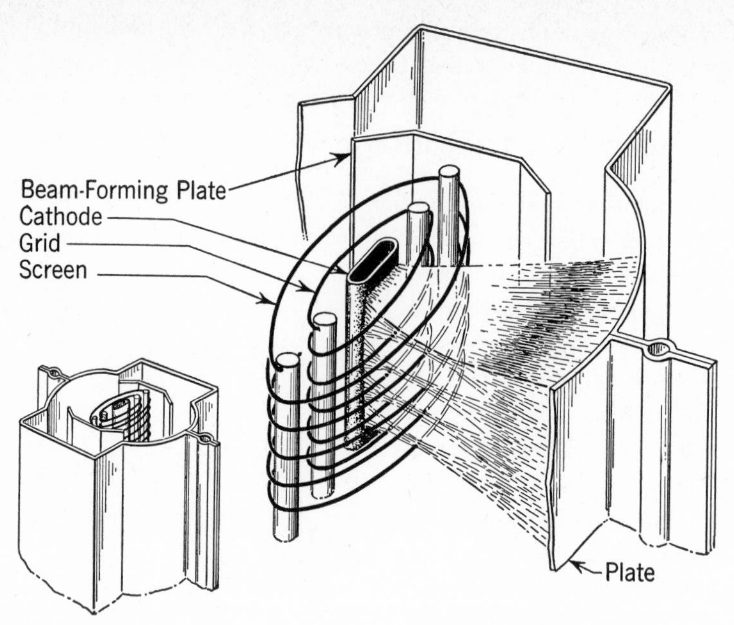

FIG. 5-16. Arrangement of electrodes in a beam-power tube. (Courtesy Radio Corporation of America.)

Fig. 5-16. The cathode is large, is indirectly heated, and has the shape indicated. The wires of the screen grid are *exactly the same* size *and exactly behind* the wires of the control grid. The screen-grid wires may be mentally pictured as being in the "shadow" of the control grid. The beam-forming, or beam-confining, plate is at zero potential, being connected to the cathode.

Because the highly positive screen grid is quite effectively enclosed by the beam-forming zero-potential confining plates, a region of zero potential exists [8,9] somewhat as shown in Fig. 5-17. Secondary electrons emitted from the plate are not drawn back to the highly positive screen grid. At such a region of zero potential a **virtual cathode** [8] (region of zero potential gradient and zero absolute potential) exists. In this connection it should be remembered, however, that at all times the *control grid* is actually determining the electron flow to the plate.

Because the screen-grid wires are immediately behind those of the control grid, the initial repulsion given the electrons by the negative control-grid wires prevents the electrons from striking the positive screen-grid wires, as Fig. 5-16 illustrates. The current taken by the screen grid is, therefore, very small except at low plate potentials. Furthermore, little secondary emission can occur from the screen grid because few electrons hit it. And, as previously explained, the zero potential gradient (virtual cathode or space charge if desired) prevents secondary emission from the plate from causing distortion.

Characteristics. The relation between the plate voltage and plate and screen-grid currents is shown in Fig. 5-18. A load line for a 2500-ohm load is indicated. This line, the nature of which will be discussed in Chapter 8, has been corrected for the rectification occurring with large signals. For these curves the screen-grid voltage was +250 volts. The curves were taken at various control-grid direct voltages. The manner in which the plate current changes with control-grid variation is shown in Fig. 5-19. As previously mentioned, the knees of these curves are *quite definite* and occur at *low values of plate voltage.* This makes it possible for the tube to handle large amounts of power with low distortion and with low plate voltage. Characteristics of a typical beam-power tube as a linear class-A amplifier (Chapter 8) are shown in Fig. 5-20.

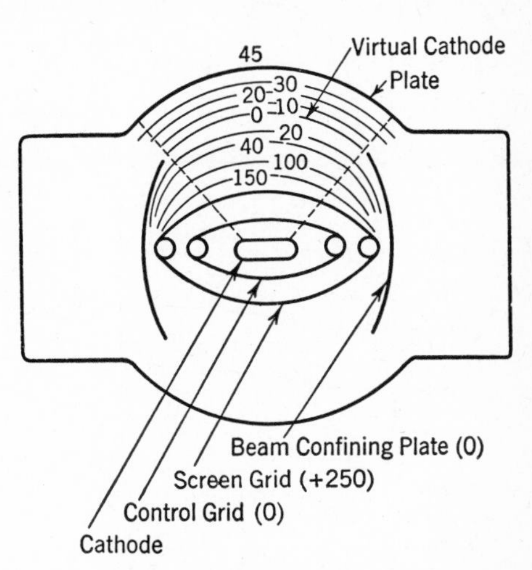

FIG. 5-17. Top view of a beam-power tube, showing relative potentials at several points in the beam. The instantaneous condition depicted is when the grid has been driven to zero by the applied signal voltage and the plate current has increased, causing the instantaneous plate voltage to fall to +45 volts because of the drop in the load resistor. (Courtesy Radio Corporation of America.)

5-15. *CO-PLANAR GRID THERMIONIC VACUUM TUBE*

This tube has four electrodes and hence is a tetrode, but (like the beam-power tube) it is considered at this point because the co-planar grid tube has characteristics like those of the power-output pentode.[2,3] The electrodes are arranged as shown in Fig. 5-21, with the two grids in the same plane. One grid, sometimes called a net, is made positive with respect to the cathode, and accordingly is a space-charge reducing, or accelerating, grid. Co-planar grid tubes chiefly are of theoretical interest; discussion of the theory will be found in references 2 and 3.

5-16. *VARIABLE-MU REMOTE CUTOFF TUBES*

The amplification factors of the tubes that have been previously considered are substantially constant, or at least of about the same magnitude, over

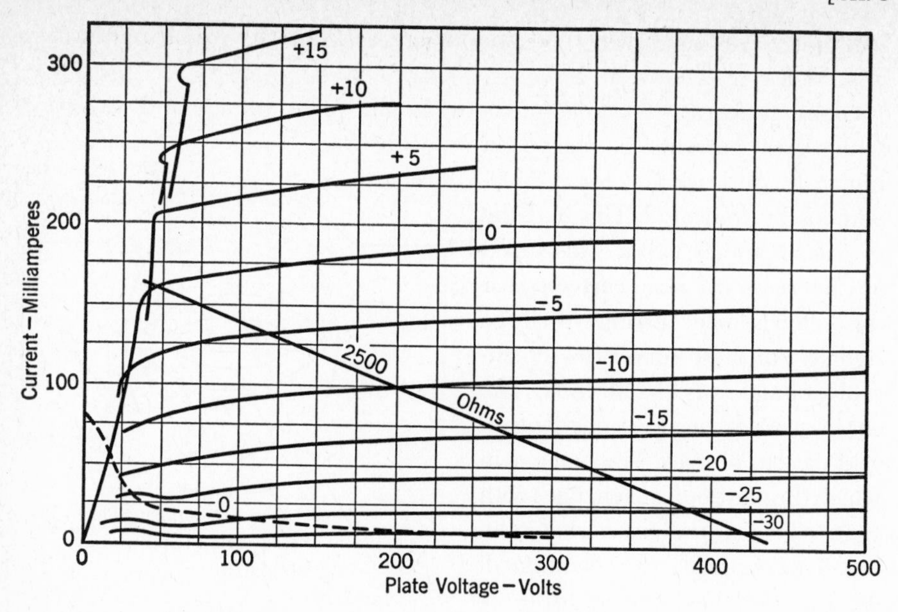

FIG. 5-18. Characteristics of a beam-power tube. Full lines are plate-current curves; broken line is screen-grid current when control-grid voltage is zero. Screen grid at +250 volts. Control-grid voltages shown on curves. (From reference 8.)

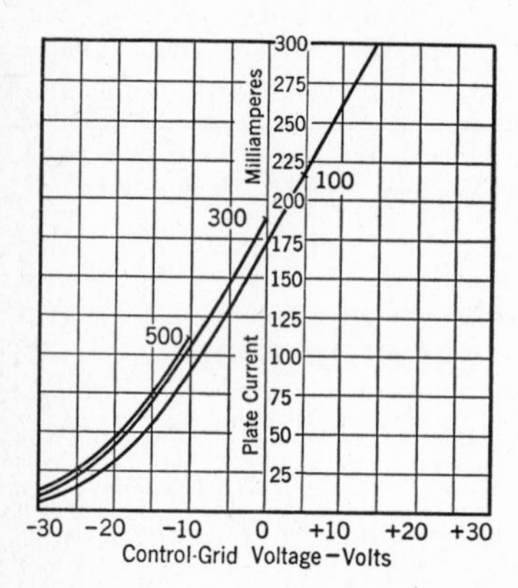

FIG. 5-19. Variations in plate current with control-grid voltage for the beam-power tube of Fig. 5-18. Plate voltages in volts shown on curves.

the normal operating ranges. Tubes with variable amplification constants are desirable for some purposes. Such tubes have been developed [10] and are called **variable-mu tubes.**[1] They have been called **super-control tubes,** and **remote cutoff tubes.** Both screen-grid tetrodes and suppressor-grid pentodes are made in this form. These are small-signal tubes and not power-output tubes.

The characteristics of a conventional screen-grid tetrode and of a variable-mu tube are shown for comparison in Fig. 5-22. As is indicated, a large negative control-grid voltage is required to cause cutoff for the variable-mu tube; that is, the cutoff point is remote from zero.

Because of the *gradually* varying characteristic curve, this tube may be operated as an amplifier at any point along the curve with a reasonably large applied signal. Various amounts of amplification are possible, depending on the point of operation. Thus if a variable-mu tube is biased to −5 volts, operation is on the steep portion of the characteristic curve, and a large plate-current change is produced by a small alternating signal voltage between grid and cathode. This gives high amplification. If the tube is biased to −30 volts, however, the same alternating grid voltage will produce but a small plate-current change. This gives low amplification.

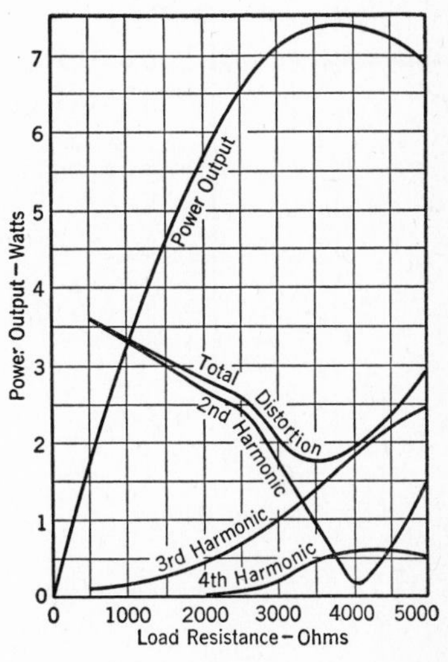

FIG. 5-20. Variations in power output and percentage distortion with load resistance for a beam-power tube. (From reference 8.)

Of course this action is also true, in a sense, for a conventional screen-grid tetrode or suppressor-grid pentode. Referring to Fig. 5-22 it is plain that the slope of the curve for the conventional screen-grid tetrode changes, and that various degrees of amplification are available. There is, however, a marked distinction between the curves for the variable-mu tube and for the conventional tube. With the variable-mu tube, the change of slope is so gradual that at *any point* on the curve a reasonably large input alternating signal voltage can be impressed, and the output current will not be *excessively* distorted. The conventional tube can be operated with appreciable input signal voltage *only* over the straight region having essentially constant slope and constant amplification. If the conventional tube were operated on any other part of the curve, and *if appreciable alternating signal voltage were used*, then excessive distortion would result.

Various combinations [10] of nonuniformly wound and nonuniformly spaced control and screen grids will give the remote-cutoff characteristic. In one type the control grid is uniformly wound at the ends, but the turns at the center are omitted. As the bias on the control grid is made progressively more negative, cutoff first occurs at the portions of the structure where the control-grid wires exist. This reduces the plate current along the steep part of the curve of Fig. 5-22. As the grid is made negative, the gap in the control grid is in effect closed by the strong electric field and the plate current gradually is forced to zero. Thus at low values of control-grid bias the current through the variable-mu tube is controlled by the high-amplification portions of the tube. At high values of control-grid bias, the current through the variable-mu tube is controlled by the low-amplification portion of the tube.

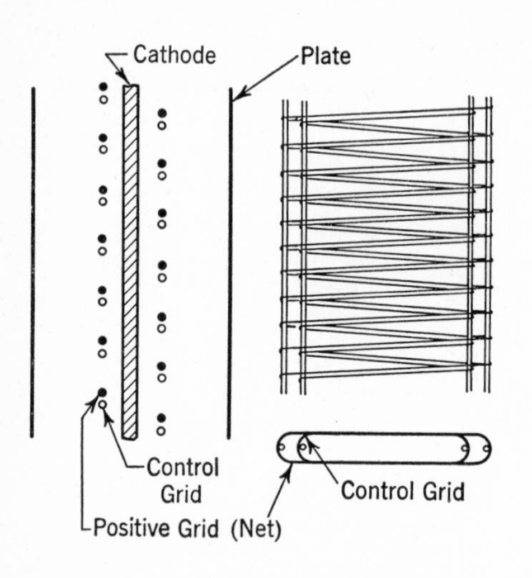

FIG. 5-21. Arrangement of electrodes in a co-planar grid tube. (From reference 2.)

Variable-mu tubes are extensively used in automatic volume-control circuits of radio receiving sets, Chapter 13. When a strong signal is being received, the tube is biased so that operation is on the lower part of the curve where the amplification is low. When the transmission path from the distant radio broadcasting station changes and a weak signal is received, the bias automatically is reduced, and the tube is operated on the portion of the curve where the amplification is high.

5-17. *MULTIELECTRODE THERMIONIC GAS TUBES*

Tubes containing gas and mercury vapor generally are used for rectification and special switching functions, and are not used for speech amplification. An exception now will be considered.

Multielectrode Thermionic Gas Amplifying Tubes. Experimental work [11,12] on gas amplifier tubes resulted in the development of tubes satisfactory for speech power-output purposes, one tube being the RK-100. This tube contains mercury vapor, but inert gas may be used to minimize temperature variations. A hot cathode supplies electrons. These electrons are drawn over to a gridlike electrode at about +10 volts with respect to the cathode, Fig. 5-23. This action causes ionization by collision between the cathode and this first electrode, which is called a **cathanode.** The cathanode and the control grid are close to the plate.

Because of this construction and other factors, ionization is largely prevented from occurring in the region between the cathanode and plate; hence, a positive-ion sheath *does not* form about the negative control grid as in Thyratrons. Thus, the control grid can regulate the flow of electrons from the cathanode to the plate, and speech amplification is possible. Ionization by collision in the region between the cathode and the cathanode results in space-charge neutralization in this region, and large electron currents flow to the *cathanode.* Many electrons pass through the spaces in this electrode and flow from the *cathanode* toward the positive plate, the instantaneous current being regulated by the control grid. What little ionization occurs is effective in reducing negative space-charge effects, but the grid does not lose control.

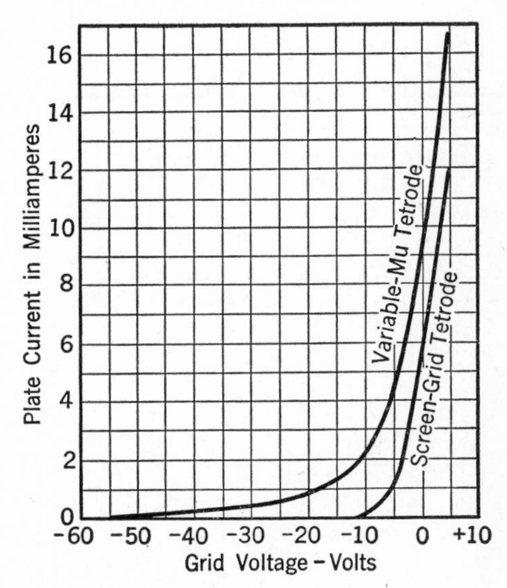

FIG. 5-22. Comparison of a sharp-cutoff tube and a variable-mu tube.

Multielectrode Thermionic Gas Rectifier Tubes. The grid-controlled gas rectifier tubes, or Thyratrons, discussed in Chapter 4 contain a thermionic cathode, grid, and anode sealed in an enclosure containing an inert gas or mercury vapor at low pressure. In these tubes the voltage between grid and cathode determines the anode voltage at which the "control" grid loses control and the tube becomes conducting.

Although the three-electrode tube just described has been satisfactory, the addition of a fourth electrode, in the form of a shield around the grid, improves operation [13,14] as follows: *First,* less grid current is passed with the shielded construction both before and after breakdown. The grid current before breakdown is important, because in high-impedance control circuits, the drop in voltage caused by the grid-current flow decreases the voltage (from a control device such as a phototube) impressed on the grid. *Second,* the shield isolates the grid from electric charges that accumulate on the walls of the tube. *Third,* the grid is shielded from material sputtered and evaporated from the cathode. *Fourth,* the grid is heat-shielded and lower grid emission occurs. *Fifth,* and perhaps most important, a shield around the grid makes possible variations in the characteristics of the tube. Referring to Fig. 4-22 for a three-electrode Thyratron, and assuming that the gas temperature is

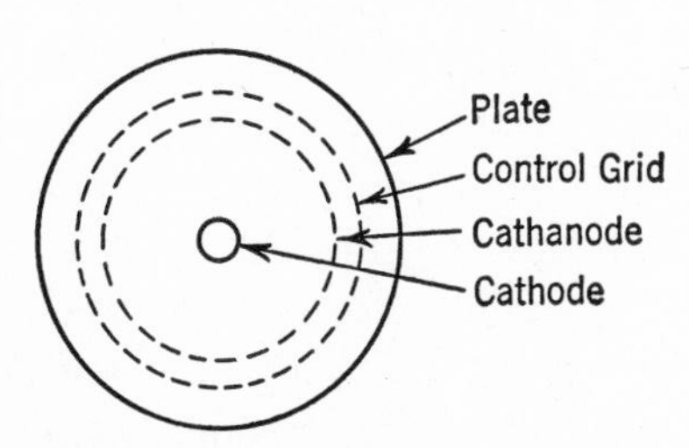

FIG. 5-23. Arrangement of electrodes in a thermionic *gas* tube once used for audio power-output purposes.

constant, it follows that this tube has *one set* of characteristics only. That is, for the 30° temperature curve, if the grid is 1.0 volt negative and the anode is 150 volts positive, the tube will break down. But, it might be very desirable to have a tube that would break down at −0.5 volt on the grid and +150 volts on the anode. With the multielectrode Thyratron containing a shield around the control grid, varying the potential of the shield will alter the breakdown characteristics of the tube.

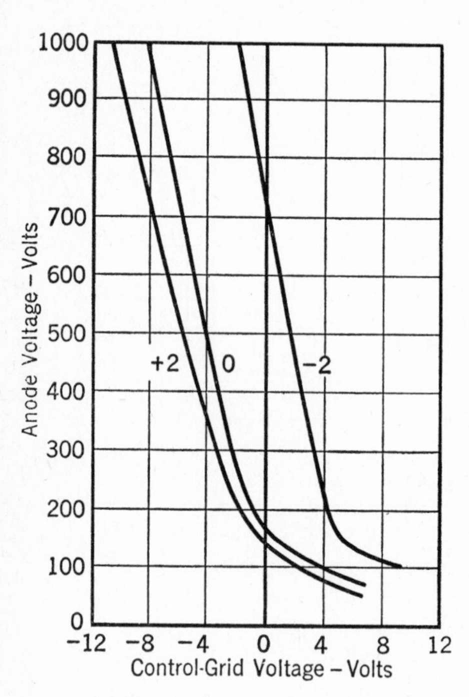

FIG. 5-24. Characteristics of a multielectrode thermionic gas rectifier tube (shield-grid Thyratron). Numbers on curves indicate voltage in volts from shield grid to cathode.

These changes in characteristics for the shield-grid Thyratron are indicated in Fig. 5-24. With zero voltage between the shield grid and the cathode, the tube has certain characteristics as shown. If, now, the shield is made slightly negative, the control grid will have to be positive before it can initiate a current flow that will break down the tube. If, however, the shield is positive, it is attempting to initiate a current flow, and the grid must be made quite negative to prevent breakdown and current flow to the anode. It is apparent that such a tube would be more flexible than one without a shield grid.

Multielectrode Ignitrons. The Ignitron, using a mercury pool as a source of electrons, was considered in Chapter 3. The advantages of the mercury pool over thermionic cathode as a source of electrons where high currents are to be handled, as in electric power devices, was stressed. Also, in that same chapter the use of shields around the anodes of such tubes was explained.

The advantages of the mercury-pool cathode and the shielded anode led to the development of a multielectrode Ignitron called a **pentode Ignitron.**[15] This tube contains a mercury pool, an Ignitor, three grids, and an anode. Next to the anode is the first grid, called the **intermediate anode.** In one circuit it is used to divide potential gradients during nonconducting periods. The second or center grid is the **control grid,** used to determine the time of starting and to reduce the deionization period at the end of conduction. The third or outer grid is a **shield grid** and acts to pick up the arc at the beginning of the conduction period and to shield the rest of the grid-anode structure from residual ionization after conduction.[15]

REFERENCES

1. Institute of Radio Engineers. *Standards on Electron Tubes: Definitions of Terms,* 1950, Proc. I.R.E., Vol. 38, No. 4, April 1950; *Methods of Testing,* 1950, Vol. 38, Nos. 8 and 9, Aug. and Sept., 1950. *Standards on Abbreviations, Graphical Symbols, Letter Symbols, and Mathematical Signs,* 1948; *Standards on Graphical Symbols for Electrical Diagrams,* Proc. I.R.E., Vol. 42, No. 6, June 1954.
2. Pidgeon, H. A. *Theory of multielectrode vacuum tubes.* Electrical Engineering, Nov. 1934, Vol. 53. Also Bell System Technical Journal, Jan. 1935, Vol. 14, No. 1.
3. Pidgeon, H. A., and McNally, J. O. *The output power from vacuum tubes of different types.* Proc. I.R.E., Feb. 1930, Vol. 18.
4. Hull, A. W., and Williams, N. H. *Characteristics of shielded-grid Pliotrons.* Physical Review, Apr. 1926, Vol. 27, No. 4.
5. Wallis, C. M. *Space-current division in the power tetrode.* Proc. I.R.E., April 1947, Vol. 35, No. 4.
6. Doolittle, H. D. *Design problems in triode and tetrode tubes for U.H.F. Operation.* Cathode Press, Fall 1951, Vol. 8, No. 4, Machlett Laboratories, Inc.
7. Henney, K. *Two kinds of pentodes.* Electronics, Apr. 1930, Vol. 1, No. 4.
8. RCA Manufacturing Co. *Application note on the operation of the 6L6.* Application Note No. 60, June 10, 1936.
9. Dreyer, J. F. *The beam power tube.* Electronics, Apr. 1936, Vol. 9, No. 4.
10. Ballantine, S., and Snow, H. A. *Reduction of distortion and crosstalk in radio receivers by means of variable-mu tetrodes.* Proc. I.R.E., Dec. 1930, Vol. 18, No. 12.
11. *Tubes with cold cathodes.* Electronics, Jan. 1933, Vol. 6, No. 1.
12. LeVan, J. D., and Weeks, P. T. *A new type gas-filled amplifier tube.* Proc. I.R.E., Feb. 1936, Vol. 24, No. 2.
13. Maser, H. T., and Livingston, O. W. *Shield grid Thyratrons.* Electronics, Apr. 1934, Vol. 7, No. 4.
14. Westinghouse Electric Corp. *Industrial Electronics Reference Book.* John Wiley & Sons.
15. Steiner, H. C., Zehner, J. L., and Zuvers, H. E. *Pentode Ignitrons for electronic power converters.* Electrical Engineering, Oct. 1944, Vol. 63, No. 10.

QUESTIONS

1. Explain the differences between multielectrode and multiple-unit tubes.
2. Why have not multiple-unit tubes been considered in detail in this chapter?
3. Discuss the effects of negative space charge in multielectrode thermionic vacuum tubes.

4. What is meant by the term virtual cathode? Could a virtual cathode exist in a gas tube?
5. Discuss the thermionic vacuum space-charge-grid tetrode, and explain why its use is limited.
6. Discuss the thermionic vacuum screen-grid tetrode, explaining the cause of, and the effects of, secondary emission.
7. Does secondary emission ever occur from the screen grid in the thermionic vacuum screen-grid tetrode?
8. Over what portion of the characteristic curve is a thermionic vacuum screen-grid tube operated?
9. Are both small-signal and power-output tetrodes available? Which type has been more widely used?
10. How do the coefficients of a typical small-signal screen-grid tetrode compare with those of a typical small-signal triode?
11. What is meant by large capacitors at the close of section 5-5?
12. Why are not the circuits used for finding the coefficients of triodes satisfactory for use with tetrodes and pentodes?
13. Discuss the division of the space current in a thermionic vacuum screen-grid tetrode operated in the normal manner with control grid negative, and with screen and plate positive. Does this division cause noise? Explain.
14. What system is used to designate the potentials of the various grids in a screen-grid tetrode? In a suppressor-grid pentode?
15. Is it always necessary to use number subscripts to designate the control grid? Explain.
16. What is the purpose of the capacitor in Fig. 5-5a, and of the mutual inductance of Fig. 5-5c?
17. Explain how to operate a triple-grid pentode as a space-charge-grid pentode, or as a suppressor-grid pentode. What will be the general characteristics of each?
18. Why is the magnitude of the mutual conductance similar for triodes, screen-grid tetrodes, and suppressor-grid pentodes of the small-signal type?
19. Explain why the amplification factors of comparable triodes, tetrodes, and pentodes are different.
20. How do the small-signal pentode and power-output pentode differ, both physically and as to their coefficients?
21. Referring to Fig. 5-12, it is sometimes observed that the plate-current curves bend up slightly at high plate-voltage values. What would be the cause?
22. Does the suppressor grid prevent secondary emission? Explain its action.
23. How is secondary emission controlled in a beam-power tube?
24. How is a variable amplification factor obtained?
25. Does an ordinary small-signal pentode have a variable amplification factor? If your answer is yes, explain why a special variable-mu tube is necessary.

PROBLEMS

1. Prepare a table showing the numerical values of μ, r_p, and g_{pg} for typical small-signal triodes, tetrodes, and pentodes, and a second table for corresponding power-output tubes.

2. A screen-grid tetrode has a $\mu = 500$, and $r_p = 500{,}000$ ohms. It is operated as an amplifier with 250,000 ohms as a plate load resistor. Find the voltage gain of the stage. If a signal voltage of 0.05 volt is impressed on the control grid, what signal voltage will appear across the load resistor?
3. A screen-grid tetrode has a $g_{pg} = 1000$ micromhos. It is biased to operate as an amplifier, but does not have a plate load resistor. If a signal voltage of 0.01 volt is impressed on the control grid, calculate the alternating-current component that will flow in the plate circuit.
4. Referring to problem 2, if the voltage between the plate and cathode is 115 volts, and the plate current is 1.18 milliamperes, what must be the voltage of the plate power supply? How much power will the power-supply deliver? Where will it be dissipated?
5. A small-signal suppressor-grid pentode has a $\mu = 1750$ and a plate resistance of 2,000,000 ohms. It is operated as an amplifier with 250,000 ohms as a plate load resistor. If a signal voltage of 0.0075 volt is impressed on the control grid of the tube, what signal voltage will appear across the plate load resistor?
6. Referring to problems 2 and 5, what per cent of the theoretically available voltage appears across the load resistors in each problem?
7. A power-output pentode having characteristics similar to those of Fig. 5-13 is operated as an amplifier. It has an amplification factor of 150 and a plate resistance of 60,000 ohms. If a peak alternating signal voltage of 12.5 volts is impressed on the tube, what will be the voltage across a load resistor of 6000 ohms?
8. Referring to problem 7, what will be the approximate power output?
9. What per cent of the theoretically available output voltage and power is obtained for the tube of problems 7 and 8?
10. A beam-power tube has a $\mu = 135$, $r_p = 22{,}500$ ohms, and is operated with a plate load resistor of 2500 ohms. Calculate the approximate power output to the load resistor when the peak alternating signal voltage impressed on the control grid is 10 volts. What per cent is this of the theoretically possible power output?

CHAPTER 6

ELECTRON-TUBE RECTIFIERS

A rectifier is defined [1] as a "device having an asymmetrical conduction characteristic which is used for the conversion of an alternating current into a current having a unidirectional component." The rectifying action depends on its ability to offer low resistance to current flow in one direction, but to offer high resistance to flow of current in the opposite direction.

Rectifiers of many types have been devised. These may be broadly classified under two headings: First, **mechanical rectifiers**; and second, **electrical rectifiers.** The commutator on an ordinary direct-current machine is an example of a mechanical rectifier. Only *electrical rectifiers* will be considered in this chapter.

The **rectifier unit** includes the rectifier with its auxiliaries and the rectifier transformer equipment.[2] Accordingly, in the discussions that follow, the term *rectifier* will designate the device that changes the wave form because of its asymmetrical (or nonsymmetrical) conducting properties. The term *rectifier unit* will designate the entire device including the rectifier itself and all the associated circuits.

An *ideal* rectifier is a device that offers *zero* resistance to current flow in one direction, and *infinite* resistance to current flow in the opposite direction. Thus, the ideal rectifier merely is a "switch" that opens and closes the circuit as the voltage impressed on the circuit changes. Two-electrode electron tubes, or diodes, are extensively used as rectifiers. To an important extent, tubes are being supplanted by semiconductor devices (Chapter 10).

6-1. *ANALYSIS OF RECTIFIED WAVES*

In one sense, rectification is a process of distortion. This is based on the definition of distortion [1,2] as "a change in wave form." There is a growing tendency, however, to use the term distortion only when the change in wave form is undesired.[1,3] If a sine-wave voltage is impressed on a rectifier, the rectified output current is not a sine wave. The shape of the output wave has been changed by the process of rectification, and *desired* distortion has resulted.

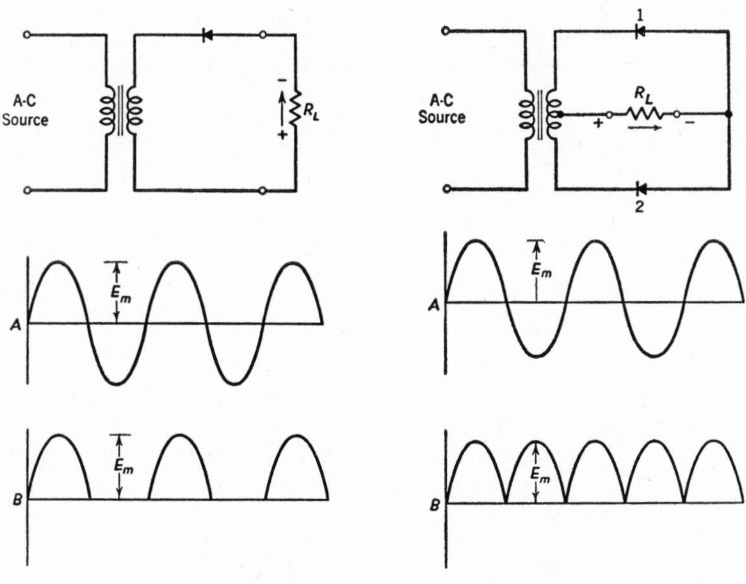

FIG. 6-1. Input voltage A and output current and voltage B for a half-wave rectifier with resistance load.

FIG. 6-2. Input voltage A and output current and voltage B for a full-wave rectifier with resistance load.

If the rectified output currents or voltages of Figs. 6-1 and 6-2 are examined, it will be apparent that these are unidirectional, and that they contain a direct component. What is not so apparent, however, is the fact that they contain *sinusoidal alternating components* produced by the process of rectification.

Thus, if an ideal rectifier is used in the half-wave rectifier circuit of Fig. 6-1 to rectify the sine-wave voltage of maximum value E_m and frequency f, the rectifier voltage will be as indicated by B, and can be represented[4] by the following:

$$e = \frac{E_m}{\pi} + \frac{E_m}{2} \sin \omega t - \frac{2E_m}{3\pi} \cos 2\omega t - \frac{2E_m}{15\pi} \cos 4\omega t \cdots \qquad 6\text{-}1$$

The first of these terms is the direct component since it contains *no* frequency term ($\omega = 2\pi f$). The other terms represent the sinusoidal alternating components. For the half-wave rectifier the first of these alternating terms would be the *same frequency as the alternating voltage being rectified*; the second term would be twice the frequency; and the third would be four times the frequency. The first term of equation **6-1** would be the voltage indicated by a direct-current voltmeter connected across the load resistance R_L of Fig. 6-1, and the second, third, and fourth terms would be as measured by a wave analyzer when set on these various frequencies.

Similarly, if ideal rectifiers are used in the full-wave rectifier circuit of Fig. 6-2, and if the sine-wave voltage shown by curve A is impressed, the rectified voltage will be as shown by B, and can be represented by the following:

$$e = \frac{2E_m}{\pi} - \frac{4E_m}{3\pi} \cos 2\omega t - \frac{4E_m}{15\pi} \cos 4\omega t - \frac{4E_m}{35\pi} \cos 6\omega t \cdots \qquad 6\text{-}2$$

The first term is the direct component that would be measured by a direct-current voltmeter connected across the load resistance R_L of Fig. 6-2. The other terms are the alternating components that would be measured by a wave analyzer connected across R_L. These statements justify the viewpoint that *rectification is a process of desired distortion.*

6-2. *RECTIFIER UNITS*

A simple rectifier unit is shown in Fig. 6-1. The transformer increases or decreases (as desired) the alternating voltage impressed on the rectifier. This is connected to a load represented by resistor R_L. Current passes through the rectifier *only* in the direction (conventional direction) of the arrow. That is, the resistance of the rectifier is assumed to be *zero* for current in the direction of the arrow, and *infinite* in the opposite direction. Thus, if the alternating voltage of curve A is induced in the secondary of the transformer (assumed to be ideal), then the current through the load R_L and the voltage across R_L will be as shown by B. This type of rectifier unit is known as a **half-wave rectifier.**[2]

A simple rectifier unit employing *two* rectifiers is shown in Fig. 6-2. When the upper end of the input transformer secondary is *negative* so that current *passes* through rectifier 1, the lower end of the transformer will be *positive* and *no current* will pass through rectifier 2. On the next half-cycle of the impressed alternating voltage, the *lower* end of the transformer will be negative and rectifier 2 instead of rectifier 1 will conduct. In this manner, *both* halves of the impressed alternating impulses will be rectified. With an alternating voltage A induced in the secondary of the transformer, the current through the resistor R_L and the voltage across the resistor R_L will be as shown by B of Fig. 6-2. This unit is a **full-wave rectifier.**[2] The circuit of Fig. 6-2 sometimes is called a **transformer bridge circuit.**

The circuit of another rectifier unit is shown in Fig. 6-3; this is called a **Graetz bridge** or a **rectifier bridge circuit.** Four rectifiers are used in a balanced bridge arrangement. When the *upper* bridge terminal is made positive by the applied alternating voltage, current will flow down through rectifier 1, across through R_L, and down through rectifier 2. Theoretically no current will flow through rectifiers 3 and 4. When the applied alternating voltage reverses direction on the next half-cycle, and the *lower* end is positive, current will flow up through rectifier 3, across through R_L and up through rectifier 4. Hence, current flows through R_L in the same direction for *each half-cycle* of the applied voltage, giving a *full-wave* rectifier unit and a rectified current such as B of Fig. 6-2.

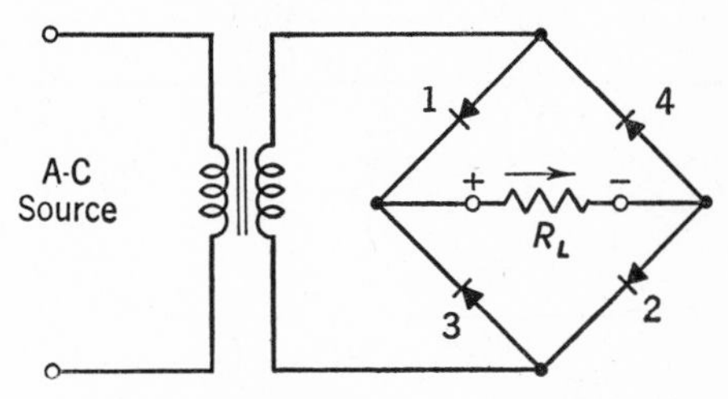

FIG. 6-3. The Graetz, or rectifier bridge, full-wave rectifier circuit.

The *rectifier unit* is composed of elements illustrated by Fig. 6-4. The transformer increases or decreases the supply voltage, depending on whether high direct voltage or low direct voltage is desired. The transformer also isolates the alternating power-supply system from the rectifier, filter, and load. The purpose of the filter is to *suppress the alternating components* of the rectified wave as listed in equations 6-1 and 6-2. For battery-charging uses, often no filter is required; for most voltage supply purposes, such as to vacuum tubes, adequate filtering is necessary.

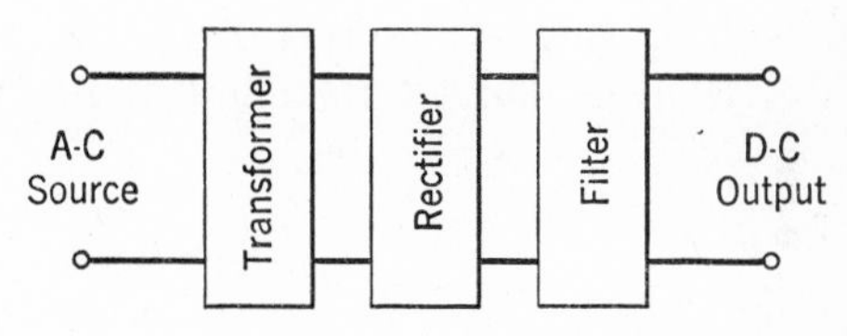

FIG. 6-4. The elements of a rectifier unit.

6-3. *RECTIFIER FILTERS*

An **electric wave filter** is [2] a "selective network which transmits freely electric waves having frequencies within one or more frequency bands and which attenuates substantially electric waves having other frequencies." Many types of filters are possible. One is [2] the **low-pass filter,** "a wave filter having a single transmission band extending from zero frequency up to some critical or cutoff frequency, not infinite." The **ripple filter** is [2] a "low-pass filter designed to reduce the ripple current, while freely passing the direct current, from a rectifier or generator."

It is convenient, and usually satisfactory, to assume that ripple filters are composed of inductors and capacitors that cause no heat losses for alternating voltages and currents. Such inductors and capacitors have zero equivalent-series effective resistance, and hence are dissipationless. The upper circuit of Fig. 6-5 is an inductor-input filter (also called choke-input filter), and is a T section. The lower circuit is a capacitor-input filter (also called condenser-input filter), and is a π section.

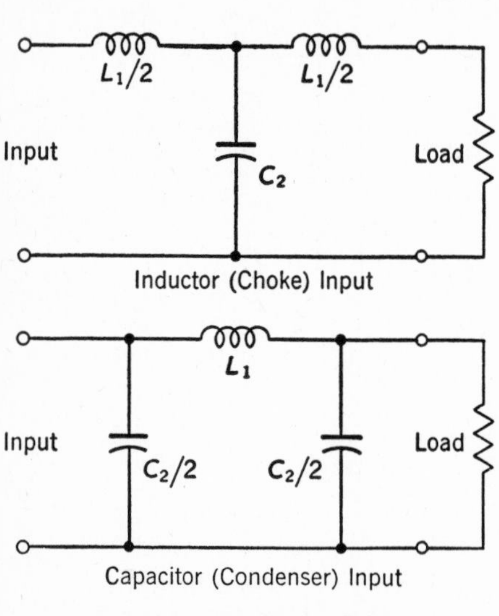

FIG. 6-5. **Low-pass filters.**

The theoretical **cutoff frequency** [2] is the frequency at which the attenuation of a dissipationless filter theoretically becomes infinite. Thus, in Fig. 6-5, if equal voltages of different frequencies are impressed across the input terminals, current will flow to the load, and power will be delivered to the load, for frequencies from zero up to the cutoff frequency. If the cutoff frequency is well below the lowest alternating component of equations 6-1 and 6-2, the output of a rectifier will be filtered, and only direct current will flow to the load. For both the T and π sections of Fig. 6-5, the cutoff frequency is approximately

$$f_c = \frac{1}{\pi\sqrt{L_1 C_2}} \tag{6-3}$$

where f_c is in cycles per second when L_1 and C_2 are in henrys and farads.

The attenuation, or loss, of a low-pass filter section of Fig. 6-5 is low for frequencies close to the cutoff frequency; however, the loss is high for frequencies considerably beyond the cutoff frequency. It is possible to design filter sections containing resonant branches that have high loss at frequencies close to cutoff. If these two types of sections are combined as in Fig. 6-6, the resonant section at the left with the section at the right will provide high loss for all frequencies above the cutoff frequency. A combination such as Fig. 6-6 sometimes is used as a rectifier filter, and often is called a **tuned-input filter.**

6-4. *OPERATION OF RECTIFIER AND FILTER*

If certain assumptions are made, the operation of a rectifier and filter can be explained on the basis of elementary circuit theory. A simple full-wave rectifier directly connected to a resistance load is shown in Fig. 6-2. The shapes of the output voltage across the resistor, and of the current through the resistor, are shown, and the mathematical relation applying is given by equation 6-2. This equation represents the output voltage of the full-wave rectifier, and the effect of each component can be considered separately. The first term of equation 6-2 is $2E_m/\pi$, and this is the direct voltage of the rectifier. This direct-voltage component forces the direct-current component through the resistance of the circuit.* The second term of equation 6-2 is $-(4E_m/3\pi)\cos 2\omega t$. It may be considered that this is an independent alternating voltage of magnitude $4E_m/3\pi$, and that it forces an alternating-current component of twice the frequency of the alternating supply voltage through the impedance of the circuit. This same method can be applied to each component of the rectifier voltage wave.

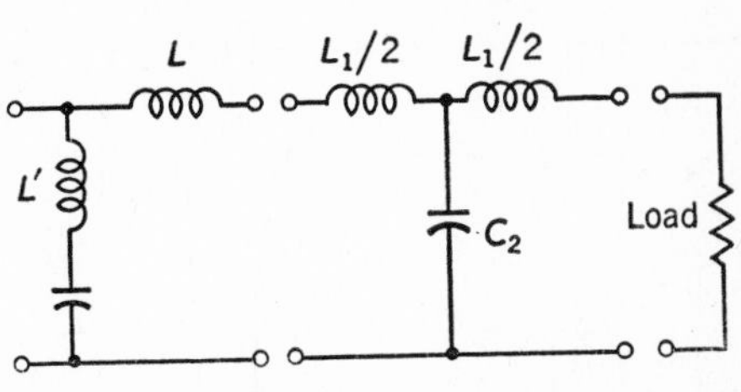

FIG. 6-6. Illustrating the design of a "tuned-input" filter.

Based on this analysis, it is possible to represent a rectifier unit composed of a rectifier, a filter, and a load resistance by the circuit of Fig. 6-7. The rectifier is replaced by a battery having an electromotive force of the same magnitude as the first component of equation 6-2, and several generators, each producing a voltage of magnitude and frequency corresponding to a term of this equation. The impedance Z_i represents the internal impedance of the transformer secondary and of the rectifier units (such as the plate resistance of tubes). For the direct-current component, Z_i is resistance only. The battery and the generators of Fig. 6-7 acting together produce exactly the same voltage wave, and have the same internal impedance as the rectifier and the transformer.

If the magnitudes and the frequencies of the various components of the voltage wave impressed by the rectifier on the filter and connected load resistance are known, then the magnitudes of the various current components that flow through the resistor can be computed. The purpose of the filter is

* In section 6-9 it is explained that under certain conditions this simplified theory will not apply.

to permit the direct component of the rectifier output voltage to force the desired direct current through the load resistor, and to *prevent* the alternating components from forcing appreciable undesired alternating current through the load resistor. If this is done, then the so-called "ripple" will be removed from the rectifier output, and the current and voltage waves at the load will be "smoothed," to use a popular term.

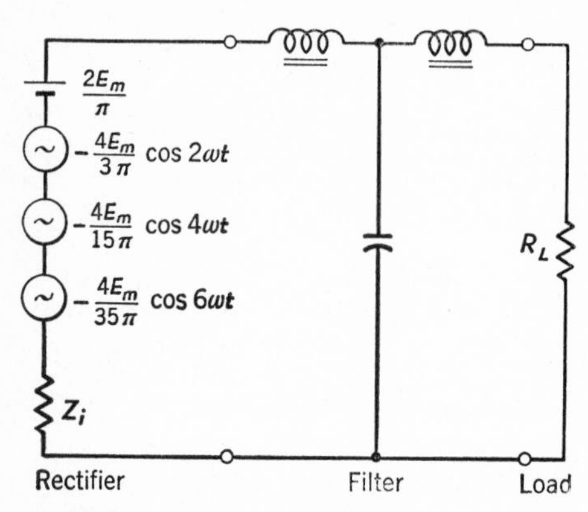

FIG. 6-7. Equivalent circuit for a rectifier unit consisting of a full-wave rectifier, a filter, and resistance load R_L.

If the cutoff frequency of the low-pass filter section is well below the lowest frequency component of the rectified wave, and if the attenuation of the section is sufficiently great, then but little alternating current will flow through the load resistance, and but little alternating current hum would result if, for instance, a speech amplifier were connected to the rectifier unit instead of to resistor R_L. A method of design based on filter theory is a possibility, but it is easier in many instances to use basic electrical theory. This will be treated in section 6-9.

From the previous discussion it appears that the design of a rectifier unit is a simple matter. Sometimes it is, sometimes it is not; it depends on the type of filter and load, and on the magnitudes of the resistances, inductances, and capacitances involved. If the equation of the rectified voltage wave that is impressed on the filter is known, then the direct- and alternating-current components flowing through the load resistor or other circuit can be found, and the percentage ripple can be computed. But, only for certain types of filter circuits and loads are the voltages known; thus, the method previously considered is not always applicable.

Before this subject is discussed further it is advisable to consider the performance of a rectifier with three different types of loads. After these have been studied, the reasons for the preceding statements will be apparent.

6-5. *RECTIFIER WITH RESISTANCE LOAD*

Suppose that the rectifier of Fig. 6-8 is connected to a resistance load as indicated. If the internal impedance to the left of terminals 1–2 is neglected, then the voltage across these terminals will be as shown by curve *B*, and represented by equation 6-2. The current flowing through resistor R_L will be as shown by curve C. Each term in the equation for the current can be found by dividing the corresponding term of the voltage equation by the resistance.

6-6. *RECTIFIER WITH INDUCTIVE LOAD*

The rectifier of Fig. 6-9 is connected to a load consisting of inductor L in series with resistor R_L. If the impressed voltage is as shown in A, the rectified voltage will be represented by B, and the current flow by C. The current wave has been "smoothed" by the action of the inductor. An explanation for this smoothing action is as follows: The rectified voltage is given by equation 6-2 and shown by curve B of Fig. 6-9. The *direct* component of the current of curve C is equal to the first term of equation 6-2 divided by the *direct-current resistance.* Each *alternating* component of the current C is equal to the voltage for that frequency component divided by *the impedance at that frequency.* The impedance increases for the higher frequency terms of equation 6-2. Also, the magnitudes of the various rectifier-output-voltage components of equation 6-2 decrease rapidly. The phase angle of each component is determined by the frequency, the resistance, and the reactance. The instantaneous sum of these components gives the current wave C. Thus, the magnitudes of the various alternating components are reduced and the phases so changed that a "smoothed current" flows. It should be noted that the peak instantaneous current *does not* rise to a high value, and that rectified current flows *from the rectifiers at all times.*

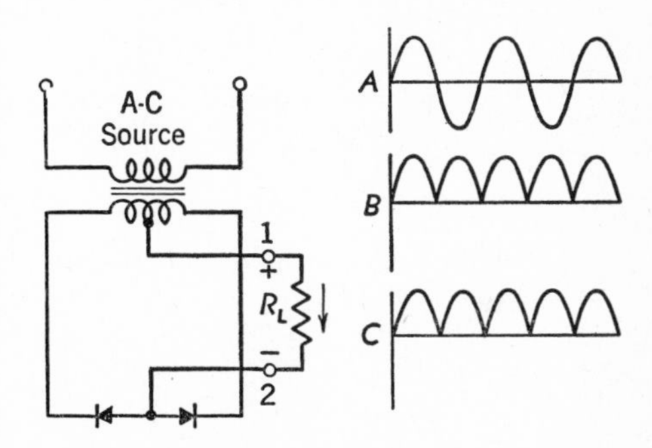

FIG. 6-8. When a sinusoidal voltage A is impressed on a full-wave rectifier circuit with a *resistance* load, the voltage across the load and the current through the load will be as shown by B and$C$.

6-7. *RECTIFIER WITH CAPACITIVE LOAD*

The load circuit of Fig. 6-10 consists of capacitor C connected across resistor R_L. The operation of the circuit is as follows: Suppose that the shape of the voltage wave *to be rectified* is as shown by curve A of Fig. 6-10. The *rectified* voltage will be impressed on capacitor C and load resistor R_L in parallel. The capacitor will draw a *high peak charging current* as indicated by i_C of curve B. (In this case it is assumed that the capacitor was partially charged, as would be true after the rectifier had been in operation a few cycles.) Current i_R also will flow to the resistor as indicated.

When the rectified voltage reaches its peak value and starts to decrease, the current through the capacitor will pass through zero, reverse in direction and the capacitor will discharge as indicated by i_C of curve B. The current i_R to the resistor R_L is now composed of that flowing from the rectifier and that supplied by the discharging capacitor. This current i_R must be maximum when that through the capacitor is zero, because this is the instant of maximum rectified impressed voltage.

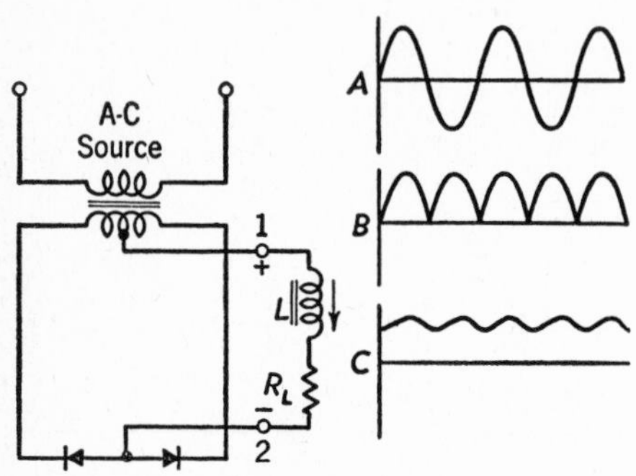

FIG. 6-9. When a sinusoidal voltage A is impressed on a full-wave rectifier with an inductive and resistive load, the voltage across the load and the current through the load will be as shown by curves B and C respectively.

The capacitor can discharge only through R_L, because *the rectifiers prevent reversed current flow*. Thus, as indicated by i_T, the total, or rectifier current, drops to zero at an angle θ_2. The capacitor then discharges until θ_1 is reached, at which time the rectified voltage again starts to charge the capacitor. The current flowing through the rectifier must be the instantaneous sum of i_C and i_R as given by i_T. This current is even more peaked than the capacitor charging current alone. Note in particular that the rectifier current i_T is zero for a large part of the cycle.

The product of the instantaneous current i_R and the load resistance R_L is equal to the voltage across terminals 1–2. Since the load resistance is constant, then this voltage must have the same shape as the current i_R. This is indicated in Fig. 6-10, part C. Since the shape of the actual output wave as shown by the *heavy* line is *not* the same as that of the dotted wave but includes a greater area under it (since it never reaches zero) equations 6-1 and 6-2 *do not apply* to the voltage wave for a rectifier with a capacitor-input filter. The *direct-voltage component* is higher in the rectifier of Fig. 6-10 with a capacitor-input filter than in the other circuits previously considered.

The angle θ_2 at which the rectifier *ceases* to conduct is referred to as the **cutout point.** Similarly, the angle θ_1, at which the tube *starts* to conduct, is called the **cutin point.** The calculation of these points depends on the circuit constants.[5]

6-8. COMPARISON OF RECTIFIER CIRCUITS

Suppose that the two filter circuits of Fig. 6-11 are considered. These sometimes are called choke-input filters and condenser-input filters in practice. Both are low-pass filters. With the usual values of inductance and capacitance, the cutoff frequencies of the filters are below the lowest alternating components of equations 6-1 and 6-2.

In the preceding sections it was shown that two fundamental types of rectifier circuits exist; *first*, those in which current flows continuously from the rectifiers, and *second*, those in which cutout occurs. The second type of circuit was found to give a higher rectified voltage, but also was shown to draw a high peak current, a serious objection with gas rectifier tubes.

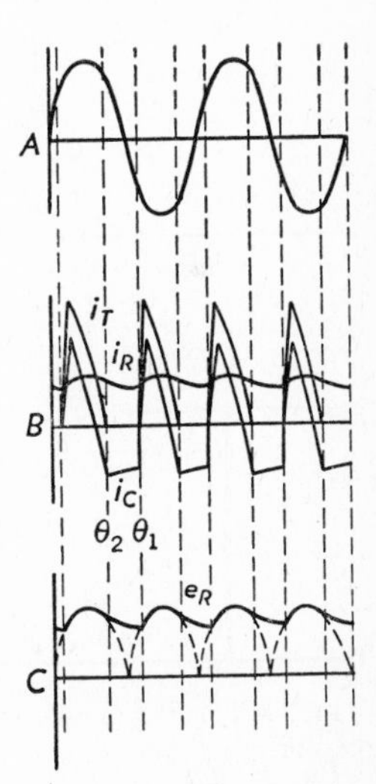

FIG. 6-10. When the sinusoidal voltage *A* is impressed on a full-wave rectifier with capacitive and resistive load, the instantaneous currents flowing will be as in *B*, and the voltage across the load as in *C*. Furthermore, curves such as these apply to any rectifier circuit in which cutout occurs.

From the similarity of the filters of Fig. 6-11 with the circuits of Figs. 6-9 and 6-10, it might be inferred that cutout *never* occurs for the inductor-input filter, and *always* occurs for the capacitor-input filter. This is not true, however.[5] Cutout of the rectified current is caused by energy storage of a capacitor somewhere in the circuit; it is not necessary that the capacitor be directly across the input terminals. Capacitor input does not always produce cutout of the rectified current. Furthermore, the capacitor across the input terminals produces *no* filtering or "smoothing" effect *unless* its capacitance is sufficiently large to produce cutout. If cutout is not caused, the capacitor will be ineffective in changing the shape of the voltage wave as in part *C* of Fig. 6-10.

The high inductance in the inductor-input filter keeps the peak value of the rectified current low. Thus, for the same peak currents, higher alternating voltages (for rectification) can be impressed on the plates of rectifiers with inductor-input filters than for rectifiers with capacitor inputs. For *equal* alternating voltages, the advantage of the capacitor-input filter in maintaining a higher *direct* voltage at the filter-input terminals (and hence across the load) is evident from Fig. 6-12.

Filters with *inductor input* are almost universally used with gas thermionic tubes because the input inductor prevents the flow of high instantaneous peak currents, which will damage these tubes. Filters with *capacitor input* are extensively used with vacuum tubes where a high instantaneous current is permissible.

6-9. *DESIGN OF RECTIFIER FILTERS* [5,6,7,8,9]

Rectifier filter systems are usually "designed" by approximate methods. The values of capacitance and inductance selected are sufficiently large to prevent objectionable hum. The magnitude of the voltage of the secondary of the transformer is made sufficiently high for the required output voltage to be obtained. A logical procedure is to start with the direct current and voltage requirements of the load, and then proceed to design the filter and the transformer, making use of curves such as Fig. 6-12 and data given by charts as in references 10 and 11.

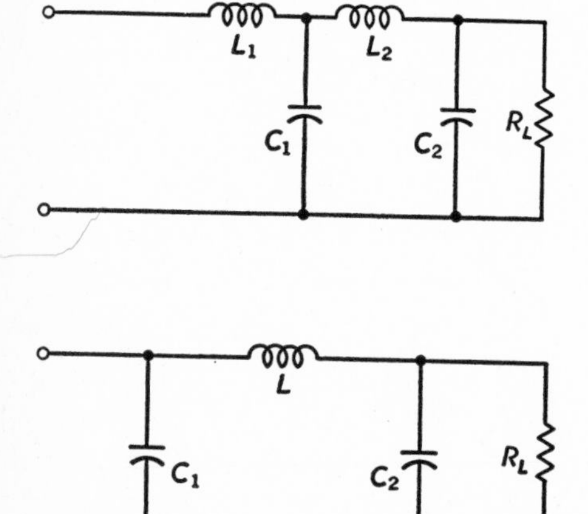

FIG. 6-11. Choke-input (above) and capacitor-input filters with resistance loads.

As an illustration, suppose that a device must be supplied with a direct voltage of 320 volts and a current of 50 milliamperes. The voltage input to the filter must exceed this amount by the voltage drop in the filter coils. A typical coil has an inductance of 12 henrys at 75 milliamperes, and a direct-current resistance of 290 ohms.

If the *vacuum* tube of Fig. 6-12 is to be used, then the filter would probably have capacitor input and would use one choke coil as in Fig. 6-11. If the current is 50 milliamperes and the coil has a resistance of 290 ohms, the direct voltage drop in the filter will be $0.05 \times 290 = 15$ volts (approximately). The direct voltage at the filter input must be $320 + 15 = 335$ volts. From Fig. 6-12 a voltage of 335 volts and a current of 50 milliamperes at the filter input would require an effective or rms value of about 315 volts. Thus, for a full-wave rectifier using the tubes of Fig. 6-12 a transformer with a 630-volt center-tapped secondary would be needed. If this size were not available, a larger size would be selected and sufficient series resistance added to reduce the voltage at the load to the amount desired. The capacitors selected

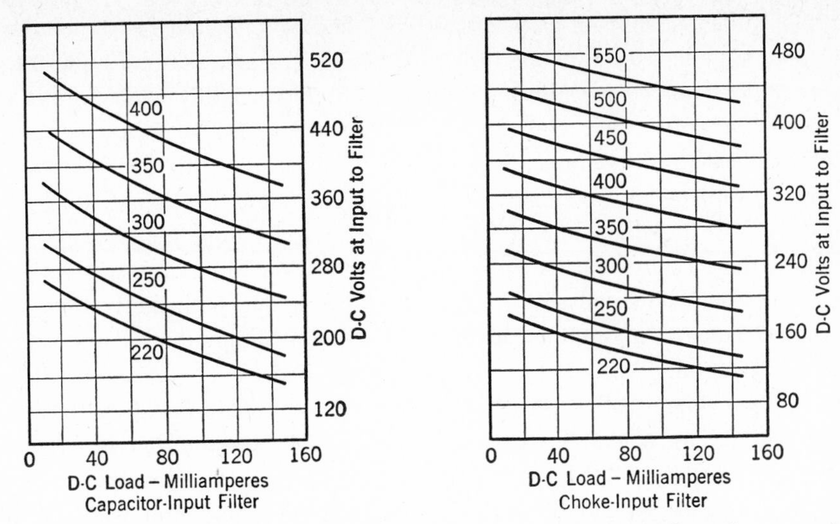

FIG. 6-12. As these curves indicate, the direct voltage available with a capacitor-input filter (with which cutout occurs) is greater than that available with a choke-input filter. Numbers on curves are effective value of the alternating voltage impressed for rectification between the plates and cathodes of a full-wave rectifier using vacuum diodes. (Data from RCA Tube Manual.)

would probably be 450-volt 8-microfarad or 16-microfarad electrolytic capacitors.

To summarize, a rectifier unit can be designed with the degree of accuracy usually required by the use of curves such as Fig. 6-12. If the filter is of the choke-input type, and cutout does not occur, then an analysis can be made using equations 6-1 and 6-2. When capacitor-input filters are used, and if cutout occurs, either a cut-and-try solution [5] or a graphical solution [9] is advisable.

If the current drawn by the load is reasonably constant, and if a light and cheap rectifier is desired, then L_1 and L_2 of Fig. 6-13 can be replaced by re-

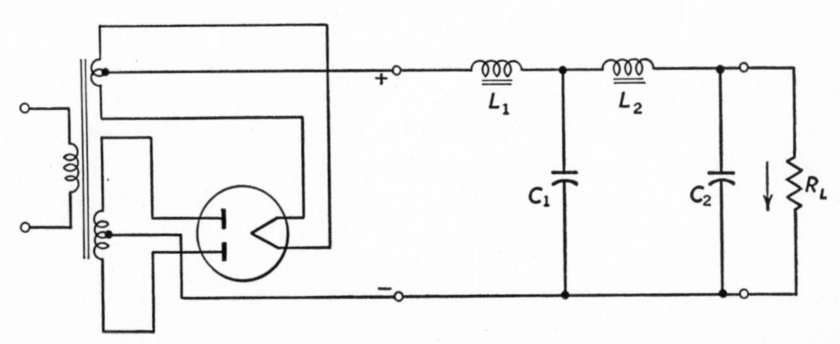

FIG. 6-13. A typical full-wave rectifier unit with the inductors in the *positive* lead.

sistors of sufficient current-carrying capacity and of sufficient resistance. In fact, for some purposes only one resistor and one capacitor may be sufficient. If cutout does not occur, filtering action is obtained in this way: Suppose that a voltage wave such as given by equation 6-1 or 6-2 is impressed and maintained across the + and − terminals of Fig. 6-13, with the inductors L_1 and L_2 replaced by resistors. Then, large capacitors at C_1 and C_2 will offer very low impedance to the alternating-current components, and the impressed harmonic voltages will largely be dissipated in the two series resistors replacing the coils. *Without* these resistors, and with the voltage of equation 6-1 or 6-2 maintained at the input, the capacitors would produce no filtering action.

It has been shown [12] that for filters of the types shown in Fig. 6-13 the inductors should be placed in the *positive lead* as indicated; that is, in the connection from the load to the cathode and not in the lead between the anode transformer and load. If this is not done, residual hum will result because of current that flows through the distributed capacitance between the secondary winding and ground, and back to the cathode without passing through, and being impeded by, the inductors.

As has been mentioned, an inductor-input filter ordinarily is used with gas tubes. This first inductor is sometimes called the **input choke,** and often is of the "swinging" variety; that is, its inductance varies with the amount of direct current being carried. Inductance values of 10 henrys at 100 milliamperes and 15 henrys at zero current are typical.

The second inductor is often called a **smoothing choke.** The ratings on these have, in the past, been misleading. Since the inductance of interest is the *incremental* inductance (Chapter 7), the value of direct current for the rated inductance should always be specified. A value of 20 henrys at 120 milliamperes is typical. The terms *input* choke and *smoothing* choke are misleading, because an analysis of filter action shows that the first inductor contributes greatly to the filtering.

A common requirement of a rectifier and associated filter is that the **regulation** is good; that is, that the voltage does not rise greatly when the load is decreased or entirely removed. The direct voltage rises on a circuit when cutout occurs, and this *can occur with choke-input as well as capacitor-input filters.*[5] The voltage rises because when cutout occurs the shape of the rectified impressed voltage wave changes from that of the broken line in part C of Fig. 6-10 to that shown by the solid line. This latter wave shape contains a larger direct-voltage component.

The size of the filter input choke that is sufficiently large to prevent cutout can be computed as follows: If the maximum magnitude of the second harmonic current flowing through choke coil L_1 of Fig. 6-13 exceeds in value the direct current, then over a part of the cycle the current furnished by the tube will be zero, cutout will occur, and the direct voltage will rise. The critical value of L_1 at which this will occur can be calculated with the required degree of accuracy if certain assumptions are made. All reactances in the circuit except that of coil L_1 and that of capacitor C_1 will be neglected. Also, all resistances except that of the load resistor R_L will be neglected. And in addition, only the effect of the second harmonic will be considered.

Referring to equation 6-2, the value of direct current through the choke coil L_1 of Fig. 6-13 will be

$$I_{dc} = (2E_m/\pi)/R_L \qquad 6\text{-}4$$

and the maximum value of the second harmonic current will be

$$I_m = (4E_m/3\pi)/(2\pi \times 2 \times 60L_1) \qquad 6\text{-}5$$

when 60-cycle power is being rectified. Equating these two expressions and solving gives

$$L_1 = \frac{R_L}{1130} \qquad 6\text{-}6$$

According to this relation, if the inductance in *henrys* of the first inductor of an inductor-input filter is less than the magnitude in *ohms* of the load resistor divided by 1130, then cutout will occur, the voltage will rise, and the regulation will be poor. The value of inductance early was reported [13] to be

$$L_1 = \frac{R_L}{1000} \qquad 6\text{-}7$$

In view of the many assumptions made, it is advisable to use equation 6-6. In both instances the units are the same.

From these equations it is evident that when the load resistance R_L is low, and a large direct current is being drawn, the value of the input inductor L_1 may also be low. But, when R_L is high, then the direct current will be small, and L_1 must be large to prevent cutout. An iron-cored coil, the core of which saturates for the direct-current values used, will have variable incremental inductance. As an example, a typical coil has 13.5 henrys inductance with zero direct current, and 9.5 henrys at 110 milliamperes. Such a coil is often called a **swinging choke.** It is fortunate that it is not necessary to have constant inductance, because coils having such properties would be more expensive than those in which the inductance varies with the direct current flowing to the load.

Referring to the *inductor-input filter* of Fig. 6-11, approximate design equations are easily derived. For this figure, the ratio of the *alternating* voltage across the load resistor R_L to the *alternating* voltage (such as the components of equation 6-2) impressed across the filter-input terminals is

$$\text{Voltage ratio} = \frac{1}{\omega^4 L_1 L_2 C_1 C_2} \qquad \text{6-8}$$

If L_2 and C_2 are omitted from the filter, giving a simpler type composed only of an inductor L_1 and a capacitor C_1 shunted across the load R_L, the corresponding equation is

$$\text{Voltage ratio} = \frac{1}{\omega^2 L_1 C_1} \qquad \text{6-9}$$

In deriving these equations it is assumed that the inductive reactances are very large at the frequency ($\omega = 2\pi f$) under consideration, and that the capacitive reactances are small and for certain purposes negligible. It is also assumed that the load resistance R_L is large and that the resistance of the inductors is negligible.

These equations give the *ratio* of the alternating voltage across the load, at each frequency component, to that existing at the same frequency in the impressed rectified voltage wave. They do *not* give the **percentage ripple** at the load. This is computed as the ratio of the effective value of the alternating component of the voltage (or current) at the load to the direct voltage (or current) value at the load. These equations can be derived using the same assumptions as for equations 6-8 and 6-9. The results must be multiplied by 100 to convert to per cent. The inductances and capacitances are expressed in henrys and farads, respectively.

The power-output tubes of amplifiers (Chapter 8) often do not need their plate supply so well filtered as is required by the preceding amplifying tubes operating at low signal level. Thus, plate-supply voltages for power-output tubes are often tapped off ahead of the last choke.

The filter capacitors used are of two general types: paper capacitors and electrolytic capacitors. As the name implies, the dielectric of the first is paper, which may be dry, but usually is impregnated with oil, paraffin, or other substance. The dielectric of the second type is an oxide layer. Two precautions are necessary when using the latter: *First,* a direct polarizing voltage must be used on the electrolytic type to form the oxide layer. This means that these capacitors cannot be used in alternating-current circuits in their simple form. The *second* precaution is that electrolytic capacitors should not be operated in series unless a resistor of high resistance is connected in parallel with each to equalize the voltage across them. Otherwise they may divide the voltage unequally, and will be damaged if the voltage is sufficiently high.

In many rectifier units voltage dividers are connected across the output terminals of the rectifier. These are usually resistors, often of about 25,000 ohms. Fixed or adjustable taps are available so that the desired voltages can be obtained. In addition to serving in this manner, the voltage divider draws a direct current and this tends to prevent the terminal voltage from increasing when the load is removed, thus improving the regulation. Such resistors are often called **bleeders.** In many instances it is necessary to connect large capacitors across the voltage-divider elements to reduce feedback due to common impedances (Chapter 8).

A voltage divider or bleeder connected across a filter performs another important function. The filter capacitors of a rectifier will hold a charge for a considerable period of time if they have high leakage resistance. Many deaths occur because operators shut off rectifier units, but forget that the capacitors are charged. For this reason, if for no other, it is good practice to place a resistor of high value across the filter.

6-10. *VOLTAGE REGULATORS USING GAS TUBES* [14]

The purpose of a rectifier and associated filter is to supply direct-current power to a device. For many purposes in electronics it is desired that the voltage remain essentially constant for all operating conditions. **Voltage regulators,** or **voltage stabilizers,** are used to maintain voltages constant. Furthermore, because they do hold the voltage constant, voltage regulators reduce the "ripple" of rectified power, and hence reduce hum.

A simple and effective voltage regulator using a cold-cathode gas diode is shown in Fig. 6-14. The principle of operation can be explained from the characteristics of the tube shown in general form by Fig. 3-13. As explained for this figure, the normal glow occurs between points C and D, and if the current increases, the area of the cathode covered by the glow increases, and the voltage drop across the tube remains essentially constant. A typical tube initially breaks down at about 115 volts, and then the current can vary from about 5 to 40 milliamperes, but the voltage drop across the tube remains essentially at 105 volts. Tubes with different constant voltages such as 75 or 150 volts are available.

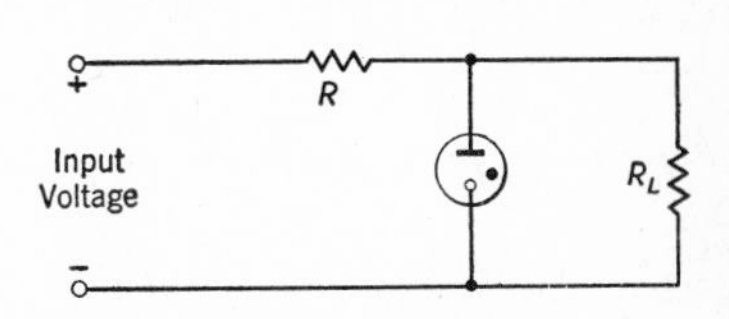

FIG. 6-14. A cold-cathode gas diode used to stabilize the voltage across the load R_L.

The current through the dropping resistor R is the sum of the current taken by the load resistor R_L and the current taken by the gas regulator tube. The drop in voltage across resistor R must equal the difference between the impressed voltage (perhaps from a rectifier filter) and the voltage across the tube and load resistor which are in parallel. When the circuit initially is energized, the instantaneous voltage across the tube must be such that it will break down, but when this occurs, the value of resistor R must be such that the current through the tube does not exceed the rated value.

During operation with the load connected, if the load resistance is reduced and more current is taken, the gas tube takes less current. If the resistance of the load is increased so that it takes less current, then the gas tube takes more current, the combined current always causing an IR drop equal to the difference between the impressed voltage and the voltage across the load that is to be maintained constant. In this way the tube compensates for load changes.

The circuit of Fig. 6-14 also will compensate for changes in the source voltage. Thus if the impressed voltage rises, the gas regulator tube takes more current, causes an increased IR drop, and maintains the voltage essentially constant at the load. If, on the other hand, the voltage of the source falls, then the gas tube takes less current, there is less IR drop across the series resistor, and the voltage at the load is maintained essentially constant.

A gas tube will maintain the direct voltage across a device essentially constant at the rated voltage of the tube, provided the current through the tube remains within the rated limits. If the current through the tube falls too low, the tube ceases to function; if it rises too high, the tube may be permanently damaged. Several tubes can be operated in series to control voltages of value greater than that of a single tube.

6-11. *VOLTAGE REGULATORS USING VACUUM TUBES* [15]

Voltage-regulating circuits employing vacuum tubes as the variable regulating element are extensively used. In many of these, a gas tube is used as a constant-voltage reference. The basic principle of a common type is shown in Fig. 6-15. The slider on the high-resistance voltage divider R is adjusted until the grid-battery voltage is overcome and the grid is negative. The tube acts like a series resistor in that it causes a voltage drop between input voltage and the load device R_L across which the voltage is to be held constant.

If the voltage across the load R_L rises, then the voltage across the voltage divider R will be greater, the grid will be more negative with respect to the cathode, the current passed by the triode will be reduced, and the current to the load and the voltage across the load returns to the original value. If the voltage across the load R_L falls, then the voltage across divider R will be less, the grid will be less negative, the current passed by the triode will increase, and conditions at the load will return almost to the original value. This circuit will maintain the voltage at the load essentially constant within limits because of voltage changes caused by a fluctuating load, or because of voltage changes caused by a fluctuating source.

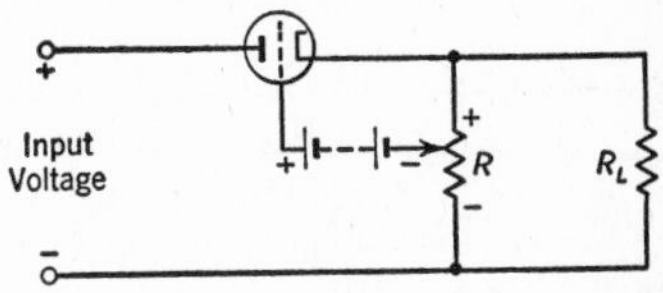

FIG. 6-15. A vacuum triode used to stabilize the voltage across load R_L.

In practice, the circuit of Fig. 6-16 instead of Fig. 6-15 commonly is used. Tube 1 is a vacuum power-output tube having sufficient current-carrying capacity and low voltage drop. Tube 2 is a small-signal tube, often a pentode. Tube 3 is a gas diode that provides a constant reference voltage between the cathode of tube 2 and the negative side of the circuit, often at ground potential. Current for gas tube 3 flows through resistor R_3. The current magnitude is such that a normal glow exists and the gas tube acts as an essentially *constant voltage source* of, for instance, 105 volts. The magnitude and sign of the bias voltage between grid and cathode of tube 2 is determined by the voltage drop across the lower part of voltage divider R_1 and the essentially constant voltage across the gas tube. This bias voltage controls the current through tube 2, and this current varies the voltage drop across resistor R_2, which in turn controls the current through the series-connected power-output tube 1.

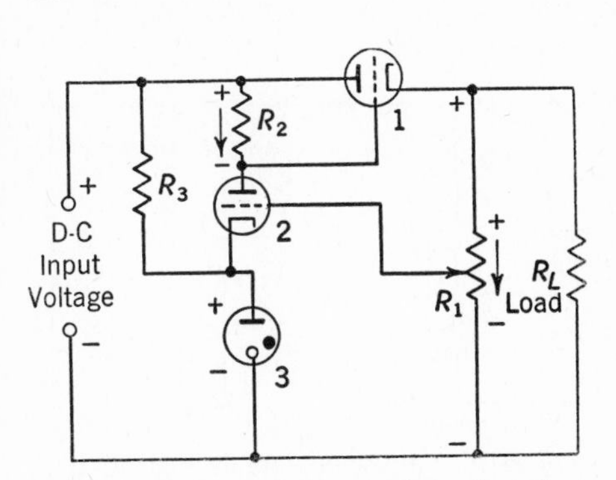

FIG. 6-16. A voltage regulator based on Fig. 6-15 using a cold-cathode gas diode instead of a battery to obtain a reference voltage. This sometimes is called a degenerative-amplifier regulator.

To explain the operation of Fig. 6-16, assume that voltage divider R_1 is adjusted so that the grid of small-signal tube 2 is a few volts negative. If the voltage across the load R_L were to rise, the increased voltage drop across R_1 would cause the bias voltage of tube 2 to fall. Then, the plate current of tube 2 would rise, increasing the voltage drop across R_2, and hence the bias between grid and cathode of tube 1. This would reduce the plate current of tube 1, and hence the voltage across load resistor R_L, thus maintaining the output voltage across R_L essentially constant. If the voltage across the load R_L were to fall, action opposite to that just explained would occur. The circuit of Fig. 6-16 is called a **degenerative-amplifier regulator.**

A circuit is shown in Fig. 6-17 that will compensate only for changes in the impressed voltage, and that is, accordingly, best suited for use where the load is essentially constant. The gas tube maintains the voltage of the cathode of the high-vacuum tube at an essentially *constant positive voltage* with respect to the negative side of the impressed voltage source. The resistors R_1 and R_2 are of high resistance and so proportioned that the IR_1 drop is less than the voltage drop across the gas tube, and thus the grid of the high-vacuum tube is negative with respect to the cathode.

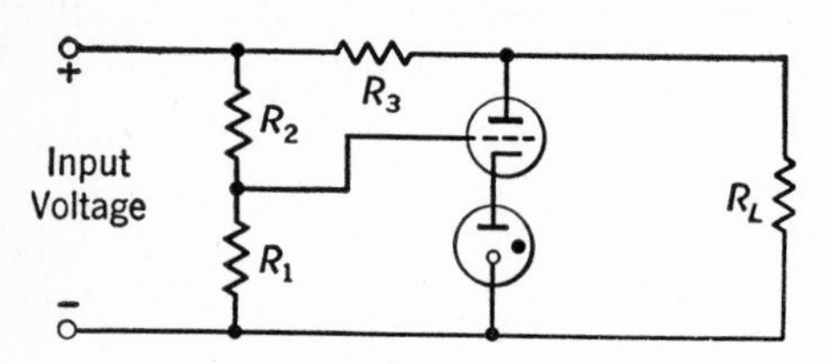

FIG. 6-17. The vacuum triode and the cold-cathode gas diode are used to regulate the voltage across load R_L. This sometimes is called a transconductance-bridge regulator.

Now suppose that the voltage of the source rises slightly; when this occurs, the IR_1 drop will be greater, and the grid of the tube will be *less* negative with respect to the cathode. This will cause an increase in the plate current taken by the triode (and also flowing through the gas tube, which is a constant-voltage device). This increase in current taken by the triode causes an increase in the IR_3 drop, and with proper adjustments, the original voltage rise is absorbed. If the impressed voltage falls, the action is opposite. This circuit is sometimes called a **transconductance-bridge regulator.**

Another voltage-regulator circuit is shown in Fig. 6-18. The battery is adjusted so that the voltage between the grid and cathode is negative the desired amount when the impressed voltage is normal. For this to be true, the voltage of the battery in the grid circuit is less than the IR_1 drop. Now suppose that the impressed voltage rises; then, the IR_1 drop will be greater, and the grid will be more negative with respect to the cathode. This will tend to *reduce* the current passed by the tube, and flowing through resistor R_3. But, the increase

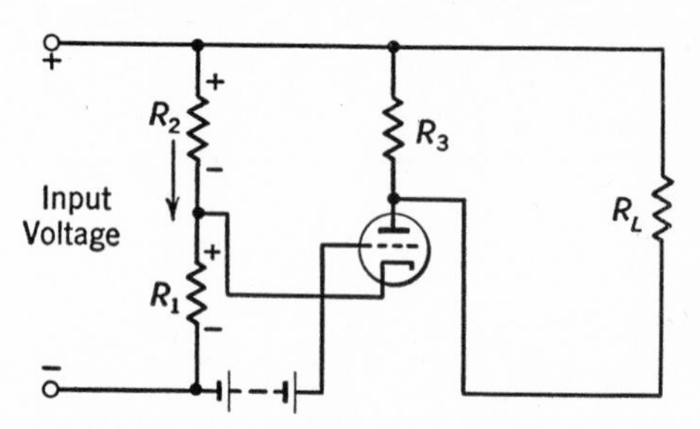

FIG. 6-18. A vacuum triode used to stabilize the voltage across load R_L. This sometimes is called an amplification-factor bridge regulator. In practice a cold-cathode gas diode probably would be used instead of the battery.

in the impressed voltage increases the potential between the plate and cathode, and this tends to *increase* the current through the tube. If the change in voltage of the source is in the opposite direction, the effects just discussed will be reversed.

For some ratio of R_2 to R_1, the plate current through R_3, and hence the output voltage, is maintained constant. This is sometimes called an **amplification-factor bridge-regulator.** This circuit, just as for the transconductance regulator, will not compensate for changes caused by variations in the load, or output, circuit. The circuit of Fig. 6-18 can be changed to use a gas tube as a source of constant potential instead of the battery in the grid circuit.

6-12. *VOLTAGE-MULTIPLYING RECTIFIERS* [16,17,18,19]

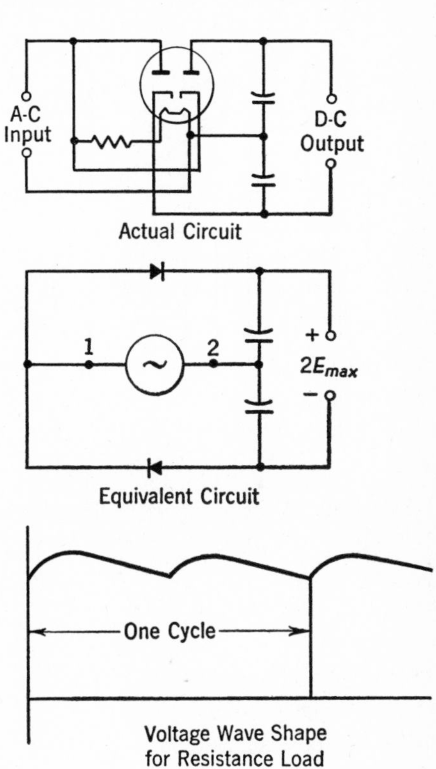

FIG. 6-19. The voltage-doubling rectifier.

A voltage-doubling circuit, such as is employed in certain small radio receiving sets, is shown in Fig. 6-19. A simplified diagram also is included. The manner in which this operates to produce a direct voltage approximately twice the maximum value of the impressed alternating voltage is essentially as follows: During the part of the alternating cycle that generator terminal 1 is positive and 2 is negative, *conventional* current will pass through the upper rectifier and charge the upper capacitor. On the negative half-cycle, when terminal 1 is negative and 2 is positive, the lower rectifier will conduct, and the corresponding capacitor will be charged. Thus, when no direct-current power is taken, the voltage across the output terminals will be approximately $2E_{max.}$ as indicated, where $E_{max.}$ is the maximum value of the applied alternating voltage.

If a resistor is connected across the output terminals, a rectified or direct current will flow. The shape of this current will be as indicated in Fig. 6-19; it is seen to be comparatively smooth. Some filtering must be accomplished, however, before this voltage can be applied to the plates of receiver tubes. The manner in which the output voltage varies for given load currents is shown in Fig. 6-20. The advantage of large capacitors is apparent. Voltage-tripling and voltage-quadrupling (etc.) circuits have been devised.[19]

6-13. *RECTIFIER TUBES*

The type of tube for a given rectifier unit will, of course, be determined by the use for which the unit is designed. Both vacuum and gas (or vapor) thermionic diodes are used. These were discussed in Chapter 3. Vacuum rectifier tubes are usually employed in radio receivers and in many small sound-system amplifiers, because the power involved is small, and efficiency is of little importance. Furthermore, gas (or vapor) tubes cause radio interference and require special shielding and filtering precautions in radio receiving sets. Gas tubes are used in plate-supply rectifiers for large radio transmitters.

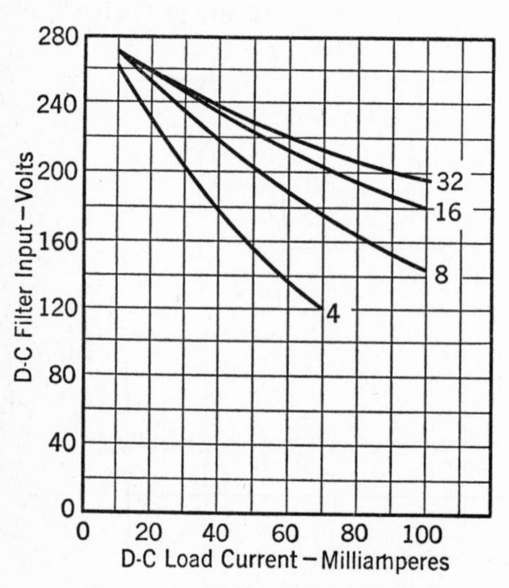

FIG. 6-20. Voltage inputs to filter for a voltage doubler using vacuum diodes. Numbers on curves refer to capacitance in microfarads of each capacitor of Fig. 6-19. (Data from RCA Tube Manual.)

As mentioned in Chapter 3, the peak current and the peak inverse anode voltage are factors limiting the life and hence the rating of tubes. The peak current is especially important in gas (or vapor) tubes, because if a certain value is exceeded, then the safe internal voltage drop will be too large and the tube will be damaged. If the safe peak inverse anode voltage is exceeded, the tube may arc back.

It is of interest to examine the voltages between electrodes during the normal operation of a tube. This is illustrated, for the full-wave rectifier, by Fig. 6-21, in which the tube is replaced by a generalized rectifier unit. When the upper end of the transformer is negative, a conventional current will flow as indicated by the arrow, but *no current* will flow through the *lower* unit. When the upper unit conducts, the resistance of this unit, and the voltage drop across it, become very low. Hence, point p assumes a potential approaching as a limit that of the upper end of the transformer winding.

Now at the same instant the upper end of the transformer secondary is negative, the lower end is positive. Thus, substantially the *entire* secondary voltage appears as an *inverse voltage* across the lower rectifier unit, which is not conducting at the instant under consideration. This inverse voltage tries to force a current through, and break down, the lower unit. On the next half-cycle the action is reversed.

6-14. *HIGH-VOLTAGE LOW-CURRENT POWER SUPPLIES* [20,21,22]

Power supplies that provide several thousand volts and a few hundred microamperes sometimes are needed. Conventional power supplies may be used, or a circuit such as Fig. 6-22 may be employed. It will be noted that this circuit does not use a transformer, which is expensive and bulky. The rectifier unit of Fig. 6-22 operates as follows: The oscillator produces a voltage of, perhaps, 100 kilocycles, and this voltage is impressed across a series resonant circuit composed of C, L, and C'. If these elements are tuned to the frequency of the oscillator, then a large voltage will exist across capacitor C and inductor L. Capacitor C' is so large that its reactance is negligible. The cathode of the tube shown is connected, in effect, in parallel with the inductor and will have the high voltage impressed between cathode and anode. Rectification will occur, and a relatively high direct voltage will exist at the filter output.

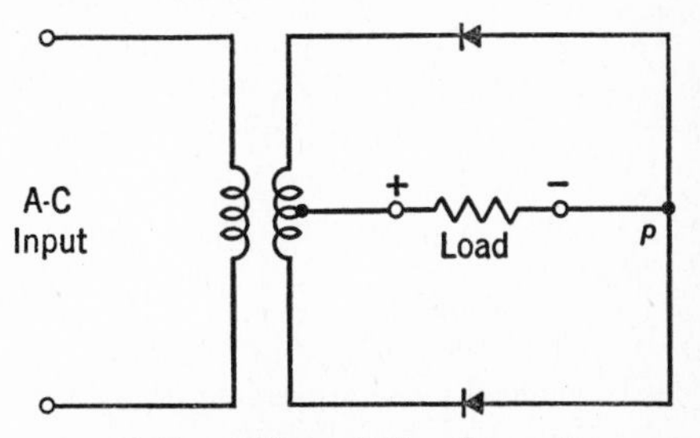

FIG. 6-21. Substantially the entire secondary voltage appears as an inverse voltage across the rectifier that is not conducting.

6-15. *POLYPHASE RECTIFIERS* [23,24,25]

Single-phase rectifiers were considered in the preceding pages. For rectifying large amounts of power, above about one kilowatt for instance, polyphase rectifiers commonly are used. An example is in the plate power supply of large radio transmitters.[26] One important reason for using polyphase rectifiers is that the "ripple frequency," or the frequency of the first alternating component, is greater with polyphase rectifiers than with single-phase types. Thus, the single-phase, half-wave rectifier has a lowest-frequency component that is the same as that of the voltage being rectified; for the single-phase, full-wave rectifier, the ripple frequency is twice that of the voltage being rectified; for the three-phase half-wave rectifier, the ripple frequency is three times that of the wave being rectified; etc. For given inductors and capacitors, filtering is more effective for the higher frequencies because the reactance of the series inductors is greater, and the reactance of the shunt capacitors is less.

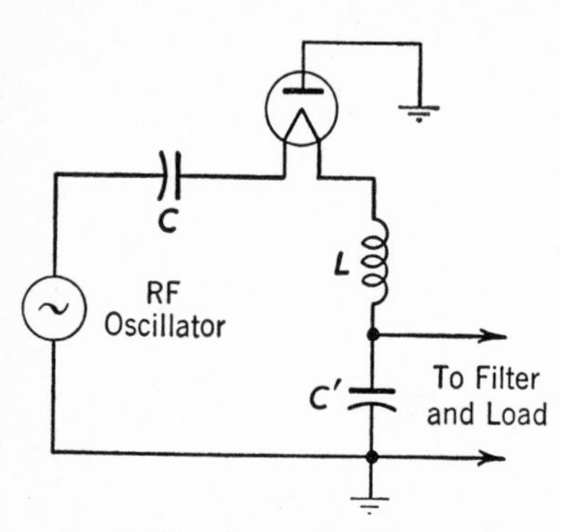

FIG. 6-22. A rectifier for supplying a voltage of several thousand volts and a direct current of a few microamperes. The high voltage to be rectified is obtained by series resonance.

Polyphase rectification is a special subject, and only the three-phase rectifier of Fig. 6-23 can be considered here. If several factors of importance in practice[24,25] are neglected, the individual rectifier, if acting alone, would impress voltages across the resistance load as shown by the light lines. The combined action in the three-phase circuit is, however, as shown by the heavy line. Because one phase after another is effective in maintaining the voltage across the load at a relatively high value, the direct voltage and current components of the resultant rectified wave are high. Polyphase rectifiers using a large number of phases have been developed.[24,25]

Rectifiers may seriously interfere with the operation of communication systems. Because of the nature of the current passed by a rectifier, the wave shape of the current drawn from the supply system may be badly distorted and may contain harmonics. These harmonics may induce noise-producing signals into paralleling telephone systems, causing serious interference with normal operation. This is called **inductive interference.** Also, the rectified direct-current output of a rectifier contains a "ripple," which is in reality composed of harmonics of the impressed power-system frequency. The output circuit of a rectifier may cause inductive interference if these harmonics are induced into communication circuits. When large rectifiers are designed, installed, or operated, it is well to give thought to inductive coordination with communication systems.[27]

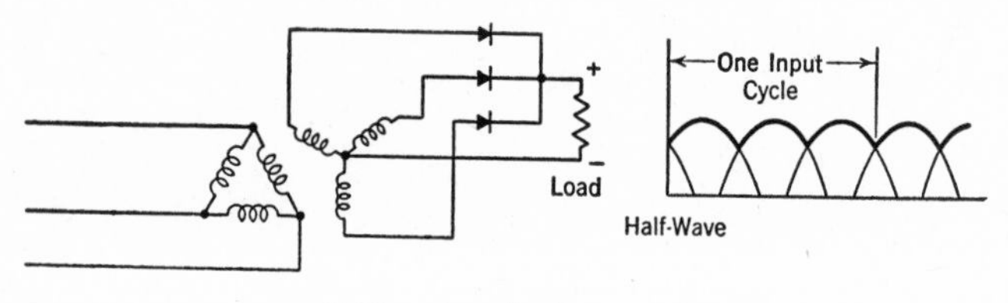

FIG. 6-23. A polyphase rectifier.

6-16. *THYRATRONS AS CONTROLLED RECTIFIERS* [24,25,28,29]

Rectifiers and associated equipment have been considered in the preceding pages. In general it can be said that the direct-voltage component forces a direct current through the resistance of the load device. It is important to note that for given conditions, the amount of direct current drawn by the load device is largely determined by the resistance of the device; it cannot be controlled at will at the rectifier, except by controlling the magnitude of the alternating voltage impressed on the rectifier. Of course, the direct current can be varied by changing the resistance of the load, or by changing the resistance of a rheostat connected in series with the load. This is satisfactory in some applications, but in many circuits the load resistance is fixed, and a series rheostat is often cumbersome and dissipates power. This power may be unimportant in low-power circuits, but it is of much importance where large amounts of power are to be controlled. Also, a rheostat must be adjusted manually, or by motor drive.

The Thyratron, discussed in Chapters 4 and 5, is a grid-controlled gas rectifier tube. By varying the potential on the control grid, the Thyratron can be made to fire and conduct at will. The Thyratron can be constructed so that it can rectify large currents, and can stand reasonably high peak inverse anode voltages. Thus, a Thyratron can be used as a **controlled rectifier,** and when so used the direct current supplied to a load device having constant resistance can be varied as desired at the rectifier, and without a

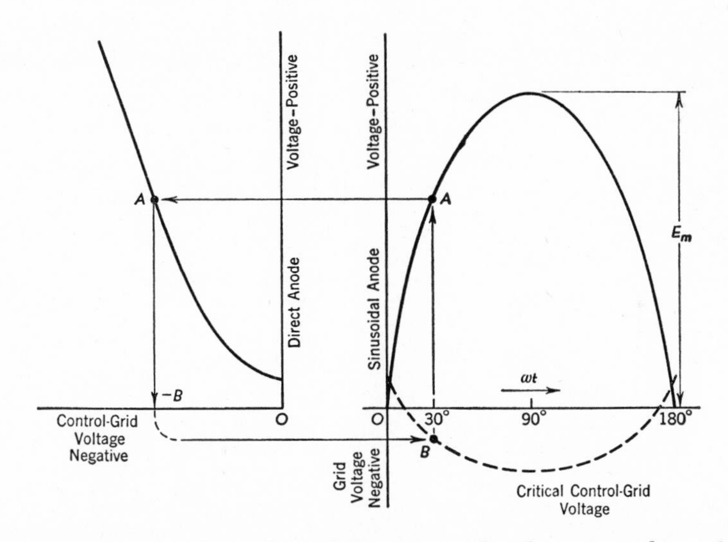

FIG. 6-24. Illustrating the method of determining the alternating-voltage characteristics of a Thyratron from the direct-voltage characteristics.

series rheostat. Furthermore, because a grid is used to vary the direct current flowing, control can be accomplished with negligible power loss compared to the amount of power controlled. This is in marked contrast with a series rheostat. Also, the grid control is simple and flexible.

To understand the operation of a Thyratron as a rectifier, and to be able to predict its performance, it is necessary to know its characteristics with alternating voltages on the control grid and on the anode. The breakdown characteristics with alternating voltages on the electrodes can be predicted from the characteristics obtained with direct voltages, such as those of Fig. 4-22. Thus, suppose that the 60°C. curve of this figure is representative of operating conditions and is reproduced as in Fig. 6-24. Only the *positive* half-cycle of the alternating voltage on the anode need be considered because the tube will not conduct when the anode is negative.

In Fig. 6-24, the maximum value of the positive half-cycle is E_m volts. Values of the angle ωt on the X axis are selected and then projected upward until they strike the alternating half-cycle voltage curve. These voltage values are then projected to the left until they intercept the direct-voltage characteristic curve and are then carried downward to determine the critical grid-voltage values for breakdown. These critical grid-voltage values are then plotted beneath the half-cycle voltage curve, giving the critical grid-voltage curve shown by the broken line of Fig. 6-24. This curve is negative because a negative-grid Thyratron is under consideration. This procedure will be illustrated. When the instantaneous anode voltage is at 30° on the half-cycle, the value is A volts. Referring to the direct-voltage characteristic curve at the left, the tube will break down and fire if the control-grid voltage is less in magnitude than $-B$ volts. This value is then plotted as the critical control-grid voltage value at 30°. The complete critical control-grid curve is obtained in this way. It is important to note that a different control-grid curve must be made for each different magnitude of applied alternating anode voltage.

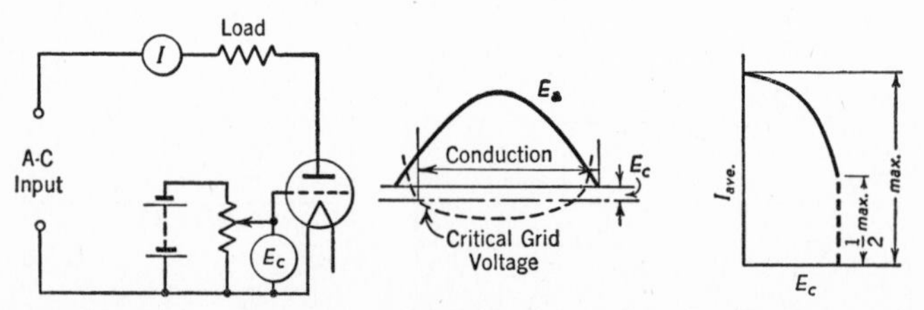

FIG. 6-25. With this type of amplitude control, current flows whenever the grid bias E_c is less negative than the critical grid-voltage value. The average anode current is controllable from maximum to one-half maximum value as indicated. For safety, a resistor should be placed in series with the grid. (Adapted from reference 30.)

Amplitude Control. The magnitude of the rectified current passed by a thermionic grid-controlled gas tube, or Thyratron, can be varied by changing the *magnitude,* or *amplitude,* of a direct voltage impressed on the control grid. This is known as amplitude control. A simple circuit arrangement is shown in Fig. 6-25. If the direct negative grid-bias voltage E_c is less in magnitude than the critical grid-voltage curve, the tube will fire. Once the tube fires, it will continue to conduct throughout the remainder of the cycle until the alternating anode potential falls to zero, when the tube deionizes and conduction ceases. When the anode again becomes positive, during the next positive half-cycle, the tube will again fire. The shape of the current wave is shown by the "conduction" portion of Fig. 6-25. In this curve and in those that follow, it is common practice to assume that the magnitudes of the *Y*-axis scales for voltage and current are such that half-cycles coincide.

A study of the center diagram of Fig. 6-25 will indicate that control is available only for angles from zero to 90°, or one-fourth cycle. This is because once the grid loses control and the tube fires, the grid cannot again take over until the tube deionizes when the anode voltage passes through zero. Because of this, varying the control-grid voltage can only control the rectified average or direct current between the maximum value, and one-half the maximum value, as Fig. 6-25 shows.

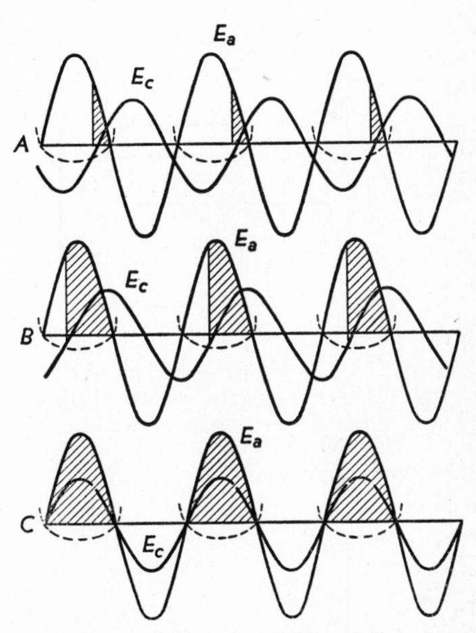

FIG. 6-26. Illustrating phase-shift control of a Thyratron. Current starts when the anode voltage E_a is positive, and when the voltage on the grid is less negative than the critical grid voltage shown by the broken lines. The shaded portions indicate the shapes of the current pulses.

Phase-Shift Control. This method is based on the curves of Fig. 6-24. The theory underlying phase-shift control is illustrated by Fig. 6-26, in which the dotted lines indicate the critical grid-voltage values. If the grid voltage (at any instant) is less negative than this critical value, the tube will start to conduct and will continue to conduct until the plate voltage falls to zero and deionization occurs. Thus, when the grid voltage and plate voltage are almost 180° out of phase, only a small amount of current will flow, and this will be as indicated by the peaked shaded area of curve *A*. When the grid voltage and plate voltage are more nearly in phase, the current will flow as shown by the shaded area of *B*, and when they are in phase, the current flows for the entire cycle and the magnitude is limited by the resistance in the anode circuit.

Several methods [24,28,29,30,31,32,33] are available for shifting the phase of the grid voltage to give the desired control. One is as shown in Fig. 6-27. Varying the value of R changes the phase relation of the grid voltage so that the average rectified current can be controlled from a maximum value (limited largely by the load resistor R_L) to zero value. The circuit of Fig. 6-27 can be modified by substituting either a fixed or a variable capacitor for the inductor in the grid circuit. The diagram of Fig. 6-28 is

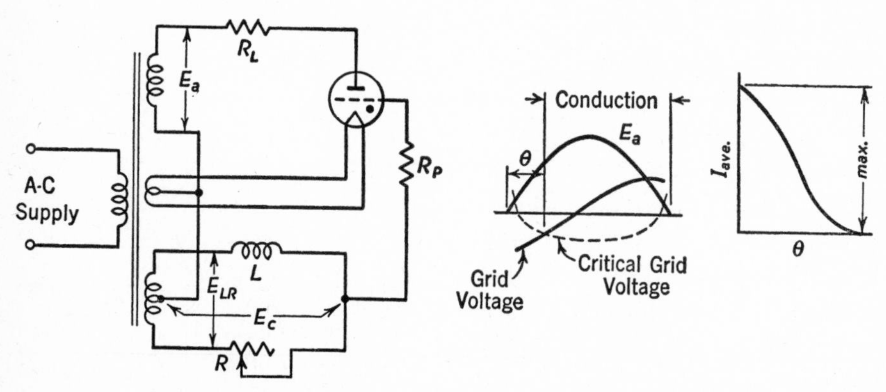

FIG. 6-27. Illustrating a method of controlling the rectified output of a Thyratron by phase-shifting. Resistor R_L is the load. Resistor R_p is to limit the grid current. The angle θ between the anode voltage E_a and the grid voltage E_c may be varied through approximately 180° by changing R. The average rectified current flowing through R_L can be varied from a maximum value to zero.

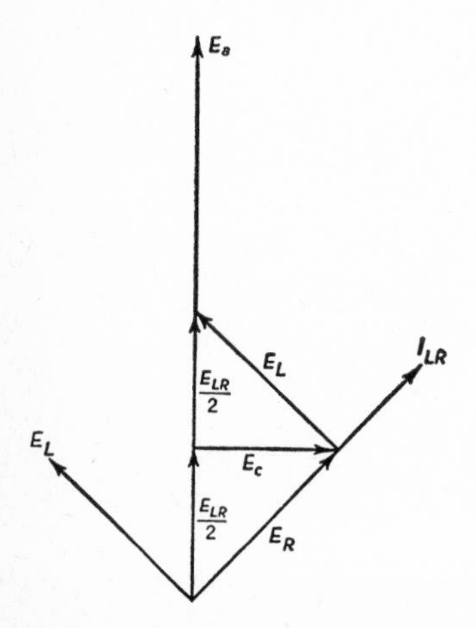

FIG. 6-28. Phasor diagram for the phase-shifting circuit of Fig. 6-27. For the condition shown, the reactance of coil L equals in magnitude the resistance of R. The anode to cathode voltage is E_a, and the grid to cathode voltage is E_c.

drawn to represent conditions when the circuit of Fig. 6-27 is adjusted (by varying either the inductor or the resistor) so that the inductive reactance of coil L and the resistance of resistor R are of the same magnitudes. E_a represents the voltage impressed on the anode. E_{LR} represents the voltage impressed on the coil L and resistor R in *series*. Because the inductive reactance and the resistance are assumed equal, the current I_{LR} flowing through the coil and resistor will lag the voltage E_{LR} by about 45°. This current will cause a voltage drop E_L across the coil, and a voltage drop E_R across the resistor as shown.

The sum of E_R and E_L must equal the secondary voltage E_{LR}. The net voltage impressed between the control grid and cathode of the Thyratron is from the center tap of the transformer winding producing voltage E_{LR} and the contact between coil L and resistor R. This is shown by E_c of Fig. 6-28. Thus, for the condition that $X_L = R$, the grid voltage E_c is 90° out of phase with the anode voltage.

To explain how 180° phase shift is possible with the circuit of Fig. 6-28, suppose that the resistance of R is reduced and approaches zero. Then, the current I_{LR} lags anode voltage E_A greater than 45°, and approaches an angle of lag of 90° as a limit. E_R becomes smaller, and E_L becomes larger. E_c lags by an angle greater than 90°, and when R is zero, the lag is 180°. But, if the inductance of the coil is made *less*, or if the value of resistor R is made *greater* than the inductive reactance of coil L, then the voltages change in the opposite direction, and the voltage E_c lags by an angle that approaches zero as a limit. If R can be varied over a wide range, then the coil L (or a capacitor) can be fixed in value, and still a phase shift of essentially 180° is possible. When a 180° phase shift is available, then the value of the average rectified current flow, which is the direct current, can be varied as desired from zero to the maximum value.

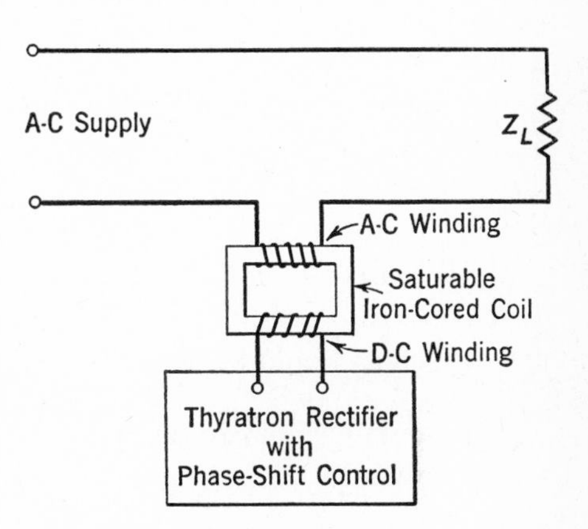

FIG. 6-29. A saturable-core reactor and a controlled rectifier can be used to vary the current to a load of impedance Z_L. The reactance of the coil in series with load Z_L varies because the rectified current in the winding varies. The direct current controls the point of operation on the saturation curve of the iron core.

In industry there are many applications of the phase-shift grid-control principle. For instance, large amounts of alternating-current power can be controlled by a circuit such as Fig. 6-29. The amount of direct current flowing from the Thyratron to the **saturable-core reactor** is controlled by a phase-shifting circuit. The series impedance (largely reactance) of the reactor will be different for different degrees of saturation, because the design of the coil and core is such that the core saturates. Thus, the alternating current flowing to the load impedance Z_L is controlled by the reactor and can be regulated over wide limits. Such a system is used to dim theater lights, and for many other purposes. An operator, located at any convenient point, can by the adjustment of a small resistor, control a large amount of power.

6-17. *IGNITRONS AS CONTROLLED RECTIFIERS* [24,28,29,34]

Ignitrons have mercury-pool cathodes instead of thermionic cathodes, an advantage when large currents are to be rectified. Ignitrons are fired, and caused to conduct, on any part of the impressed cycle, by passing a small current through the Ignitor electrode (section 3-21). Ignitrons are sturdy, and will pass large rectified currents with a low tube voltage drop. Thyratrons and Ignitrons work well together. The phase-shifting circuit of Fig. 6-27, or modifications of it, can be used to fire a Thyratron at any part of the impressed cycle. The current controlled by the Thyratron can be passed into the Ignitor electrode and to the mercury pool, thus firing the Ignitron. With such a combination, it is possible to regulate the current within close limits. It is also possible to effect this regulation without the dissipation of power such as would be caused by a rheostat.

6-18. *THYRATRONS AND IGNITRONS AS ELECTRONIC SWITCHES* [24,28,29,30]

These tubes can be used as switches to connect and disconnect a device from a source of alternating power. This is an important function; for instance, a device can be connected at any part of a cycle, can be energized for as long as desired, and then can be disconnected. The basic switching circuit consists of two rectifier tubes so arranged that each passes one half of the alternating current. Either Thyratrons or Ignitrons can be used, but in many circuits Thyratrons are used to control Ignitrons as explained in section 6-17.

The two Ignitrons of Fig. 6-30 constitute the electronic switch; the two Thyratrons are part of the switch-closing mechanism. At a given instant the polarity of the applied voltage will be as indicated, with the upper input terminal positive and the lower terminal negative. At this instant, the anode of Ignitron 2 will be positive, the mercury-pool cathode will be negative, and Ignitron 2 will be *ready* to fire. At this same instant the anode of Ignitron 1 is negative, and it cannot conduct, even if fired. During the *next half-cycle* of the input voltage, the anode of Ignitron 1 is positive, its mercury-pool cathode is negative, and Ignitron 1 is *ready* to fire. Neither of these Ignitrons will fire and connect the alternating-current input to the primary of the arc-welding transformer unless the Thyratrons are initially conducting and pass current through the Ignitor electrodes. The operation of the Thyratrons can be from some mechanical contacting arrangement such as a rotating switch or a foot switch. Or, the system can be entirely automatic and electrical.

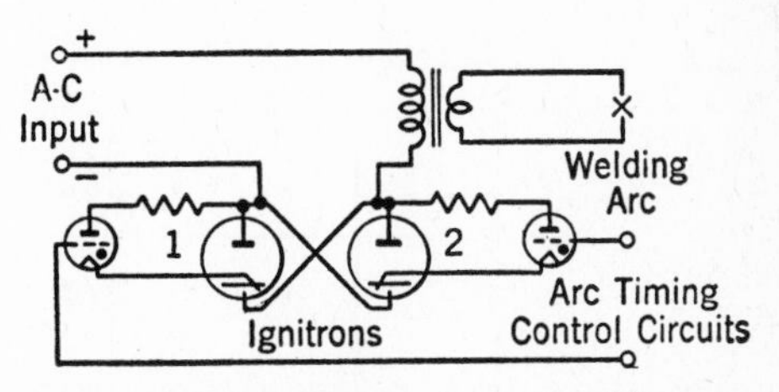

FIG. 6-30. Ignitrons may be used for current control in welding. The firing of the Ignitrons is controlled by Thyratrons. (Adapted from reference 20.)

The operation of this electronic switch is so precise that the circuit can be closed on the zero part of the cycle and in this way switching transients are negligible. This is important in certain types of welding, such as on thin metal sheets, where precise control, and operation for but a few cycles, is desired. If the ordinary type of mechanical switch were used instead of an electronic switch, the transient might be so bad that before the steady-state condition was reached, it would be time to open the circuit. This erratic operation would result in poor welds. An electronic switch using Ignitrons is without moving parts, can be remotely operated, will open and close circuits passing thousands of amperes, and can be made to close on any part of the cycle and remain closed for as many cycles as desired.

6-19. *INVERTERS* [24,28,29,31]

These are devices employing *gas* or *vapor* tubes for converting direct-current power into alternating-current power. When this is done with *vacuum tubes,* the devices are called oscillators. Inverters use Thyratrons and Ignitrons. The efficiency of an inverter is high compared with an oscillator because of the low internal loss in gas tubes. Inverters are, therefore, used where large amounts of power are to be converted.

Several types of inverters have been developed.[24] A simple single-tube inverter is shown in Fig. 6-31. The operation of this circuit is as follows: At the instant the starting switch S is closed, the voltage between the anode and cathode is zero. As capacitor C charges through R, the voltage across the capacitor increases, and the voltage between anode and cathode of the tube also increases. When this voltage is sufficient, the tube breaks down and passes current through the inductance L. When this occurs, the voltage across the capacitor cannot exceed that of the drop across the tube, which is about 15 volts. The capacitor therefore discharges through L. When the magnitude of the voltage of the capacitor reaches that of the drop across the tube, current flow is maintained by the energy stored in the coil L. This charges the capacitor in the opposite direction, and permits the anode to be driven negative so that deionization occurs. When the arc ceases, the charge across the capacitor builds up as before, and the operation is repeated. Since the current through L starts from zero, increases to a maximum, and then decreases to zero, an alternating voltage will be induced in the secondary coil magnetically coupled to it. It has been proposed to use inverters to convert from direct to alternating power in direct-current power transmission systems.[35,36]

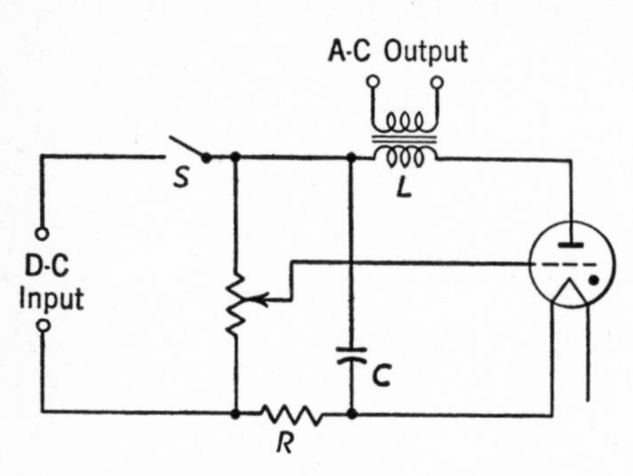

FIG. 6-31. Basic circuit of a Thyratron inverter for converting from direct current to alternating-current power.

REFERENCES

1. Institute of Radio Engineers. *Standards on Antennas, Modulation Systems, Transmitters. Definitions of Terms.* 1948.
2. American Institute of Electrical Engineers. *American Definitions of Electrical Terms.* 1941.
3. Institute of Radio Engineers. *Standards on Receivers: Definitions of Terms,* 1952. Proc. I.R.E., Dec. 1952, Vol. 40, No. 12.
4. Jolley, L. B. W. *Alternating Current Rectification.* John Wiley & Sons.
5. Stout, M. B. *Analysis of rectifier filter circuits.* Electrical Engineering, Sept. 1935, Vol. 54, No. 9.
6. Wallis, C. M. *Full wave rectifier analysis.* Electronics, March 1940, Vol. 13, No. 3.
7. Panofsky, W. K. H., and Robinson, C. F. *Graphical solution of rectifier circuits.* Electronics, April 1941, Vol. 14, No. 4.
8. Schade, O. H. *Analysis of rectifier operation.* Proc. I.R.E., July 1943, Vol. 31, No. 7.
9. Waidelich, D. L. *Analysis of full-wave rectifier and capacitive-input filter.* Electronics, Sept. 1947, Vol. 20, No. 9.
10. Lee, R. *Choke-input filter chart.* Electronics, Sept. 1949, Vol. 22, No. 9.
11. Glaser, J. L. *Monograph aids filter designers.* Electronics, Sept. 1952, Vol. 25, No. 9.

12. Terman, F. E., and Pickles, S. B. *Note on a cause of residual hum in rectifier filter systems.* Proc. I.R.E., Aug. 1934, Vol. 22, No. 8.
13. Dellenbaugh, F. S., and Quimby, R. S. *A series of articles on rectifier design.* Q.S.T., Feb., March, April, 1932.
14. Miles, R. G. *How to design VR tube circuits.* Electronics, Oct. 1952, Vol. 25, No. 10.
15. Koontz, P., and Dilatush, E. *Voltage-regulated power supplies.* Electronics, July 1947, Vol. 20, No. 7.
16. Waidelich, D. L., and Gleason, C. H. *The half-wave voltage-doubling rectifier circuit.* Proc. I.R.E., Dec. 1942, Vol. 30, No. 12.
17. Waidelich, D. L., *The full-wave voltage-doubling rectifier.* Proc. I.R.E., Oct. 1941, Vol. 29, No. 10.
18. Waidelich, D. L., and Shackelford, C. L. *Characteristics of voltage-multiplying rectifiers.* Proc. I.R.E., Aug. 1944, Vol. 32, No. 8.
19. Waidelich, D. L., and Taskin, H. A. T. *Analysis of voltage-tripling and -quadrupling rectifier circuits.* Proc. I.R.E., July 1945, Vol. 33, No. 7.
20. Mautner, R. S., and Schade, O. H. *Television high voltage R-F supplies.* R.C.A. Review, March 1947, Vol. 8, No. 1.
21. Mathers, G. W. C. *Some additions to the theory of radio-frequency high-voltage supplies.* Proc. I.R.E., Feb. 1949, Vol. 37, No. 2.
22. Sulzer, P. G. *Series-resonant high-voltage supply.* Electronics, Sept. 1952, Vol. 25, No. 9.
23. Terman, F. E. *Radio Engineers' Handbook.* McGraw-Hill Book Co.
24. Westinghouse Electric Corp. *Industrial Electronics Reference Book.* John Wiley & Sons.
25. Knowlton, A. E. *Standard Handbook for Electrical Engineers.* McGraw-Hill Book Co.
26. Armstrong, R. W. *Polyphase rectification special connections.* Proc. I.R.E., Jan. 1931, Vol. 19, No. 1.
27. McDonald, D. J. *Rectifiers and inductive coordination.* Electrical Engineering, Feb. 1945, Vol. 64, No. 2.
28. Benz, W. I. *Electronics for Industry.* John Wiley & Sons.
29. Cage, J. M. *Theory and Application of Industrial Electronics.* McGraw-Hill Book Co.
30. McArthur, E. D. *Electronics and Electron Tubes.* John Wiley & Sons.
31. Cockerall, W. D. *Industrial Electronic Control.* McGraw-Hill Book Co.
32. Coroniti, S. C. *Analysis and characteristics of vacuum-tube Thyratron phase-control circuit.* Proc. I.R.E., Dec. 1943, Vol. 31, No. 12.
33. May, J. C., Reich, H. J., and Skalnik, J. G. *Thyratron phase-control circuits.* Electronics, July 1948, Vol. 21, No. 7.
34. Sonbergh, G. *AC Resistance welding control.* Electronic Industries, May 1943, Vol. 2, No. 5.
35. Willis, C. H., Bedford, B. D., and Elder, F. R. *Constant current d-c transmission.* Electrical Engineering, Jan. 1935, Vol. 54, No. 1.
36. Bedford, B. D., Elder, F. R., and Willis, C. H. *Power transmission by direct current.* General Electric Review, May 1936, Vol. 39, No. 5.

QUESTIONS

1. What is the distinction between a rectifier and a rectifier unit?
2. Why may a rectifier be termed a distorter?
3. What is the purpose of a rectifier filter?
4. Under what conditions do equations 6-1 and 6-2 apply?
5. What causes cutout?
6. What effect does cutout have on regulation?
7. How does a resistance-capacitance filter function?
8. Can cutout occur if a resistance-capacitance filter is used?
9. Why should resistors always be connected across the capacitors of a rectifier filter?
10. How is equation 6-6 used in filter design?
11. What is meant by a "swinging choke" and what are the advantages of its use?
12. The first inductor of a filter rectifier sometimes is called an "input choke" and the second a "smoothing choke." Is this advisable?
13. How does a "bleeder" improve regulation?
14. Why should a gas regulator tube be operated with normal glow?
15. Why may the circuit of Fig. 6-16 be called a degenerative-amplifier regulator?
16. Why may the circuit of Fig. 6-17 be called a transconductance-bridge regulator?
17. Why may the circuit of Fig. 6-18 be called an amplification-factor bridge regulator?
18. How can the battery of Fig. 6-18 be replaced with a gas tube?
19. Why are resonant circuits sometimes used in rectifier units?
20. Why is filtering easier for a ripple frequency of 180 cycles than for a frequency of 60 cycles? Would your answer hold for a filter composed of resistance and capacitance?
21. What are the important advantages of using Thyratrons and Ignitrons as controlled rectifiers?
22. Why is only the positive half-cycle shown in Fig. 6-24?
23. What would be an important advantage of using an Ignitron instead of a Thyratron in a rectifier?
24. What are the important characteristics of an electronic switch?
25. Can an electronic switch be conveniently arranged to open on any part of a cycle?

PROBLEMS

1. A filter arranged as a π section has one 12-henry choke coil and two 8-microfarad capacitors. Calculate the cutoff frequency.
2. Suppose that the wave form of the voltage produced by the series combination of the battery and generator of Fig. 6-7 is represented by equation 6-1. Suppose that the magnitude of the direct component is 350 volts; that each coil has a direct-current resistance of 210 ohms and an inductance of 12

henrys under operation conditions; that the value of Z_i is 250 ohms direct-current resistance; that the capacitor is 16 microfarads; and that the R_L is 7500 ohms. Calculate the value of the direct current that would flow in the circuit, the value of the first alternating-current term that will flow in the load resistor, and the per cent ripple.

3. Repeat problem 2, assuming that the wave shape is as given by equation 6-2.
4. Referring to Fig. 6-8, there is a voltage of 750 volts 60 cycles effective value between the two outside transformer secondary terminals. The load resistor R_L is 10,000 ohms. Neglect all other resistances and reactances, and calculate the current through the resistor for each term of equation 6-2. Calculate the per cent ripple, assuming that the lowest-frequency term is the only alternating-current component present.
5. Referring to Fig. 6-9, assume that all conditions of problem 4 apply and that the choke coil L has an inductance under operating conditions of 10 henrys and a direct-current resistance of 180 ohms; and repeat the requirement of problem 4.
6. Repeat the calculations in section 6-9, assuming that a choke-input filter having two of the same coils had been used.
7. A choke-input filter is connected to a resistance load that varies from 1000 to 15,000 ohms. What is the least value and the greatest value of inductance that the input choke must have?
8. Derive equations 6-8 and 6-9.
9. Referring to Fig. 6-14, the input voltage is 125 volts, and the constant-voltage drop across the tube is 105 volts. What must be the value of the protective resistance R?
10. Referring to Figs. 6-27 and 6-28, draw a phasor diagram to represent conditions when a capacitor instead of inductor L is used.

CHAPTER 7

VACUUM-TUBE VOLTAGE AMPLIFIERS

The first extensive use of the vacuum tube for amplification was by the Bell System in repeaters installed at intervals of several hundred miles in the transcontinental telephone lines opened for service about 1915. For some thirty-five years the thermionic vacuum tube was the only device in general use for amplification; however, the transistor (Chapter 10) now is available as a worthy competitor.

Amplification is [1] a "term used to denote an increase of signal magnitude." The signal to be increased in magnitude often is a telephone message attenuated in transmission from a distant station, or the weak radio signal induced in a receiving antenna.

An **amplifier** has been defined [2] as a "device whose output is an enlarged reproduction of the essential features of an input wave and which draws power therefor from a source other than the input signal." Three important characteristics of an electric signal are its voltage, current, and power. This chapter largely will be devoted to amplifiers that are designed primarily to *increase the voltage* of an electric signal wave.

7-1. *VACUUM-TUBE VOLTAGE AMPLIFIERS*

If thermionic vacuum tubes are operated so that the grid is negative with respect to the cathode at all times, even when the input signal is impressed, then for most purposes at frequencies below perhaps 50 megacycles, it is satisfactory to assume that the grid input circuit of the tube draws negligible power. Because they draw negligible power, it is permissible to classify these devices as being **voltage amplifiers,** as distinguished from power amplifiers, to be considered in the following chapter. As an example of the distinction, consider an audio amplifier for amplifying the output from a microphone until the signal magnitude is sufficient to operate several loud-speakers. When actuated by speech sounds, the microphone generates a feeble signal voltage that is impressed on the amplifier. The first several tubes are small-signal types, operated as voltage amplifiers. They take negligible input power, and amplify the signal voltage until this voltage is of magnitude sufficient to drive a power-output tube. This final tube furnishes the driving power to the loud-speakers.

The voltage amplification produced by a vacuum-tube amplifier is the "ratio of the magnitude of the voltage across the input of the transducer * to the magnitude of the voltage delivered to a specified load impedance connected to the transducer." Voltage amplification is commonly expressed as a ratio, or in decibels. The term "gain" is not advocated for expressing voltage amplification.

7-2. *THE DECIBEL*

This unit is used in rating amplifiers. The decibel is based on the **bel**, defined [4] as follows: "The bel is the fundamental division of a logarithmic scale for expressing the ratio of two amounts of power, the number of bels denoting such a ratio being the logarithm to the base 10 of this ratio." Thus, the number of bels is given by the equation,

$$N = \log_{10}\left(\frac{P_1}{P_2}\right) \tag{7-1}$$

where P_1 and P_2 represent two amounts of power.

The **decibel** (abbreviated **db**) is defined [4] as "one tenth of a bel, the number of decibels denoting the ratio of two amounts of power being 10 times the logarithm to the base 10 of this ratio." The number of decibels is given by the equation

$$n = 10 \log_{10} \frac{P_1}{P_2} \tag{7-2}$$

From the theory of logarithms it follows that

$$\frac{P_1}{P_2} = 10^{0.1 \times n} \tag{7-3}$$

* A **transducer** [1,3] is a device capable of being actuated by one or more transmission systems or media and of supplying related waves to one or more other transmission systems or media. The vacuum-tube amplifier is an **active transducer** [1,3] because its "output waves are dependent upon sources of power, apart from that supplied by any of the actuating waves, which power is controlled by one or more of these waves."

where P_1 and P_2 represent two amounts of power, and n is the number of decibels expressing their ratio.

Power is equal to $EI \cos \theta$ or to I^2R. It is possible to expand equations 7-1 and 7-2 so that voltage or current measurements may be used and power need not be measured. Expanding equation 7-2,

$$n = 10 \log_{10} \left(\frac{P_1}{P_2}\right) = 10 \log_{10} \frac{E_1 I_1 \cos \theta_1}{E_2 I_2 \cos \theta_2} = 10 \log_{10} \frac{E_1^2 Z_2 \cos \theta_1}{E_2^2 Z_1 \cos \theta_2}$$
$$= 20 \log_{10} \frac{E_1}{E_2} + 10 \log_{10} \frac{Z_2}{Z_1} + 10 \log_{10} \frac{\cos \theta_1}{\cos \theta_2} \quad \text{7-4}$$

In terms of current values,

$$n = 10 \log_{10} \left(\frac{P_1}{P_2}\right) = 10 \log_{10} \left(\frac{I_1^2 R_1}{I_2^2 R_2}\right)$$
$$= 20 \log_{10} \left(\frac{I_1}{I_2}\right) + 10 \log_{10} \left(\frac{R_1}{R_2}\right) \quad \text{7-5}$$

From equations 7-4 and 7-5 it is seen that if $Z_1 = Z_2$, $\theta_1 = \theta_2$, and $R_1 = R_2$, then the number of decibels expressing the power ratio can be found *under these special conditions* from the voltage relation

$$n = 20 \log_{10} \left(\frac{E_1}{E_2}\right) \quad \text{7-6}$$

It follows that

$$\frac{E_1}{E_2} = 10^{0.05 \times n} \quad \text{7-7}$$

where E_1 and E_2 are the input and output voltages and n is the number of decibels expressing the power ratio. Also, the number of decibels can be found *under the special conditions* from the relation

$$n = 20 \log_{10} \left(\frac{I_1}{I_2}\right) \quad \text{7-8}$$

It follows that

$$\frac{I_1}{I_2} = 10^{0.05 \times n} \quad \text{7-9}$$

where I_1 and I_2 are the input and output currents and n is the number of decibels expressing the power ratio.

As previously mentioned, equations 7-6 and 7-8 hold only under special conditions, which are that the *magnitude* and the *angle* of the impedances of the output circuit under test and the terminating load are identical.

The theory just discussed is illustrated by Fig. 7-1. If a voltmeter is connected across the terminals 1–2 and another is connected across the terminals 3–4, then equation 7-6 can be used to give the *power loss* or *gain* of this circuit only when the impedance measured into the circuit at terminals 1–2 with the driving oscillator disconnected is identical in magnitude and angle to the load impedance. If this is not true, equations 7-6 and 7-8 do not give the *power* loss or gain in decibels in a circuit as defined by equation 7-2.

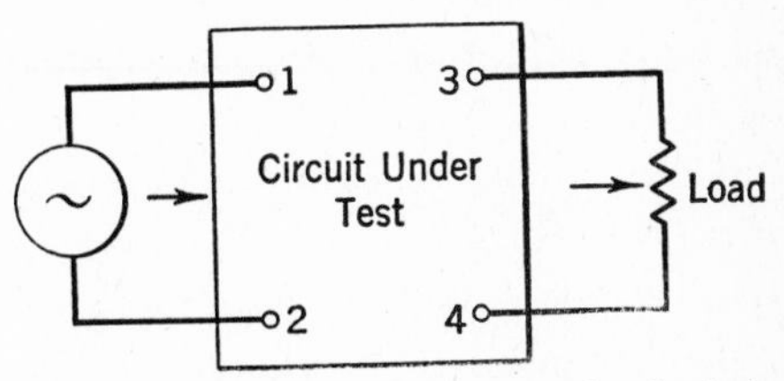

FIG. 7-1. Circuit for explaining the decibel.

If E_1 is the voltage input and E_2 is the voltage output of an amplifier, equation 7-6 gives the voltage amplification in decibels; or, if equation 7-8 is used, current amplification is obtained. The input circuit of an amplifier may contain a step-up transformer, particularly if it is to be used with low-impedance signal sources. Furthermore, the output circuit of an amplifier often contains a step-down transformer, especially if the amplifier is to be used with low-impedance loads such as the common radio loud-speaker. Thus it is possible for the output voltage and the input voltage to be about the same, and for an amplifier to exhibit little *over-all* voltage amplification, yet have much amplification within.

The decibel is used in specifying the **power level** in an amplifier or other electric circuit. Thus, the power level is regarded [1] as "the ratio of the steady-state power at some point in a system to an arbitrary amount of power chosen as a reference." Sometimes this arbitrary amount of power is the input power, or it may be the power at some other point in a circuit. The arbitrary chosen amount of power is termed the **reference level,** or **zero level.** Although other values are used, 0.001 watt, or 1.0 milliwatt, is a common reference level. The abbreviation **dbm** is used to signify that the power level is specified in decibels with respect to 1.0 milliwatt as a reference. In fact, the decibel has been used (and misused) in so many ways that the proposal has been made that the decibel be used as originally intended for expressing the logarithm of electric power ratios, and that a new term **decilog (dg)** be used for expressing the logarithm of the ratio of *any* two quantities.* The **volume unit (vu)** is used to measure the volume of a complex audio-frequency wave.[1]

* See *The Decilog: A Unit for Logarithmic Measurement,* by E. I. Green. Electrical Engineering, July 1954, Vol. 73, No. 7.

7-3. *DISTORTION*

An amplified output signal wave is not an *exact* replica of the input. Distortion is defined [5] as "an undesired change in wave form." Earlier standards did not include the word "undesired." Suppose that a steady-state complex wave is impressed on the input of an amplifier. The rectified wave of Fig. 6-1 and equation 6-1 is an example. Each component of different frequency has an amplitude that possesses a definite ratio with respect to the amplitudes of the others. Also, each component of different frequency has a time-phase angle that bears a definite relationship to the time-phase angles of the others. For no distortion to occur during amplification, new components must not be created, and the amplitude ratios and time-phase angles of the components must not be altered by the amplifier. The various Standards list many types of distortion, and name and define them differently. The definitions that follow are those deemed most important to an elementary discussion of amplifiers in this chapter and those that follow.

Nonlinear distortion [4] is the type of distortion "which occurs in a system when the ratio of voltage to current therein (or analogous quantities in other fields) is a function of the magnitude of either." If nonlinear distortion occurs, new components, of frequencies not present in the original signal wave, are created.

Attenuation distortion [4] is the type of distortion "which results from failure to amplify or attenuate uniformly over the frequency range required for transmission." Distortion of this type results in a change in the relative magnitudes of the components of different frequencies in a signal wave, provided that nonlinear distortion does not cause the change. This has been called frequency distortion.

Phase distortion [4] is the type of distortion "which results from lack of direct proportionality of phase shift to frequency over the frequency range required for transmission." This definition neglects the possibility of intercept distortion * occurring, an assumption usually justified. Thus, various time-phase relations exist among the several components of a complex wave. This is illustrated by the various time-phase angles of equation 6-1. The same time-phase angles must exist in the output if a complex wave is amplified. Thus, if an amplifier causes a phase shift of θ degrees for the fundamental, the phase shift must be 2θ for the second harmonic component, etc.; that is, if the output is to be a replica of the input wave, a linear relation must exist between phase shift and frequency. This principle sometimes is stated: The rate of change of phase shift with frequency must be constant

* See *Communication Networks,* Vol. 2, by E. A. Guillemin, John Wiley & Sons.

over the frequency range of interest if no phase distortion (often called delay distortion) is to occur.

Considerable phase distortion of speech and music can be tolerated because the ear is not sensitive to phase distortion. In facsimile transmission and in television, however, this would result in a distorted picture or image.

7-4. *CLASSIFICATION OF AMPLIFIERS*

Several methods of classification are possible; the one that follows is based on the method of operation of the tubes, and is given in reference 4.

Class-A Amplifier: An amplifier in which the grid-bias and alternating grid voltages are such that plate current in a specific tube flows at all times.*

Class-AB Amplifier: An amplifier in which the grid bias and alternating grid voltages are such that plate current in a specific tube flows for appreciably more than half but less than the entire electrical cycle.

Class-B Amplifier: An amplifier in which the grid bias is approximately equal to the cutoff value so that the plate current is approximately zero when no exciting grid voltage is applied, and so that plate current in a specific tube flows for approximately one half of each cycle when an alternating grid voltage is applied.

Class-C Amplifier: An amplifier in which the grid bias is appreciably greater than the cutoff value so that the plate current in each tube is zero when no alternating grid voltage is applied, and so that plate current flows in a specific tube for appreciably less than one half of each cycle when an alternating grid voltage is applied.

Amplifiers also are classified as to the frequencies they amplify effectively. **Audio-frequency (AF)** amplifiers operate in the range of about 30 to 15,000 cycles. **Radio-frequency (RF)** amplifiers amplify bands of frequencies within the limits of about 15,000 cycles to thousands of megacycles. Radio-frequencies of particular interest are classified as medium frequency (MF, 0.3 to 3 megacycles); high frequency (HF, 3 to 30 megacycles); very high frequency (VHF, 30 to 300 megacycles); ultrahigh frequency (UHF, 300 to 3000 megacycles); and superhigh frequency (SHF, 3000 to 30,000 megacycles).

* To denote that the grid is never driven positive, and that as a result grid current never flows, the suffix 1 is used with the letter identification; for example, class-A1. To denote that the grid is driven positive and that grid current flows during a part of the input cycle, the suffix 2 is used; for example, class-A2. Such designations are only used when it is important to distinguish between the two modes of operation. They are used also with class-AB, class-B, and class-C amplifiers.

7-5. AUDIO-FREQUENCY CLASS-A1 VOLTAGE AMPLIFICATION USING TRIODES [6]

The basic theory of amplification with triodes was considered briefly in section 4-15. For class-A1 amplification the grid is negative at all times, and at audio-frequencies negligible grid input driving power is required. For small-signal operation the grid is biased for operation on the essentially straight part of the grid-voltage–plate-current characteristic, and the signal voltage to be amplified is impressed between grid and cathode. The resulting plate-current variations are a close replica of the grid-signal voltage. This current commonly is passed through a load resistor. The voltage drop across this resistor provides an amplified signal voltage. As will be explained in subsequent sections, an impedance such as an inductor, or the primary of a transformer, sometimes is substituted for the plate load resistor. From section 4-15 the amplified alternating signal voltage across the load resistance is

$$E_L = \frac{\mu E_g R_L}{r_p + R_L} \tag{7-10}$$

and the voltage amplification expressed as a ratio is

$$A_v = \frac{E_L}{E_g} = \frac{\mu R_L}{r_p + R_L} \tag{7-11}$$

In these equations effective values of voltages commonly are used, μ is the amplification factor of the small-signal triode (values from about 10 to 100 are common), r_p is the plate resistance in ohms (values from about 10,000 to 100,000 ohms being typical), and R_L is the plate load resistance of from 3 to 5 times r_p for triodes having plate resistances in the range 10,000 to 20,000 ohms. If plate load impedance is substituted for R_L, then both E_L and A_v will have both magnitudes and angles. Pentodes with all electrodes connected to base pins often are operated as triodes by connecting the screen and suppressor grids to the plate.

7-6. VOLTAGE AMPLIFICATION USING TETRODES [6]

Tetrodes seldom are used for voltage amplification; for this purpose they largely have been supplanted by pentodes. Some information on amplification with tetrodes was given in section 5-6. The coefficients of a small-signal tetrode are numerically similar to those of a small-signal pentode, and many of the statements to be made for pentode amplifiers apply if tetrodes are used. If all electrodes are connected to base pins, a pentode may be operated as a tetrode by connecting the suppressor grid to the plate, or by connecting the suppressor to the screen.

7-7. *AUDIO-FREQUENCY CLASS-A1 VOLTAGE AMPLIFICATION USING PENTODES*

For audio-frequency voltage amplification the small-signal pentode is operated with the grid at all times negative, and with resistance in the plate circuit. The equivalent circuit and the equations of section 4-15 apply. Coefficients for a typical pentode are approximately $\mu = 1500$, $r_p = 1.5$ megohms, and $g_{pg} = 1000$ micromhos. A typical load resistor is $R_L = 0.25$ megohms.

From equation 7-11, the voltage amplification is $A_v = 1500 \times 250{,}000/(1{,}500{,}000 + 250{,}000) = 215$. Tube manuals often do not list the amplification factor and plate resistance of pentodes, but do give the control-grid–plate transconductance. From section 4-6, the product of grid-signal voltage and grid–plate transconductance gives the plate signal current with no plate load resistor. Thus, neglecting R_L, which is one-sixth of r_p (preceding paragraph), the plate signal current is approximately $I = E_g g_{pg}$, the output voltage across the plate load resistor will be $E_L = IR_L = E_g g_{pg} R_L$, and the voltage amplification will be $A_v = E_L/E_g = g_{pg} R_L$. Using the numerical values of the preceding paragraph, the voltage amplification is approximately $A_v = 1000 \times 10^{-6} \times 250{,}000 = 250$. This value is too high because r_p was neglected. Nevertheless, the method is used, and is satisfactory for many purposes, because the tube coefficients listed in manuals are for an "average tube," from which a specific tube may differ considerably.

7-8. *PHASE RELATIONS FOR AMPLIFIERS WITH RESISTANCE LOADS*

The phase relations for a *triode* with resistance load were considered in section 4-17. The phase relations are the same for vacuum triodes, tetrodes, and pentodes operating as class-A amplifiers into resistance loads. This statement applies at audio-frequencies, and at the lower radio-frequencies where interelectrode and stray capacitances are negligible, and where the transit time (section 8-23) need not be considered.

As explained in section 4-17 and associated figures, and as shown in Fig. 7-2, for a class-A amplifier with resistance load the positive half-cycle of the signal voltage applied between grid and cathode causes the plate current to increase. This increases the instantaneous voltage drop across the load re-

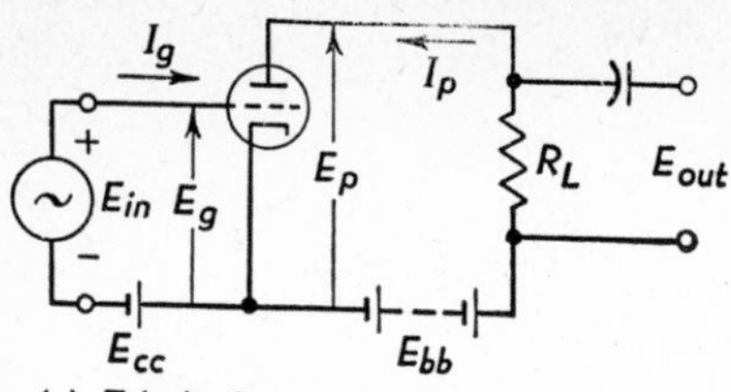

(a) Triode Operated as Class A-1 Voltage Amplifier

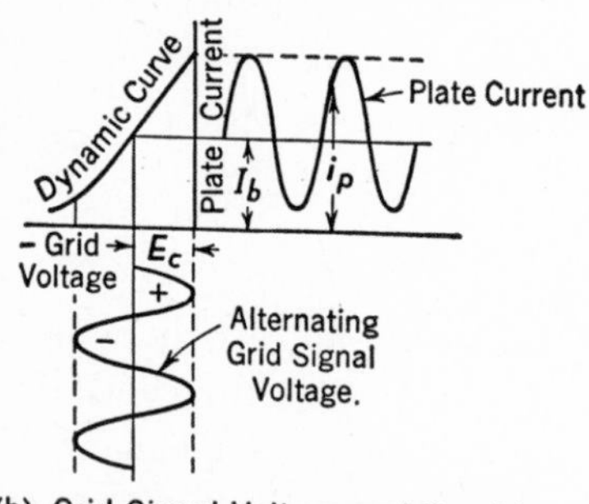

(b) Grid Signal Voltage and Resulting Plate-Current Variations

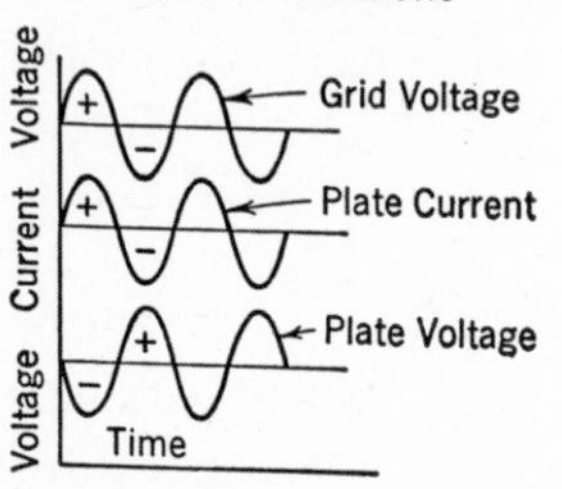

(c) Grid Signal Voltage and Resulting Plate-Current and Plate-Voltage Variations

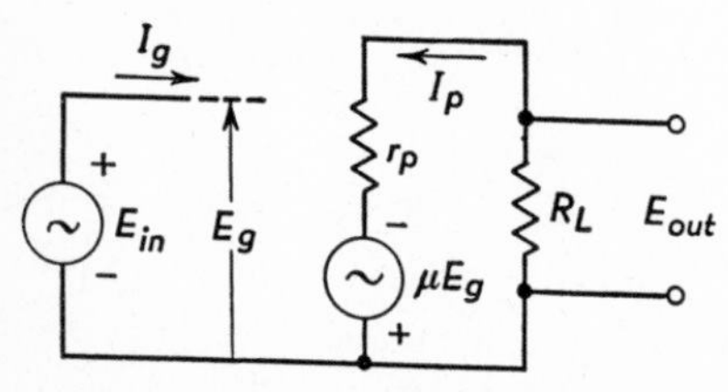

(d) Equivalent Amplifier Circuit

E_{out} E_{in} $I_p r_p$
E_p E_g $I_p R_L$ μE_g I_p

(e) Phasor (Vector) Diagram for Amplifier

FIG. 7-2. Current and voltage relationships in a vacuum-tube amplifier with resistance load.

sistor, and hence the voltage between the plate and cathode must fall. The negative half-cycle of applied grid-signal voltage causes the plate current to fall, and this decreases the instantaneous voltage drop across the load resistor, allowing the voltage between plate and cathode to rise. Thus, grid-voltage variations, plate-current changes, and the resulting voltage drops across the load resistor are *in phase*, but the plate-to-cathode voltage variations are *180° out of phase.*

These relations are shown in Fig. 7-2. Voltage E_g, the alternating signal voltage impressed between the grid and cathode, is taken as the base (X axis) in drawing this diagram. Amplified signal voltage μE_g forces signal (plate) current I_p through the circuit. Because the circuit is considered to be completely resistive, the plate current I_p is in phase with the driving voltage μE_g. Plate current I_p causes an internal voltage drop $I_p r_p$ and an external voltage drop $I_p R_L$, and these drops must be in phase with the plate current. The alternating plate voltage measured between plate and cathode is the sum of $-\mu E_g$ and $+I_p r_p$, or is $-I_p R_L$. This follows from equation 4-18 and from Figs. 4-13 and 7-2. Thus

$$-\mu E_g + I_p r_p + I_p R_L = 0 \quad 7\text{-}12$$

$$-\mu E_g + I_p r_p = -I_p R_L \quad 7\text{-}13$$

Since the output voltage is $-\mu E_g + I_p r_p$, if the load resistor is made to approach infinity in magnitude, plate current I_p will approach zero, and for this condition the output voltage will approach $-\mu E_g$. This indicates a 180° shift in phase between the alternating signal input voltage impressed between grid and cathode, and the amplified output signal voltage between plate and cathode for a resistance load.

7-9. *INPUT IMPEDANCE OF VACUUM TUBES*

The direct current drawn by the grid of a thermionic vacuum tube was discussed in section 4-11. With the grid negative the positive ion current is very small and in most instances is negligible. If, however, a very large value of resistance R_g is connected between the grid and cathode, and in series with the grid-bias source, then this grid current, and the I_cR_g voltage drop, are important. This drop will subtract from the applied grid bias, causing the actual voltage from grid to cathode to be less than the applied value. This is one factor limiting the size of the grid resistor.

In the discussions of vacuum-tube amplifiers given in the preceding pages it has been indicated that when an alternating signal voltage is impressed between grid and cathode, no alternating current flows. This implies that the grid-input impedance is infinite. There are, however, instances where the actual magnitude of the grid-input impedance must be considered, and this subject will now be discussed. Triodes will be considered separately from tetrodes and pentodes, because the input impedances are different.

Input Impedance of Triodes.[7] At low frequencies, such as at audio-frequencies, the input impedance of triodes as operated in ordinary amplifier circuits is very large, and the current drawn is usually negligible. At radio-frequencies, however, the distributed capacitances between the tube electrodes affects the grid-input impedance and has a pronounced influence on the operation of an amplifier.

In a triode the three electrode capacitances are the **grid-cathode capacitance** C_{gk}, **the grid-plate capacitance** C_{gp}, and the **plate-cathode capacitance** C_{pk}. These exist between the electrodes of the tubes, but for convenience are indicated as in Fig. 7-3a. These values of capacitance include the wires in the tube connecting to the electrodes. Considering the tube as a generator of voltage, μE_g, the equivalent circuit becomes that of Fig. 7-3b.

It is possible to make two further simplifications. Referring to Fig. 7-3b, if the load resistance R_L is reasonably low (say 50,000 ohms or less, as is usually true for a triode), then the parallel reactance of the plate-cathode capacitance C_{pk} is so large as to be negligible in comparison with R_L. The other simplification is that the effect of the grid-cathode capacitance C_{gk} can be excluded until the final equation for the grid-input impedance is written. These modifications result in the simplified circuit of Fig. 7-3c. The magnitude of the input impedance of the grid circuit of a tube is equal to the applied alternating grid signal voltage divided by the alternating current that flows. The angle of the impedance will be the phase angle between the grid voltage and grid current.

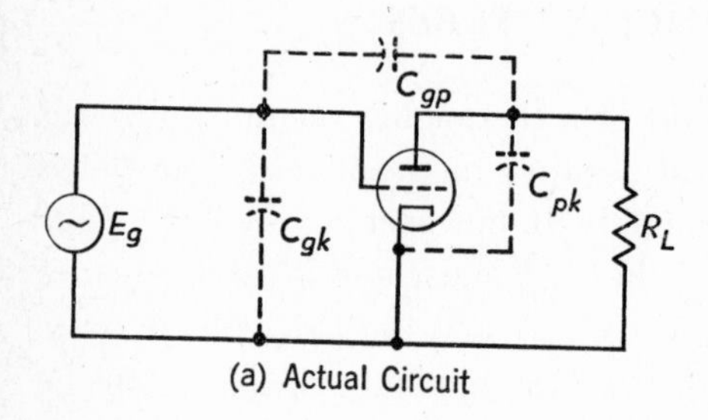

(a) Actual Circuit

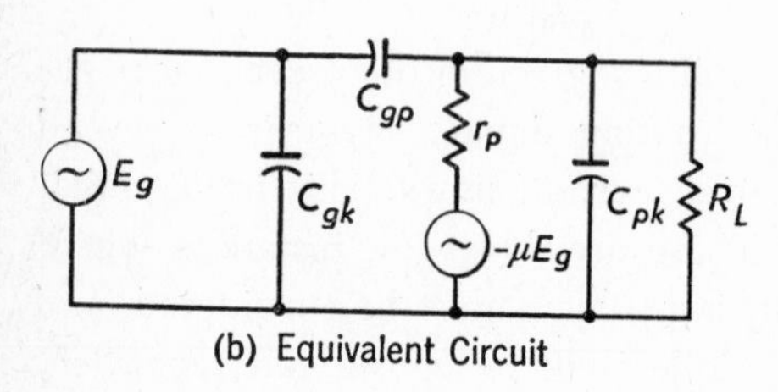

(b) Equivalent Circuit

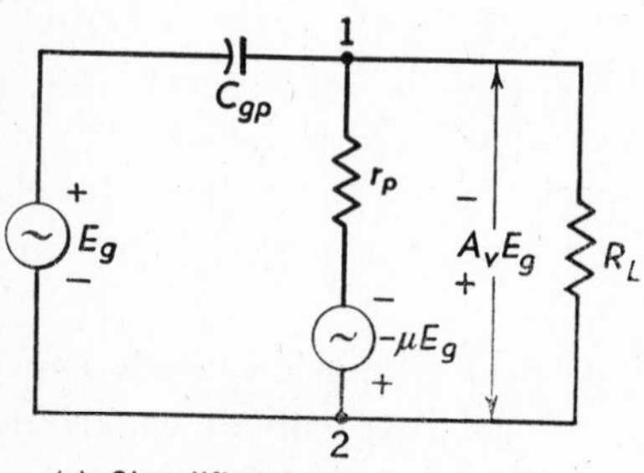

(c) Simplified Equivalent Circuit

FIG. 7-3. Circuits for studying the grid-input impedance of a triode.

When the tube is amplifying and has an alternating signal voltage of E_g volts impressed on the grid, there will be an amplified output voltage of A_vE_g across the resistor R_L in accordance with equation 7-11. The amplified output voltage across R_L also exists across points 1–2 of Fig. 7-3c. Also, the amplified voltage A_vE_g existing across points 1–2 will be 180° out of phase with the impressed voltage E_g. Thus when the polarity of the impressed grid-signal voltage E_g is as indicated, the polarity of the amplified output voltage A_vE_g will be as shown, and the current that flows through the grid-plate capacitance C_{gp} will be proportional to the combined voltages $E_g + A_vE_g$ or to $E_g(1 + A_v)$. Assuming that the capacitive reactance of C_{gp} is large compared to the internal impedance of the source E_g, this current through E_g will lead the voltage across the grid input terminals by approximately 90°.

From this it follows that the *apparent capacitance* of a triode is *increased* by the factor $(1 + A_v)$. Thus the input capacitance of the circuit of Fig. 7-3c is $C_{gp}(1 + A_v)$. When this is added to the grid-cathode capacitance C_{gk} of the tube that is effectively in parallel with it, the total equivalent grid-input capacitance is approximately equal to

$$C_g = C_{gk} + C_{gp}(1 + A_v) \tag{7-14}$$

For a typical small-signal amplifying triode, $C_{gp} = 3.2 \times 10^{-12}$ farad, $C_{gk} = 3.2 \times 10^{-12}$ farad, and $C_{pk} = 2.2 \times 10^{-12}$ farad. This increase in capacitance of a triode can be verified experimentally by measuring with a bridge the grid-input capacitance of a triode amplifier with and without the cathode heated. This increase in grid-input capacitance of a *triode* operating as a class A amplifier with a resistance load is sometimes called the **Miller effect.** The preceding analysis is for a triode *with resistance load.* Many triode amplifiers are operated this way, or are operated in circuits that are largely equivalent to resistance, and so equation 7-14 applies.

If a generalized study[7] of the input impedance of a triode is made for three conditions of load impedances consisting of pure resistance, resistance and capacitance, and resistance and inductance, the following grid-input impedances will exist. *First,* when the plate load impedance is pure resistance, then the grid-input impedance will be pure capacitance. *Second,* when the plate load impedance is resistance and capacitance, the grid-input impedance will be resistance and capacitance, and the resistance is usually positive (that is, an increase in current results in an increase in voltage drop). *Third,* when the plate load impedance is resistance and inductance, then the grid-input impedance is resistance and capacitance, but under certain conditions the *resistance component of the grid-input impedance is negative* (that is, an increase in current results in a *decrease* in voltage drop).

Input Impedance of Tetrodes and Pentodes. The screen grid in the tetrode is placed between the control grid and the plate. If the screen grid is connected to the cathode through a sufficiently large capacitor (the size depending on the frequency) then plate voltage variations cannot greatly affect the control grid, and no increased input capacitive effect due to amplifying action results. For all audio-frequencies, and all but very high radio-frequencies, the control-grid input capacitance is essentially the same with the cathode hot or cold, and is but a few micromicrofarads. In the suppressor-grid pentode the control grid is shielded by the grounded (to cathode) suppressor grid in addition to the screen grid. The grid-input capacitance of a pentode is also very low, being but a few micromicrofarads, and is essentially the same if the cathode is hot or cold.

7-10. *RESISTANCE-COUPLED AUDIO VOLTAGE AMPLIFIERS USING TRIODES*[8]

More than one tube often is needed to provide the required voltage amplification. Hence, an amplifier commonly contains several tubes operated in

"cascade," with one amplifier "stage" driving the next through a suitable coupling network. In the **resistance-coupled amplifier** the coupling network consists of a capacitor and resistors as in Fig. 7-4.

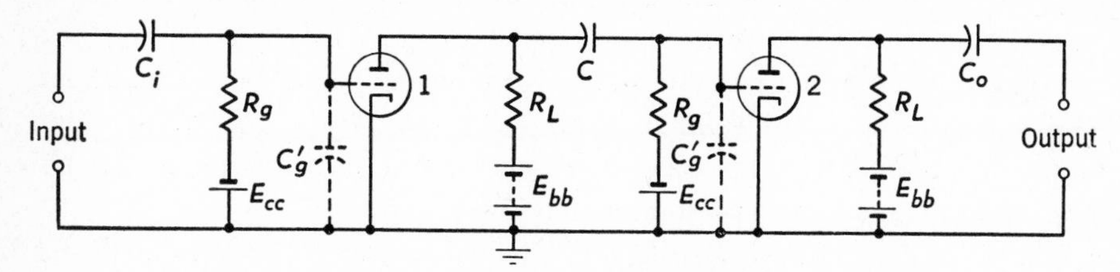

FIG. 7-4. Two-stage resistance-coupled amplifier using triodes.

Direct-Current Aspects. The input capacitor C_i isolates the signal-input source and amplifier from direct voltages present in either. The **grid resistor** R_g connects the negative biasing voltage E_{cc} to the grid. For class A1 operation, negligible grid current flows through R_g, and hence the direct voltage between grid and cathode is essentially E_{cc}. The plate-supply voltage E_{bb} is impressed between plate and cathode through **load resistor,** or **coupling resistor,** R_L. The plate current I_b flows through R_L, and hence the actual voltage between plate and cathode is $E_b = E_{bb} - I_bR_L$. Consideration of this equation *alone* indicates that R_L should be small so that E_{bb} need not be excessive. Coupling capacitor C isolates the two stages, so that the plate of the first tube and the grid of the second may be at different voltages. This capacitor must have negligible leakage. Resistors R_g and R_L of the second stage operate as explained. Capacitor C_o isolates the amplifier and load from a direct-voltage standpoint.

Alternating-Current Aspects. The alternating signal input voltage forces signal current through input capacitor C_i and the parallel combination of R_g and C_g' of Fig. 7-4. Capacitor C_g' represents the sum of the input capacitance (equation 7-14) of the tube, and "stray" wiring capacitance. The reactance of C_i should be negligible at all frequencies to be amplified; then, the entire signal input voltage will be between grid and cathode for amplification. The signal voltage between grid and cathode will cause alternating signal plate current to flow through the coupling network at the right of tube 1. The voltage drop E_g' of Fig. 7-5 that results across R_g and C_g' (in parallel) is the signal that tube 1 impresses on tube 2 for further amplification. If the *direct* plate voltage is maintained constant, then the greater R_L, the greater will be the signal voltage impressed on tube 2. But, as explained in the preceding paragraph, R_L should be low. The selection of R_L is a compromise, values of

R_L usually being from three to five times the plate resistance r_p of the first tube *when triodes* are used.

Equivalent Circuits of Triode R-C Amplifier. The equivalent circuit of tube 1 of Fig. 7-4 and the coupling network between tube 1 and tube 2 are shown in Fig. 7-5. The alternating signal voltage E_g impressed between cathode and grid of tube 1 is amplified to a value μE_g and is impressed through the plate resistance r_p onto the network composed of R_L, C, R_g, and C_g'. It is possible to write one equation for the circuit of Fig. 7-5 that will give voltage E_g' for all signal frequencies. From such an equation all that usually needs to be known about a small-signal amplifier can be determined. The information usually desired is the voltage amplification at various frequencies, and sometimes the phase shift.

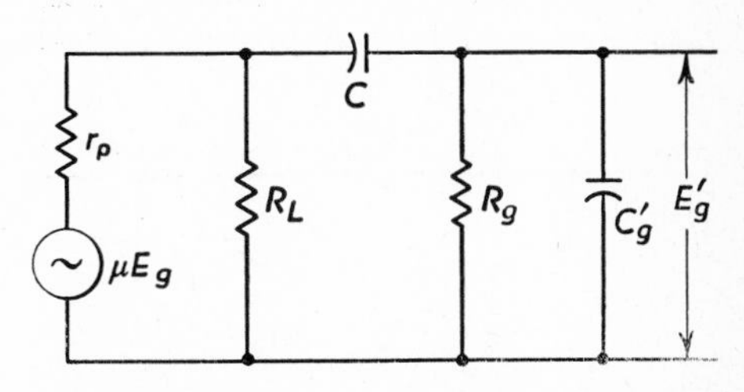

FIG. 7-5. Equivalent circuit for one stage of Fig. 7-4.

The voltage amplification is the ratio of signal voltage E_g' to the input signal voltage E_g, and can be expressed as a ratio, or in decibels (section 7-2). The phase shift is the difference in phase, expressed in degrees, between input signal voltage E_g and voltage E_g'. The phase shift is composed of two parts, that produced by the tube, and that produced by the coupling network. That is, with *only* resistance in the plate circuit, there is a 180° shift in phase for a single tube caused by the fundamental amplifying action of the tube (see Fig. 7-2). To this must be added or subtracted the shift in phase caused by the resistance-capacitance coupling circuit. Rather than write one equation for Fig. 7-5, the practice is to consider the circuit at three different frequency regions and in the following order: midfrequency, low frequency, and high frequency.

Amplification at Mid-Audio-Frequencies in Triode R-C Amplifier. In a typical resistance-coupled voltage amplifier using small-signal triodes, $R_L = 50{,}000$ ohms, $C = 0.05$ microfarad, $R_g = 250{,}000$ ohms, $\mu = 13.8$, $r_p = 12{,}000$ ohms, $C_{gp} = 2.2$ micromicrofarads, and $C_{gk} = 3.4$ micromicrofarads. Assuming that a single stage will have a voltage amplification of 10, from equation 7-14 the input capacitance to the grid of the second tube will be about $C_g = 3.4 + 2.2(1 + 10) = 3.4 + 24.2 = 27.6$ micromicrofarads. Estimating that there will be about 25 micromicrofarads capacitance in the wiring in the grid circuit, then the total input capacitance C_g' of Fig. 7-5 will equal approximately 50 micromicrofarads.

The audio-frequency voltage amplifier under consideration would be designed to amplify from about 30 to 15,000 cycles per second. A frequency of 1000 cycles will be considered in the mid-audio-frequency region. At this frequency the capacitive reactance of the coupling capacitor C is $X_C = 1/(2\pi fC) = 1/(6.28 \times 1000 \times 0.05 \times 10^{-6}) = 3180$ ohms. At this same frequency the capacitive reactance offered by the total input capacitance C_g' is $X_{Cg}' = 1/(6.28 \times 1000 \times 50 \times 10^{-12}) = 3{,}180{,}000$ ohms.

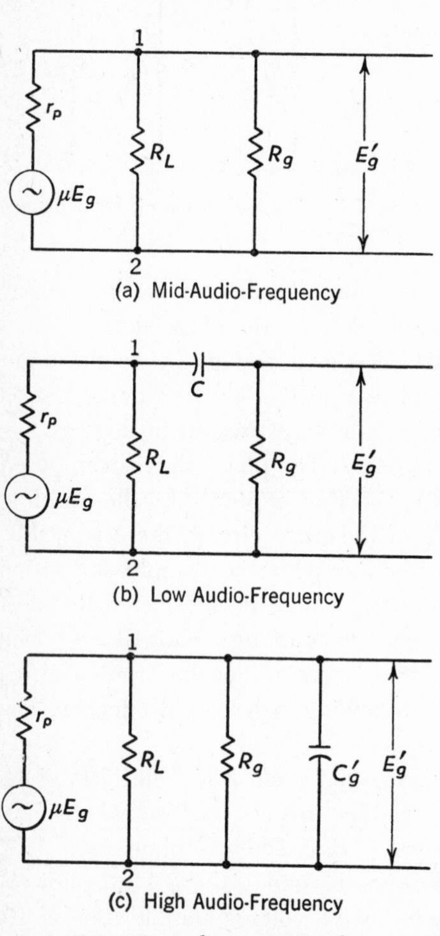

FIG. 7-6. Equivalent circuits for a resistance-coupled voltage amplifier.

The values of the equivalent circuit of Fig. 7-5 now are known: $R_L = 50{,}000$ ohms, $R_g = 250{,}000$ ohms, $X_C = 3180$ ohms, and $X_{Cg}' = 3{,}180{,}000$ ohms. For an approximate analysis, the effects of both C and C_g' can be neglected at 1000 cycles, and *for this mid-audio-frequency* the equivalent circuit of Fig. 7-5 becomes Fig. 7-6a. The voltage amplification of this equivalent circuit can be computed by equation 7-11 if R_L of this equation is replaced by R_e, where $R_e = R_L R_g/(R_L + R_g)$. Using the values previously stated, $R_e = (50{,}000 \times 250{,}000)/(50{,}000 + 250{,}000) = 41{,}700$ ohms. The voltage amplification at mid-audio-frequencies is $A_v = \mu R_e/(r_p + R_e) = (13.8 \times 41{,}700)/(12{,}000 + 41{,}700) = 10.71$. Because at mid-audio-frequencies such as 1000 cycles the equivalent circuit (Fig. 7-6a) consists of resistance only, the coupling network will cause negligible phase shift at this frequency. However, there will be a phase shift of 180° caused by the tube (section 7-8).

Amplification at Low Audio-Frequencies in Triode R-C Amplifier. At 30 cycles the capacitive reactance of coupling capacitor C is about 106,000 ohms, and cannot be neglected. The reactance of the input capacitance C_g' has increased to about 106,000,000 ohms, and is negligible. Thus, for low audio-frequencies the equivalent circuit of Fig. 7-5 can be simplified to Fig. 7-6b. To find the amplification and phase shift of the stage at low frequencies, the magnitude and angle of the voltage E_g' across R_g must be computed.

The impedance between points 1–2 of Fig. 7-6b will be the impedance of the parallel circuit composed of R_L as one branch, and the coupling capacitor C and the grid resistor R_g as the other branch. That is, from the general formula $Z_e = Z_1Z_2/(Z_1 + Z_2)$,

$$Z_{12} = \frac{R_L(R_g - jX_C)}{R_L + (R_g - jX_C)} \qquad 7\text{-}15$$

The total impedance Z_t in series with the amplified voltage μE_g will be the sum of the plate resistance r_p and Z_{12}, the impedance of the parallel branch. Therefore

$$Z_t = r_p + Z_{12} = r_p + \frac{R_L(R_g - jX_C)}{R_L + (R_g - jX_C)} \qquad 7\text{-}16$$

The current flowing because of the amplified voltage of magnitude μE_g will be

$$I = \frac{\mu E_g}{Z_t} = \frac{\mu E_g}{r_p + Z_{12}} \qquad 7\text{-}17$$

The voltage across the parallel impedance Z_{12} of the coupling circuit will be

$$E_{12} = IZ_{12} = \frac{\mu E_g Z_{12}}{r_p + Z_{12}} \qquad 7\text{-}18$$

The current flowing through R_g will equal the voltage across points 1–2 divided by the series impedance of the coupling capacitor C and the grid resistor R_g,

$$I_{Rg} = \frac{E_{12}}{R_g - jX_C} \qquad 7\text{-}19$$

The voltage across E_g' across R_g, which is the amplified output voltage that is available to drive the grid of the next tube, is

$$E_g' = I_{Rg}R_g = \frac{E_{12}R_g}{R_g - jX_C} \qquad 7\text{-}20$$

The voltage amplification and phase shift can be found by making the indicated substitutions and solving for the ratio of the output voltage E_g' across the grid resistor R_g to the impressed input voltage E_g. Thus

$$A_{v(\text{low})} = \frac{E_g'}{E_g} = \frac{\mu R_L R_g}{(r_p R_g + R_L R_g + r_p R_L) - j\left[\dfrac{R_L + r_p}{2\pi f C}\right]} \qquad \text{7-21}$$

Also to be considered is the 180° shift in phase caused by the tube; this will be done later.

Substituting in equation 7-16 the values given for the amplifier under consideration, and at a frequency of 30 cycles per second, gives

$$Z_t = 12{,}000 + \frac{50{,}000(250{,}000 - j106{,}000)}{50{,}000 + (250{,}000 - j106{,}000)}$$

$$= 12{,}000 + (42{,}700 - j2610) = 54{,}700 - j2610 \text{ ohms}$$

In polar notation this impedance is about $54{,}750 \,\underline{/-2.7^\circ}$ ohms.

Assume that an effective value of 1.0 volt is impressed between the grid and cathode of the tube, and that the fictitious amplified voltage μE_g between point 2 and r_p of Fig. 7-6b is taken as a reference. Then, from equation 7-17, the current flowing from the tube will be

$$I = \frac{13.8 \times 1.0}{54{,}750 \,\underline{/-2.7^\circ}} = 0.000253 \,\underline{/+2.7^\circ} \text{ ampere}$$

From a preceding calculation the impedance of the parallel portion $Z_{12} = 42{,}700 - j2610$ ohms, or $Z_{12} = 42{,}750 \,\underline{/-3.5^\circ}$ ohms. The voltage drop across this parallel portion is given by equation 7-18 and is

$$E_{12} = (0.000253 \,\underline{/+2.7^\circ})(42{,}750 \,\underline{/-3.5^\circ}) = 10.8 \,\underline{/-0.8^\circ} \text{ volts}$$

The impedance of the path provided by the coupling capacitor C and the grid resistor R_g is $250{,}000 - j106{,}000 = 272{,}000 \,\underline{/-23^\circ}$ ohms. The current through R_g can be found from equation 7-19 and equals

$$I_{Rg} = \frac{10.8 \,\underline{/-0.8^\circ}}{272{,}000 \,\underline{/-23^\circ}} = 0.0000397 \,\underline{/+22.2^\circ} \text{ ampere.}$$

The output voltage E_g' is given by equation 7-20 and equals

$$E_g' = 250{,}000 \times 0.0000397 \,\underline{/+22.2^\circ} = 9.9 \,\underline{/+22.2^\circ} \text{ volts.}$$

Thus, the voltage amplification $A_{v(\text{low})} = 9.9/1 = 9.9$, and the total phase shift for the tube and R-C coupling network is $180^\circ + 22.2^\circ = 202.2^\circ$ as indicated in Fig. 7-7.

The voltage amplification and phase shift can also be computed directly from equation 7-21 without going through each step as indicated. Thus,

$$A_{v(\text{low})} = \frac{13.8 \times 50{,}000 \times 250{,}000}{(12{,}000 \times 250{,}000) + (50{,}000 \times 250{,}000) + (12{,}000 \times 50{,}000) - j\left(\dfrac{50{,}000 + 12{,}000}{6.28 \times 30 \times 0.05 \times 10^{-6}}\right)} = 9.9$$

and the phase angle of the coupling circuit is +22.2°, giving a total phase shift of 180° + 22.2° = 202.2°.

It is common practice to assume that an amplifier ceases to be useful when the low-frequency voltage amplification has dropped to about 70 per cent of the amplification at midfrequencies. (The value is 0.707, a drop in voltage amplification of 3 decibels.) The low frequency at which this arbitrary limit is reached can be determined from equation 7-21, and occurs at the frequency at which the real and imaginary parts of the denominator are equal. That is, when

$$f_{\text{low}} = \frac{R_L + r_p}{2\pi C(r_pR_g + R_LR_g + r_pR_L)} \qquad 7\text{-}22$$

In a given circuit *the low-frequency performance largely is determined by the size of the coupling capacitor C.* At low frequencies, where the reactance of this capacitor is high, the amplification is low, and the total phase shift approaches 270°.

Amplification at High Audio-Frequencies in Triode R-C Amplifier.[9] The high audio-frequency of interest is assumed to be 15,000 cycles. The capacitive reactance of the 0.05-microfarad coupling capacitor C of Fig. 7-5 is 212 ohms at this frequency and can be neglected. But, at this same upper frequency the capacitive reactance of the grid-input capacitance C_g', assumed to be 50 micromicrofarads, is 212,000 ohms, and since it is in parallel with R_g of 250,000 ohms, the effect of C_g' must be considered. These facts make possible the representation of the resistance-capacitance amplifier circuit at high frequencies by Fig. 7-6c.

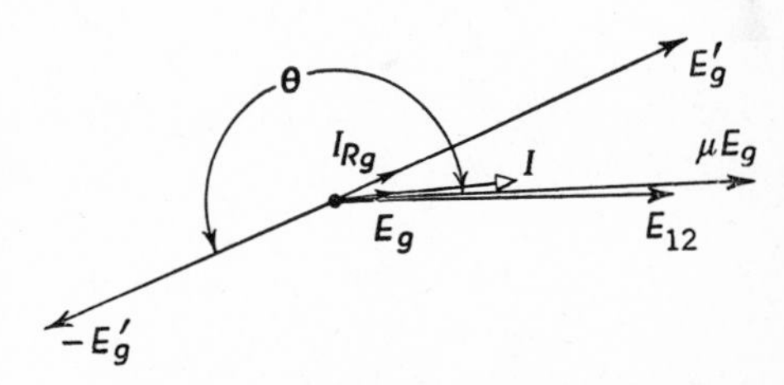

FIG. 7-7. Approximate diagram for resistance-coupled voltage amplifier at *low* frequencies. The angle by which output voltage $-E_g'$ leads input voltage E_g is indicated by θ.

The ratio of E_g' to E_g and the phase angle between them can be determined much as for the low-frequency case, but with more ease. The amplified output voltage E_g' is of course equal to the voltage drop across the parallel branch composed of R_L, R_e, and C_g'. Because the load resistor R_L and the grid resistor R_g are in parallel, the first step is to solve for their equivalent resistance, $R_e = R_LR_g/(R_L + R_g)$.

The equivalent resistance R_e is in parallel with C_g', and thus the equivalent impedance between points 1–2 is

$$Z_{12} = \frac{-jX_{Cg}'R_e}{R_e - jX_{Cg}'} \qquad 7\text{-}23$$

The total impedance Z_t in series with the amplified voltage μE_g is the sum of the plate resistance r_p and Z_{12}. That is

$$Z_t = r_p + Z_{12} = r_p + \frac{-jX_{Cg}'R_e}{R_e - jX_{Cg}'} \qquad 7\text{-}24$$

The current supplied by the amplified voltage of magnitude μE_g will be

$$I = \frac{\mu E_g}{Z_t} \qquad 7\text{-}25$$

The voltage across the parallel branch Z_{12} is the desired amplified output voltage E_g'. That is,

$$E_g' = IZ_{12} \qquad 7\text{-}26$$

Making the necessary substitutions gives the following equation for the voltage amplification at high frequencies

$$A_{v\,(\text{high})} = \frac{\mu R_L R_g}{(r_p R_g + R_L R_g + r_p R_L) + j(2\pi f C_g' R_L R_g r_p)} \qquad 7\text{-}27$$

This equation gives both the magnitude of the voltage amplification and the phase shift (caused by the coupling circuit) between the input and the output voltages as will be illustrated by a numerical solution. It does not consider the 180° shift caused by the tube.

Substituting in equation 7-24 the values given for the amplifier under consideration, and at a frequency of 15,000 cycles, gives for the total impedance

$$Z_t = 12{,}000 + \frac{-j212{,}000 \times 41{,}700}{41{,}700 - j212{,}000} = 12{,}000 + (40{,}200 - j7890)$$

$$= 52{,}200 - j7890 \text{ ohms,}$$

or

$$Z_t = 52{,}800 \,\underline{/-8.6^\circ} \text{ ohms}$$

From equation 7-24, impedance Z_{12} of Fig. 7-6c is $41{,}000 \,\underline{/-11.1^\circ}$ ohms. Assuming that an effective value of 1.0 volt is impressed on the grid of the tube, and that the amplified voltage μE_g between point 2 and r_p of Fig. 7-6c is taken as a reference, then the current flowing through the tube will be

$$I = \frac{13.8 \times 1.0}{52{,}800 \,\underline{/-8.6^\circ}} = 0.000261 \,\underline{/+8.6^\circ} \text{ ampere}$$

The desired output voltage E_g' is given by equation 7-27; thus

$$E_g' = (0.000261 \,\underline{/8.6^\circ})(41{,}000 \,\underline{/-11.1^\circ}) = 10.7 \,\underline{/-2.5^\circ} \text{ volts}$$

If equation 7-27 is used to calculate the high audio-frequency voltage amplification and the phase shift, then,

$$A_{v(\text{high})} = \frac{13.8 \times 50{,}000 \times 250{,}000}{[(12{,}000 \times 250{,}000) + (50{,}000 \times 250{,}000) + (12{,}000 \times 50{,}000)] + j[6.28 \times 15{,}000 \times 50 \times 10^{-12} \times 50{,}000 \times 250{,}000 \times 12{,}000]} = 10.7$$

and the phase angle is $-2.5°$, giving a total phase shift for the tube and load of $180° - 2.5 = 177.5°$ as shown in Fig. 7-8. Thus, either the step-by-step method or the use of equation 7-27 gives identical results. It is noted that the amplification at 15,000 cycles is the same as at 1000 cycles, and that the phase shift caused by the coupling circuit is negligible.

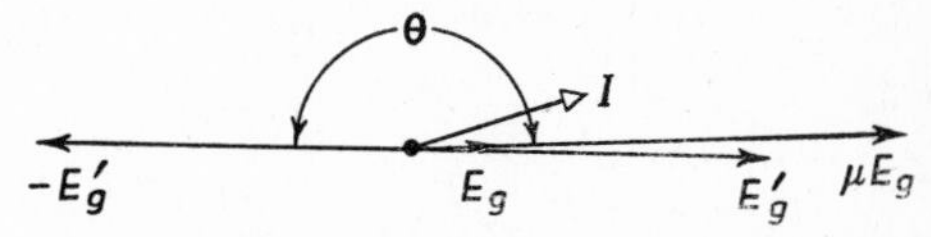

FIG. 7-8. Approximate diagram for resistance-coupled voltage amplifier at *high* frequencies. The angle by which output voltage $-E_g'$ leads input voltage E_g is indicated by θ.

The value of the high frequency at which an amplifier ceases to be useful depends on the arbitrary limit assigned. As for the low-frequency case, it is common to assume that an amplifier ceases to be useful when the high-frequency voltage amplification has dropped to about 70 per cent (actually, 0.707) of the intermediate-frequency value. The high frequency at which this limit is reached obtains when the real and imaginary components of equation 7-27 are equal. That is, when

$$f_{\text{high}} = \frac{(r_pR_g + R_LR_g + r_pR_L)}{2\pi C_g'R_LR_gr_p} \tag{7-28}$$

7-11. *RESISTANCE-COUPLED AUDIO VOLTAGE AMPLIFIERS USING PENTODES*

The resistance-coupled amplifier considered in the preceding pages used triodes. Small-signal pentodes are used more extensively in such amplifiers. Figure 7-5 represents also the equivalent circuit of a stage of a resistance-coupled voltage amplifier using a small-signal pentode. Because of the shielding between the control grid and plate afforded by the suppressor and screen grids, the grid-to-plate capacitance can be neglected. The capacitance caused by the control-grid wiring, and the capacitance from the control grid to cathode must be considered. This total may be about 20 micromicrofarads. A typical small-signal pentode will have an amplification factor of about 1500, a plate resistance of about 1,500,000 ohms, and a control-grid–plate transconductance (mutual conductance) of about 1000 micromhos, or 0.001 mho. Values for Fig. 7-5 of $R_L = 250{,}000$ ohms, $R_g = 500{,}000$ ohms, and $C = 0.005$ microfarad are typical.

The equivalent circuits of Fig. 7-6 apply to the pentode resistance-coupled voltage amplifier. It follows, therefore, that equations 7-11, 7-21, and 7-27 give the amplification at mid, low, and high audio-frequencies.

Amplification at Mid-Audio-Frequencies in Pentode R-C Amplifier. At a midfrequency assumed to be 1000 cycles, at which Fig. 7-6a applies, the equivalent parallel resistance of R_L and R_g will be $R_e = 250{,}000 \times 500{,}000/(250{,}000 + 500{,}000) = 167{,}000$ ohms. The voltage amplification will be, from equation 7-11 and section 7-10,

$$A_v = \frac{\mu R_e}{r_p + R_e} = \frac{1500 \times 167{,}000}{1{,}500{,}000 + 167{,}000} = 150 \qquad \text{(approx.)}$$

It is important to note that the very high plate resistance of the pentode renders it inefficient (from a theoretical standpoint) as a voltage amplifier. In this circuit a large percentage of the total theoretically available μE_g is lost as an $I_p r_p$ drop inside the tube, and only a small percentage is delivered for amplification by the second tube. It is important to note that, although the pentode as here used is relatively inefficient, a *large* increase in voltage (about 150 times) is possible. Only the 180° shift in phase caused by the tube with resistance load exists at this frequency.

Amplification of Low Audio-Frequencies in Pentode R-C Amplifiers. This can be represented by Fig. 7-6b, and the voltage ratio and phase shift can be calculated by equation 7-21. Thus, at the low frequency of 30 cycles,

$$A_{v(\text{low})} = \frac{1500 \times 250{,}000 \times 500{,}000}{[(1{,}500{,}000 \times 500{,}000) + (250{,}000 \times 500{,}000) + (1{,}500{,}000 \times 250{,}000)] - j\left[\dfrac{250{,}000 + 1{,}500{,}000}{6.28 \times 30 \times 0.005 \times 10^{-6}}\right]}$$

$$= 83.8 \underline{/+56.1^\circ}$$

This indicates that at 30 cycles the voltage amplification has dropped from 150 to 83.8 (or to 56.8 per cent of the 1000-cycle value) and that the phase shift caused by the coupling network has changed from zero (as at 1000 cycles) to $+56.1^\circ$, giving a total phase shift of the tube and network of $180^\circ + 56.1^\circ = 236.1^\circ$.

The voltage amplification will drop to about 70 per cent that at mid-audio-frequencies at a frequency given by equation 7-22. For the pentode voltage amplifier now under consideration, the value given by equation 7-22 is 44.5 cycles. Using this frequency in the solution of the preceding paragraph gives a voltage ratio of 106, and an angle of $+45^\circ$.

Amplification at High Audio-Frequencies in Pentode R-C Amplifiers. The equivalent circuit is shown in Fig. 7-6c, and the voltage amplification and phase shift can be calculated from equation 7-27. Thus, at the upper audio-frequency of 15,000 cycles,

$$A_{v(\text{high})} = \frac{1500 \times 250{,}000 \times 500{,}000}{[(1{,}500{,}000 \times 250{,}000) + (250{,}000 \times 500{,}000) + (1{,}500{,}000 \times 250{,}000)] + j[6.28 \times 15{,}000 \times 20 \times 10^{-12} \times 250{,}000 \times 500{,}000 \times 1{,}500{,}000]} = 144 \underline{/-15.8^\circ}$$

This indicates that at 15,000 cycles per second the voltage amplification has dropped from 150 to 144 (or to 96 per cent of the 1000-cycle value) and that the phase shift caused by the coupling network has changed from zero to $-15.8°$, giving a total phase shift of tube and network of $180° - 15.8° = 164.2°$.

The voltage amplication will drop to about 70 per cent of the value at mid-audio-frequencies at a frequency given by equation 7-28. For the amplifier being considered, the value given by equation 7-28 is 53,000 cycles. Using this frequency in the solution of the preceding paragraph gives a voltage gain of 106, and an angle of $-45°$.

Alternate Method of Computing Amplification. In the previous pages a vacuum tube in an amplifying circuit has been considered to be a generator, or source, of amplified voltage of magnitude μE_g having an internal resistance r_p connected through a resistance-capacitance coupling circuit to the grid of the following tube. If the input impedance of this entire coupling circuit is represented by Z_L, then the circuit arrangement is as shown by Fig. 7-9a. The equation for the voltage ratio is

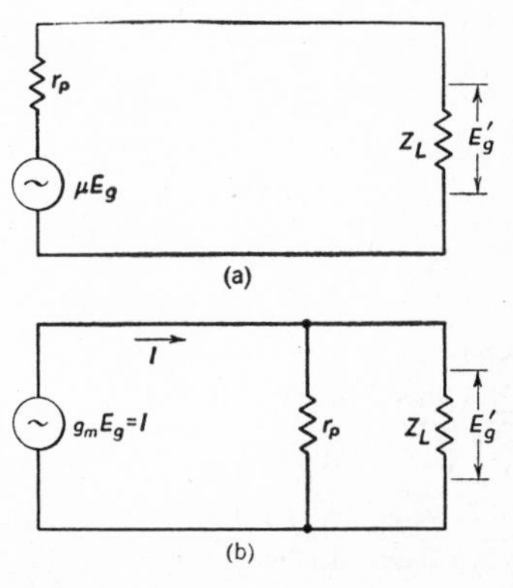

FIG. 7-9. Amplifier equivalent circuits; (*a*) in accordance with Thévenin's theorem; (*b*) in accordance with Norton's theorem.

$$A_v = \frac{E_g'}{E_g} = \frac{\mu Z_L}{r_p + Z_L} \qquad \text{7-29}$$

in accordance with equation 7-11. Equation 7-29 may be written

$$A_v = \frac{E_g'}{E_g} = \frac{\mu Z_L}{\dfrac{r_p^2 + r_p Z_L}{r_p}} = \frac{\mu r_p Z_L}{r_p(r_p + Z_L)} = \frac{\mu}{r_p} \cdot \frac{r_p Z_L}{r_p + Z_L} = g_{pg} Z_e \qquad \text{7-30}$$

where $g_{pg} = \mu/r_p$ and is the grid-plate transconductance, or mutual conductance, of the tube in mhos, and Z_e is the equivalent impedance $Z_e = r_p Z_L/(r_p + Z_L)$ of r_p and Z_L in parallel.

From this discussion it follows that either equations 7-29 or 7-30 can be used to compute voltage amplification. The circuit of Fig. 7-9a that has been used in the preceding pages represents a *constant-voltage generator* of voltage μE_g and internal impedance r_p, connected to a load of impedance Z_L. The circuit of Fig. 7-9b represents a *constant-current generator* of current $I = g_{pg}E_g$, the current flowing into a circuit composed of r_p and Z_L in parallel. Figure 7-9a and equation 7-29 are in accordance with Thévenin's theorem; Fig. 7-9b and equation 7-30 are in accordance with Norton's theorem.*

Either of the methods just discussed can be used, and in fact both are used, to design vacuum-tube voltage amplifiers. In practice the constant-voltage system (Thévenin's theorem) is generally used for circuits containing triodes, and the constant-current system (Norton's theorem) is more often used for circuits containing pentodes.

Perhaps the most important reason the constant-current method is used is its simplicity. Thus, assume that a pentode voltage amplifier is being considered at a frequency of 1000 cycles, at which the effects of the series coupling capacitor C and of the shunt capacitance C_g' of Fig. 7-5 are negligible. This gives the circuit of Fig. 7-6a. Also, assume that the effect of R_g is negligible. The current that flows from the generator is $I = \mu E_g/r_p$, if the effect of the load resistance (which *in pentode amplifiers* is low compared to the plate resistance) is neglected. Then, this current is considered to flow through the load resistance R_L, and the IR_L drop is the approximate amplified output voltage, or $E_g' = (\mu/r_p)(E_gR_L) = g_{pg}E_gR_L$, and $A_v = g_{pg}R_L$. The approximation can be made more exact in an actual amplifier by using the equation

$$A_v = g_{pg}R_e \qquad \text{7-30a}$$

where $R_e = R_LR_g/(R_L + R_g)$. *Equation 7-30a should not be used with triode*

* Brief statements of Thévenin's and Norton's theorems as they apply in this instance are as follows: *Thévenin's theorem:* The current that flows through an impedance Z_L connected to two terminals of a generator equals the open-circuit voltage E_{oc} at the terminals divided by the sum of the internal generator impedance plus the connected load impedance. That is, $I_L = E_{oc}/(Z_g + Z_L)$. It is assumed that the generated electromotive force (which equals E_{oc}) remains constant. *Norton's theorem:* The current that flows through an impedance Z_L connected to two terminals of a generator of internal impedance Z_g is the same as if the impedance Z_L were connected to a constant-current generator whose generated current is the same as the current that flows through the generator terminals when they are short-circuited, the constant-current generator being in parallel with an impedance equal to the internal impedance Z_g. Using a Norton's theorem circuit (as defined), $E_L = I_{sc}Z_e = \frac{E_{oc}}{Z_g} \cdot \frac{Z_gZ_L}{Z_g+Z_L} = \frac{E_{oc}Z_L}{Z_g+Z_L}$ The load current $I_L = \frac{E_L}{Z_L} = \frac{E_{oc}Z_L}{Z_g+Z_L} \cdot \frac{1}{Z_L} = \frac{E_{oc}}{Z_g+Z_L}$, which is the same result as obtained by Thévenin's theorem. Referring to equation 7-30, $A_v = \frac{E_g'}{E_g} = g_{pg}Z_e$, and $E_g' = E_gg_{pg}Z_e$. In this relation, $I = g_{pg}E_g$ is the alternating current that will flow through a tube with no external load in the plate circuit; that is, it is the short-circuit current of a tube.

resistance-coupled amplifiers. With triodes, the effect of the load resistance on the alternating plate current *cannot* be neglected.

7-12. *RESISTANCE-COUPLED AMPLIFIER CHARACTERISTICS*

The calculations made for resistance-coupled amplifiers using small-signal *triodes* indicate that the voltage amplification is almost constant over the audio-frequency range and beyond. Also, the calculations show that little phase shift is caused by the coupling networks over a wide range of frequencies. Typical values of the load resistor R_L, the coupling capacitor C, and the grid resistor R_g used at audio-frequencies were given. The effects of variations in the magnitudes of these values are readily determined by the equations. In general, with *triodes,* R_L is three to five times r_p, and R_g is about five times R_L. Satisfactory results are obtainable even with wide variations in these values.

The calculations made for resistance-coupled amplifiers using typical small-signal *pentodes* (similar calculations apply to tetrodes) indicate that the voltage amplification is reasonably constant over the audio-frequency range. The useful frequency band is not so wide as with triodes; but the amplification is large in comparison. It was shown that the phase shift is not excessive over the audio-frequency range. For pentodes, R_L is usually about one-fifth of r_p, and R_g is about twice R_L. Wide variations in these values give satisfactory results.

Attention is called to the fact that the resistance-capacitance coupling network causes little phase shift in the midfrequency region where the amplification is essentially constant. If a curve for voltage amplification versus frequency is plotted, there is little phase shift over the "flat" portion of the curve. Over this region the equivalent coupling circuit for the amplifier is essentially resistance, which will cause no phase shift. It is important to remember that the tube with resistance load causes a phase shift of 180°. The phase shift at *low* frequencies caused by the coupling network is *added* to this 180° shift to determine the total phase shift of the stage. The phase shift at *high* frequencies caused by the coupling network is *subtracted* from this 180° shift to determine the total shift of a stage. At the *low*-frequency limit the amplified output voltage E_g' *leads* the input voltage E_g by a total maximum angle of 270°. At the *high*-frequency limit the amplified output voltage E_g' *leads* the input voltage E_g by a total minimum angle of 90°.

A peculiar type of distortion may occur in resistance-coupled amplifiers if the grid of a tube is driven positive by a momentary overload. The positive grid may draw a large electron current from the space-charge region, and this will charge negatively the coupling capacitor C of Fig. 7-4. This charge will but slowly "leak" off to the cathode through R_g. Thus the grid may be maintained at a high negative potential for some time after the momentary overload has passed. This may "block" the tube, making it inoperative for a brief interval of time, and may cause noticeable distortion. Low values of C and R_g are of advantage in reducing this effect.

7-13. *IMPEDANCE-COUPLED AUDIO VOLTAGE AMPLIFIERS*

Such an amplifier uses an inductor of 50 to 100 henrys as a load for the tubes, replacing load resistor R_L of Fig. 7-4. An advantage of the impedance-coupled amplifier is that the direct-current resistance of the coupling inductor is low; almost the entire supply voltage E_{bb} is impressed between the plate and cathode. The operation of this amplifier is similar to the operation of the resistance-coupled type. The inductor offers high impedance to the alternating component of the plate current, and a large IZ_L signal-voltage drop will exist across this inductor. The voltage amplification and other characteristics can be calculated in the same general way as for resistance-coupled tubes. As an approximation, R_L may be replaced in equations such as 7-11 by the reactance $j\omega L$ of the inductor. The resistance and distributed capacitance of the inductor may be neglected. The impedance-coupled amplifier is not extensively used. By connecting a capacitor across the inductor, the amplifier can be made selective so that it will amplify well over only a narrow band.

7-14. *AUDIO-FREQUENCY TRANSFORMER-COUPLED VOLTAGE AMPLIFIERS*

The vacuum tubes of an amplifier sometimes are connected with transformers as in Fig. 7-10 so that the output of one tube is fed into the following tube for further voltage amplification. The tubes are operated in class A1; hence, negligible grid current flows. The **input transformer** usually steps up the voltage impressed on it, and isolates the first tube from direct volt-

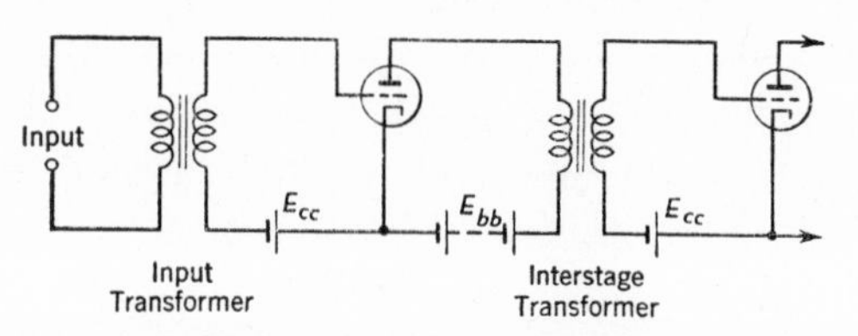

FIG. 7-10. A two-stage transformer-coupled class-A audio-frequency amplifier.

ages the signal source may contain. The turns ratio is determined by the type of signal source, and may be one primary to 20 secondary turns for low-impedance low-voltage sources.

The **interstage transformer** steps up the voltage impressed across the primary and impresses the voltage between grid and cathode of the following tube for further amplification. The interstage transformer also isolates the grid of the second tube from the direct plate voltage of the first. A ratio of one primary turn to three secondary turns is common. Although audio-frequency transformer-coupled voltage amplifiers once were widely used, resistance-coupled amplifiers usually are employed for this purpose.

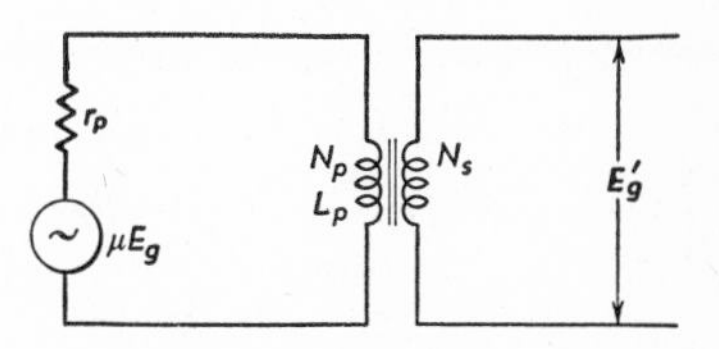

FIG. 7-11. Circuit for computing the approximate voltage amplification per stage of Fig. 7-10.

Amplification of AF Transformer-Coupled Amplifiers. Because negligible grid current flows, the grid-bias voltage equals E_{cc}. Because the resistance of the primary of interstage transformers is low, the plate-to-cathode voltage is E_{bb}, approximately. Although batteries are indicated in Fig. 7-10, these voltages may be obtained by self-biasing means (section 8-10) and from a rectifier-filter combination.

If the input impedance of the grid circuit connected to the secondary is assumed to be infinite, if the transformer is assumed to dissipate no energy, if distributed capacitances are neglected, and if the turns ratio is $N = N_s/N_p$, a stage of audio amplification can be represented by Fig. 7-11. The current magnitude in the primary is $I = \mu E_g/\sqrt{r_p^2 + (\omega L_p)^2}$. The voltage across the primary is $E_p = IX_L = I\omega L_p$. The voltage induced *in series* in the secondary and appearing as an open-circuit voltage E_g' between the grid and cathode of the next tube is $E_g' = I\omega LN$. The voltage amplification is

$$A_v = \frac{E_g'}{E_g} = \frac{\mu\omega L_p N}{\sqrt{r_p^2 + (\omega L_p)^2}} \qquad 7\text{-}31$$

Transformer-coupled audio voltage amplifiers usually employ triodes having a plate resistance of about 10,000 ohms, and an amplification factor of about 10. Cores of excellent magnetic characteristics [10] are used. The primary self-inductance is high, and at the mid-audio-frequency range of about 1000 cycles, it may be assumed that r_p is negligible compared to ωL_p, in which instance the voltage amplification is $A_v = \mu N$, approximately.

An extensive consideration of audio-frequency transformer-coupled voltage amplifiers has been given in preceding editions of this book, and in other books.[11] Such treatments are based on the equivalent circuit of a transformer.[12,13,14] The measurements of the circuit constants must be made in accordance with approved procedures,[15,16] one important reason being that cores of ferromagnetic materials are used. The present limited use of this voltage amplifier does not justify further consideration.

7-15. *RADIO-FREQUENCY VOLTAGE AMPLIFIERS*

In the preceding pages *audio-frequency* voltage amplifiers were considered; the following pages will treat *radio-frequency* voltage amplifiers. These are of two general types: first, *untuned* amplifiers for amplifying signals distributed over a wide frequency range; second, *tuned* amplifiers for amplifying only those signals within a desired frequency band, which often is, relatively speaking, narrow in width.

At audio-frequencies the interelectrode capacitances of a tube may be neglected in many instances. At radio-frequencies of, perhaps, 100,000 cycles and above, the interelectrode capacitances are troublesome. One effect is that these capacitances may cause the amplifier to oscillate. Unwanted oscillations in radio-frequency amplifiers may be prevented, or at least minimized, in several ways. First, if triodes are used, they are neutralized (Chapter 9) by special circuit arrangements, or they are operated with grounded grids instead of with grounded cathodes, as is so common. Second, tetrodes and pentodes are used instead of triodes to avoid oscillations. The screen grid in the tetrode, and the screen and suppressor grids in the pentode, shield the control grid from the signal variations in the plate circuit. Furthermore, these grids usually are connected to the cathode directly, or through capacitors of low reactance at the frequencies involved; hence, the instantaneous potential of these intervening electrodes cannot vary at a radio-frequency rate. As a result of the shielding and "grounding" effects of these electrodes, signal feedback from the plate to the control-grid circuit is so small that oscillations usually do not occur.

Tetrodes and pentodes are used so widely for radio-frequency amplification that amplifiers using these tubes will be considered more completely than radio-frequency amplifiers using triodes.

7-16. *RADIO-FREQUENCY RESISTANCE-COUPLED VOLTAGE AMPLIFIERS*

Resistance-coupled voltage amplifiers are used with signals classified as radio-frequencies (section 7-4). **Video amplifiers** are used in television to amplify the so-called **video signal** containing the image-signal information produced by a television camera, and for amplifying the received video signal for driving the cathode-ray tube in a television receiver. A video band may extend from about 30 cycles to 4.5 megacycles. The preceding discussions for resistance-coupled amplifiers apply. However, such amplifiers do not amplify satisfactorily over the entire video-signal band. For video-signal voltage amplification the basic resistance-coupled amplifier must be "compensated" at both low and high video frequencies, and are referred to as **compensated amplifiers**, and as **wide-band amplifiers.**[17,18,19,20]

Small-signal pentodes of special design [21] to give low electrode capacitances and high transconductance often are used. The circuits of Figs. 7-4, 7-5, and 7-6 apply, but it must be remembered that pentodes having plate resistances of about one megohm are used in the wide-band amplifiers now being considered. Because of the high plate resistance, the constant-current viewpoint given at the close of section 7-11 may be used to advantage. The voltage amplification at the "middle" of the video band (say at about 50,000 cycles) is calculated readily by this method.

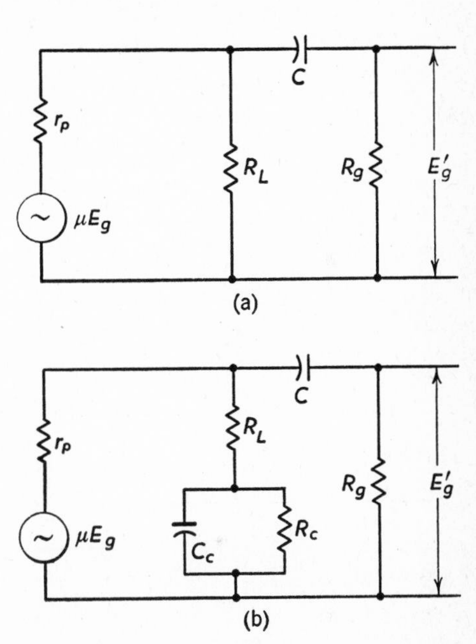

FIG. 7-12. Low-frequency compensation in a wide-band amplifier. The amplification at low frequencies is increased by the use of the parallel circuit C_c–R_c.

Low-Frequency Compensation. Equations 7-21 and 7-22 apply at low frequencies to the amplifier equivalent circuit of Fig. 7-12a *without* compensation. A study of the circuit and the equations indicates that for good low-frequency response C should have high capacitance and R_L high resistance. The insulation resistance of C should be very high, the inductance of the leads and electrodes should be low, and consideration should be given to blocking (section 7-12). As will be explained later, good high-frequency response dictates that R_L should be of low resistance compared to the value that ordinarily would be used in a pentode R-C amplifier; assume that calculations [20] indicate that R_L should be 10,000 ohms (an approximate figure only).

The circuit of Fig. 7-12b is arranged for low-frequency compensation by the parallel C_c–R_c circuit. If R_c is about 10,000 ohms, and if the capacitance of C_c is such that its reactance is large at the low frequencies at which compensation is desired, then the effective load resistance at very low frequencies is $R_L + R_c$. Since the plate resistance of r_p is so great that constant plate signal current may be assumed, at very low frequencies the increased effective load will *tend* to cause a greater percentage of the plate signal current to flow through capacitor C. As has been explained in preceding pages, the increase in the reactance of capacitor C is an important factor in reducing the voltage amplification at low frequencies; hence, the C_c–R_c combination produces compensation, the desired characteristics being calculable.[20] At all but very low frequencies, the reactance of C_c is negligible, and hence R_c is in effect short-circuited, the plate load resistance then becoming R_L.

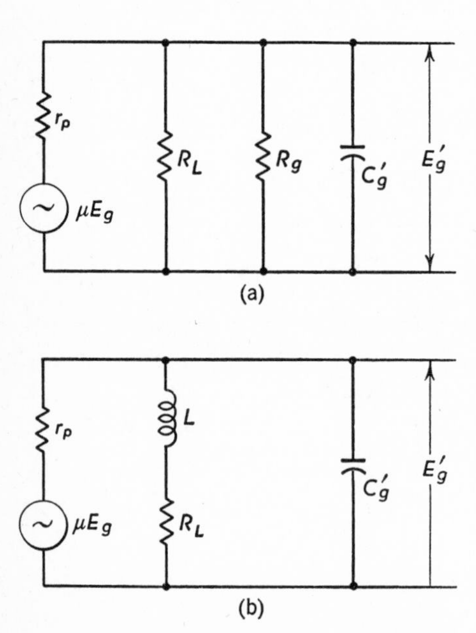

FIG. 7-13. High-frequency compensation in a wide-band amplifier. The amplification at high frequencies is increased by the addition of coil L. The effect of R_g is negligible.

High-Frequency Compensation. Equations 7-27 and 7-28 apply at high frequencies to the amplifier equivalent circuit of Fig. 7-13a *without* compensation. Assuming constant plate signal current, and recalling that the reactance of C_g' will decrease as the frequency is increased, it follows that the lower the value of R_L, the less important will be the *effect* of C_g' at the high video frequencies. Of course a low value of R_L will result in low overall amplification, but the response curve of amplification versus frequency will be "flatter" than the curve with R_L of greater resistance.

To obtain reasonable amplification at midfrequencies, and to increase the amplification at high frequencies, a coil L of low inductance is connected in series with R_L of Fig. 7-13b. At the high video frequencies the inductive reactance of this coil is relatively large, and this increases the impedance of the load composed of R_L and L in series. Again assuming constant plate signal current, the effect of L at high frequencies is to cause a larger percentage of the signal current through capacitor C_g', and thus to offset the decrease in reactance of C_g' at these high video frequencies. As previously explained, the decrease in the reactance of C_g' at high frequencies is an important factor in determining the high-frequency response. An alternate viewpoint is that the branch of Fig. 7-13b containing L and R_L and the branch containing C_g' become parallel resonant at the high video frequencies, thus sustaining the amplification. Calculations can be made [20] to determine the correct values to use to meet desired amplification and phase-shift conditions.

Other methods of high-frequency compensation are possible.[22,23,24] For instance, a small inductor may be inserted in series with the coupling capacitor C (Fig. 7-12a), and hence effectively in series with C_g' of Fig. 7-13b. The inductance of L is of such value that series resonance between L and C_g' occurs at the high video frequencies. Then, the impedance of this branch will fall, the current will rise, and this rise will tend to increase the voltage drop across C_g', which is the voltage output of the stage. The tendency for the voltage to rise will be offset by the decrease in the reactance of C_g' at the higher video frequencies, and with the proper selection of components, the high-frequency response can be extended. Overcompensation and resulting distortion must be avoided.

7-17. *RADIO-FREQUENCY IMPEDANCE-COUPLED VOLTAGE AMPLIFIERS*

If a suitable inductor is substituted for the load resistors R_L of a resistance-coupled radio-frequency amplifier, an impedance-coupled amplifier results. The amplifier may be *untuned,* or may be *tuned* by connecting a capacitor across the inductor. It should be mentioned that an inductor without a capacitor connected in parallel may exhibit resonance characteristics at certain frequencies because of the distributed capacitance between turns. The inductor may be without a core, or may have a ferromagnetic core of special material for use at radio frequencies.[25,26,27]

7-18. *RESONANT PARALLEL CIRCUITS*

Because resonant parallel circuits are extensively used in radio-frequency voltage amplifiers, the characteristics of these circuits now will be reviewed. Thus, in Fig. 7-14 are shown the equivalent-series input reactance and impedance that would be measured with a radio-frequency impedance bridge when connected at points 1–2. At frequencies considerably *below resonance,* the inductive reactance is low, and the capacitive reactance is high. For a typical inductor, the current through the inductor will lag by almost 90° *the voltage* across points 1–2. The current through the capacitor will lead by 90° the voltage across points 1–2 and will be less in magnitude than the inductor current. The resultant current flowing into the circuit from a signal source at 1–2 will be lagging, and the input impedance of the circuit will be inductive. At frequencies considerably *above resonance,* the 90° leading current through the capacitor will exceed in magnitude the

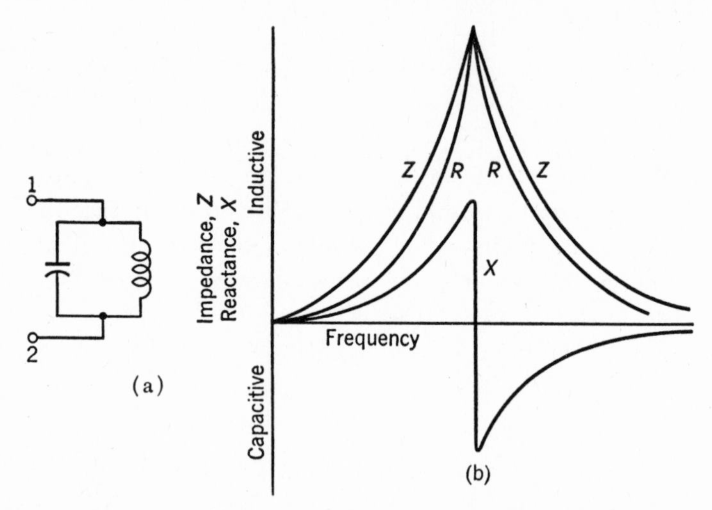

FIG. 7-14. Parallel resonant circuit (*a*), and characteristics (*b*).

current through the inductor that lags the impressed voltage by about 90°, the resultant current will be leading, and the input impedance will be capacitive. At parallel resonance (defined [4] as the frequency at which the impressed voltage and input current are in phase), the resultant current is small, and since it is in phase, the equivalent input impedance is a high value of resistance, as Fig. 7-14 shows.

Air and mica commonly are used for dielectrics in radio-frequency capacitors. For usual operation at medium and high radio-frequencies, the energy dissipation in such capacitors may be neglected, and the capacitors considered to have reactance only. At these same frequencies, inductors cause dissipation that may not be negligible, and hence as circuit elements they act like inductance and resistance in series. For these reasons it was stated in the preceding paragraph that for the capacitor the current led the voltage by 90°, and that for an inductor the current lagged by *almost* 90°. For an inductor, the ratio of inductive reactance ωL to effective resistance R gives a **quality factor Q.** Assuming that a parallel-resonant circuit contains a lossless capacitor, and that the inductor has little loss, the impedance of Fig. 7-14 will at resonance rise to a very high value of resistance, and the curve will be sharp. For such a circuit, $Q = \omega L/R$ will be a large number, and the circuit will be termed a **sharply tuned, or high-Q, circuit.** However, if the losses in the inductor are high, or if the circuit radiates or dissipates considerable energy, the equivalent impedance will be a lower value of resistance at resonance, the circuit will *not* be sharply tuned, and it will be termed a **low-Q circuit.**

If the inductor of a resonant parallel circuit is coupled *inductively* to a secondary closed circuit in which current may flow, the effect of the secondary current on the primary circuit is as if an impedance

$$Z_{reflect} = (\omega M)^2/Z_s \qquad 7\text{-}32$$

were connected in series with the primary coil. In this equation $\omega = 2\pi f$, M is the mutual inductance (henrys) between the two coils, and Z_s is the *total* series impedance of the *entire* secondary circuit, expressed in both magnitude and angle.

7-19. *RADIO-FREQUENCY TRANSFORMER-COUPLED VOLTAGE AMPLIFIERS*

Untuned Primary and Secondary. The small-signal pentode of Fig. 7-15 is connected to a radio-frequency transformer with the primary and secondary untuned. If wiring capacitances and distributed capacitances between turns are neglected, the magnitude of the primary alternating signal current will be $I_p = \mu E_g/\sqrt{r_p^2 + (\omega L_p)^2}$. The magnitude of the voltage induced in series in the secondary is $E_g' = I_p\omega M$, and when the equation for I_p is substituted, the voltage amplification ratio is

$$\frac{E_g'}{E_g} = \frac{\mu\omega M}{\sqrt{r_p^2 + (\omega L_p)^2}} \qquad 7\text{-}33$$

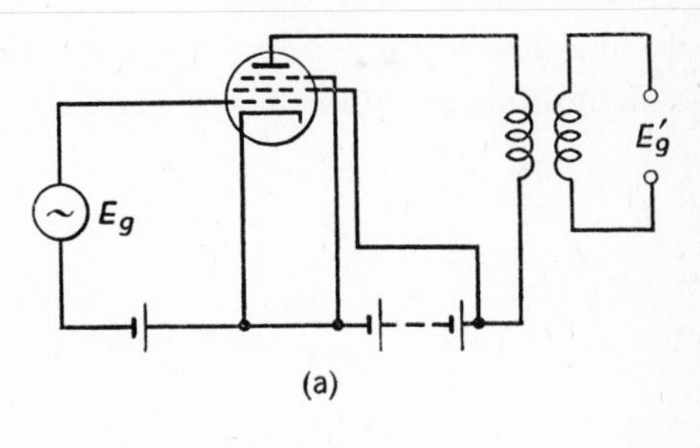

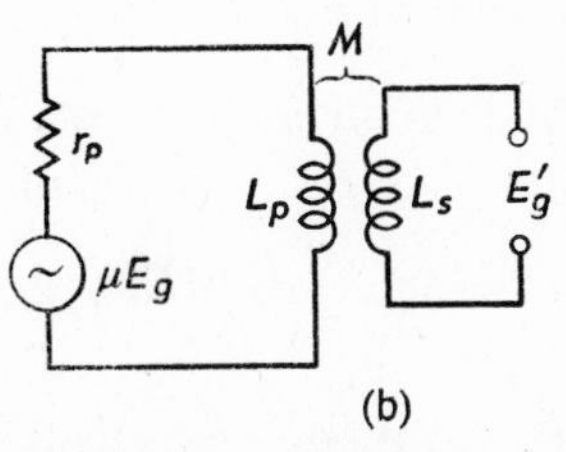

FIG. 7-15. Actual and equivalent circuits for an *untuned* radio-frequency transformer-coupled amplifier.

Untuned Primary and Tuned Secondary. The actual and equivalent circuits are shown in Fig. 7-16. Variable capacitor C includes all secondary-circuit capacitances. The amplified output voltage E_g' will equal the product of the secondary current I_s and the reactance X_C of the capacitor. The value of the secondary current I_s will equal the voltage induced in *series* in the secondary windings divided by total secondary impedance. Since the secondary will be adjusted to series resonance so that the amplifier will be selective, the series impedance of the secondary will approximately equal R_s, the resistance of the secondary. Thus

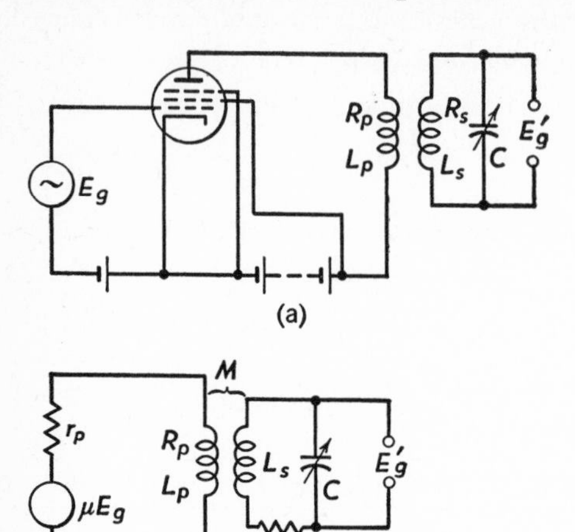

FIG. 7-16. Actual and equivalent circuits for a radio-frequency transformer-coupled voltage amplifier with tuned secondary.

$$E_g' = I_s X_C = \frac{E_s}{R_s} X_C = \frac{\omega M I_p X_C}{R_s} \qquad 7\text{-}34$$

where the magnitude of the voltage induced in the secondary is $E_s = \omega M I_p$.

The value of the primary current I_p equals the amplified voltage μE_g divided by the total impedance of the primary circuit. In calculating this total impedance, the impedance of the primary of the coil is assumed negligible in comparison with the plate resistance and reflected impedance. The impedance reflected into the primary by the tuned secondary will be a value of resistance from equation 7-32 equal to $(\omega M)^2/R_s$. Making these substitutions in equation 7-34 gives

$$E_g' = \frac{\mu E_g M}{(R_s r_p + \omega^2 M^2)C} \qquad \text{or} \qquad E_g' = \frac{\mu E_g \omega^2 M L_s}{R_s r_p + \omega^2 M^2} \qquad 7\text{-}35$$

where $X_L = \omega L_s$ is substituted for its equal (at series resonance) $X_C = 1/(\omega C)$. The voltage amplification is

$$A_v = \frac{E_g'}{E_g} = \frac{\mu M}{(R_s r_p + \omega^2 M^2)C} \qquad \text{or} \qquad A_v = \frac{\mu \omega^2 M L_s}{R_s r_p + \omega^2 M^2} \qquad 7\text{-}36$$

A further simplification may be made because $\omega^2 M^2$ usually is negligible compared to the plate resistance r_p of a pentode. Thus

$$A_v = \frac{\mu \omega^2 M L_s}{r_p R_s} = \frac{g_{pg} \omega^2 M L_s}{R_s} = g_{pg} \omega M Q \qquad 7\text{-}37$$

where g_{pg} is the grid-plate transconductance and equals μ/r_p, and where $Q = \omega L_s/R_s$. The circuit will be tuned to one frequency, but will amplify at adjacent frequencies, the band width depending on the sharpness of tuning.

Resistance may be added in series, or in parallel with the secondary to widen the band passed.

Tuned Primary and Tuned Secondary. This is the most common type of radio-frequency transformer-coupled voltage amplifier. It is highly selective, yet has good band width. The voltage ratio of the transformer itself is not high, because voltage-amplifying pentodes are used and these have high amplification factors, and because it is better to design the transformers for high selectivity and good band width than for high voltage ratio. The circuits are tuned by varying the small "trimmer" capacitors indicated in Fig. 7-17, or by varying the position of the core, if coils with ferromagnetic cores are used. Capacitors C_p and C_s include all stray, wiring, and input capacitances.

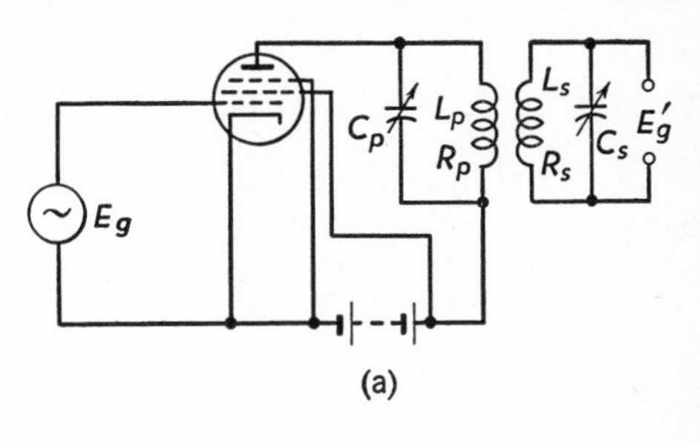

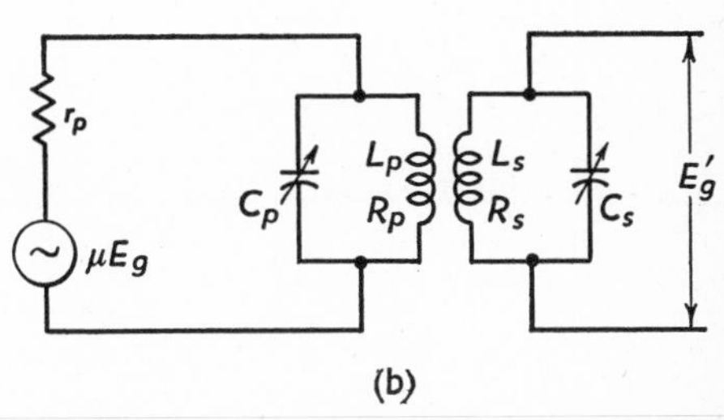

FIG. 7-17. A radio-frequency transformer-coupled voltage amplifier with tuned primary and secondary.

The amplified output signal voltage will equal the secondary current I_s multiplied by the reactance X_{Cs} of the capacitor tuning the secondary. This secondary current I_s will equal the voltage induced in *series* in the secondary winding of the transformer divided by the total secondary impedance, which will equal R_s when the secondary is tuned to resonance. The voltage induced in the secondary will equal $\omega M I_{pL}$, where I_{pL} is the current in the primary of the transformer. That is,

$$E_g' = I_s X_{Cs} = \frac{\omega M I_{pL} X_{Cs}}{R_s} \qquad 7\text{-}38$$

The plate current I_p flowing through the source of voltage μE_g can be assumed to be limited only by the plate resistance r_p of the tube because the plate resistance of a pentode is so high. This current will cause a voltage drop $I_p Z_e$, where Z_e represents the equivalent *parallel* impedance of the tuned transformer primary. This tuned circuit consists of impedance Z_{Cp} of capacitor C_p in parallel with Z_p, where Z_p is the impedance of the transformer primary *plus* the impedance reflected into the primary by the presence of the secondary, $(\omega M)^2/R_s$. This parallel impedance can be found readily by the usual relation $Z_e = Z_{Cp}Z_p/(Z_{Cp} + Z_p)$. The current I_{pL} in the transformer primary will equal the voltage I_pZ_e divided by Z_p, the impedance of the transformer primary *plus* the reflected secondary impedance. Thus,

$$I_{pL} = \frac{I_pZ_e}{Z_p} = \frac{\mu E_gZ_e}{r_pZ_p} \qquad \text{7-39}$$

Substituting this expression in equation 7-38 gives

$$E_g' = \frac{\omega MX_{Cs}\mu E_gZ_e}{R_sr_pZ_p} \qquad \text{7-40}$$

and the voltage amplification is

$$A_v = \frac{E_g'}{E_g} = \frac{\mu\omega MX_{Cs}Z_e}{r_pR_sZ_p} \quad \text{or} \quad A_v = \frac{g_{pg}\omega MX_{Cs}Z_e}{R_sZ_p} \qquad \text{7-41}$$

These equations can be simplified by assuming identical primary and secondary windings.

7-20. *VOLTAGE AMPLIFICATION AND BAND WIDTH*

In addition to amplifying, tuned radio-frequency amplifiers provide selectivity. They amplify well only those frequencies within the band to which they are tuned, and hence select these frequencies and reject all others. As will be explained in Chapter 12, a typical amplitude-modulated radio-broadcast transmitter radiates a signal having a band width of about 10,000 cycles centered around a carrier frequency of, say, 1,000,000 cycles. An amplifier to select and amplify this signal, and reject all other signals, should have the ideal characteristics of Fig. 7-18a. However, an actual tuned radio-frequency amplifier would have characteristics more like Fig. 7-18b.

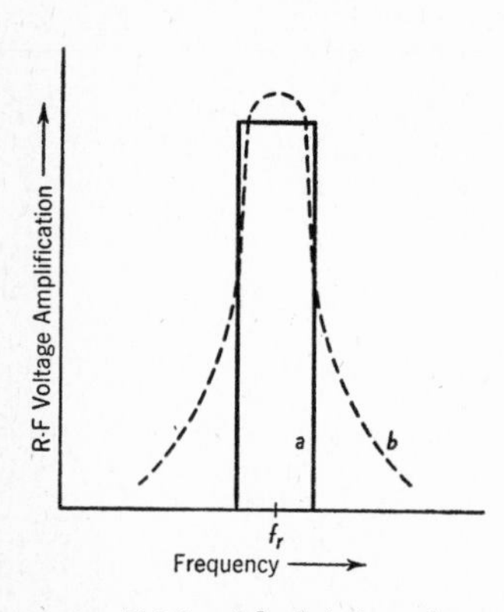

FIG. 7-18. Ideal (a) and actual (b) approximate characteristics of an amplifier for typical amplitude-modulated radio-frequency signals. (Scales approximate.)

The **band width** of an amplifier (usually and arbitrarily) is assumed to be bounded by those frequencies at which the power output is one-half of the output at the midfrequency such as f_r. From section 7-2, this corresponds to a 3-decibel decrease. Because power is proportional to voltage squared, the corresponding voltage decrease would be 0.707 of the midfrequency voltage output. An amplifier that will amplify a definite band of frequencies only is called a **band-pass amplifier.** Wide-band tuned radio-frequency amplifiers that will have characteristics better than those depicted by Fig. 7-18b often are desired. As mentioned in the preceding section, connecting resistance across the tuned circuits, or in series with them, improves the band width, but at the expense of amplification.

Overcoupling can be used to increase band width. A complete analysis of this is lengthy, and the subject must be summarized.[22,23] Equation 7-32 shows that the magnitude and the angle of the impedance reflected by the secondary into the primary for two coupled circuits is determined by the frequency. An examination of this equation discloses that *below* the resonant frequency f_r, where the total secondary impedance Z_s will be capacitive, the impedance reflected into the primary will be inductive. Also, equation 7-32 shows that *above* the resonant frequency f_r, where Z_s is inductive, the impedance coupled into the primary is capacitive. Therefore, at some frequency below f_r (the resonant frequency for the primary and secondary individually) where the inductive reactance of the primary alone is too *small* for resonance to occur, the reflected inductive reactance will add to the primary inductive reactance and will permit the phenomenon of resonance to occur. Also, at some frequency above f_r where the inductive reactance of the primary alone is so large that resonance cannot occur, the capacitive reactance coupled by the secondary into the primary will neutralize some of the excess primary inductive reactance and permit resonance to occur. Thus, theoretically, coupled circuits have three points of resonance. If the coupling is small, the effect is not noticeable, and the secondary currents and the output voltages for an amplifier using such coupling would appear *somewhat* as in Fig. 7-19a. If the coupling is critical,[4,22,23] the characteristics of the amplifier will be broadened somewhat as shown in Fig. 7-19b. If the two circuits are overcoupled,[22,23] the three resonance points are evident, and the output voltage would appear as in Fig. 7-19c, giving an amplifier with wide-band characteristics. **Critical coupling** is the degree of coupling at which maximum energy transfer occurs between two separate circuits tuned to the same frequency. For **loose coupling** the coupling is less than the critical value, and for **close coupling** it is greater.

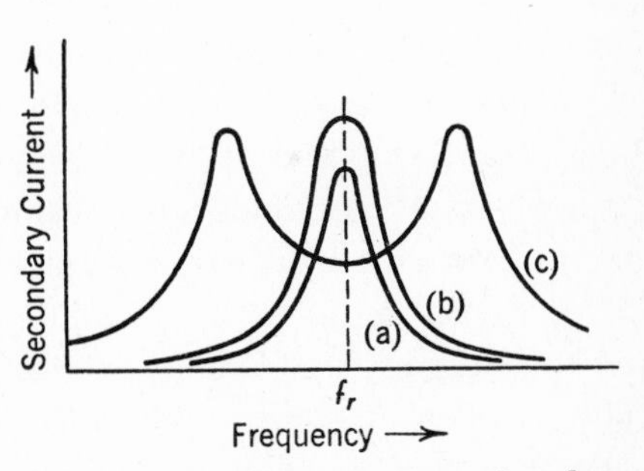

FIG. 7-19. Characteristics of inductively-coupled resonant circuits for (a) loose, (b) critical, and (c) close coupling. (Scales approximate.)

Staggered tuning [23,28,29,30] also is used to provide sufficient band width for signals such as in television. With this method, two or more stages connected in tandem (or cascade) are tuned to different frequencies, the over-all characteristics being the composite of the separate stages.

The **gain-band product**,[31] or **gain-band-width product**,[32] is an important criterion when considering the performance of an electron device such as an amplifier. Thus, a given tube in a sharply tuned circuit will provide much signal amplification over a narrow band. Or, the same tube in a broadly tuned circuit will provide amplification over a wide band, but with reduced gain. In general, high amplification is obtained at the expense of band width, and vice versa.

7-21. *NOISE* [15]

A completely shielded amplifier, with no input voltage applied, will produce random output voltages that cause noise to exist in the output of a telephone receiver connected to the amplifier. **Noise** is defined [1] as an "undesired disturbance within the useful frequency band." **Interference** is defined [1] as "undesired disturbances within the useful frequency band produced by other services." **Electrical noise** is defined [1] as "unwanted electrical energy other than crosstalk present in a transmission system." The common sources of electrical noise in vacuum-tube amplifiers will be considered.

Thermal noise (section 1-26) is an electrical noise usually attributed to the random motion of electrons in a circuit element such as a resistor.

Tube noise is a random electrical noise generated within the vacuum tubes used in an amplifier. Tube noises are caused by (1) shot-effect (section 2-12), consisting of random variations in thermionic emission; (2) **partition noise** resulting from random variations in the division of electron current between two or more positive electrodes such as the plate and screen grid of a pentode; (3) induced grid noises attributed to random fluctuations in the number of electrons passing close to a grid; (4) random variations in secondary emission; and (5) ionization of residual gas.

The electrical noise caused by an amplifier is an important factor in determining the weakest signal that can advantageously be amplified. For instance, if the first stage of an amplifier produces a certain amount of electrical noise in the same frequency band as the signal, and if the impressed signal magnitude is such that the strength of the amplified signal output is about the same as the noise, then, after amplification, the output composed of equal amounts of noise and signal would be useless for most purposes. Generally, it is important to keep the **signal-to-noise ratio** high. If the generated noise cannot be reduced, then the magnitude of the input signal must be increased, perhaps by radiating more signal power at a radio transmitter, or perhaps by installing a better receiving antenna.

The **noise figure** of an amplifier is a measure of the reduction in signal-to-noise ratio caused by noise generated within the amplifier.[15,33] Standardized means of measuring the noise figure have been developed.[15,33,34]

7-22. *DIRECT-COUPLED AMPLIFIERS*

The amplifiers considered in the preceding pages have low-frequency and high-frequency limits beyond which they will not amplify effectively. Sometimes it is desired to amplify signals of a few cycles, even as low as zero cycles; that is, direct voltages and currents. As explained in section 7-10, the low-frequency cutoff is caused by the coupling capacitor *C* of Figs. 7-4, 7-5, and 7-6b. In fact, if capacitor *C* is omitted, the circuit becomes a **direct-coupled amplifier,** which will have a cutoff at zero cycles, and which will amplify direct voltages and currents. In fact, if it were not for stray inductance and capacitance, and other such effects, the amplifier would, theoretically, amplify signals of any frequency. If the coupling capacitor *C* is removed, the plate and grid of the two tubes will be connected together, and some provision must be made to ensure that the grid of the second tube is held at the correct direct potential with respect to its cathode for class-A1 operation.

The basic circuit arrangement is shown in Fig. 7-20. Direct voltage E_{c2} must be less in magnitude than direct voltage $I_b R_{L1}$ by an amount equal to

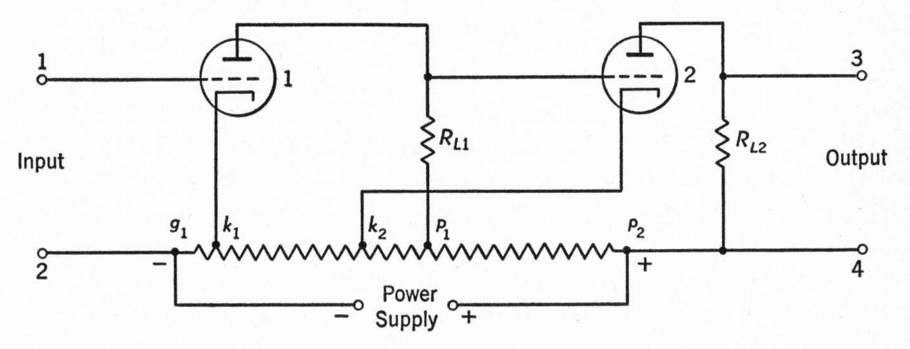

FIG. 7-20. A direct-coupled direct-current amplifier.

the desired grid-bias value. Then, if the amplifier is driven by a small low-frequency voltage of effective value E_{g1} volts, the magnitude of the signal current in the plate circuit of the first tube will be

$$I_p = \frac{\mu E_{g1}}{r_p + R_L} \tag{7-42}$$

and the amplified signal voltage across load resistor R_{L1} will be

$$E_{L1} = I_p R_{L1} = \frac{\mu E_{g1} R_{L1}}{r_p + R_{L1}} \tag{7-43}$$

and the voltage ratio of the first stage will be

$$A_v = \frac{E_{L1}}{E_{g1}} = \frac{\mu R_{L1}}{r_p + R_{L1}} \tag{7-44}$$

The voltage ratio of the second stage may be computed in a similar manner. The phase shift will be, theoretically, 180° per stage.

A direct-coupled amplifier may be arranged for operation from a power supply, an example being shown in Fig. 7-20. This is known as the Loftin-White amplifier,[35] after its originators. The several electrodes are connected to different points on the voltage divider as indicated. When no direct voltage is applied between input terminals 1–2 (such as would be the case if they were short-circuited), the grid of tube 1 is held at a negative bias equal to the voltage between points g_1 and k_1 of the voltage divider. A direct current I_{b1} will flow in the plate circuit of tube 1, and through the load or coupling resistor R_{L1}. The potential between the plate and cathode will be the voltage between k_1 and p_1 minus the $I_{b1}R_{L1}$ voltage drop.

If the cathode of tube 2 were connected to point p_1, then the grid of tube 2 would be negative by the voltage $I_{b1}R_{L1}$, but it would be *too far negative, probably beyond cutoff*. For this reason the cathode of tube 2 is connected to point k_2, making the cathode negative by such a value that $I_{b1}R_{L1}$ minus the potential between points p_1 and k_2 gives the correct negative bias for the grid of tube 2. The plate of tube 2 is positive with respect to its cathode by the voltage between points k_2 and p_2 minus the $I_{b2}R_{L2}$ drop in the resistor in the plate circuit, or other device if a resistor is not used. It will be noted that *the cathodes of the two tubes are at different potentials.* These cathodes must, therefore, be isolated from each other by transformers or by using separate filament batteries. If indirectly heated tubes with separate cathodes are used then a common filament supply can be employed. If this is done, the safe difference of potential between the cathode and heater must not be exceeded. This safe value of voltage depends on the type of tube.

A direct-coupled amplifier will amplify changes in supply voltage. The characteristics of such amplifiers tend to "drift." Balanced circuits and feedback are used to stabilize direct-coupled amplifiers.[36,37,38,39] Stray inductances and capacitances in wiring and tubes limit the high-frequency amplification, and the response to rapid changes in input voltage. To avoid the use of direct-coupled amplifiers a "chopper" sometimes is used to convert weak "direct-current" signals to "alternating-current" signals so that they may be amplified in a conventional resistance-coupled amplifier to avoid the use of a direct-coupled amplifier.[40]

7-23. *DISTRIBUTED AMPLIFIERS*

For the amplifiers that have been considered, the voltage ratios per stage are *multiplied* together to obtain the total gain of a multistage amplifier. For the so-called distributed amplifier,[41,42] the amplifications per tube are *added* to obtain the total amplification. A circuit is shown in Fig. 7-21. This device

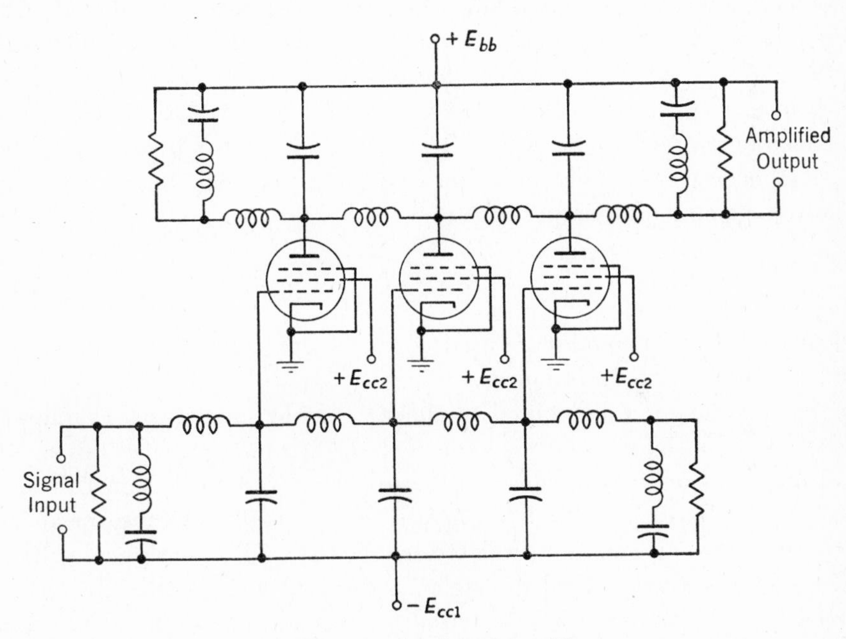

FIG. 7-21. A distributed amplifier.

will amplify from zero to many millions of cycles. One amplifier [42] operated from 20 to 30 megacycles with a variation of less than one decibel from a midfrequency gain of about 8 decibels. Other such amplifiers have been reported that have gains of 10 decibels from 10 kilocycles to 200 megacycles,[43] and 8 decibels from 10 to 360 megacycles.[44] Theory predicts [42] that three tubes, with a maximum total voltage amplification of ϵ, or 2.72 approximately, is the optimum amplification per unit for this amplifier. To obtain a greater amplification, several units (of three tubes each) in tandem, or cascade, may be used, in which event the amplification per unit should be multiplied to obtain the over-all amplification. Of course, if the amplification of each tube is in decibels, then addition should be used to obtain the over-all amplification in decibels.

The importance of adding amplifications per stage is as follows: Suppose that an amplifier composed of separate stages is constructed to have a *very* wide response, and as a result, the ratio of output to input voltage is only 0.9. If three such stages are used, the total amplification will be the product of the amplification of each stage, or 0.73. However, if three tubes are used in a distributed amplifier and the amplification is 0.9 for each tube, the total amplification will be 2.7.

As Fig. 7-21 shows, the tubes are distributed along an artificial transmission line that is in reality a low-pass constant-*k* filter, terminated at each end in *m*-derived half-sections, and resistors equal to the **iterative impedance** of the network.[11,13,45] A typical value of iterative impedance is 200 ohms resistance. Each tube works into a load which is one-half the iterative impedance because of a termination at each end. The cutoff frequency of the filter is made considerably above the maximum frequency to be amplified. The voltage amplification per tube is

$$A_v = \frac{g_{pg}Z}{2} \qquad 7\text{-}45$$

when small-signal pentodes are used, g_{pg} is the grid-plate transconductance, and Z is the iterative impedance.

The operation of a distributed amplifier [42] may be summarized as follows. When an input signal voltage is impressed, the first tube amplifies this signal and impresses an output signal on the upper network. This signal divides equally, part passing to the left where it is dissipated in the resistor at the left end of the line, and part going toward the right. The input signal voltage wave is propagated toward the right on the grid line, and the action explained for the first tube is repeated at the second tube. Because the input and output networks are identical, and the tubes are identical, the signal that the second tube starts toward the right adds in phase with the signal that the first tube previously sent to the right. In this way the signals from successive tubes add to give the final output voltage across the output terminals. The action here discussed has been for voltage amplification, but the distributed amplifier can be designed primarily for power-output purposes.[41,42] The invention of the distributed amplifier is attributed [46] to Percival.

REFERENCES

1. Institute of Radio Engineers. *Standards on Audio Techniques: Definitions of Terms,* 1954. Proc. I.R.E. July 1954, Vol. 42, No. 7.
2. Institute of Radio Engineers. *Standards on Antennas, Modulation Systems, and Transmitters.* 1948.
3. Institute of Radio Engineers. *Standards on Transducers: Definitions of Terms,* 1951. Proc. I.R.E., Aug. 1951, Vol. 39, No. 8.
4. American Institute of Electrical Engineers. *American Standard Definitions of Electrical Terms.* 1941.
5. Institute of Radio Engineers. *Standards on Receivers: Definitions of Terms,* 1952. Proc. I.R.E., Dec. 1952, Vol. 40, No. 11.
6. Pidgeon, H. A., *Simple theory of the three-electrode vacuum tube.* Journal of the Society of Motion Picture Engineers, Feb. 1935, Vol. 24; also, *Theory of multi-electrode vacuum tubes.* Electrical Engineering, Nov. 1934, Vol. 53.
7. Glasgow, R. S. *Principles of Radio Engineering.* McGraw-Hill Book Co.
8. Diamond, J. M. *Maximum output from a resistance-coupled triode voltage amplifier.* Proc. I.R.E., April 1951, Vol. 39, No. 4.
9. Sauber, J. W. *High-frequency characteristics of resistance-coupled amplifiers.* Proc. I.R.E., Jan. 1952, Vol. 40, No. 1.
10. Elmen, G. W. *Magnetic alloys of iron, nickel, and cobalt.* Electrical Engineering, Dec. 1935, Vol. 54, No. 12.
11. Everitt, W. L. *Communication Engineering.* McGraw-Hill Book Co.
12. Lee, R. *Electronic Transformers and Circuits.* John Wiley & Sons.
13. Creamer, W. J. *Communication Networks and Lines.* Harper & Brothers.
14. Lord, H. W. *The design of broad-band transformers for linear electronic circuits.* Trans. A.I.E.E., 1950, Vol. 69.
15. Terman, F. E., and Pettit, J. M. *Electronic Measurements.* McGraw-Hill Book Co.
16. Hague, B. *Alternating Current Bridge Methods.* Pitman & Sons.
17. Wheeler, H. A. *Wide-band amplifiers for television.* Proc. I.R.E., July 1939, Vol. 27, No. 7.
18. Bedford, A. V., and Fredendall, G. L. *Transient response of multistage video-frequency amplifiers.* Proc. I.R.E., April 1939, Vol. 27, No. 4.
19. Bereskin, A. B. *Improved high-frequency compensation for wide-band amplifiers.* Proc. I.R.E., Oct. 1944, Vol. 32, No. 10.
20. Fink, D. G. *Principles of Television Engineering.* McGraw-Hill Book Co.
21. Whyte, J. R. *Choosing pentodes for broad-band amplifiers.* Electronics, April 1952, Vol. 25, No. 4.
22. Terman, F. E. *Radio Engineering.* McGraw-Hill Book Co.
23. Terman, F. E. *Radio Engineers' Handbook.* McGraw-Hill Book Co.
24. Bereskin, A. B. *Cathode-compensated video amplification.* Electronics, June 1949, Vol. 22, No. 6.
25. Shea, H. G. *Magnetic powders.* Electronic Industries, Aug. 1945, Vol. 4, No. 8.
26. Martowicz, C. T. *Powdered iron cores.* Electronic Industries, June 1945, Vol. 4, No. 6.

27. Beller, H., and Altman, G. O. *Radio frequency cores of high permeability.* Electronic Industries, Nov. 1945, Vol. 4, No. 11.
28. Wallman, H. *Stagger-tuned amplifier design.* Electronics, May 1948, Vol. 21, No. 5.
29. Easton, A. *Stagger-peaked video amplifiers.* Electronics, Feb. 1949, Vol. 22, No. 2.
30. Wittenberg, R. C. *Broad-banding by stagger tuning.* Electronics, Feb. 1952, Vol. 25, No. 2.
31. Morton, J. A., and Ryder, R. M. *Design factors of the Bell Telephone Laboratories 1553 triode.* Bell System Technical Journal, Oct. 1950, Vol. 29, No. 4.
32. Young, C. W. *Gain-bandwidth nomograph.* Electronics, Nov. 1950, Vol. 23, No. 11.
33. Institute of Radio Engineers. *Standards on Radio Receivers: Methods of Testing Amplitude-Modulation Broadcast Receivers.* 1948.
34. Federal Telephone and Radio Corp. *Reference Data for Radio Engineers.* 1949.
35. Loftin, D. H., and White, S. Y. *Cascade direct-coupled tube systems operated from alternating current.* Proc. I.R.E., April 1930, Vol. 18, No. 4.
36. Artzt, M. *Survey of d-c amplifiers.* Electronics, Aug. 1945, Vol. 18, No. 8.
37. Williams, A. J., Tarpley, R. E., and Clark, W. R. *D-c amplifier stabilized for zero gain.* Trans. A.I.E.E., 1948, Vol. 67.
38. Van Scoyoc, J. N., and Warnke, G. F. *A d-c amplifier with cross-coupled input.* Electronics, Feb. 1950, Vol. 23, No. 2.
39. Goldberg, E. A. *Stabilization of wide-band d-c amplifiers for zero and gain.* R.C.A. Review, June 1950, Vol. 11, No. 2.
40. Schafer, C. R. *D-c amplifier using air-coupled chopper.* Electronics, March 1950, Vol. 23, No. 3.
41. Ginzton, E. L., Hewlett, W. R., Jasberg, J. H., and Noe, J. D. *Distributed amplification.* Proc. I.R.E., Aug. 1948, Vol. 36, No. 8.
42. Copson, A. P. *A distributed power amplifier.* Electrical Engineering, Oct. 1950, Vol. 69, No. 10.
43. Kennedy, F., and Rudenberg, H. G. *200-mc traveling wave chain amplifier.* Electronics, Dec. 1949, Vol. 22, No. 12.
44. Scharfman, H. *Distributed amplifier covers 10 to 360 mc.* Electronics, July 1952, Vol. 25, No. 7.
45. Albert, A. L. *Electrical Communication.* John Wiley & Sons.
46. Payne, D. V. *Distributed amplifier theory.* Proc. I.R.E., June 1953, Vol. 41, No. 6.

QUESTIONS

1. What is the fundamental difference between voltage amplifiers and power amplifiers?
2. For what purposes is the decibel used? What changes are contemplated?
3. What is distortion, and what are the different types?
4. Referring to the definition of distortion, can you name instances where changes in wave form may be desired?
5. In what applications is phase distortion serious?

6. What is meant by the terms class A, B, and C amplifiers?
7. What is meant by voltage amplification? In what units is it measured?
8. What is the phase relation between the voltage impressed on the grid of a tube and the voltage across a resistance load?
9. Is the input impedance of a vacuum tube independent of frequency?
10. Does the input impedance of a pentode depend on the load impedance in the plate circuit?
11. Referring to section 7-9, what is the significance of a negative resistance?
12. Why are capacitors C_i and C_o used in Fig. 7-4?
13. What important characteristic should be had by capacitor C of Fig. 7-4?
14. If two stages of amplification are "in cascade," how would you find the combined voltage amplification as a ratio, and in decibels?
15. What is meant by the phase shift of an amplifier? What causes phase shift?
16. For a resistance-coupled amplifier, what is the relation between phase shift and frequency distortion?
17. Is it more difficult to obtain in resistance-coupled amplifiers a wide frequency response with a triode or with a pentode?
18. Is the larger percentage of the amplified voltage μE_g available in the output circuit with a triode or a pentode?
19. Why are pentodes so extensively used at radio-frequencies?
20. In section 7-11 it is stated that the voltage amplification has dropped from 150 to 83.8. What would this drop be in decibels?
21. For the resistance-coupled amplifier, what circuit element largely determines the low-frequency response? The high-frequency response?
22. What are the differences between Thévenin's theorem and Norton's theorem?
23. Why should not equation 7-30a be used with triodes?
24. What factors limit the magnitude of the grid resistor R_g that should be used in resistance-coupled amplifiers?
25. In section 7-13 it is stated that an impedance-coupled amplifier can be made selective by connecting a capacitor across the coil in the plate circuit. Why is this true?
26. What are the advantages and disadvantages of transformer-coupled audio-amplifiers?
27. How can the low-frequency response and high-frequency response of a video resistance-coupled amplifier be increased?
28. Referring to section 7-18, what relation exists between band passed and sharpness of tuning?
29. Referring to Fig. 7-20, how does the amplifier operate to amplify the feeble current from a phototube?
30. For the distributed amplifier, why are the amplifications of each tube added to obtain the total amplification?

PROBLEMS

1. A large triode operated as a class C radio-frequency power amplifier (Chapter 8) requires 400 watts grid-driving power, and has an output of 10 kilowatts. Calculate the power gain as a ratio, and expressed in decibels.
2. A voltage-amplifying audio-frequency amplifier using small-signal pentodes

has a voltage amplification per stage of 120, and an output voltage of 29 volts. Express the ratio in decibels, and calculate the impressed grid signal voltage.

3. Referring to Fig. 7-2, if $\mu = 13.8$, $g_{pg} = 1450$ micromhos, $R_L = 35{,}000$ ohms, and $E_g = 3.5$ volts maximum value, calculate the effective value of the alternating current through R_L, the effective value of the alternating voltage across R_L, and the signal power delivered to R_L. Calculate the voltage amplification in decibels.
4. For a resistance-capacitance coupled amplifier stage using a triode, plot curves showing how the direct-voltage drop across load resistor R_L, and the alternating-voltage drop across R_L, vary with the magnitude of R_L, and prove that R_L usually should be three to five times r_p.
5. A small-signal pentode has a plate resistance of 1.0 megohm, a control-grid–plate transconductance of 1185 micromhos, and an amplification factor of 1185. It is operated with a plate load resistor of 250,000 ohms. If the grid input signal voltage has a maximum value of 0.054 volt, what will be the effective value of the amplified output voltage appearing across the load resistor? What percentage is this of the maximum theoretical voltage output? Express the actual and theoretical voltage ratios in decibels.
6. The tube used in problem 3 has a grid-plate capacitance of 3.2, a grid-cathode capacitance of 3.2, and a plate-cathode capacitance of 2.2, all values being in micromicrofarads. Calculate the approximate grid-input capacitance when operated as in problem 3. Estimate the total input capacitance (when part of a multistage amplifier).
7. A resistance-coupled voltage amplifier using triodes has the following values: $R_L = 50{,}000$ ohms, $R_g = 0.25$ megohm, $C = 0.025$ microfarad, $\mu = 20$, $r_p = 20{,}000$, $C_{gp} = 2.8$ micromicrofarads, $C_{gk} = 3.4$ micromicrofarads, and $C_{pk} = 4.8$ micromicrofarads. Calculate the voltage amplification per stage and total phase shift at 50, 1000, 10,000, and 50,000 cycles.
8. A resistance-coupled voltage amplifier using pentodes has the following characteristics: $R_L = 100{,}000$ ohms, $R_g = 250{,}000$ ohms, $C = 0.01$ microfarad, $\mu = 1185$, $r_p = 1.0$ megohm, input capacitance 7 micromicrofarads, and output capacitance 12 micromicrofarads. Calculate the amplification per stage and phase shift at 50, 1000, 10,000, and 50,000 cycles.
9. An audio-frequency transformer-coupled voltage amplifier uses triodes having an amplification factor of 13.8, a plate resistance of 12,000 ohms, and a mutual conductance of 1150 micromhos. It is connected with high-quality audio-frequency interstage transformers having a step-up ratio of 1 to 2.6. Calculate the voltage amplification for two stages, and express the results as a ratio, and in decibels.
10. A wide-band amplifier uses a pentode having a plate resistance of 1.0 megohm, and a control-grid–plate transconductance of 9000 micromhos. Referring to Figs. 7-12 and 7-13, $C = 0.1$ microfarad, $R_L = 1500$ ohms, $R_g = 250{,}000$ ohms, $C_c = 20$ microfarads, $R_c = 5000$ ohms, $C_g' = 20 \times 10^{-12}$ farad, and $L = 30 \times 10^{-6}$ henry. Calculate the approximate voltage amplification at 30, 100,000, and 3,000,000 cycles. If these values do not give satisfactory response, modify them so that they do.

CHAPTER 8

VACUUM-TUBE POWER AMPLIFIERS

Audio-frequency and radio-frequency *voltage* amplifiers were discussed in the preceding chapter. These are used to increase the weak alternating signal voltage, such as the output voltage of a microphone, until this voltage is sufficiently large to drive a *power* amplifier. Voltage amplifiers are of the class-A type (section 7-4) because voltage amplifiers must not appreciably distort the signal.

The purpose of the *power* amplifier is to furnish the power to drive a device such as a loud-speaker or a radio transmitting antenna. Class-A amplifiers are used as power amplifiers, but their plate-circuit efficiency is low because a large direct plate current flows even when no alternating signal voltage is impressed. Class-AB, -B, and -C amplifiers are used for power amplifiers, because the efficiency of these amplifiers is higher than for the class-A type.

Audio-frequency power amplifiers will be discussed first, and then radio-frequency power amplifiers will be treated. It is of interest to note that class A1 "power" amplifiers draw negligible power from the source of driving signal and thus are not *power* amplifiers in the strict sense. Power tubes usually are connected through impedance-transforming networks to the devices they are to drive (such as a loud-speaker) so that maximum (essentially undistorted) power transfer can be obtained. A power tube therefore delivers power to a reflected fictitious resistance. If a power tube actually worked into a resistor, merely the development of heat would result.

8-1. *TRANSFORMERS AS IMPEDANCE CHANGERS*

The common transformer is an impedance-transforming network. A transformer consists of a primary winding and a secondary winding on a core of ferromagnetic material such as laminated iron.* The iron is of excellent magnetic quality, and it may be assumed that all the magnetic lines of force produced by the primary windings N_p link with each of the secondary windings N_s. It will also be assumed that the losses are negligible. In such a transformer, the ratio of primary and secondary voltages varies directly as the ratio of turns. The currents, however, vary inversely as the ratio of turns. That is,

$$\frac{E_s}{E_p} = \frac{N_s}{N_p} \quad \text{and} \quad \frac{I_s}{I_p} = \frac{N_p}{N_s} \qquad \text{8-1}$$

The transformer of Fig. 8-1 is connected to a load of Z_L ohms and a current of $I_s = E_s/Z_L$ will flow. The *current* in the *primary* will be $I_p = I_sN_s/N_p$, and (in terms of the secondary voltage) the *voltage* across the *primary* will be $E_p = E_sN_p/N_s$. Now the impedance measured at the transformer primary will be $Z_p = E_p/I_p$, and hence

$$Z_p = \frac{E_p}{I_p} = \left(\frac{E_sN_p}{N_s}\right)\left(\frac{N_p}{I_sN_s}\right) = \frac{E_sN_p^2}{I_sN_s^2} = Z_L\left(\frac{N_p}{N_s}\right)^2 \qquad \text{8-2}$$

This equation shows that the impedance that would be measured (with the generator disconnected) at the primary terminals 1–2 of an *ideal* transformer would be the load impedance Z_L times the ratio of the turns squared. Thus, the transformer may be regarded as an *impedance changer* and may be used to match circuits and obtain maximum power transfer (section 4-21). An *ideal* transformer can change the magnitude of an impedance, *but it cannot change the phase angle.* Thus in Fig. 8-1, if Z_L is pure resistance, then for the ideal case, Z_p measured at points 1–2 also will be a pure resistance, but larger or smaller, depending on the turns ratio. An actual transformer will change the phase, the amount depending on the deviation from the ideal transformer.

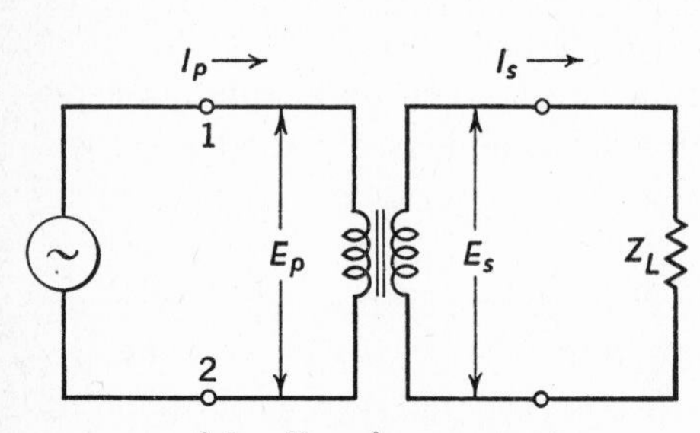

FIG. 8-1. Transformer circuit.

* It is common and convenient to speak of the *iron* core of a transformer. Actually the core would be of silicon-steel laminations or of other magnetic material such as compressed powdered Permalloy, a material composed largely of iron and nickel. See reference 10 of preceding chapter.

8-2. *POWER-AMPLIFIER FUNDAMENTALS* [1]

As defined in section 7-4 for a tube operated as a class-A1 amplifier the negative direct grid-bias voltage is such that plate current flows through the tube at all times. The magnitude of the alternating signal voltage applied between grid and cathode should be such that the grid *never* is driven positive. The positive half-cycle of the signal voltage *subtracts* from the direct grid bias, and *more* plate current flows. The negative half-cycle of the negative signal voltage *adds* to the negative grid bias, and *less* plate current flows. Alternating signal-current and signal-voltage components then exist in the plate circuit, and these drive the loud-speaker or other device connected to the amplifier.

These explanations were discussed in section 4-17, and illustrated by Fig. 4-18. As there shown, the tube operates on the *dynamic* rather than the static curve and works into a resistance load. The way in which the output signal is distorted because of nonlinearity of the dynamic curve was illustrated. It may appear that since the power amplifier tube often works into an output transformer with a load on the secondary instead of working directly into a resistor, the dynamic curve would *not* be as in Fig. 4-18, but this is not true.

In Fig. 8-2 is indicated a triode driven by an alternating signal voltage E_g and connected through an output transformer to a load Z_L'. Assume that Z_L' represents the moving voice-coil of a dynamic loud-speaker. Then, the transformer will have a step-up ratio to increase the load impedance to the desired value as "viewed" by the tube connected to the primary. The load on the tube is the impedance measured (with the tube disconnected) across the primary of the transformer at points 1–2 with the voice coil Z_L' connected to the secondary. If Z_L' is largely resistance, then the load impedance measured at points 1–2 will be largely resistance. Any difference between the phase angle of Z_L' and that measured at the primary will be the result of imperfections *in the transformer* and errors in measurement.

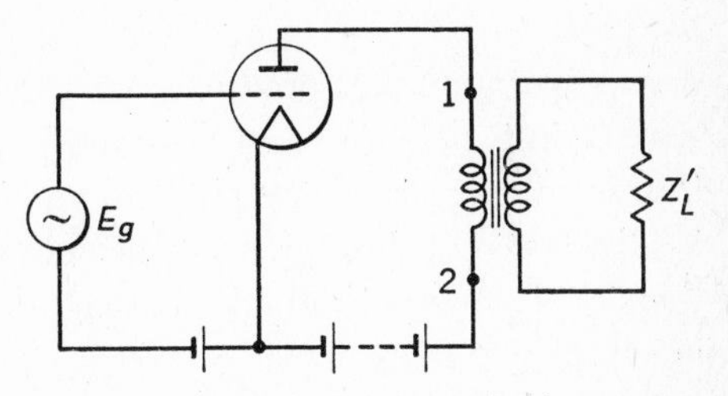

FIG. 8-2. For an ideal transformer, the load impedance reflected into the plate circuit equals Z_L' multiplied by the turns ratio squared.

To check these statements, measurements were made of the impedance of the voice coil of a dynamic loud-speaker of a radio receiving set. The impedance at 1000 cycles was $8.0 + j3.7$ or $8.8 \underline{/+24.8^\circ}$ ohms. Measurements were then made across the primary of an inexpensive impedance-matching transformer with the voice coil connected as indicated in Fig. 8-2. Again using 1000 cycles, the load impedance as measured at the primary (section 7-18) was 2065 ohms resistance and 1220 ohms inductive reactance, or $2400 \underline{/+30.5^\circ}$ ohms.

8-3. *POWER AMPLIFIER GAIN* [2,3,4]

In general, the purpose of a power amplifier is to increase the power level in a system. The **power level** at a given point in a transmission system is [2] "the ratio of the power at that point to some arbitrary amount of power chosen as a reference." The power level ratio commonly is expressed in decibels with one milliwatt as a reference (**dbm**), or in decibels with one watt as the reference (**dbw**).

A power amplifier is classed as an **active electric transducer,** which is an electric network by means of which energy may flow from one part of a system to another, and containing one or more sources of energy.[2]

In the class-A1 power amplifier of the preceding section, negligible power is required to drive the amplifier. Other power amplifiers require **grid driving-power,** defined [2] as "the average product of the instantaneous values of grid current and of the alternating component of grid voltage over a complete cycle. This comprises the power supplied to the biasing device and to the grid."

A power amplifier produces a power gain in an electric transmission system. **Gain** is defined [2] as a term used to "denote an increase in signal power in transmission from one point to another." The **power gain** of an amplifier is [2] the "ratio of the signal power delivered to a specified load impedance to the signal power absorbed by its input." Both gain and power gain usually are expressed in decibels.

Of much interest is the power that may be drawn from an amplifier. Summarized, the **available power** [3] is the power that an amplifier (or generator) will deliver to a load impedance that is the conjugate of the amplifier internal impedance measured at the output terminals with the load disconnected. This power is *available,* even if not taken from the generator.

The **available power gain** [3] is the ratio of the available power from the output terminals of the amplifier to the available power from the driving generator.

The **insertion power gain** [3] is the ratio of the power supplied an external termination (such as a load) when the amplifier is inserted between the signal source and the external termination, to the power delivered to the termination when the signal source is connected directly to the external termination.

8-4. *AUDIO-FREQUENCY CLASS-A1 SINGLE TRIODE POWER AMPLIFIERS*

In accordance with section 8-2, the circuit of Fig. 8-2 is equivalent to Fig. 8-3, and from section 4-15, the power output in watts of the tube of Fig. 8-2 is

$$P = \frac{\mu^2 E_g{}^2 R_L}{(r_p + R_L)^2} \qquad 8\text{-}3$$

where R_L is the load resistance *reflected into the primary* (section 8-2) and E_g is the *effective* value of the impressed grid signal voltage. This equation applies only if μ and r_p are substantially constant.

As previously discussed, when a tube has resistance in its plate circuit (as it effectively has when connected as in Fig. 8-2), the tube operates on the load, or dynamic, curve rather than the static curve (section 4-16). If the distortion is to be kept low, the tube must be operated on the essentially *straight portion* of this curve. For *graphically* determining the optimum load resistance, for calculating the power output, and for finding percentage distortion, **load lines** plotted on a family of static curves with *plate voltage* on the X axis and *plate current* on the Y axis are used.[5] The use of load lines is better suited to quantitative analysis. Load, or dynamic, curves and load lines represent the same data plotted on different axes.

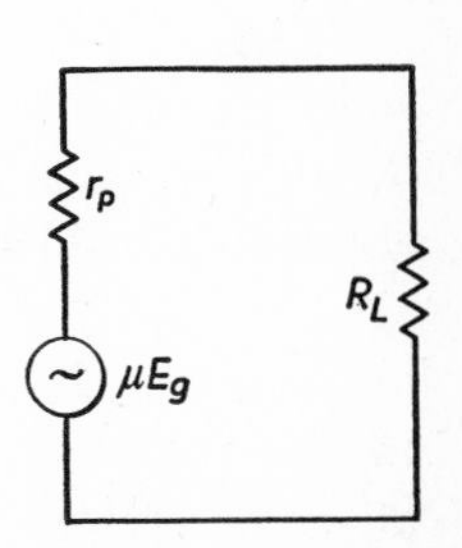

FIG. 8-3. Equivalent circuit for Fig. 8-2 when Z_L' is a resistor.

The problem generally is as follows: To find the value of load resistance into which a tube should work so that the maximum power output is obtained without the percentage of second-harmonic distortion (which is usually the most troublesome harmonic), exceeding 5 per cent, a value usually acceptable. In the solution of the problem, the static plate-voltage–plate-current family of curves such as Fig. 8-4 is used. The plate voltage E_b to be applied is determined by the rating of the tube, by the type of amplifier, and by the power supply. The load resistance assumed for the first trial calculation usually is about twice the plate resistance of the tube. The value of the grid-bias voltage can be determined by the relation

$$E_c = \frac{0.7 E_b}{\mu} \qquad 8\text{-}4$$

Factors other than 0.7 are used,[5] but this value usually is satisfactory. For Fig. 8-4 these values are +250 volts, −50 volts, and 3900 ohms. The point M is where the plate voltage is 250 volts and the plate current zero, and point Q is where the plate current equals 250/3900 = 64 milliamperes. Line X–Y is parallel to A–B and through point P corresponding to the voltage values selected. Line X–Y is the trial load line. It is assumed that the maximum alternating signal voltage applied to the grid *will be sufficient to drive the grid to zero and to twice the direct bias voltage*. Thus, line X–Y terminates on the $E_c = 0$ and $E_c = -100$-volt curves. These points should be extended to the plate-current axis as indicated, giving the $I_{max.}$ and $I_{min.}$ lines shown. Also, the I_b, $E_{max.}$, and $E_{min.}$ lines should be drawn. Under these conditions,

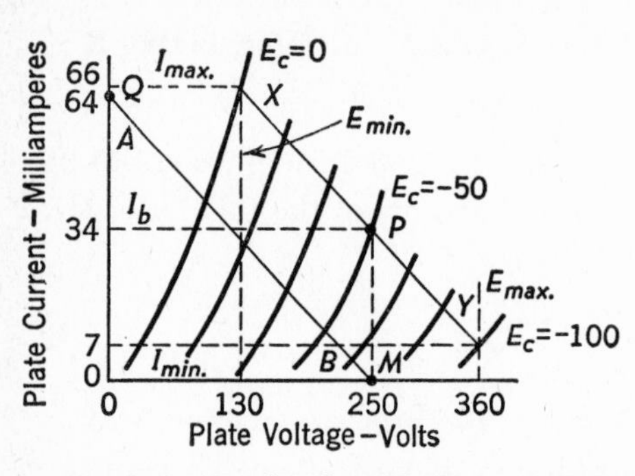

FIG. 8-4. Graphical method for finding the power output and percentage second-harmonic distortion for a triode. (Adapted from reference 5.)

$$\text{Power output} = \frac{(I_{max.} - I_{min.})(E_{max.} - E_{min.})}{8} \qquad 8\text{-}5$$

For Fig. 8-4 this equals $[(66 - 7)(360 - 130)]/8 = 1700$ milliwatts.

The percentage *second-harmonic* distortion is determined approximately by the relation

$$\text{Second-harmonic distortion} = \frac{\dfrac{(I_{max.} + I_{min.})}{2} - I_b}{I_{max.} - I_{min.}} \times 100 \qquad 8\text{-}6$$

For Fig. 8-4 this becomes

$$\frac{\dfrac{(0.066 + 0.007)}{2} - 0.034}{0.066 - 0.007} \times 100 = 4.2 \text{ per cent}$$

Since this value does not exceed 5 per cent, the conditions for operation assumed would be satisfactory; otherwise, new operation conditions would have to be selected and a new check made. Usually 5 per cent distortion is not exceeded when R_L equals about twice r_p, and when the amplifier is not driven beyond the limits $e_c = 0$ and e_c equals twice the bias value. The derivation of equation 8-5 is as follows: Power in watts is given by the relation EI, where E is the *effective* voltage, I is the *effective* current, and the circuit is of pure resistance. Thus, $(I_{max.} - I_{min.})/(2\sqrt{2})$ is the *effective* current I; also, $(E_{max.} - E_{min.})/(2\sqrt{2})$ is the *effective* voltage E. When these values are substituted in the relation $P = EI$, equation 8-5 results.

The way in which distortion is caused in a triode because the dynamic characteristic curve is not a straight line was illustrated by Fig. 4-18. This distorted plate current is reproduced in Fig. 8-5. It will be noted that the wave is nonsymmetrical with respect to the X axis. A steady-state nonsinusoidal wave consists of a fundamental and harmonics. The wave of Fig. 8-5 is composed largely of a fundamental and a second harmonic as shown by the dotted lines.

The reason the axis of the second harmonic does not coincide with that of the fundamental is as follows: When distortion occurs in a triode operated in class A, rectification results because the positive and negative halves of the cycle are not the same. If the distortion is considerable, a noticeable increase in plate current occurs when the signal voltage is impressed. This increase in current is equal to the *amplitude of the second harmonic.*

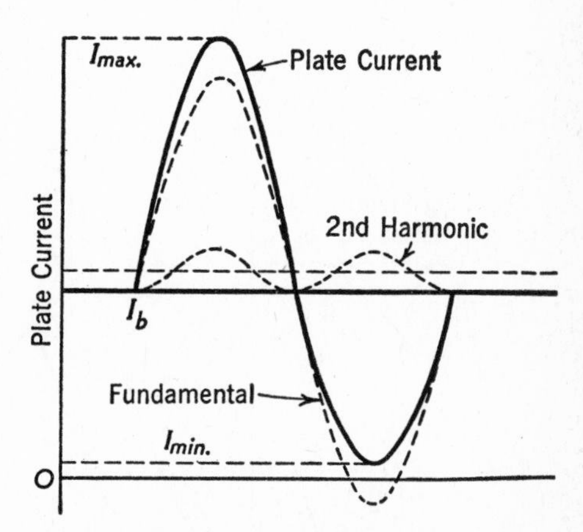

FIG. 8-5. The second-harmonic component causes the plate current to be nonsymmetrical about the X axis, I_b.

As Fig. 8-5 shows, with no alternating signal voltage on the grid, the plate current is I_b. With a signal, the *average* value of the plate current is $(I_{max.} - I_{min.})/2$, and this also equals the amplitude of the fundamental component. The amplitude of the second harmonic will be $(1/2)\{[(I_{max.} + I_{min.})/2] - I_b\}$, and this value, divided by the amplitude of the fundamental and multiplied by 100, gives the percentage second harmonic of equation 8-6. These statements can be proved by writing equations that $I_{max.}$ equals I_b plus the maximum value of the fundamental plus twice the maximum value of the second harmonic, and that $I_{min.}$ equals I_b minus the maximum value of the fundamental plus twice the maximum value of the second harmonic. The equations are then solved simultaneously for the fundamental, and for the second harmonic. Much information exists on the graphical analysis of vacuum-tube circuits.[6,7,8]

In previous discussions of amplifiers (sections 4-20 and 7-8), constant *plate-voltage* curves were used. Grid-voltage values were plotted on the X

axis, and plate-current values on the Y axis. These curves are useful in explaining amplifier operation from the qualitative standpoint. It also is possible to study amplifier operation using constant *grid-voltage* curves with plate voltage on the Y axis, and plate current on the X axis. These are particularly useful because the areas under such curves, being the product of voltage and current, represent *power*. Thus, Fig. 8-4 has been replotted as Fig. 8-6. A load (or dynamic) curve represents the effect of the load resistance plotted on constant plate-voltage curves. Load lines represent the effect of the load resistance plotted on constant grid-voltage curves.

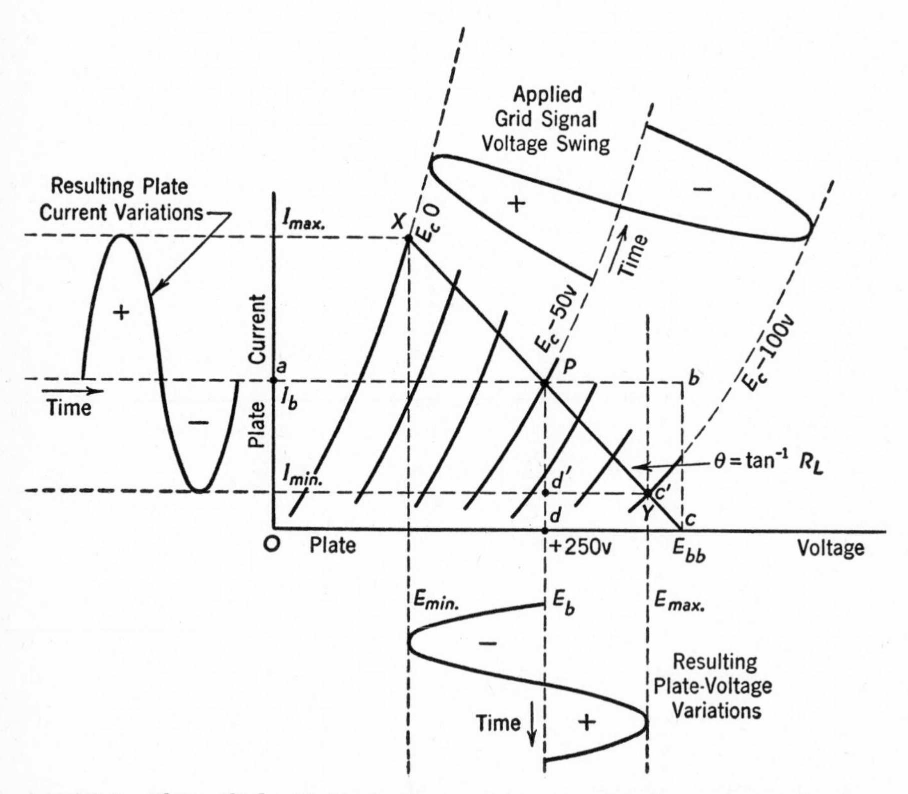

FIG. 8-6. The applied grid-signal voltage swing causes the plate current and voltage to vary as shown. The total power delivered by the plate supply source to the tube and resistor R_L in the plate circuit is represented by the area *Oabc*. The power delivered to the tube alone is given by the area *OaPd*. The alternating signal power output of the tube is shown by the area of the triangle *d'Pc'*.

The tube having the characteristics of Fig. 8-6 is biased to −50 volts, and the alternating grid signal voltage, or grid swing, drives the grid between the limits of zero and −100 volts. The significance of the load line X–Y now will be discussed. Suppose there is *no load* in the plate circuit and that the direct voltage between plate and cathode is +250 volts. When the grid is driven by the signal to zero, the plate current theoretically will rise to a very high value, determined by extending line *d–P* of Fig. 8-6 until it intersects an extended $E_c = 0$ plate-current curve. When the impressed signal drives the grid to −100 volts, the plate current falls to zero. Now suppose that there is a *very high-resistance load* in the plate circuit and that the bias is −50 volts and the direct plate voltage is +250 volts as before. Because of the high resistance, the same grid swing will produce little change in plate current but will produce a large change in plate voltage. Operation is as if the load line were almost horizontal. It follows that for a typical load of $R_L = 2r_p$ approximately, the load line will be line *X–Y* of Fig. 8-6 passing through operating point *P*, and will lie between lines *c–b* and *c–d* of Fig. 8-6.

The slope of load line *X–Y* is determined by the resistance of the load resistor. If θ is defined as the angle between the load line *X–Y* and *the vertical*, then the slope of the line and the tangent of the angle equal plate voltage in volts divided by plate current in amperes.

If it is assumed that Fig. 8-6 represents a tube working into a resistor (and not into a transformer that *reflects* load resistance into the plate circuit) then *both* the direct and alternating components of the plate current flow through the resistor. If, on the other hand, the tube is working into a transformer connected to a load (such as a loud-speaker) and *resistance is reflected* into the plate circuit, then the direct component of the plate current flows through only the resistance of the primary of the transformer, but the alternating component flows through only the reflected resistance.

Referring to Fig. 8-6, and assuming that there is little plate-current distortion so that I_b represents the direct component both *before* and *after* the grid signal voltage is applied, then the total direct-current power supplied by the source of voltage and delivered to the tube and circuit (excluding cathode heating power) is

$$P_i = E_{bb} \times I_b = \text{area } Oabc \qquad 8\text{-}7$$

The average power delivered to the tube alone

$$P_{iT} = E_b \times I_b = \text{area } OaPd \qquad 8\text{-}8$$

and this is dissipated in the plate circuit of the tube. The voltage across the load resistance rises and falls in accordance with the $E_{\text{max.}}$ to $E_{\text{min.}}$ variations, and the current through the load resistance rises and falls in accordance with the $I_{\text{max.}}$ to $I_{min.}$ changes. Assuming the variations are sinusoidal, *one-half* of each of these variations represents the peak or maximum value of the respective wave. The alternating power dissipated in the load resistance will be the alternating-current power output of the tube, or

$$P_o = E \times I = \frac{E_{max.}}{\sqrt{2}} \times \frac{I_{max.}}{\sqrt{2}} = \frac{E_{max.}I_{max.}}{2} \qquad 8\text{-}9$$

and is shown by the triangular area $d'Pc'$. This figure and these areas show that the **plate-circuit efficiency,** or ratio of alternating power output of a tube to direct power input to the plate of a tube is *low* for a class-A1 amplifier. However, the distortion is low for a well-designed circuit.

8-5. *AUDIO-FREQUENCY CLASS-A1 PARALLEL POWER AMPLIFIERS*

Two or more triodes may be operated in parallel to obtain an output greater than one tube can deliver. Corresponding electrodes of two or more tubes of the same type are connected together. For the same driving signal voltage the output is twice that of a single tube. Because the plates are in parallel, the equivalent plate resistance for Fig. 8-3 is one-half the plate resistance of a single tube. The equivalent generator voltage is μE_g. The same reasoning as for a single tube indicates that the resistance of the load should be about the same as the plate resistance of a *single* tube (twice the equivalent plate resistance of the *parallel* combination). As an illustration of the problems of early radio, for the first transatlantic radio *telephone* tests, several hundred tubes, each of about 25 watts capacity, were operated in parallel in the final stage.

8-6. *AUDIO-FREQUENCY CLASS-A1 PUSH-PULL POWER AMPLIFIERS* [9,10]

A circuit with two triodes in push-pull for audio-frequency power amplification is shown in Fig. 8-7. The two tubes and corresponding transformer windings are assumed to be identical so that the amplifier is a *balanced* transducer. The grids of the two tubes are biased to *about* one-half the cutoff value. One-half the alternating signal voltage induced in the secondary of the input transformer is impressed on the grid of *each* tube. The grids never are driven positive for class-A1 operation. With no signal impressed, and with identical tubes, the plate current drawn by each is the same. Since direct current flows in opposite directions through each of the identical halves of the primary of the output transformer, no resultant magnetizing effect is produced. Smaller magnetic cores may, therefore, be used because magnetic saturation and resulting distortion are avoided.

When an alternating signal voltage is impressed on the two grids, they are driven alternately less negative and more negative. Thus, as the plate current in one tube *increases,* that in the other tube *decreases.* The net magnetizing effect in the transformer is the same as if an alternating current of *twice the value of either alternating plate-current component flowed through one half of the primary winding.* The power output of two tubes in class-A1 push-pull may be considerably greater than twice that of a single tube in class-A1.

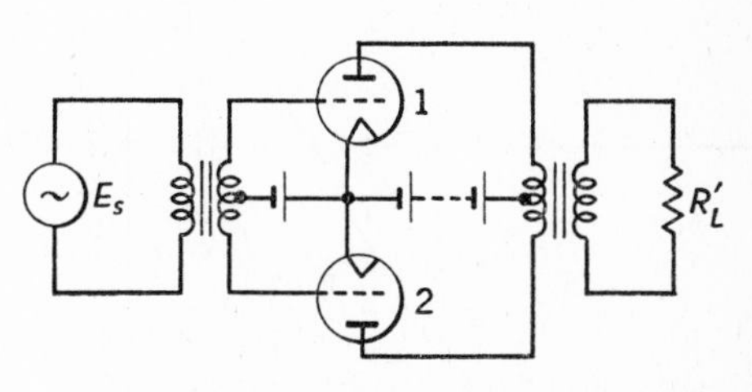

FIG. 8-7. Two triodes connected for push-pull class-A operation. The turns ratio of the input transformer is assumed unity in the discussion in section 8-6.

Operation of Push-Pull Amplifiers. This is illustrated by Fig. 8-8. The characteristics are plotted with the curve for tube 2 inverted. The current variations in each tube when an alternating signal voltage is impressed on the grids are shown by the curves: plate current, tube 1, and plate current, tube 2. As illustrated, these two plate currents are distorted because the tubes are operated with alternating signal voltages which are so large that operation is over *more* than the (essentially) linear part of the dynamic curves. As was indicated by Fig. 8-5, this will be largely a second-harmonic distortion. As Fig. 8-8 shows, however, these second-harmonic components existing in the plate current of each tube will produce magnetizing effects that *are balanced out in the transformer.* The characteristics of tube 2 are inverted and the second-harmonic components are *in phase.* Since they pass through identical primaries, *but in opposite directions,* their effective magnetomotive forces cancel, no second-harmonic magnetization is produced, and no second-harmonic voltage is induced in the secondary. This is *one reason* that the power output of two tubes in push-pull may be greater than twice the output of one tube; *the second-harmonic distortion is balanced out.*

As indicated in a preceding paragraph, the net magnetizing effect of the alternating plate-current components is the same as if an alternating current of twice the value of either component flowed through one half of the primary winding. This is equivalent to saying the net magnetizing effect is as if a combined resultant current flowed as shown in Fig. 8-8 *from a source which had a dynamic curve which was the resultant of the dynamic curves of the two separate tubes.* If this resultant dynamic curve were straight as shown by the broken line, then the resultant current for the two tubes would be a sine wave composed only of a fundamental. The resultant dynamic

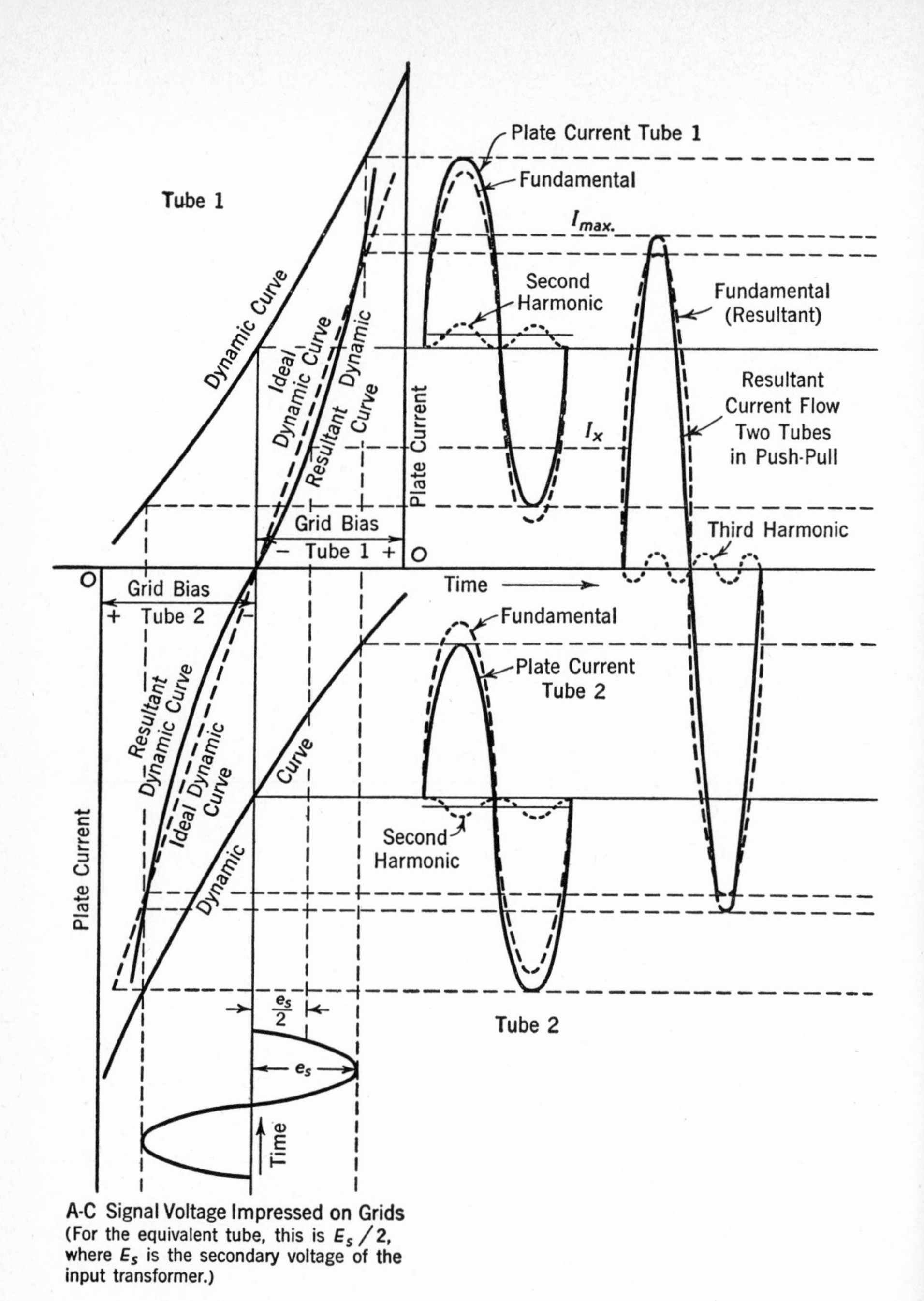

FIG. 8-8. Diagram for analyzing the class-A push-pull amplifier of Fig. 8-7.

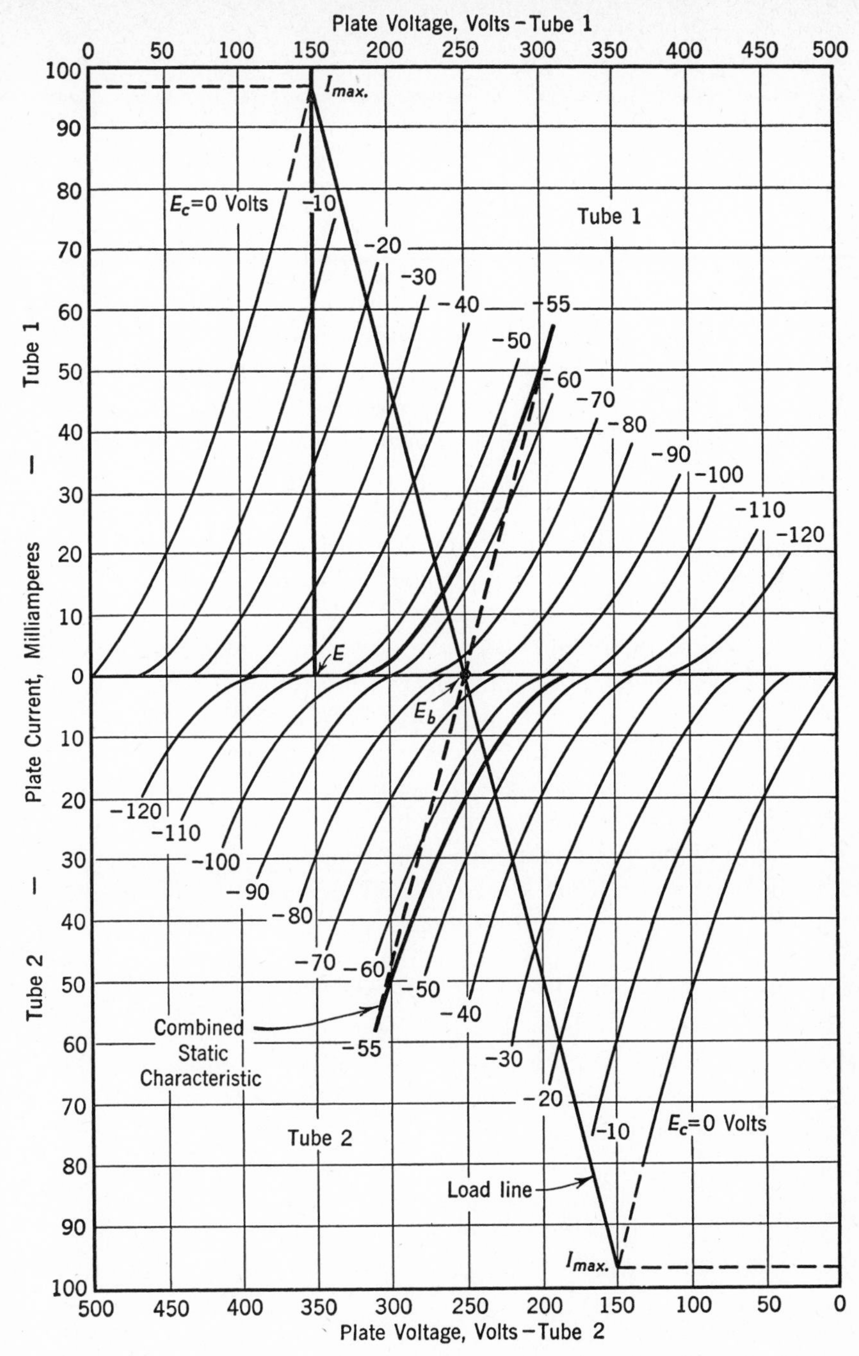

FIG. 8-9. Static curves for a push-pull amplifier. The value of E is 0.6 E_b.

characteristic is slightly curved, as Fig. 8-8 shows, and therefore a third-harmonic distortion (and under overloads also a fifth harmonic) exists in the output, as this figure indicates.

Equivalent Circuits for Push-Pull Triodes. A push-pull audio-frequency power amplifier could be designed from the *dynamic* curves of Fig. 8-8. Much time would be consumed in plotting these dynamic curves, and thus the static curves commonly are used.[9] The static curves are plotted as shown in Fig. 8-9, with the curves of one tube inverted as before. The combined static curves—only one of which is shown by the broken line—represent the characteristic of *an equivalent, or composite, tube.* The plate resistance of each tube is the reciprocal of the slope of the static curve at the point of operation (section 4-5). Hence, the *apparent* plate resistance of triodes in push-pull would be the reciprocal of the slope of the *combined static curve shown by the broken line.* For Fig. 8-9 this value is about 1000 ohms. The corresponding value of the tube plate resistance determined from the slope of an *individual* static curve (not the combined curve) is about 1700 ohms. It is possible to avoid plotting these curves in detail.[11]

The method of determining the value of the load for a push-pull circuit will now be considered. For a *single class-A tube,* it may be necessary to try several values of load resistance before one is found with which the second-harmonic distortion is less than 5 per cent. This value of load resistance which gives maximum *undistorted* power output (but not necessarily maximum power output) for a *single* class-A tube is usually about twice the plate resistance. With push-pull tubes, the second-harmonic distortion is balanced out in the transformer. Thus, a value of load resistance may be selected that is *equal* to the apparent plate resistance, and thus the *maximum possible* power output can be obtained. This is the *second important reason* for the high output of push-pull tubes.

It now is possible to draw the equivalent circuit of Fig. 8-10. The two tubes in push-pull class A act like one fictitious, or composite, tube that has a static curve as shown by the dotted line of Fig. 8-9, giving a plate resistance of 1000 ohms. This has been called r_p' in Fig. 8-10. The signal voltage impressed on the grid of this fictitious tube is one-half the signal voltage E_s induced in the input transformer secondary; thus the voltage on the grid of the fictitious tube will be $E_g' = E_s/2$. This fictitious tube was shown to have an *apparent* plate resistance r_p' of about 1000 ohms. In accordance with the statements of the preceding paragraph, the selected load R_F for the fictitious tube should also be 1000 ohms.

This value of R_F is not the *plate-to-plate* load that should be used with the two tubes in push-pull. The value of the actual plate-to-plate load equals

$4R_F$. The reason is as follows: The fictitious tube from which the equivalent current of Fig. 8-8 flows operates into *one half* of the primary. Thus, for an output transformer of 1 to 1 over-all ratio, if one half of the primary turns is used, the ratio becomes 1:2, and an impedance in the secondary will be reflected as *one-fourth* the value (section 8-1) when measured across *one half* of the primary in the manner the fictitious tube operates. Thus, the actual plate-to-plate load connected across the secondary of the transformer of unity turns ratio is $4R_F$ as previously mentioned (section 8-2).

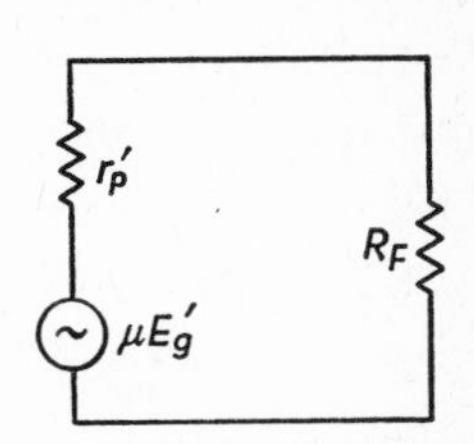

FIG. 8-10. One possible equivalent circuit for a class-A push-pull amplifier.

The approximate power output of a push-pull amplifier can be calculated using the circuit of Fig. 8-10. Assuming an alternating signal voltage of $E_s = 110$ volts peak or 77.7 volts effective value, a bias of -55 volts, and a $\mu = 3.5$, the output signal current is $I = 77.7 \times 3.5/(2 \times 2000) = 0.068$ ampere, and the power delivered to the load is $I^2R_F = 0.00462 \times 1000 = 4.62$ watts. The equivalent circuit of Fig. 8-10 and the calculations just made are of value in explaining push-pull amplifier operation, rather than for design purposes.

Push-pull amplifiers can be designed from simple calculations. Using the static curves for the tube (such as the upper set of curves of Fig. 8-9), erect a vertical line at $E = 0.6E_b$, where E_b is the operating plate voltage.[5] Both E and E_b are shown on Fig. 8-9. Draw a load line from $I_{max.}$ to E_b. The tangent of the angle (that is, E_b/I_b) that this line makes with the vertical when *multiplied by four* is the value of the plate-to-plate load resistance to be used in accordance with the explanations previously given. For Fig. 8-9 this is $100/0.096 = 1040$ ohms. (In the preceding discussion, a resistor of 1000 ohms was used for R_F of Fig. 8-10). The power output of the two tubes in class-A push-pull is [5]

$$P = \frac{I_{max.}E_b}{5} \qquad 8\text{-}10$$

where the power is in watts, where $I_{max.}$ is in amperes and E_b is in volts.

Distortion in Class-A Push-Pull Amplifiers. As explained, the second harmonics are balanced out in the transformer primary in a push-pull stage. Likewise, power hum introduced into the plate circuit *from the power supply* is balanced out. Some third-harmonic distortion, and a small amount of fifth-harmonic distortion, sometimes occur.

Push-pull amplifiers are used for *voltage* amplification in high-quality equipment. A push-pull circuit is *balanced* with respect to ground.[12] Much vacuum-tube equipment is designed so that during operation one side is grounded, or may be grounded. Such equipment is *unbalanced* with respect to ground. Unbalanced amplifiers sometimes are referred to as single-ended amplifiers; balanced amplifiers sometimes are called double-ended amplifiers. A single-ended push-pull audio amplifier that does not use transformers has been described.[13]

8-7. *AUDIO-FREQUENCY CLASS-A1 PENTODE POWER AMPLIFIERS*

When the power-output pentode is used in a power amplifier, a large power output is obtained with small alternating signal voltage applied to the grid as compared with a triode. Fig. 8-11 shows a suppressor-grid pentode as a power amplifier. An equivalent circuit for this tube also is shown in Fig. 8-11, in which r_p is the plate resistance and R_L is the load resistance R_L' reflected into the primary. In accordance with the theory in section 8-1, $R_L = N^2 R_L'$, where N is the turns ratio of the (ideal) transformer.

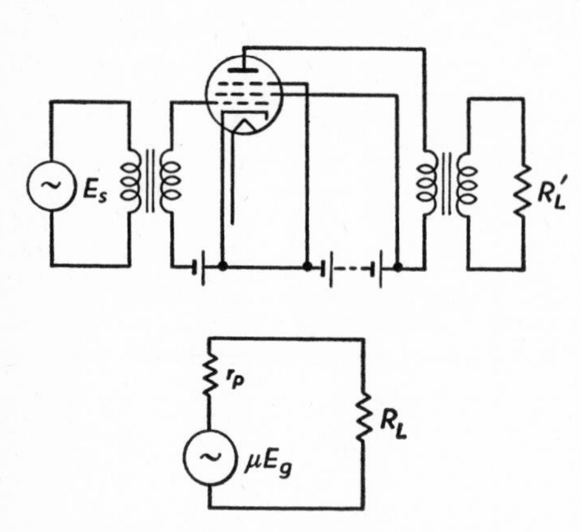

FIG. 8-11. Actual and equivalent circuits for a class-A power pentode.

It is evident from Fig. 5-13 that if a large value of load resistance is used, the dynamic curve will be very badly curved at the upper end. For low distortion, the value of the load resistance chosen should be such as to give a straight dynamic curve on which to operate. For *triodes*, the value of load resistance selected is as close as practicable to the plate resistance of the tube to make the power output large. As discussed in section 4-21, the power output is maximum when the load resistance equals the plate resistance, but with triodes the value of load resistance actually selected is about twice the plate resistance, to reduce distortion. Owing to the excessive bending of the *dynamic curve for the pentode*, the value of the load resistance selected must be very much *lower* than the plate resistance of the tube. For the tube of Figs. 5-12 and 5-13, the load resistance chosen would be about 6000 ohms, although from Fig. 5-12 the plate resistance of the tube is 50,000 ohms. This mismatch causes the power output to be lower than the value theoretically obtainable (which would result in excessive distortion).

The power output under these conditions can be obtained from equation 8-3. The suppressor-grid power-output tube is widely used however, because, as Fig. 8-12 shows, only a relatively small alternating signal voltage need be applied to the control grid to drive the tube.

Characteristics of Class-A Power Pentodes. As shown in Fig. 5-13, even at the low load resistance of 6000 ohms the dynamic curve of a suppressor-grid power pentode is not a straight line. This curve has been replotted as

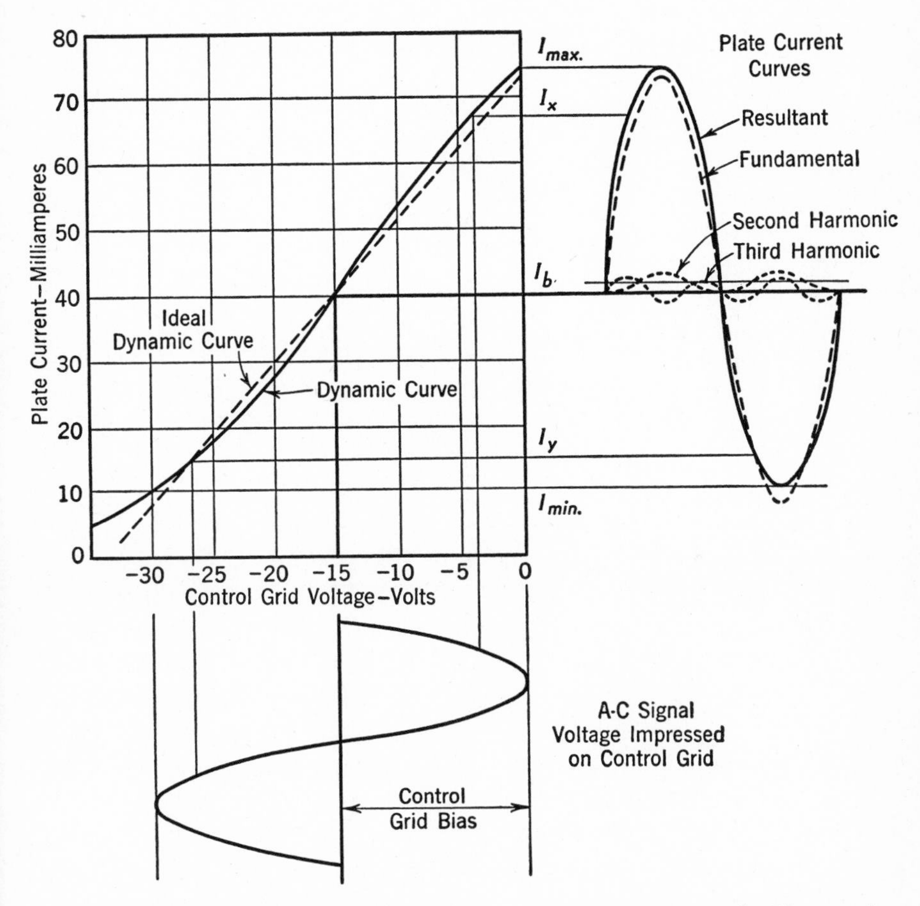

FIG. 8-12. Diagram for analyzing the performance of a power pentode. The meanings of I_x and I_y are given in the text.

Fig. 8-12. When points on this dynamic curve corresponding to given values of the alternating grid signal voltage are projected, the resultant plate-current curve is obtained. If the points are projected from the *ideal* dynamic curve, a sine wave is obtained corresponding to the fundamental. The resultant wave has further been broken down into second and third harmonics.

The power output and the harmonic distortion for suppressor-grid pentodes can be calculated from equations based on Fig. 5-12 and derived from Fig. 8-12 in which I_x equals the plate current flowing at $0.293E_g$ and I_y is the current at $1.707E_g$. The voltage E_g is the maximum value of the alternating sinusoidal signal voltage which when applied to the grid will just drive it to zero potential. The power output for a suppressor-grid pentode or a beam-power tube (which has a similar family of curves, as shown in Fig. 5-18) is [5]

$$\text{Power output} = [I_{max} - I_{min.} + 1.41(I_x - I_y)]^2 \frac{R_L}{32} \qquad 8\text{-}11$$

This value will be microwatts if the current is in milliamperes and the load resistance R_L is in ohms. The value R_L is the load resistance reflected into the primary for a transformer-coupled amplifier. From Fig. 5-12 this load resistance is equal to $(E_{\text{max.}} - E_{\text{min.}})/(I_{\text{max.}} - I_{\text{min.}})$.

The percentage distortion can be found from Fig. 8-12 by the following equations

$$\text{Second harmonic distortion} = \frac{I_{\text{max.}} + I_{\text{min.}} - 2I_b}{I_{\text{max.}} - I_{\text{min.}} + 1.41(I_x - I_y)} \times 100 \qquad 8\text{-}12$$

$$\text{Third harmonic distortion} = \frac{I_{\text{max.}} - I_{\text{min.}} - 1.41(I_x - I_y)}{I_{\text{max.}} - I_{\text{min.}} + 1.41(I_x - I_y)} \times 100 \qquad 8\text{-}13$$

The total percentage distortion—that is, the sum of the second and third harmonic distortion—is equal to the square root of the sum of the percentage second harmonic squared and the percentage third harmonic squared. In Fig. 5-15 is shown the manner in which the harmonic output varies with load resistance and input voltage.[14] In selecting the best value of load resistance to use with a suppressor-grid power pentode or a beam-power tube (section 5-13) several trial load lines for different resistance values are drawn on the family of curves such as in Fig. 5-12. For the curve giving low distortion, the distance I_b to $I_{\text{max.}}$ will approximately equal the distance I_b to $I_{\text{min.}}$

Pentodes in Push-Pull.[15] Suppressor-grid power pentodes are operated in push-pull circuits similar to that of Fig. 8-7. The third-harmonic distortion in pentodes is much higher than in triodes under comparable conditions (Fig. 5-15). Odd harmonics are *not* canceled in push-pull circuits. Furthermore, in properly designed single-tube circuits, the second-harmonic output is fairly low. Thus from the standpoint of distortion *produced by the tubes* *themselves,* little advantage results from push-pull *pentode* operation. As in all push-pull operation, there are certain other advantages, such as the reduction in hum in the output transformer, thus requiring less filtering in the power-supply rectifier.

8-8. *AUDIO-FREQUENCY CLASS-AB1 POWER AMPLIFIERS*

In such amplifiers (section 7-4) two tubes (often triodes) are connected in a push-pull circuit such as Fig. 8-7. A *higher* negative grid bias is used than for class-A operation, and higher plate voltage (and screen voltage if a pentode) also is used. The design follows the general method given in section 8-6 for the class-A push-pull amplifier. For example, on Fig. 8-9 the curves could be drawn for a plate voltage of 275 volts and a grid-bias voltage of −70 volts. For given tubes, higher output and efficiency are obtainable with but little greater distortion than with class-A operation.

8-9. *AUDIO-FREQUENCY CLASS-B2 POWER AMPLIFIERS*

In accordance with definitions (section 7-4), a tube operated in class B is biased to about cutoff, and plate-current is almost zero with no alternating signal voltage impressed between grid and cathode. In audio power amplifiers two tubes are connected essentially as in Fig. 8-7, and with impressed signal each tube passes plate current for approximately one-half cycle. The operation is, in a sense, "push-push," a term seldom used. Tubes that operate close to plate-current cutoff with *no* grid bias have been developed for class-B audio amplifiers. The *load* (*or dynamic*) characteristics for two such tubes are shown in Fig. 8-13. Ideal characteristics are shown by the broken line.

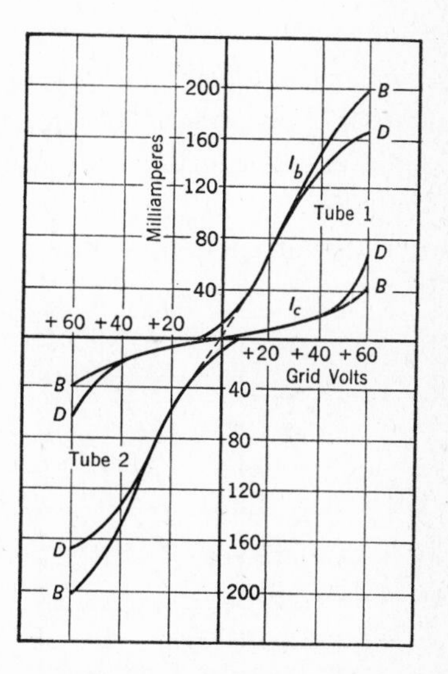

FIG. 8-13. Dynamic transfer characteristics for two tubes operated as class-B triodes. Plate current I_b and grid current I_c for plate-to-plate loads per tube of B, 1450 ohms, and D, 2000 ohms. $E_p = 400$ volts. (Data from reference 5.)

Operation. Each tube passes plate current to the load for the half-cycle of signal voltage that the grid-cathode voltage is positive. For this half-cycle much grid current flows and operation is class-B2. This method commonly is referred to as class-B, it being understood that grid current flows. One tube, and then the other, passes plate current, and the effect is as if one tube worked on *both* half-cycles into one half of the primary of the output transformer. If the tubes are to be operated in accordance with curve *B* of Fig. 8-13, the effective plate-to-plate load should be $1450 \times 4 = 5800$ ohms (section 8-6). If the secondary of an ideal unity-ratio transformer is terminated in 5800 ohms resistance, this appears as 1450 ohms to a tube connected to one-half of the primary.

The point of operation of Fig. 8-13 about which the grid-signal voltage varies is zero grid volts. Because one tube conducts, and then the other conducts, for a sinusoidal grid-signal voltage *each* tube will pass a plate current essentially as shown by Fig. 6-1. From equation 6-1, this current consists of a direct component, a component of the same frequency as the fundamental signal frequency impressed on the grid, and even harmonic terms. Because the direct components from the two tubes in class-B flow through identical halves of the primary of the output transformer in *opposite* directions, negligible magnetic effects result from these components. The phases of the fundamental signal components are such that their magnetic effects *add,* and thus an amplified sinusoidal signal is passed to the output terminals of the output transformer. The phases of even harmonics are such that their magnetic effects cancel in the transformer primary as in the push-pull class-A amplifier.

Plate-current–plate-voltage *static* characteristic curves for the tube of Fig.

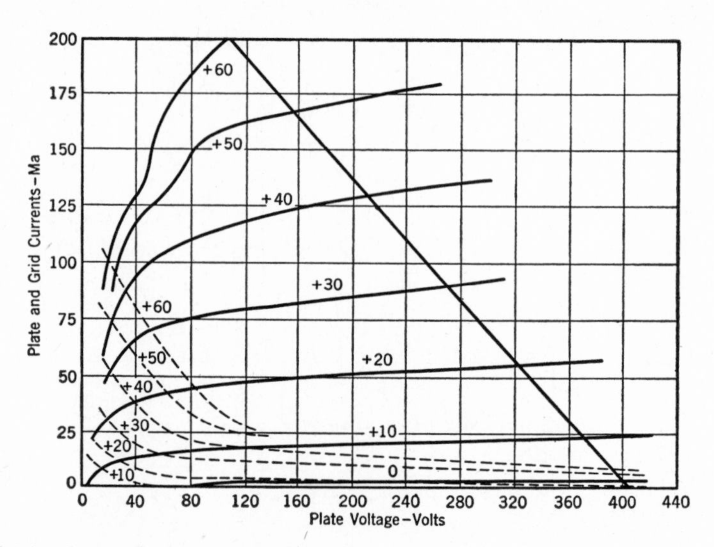

FIG. 8-14. Static plate current (solid) and grid current (broken) curves for the same tubes as Fig. 8-13. Values on curves are grid voltages in volts. Load line for 1450 ohms shown. (Data from reference 5.)

8-13 are given in Fig. 8-14. A load line corresponding to 1450 ohms is shown. The intercepts on the X axis in volts divided by those on the Y axis in amperes equals the load resistance. Because one tube only operates at a time, it is not necessary to show inverted characteristics as for the class-A push-pull stage of Fig. 8-9. For the tube shown, the direct plate voltage is +400 volts and the grid-bias voltage is zero.

Power Output. The power output of this amplifier with large alternating signals impressed on the grid can be approximately determined in several ways. Perhaps the easiest is as follows: Referring to either Fig. 8-13 or 8-14, for a maximum grid-signal voltage on the tube of 50 volts, the corresponding peak value of plate current is about 170 milliamperes for a 1450-ohm *tube* load (not plate-to-plate). The alternating signal power delivered by the two tubes would be $P = I^2R_L$ or $P = (0.707 \times 0.170)^2 \times 1450 = 20.9$ watts (approximately).

An equivalent circuit for a class-B audio amplifier is shown in Fig. 8-15. In accordance with preceding statements, it is assumed that a single tube has one-half the signal voltage applied, that the tube passes plate current for the entire cycle, and that R_L is 1450 ohms. For this tube $r_p = 17{,}500$ ohms and $\mu = 70$, the values being approximate. If the maximum grid-to-grid sinusoidal alternating signal voltage is 100 volts, then the effective value is 70.7 volts, and $\mu E_g/2$ is $70 \times 70.7 \times 0.5 = 2470$ volts. The alternating component of plate current is $I = 2470/(17{,}500 + 1450) = 0.13$ ampere, and the power delivered to the load is $P = (0.13)^2 \times 1450 = 24.7$ watts. These calculations, based on the equivalent circuit of Fig. 8-15, are made for illustrative purpose; the calculations of the preceding paragraph are more reliable.

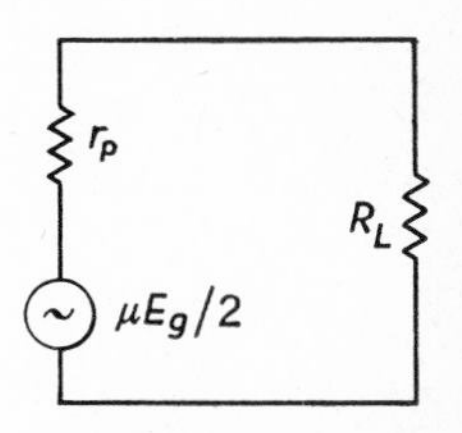

FIG. 8-15. Equivalent circuit for two tubes in class-B.

Input Power.[16] As indicated in Fig. 8-13, the grids draw much current. Alternating signal power must, therefore, be supplied by the preceding stage to the grid circuit of a class-B2 amplifier. For example, 650 milliwatts are required to drive the tubes here considered.[5] Care must be taken in designing the preceding driving, or input, stage so that it can supply the required power without distortion.

8-10. *AN AUDIO-FREQUENCY AMPLIFIER*

The amplifier of Fig. 8-16 is to be used solely to illustrate certain principles treated in preceding chapters. It is not to be considered the latest in

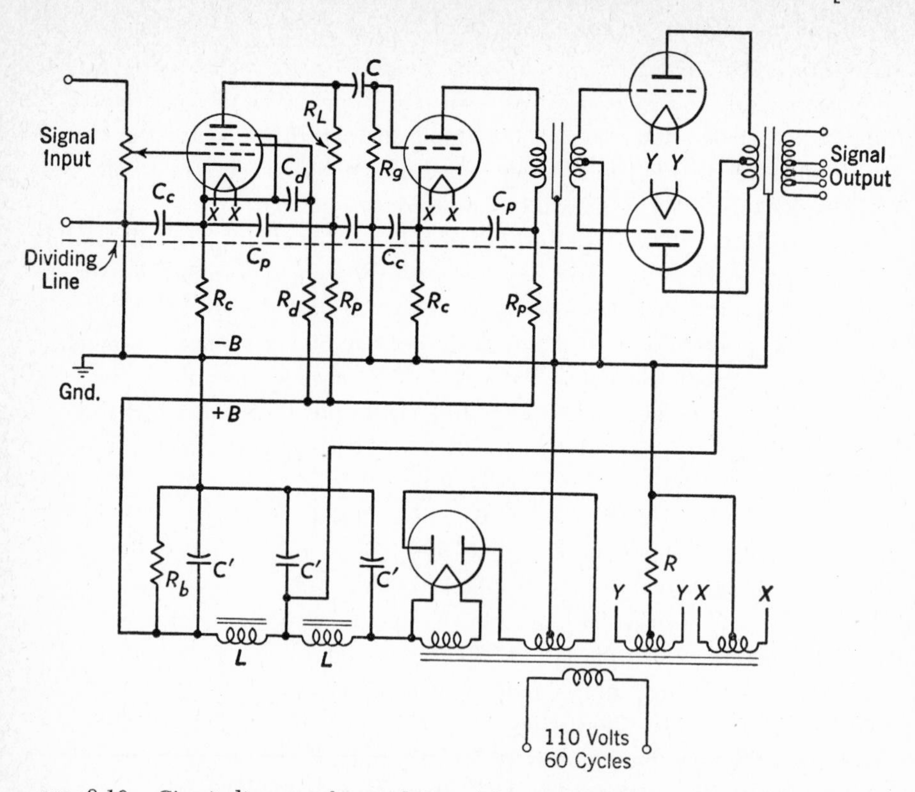

FIG. 8-16. Circuit diagram of an audio amplifier. Capacitor C and resistors R_L and R_g are the usual coupling units. Capacitors C_c and resistors R_c serve to maintain the correct grid bias. Resistor R_d carries the direct screen-grid current. The voltage drop in this resistor subtracts from the rectifier voltage and provides the correct positive voltage for the screen grid. Capacitor C_d grounds the screen grid from an alternating voltage standpoint. Capacitors C_p are called decoupling capacitors and provide a return path to the cathode for alternating signal components. Resistors R_p are decoupling and dropping resistors. They carry the direct component of the plate current. In these resistors there is an IR drop that subtracts from the power-supply voltage, thus providing the correct positive voltage for the plate. Resistors R_p also increase the impedance of the path for alternating current that might otherwise flow into the power supply. Capacitors C' and coils L are the filter across which is the bleeder R_b. The push-pull output stage is connected to the power supply as indicated because complete filtering is not necessary. Resistor R is a biasing resistor for the push-pull power-output stage, and in some circuits may need a by-passing capacitor. The first two stages are voltage amplifiers. Values of resistance, inductance, capacitance, etc., can be found in tube manuals.

amplifier circuitry. Above the broken dividing line is the path from tube to tube for the alternating signal being amplified; all signal components *must* be kept *above* this line. The capacitors immediately above the dividing line, and the resistors just below the line, serve to separate the alternating signal components from the direct components furnished by the power supply at the bottom. The functions of the various elements that are not apparent from preceding discussions on amplifiers are explained beneath Fig. 8-16 or in the following paragraphs.

In early amplifiers "C" batteries were used to bias the tubes. Later, a small "biasing cell" was developed. With separate-heater tubes the biasing voltage can be obtained by having the direct plate current flow through a cathode resistor such as R_c of Fig. 8-16. This is termed **self-biasing.** The direction of the direct plate-current component is such that the cathode is made positive with respect to ground, which is equivalent from a direct-voltage standpoint to making the grid negative with respect to the cathode. If the alternating-current component of the plate voltage passes through a cathode resistor, negative current feedback causing degeneration will result (section 8-19). This might seriously affect operation as an amplifier. Referring to Fig. 8-16, capacitors C_c and C_p provide a path of low impedance between stages for signal components. In some circuits cathode capacitor C_c would be connected directly across cathode resistor R_c. The reactance of the capacitor should be low—say less than 10 per cent—in comparison with the resistance of the resistor for the lowest frequencies to be amplified.

8-11. *COMMON IMPEDANCES*

As has been mentioned, the alternating components should be kept above the dotted line of Fig. 8-16. *It is very important that in amplifiers alternating-current components do not flow through impedances that are common to two or more tubes.* If this occurs, one tube may affect other tubes to the extent that the stages will oscillate, usually at a low frequency; this is called **motor boating.** One possible cause of motor boating will now be explained.

Referring to Fig. 8-17, the impedance of the input of the power supply is common to the plate circuits of *both* tubes. Because of the intervening amplifying stages, the signal level is large in the plate circuit of the second tube. Thus, a large signal alternating-current component will flow through the input impedance of the power supply, and an alternating voltage drop will exist across this input impedance which is *common* to all stages. This is a case of feedback in a sense, because any output signal will produce a voltage variation in a preceding stage of amplification. If the phase relations are correct, oscillations may result. In the circuit under study this is most likely to occur at very low frequencies such that the reactance of filter capacitors is appreciable.

If resistors, as shown dotted, are connected in series with the plates of the tubes and if capacitors, also shown dotted, are connected from the plate circuits to the cathodes, then the stages are said to be **decoupled.** This means

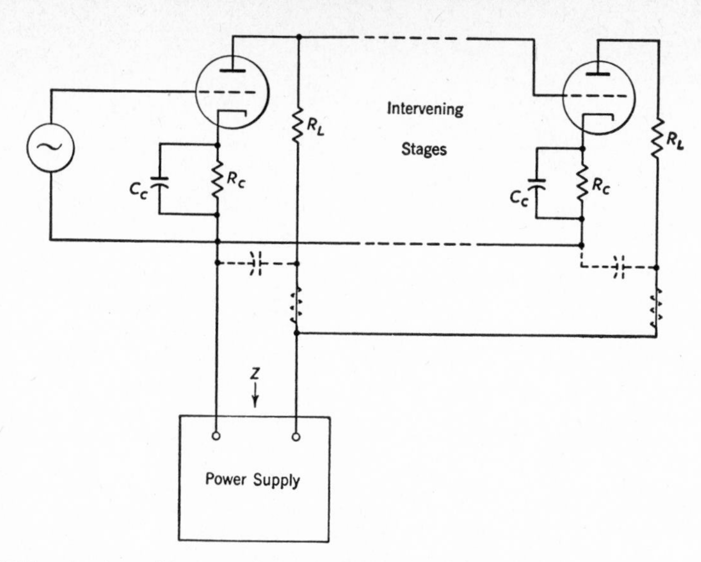

FIG. 8-17. An impedance, such as the input impedance of a power supply that is common to several amplifier stages, may cause oscillations.

that the impedance of the paths leading to any common impedance (such as the input of the power supply) is made high with the resistors, and also, the paths have been shunted with the low reactance of the large capacitors. Thus, variations in the plate circuit of the output tube largely are prevented from affecting preceding tubes, and the tendency to oscillate or motor boat is reduced. Of course, connecting capacitors directly across common impedances is effective in some instances, but a power supply already has capacitors in it, and yet motor boating may result. Motor boating is a low-frequency phenomenon, and even a large capacitor has considerable reactance at such frequencies.

8-12. *PHASE INVERTERS*

Referring to Fig. 8-16, an interstage transformer is used to connect from the "single-ended" voltage amplifier to the push-pull power stage. The split secondary of this interstage transformer provides a voltage for the grid of the lower tube that is 180° out of phase with the grid voltage impressed on the upper tube. Sometimes it is desired to obtain a signal voltage 180° out of phase without a transformer. A circuit that accomplishes this is called a **phase inverter.**

A circuit for phase inversion is shown in Fig. 8-18. The upper tube is a resistance-coupled voltage amplifier in the usual sense. The amplified output voltage existing across R_L and R_g is essentially 180° out of phase with the

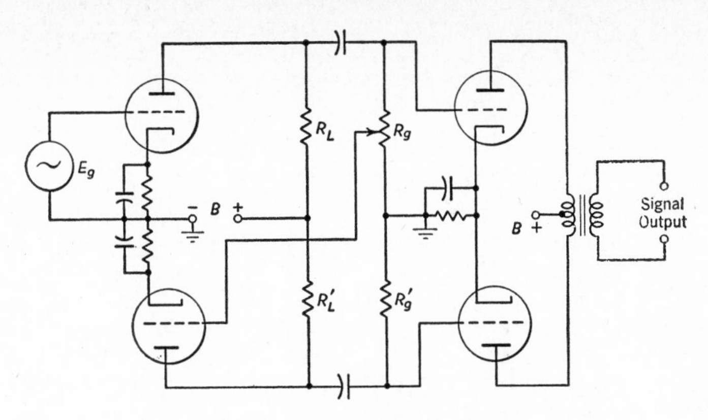

FIG. 8-18. A phase-inverting circuit.

impressed signal voltage E_g. A portion of this amplified output voltage is selected from resistor R_g, impressed on the grid of the lower tube, and is then amplified. Thus the signal voltages impressed on the grids of the two tubes are essentially 180° out of phase, and the amplified signal voltages across R_L and R_L' in the plate circuits will be 180° out of phase. The two grid input signal voltages across R_g and R_g' are also essentially 180° out of phase, and for proper adjustment of the voltage divider R_g can be made equal.

8-13. *RADIO TRANSMITTERS*

Audio-frequency power amplifiers were treated in the preceding pages. The following pages will be devoted to *radio-frequency* power amplifiers. Before considering these, however, it is advisable to discuss radio transmitters. In an amplitude-modulated radio-telephone transmitter the speech signals are used to modulate (Chapter 12) a radio-frequency carrier so that the information may be transmitted through space at radio-frequencies. There are many reasons why a frequency shift, or translation, is necessary. Two reasons are as follows: Energy cannot be radiated effectively from ordinary antennas at audio-frequencies; also, if translations to various radio-frequency channels were not accomplished, all stations would operate at *the same frequency band* without the possibility of receiving one station and excluding others, at least by present means.

As will be shown in Chapter 12, in the process of modulation, the information is translated from audio-frequencies to radio-frequencies. If a broadcast station operates at a carrier frequency of 1,000,000 cycles, and if the voice frequencies vary from 100 to 5000 cycles, then the signal radiated by the antenna will contain frequency components as follows: The carrier plus the voice, varying between the limits of 1,000,100 and 1,005,000 cycles; the carrier, of 1,000,000 cycles; and the carrier minus the voice, varying between the limits of 999,900 and 995,000 cycles. The first group of these is called the **upper sideband,** and the last the **lower sideband.** The radiated wave containing these three components appears as in Fig. 8-19.

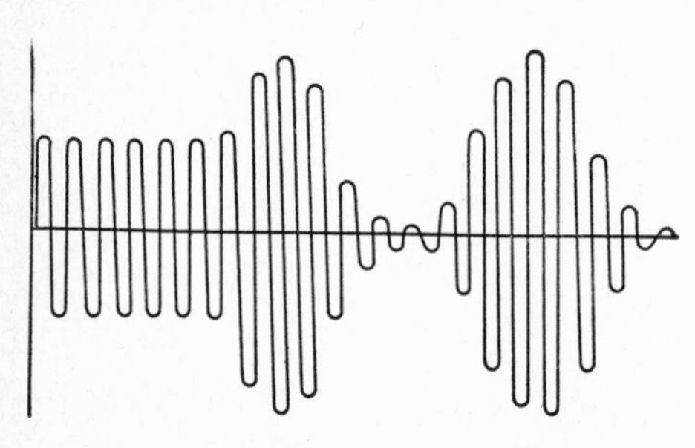

FIG. 8-19. Showing the shape of a wave before and after modulation. The unmodulated portion at the left contains only the carrier; the modulated portion contains *three* components: The carrier frequency; the carrier plus the voice (upper sideband); the carrier minus the voice (lower sideband). The component of carrier-frequency *does not* vary in amplitude in the modulated portion. The *apparent* change in amplitude is caused by the presence of the sidebands.

There are two general types of radio transmitters. In one, modulation occurs at high-power level in the last stage (before the antenna); this is called **high-level modulation.** In the other, modulation occurs at low-level, and stages of amplification intervene between the modulating stage and the antenna; this is called **low-level modulation.** The two systems are shown in Fig. 8-20.

A *high-level transmitter* functions as follows: The crystal-controlled oscillator (Chapter 9) generates the carrier, a single-frequency radio wave. The output of the oscillator drives a class-C radio-frequency power amplifier. The audio-frequency amplifier is driven by a microphone. The audio amplifier may be class-A voltage-amplifying stages driving class-AB or class-B power-output stages. The radio-frequency amplifier and the audio-frequency amplifier together drive the **modulator,** defined [17] as "a device to effect the process of modulation." The modulator is the stage in which the information is translated from audio-frequencies to the radio-frequency sidebands. The audio-frequency *driving stage* sometimes is incorrectly called a modulator. In high-level modulation the modulator is connected directly to the transmitting antenna.

A *low-level transmitter* functions as follows: The carrier is generated in the crystal oscillator and the output is amplified, but not to a high-power level as in the preceding transmitter. Similar statements apply to the audio-frequency amplifier. Modulation is effected at low-power level as compared to the type previously treated. Class-B radio-frequency stages are used to increase the power in the modulated signal to the level desired for radiation. Class-A *power* amplifiers are not used at radio-frequencies because they are less efficient than class-B and class-C power amplifiers.

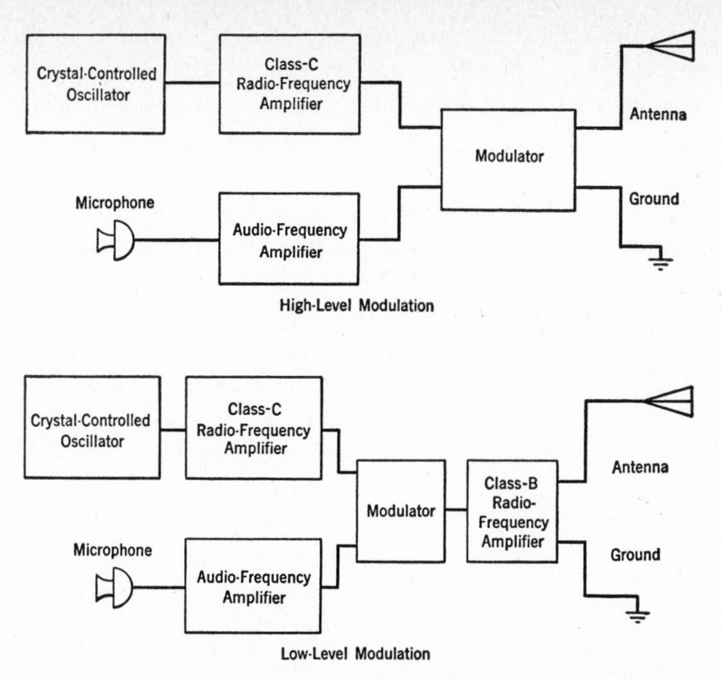

FIG. 8-20. Basic types of amplitude-modulation radio transmitters.

8-14. *RADIO-FREQUENCY CLASS-B POWER AMPLIFIERS*

These amplifiers commonly operate in class-B2 with grid-current flowing. As explained in section 8-9, two tubes must be used in *audio-frequency* class-B amplifiers so that even-numbered harmonics will be canceled in the output transformer. Two tubes per stage sometimes are used so that a class-B *radio-frequency* amplifier will be balanced with respect to "ground"; however, if an unbalanced amplifier is satisfactory, one tube per stage may be used, as will be explained.

In Fig. 8-21 an amplitude-modulated signal containing a carrier component and upper and lower sidebands is shown driving a tube biased to cutoff and operating in class-B. It is assumed that the wave shown *has been* modulated by a sine wave of, for example, 1000 cycles per second. The load in the plate circuit is so chosen that the dynamic characteristic curve is approximately a straight line. As Fig. 8-21 indicates, the impressed modulated signal wave will be distorted. The tuned load circuit largely suppresses the undesired components that are created by the distortion, but accepts and passes the desired signal.

As indicated in Fig. 8-22, a parallel circuit, resonant to the carrier frequency, is in the plate circuit. *To this frequency,* and *to the sidebands* which also are being amplified, this resonant circuit is essentially a *pure resistance* of the correct value to load the tube properly. To the harmonics the load is a low value of reactance; hence, these unwanted components are amplified little. The term R_L includes the reflected impedance of the connected load resistor that represents the antenna or the following amplifier stage to which the circuit shown is supplying power. Because the load essentially is pure resistance to the carrier and sidebands, the amplifier can be represented by the equivalent circuit of Fig. 8-22. Load resistance R_L *presents itself only to the carrier and sideband alternating components.* To the direct component, the coil has negligible resistance.

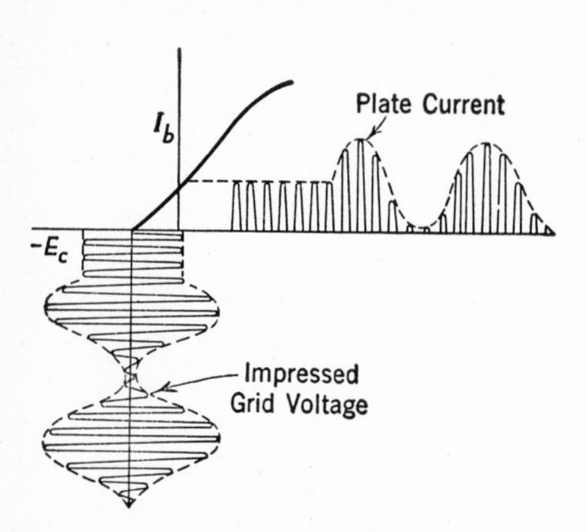

FIG. 8-21. In the class-B amplifier the plate current is badly distorted. The resonant tuned load circuit selects only the desired frequency components and does not transmit the undesired products of distortion.

In Fig. 8-22 a sinusoidal signal voltage is impressed between grid and cathode of a class-B amplifier tube. The instantaneous plate current i_b will flow for a time equal to about one-half cycle and represented by the angle θ, sometimes called the **angle of flow.** Although this plate current appears to be "spurts," it is a recurring phenomenon and can be analyzed into a fundamental (of the same frequency as the alternating signal voltage applied to the grid), and a series of harmonics. Thus, the *instantaneous voltage* e_b between the plate and cathode is composed of the constant direct plate voltage E_b and the value of the fundamental alternating voltage component at that instant. The design of a class-B radio-frequency amplifier is similar to that of the class-C amplifier to be considered in the following pages.*

8-15. *RADIO-FREQUENCY CLASS-C POWER AMPLIFIERS*

The basic circuit for a class-C amplifier is the same as Fig. 8-22 for a class-B amplifier. The grid of the class-C tube is biased with direct voltage to considerably beyond cutoff, and plate current flows only for a small part of the cycle, as Fig. 8-23 indicates. Since the angle of flow is so small, the distortion is great. The class-C amplifier is not used for amplifying ampli-

* W. H. Doherty in Proc. I.R.E., Sept. 1936, announced a high-efficiency power amplifier for modulated waves. This special class-B amplifier has been used extensively in radio-broadcast transmitters.

tude-modulated waves such as those shown in Fig. 8-21. As Fig. 8-20 indicates, the class-C amplifier is used in radio-telephone transmitters for amplifying a *single frequency,* such as the carrier wave. It also is used as an amplifier in radio-telegraph transmitters. The signal is distorted badly, and harmonics are produced. The plate load circuit is resonant to the fundamental frequency, however, and the harmonics are not effectively amplified. It should be remembered that any continuously recurring current can be analyzed into a fundamental and harmonic components.

If the grid of the tube is driven sufficiently positive, a "dip" may occur in the *plate current* as in Fig. 8-23. This is caused by reaching the region of saturation (section 3-3), and because the highly positive grid is drawing many of the available electrons. The *grid current* may show a similar dip because of secondary emission from the grid (section 4-11). The grids of large power tubes are specially constructed so that thermionic and secondary grid emission is minimized; also, so that the grids can handle the currents encountered.

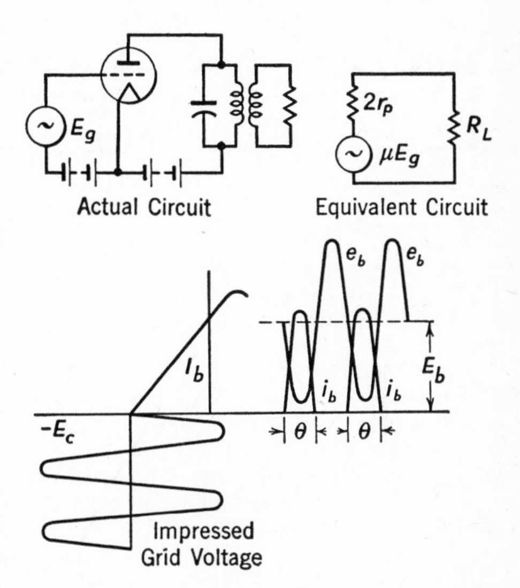

FIG. 8-22. Actual and equivalent circuit and diagram for a class-B radio-frequency amplifier showing the shape of the plate current and instantaneous plate voltage. The effect of the tuned resonant plate load causes the instantaneous plate-voltage variations to exceed the plate-supply voltage E_b. The internal resistance of the fictitious generator is taken as $2r_p$. (See W. L. Everitt's *Communication Engineering,* second edition; McGraw-Hill Book Co.)

The class-C amplifier operates into a resonant circuit tuned to the fundamental which it is desired to amplify. In the following explanations the use of the term "tank circuit" to describe the parallel resonant load circuit will be avoided. Also, no reference will be made to "flywheel action." The tuned parallel load circuit of a class-C amplifier operates just as any parallel resonant circuit (section 7-18).

8-16. *DESIGN OF A CLASS-C AMPLIFIER*

The problem usually is to find conditions of operation and circuit constants such that the maximum amount of power and high efficiency can be obtained with a given tube. The first cost and operating charges for a large tube and associated apparatus are high; thus a saving is effected by using the smallest tube that will carry the load. The basic circuit for the class-C amplifier is shown in Fig. 8-24. The tube is biased considerably beyond cutoff, and a large alternating signal voltage is impressed between grid and cathode. The angles of flow θ_p for the plate current and θ_g for the grid current are considerably less than 180°.

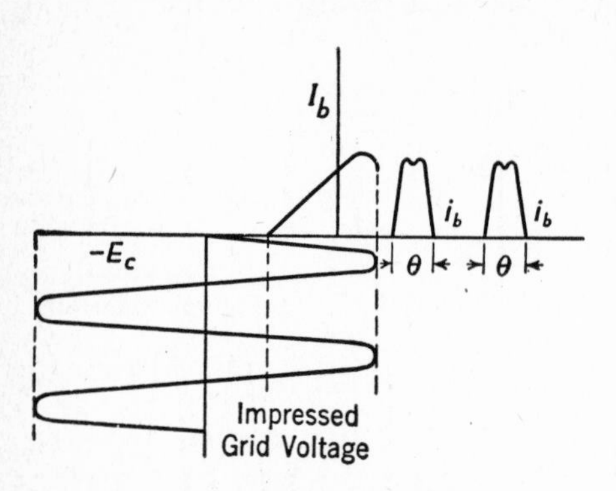

FIG. 8-23. For class-C amplification the plate current may show "dips" if the tube is driven with a large grid voltage as indicated. The angle θ is less than in Fig. 8-22 for the class-B amplifier. The instantaneous plate-voltage variations are similar to those shown in Fig. 8-22.

The purpose of the stage is to deliver power to some device such as a radio transmitting antenna. This power-absorbing circuit is represented by R of Fig. 8-24a, and this value of R will be reflected into the parallel resonant circuit. At the frequency at which this parallel resonant circuit is tuned, its input impedance will be R' as shown in Fig. 8-24b. This resistance is offered *only* to the fundamental alternating component of the plate current. The phase relations of the grid current, plate current, and plate voltage are shown in Fig. 8-24d. Because of the impedance R', the plate voltage rises above and falls below the supply voltage E_{bb}.

Triodes are used in class-B and class-C where large amounts of power are to be handled. With the cathode heating power normal, the design of a class-C stage must consider plate voltage, grid voltage, plate current, and grid current. In the preceding chapters the characteristics of a tube were plotted either as constant plate-voltage curves or as constant grid-voltage curves. In the design of class-B and class-C radio-frequency amplifiers, it is convenient to use constant plate-current and constant grid-current curves, with plate voltage and grid voltage as variables. Such a set of curves for a triode is shown in Fig. 8-25. Curves of this type are furnished by tube manufacturers or can be obtained experimentally (section 4-12). The design of a class-C amplifier is a cut-and-try process,[18] the number of attempts that are necessary depending largely on the experience of the designer and familiarity with the tube or similar tubes.

The first step is to select a so-called **Q point,** or quiescent (no-signal) electrode potential. In Fig. 8-25, this point has been selected at a direct plate

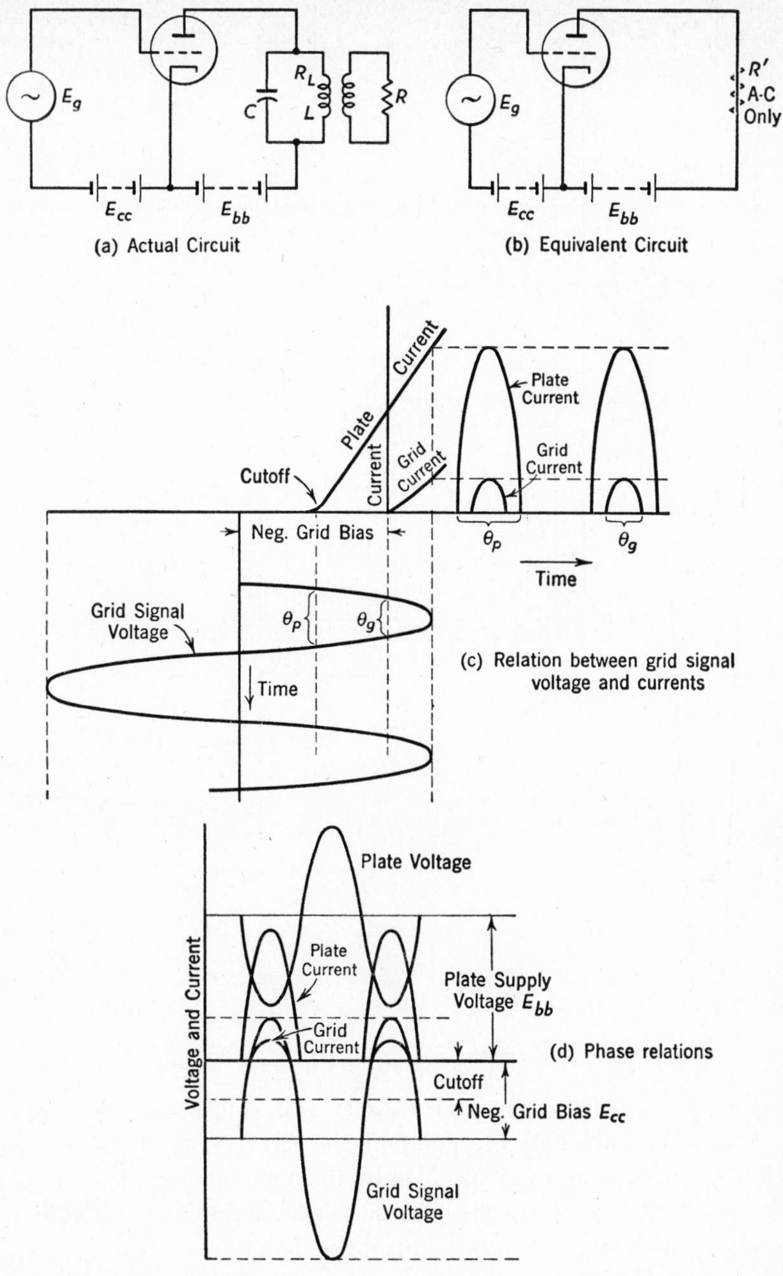

FIG. 8-24. Circuits for class-C radio-frequency power amplifier, and instantaneous current and voltage variations.

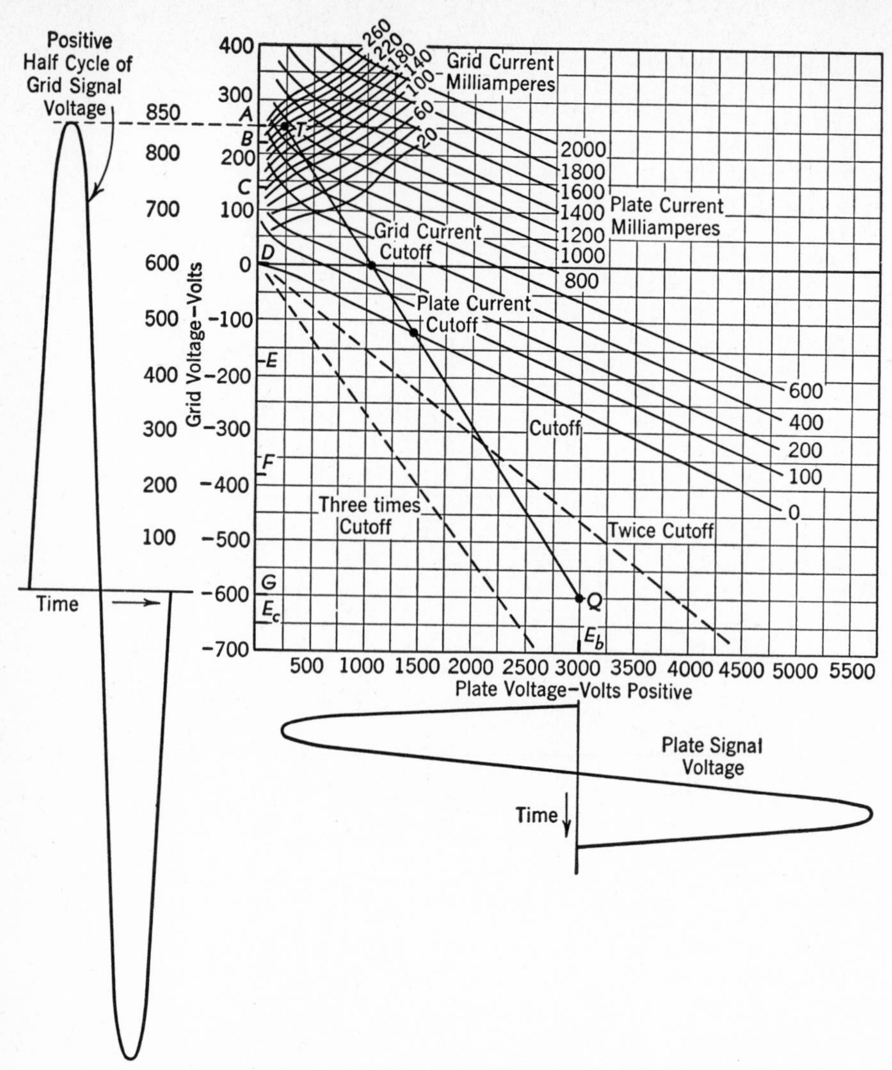

FIG. 8-25. Constant-current curves and details of class-C amplifier design.

voltage of $E_b = +3000$ volts, and a direct grid voltage of -600 volts. To ensure long life and safety of operation, the direct plate potential selected is below the maximum value specified by the manufacturer. The direct grid voltage is selected at -600 volts because this is 2.5 times the cutoff bias that experience has shown to be a good operating point.

The second step is to assume **a tip point** or ***T* point.** This point determines the maximum positive value to which the grid is driven, and the minimum positive potential to which the plate falls (Fig. 8-25). The T point assumed on Fig. 8-25 is $e_c = +250$ volts and $e_b = +250$ volts, small letters e_c and e_b being used to represent *total instantaneous values.* It will be noted that the plate voltage does not fall below the grid voltage. This is common practice. The T point determines the magnitude of the alternating signal grid-voltage variations or swing that can be impressed on the tube; for Fig. 8-25 this will be a maximum voltage value of 850 volts from the −600-volt value to the +250-volt value. The T point also fixes the maximum value of the alternating plate-voltage value. When a signal voltage of 850 volts maximum value is impressed on the grid, the maximum plate-voltage swing will be from +3000 down to +250 volts, or a maximum signal voltage of 2750 volts. A line connecting the Q and T points determines the plate-current cutoff point and the grid-current cutoff point as shown in Fig. 8-25. This will determine the angle of flow for both the plate current and grid current.

The third step is to determine maximum and the average values of the plate and grid currents that will flow. There are various means of accomplishing this; one of the simplest was developed by Chaffee. This method[19] consists of evaluating the maximum and average values of the plate-current and grid-current pulses such as shown in Fig. 8-24. The average value of current is found from the relation

$$I_{\text{av.}} = 0.0833(0.5A + B + C + D + E + F) \qquad \text{8-14}$$

and the maximum value of the fundamental component from the relation

$$I_{\text{max.}} = 0.0833(A + 1.93B + 1.73C + 1.41D + E + 0.052F) \qquad \text{8-15}$$

where A, B, C, D, E, and F are the total instantaneous values of the plate and grid currents flowing at 15° increments of grid driving voltage. The values A, B, C, etc., are marked on Fig. 8-25. The evaluation can best be illustrated by Table 8. From this table, the direct or average value of plate current is 0.181 ampere, the maximum value of the *fundamental* component of the plate current is 0.331 ampere, the direct or average value of the grid current is 0.0226 ampere, and the maximum value of the *fundamental* component of the grid current is 0.0434 ampere.

The approximate grid dissipation is calculated as follows: First find the grid-signal driving power. This equals the effective value of the grid driving voltage multiplied by the effective value of the grid alternating-current fundamental component or

$$\text{Grid signal driving power} = \frac{E_{c(\text{max.})}}{\sqrt{2}} \times \frac{I_{g(\text{max.})}}{\sqrt{2}} = \frac{850}{\sqrt{2}} \times \frac{0.0434}{\sqrt{2}} = 18.5 \text{ watts}$$

This will be the power supplied to the grid circuit by the source of grid driving voltage. When this alternating signal power is put into the grid, the grid

Table 8

EVALUATION OF CURRENT COMPONENTS FOR CLASS-C OPERATION

(Method from reference 19) (Currents in amperes)

ANGLE θ (DEGREES)	COS θ	$E_{g(\max.)}$ TIMES COS θ	POINT	PLATE CURRENT	I_b FROM EQ. 8-14	$I_{p(\max.)}$ FROM EQ. 8-15	GRID CURRENT	I_c FROM EQ. 8-14	$I_{g(\max.)}$ FROM EQ. 8-15
0	1.0	850	*A*	0.92	0.46	0.92	0.187	0.094	0.187
15	0.966	821	*B*	0.86	0.86	1.66	0.140	0.140	0.270
30	0.866	736	*C*	0.62	0.62	1.07	0.037	0.037	0.064
45	0.707	600	*D*	0.23	0.23	0.32	0	0	0
60	0.500	425	*E*	0	0	0	0	0	0
75	0.259	220	*F*	0	0	0	0	0	0
90	0.0	0	*G*	0	0	0	0	0	0
				Sums	2.17	3.97		0.271	0.521
				0.0833 × Sums	0.181	0.331		0.0226	0.0434

circuit acts as a rectifier, producing a direct current of 0.0226 ampere. This current flows in the grid circuit in a direction *against* the grid-bias voltage. Thus,

$$\text{Power dissipated in grid circuit external to tube} = E_c \times I_c = 600 \times 0.0226 = 13.6 \text{ watts}$$

If the source of grid bias were a battery, this power would be dissipated in charging the battery. The approximate grid power dissipated *within* the tube is then equal to the difference between alternating power input and the rectified power output, or 18.5 − 13.6 = 4.9 watts.

The approximate plate dissipation is calculated as follows: First find the power input to the plate circuit from the source of plate voltage. This will be the product of the direct plate voltage and the direct plate current, or

$$\text{Plate power input} = E_b \times I_b = 3000 \times 0.181 = 543 \text{ watts}$$

Next find the plate signal output power. This equals the product of the effective value of the alternating plate signal voltage and the effective value of the alternating component of plate current, or

$$\text{Plate signal output power} = \frac{E_{p(\max.)}}{\sqrt{2}} \times \frac{I_{p(\max.)}}{\sqrt{2}} = \frac{2750}{\sqrt{2}} \times \frac{0.331}{\sqrt{2}} = 455 \text{ watts}$$

The approximate plate power dissipated *within* the tube is then equal to the difference between the direct power input and the alternating power output, or 543 − 455 = 88 watts.

The plate efficiency equals the power output divided by the power input, or

$$\text{Plate efficiency} = \frac{P_o}{P_i} = \frac{455}{543} \times 100 = 83.8 \text{ per cent}$$

The power amplification is the ratio of the signal power output from the plate to the signal power input to the grid, or 455/18.5 = 24.6.

The power input, power output, grid dissipation, and plate dissipation, and other values calculated should be checked against the manufacturer's ratings for the tube under consideration. Before deciding on a final Q point and T point, it may be advisable to make the preceding calculations at several other points to determine the optimum values. Once these are selected and the preceding values are calculated, the output circuit must be calculated. The output circuit, in reality, is the circuit that draws power from the tube, and if the tube is to operate between certain Q and T points, and give a specified output and efficiency, then the plate output circuit must have given characteristics. Before attempting to design the output circuit it will be well to review its function.

As is evident from the preceding discussions, and as indicated by Fig. 8-24, the plate current in a class-C tube is distorted badly. This distorted current will, however, contain a fundamental component which is the signal frequency being amplified. It is the function of the parallel resonant circuit in series with the plate of the tube to load the tube correctly, and to select the fundamental signal component and draw power from the tube at this frequency. The power must not be dissipated in the parallel circuit itself, but must be passed on to the device that is to be driven by the class-C stage. To draw the correct amount of power from the tube, this parallel circuit must have a certain input resistance. To select the fundamental component and reject the undesired harmonics, the tuning of the parallel circuit must be reasonably "sharp," so that little impedance is offered by it to the harmonics. Furthermore, the correct input resistance must exist when the device to be driven is coupled into the parallel circuit (section 7-18). Experience has shown that for good results the effective Q, which is the ratio of reactance to resistance with the load reflected, should be about 12.

The equivalent input resistance of the parallel circuit with the reflected or coupled loading device must equal the alternating output power divided by the effective value of the current squared, or

$$R_e = \frac{455}{(0.331 \times 0.707)^2} = 8320 \text{ ohms}$$

Because the resonant circuit must be adjusted experimentally for correct operation, the calculations to determine the size of the coil and capacitor may be approximate. When adjusted for resonance so that the input impedance is pure resistance, and neglecting losses in the capacitor, the equivalent input resistance is equal to the reciprocal of the conductance of the branch composed of the coil of inductance L, and effective and reflected resistance R_L. That is,

$$R_e = \frac{R_L^2 + X_L^2}{R_L} = R_L + \frac{\omega^2 L^2}{R_L} = \frac{\omega^2 L^2}{R_L} \tag{8-16}$$

where R_L is negligible in comparison to the term $\omega^2 L^2/R_L$. From this equation,

$$R_e = Q_L \omega L \qquad \text{and} \qquad L = \frac{R_e}{\omega Q_L} \tag{8-17}$$

where R_e is the equivalent input resistance at resonance and must be of such value as correctly to load the tube, Q_L equals $\omega L/R_L$ and has a value of about 12, and R_L is the effective coil resistance *plus* the resistance reflected into the coil by the loading device. For an approximate solution, the capacitance of the capacitor can be found when the relation $\omega L = 1/(\omega C)$, and

$$C = \frac{1}{\omega^2 L} \tag{8-18}$$

In the above equations all values are in ohms, henrys, and farads.

The angle of flow for both the plate current and grid current can be found from Fig. 8-25. Assume that the impressed grid signal voltage is sinusoidal. Plate current will start to flow when the alternating grid signal voltage of 850 volts maximum first drives the grid beyond the plate-current cutoff point. Using the line Q–T of Fig. 8-25, we see that plate current first flows when the grid is 125 volts negative, or when the *alternating* grid signal has gone from 0 to +475 volts. The sine of the angle at which plate current starts is $\sin\theta = 475/850 = 0.559$, and $\theta = 34°$. Plate current then flows until the angle is 146°, giving an angle of flow $\theta = 112°$. This is a small angle of flow, an angle of 120° to 140° being more common.

The grid-current angle of flow can be found as follows: Grid current will flow when the grid signal voltage drives the grid beyond grid-current cutoff. From Fig. 8-25 this will occur approximately where the +850 maximum grid driving voltage exceeds the −600-volt bias value. The angle at which grid current starts to flow is, accordingly, $\sin\theta = 600/850 = 0.75$, and $\theta = 49°$ approximately. Grid current starts to flow, therefore, when the angle of the applied grid signal voltage is 49°, and flows until the angle is 131°; that is, it flows for about 82°.

In this discussion the angle of flow is taken as the *entire* angle that current flows. There seems to be no general agreement as to this, however, and the angle of flow is often specified as one-half the angle that current flows. In the preceding calculations the class-C tube is in effect operating into a resistance load. Calculations for class-C tubes with reactive loads also may be made.[20]

8-17. *FREQUENCY MULTIPLIERS*

When a tube is operated in class-C, the plate current flows in pulses, somewhat as in a half-wave rectifier. Such a current contains a direct component, and alternating components. The alternating components consist of a fundamental of the same frequency as the grid-driving voltage, and various harmonics. If the resonant parallel load circuit is tuned to the fundamental, then the output will be of the same frequency as the grid input signal, and the device is a class-C amplifier. If the resonant circuit in the plate of the tube is tuned to one of the harmonics, then the output will be at the frequency of the harmonic, and the device is a **frequency multiplier.** Such operation is common for increasing frequency.

The shape of the current pulses, and thus the harmonic content, depend on the angle of flow of the plate current. For this reason, there is an optimum angle of flow for a frequency multiplier, depending on the particular frequency component desired. An approximate rule to follow is that the plate-current angle of flow for a frequency multiplier is

$$\theta_p = \frac{180^\circ}{n} \qquad \text{8-19}$$

where n is the ratio of the desired output frequency to the grid driving frequency, and θ_p is the *entire* angle during which plate current flows. Thus, for a frequency tripler the angle of flow should be about 60°. The alternating power output of a frequency multiplier is about $1/n$ of that of a similar class-C amplifier.

8-18. *NEUTRALIZATION*

Triodes are used in class-B and class-C radio-frequency *power* amplifiers, and in frequency multipliers. Because radio-frequencies are being amplified, considerable signal is fed back into the grid circuit from the plate circuit through the grid-plate capacitance. This feedback may cause the amplifier to **oscillate** or **regenerate.** In radio-frequency amplifiers, special precautions must be taken to prevent oscillations from occurring, and this is known as

neutralization. Proper neutralization is important, and radio-frequency power amplifiers using triodes must be adjusted carefully so that oscillation does not occur. Doherty has made an excellent analysis of neutralization.[21] Because of the shielding action of the screen grid in the tetrode, and the screen grid and suppressor grid in the pentode, these tubes have far less tendency to oscillate than do triodes, and neutralization often is unnecessary.

Reason for Feedback. A simplified circuit of class-C *unneutralized* radio-frequency power amplifier using a single triode is shown in Fig. 8-26. The

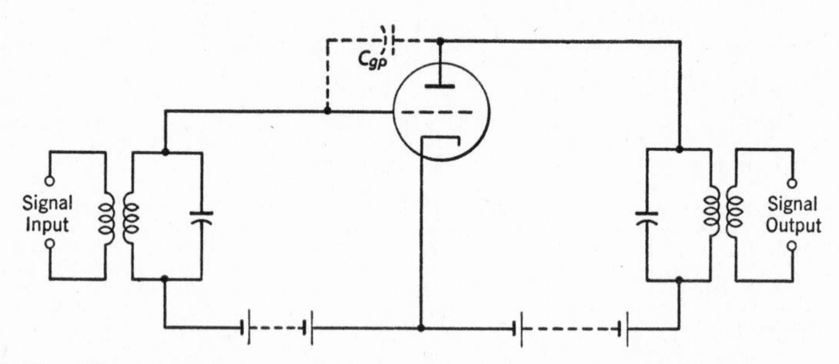

FIG. 8-26. The capacitance within a triode between the grid and plate electrodes can be conveniently represented by the external capacitor C_{gp}. Signal feedback from the plate to the grid circuit through this capacitance may cause a triode to oscillate.

grid-plate capacitance that exists *within* the tube is shown externally as a dotted capacitor connected between the plate and grid. When the input and output circuits are tuned to resonance, they are equivalent electrically to resistors connected between the grid and plate circuits. In accordance with section 7-8, the input signal alternating voltage across the tuned input circuit will be 180° out of phase with the output signal voltage across the tuned output circuit, and because of this, the alternating voltage across the capacitance C_{gp} will be numerically the *sum* of these input and output voltages. This combined voltage will cause a current flow through the interelectrode capacitance C_{gp}. This current will cause a voltage drop across the resonant circuit connected between grid and cathode. For the 180° phase relation mentioned above, the voltage drop in the grid circuit caused by the current fed back through the grid-plate capacitance will *not* add directly to the impressed grid voltage and will not tend to cause oscillations or feedback.

It is, however, difficult to obtain perfect adjustment of the parallel resonant circuits. Also, even if they were adjusted to perfect resonance, they could not offer zero impedance to other frequencies, and thus would cause *some* amplification for frequencies other than that frequency to which the circuits were tuned. The plate voltage and grid voltage relations for any frequency to which the plate load circuit is slightly inductive are such that the tube may oscillate. At such a frequency the voltage drop in the grid circuit adds in part to the impressed grid signal voltage and increases the net signal voltage between grid and cathode, thus tending to cause oscillation. Although there are several methods of obtaining neutralization, the system most widely used in radio-frequency amplifiers is intentionally and by controlled means to feed back into the grid circuit a signal of such magnitude and phase that it cancels the effect of the signal fed back through the grid-plate capacitance of the tube.

Coil Neutralization. This is a simple and effective means of preventing unwanted oscillations. Referring to Fig. 8-26, if a coil of the proper size is connected from the plate to the grid, then a current of the same magnitude as that flowing through the grid-plate capacitance will flow through the coil. These two currents will be substantially 180° out of phase, and the resultant current flowing back into the grid circuit from the plate will approach zero; thus, the plate and grid circuits are effectively isolated. It may be considered that the neutralizing coil in parallel with the grid-plate capacitance is in effect a parallel resonant circuit, the impedance of which approaches infinity as a limit. Then, the plate and grid circuits would be isolated and no feedback could occur. It is necessary to connect a capacitor in *series* with the coil so that the grid and plate may be at different direct potentials.

Capacitor Neutralization. It is common practice to feed back current through a capacitor into the grid circuit to offset the effect of the current fed back through the grid-plate capacitance. For such action to be effective, it is necessary to connect the capacitor to a voltage that is 180° out of phase with the plate voltage; otherwise, the two currents fed back into the grid circuit would be in phase and no neutralization would result. There are many methods of obtaining the required 180° out-of-phase voltage. But as Doherty has shown,[21] some of these methods, although they have been used for years, do not achieve true neutralization and are only partially effective.

The circuit of Fig. 8-27 provides the 180° out-of-phase voltage required to produce neutralization if the neutralizing current is fed back through a capacitor. The Circuit L–C_1 is adjusted so that inductive reactance predominates and the current through it will lag by 90°. This current will cause an IX_C drop across capacitor C_1 and this voltage will lag the current producing it by 90°. This produces a voltage across capacitor C_1 that is 180° out of phase with the plate voltage that is forcing a current through the grid-plate capacitance. When the neutralizing capacitor C_N is correctly adjusted, a current flows into the grid circuit that is equal in magnitude and opposite in phase to the current that flows through the grid-plate capacitance. In this way,

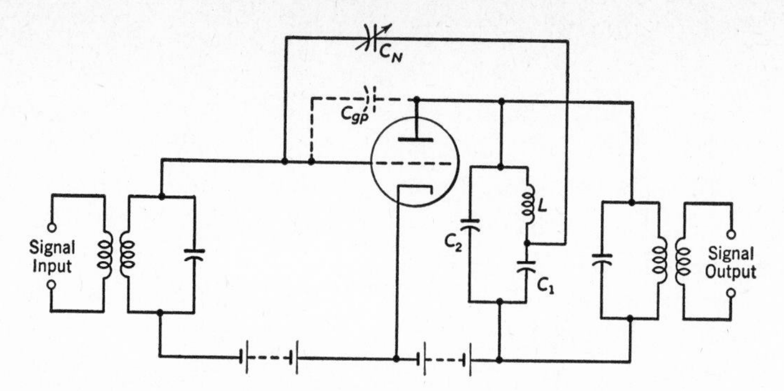

FIG. 8-27. The neutralizing capacitor C_N is adjusted to feed back signal 180° out of phase with that fed back by the interelectrode capacitance C_{gp}.

neutralization is accomplished. Capacitor C_2 tunes the circuit L–C_1–C_2 to resonance so that it offers a high impedance from plate to cathode and hence has but little effect on the amplifier as a whole.

Neutralization in Push-Pull Amplifiers. If two tubes are used in a push-pull, or balanced, radio-frequency amplifier using triodes such as Fig. 8-28, neutralization is simple. If identical tubes are used and the circuits are symmetrical throughout, then the two plate voltages will be 180° out of phase. The two cross-connected neutralizing capacitors C_N when adjusted to the correct value will then introduce neutralizing currents of the proper magnitude into each grid circuit, and oscillations will be prevented.

Adjustments for Neutralization. A simple means of adjusting the neutralizing circuits is to remove the plate voltage from the tube to be neutral-

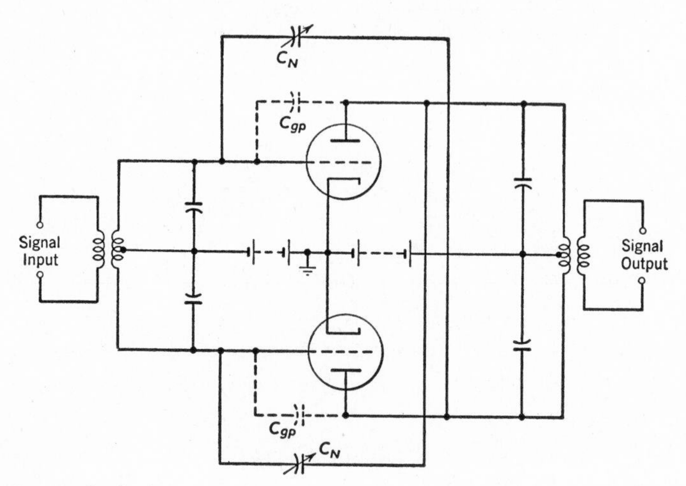

FIG. 8-28. Method of neutralizing a push-pull, or balanced, radio-frequency amplifier with capacitors C_N.

ized, and to apply an exciting voltage to the grid circuit. The resonant plate load circuit is then adjusted to resonance. If the circuit is originally partially neutralized, it may be necessary to vary the neutralizing capacitors until they are out of adjustment. Resonance in the parallel load circuit may be conveniently determined with a wave meter or a grid-dip meter. After the circuit is adjusted to resonance, the neutralizing capacitor is varied until the wave meter coupled to the load circuit shows a minimum or zero reading. The circuit is now effectively neutralized because if a signal in the grid circuit produces negligible signal in the plate circuit, then a signal in the plate circuit will induce negligible signal into the grid circuit.

8-19. *CONTROLLED FEEDBACK*

As explained in the preceding pages, feedback from the plate circuit to the grid circuit through the interelectrode capacitance may cause triodes to oscillate. Such feedback in amplifiers is undesired. However, there are instances where feedback has beneficial effects, provided that it is of the correct type, and is controlled. There are two types of feedback; one is **positive feedback,** or **regeneration,** in which the signal fed back combines with the signal to be amplified and *increases* the signal actually applied between grid and cathode. This results in an unstable amplifier and may cause oscillations. The other type is **negative feedback,** or **degeneration,** in which the signal fed back combines with the signal to be amplified, and *decreases* the signal actually applied between grid and cathode. This results in a stable amplifier having excellent characteristics for many applications.

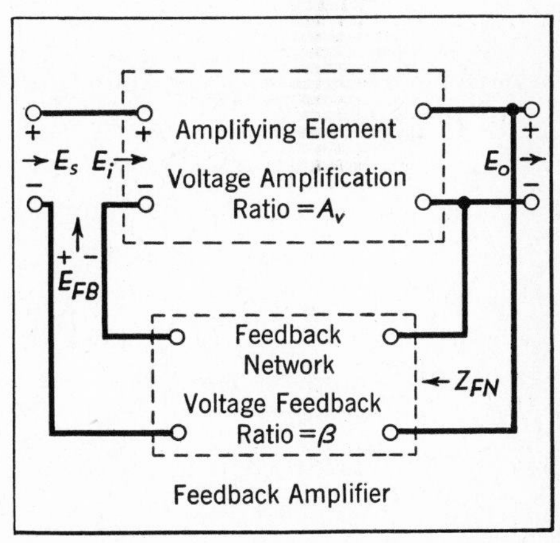

FIG. 8-29. An amplifier with feedback.

An amplifier arranged for controlled feedback is shown in Fig. 8-29. The complete **feedback amplifier** is enclosed by the rectangle of light solid lines. The amplifying tubes and circuit (the amplifying element) are enclosed by the broken lines. The voltage input to the amplifying element is E_i, the voltage output is E_o, and the voltage amplification of the amplifying element is $A_v = E_o/E_i$. Attention is called to the fact that the load impedance to be used in calculating A_v must include the input impedance of the feedback network, marked Z_{FN} on Fig. 8-29. The assumed positive directions for the voltages E_i and E_o are in accordance with convention (section 4-14). The feedback network impresses a voltage $E_{FB} = \beta E_o = \beta E_i A_v$ in series with the signal voltage E_s impressed on the feedback amplifier. A voltage equation for the input circuit is $E_i = E_s + E_{FB}$, which may be written $E_i = E_s + E_i \beta A_v$. For the amplifying element alone, $A_v = E_o/E_i$, then $E_i = E_o/A_v$, and it can be written that $E_o/A_v = E_s + E_o\beta$, from which

$$A_{FB} = \frac{E_o}{E_s} = \frac{A_v}{1 - A_v\beta} \qquad \text{8-20}$$

where A_{FB} is the over-all amplification of the entire feedback amplifier of Fig. 8-29. All terms in this equation may have both magnitudes and angles, and may be written as complex numbers. Also, the amplification ratio A_v and the feedback voltage ratio β in general are functions of frequency. For this reason, $A_v\beta$ of equation 8-20 may cause the over-all amplification A_{FB} to be greatly different at different frequencies. Also, the nature of A_v and β will depend on the type of amplifying element and the type of feedback network. Several values of $A_v\beta$ now will be considered.

Feedback: $A_v\beta$ Real, Positive, < 1.0. Assume that the *product* of the voltage amplification A_v of the amplifying element (enclosed by broken lines of Fig. 8-29) and the feedback voltage ratio β of the feedback network is a real positive number such as 0.5. Inserting this in equation 8-20 gives an over-all amplification of the feedback amplifier as $A_{FB} = 2A_v$. When βA_v is a real positive number less than unity, the voltage βE_o fed back is in phase with the applied signal voltage E_s, and E_i of Fig. 8-29 will be increased by the feedback, which is *positive*, or *regenerative*. From equation 8-20, $E_o = A_{FB}E_s = 2A_vE_s$. Hence, for the condition under consideration of $A_v\beta = 0.5$, the over-all gain of the feedback amplifier is twice that of the amplifying element alone, but if there is no input signal voltage, E_s is zero and the output signal E_o is zero. There is regeneration, but no oscillations. However, should the product $A_v\beta$ approach unity, then the denominator of equation 8-20 will approach zero, and the amplification of the feedback amplifier will approach infinity. This means that the feedback amplifier is in an unstable state, and if $A_v\beta$ is equal to or greater than unity, then an output voltage E_o can exist with no input signal voltage E_s. If the input terminals at E_s are connected together, sustained oscillations will result.

Feedback: $A_v\beta$ Real, Negative. For any real *negative* value of the product $A_v\beta$, the denominator of equation 8-20 will exceed unity, and A_{FB} will be less than A_v. The voltage βE_o fed back is in phase *opposition* to the applied signal voltage E_s. The feedback is *negative,* or *degenerative.* From equation 8-20, if $A_v\beta = -0.5$, then $A_{FB} = A_v/1.5$ or $A_{FB} = 0.67A_v$. The over-all voltage amplification with *negative* feedback is less than the amplification of the amplifying element alone. This reduction in amplification results in stable over-all amplification. The principle was used in the negative feedback amplifier developed by Black.[22]

Feedback: $A_v\beta$ Complex. The circuit of Fig. 8-29 has been investigated when $A_v\beta$ is real and positive, and when $A_v\beta$ is real and negative. The voltage amplification ratio A_v is real for midfrequencies, but is complex, having both magnitude and angle, at low and at high frequencies (section 7-12). If the feedback network is of resistors only, the voltage feedback ratio β will be real, but if it contains reactors, β may be complex. Also, the voltage amplification A_v may be complex at low frequencies, real at midfrequencies, and again complex at high frequencies. Hence, feedback conditions must be investigated when the product $A_v\beta$ is complex. Such studies are based on the following rule given [23] by Nyquist: "Plot plus and minus the imaginary part of $A_v\beta$ against the real part for all frequencies from zero to infinity. If the point $1 + j0$ lies completely outside this curve the system is stable; if not, it is unstable."

The method of plotting $A_v\beta$ is illustrated in Fig. 8-30. The magnitude and angle of A_v and of β must be calculated, or measured,[24] at frequencies from

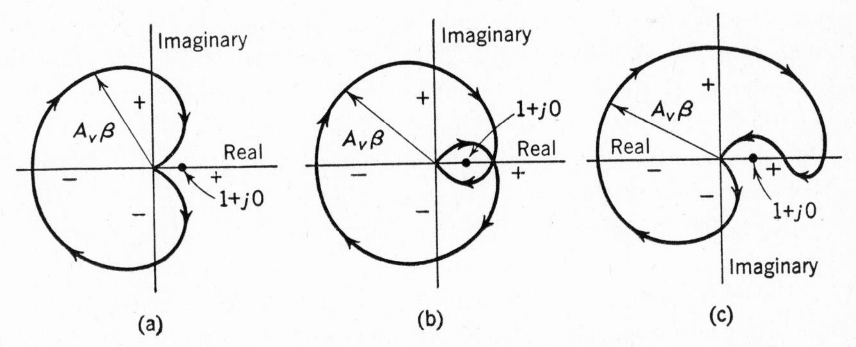

FIG. 8-30. Feedback characteristics obtained by plotting $A_v\beta$. The amplifier represented in (a) is stable, in (b) is unstable, and in (c) is conditionally stable.

zero to infinity. Then, the product $A_v\beta$ is determined, and this complex number is plotted on the real and imaginary axes.[25] If desired, it may be plotted in polar form. If the locus of the term $A_v\beta$ at frequencies from zero to infinity does not form a closed curve (as for an amplifier that will operate at zero frequency) the locus of the conjugate of $A_v\beta$ also should be plotted.[25]

If the curve does *not* pass through or enclose the $1 + j0$ point as in Fig. 8-30a, the amplifier with feedback $A_v\beta$ theoretically will be stable. From a practical standpoint, the curve should not approach $1 + j0$ too closely. If this point is enclosed as in Fig. 8-30b, then the amplifier is unstable because feedback is such that oscillations may occur. If a curve such as Fig. 8-30c is obtained, then the amplifier is *conditionally* stable, and will not oscillate. However, this amplifier may oscillate if A_v is small, as it may be momentarily when an amplifier first is energized. Then, the product $A_v\beta$ may be such that point $1 + j0$ is enclosed. Sometimes the curve resulting from plotting $A_v\beta$ is such that a question may arise whether or not point $1 + j0$ is enclosed. Nyquist gives a method for determining this fact.[23] Even for an amplifier as simple as a resistance-coupled amplifier, it is bothersome to use one equation to determine the voltage-amplifying ratio A_v at all frequencies from zero to infinity. It is simpler to make calculations of A_v using separate low-frequency, midfrequency, and high-frequency equivalent circuits.

8-20. *NEGATIVE FEEDBACK AMPLIFIERS*

Amplifiers with negative feedback are in wide use. Also, the feedback principle has been applied to many other electrical devices (loud-speakers, phonograph pickups, servo devices, etc.), and also to mechanical systems.

Stability. One reason is indicated by examining equation 8-20. With negative feedback making $A_v\beta$ a large negative number, the over-all voltage amplification A_{FB} approaches the limiting value $1/\beta$. For large values of negative feedback the over-all amplification A_{FB} becomes almost independent of the amplification A_v of the amplifying element. This means that the amplification ratio A_{FB} of a feedback amplifier with negative feedback tends to be independent of the factors such as aging of tubes, and plate voltage fluctuations that influence A_v. Thus, feedback amplifiers may be very stable in operation.

Nonlinear Distortion. Negative feedback can be used to reduce the over-all distortion of an amplifier. In a sense, and within limits, if the amplified output signal is distorted, the feedback network will introduce some of this distortion into the input. Thus, the input signal impressed on the amplifying element is predistorted, so that the output wave is less distorted than if there were no feedback. Suppose d represents the nonlinear distortion voltages actually produced by the amplifying element, which would appear in the output *without* feedback. Let D be the nonlinear distortion output voltages *with* feedback. The feedback network will feed back a voltage $D\beta$, and this will be amplified to $A_vD\beta$ at the output. The *total* distortion *with feedback* is $D = d + A_vD\beta$, and the distortion with feedback is

$$D = \frac{d}{1 - A_v\beta} \qquad 8\text{-}21$$

This equation shows that with a large amount of negative feedback $A_v\beta$, the nonlinear distortion D will be less than d, the distortion without feedback.

Other Characteristics.[22,24,25] Negative feedback is effective in increasing the signal-to-noise ratio of an amplifier output, provided that the noise is not introduced by the input circuit, such as by "hum" induced in an input transformer. Negative feedback can be used to vary the band width of an amplifier, and also to control the input and output impedances.

Examples.[26] Simple examples of feedback amplifiers are shown in Fig. 8-31. The upper figure illustrates negative **voltage feedback.** With *no* feedback, the impressed signal voltage E_s will be amplified, producing an output voltage E_o, the 180° phase shift being indicated by the positive and negative signs. Assuming that capacitor C has negligible reactance, the current flowing in R_1 and R_2 because of output voltage E_o also will be 180° out of phase with E_s. Hence, the feedback voltage across R_2 is 180° out of phase with E_s. The magnitude of the feedback voltage ratio is

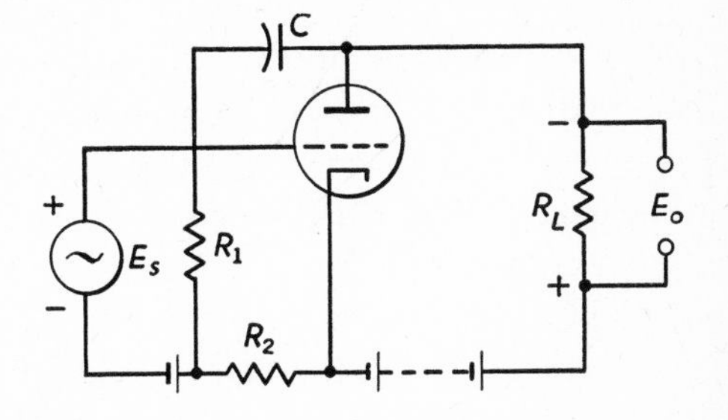

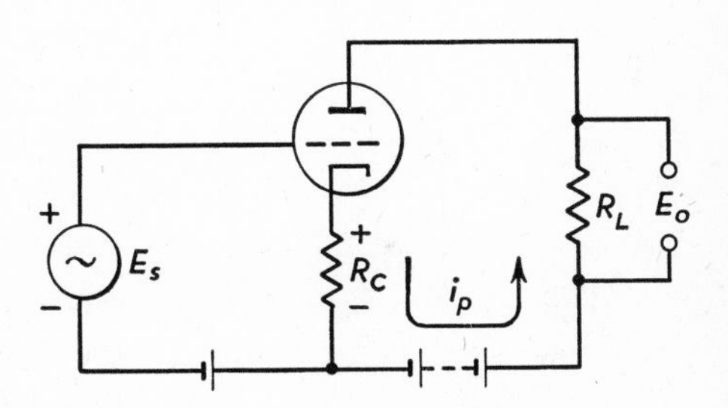

FIG. 8-31. Basic feedback circuits.

$$\beta = \frac{E_2}{E_o} = \frac{R_2}{R_1 + R_2} \qquad 8\text{-}22$$

The lower circuit of Fig. 8-31 illustrates negative **current feedback.** Because of the 180° phase reversal existing when a vacuum tube works into a resistance load, when the signal voltage E_s has the polarities shown, the current will be directed as indicated by i_p. This results in a voltage across R_c that will combine with E_s and *reduce* the signal voltage actually impressed between grid and cathode. This is a degenerative effect, and will occur in an amplifier with self-bias (section 8-10) if the capacitor commonly connected across R_c is omitted, is too small, or becomes "open." The feedback voltage ratio is

$$\beta = \frac{R_c}{R_L} \tag{8-23}$$

In these elementary illustrations the circuits have been considered over the midfrequency region where only 180° phase shifts occur.

8-21. *THE GROUNDED-GRID AMPLIFIER*

The amplifiers considered in the preceding chapters were operated with the cathodes connected to a common point (if more than one stage was used). This common point often is grounded. Such amplifiers are known as **common-cathode amplifiers,** or as **grounded-cathode amplifiers.**[17] When self-biasing is used (section 8-10), a resistor is placed in series with the cathode to obtain a direct biasing voltage; however, a large capacitor of low reactance is connected in parallel with this biasing resistor, and from an alternating-voltage standpoint the cathode is at the potential of the common point, or ground. Ground commonly means an alternating-voltage ground.

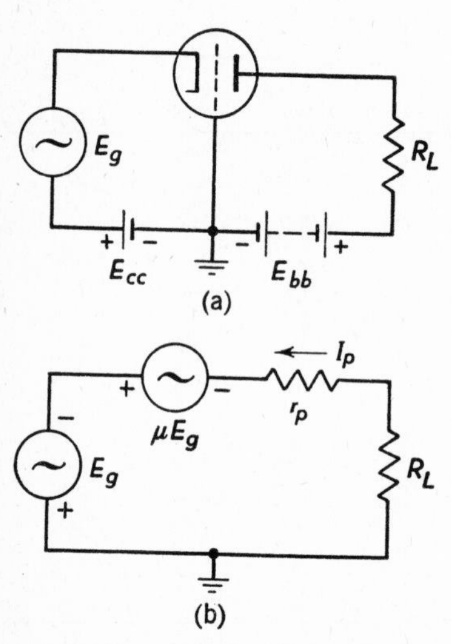

FIG. 8-32. Simplified and equivalent circuits of a grounded-grid amplifier.

There are two other methods of operating vacuum tubes in multistage amplifiers. In one the grids are connected to a common point, which often is grounded. Such amplifiers are known as **common-grid amplifiers,** or as **grounded-grid amplifiers.**[17] In the other (section 8-22), the plates are connected to a common point, which often is grounded, and these sometimes are called **common-plate amplifiers,** or **grounded-plate amplifiers.**[17] Single tubes commonly are referred to as grounded-cathode, grounded-grid, or grounded-plate amplifiers.

The grounded-grid amplifier, developed about 1940, is shown in Fig. 8-32. This has been referred to as an inverted amplifier,[27] and should not be confused with an inverted vacuum tube[28] in which the input signal is injected between the plate and cathode, and the output obtained between the grid and cathode. The equivalent circuit of Fig. 8-32 applies to the grounded-grid amplifier, but the source of alternating signal voltage E_g is in effect in series with the voltage μE_g. These two voltages act in phase because a corresponding change in the magnitude of either causes a similar change in the current flowing. If the alternating signal voltage is E_g volts and the internal impedance of the source is zero, the equation for the plate current flowing is

$$I_p = \frac{E_g + \mu E_g}{r_p + R_L} \tag{8-24}$$

and the output voltage is

$$E_L = I_p R_L = \frac{(E_g + \mu E_g) R_L}{r_p + R_L} \tag{8-25}$$

The gain per stage is

$$A_v = \frac{E_L}{E_g} = \frac{(1 + \mu) R_L}{r_p + R_L} \tag{8-26}$$

This equation, by comparison with equation 4-21, indicates that the grounded-grid amplifier acts like a generator of open-circuit voltage $(1 + \mu)E_g$ instead of μE_g for the grounded-cathode amplifier.

The outstanding advantage of the grounded-grid amplifier is that triodes may be operated in this manner at radio-frequencies without neutralization. For this reason grounded-grid triode amplifiers are used extensively, particularly at frequencies of hundreds of megacycles. Both voltage amplifiers and power amplifiers of the basic types previously considered employ grounded-grid triodes. As Fig. 8-32a shows, the grid acts as a shield between the plate-to-ground output circuit and the cathode-to-ground input circuit. Hence, stray currents that flow from plate to grid because of the plate-voltage signal variations flow to ground and back to the plate through the load, and do not flow into the input circuit. For this reason the grounded-grid triode has little tendency to oscillate. Attention is directed to the fact that the signal source driving a grounded-grid triode must supply power to the input circuit. Internal impedance in the source will cause negative feedback. The input and output currents are substantially the same in a grounded-grid amplifier. The term "current-follower circuit" has been applied to this amplifier.[29] Both small-signal tubes and power-output tubes may be operated with grounded grids, or with grounded plates as discussed in the next section.

8-22. *THE GROUNDED-PLATE AMPLIFIER* [30,31,32]

The grounded-plate amplifier of Fig. 8-33 also is called a **cathode follower.**[17] It is not an amplifier in the sense that the output voltage is greater than the input voltage. It is a current-feedback amplifier without a load resistor R_L, and with the entire "output" voltage E_o fed back in series to reduce the input voltage from E_i to $E_i - E_o$, which is the value that appears between grid and cathode for amplification. From equation 8-23, $\beta = 1$. It has been pointed out [29] that this amplifier is a "voltage-follower circuit."

Referring to Fig. 8-33, the output voltage is $E_o = I_p R_c$. The plate current flowing is

$$I_p = \frac{(E_i - E_o)\mu}{r_p + R_c} \tag{8-27}$$

The output voltage is accordingly

$$E_o = \frac{(E_i - E_o)\mu R_c}{r_p + R_c} \tag{8-28}$$

Solving for the ratio of the output voltage E_o to the impressed signal voltage E_i gives the voltage amplification. Thus

$$A_v = \frac{E_o}{E_i} = \frac{\mu R_c}{r_p + R_c(1 + \mu)} \tag{8-29}$$

The question immediately arises: What use is an amplifier having a voltage amplification *less than unity*? The answer is that this device should be regarded as an *isolating stage,* or as an *impedance transforming stage,* and sometimes as a power amplifier. It acts as an isolating stage because the grid input capacitance is *low* in comparison with the usual amplifying tube (section 7-9). For this reason, cathode-follower stages are sometimes used between wide-band amplifier stages so that the wide-band stage works into a low capacitance.[33,34] At frequencies below several million cycles, where the reactances of the electrode capacitances of the tube are negligible in comparison with the resistances involved, the internal output impedance of the tube and associated circuit is essentially *pure resistance,* and is a *low value.* The cathode follower therefore acts like an impedance transformer and can be used to connect a circuit

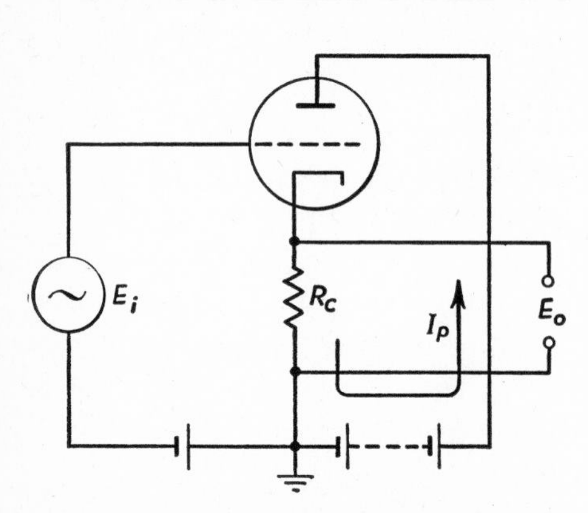

FIG. 8-33. In the grounded-plate amplifier, or cathode follower, the entire "output" voltage is fed back in series to oppose the impressed signal voltage E_i.

of high impedance to one of low impedance.[35] The internal resistance is *approximately* equal to the value

$$R_{\text{int}} = \frac{r_p}{1+\mu} \qquad \text{8-30}$$

This equation can be derived from the relation $R_{\text{int.}} = E_{oc}/I_{sc}$, where E_{oc} is the open-circuit voltage or output voltage E_o (from equation 8-29) and I_{sc} is the short-circuit current, equal to $I_{sc} = \mu E_i/r_p$. The voltage E_o of the cathode follower is *in phase* with the applied voltage E_i, and this fact is often important.

8-23. *VHF POWER AMPLIFIERS*

As explained in Chapter 7, voltage amplifiers are used to increase a weak signal voltage until it is of sufficient magnitude to drive the grid of a power-output tube. This viewpoint is satisfactory at audio-frequencies and radio-frequencies up to a few megacycles when operation is class-A. At higher frequencies the grid input circuit tends to draw considerable power from the signal source even when the grid is negative. For this and other reasons, vacuum-tube amplifiers for very high frequencies (30 to 300 megacycles) usually are power amplifiers although the tubes may be small physically.

Generally speaking, the higher the frequency, the smaller will be the signal output power obtainable, and the lower will be the efficiency, of a thermionic vacuum triode or tetrode. For example, a "50-watt" triode that delivers 100 per cent output at 30 megacycles can deliver only 50 per cent at 130 megacycles. Among the factors limiting output at very high frequencies are[36] (1) reactances of electrode leads and electrodes, (2) losses in leads and within the tube, and (3) electron transit time.

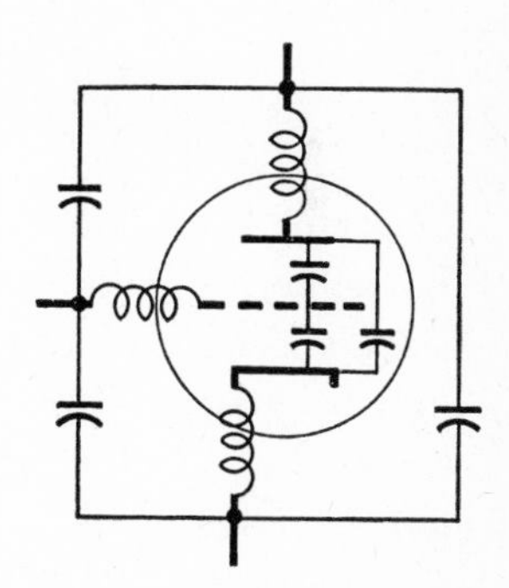

FIG. 8-34. Illustrating the internal and external inductances and capacitances associated with a thermionic vacuum tube.

Reactances of electrode leads and electrodes are caused by the series self-inductances of the leads and electrodes, and the stray and interelectrode capacitances. Such reactances are indicated by the lumped units in Fig. 8-34, but at very high frequencies these reactances are more correctly considered as being distributed. In fact, unequal voltage distributions may exist along the electrodes because of standing-wave phenomena (discussed in books on transmission-line theory). Also, because of these reactances, signal voltages impressed on the electrode leads may be less at the electrodes; furthermore, impedance mismatches and circuit resonances may result. Reactances may be controlled by changes in design as indicated in Fig. 8-35. The electrode arrangement for a power-output triode for 2 megacycles is shown at the left, and the arrangement for a 110-megacycle power-output triode of about the same capacity is shown at the right. For the high-frequency tube, connecting leads have been shortened, diameters of the electrodes have been increased, and the electrodes have been connected to metal-disk terminations that have low inductance, and which also are suited for mounting in coaxial circuit elements. Because of standing-wave phenomena, the electrode length should not exceed 1/16 wave length in the direction of wave travel, which is in the direction of the axis of the tube. For 100 megacycles this maximum length becomes *18 centimeters*.[37]

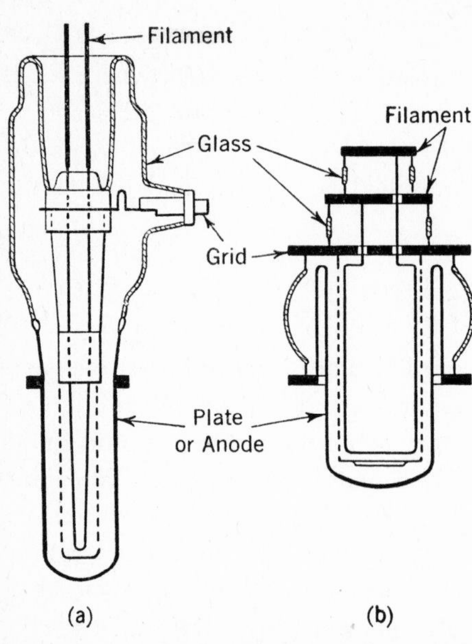

FIG. 8-35. Structures of typical large power-output triodes for 2 megacycles (left) and 110 megacycles (right). (Reference 37.)

Losses in leads and within the tube cause an increase in the signal input driving power required and lower the available power output. Among the important losses are those that occur in leads and terminations because of skin effect, radiation losses, and dielectric hysteresis losses in insulation, bases, and metal-to-glass seals. Similar losses occur within the tube. At very high frequencies, currents passed by the stray and interelectrode capacitances are increased in magnitude, and increased I^2R losses occur in metal parts conducting these currents. Furthermore, these "capacitive" currents must be emitted by the cathode, limiting its useful output. Power tubes are cooled by forcing air through metal fins in contact with the plate.

Electron transit-time effects occur if the time required for electrons to pass from cathode to plate is an appreciable portion of the time for one cycle. In an ordinary radio-receiving tube the transit time is about 10^{-9} second, which is 10 per cent of the time for one cycle at 100 megacycles.[38] Transit-time effects reduce the output of a thermionic vacuum triode in several ways: (1) Plate signal current will lag behind grid-signal voltage, causing the amplified output signal voltage and current to be out of phase.[36] (2) Debunching caused by the mutual repulsion among the negative electrons constituting the signal cathode-to-plate current will alter the signal characteristics. Debunching and spreading occurs at any frequency, but if the frequency is very high, the effect is exaggerated.[36,37,38,39] (3) The signal input impedance between grid and cathode is caused to fall to a low value, and to have an important in-phase component; accordingly, the input admittance will rise, and the conductance component will cause the input current and voltage to be partly in phase.[37] This will cause the power drawn by the grid-input circuit to increase.

Tubes for very high frequencies are designed so as to minimize the effects of reactances, power losses, and transit time. This is illustrated by Fig. 8-35b for a tube capable of handling many kilowatts. Short parallel leads and coaxial construction; special metals, glasses, ceramics, and glass-to-metal seals; close spacings of a few thousandths of an inch to reduce transit time; and careful cathode design and construction are incorporated in such tubes. Also, rugged grids that do not readily emit electrons are used, and the plates likewise are specially constructed for such tubes. Tubes for handling a few watts or less at very high frequencies commonly use parallel electrodes and sometimes are called *planar tubes*. The so-called "lighthouse" construction makes possible very small electrode spacings of several thousandths of an inch, and is suited for coaxial mounting as in the amplifier of Fig. 8-36. The short-circuited coaxial elements and the gaps associated with them act as parallel resonant circuits such as used in radio-frequency amplifiers.

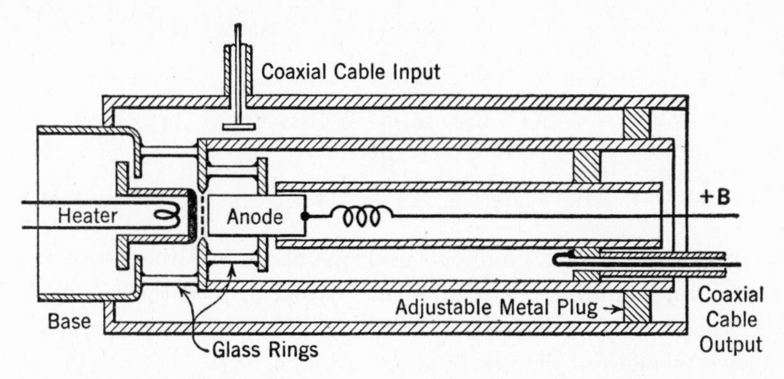

FIG. 8-36. A planar-electrode tube in a coaxial system for amplification at ultrahigh frequencies. Metal parts are shaded. At the left of the anode is the grid (broken line), and at the left of the grid is the oxide coating on the cathode (heavy black line). The gap between the cathode and tube base, and the gap between the anode and the tubular portion around it, provide capacitances for producing resonance with the coaxial elements that are shorted with metal plugs so as to make the input to the element inductive. The tube operates as a grounded-grid amplifier, with coaxial input and output.

Generally speaking, conventional transmission-line theory applies to these coaxial elements, but it must be remembered that the tube electrodes become an important part of the coaxial elements.

8-24. *UHF POWER AMPLIFIERS*

In the ultrahigh frequency range (300 to 3000 megacycles) the effects of lead and electrode reactances, power losses, and transit time become more pronounced. Thermionic vacuum tube triodes and tetrodes both are used for amplification at the lower ultrahigh frequencies, but for frequencies above about 1000 megacycles, triodes offer advantages. Two types of construction are used, coaxial electrodes and parallel electrodes.

Coaxial Electrode Construction.[40] As mentioned in the preceding section, the electrodes must be less than 1/16 wave length long and at 1000 megacycles this is 1.8 centimeters; hence, the electrodes must be very short. To obtain the area required for cathode emission, the diameters of the coaxial elements must be large in comparison to their lengths. For tubes designed to handle kilowatts of power, the plate (or anode) and the grid are rigid massive structures as compared with low-frequency tubes.[40]

Coaxial construction also is used for low-power ultrahigh frequency tubes; an example is the "pencil-type" triode [41] of Fig. 8-37. The grid is attached

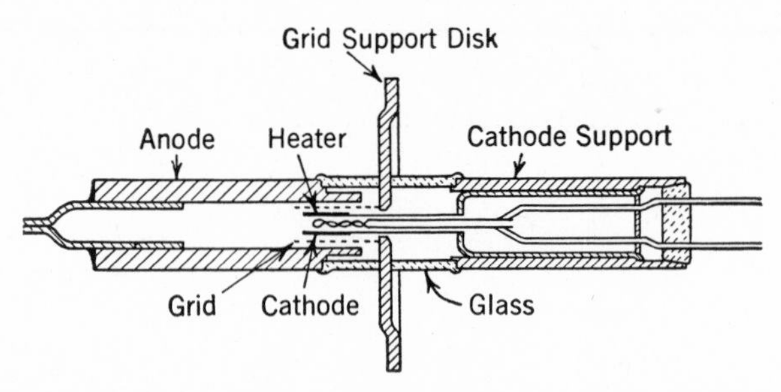

FIG. 8-37. Coaxial construction of a "pencil-type" triode. (Reference 41.)

to a supporting disk for grounded-grid operation, and the entire structure is adapted for mounting in coaxial elements that provide the tuned amplifier circuits (section 9-11). The dimensions are small, as can be judged by the fact that the length of the grid is about 0.3 inch.

Parallel Electrode Construction. The parallel, or planar, construction of a low-power triode for the ultrahigh frequencies, and above, is shown in Fig. 8-38. The oxide coating is on the upper end of the indirectly heated cathode. The grid is of wires 0.0003 inch in diameter, has 1000 wires per inch, and is spaced 0.0006 inch from the cathode. Grid-to-plate spacing is

0.012 inch.[42] As a result of precise construction and close spacing, the tube has an amplification factor of 300 and grid-plate transconductance of 50,000 micromhos, operates with a plate voltage of 200 volts and a plate current of 20 milliamperes, and has a power gain of 13 decibels at 50 milliwatts output and a band width between half-power points of 100 megacycles at a frequency of 4000 megacycles. This tube is used in the Bell System transcontinental point-to-point radio-relay system.[43] This operates between 3700 and 4200 megacycles. A simplified diagram of the planar triode in an amplifier with wave-guide input and output is shown in Fig. 8-38.

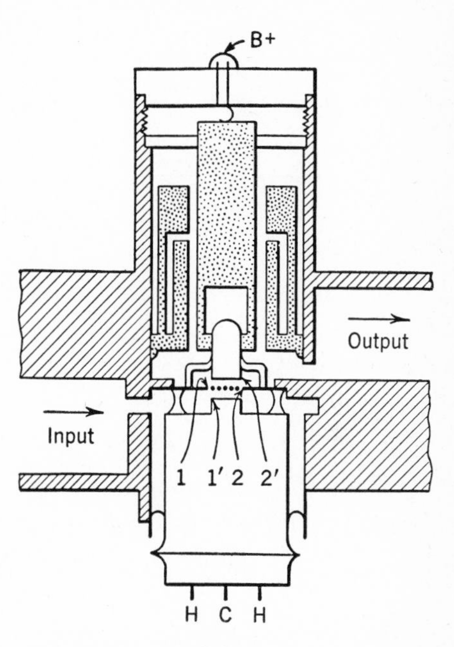

FIG. 8-38. Parallel, or planar, electrode construction of a triode for operation at ultrahigh frequencies and above. The tube is mounted in a wave-guide element for grounded-grid operation. Arrows 1 and 2 indicate the grid. Arrow 1′ indicates the indirectly-heated cathode. Arrow 2′ leads to the anode, or plate. The nonshaded portions between cathode and grid, and grid and anode, are glass. The electromagnetic signal wave enters the input cavity beneath the grid, controls the cathode to anode electron flow, and emerges as an amplified output wave. (Reference 42.)

8-25. *SPACE-CHARGE-WAVE AMPLIFIERS*

A thermionic vacuum tube having outstanding abilities as a power amplifier of ultrahigh frequencies (300 to 3000 megacycles), and superhigh frequencies (3000 to 30,000 megacycles) was introduced about 1946. Several modifications of the early tube have been made, and all are classified [44] as **space-charge-amplifier tubes.** These tubes consist [44] of four parts: (1) One or more electron guns (Chapter 17) for producing an axial flow of electrons; (2) an input section where the signal to be amplified is introduced; (3) an amplifier section where energy is added to the weak input signal; and (4) an output section where the amplified signal is fed into the outgoing transmission system. Three types of space-charge-wave tubes are the **traveling-wave tube,** the **double-beam electron-wave tube,** and the **single-beam electron-wave tube.**

The Traveling-Wave Amplifier Tube. Early work on this device was done by Kompfner and others in England,[45,46] and by Pierce and Field in the United States.[47,48] A simplified diagram of the traveling-wave tube is shown in Fig. 8-39. In a typical tube an electron gun (Chapter 17) produces an electron beam of about 10 milliamperes with electron velocities produced by an accelerating voltage of about 1500 volts. The signal to be amplified is transferred from the wave guide to the helix by the termination that acts like a receiving antenna. At the output end the amplified signal is transferred from the helix to the wave guide by the termination that acts like a transmitting antenna. The electrons in the beam have a tendency to spread because of mutual repulsion, and are confined to a beam by the action of an axial magnetic field (section 17-5) produced by a coil around the tube,[47] or by magnets.[49]

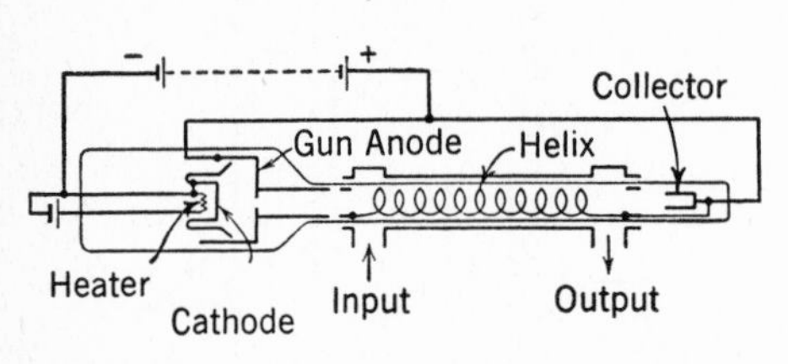

FIG. 8-39. Diagram of a basic traveling-wave amplifying tube. (Reference 47.)

The incoming signal wave starts down the helix traveling *along the wire* at almost the speed of light. But, the pitch of the helix is such that the signal wave travels *along the axis* of the tube at about 1/13 the speed of light. The electrons in the beam at the center of the helix travel slightly faster than the signal wave travels along the axis of the tube. Under these conditions the traveling signal wave absorbs energy from the electron beam, and the signal wave will increase in magnitude, or "grow," as it progresses from the input end to the output end.

The reason for this amplification has been investigated by Pierce [48,50] and others.[51] If the helix is assumed lossless, the transmission path from input to output of a traveling-wave tube propagates several signal wave components: (1) a forward component that is attenuated in passing from input to output; (2) a forward component that is unattenuated; (3) a forward component that is amplified; and (4) a backward component that is affected but little in traveling against the electron beam. The situation has been likened to a breeze blowing past ripples on the surface of water, the ripples abstract energy from the breeze and grow in size as they are blown along.[47]

An outstanding characteristic of the tube is the band width over which it will operate with good power gain. For instance, an early traveling-wave tube using a beam voltage of 1670 volts had a band width of about 800 megacycles (between −3 decibel points), power gain of 23 decibels, and 200 milliwatts output, when operating at about 3600 megacycles.[47] It is of interest to note that with the electron beam *off*, the *loss* encountered by the signal in passing along the helix was 33 decibels. Later tubes designed to have low noise figures (about 10 decibels) and to operate at lower beam-accelerating voltages of 500 volts, at 3000 megacycles gave 20 decibels gain over a band width of 500 megacycles. Traveling-wave tubes with outputs of about 10 watts are common, although tubes handling up to 1000 watts have been reported.[52] In the first-mentioned tubes,[49] the cathodes are designed and operated to minimize random emission fluctuations that cause noise, disk-loaded wave guides instead of helices sometimes are used,[53] and the inside of a tube sometimes is coated with a "lossy" material (graphite) to absorb any backward-traveling energy that might cause oscillations. Backward-wave oscillators operating up to 50,000 megacycles have been studied.[54]

The Double-Beam Electron-Wave Tube.[44,48,55,56] A tube using no structure, such as a helix, for slowing and guiding a signal wave, but using two electron beams, will amplify. The two electron guns of Fig. 8-40 produce two beams, which have slightly different electron velocities. The two beams intermingle as the electrons travel along the axis of the tube. They are prevented from spreading by axial magnetic fields. At the input end the two beams pass through a signal input helix, or wave cavity, and at the output end they again pass through an output helix or cavity. As the signal wave impressed on the electron beams travels toward the output end, the signal wave abstracts energy from the electron beams and the signal variations "grow" in magnitude. The tube sometimes is called a growing-wave tube. The absorption of energy is attributed to electromechanical reactions. Among the advantages of the double-beam tube is the fact that no helix or wave guide extending the length of the tube is necessary. This is important at superhigh frequencies.

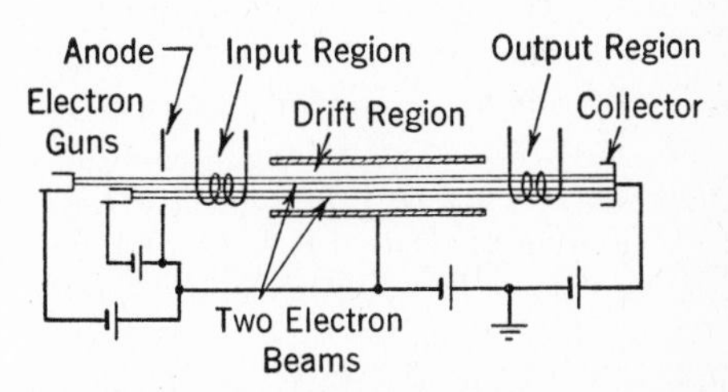

FIG. 8-40. In the double-beam electron-wave tube the two electron guns at the left produce two separate beams of electrons, which travel with slightly different velocities to the collector. (Reference 44.)

The Single-Beam Electron-Wave Tube.[44,56] Only one electron gun and a single electron beam are used in this tube. However, the electrode arrangement is such that the outer electrons of the beam are caused by the electrode potentials to travel faster than the electrons at the center of the beam. The beam passes through input and output helices, or through suitable cavities serving the same purposes. As the signal-wave variations impressed on the electron beam travel to the output end, energy is absorbed from the beam and signal amplification occurs. Other power amplifiers for ultrahigh and superhigh frequencies, such as the Klystron, will be discussed in the next chapter.

REFERENCES

1. Pidgeon, H. A., and McNally, J. O. *A study of the output power obtained from vacuum tubes of different types.* Proc. I.R.E., Feb. 1930, Vol. 18.
2. American Institute of Electrical Engineers. *American Standard Definitions of Electrical Terms.* 1941.
3. Institute of Radio Engineers. *Standards on Electron Tubes: Definitions of Terms,* 1950. Proc. I.R.E., April 1, 1950, Vol. 38, No. 4.
4. Haefner, S. J. *Amplifier-gain formulas and measurements.* Proc. I.R.E., July 1946, Vol. 34, No. 7.
5. Radio Corporation of America. *Receiving Tube Manual.*
6. Kilgour, C. E. *Graphical analysis of output tube performance.* Proc. I.R.E., Jan. 1931, Vol. 19, No. 1.
7. Pidgeon, H. A. *Simple theory of the three-electrode vacuum tube.* Journal of the Society of Motion Picture Engineers, Feb. 1935, Vol. 24.
8. Preisman, A. *Graphical Construction for Vacuum Tube Circuits.* McGraw-Hill Book Co.
9. Thompson, B. J. *Graphical determination of performance of push-pull audio amplifiers.* Proc. I.R.E., Apr. 1933, Vol. 21, No. 4.
10. Krauss, H. L. *Class-A push-pull amplifier theory.* Proc. I.R.E., Jan. 1948, Vol. 36, No. 1.
11. Houghton, E. W. *Class A push-pull calculations.* Electronics, June, 1937, Vol. 10, No. 6.
12. Offner, F. F. *Balanced amplifiers.* Proc. I.R.E., March 1947, Vol. 35, No. 3.
13. Peterson A. and Sinclair D. B. *A single-ended push-pull amplifier.* Proc. I.R.E., Jan. 1952, Vol. 40, No. 1.
14. Pidgeon, H. A. *Theory of multielectrode vacuum tubes.* Electrical Engineering, Nov. 1943, Vol. 53.
15. Radio Corporation of America. *Class AB operation of type 6F6 tubes connected as pentodes.* Application Note No. 54.
16. Pender, H., and McIlwain, K. *Electrical Engineers' Handbook—Vol. 5: Electric Communication and Electronics.* John Wiley & Sons.
17. Institute of Radio Engineers. *Standards on Antennas, Modulation Systems, and Transmitters: Definitions of Terms.* 1948.
18. Sarbacher, R. I. *Graphical determination of power amplifier performance.* Electronics, Dec. 1942, Vol. 15, No. 10.
19. Chaffee, E. L. *A simplified harmonic analysis.* Review of Scientific Instruments, Oct. 1936.
20. Cawood, D. W. *Calculations for class "C" amplifiers with a reactive load.* Sylvania Technologist, July 1951, Vol. 4, No. 3.
21. Doherty, W. H. *Neutralization of radio-frequency power amplifiers.* Pick-Ups. (Western Electric Co.), Dec. 1939. See also pp. 206-213, *The Electronic Engineering Handbook,* Blakiston Co., 1944.

22. Black, H. S. *Stabilized feedback amplifiers.* Electrical Engineering, Jan. 1934. Vol. 53.
23. Nyquist, H. Regeneration Theory. Bell System Technical Journal, Jan. 1932, Vol. 11, No. 1.
24. Terman, F. E., and Pettit, J. M. Electronic Measurements. McGraw-Hill Book Co.
25. Bode, H. W. *Network Analysis and Feedback Amplifier Design.* D. Van Nostrand Co.
26. Schulz, E. H. *Comparison of voltage-and current-feedback amplifiers.* Proc. I.R.E., Jan. 1943, Vol. 31, No. 1.
27. Strong, C. E. *The inverted amplifier.* Electronics, July 1940, Vol. 13, No. 7.
28. Terman, F. E. *The inverted vacuum tube—a voltage reducing power amplifier.* Proc. I.R.E., April 1928, Vol. 16, No. 5.
29. Jones, M. C. *Grounded-grid radio-frequency voltage amplifiers.* Proc. I.R.E., July 1944, Vol. 32, No. 7.
30. Sziklai, G. C., and Schroeder, A. C. *Cathode-coupled wide-band amplifiers.* Proc. I.R.E., Oct. 1945, Vol. 33, No. 10.
31. Schlesinger, K. *Cathode-follower circuits.* Proc. I.R.E., Dec. 1945, Vol. 33, No. 12.
32. Pullen, K. A. *The cathode-coupled amplifier.* Proc. I.R.E., June 1946, Vol. 34, No. 6.
33. Kline, M. B. *Cathode follower bandwidth.* Electronics, June 1949, Vol. 22, No. 6.
34. Baer, R. H. *Cathode follower response.* Electronics, Oct. 1950, Vol. 23, No. 10.
35. Reich, H. J. *Input admittance of cathode-follower amplifiers.* Proc. I.R.E., June 1947, Vol. 35, No. 6.
36. Spangenberg, K. R. *Vacuum Tubes.* McGraw-Hill Book Co.
37. Doolittle, H. D. *Design problems in triode and tetrode tubes for U.H.F. operation.* Cathode Press (Machlett Laboratories), Fall 1951, Vol. 8, No. 4.
38. Ferris, W. R. *Input resistance of vacuum tubes as ultra-high-frequency amplifiers.* Proc. I.R.E., Jan. 1936, Vol. 24, No. 1.
39. Terman, F. E. *Radio Engineering.* McGraw-Hill Book Co.
40. Smith, P. T. *Some new ultra-high-frequency power tubes.* R.C.A. Review, June 1952, Vol. 13, No. 2.
41. Rose, G. M., Power, D. W., and Harris, W. A. *Pencil-type UHF triodes.* R.C.A. Review, Sept. 1949, Vol. 10, No. 3.
42. Morton, J. A., and Ryder, J. M. *Design factors of Bell Telephone Laboratories 1553 triode.* Bell System Technical Journal, Oct. 1950, Vol. 29, No. 4.
43. Roetken, A. A., Smith, K. D., and Friis, R. W. *The TD-2 microwave radio relay system.* Bell System Technical Journal, Oct. 1951, Vol. 30, No. 4.
44. Hutter, R. G. E. *Space-charge-wave amplifier tubes.* Sylvania Technologist (Sylvania Electric Products), Oct. 1952, Vol. 5, No. 4.
45. Kompfner, R. *The travelling-wave valve.* Wireless World, Nov. 1946, Vol. 52, No. 11.
46. Kompfner, R. *The traveling-wave tube as an amplifier of microwaves.* Proc. I.R.E., Feb. 1947, Vol. 35, No. 2.

47. Pierce, J. R., and Field, L. M. *Traveling-wave tubes.* Proc. I.R.E., Feb. 1947, Vol. 35, No. 2.
48. Pierce, J. R. *Traveling-wave Tubes.* D. Van Nostrand Co.
49. Dodds, W. J., Peter, R. W., and Kaisel, S. F. *New developments in traveling-wave tubes.* Electronics, Feb. 1953, Vol. 26, No. 2.
50. Pierce, J. R. *Theory of the beam-type traveling-wave tube.* Proc. I.R.E., Feb. 1947, Vol. 35, No. 2.
51. Chu, L. J., and Jackson, J. D. *Field theory of traveling-wave tubes.* Proc. I.R.E., July 1948, Vol. 36, No. 7.
52. Webber, S. E. *1000-watt traveling-wave tube.* Electronics, June 1950, Vol. 23, No. 6.
53. Field, L. M. *Some slow-wave structures for traveling-wave tubes.* Proc. I.R.E., Jan. 1949, Vol. 37, No. 1.
54. Heffner, H. *Backward-wave tube.* Electronics, Oct. 1953, Vol. 26, No. 10.
55. Nergaard, L. S. *Analysis of a simple model of a two-beam growing-wave tube.* R.C.A. Review, Dec. 1948, Vol. 9, No. 4.
56. Haeff, A. V. *The electron-wave tube—A novel method of generation and amplification of microwave energy.* Proc. I.R.E., Jan. 1949, Vol. 37, No. 1.

QUESTIONS

1. What classes of amplifiers are used for voltage amplification? For power amplification?
2. What is the relation between r_p and R_L for maximum power output? For maximum "undistorted" power output?
3. In section 8-4 it is stated that the distortion should not exceed 5 per cent. What type of distortion is this?
4. Referring to Fig. 8-6, if θ is measured between the load line and the horizontal (as sometimes done), what is its value?
5. Equation 8-8 indicates that power equal to $E_b I_b$ is dissipated in the tube. Account for its loss.
6. Following equation 8-9, plate-circuit efficiency is discussed. Referring to Fig. 8-6, this is the ratio of what areas?
7. In discussing the equivalent circuit for a push-pull audio amplifier, it is stated that a load resistance equal to the plate resistance may be selected. Why?
8. How are even harmonics cancelled out in a push-pull stage?
9. Why are pentodes often used as power-output tubes?
10. What is the difference between class-B1 and class-B2 operation?
11. Why is driving power required for certain power amplifiers? What types are these?
12. What is the fundamental difference between a load (or dynamic) curve and a load line?
13. How do the class-B and class-C amplifiers differ?
14. Referring to Fig. 8-24, is the plate-circuit angle of flow indicated the actual angle that will apply with the parallel plate load circuit inserted, or would it apply only without the plate load?

15. What is meant by constant plate-voltage and constant grid-voltage characteristic curves?
16. In section 8-16 it is stated that it is common practice to keep the plate voltage from falling below the grid voltage. Why is this done?
17. In section 8-16 it is stated that the tuning must be sharp to reject the harmonics. Why?
18. How does a frequency multiplier operate?
19. Explain what causes an unneutralized triode to oscillate?
20. What is the basic principle of neutralization?
21. What is the basic principle of coil neutralization? Of capacitor neutralization?
22. What is the basic principle of controlled negative feedback?
23. Is there a fundamental difference between current feedback and voltage feedback? Why?
24. What is the cathode follower, and why is it useful?
25. Referring to Fig. 8-16, why is the plate supply for the push-pull stage connected between the two filter chokes instead of at the end of the filter?
26. What is an important reason for using grounded-grid tubes?
27. What is meant by the term "grounded"?
28. Why are cathode followers used?
29. Why are tubes for ultrahigh frequencies made with short leads, short electrodes, and close spacings?
30. What is electron transit time, and what are its effects?

PROBLEMS

1. Derive equation 8-5.
2. Derive equation 8-6.
3. A triode having a plate resistance of 1900 ohms and an amplification factor of 3.8 is operated in class-A1 with a grid bias voltage of −63 volts and a plate voltage of +350 volts. Approximately what maximum undistorted power can be delivered by this tube?
4. Figure 8-6 represents conditions when a resistor is inserted into the plate circuit. Redraw Fig. 8-6 for an amplifier with a load resistance *coupled* into the plate circuit through a transformer.
5. If two of the triodes of problem 3 are operated in parallel in class-A1, and with the same electrode potentials, etc., approximately what maximum undistorted power can be delivered?
6. If two of the triodes of problem 3 are operated in push-pull class-A1, and with the same electrode potentials, etc., approximately what maximum undistorted power can be delivered?
7. Refer to the discussion in section 8-9 regarding the cancellation of even harmonics. Draw a diagram and explain in detail how this occurs.
8. Assume curve *D* of Fig. 8-13 is a straight line, and that a plate-to-plate load resistance of 8000 ohms is used with two of these tubes in class-B2. Calculate the power output for maximum grid-driving voltage as indicated in Fig. 8-13.
9. Repeat the calculations (in section 8-16) for the class-C amplifier, but using a bias of three times cutoff.

10. If the class-C amplifier of problem 9 is operating at one million cycles, calculate the approximate values of L and C. What is the approximate effective value of the current through the capacitor? Through the coil?
11. Plot an $A_v\beta$ diagram for a typical R-C amplifier with feedback, and determine if the device will oscillate.
12. A tube operated as a cathode follower has an amplification factor of $\mu = 13.5$ and $g_{pg} = 1300$ micromhos. Calculate the apparent internal resistance. Prove that the output voltage will be in phase with the input voltage. Derive equation 8-30.
13. Figure 8-36 is a circuit for an amplifier with coaxial elements and "gap" capacitances providing the input and output circuits. Draw an equivalent circuit using lumped inductance and capacitance, proving that Fig. 8-36 is a tuned, grounded-grid amplifier.
14. It is desired that the tube of Fig. 8-36 be self-biased. Where should a resistance be connected for this, and what should be its resistance?
15. Calculate the plate resistance of the planar triode of Fig. 8-38. If the amplification factor were not given, how could it be estimated? Make such an estimate, and compare it with the value given. From the data given, determine the plate-circuit efficiency.

CHAPTER 9

VACUUM-TUBE OSCILLATORS

In communication systems a source of alternating-current power is needed for many purposes. This commonly is supplied by an **oscillator,** defined [1] as a "nonrotating device for producing alternating current, the output frequency of which is determined by the characteristics of the device." In a sense an oscillator is a converter because it changes direct-current power from the supply source to alternating-current power. Thermionic vacuum tubes are extensively used in oscillators; these will be discussed in this chapter. Gas-tube oscillators sometimes are called **inverters** (section 6-19).

There are several bases on which oscillators may be classified. From the standpoint of the generated output frequency, a convenient classification is **audio-frequency oscillators** (20 to 15,000 cycles) and **radio-frequency oscillators** (10 kilocycles to 30,000 megacycles). Oscillators will produce outputs at frequencies from a fraction of a cycle per second [2] to (1955) about 50,000 megacycles. On the basis of output frequency, radio-frequency oscillators may be classified further as **high-frequency oscillators** (3 to 30 megacycles); **very high-frequency oscillators** (30 to 300 megacycles); **ultrahigh-frequency oscillators** (300 to 3000 megacycles); and **superhigh-frequency oscillators** (3000 to 30,000 megacycles).[3]

Oscillators may be classified, on the basis of the shape of the output wave, as **sinusoidal oscillators** and **nonsinusoidal oscillators.** Furthermore, many of the sinusoidal oscillators are **tuned-circuit oscillators** employing resonant inductive-capacitive circuits (or their electrical equivalent) for determining the output frequency. Oscillators of this general type have been termed [4] harmonic oscillators. Also, some nonsinusoidal oscillators are **relaxation oscillators,** which commonly use charge and discharge characteristics of resistor-capacitor circuits to determine the fundamental frequency.

Oscillators also may be classified on the basis of the element that makes possible the oscillations, and the conversion from direct- to alternating-current power. Thus, the common type using a vacuum-tube amplifier and positive feedback (section 8-19) could be called a **feedback-amplifier oscillator.** In practice, this is known as a **feedback oscillator,** defined [1] as an "oscillating circuit, including an amplifier, in which the output is coupled in phase with the input, the oscillation being maintained at a frequency determined by the parameters of the amplifier and the feedback circuits such as *L–C, R–C,* and other frequency-selective elements." A second type, called a **negative-resistance oscillator,** employs a negative-resistance element to make oscillation possible. If once excited, a *lossless* circuit containing *only* inductance and capacitance will oscillate indefinitely. A negative resistance that *supplies* power (instead of absorbing power as does an ordinary resistor), when connected to a *L–C* circuit having loss may be able to overcome the loss and sustain oscillations. There are many types of oscillators (the Standards [1] list 32 types), but some of these differ only in detail. The procedure will be to consider feedback oscillators first, and then negative-resistance oscillators.

9-1. *TUNED-PLATE FEEDBACK OSCILLATOR*

The **tuned-plate oscillator** [1] employs a tuned parallel *L–C* circuit between plate and cathode of an amplifying tube to determine the frequency, and positive feedback from plate to grid to sustain oscillations. A simple tuned-plate feedback oscillator is shown in Fig. 9-1. The tube is assumed to be operating as a class-A1 amplifier; but, as will be explained, the grid is driven slightly positive.

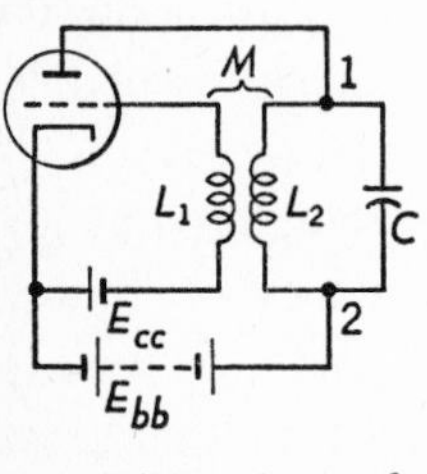

FIG. 9-1. A simple radio-frequency oscillator. Power is taken by coupling a coil with L_2.

The parallel resonant circuit L_2–C *largely* determines the frequency of oscillation. At the resonant frequency the driving-point, or input, impedance of the parallel circuit only (measured between points 1–2) is a *high* value of resistance (section 7-18). To all other frequencies the impedance between points is lower than at resonance, and if the loss in inductor L_2 is low, the impedance largely is reactive. If a nonsinusoidal current of essentially constant magnitude and containing many frequency components flows into the resonant parallel circuit L_2–C, at the resonant frequency the current in L_2 will be larger than at other frequencies. This current component of resonant frequency will induce a large voltage in the secondary L_1, and this voltage will be impressed between grid and cathode for amplification. Resonance occurs when the inductive susceptance of L_2 equals the capacitive susceptance of C. If the power losses in L_2 and C are negligible (usually a reasonable assumption), then the resonant frequency is

$$f_r = \frac{1}{2\pi\sqrt{L_2 C}} \tag{9-1}$$

If the grid goes positive and draws appreciable power from coil L_2, this is equivalent to an increase in the resistance of L_2 (section 7-18), and the resonant frequency will be changed from equation 9-1. This also occurs if appreciable power is drawn from the resonant circuit in any way, such as by coupling the resonant circuit to an antenna.

When the oscillator of Fig. 9-1 first is energized, a transient condition will exist. The component of current for which L_2–C is resonant will exceed in magnitude other current components; hence, a relatively large voltage at the frequency of resonance will be induced in L_1, and will be impressed between grid and cathode. If the phase relation of this impressed voltage is correct (section 8-19), the tube will amplify this voltage, and the process will be repeated. In this manner, oscillations at the frequency of resonance increase in magnitude until the grid is driven positive and draws appreciable current and power. When current flows, alternating voltage drops occur. Also the input impedance of the grid-cathode circuit will contain a relatively low value of resistance. Because of the mutual inductance between L_1 and L_2, this will be reflected (section 7-18) as resistance in series with L_2. In addition to changing the resonant frequency (preceding paragraph), this also will reduce the magnitude of the equivalent load resistance offered the tube by the tuned circuit, and the tube will become a less efficient amplifier. Because of this nonlinear characteristic, oscillations do not "build up" indefinitely, but soon reach a magnitude where they are constant for given conditions.

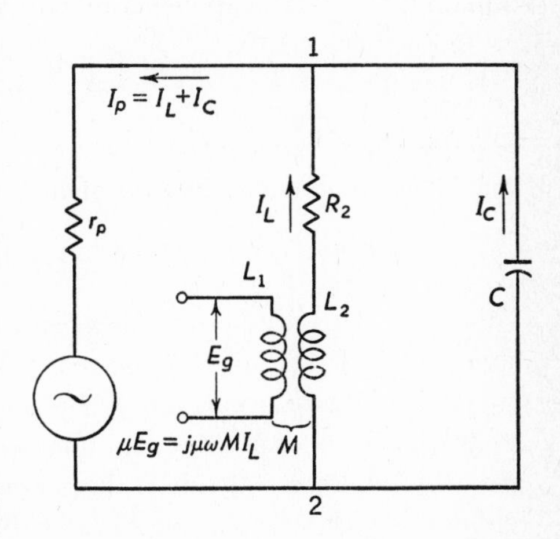

FIG. 9-2. Equivalent circuit of the oscillator of Fig. 9-1.

Theory of Tuned-Plate Feedback Oscillator. The mathematical solution of a feedback oscillator such as Fig. 9-1 early was given by van der Bijl.[5]

Others have analyzed oscillators,[4] particularly from the positive feedback standpoint (section 8-19). Following in general the procedure of van der Bijl, the circuit of Fig. 9-1 has been redrawn as Fig. 9-2, where the amplifying tube is replaced by a generator producing a voltage of magnitude μE_g and having internal resistance r_p. Resistor R_2 is the effective alternating-current resistance of L_2. It is assumed (1) that the tube operates as a class-A1 amplifier, the grid is always negative, and the grid-input impedance is infinite; (2) that interelectrode capacitances are negligible; (3) that the input impedance of parallel circuit L_2–C as measured between points 1–2 is substantially pure resistance at the frequency of oscillation; and (4) that steady-state sinusoidal voltages and currents exist. The signal voltage E_g impressed between grid and cathode to sustain oscillations is the voltage induced in L_1 by the current I_L in coil L_2, and is $E_g = -j\omega MI_L$. But, feedback connections *must be made* so that $E_g = j\omega MI_L$ if the tube is to drive itself and oscillate, and hence in Fig. 9-2, $\mu E_g = j\mu\omega MI_L$.

Steady-state voltage equations for the two loops of Fig. 9-2 are

$$(I_L + I_c)r_p + I_LR_2 + j\omega L_2I_L = j\mu\omega MI_L \qquad 9\text{-}2$$

$$\frac{I_c}{j\omega C} - j\omega L_2I_L - I_LR_2 = 0 \qquad 9\text{-}3$$

Solving for I_c in the second equation, and substituting in the first, gives

$$-\omega^2 r_pLC + r_p + R_2 + j\omega(r_pR_2C - \mu M + L_2) = 0 \qquad 9\text{-}4$$

Equating the real components to zero and solving for f gives

$$f = \frac{1}{2\pi}\sqrt{\frac{1 + R_2}{L_2Cr_p}} \qquad 9\text{-}5$$

which is the approximate frequency of oscillation of Fig. 9-1. Equating the imaginary components to zero gives

$$r_pR_2C - \mu M + L_2 = 0 \qquad \text{and} \qquad g_{pg} = \frac{\mu}{r_p} = \frac{L_2}{r_pM} + \frac{CR_2}{M} \qquad 9\text{-}6$$

These two steps are permitted because a complex expression such as equation 9-4 equals zero only if the real and imaginary components equal zero. For oscillations to occur, the grid-plate transconductance g_{pg} of the oscillator tube must be a positive real number equal to, or greater than, the value given by equation 9-6. Thus, there are circuit conditions for which a tube will not oscillate even if voltage is fed back between grid and cathode in the correct phase. The equations just given may be derived on the basis of the positive feedback amplifier of section 8-19.

Phase Relations in Tuned-Plate Feedback Oscillator. The phase relations for a vacuum-tube oscillator of Fig. 9-1 are shown in Fig. 9-3. These are similar to Fig. 8-24 for a class-C power amplifier. One difference is that μE_g and I_L must be exactly 90° out of phase, because $\mu E_g = j\mu\omega M I_L$. These diagrams assume class-A1 operation. When the tube is not oscillating, the grid is held to a negative value E_c by battery E_{cc} of Fig. 9-1. For these con-

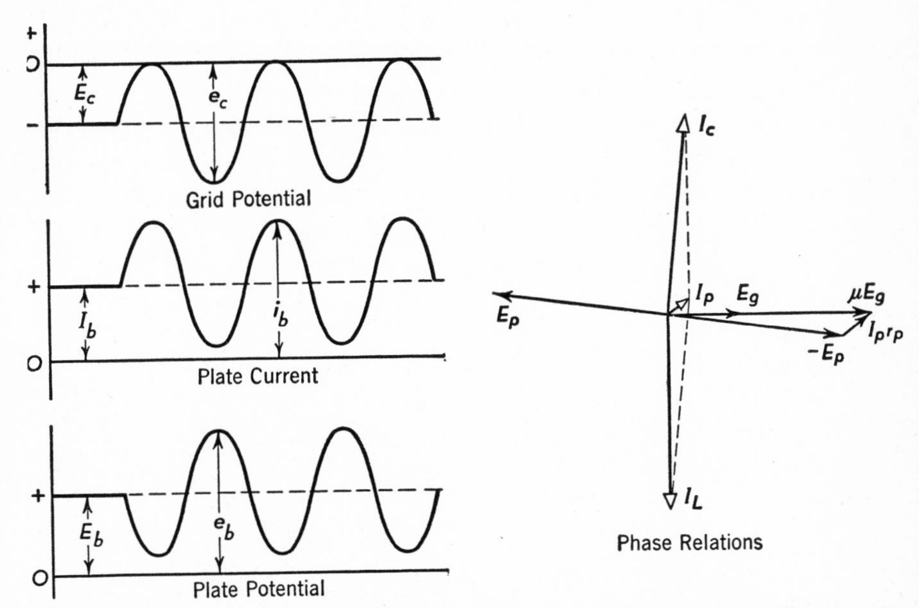

FIG. 9-3. At the left are shown theoretical instantaneous voltage and current relations for the oscillator of Fig. 9-1 when not dissipating energy in the resonant circuit and not delivering energy externally. At the right are shown relations when a small amount of energy is dissipated in the inductor, or delivered to an external circuit coupled to the inductor.

ditions, the plate current is I_b and the plate voltage is E_b. When signal voltage is fed back in the proper magnitude and phase, oscillations occur and the instantaneous variations are as shown in Fig. 9-3. Because of the low direct-current resistance of coil L_2, almost all the direct voltage E_{bb} appears as a voltage E_b between plate and cathode.

As indicated in Fig. 9-3, the instantaneous plate voltage e_b rises above the direct voltage E_b. This is because the plate current consists of both direct and alternating components. The alternating current flowing through the parallel tuned circuit L_2–C causes a voltage drop across the impedance (largely resistance) offered between points 1–2. This alternating voltage drop will combine with the direct voltage E_b, causing the instantaneous voltage e_b between plate and cathode to vary as shown. Instantaneous plate voltage variations are substantially 180° out of phase with grid voltage variations because these are the relations in an amplifier, and an oscillator such as Fig. 9-1 is an amplifier with positive feedback. If oscillation will not start, it may be that the feedback circuit is incorrectly phased, and the leads from coil L_1 to grid and source E_{cc} should be reversed.

The tuned parallel circuit L_2–C sometimes is called a **tank circuit**, defined [1] as "a circuit capable of storing electrical energy over a band of frequencies continuously distributed about a single frequency at which the circuit is said to be resonant, or tuned." Furthermore,[1] the selectivity of the tuned circuit is proportional to the Q of the circuit (section 7-18). Oscillator action sometimes is explained from the tank-circuit energy-storage viewpoint. However, it should be remembered that after the starting period an oscillator operates in accordance with conventional steady-state circuit theory, and that "tank-circuit," "flywheel action," and "circulating-current" explanations are not required.

9-2. *TUNED-GRID FEEDBACK OSCILLATOR*

The tuned-grid oscillator [1] employs a tuned parallel L–C circuit between grid and cathode of an amplifying tube to determine the frequency, and positive feedback from plate to the tuned circuit to sustain oscillations. A basic circuit is shown in Fig. 9-4. Alternating plate current flowing in coil L_2 will introduce a voltage in coil L_1. Assuming that the tube operates as a class-A1 amplifier, the voltage induced *in series* in coil L_1 will cause alternating current to flow *around* the circuit composed of L_1 and C *in series*. The current will be greatest for the frequency to which L_1–C is in *series* resonance. The voltage drop I_pX_C across capacitor C is the alternating voltage impressed between grid and cathode for amplification, and this voltage drop will be greatest, and the circuit will oscillate, at approximately the frequency

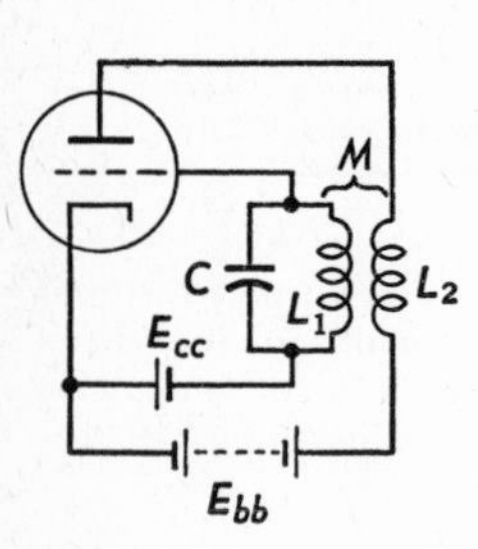

FIG. 9-4. A simple tuned-grid feedback oscillator.

$$f = \frac{1}{2\pi\sqrt{L_1C}} \tag{9-7}$$

As oscillations increase, the grid will be driven positive, grid current will flow, and the impedance between grid and cathode will become a value of perhaps several thousand ohms, depending on the tube and electrode potentials. This resistance is in effect in parallel with the tuned circuit L_1–C, and will reduce the signal voltage impressed between grid and cathode. For this reason oscillations reach a maximum value for a given set of conditions.

9-3. *TUNED-GRID TUNED-PLATE FEEDBACK OSCILLATOR*

In a **tuned-grid tuned-plate oscillator** tuned parallel *L–C* circuits are in series with both the plate and grid, and feedback is obtained by the plate-to-grid interelectrode capacitance [1] as indicated by the dotted capacitor C_{gp} in Fig. 9-5. This is in effect a triode tuned-circuit amplifier without neutralization such as was discussed in section 8-18, where it is stated that this circuit may oscillate at a frequency for which the tuned-plate circuit is inductive (see also section 7-9). If the two circuits are in parallel resonance at exactly the same frequency, for this frequency they may be considered to offer resistance only to current flow. For this frequency the voltages from plate to cathode and from grid to cathode are 180° out of phase, and current fed back through the interelectrode capacitance will *not* produce a voltage drop in the equivalent grid-circuit resistance that will result in oscillations. But if the frequency under consideration is below the resonant frequency to which the circuits are tuned, the two circuits will appear inductive, and hence the oscillator can be represented by Fig. 9-5b. This circuit commonly is used for high- and very high-frequency oscillators. At lower frequencies additional capacitance sometimes is connected from plate to grid. This oscillator will be considered further in section 9-10.

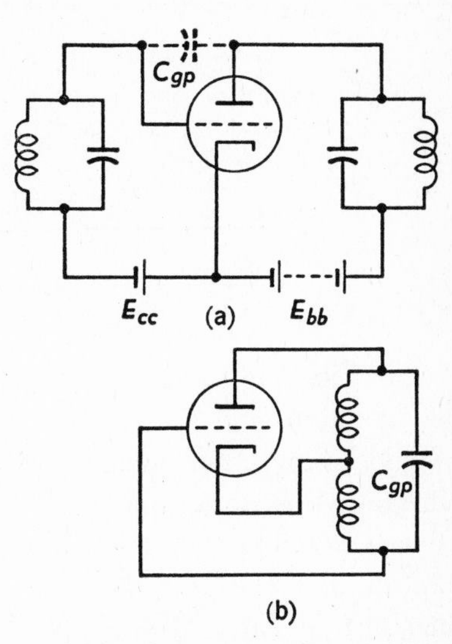

FIG. 9-5. Actual (a) and equivalent (b) circuits for a tuned-grid tuned-plate oscillator.

9-4. *AUDIO-FREQUENCY L–C FEEDBACK OSCILLATOR* [6,7]

The oscillator of Fig. 9-6 was used extensively for generating audio-frequencies, but largely has been supplanted by the oscillator to be discussed in the next section. Feedback is regulated by the *R–C′* series combination, and the name **resistance-stabilized oscillator** sometimes is applied. Tube 1 is the oscillator tube, and the frequency is essentially controlled by the L_2–C combination (equation 9-7). Tube 2 is a **buffer tube;** its purpose is to amplify, and to isolate the oscillator tube from variations in connected load, thus increasing stability. Note that the two grids are connected together. The first tube oscillates in accordance with preceding explanations. The biases, which may be fixed or may be obtained by an R–C combination in the cathode circuit (section 8-10), are such that the tubes operate essentially as class-A amplifiers with the grid of the first tube driven slightly positive.

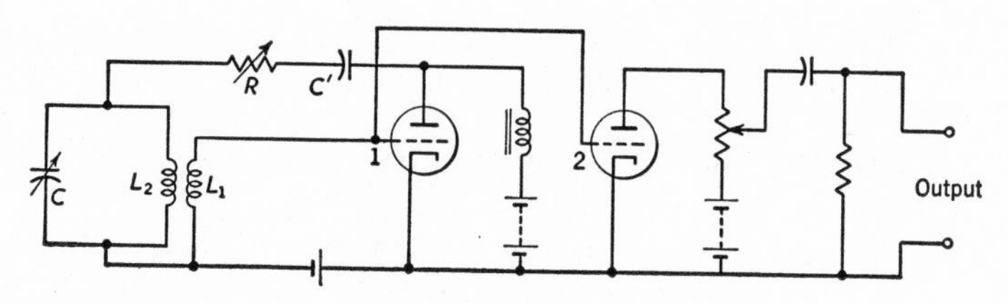

FIG. 9-6. Circuit of an audio-frequency L–C feedback oscillator. The output terminals may be connected to additional amplifying stages.

The blocking capacitor C' should have low reactance at the frequency of operation, and the reactance of the inductor in series with the battery should be high. For audio-frequencies, the coils L_2 and L_1 may be a one-to-one ratio transformer with an iron core; close coupling increases the frequency stability. The losses in the coils and the iron core should be low so that the impedance of the resonant circuit L_2–C will be high for the fundamental frequency. Also, the ratio of L_2/C should be low. If these conditions are satisfied, the tube will oscillate, if, when a voltage of value E_g is impressed on the grid, a voltage of the same value E_g is developed across the tuned parallel circuit L_2–C. The starting feedback resistance is given by the relation [7]

$$R = R_L(\mu - 1) - r_p \qquad 9\text{-}8$$

where R_L is the equivalent resistance of the tuned circuit at resonance, μ is the amplification factor of the tube, and r_p is the plate resistance of the tube. This equation can be derived on the basis that the tube generating a voltage μE_g is driving a series circuit composed of r_p, R_L, and R.

Triodes with an amplification factor or μ of 10 to 20 are satisfactory for these oscillators. With such tubes, the resistance of the plate circuit (load resistance) at resonance should be from about 10,000 to 50,000 ohms. The corresponding value of the feedback resistor R is several hundred thousand ohms. Although the discussions just given apply largely to audio-frequency oscillators, these oscillators can be made to operate up to several hundred thousand cycles.

9-5. *AUDIO-FREQUENCY R–C FEEDBACK OSCILLATOR*

The frequency of the **R–C oscillator** is controlled by resistance-capacitance elements.[1] The principle is shown in Fig. 9-7. Each tube of a resistance-coupled audio amplifier shifts the signal voltage about 180°, and hence in a two-stage amplifier the input and output voltages are in phase. Thus, if the correct portion of the output voltage is fed, without phase shift, into the input, the amplifier will oscillate under steady-state conditions. It is, of course, necessary to add a frequency-controlling arrangement if a variable-frequency sinusoidal oscillator is desired, and this is the *R–C* elements of Fig. 9-7. The blocking capacitor is assumed to be of sufficient capacitance to offer negligible reactance at the frequencies under consideration. The frequency of oscillation and the required voltage amplification now are to be determined.

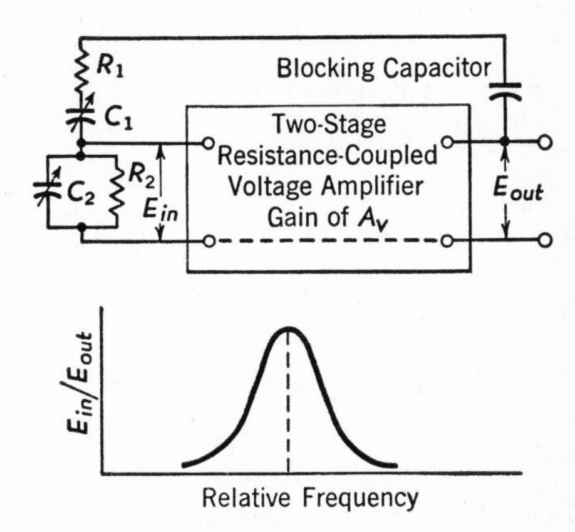

FIG. 9-7. Illustrating the principle of a *R–C* feedback oscillator.

When oscillating under steady-state conditions, the output voltage E_{out} is A_vE_{in}. The impedance of the *series* R_1–C_1 combination is $Z_1 = R_1 - jX_1$, and of the *parallel* R_2–C_2 combination is $Z_2 = -jX_2R_2/(R_2 - jX_2)$. The amplifier input impedance is assumed infinite and (neglecting the blocking capacitor) the ratio of the input voltage to output voltage equals the ratio of Z_2 to $Z_1 + Z_2$, or

$$\frac{E_{in}}{E_{out}} = \frac{E_{in}}{E_{in}A_v} = \frac{Z_2}{Z_1 + Z_2} = \frac{\dfrac{-jX_2R_2}{R_2 - jX_2}}{(R_1 - jX_1)\dfrac{-jX_2R_2}{R_2 - jX_2}} \tag{9-9}$$

This equation reduces to

$$\frac{1}{A_v} = \frac{-jX_2R_2}{(R_1 - jX_1)(R_2 - jX_2) - jX_2R_2} \tag{9-10}$$

If the indicated multiplication is performed and the real terms and imaginary terms separated, then

$$X_1X_2 - R_1R_2 = 0 \quad \text{and} \quad f = \frac{1}{2\pi\sqrt{R_1R_2C_1C_2}} \tag{9-11}$$

and

$$\left.\begin{aligned} +jX_2R_2A_v - jX_1R_2 - jX_2R_1 - jX_2R_2 = 0 \\ A_v = \frac{R_1}{R_2} + \frac{C_2}{C_1} + 1 \end{aligned}\right\} \tag{9-12}$$

In these equations when resistances are in ohms and capacitances in farads, f will be the frequency of oscillation in cycles per second, and A_v will be the required voltage gain expressed as a ratio. Note that if $R_1 = R_2 = R$, and $C_1 = C_2 = C_m$, then equation 9-11 becomes

$$f = \frac{1}{2\pi RC} \tag{9-13}$$

To explain why the oscillations are given by equation 9-11, it is convenient to consider a type [8] widely used at audio frequencies. The relationship of feedback units is $R_1C_1 = R_2C_2$, and $R_1 = R_2$ and $C_1 = C_2$. The two variable capacitors C_1 and C_2 are "ganged" so that they vary together. Several different resistor values may be selected by a tapped switch to provide various frequency ranges; yet, $R_1 = R_2$ for all settings. Assume that the resistors each are 848,000 ohms, and the capacitors each are 200 micromicrofarads. Equations 9-11 and 9-13 give an oscillation frequency of 940 cycles for this combination. If the two-stage *amplifier* is assumed to have the same gain at the frequencies to be considered, and if the ratio E_{in} to E_{out} as given by equation 9-9 is calculated at, for example, $0.1f$, f, and $10f$ (or 94, 940, and 9400 cycles), a curve such as Fig. 9-7 will result. This shows that the ratio of E_{in} to E_{out} is greatest at the frequency given by equations 9-11 and 9-13, and that the circuit will oscillate at this frequency. This fact can be verified mathematically or experimentally. The feedback through R_1–C_1, to R_2–C_2 is *positive feedback*, tending to cause an increase in the magnitudes of the oscillations.

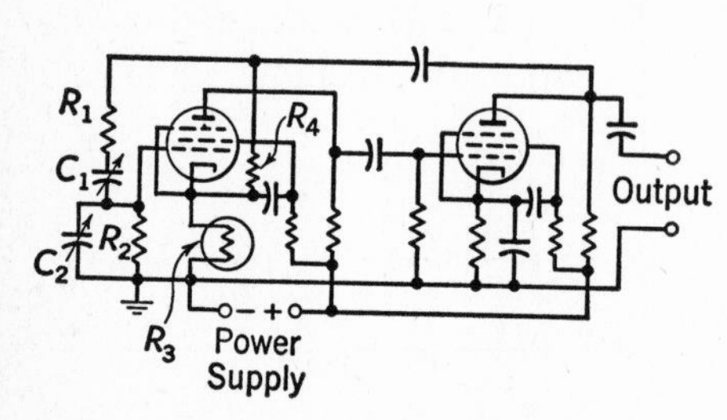

FIG. 9-8. Basic circuit of a R–C feedback oscillator. The output terminals commonly are connected to an amplifier.

For the resistors and capacitors used, equation 9-12 gives a value $A_v = 3$ for the condition of oscillation. A voltage gain of 3 is indeed small for a two-stage resistance-coupled amplifier, and in a commercial form is obtained by applying *negative feedback inside* the two-stage amplifier of Fig. 9-7. As shown in Fig. 9-8, this

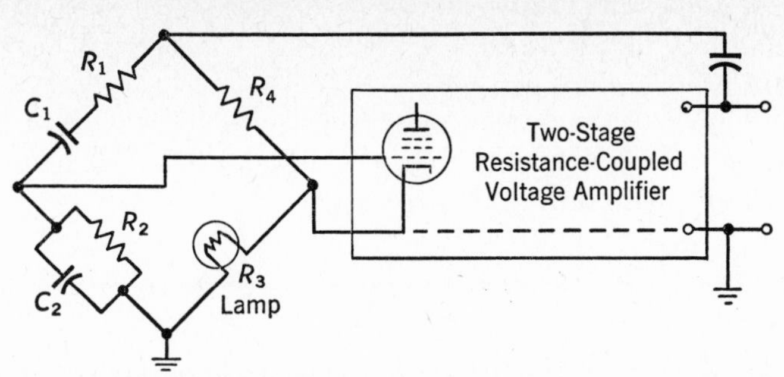

FIG. 9-9. The R–C feedback oscillator of Figs. 9-7 and 9-8 may be considered from the standpoint of the Wien bridge here shown.

negative feedback is obtained (section 8-19) by passing output signal back through the series R_4–R_3 combination, it being noted that R_3 is the filament of a small lamp. The lamp is operated at a temperature such that the resistance of the filament is nonlinear, varying with the magnitude of the current fed back. If the resistance of the filament R_3 increases, the *negative* feedback increases. This action stabilizes the magnitude of the oscillations. When properly designed and adjusted, the oscillator operates at a frequency given by equations 9-11 and 9-13 at the maximum ratio given by the curve of Fig. 9-7, and with a negative feedback *slightly less* than the positive feedback. This amount of negative feedback results in excellent wave shape. The R-C oscillator may be analyzed from the standpoint of a Wien bridge as indicated in Fig. 9-9. For the condition of oscillation the bridge is almost, but not quite, balanced.[9] The R–C oscillator is widely used as a source of audio-frequencies, and of frequencies up to about 500 kilocycles.

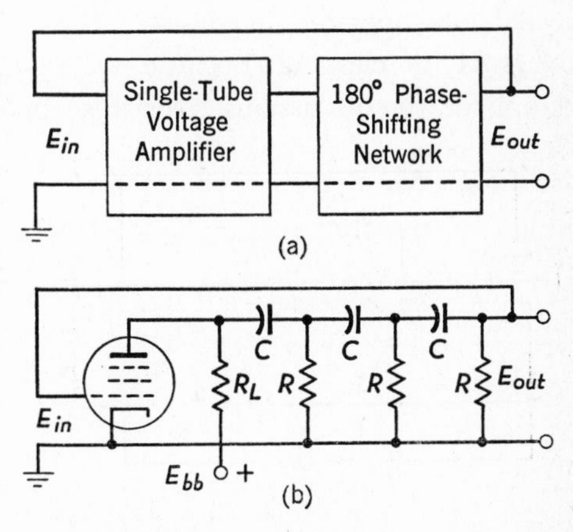

FIG. 9-10. Basic circuits of a phase-shift oscillator.

9-6. *PHASE-SHIFT FEEDBACK OSCILLATOR*

The **phase-shift oscillator** is an oscillator produced [1] by "connecting any network having a phase shift of an odd multiple of 180° (per stage) at the frequency of oscillation, between the output and the input of an amplifier." The basic circuit arrangement is shown in Fig. 9-10. Because a single amplifying tube produces a phase shift of 180°, the 180° phase-shifting network of Fig. 9-10a must be inserted between the output and input to cause a total shift of 360° and provide a feedback input voltage of the proper phase to drive the tube. One of the many possible phase-shifting networks is shown in Fig. 9-10b. This network gives the desired 180° phase shift at the frequency

$$f = \frac{1}{2\pi\sqrt{6}RC} \qquad 9\text{-}14$$

where f is in cycles per second when R and C are in ohms and farads.[10] For this frequency, oscillations will occur if the voltage gain is at least 29. In deriving equation 9-14 it is assumed [10] that the resistance of R greatly exceeds R_L.

9-7. *RADIO-FREQUENCY FEEDBACK OSCILLATOR*

Many radio-frequency oscillators are fundamentally class-C radio-frequency amplifiers with positive feedback. Class-C operation is used to obtain high efficiency. The tuned parallel resonant circuit rejects the unwanted harmonics of the distorted class-C signal output (section 8-16) and passes the desired fundamental. Class-C operation could be used at audio-frequencies if it were not for the filtering problem involved. *Simple* tuned circuits could not be used for rejecting unwanted frequencies in wide-band audio oscillators operated in class-C. Thus, audio oscillators usually operate in class-A1, or class-A2. Because a class-C oscillator is a self-excited class-C amplifier, the discussions given in sections 8-15 and 8-16 apply to the oscillator. The efficiency of a class-A oscillator is less than 50 per cent, but for a class-C oscillator it may be about 80 per cent.

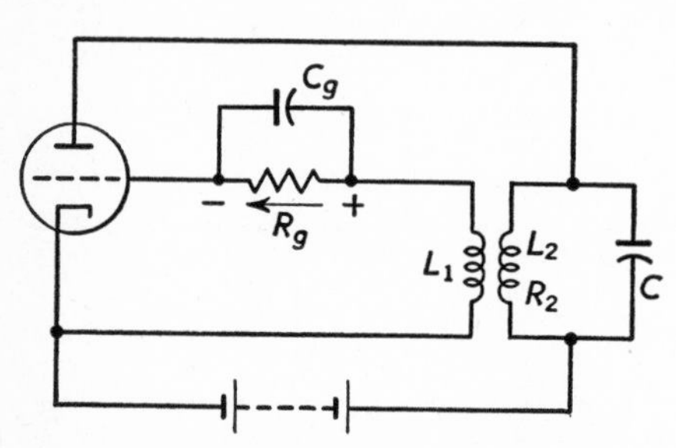

FIG. 9-11. A class-C oscillator, self-biased by the grid resistor–capacitor combination.

In the class-A oscillator considered in section 9-1, a constant grid-bias voltage, such as supplied by batteries, was used. In the class-C oscillator, however, the grid bias usually is obtained by the so-called "grid-leak-capacitor method," better known as grid-resistor-capacitor method. With this method (Fig. 9-11), when the tube is not oscillating, the bias on the grid is zero, and hence oscillations will be self-starting when the circuit is energized. If a tube is biased initially beyond cutoff for class-C operation, oscillations will not start because no plate current will flow through the tube. When the tube first is energized the bias is zero and oscillations will start. These will increase as explained in section 9-1, and the grid will be driven positive. This will cause a rectified *grid current* to flow in "spurts," and this will have a direct-current component I_c (section 6-1). This current will cause an I_cR_g drop across the resistor as shown in Fig. 9-11. Grid capacitor C_g acts as a low-impedance path for the alternating components of the grid current, but the direct current must pass through the grid resistor R_g to produce the required I_cR_g voltage drop to bias the tube.

The value of the grid capacitor C_g is not critical, but its reactance at the fundamental frequency of operation should be low in comparison with the resistance of the grid resistor. If the capacitance is too great, intermittent operation and blocking of the tube may result. Grid capacitors of 100 to 250 micromicrofarads are often used with the small tubes. The resistance of the grid resistor for a given bias can be determined if the magnitude of the direct component of the grid current is known or can be estimated. A value of 10,000 ohms for the bias resistor is often satisfactory for small tubes, but 25,000 or even 50,000 ohms may be necessary.

Experience with oscillator design has indicated that the class-C oscillator will be unstable if the ratio of $\omega L_2/R_2$ is less than about 12. The value of R_2 is largely determined by the load, usually inductively coupled to L_2. This reflected load resistance acts like a resistance in series with the coil, and thus reduces the ratio $Q = \omega L_2/R_2$ to lower values than for the coil and capacitor alone.

If either C_g, or R_g, or both of these are too large **intermittent operation** may result. This is explained as follows: After stable operation has been reached, the grid has assumed an equilibrium negative grid voltage. If now the amplitude of oscillation should be reduced by any slight change in circuit conditions, the grid bias voltage must also reduce. This cannot occur rapidly if the capacitor or resistor is too large, and the tube may stop oscillating until the grid-bias voltage across R_g–C_g reduces to the value where oscillations will start. Ordinarily, intermittent operation results only in erratic performance.

The similar phenomenon of **blocking** also causes oscillations to stop, but often with disastrous results. Thus, if the grid resistor is too large, so that the current of electrons flowing to the grid is low; if the grid is driven highly positive by the alternating voltage impressed on it; and if at this instant the plate potential is high, as may be the case if the amplitude of oscillation is small; then the *secondary emission current from the grid to the plate may exceed the electron flow from the cathode to the grid.* If this occurs, the grid current will reverse, the grid will be biased positively by this reversed-current flow through the large grid resistor, the plate current will rise to a high value, and the tube will be overheated. This will cause the evolution of occluded gases and probably ruin the tube. Blocking can be prevented by conservative design so that in operation the plate voltage on the negative half-cycle does not fall to a potential too close to that to which the grid is driven. In the design of vacuum tubes precautions are taken to reduce grid emission.[11]

The Hartley Oscillator. The Hartley oscillator as shown in Fig. 9-12a and modifications of it are used extensively in radio, and sometimes at audio-frequencies. Grid bias is obtained by the C_g–R_g combination. Direct voltage is applied to the plate through the inductor, or choke coil. The direct-current component of plate current flows through this coil. Capacitor C′ blocks the flow of direct current to the right, and isolates the parallel resonant circuit (which may require manual adjusting) from the high direct voltage. Mutual inductance commonly exists between L_2 and L_1, but is not necessary. The mutual inductance being neglected, oscillations will occur at *approximately* the frequency that the capacitive reactance of C minus the inductive reactance of L_1 equals the inductive reactance of L_2. In effect, the Hartley oscillator uses inductive voltage divider L_1–L_2 to regulate the amount of feedback.

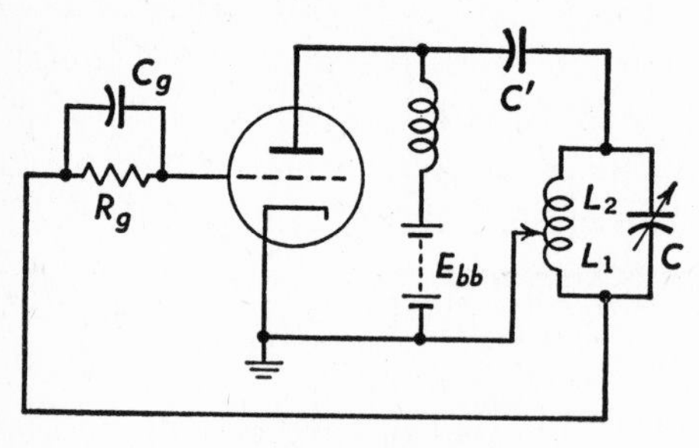

(a) Hartley Oscillator

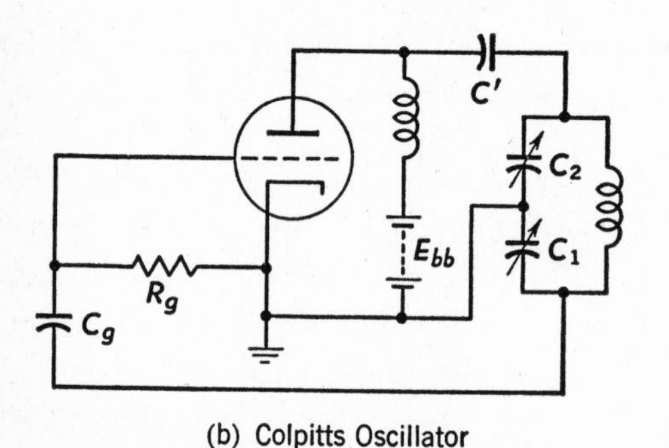

(b) Colpitts Oscillator

FIG. 9-12. Common oscillator circuits.

The Colpitts Oscillator. This is similar to the Hartley oscillator except that feedback voltage is determined by capacitive voltage divider C_1–C_2 of Fig. 9-12b instead of by an inductive voltage divider. The grid resistor R_g must be connected directly from grid to cathode, and capacitor C_g connected as shown (or omitted because of C_1). The frequency of oscillation will be *approximately* that at which the inductive reactance of L minus the capacitive reactance of C_1 equals the capacitive reactance of C_2.

9-8. *QUARTZ-CRYSTAL-CONTROLLED OSCILLATORS*

In many instances, frequency stability of vacuum-tube oscillators is very important. This subject was studied by Llewellyn [12] and others.[13] Quartz-crystal elements commonly are used to stabilize radio-frequency oscillators. These elements are based on the discovery of the piezoelectric effect by the Curies.[14] Thus, if the dimensions of certain crystals are altered by an applied mechanical force, a difference of potential will be produced between opposite crystal surfaces. Furthermore, if a direct voltage is applied to the opposite crystal surfaces, a mechanical force will exist within the crystal and its dimensions will be changed. Also, if the direction of the mechanical force is changed, the polarity of the difference of potential will be reversed; and changing the sign of the applied voltage alters the direction of action of the mechanical force.

A typical quartz-crystal element is about one-half inch square and one-sixteenth inch thick. It is sawed from a natural crystal, ground, and etched until it is resonant at exactly the frequency desired. Crystals cut from the same specimen, but in different planes, exhibit different frequency variations with temperature.[15,16] However, if they are cut and operated properly, their characteristics are substantially independent of ordinary temperature variations.[17,18]

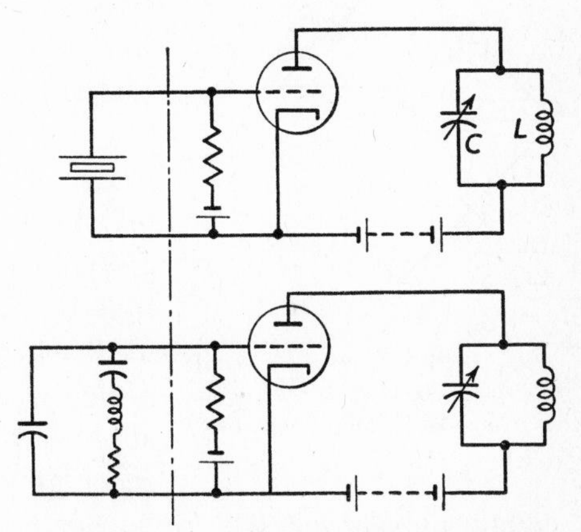

FIG. 9-13. A crystal oscillator, showing the crystal replaced by its equivalent electrical network (lower illustration). A "choke" sometimes is connected in series with the grid resistor.

When used to stabilize frequency of oscillation, the quartz-crystal element is placed in a suitable holder [19] in a circuit such as Fig. 9-13. When the crystal is not vibrating, the holder and crystal are electrically equivalent to a small capacitor. When the crystal vibrates at one of several possible modes, the crystal is electrically equivalent to the series resonant circuit shown in the lower part of Fig. 9-13. The equivalent Q of a crystal in an ordinary holder is about 10,000 or more, much greater than can be obtained by using inductors and capacitors. Under special conditions a Q of over 300,000 has been obtained, and a crystal has maintained an oscillator frequency constant within one part in a million over a 100-degree Centigrade range.[18]

The vibrating crystal and holder constitute a resonant circuit between grid and cathode of Fig. 9-13, and *L–C* constitutes a resonant plate load; thus, Fig. 9-13 is in effect a tuned-grid, tuned-plate feedback oscillator and operates as explained in section 9-3. Although a triode is shown in Fig. 9-13, tetrodes and pentodes are used also, in which event it may be necessary to connect a small capacitor from plate to grid to ensure sufficient feedback. Crystals for frequencies of a few megacycles are thin and fragile, and may shatter if permitted to vibrate too violently. It should be mentioned that synthetic crystals have been produced.[20]

9-9. *BRIDGE-STABILIZED OSCILLATORS*

As shown in Fig. 9-9, the Wien bridge is, in effect, used to control the *R–C* oscillator. The bridge principle was applied by Meacham to produce an oscillator of unusual frequency stability.[21] It has been stated [4] that this oscillator produces oscillations of the greatest frequency stability. The basic circuit is shown in Fig. 9-14. The vibrating crystal is equivalent to a series resonant circuit (Fig. 9-13), and is in effect resistance in the bridge of Fig. 9-14. In fact, a series *L–C* circuit may be used instead of a crystal. Resistor R_3 is a small lamp as in the *R–C* oscillator, and controls the amplitude of oscillation by negative feedback as indicated in the diagram. If the bridge were exactly balanced, there would be no feedback, and no oscillations. Oscillations occur at a frequency at which the bridge is slightly unbalanced. The frequency is almost completely controlled by the crystal or other resonant element.[4,21]

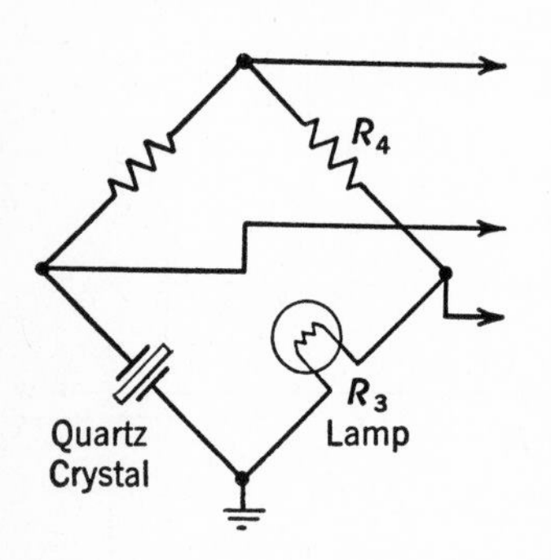

FIG. 9-14. A quartz crystal in a bridge circuit for controlling an oscillator. (See Fig. 9-9.)

9-10. *VHF AND UHF FEEDBACK L–C OSCILLATORS*

The operation of thermionic vacuum tubes as amplifiers at very high frequencies (30 to 300 megacycles) and ultrahigh frequencies (300 to 3000 megacycles) was discussed in sections 8-23 and 8-24. It was explained that at these frequencies lead inductances, interelectrode capacitances, energy losses, and electron transit time became very important. Also it was shown how these effects were minimized, and how close-spaced planar triodes were used as amplifiers up to 3000 megacycles and above. Since an amplifier with correct feedback is an oscillator, it follows that triode oscillators are possible within this range.

The basic *L–C* oscillator circuits of the preceding pages using *separate* inductors and capacitors as in Fig. 9-15 are useful up to several hundred megacycles. An oscillator using a combined *L–C* tuning unit, called a **butterfly circuit,** or **butterfly resonator,** operates from 100 to 1000 megacycles.[22] This tuned circuit, or resonator, resembles somewhat a variable air capacitor in appearance, with variable capacitance provided between stationary and movable plates; however, it is constructed so that variable inductance also is provided by a portion of the stationary structure.[22]

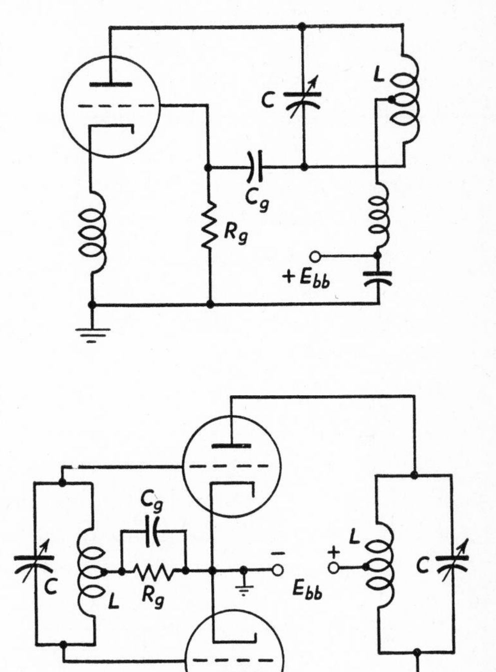

FIG. 9-15. Two circuits used for *VHF* and *UHF* oscillators. *R–F* chokes are included in the cathode and plate-supply leads of the upper circuit.

9-11. *VHF AND UHF FEEDBACK TRANSMISSION-LINE OSCILLATORS*

Sections of transmission lines early were used[23] to provide frequency control in oscillators operating above about 30 megacycles. Transmission-line control is particularly adapted to oscillators operating at about 100 to 1000 megacycles. Such oscillators will be discussed in this section. Although two-conductor transmission lines will be used in the illustrations, it should be kept in mind that in general the theory applies to oscillators using coaxial construction. Also, it should be noted that the two-conductor line is well suited for obtaining unshielded balanced signal output, and that coaxial construction is well suited for shielded unbalanced output. Many, but not all, two-conductor line oscillators are of the balanced type.

Resonant Transmission Line Theory. This subject is of much importance and is covered in articles[24] and many textbooks, references 19 and 25 being especially useful. In practice the lines often used are of closely paralleled tubes or rods of large diameter compared to ordinary wires; also, the lines are physically short, perhaps a few inches to several feet long. Because of the large size, short length, and high frequency (which tends to cause series-inductance and shunt-capacitance reactive effects to "overshadow" series-resistance and shunt-conductance loss effects), these line sections commonly are considered lossless. Of course energy losses do occur within the system, and also, some energy is radiated. Such parallel lines sometimes are called **Lecher wires.**

If a section of a *lossless* two-parallel-wire line is investigated at frequencies for which the length of the line is between zero and one-fourth wave length, the input, or driving-point, impedance will be *capacitive reactance* if the distant end of the line is *open-circuited.* Furthermore, the magnitude of the reactance and the equivalent capacitance lie between the theoretical limits of zero and infinity. However, if the distant end of the line is *short-circuited,* the input impedance will be *inductive reactance,* and the magnitude of the reactance and the equivalent inductance lie between zero and infinity. For sections longer than one-fourth wave length and less than one-half wave length, the nature of the reactance changes from capacitive to inductive, and vice versa. These short sections of lines commonly are called **resonant lines** because their input impedance varies like the impedance of L–C resonant circuits. At very high, and ultrahigh frequencies it is both convenient and economical to use short line sections in oscillators as the frequency-determining element instead of inductors and capacitors. The equivalent Q obtained by transmission-line tuning is higher than with L–C circuits and is less than with quartz crystals, but lines can be used at frequencies above those at which crystals are applicable.

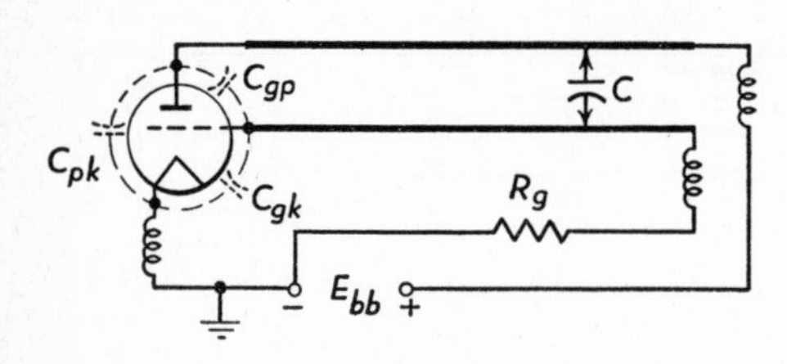

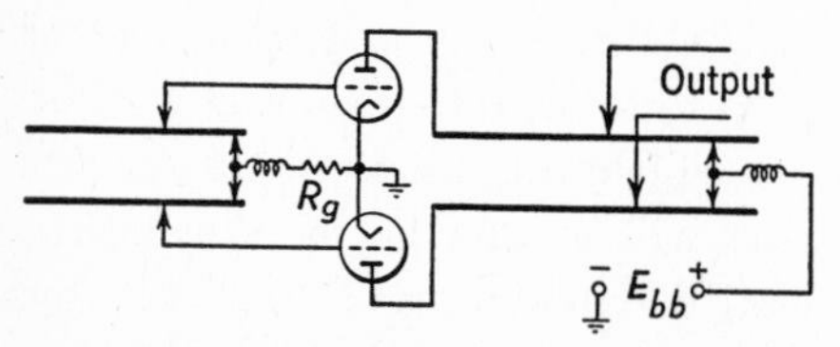

FIG. 9-16. Basic transmission-line oscillators. The upper circuit is an unbalanced oscillator, and the lower circuit is a balanced oscillator.

Parallel-Wire Transmission-Line Oscillators. An elementary unbalanced transmission-line oscillator is shown in Fig. 9-16. At the high frequencies generated, capacitor C in effect short-circuits the line. Resistor R_g is to provide direct grid-bias voltage after oscillations start and the grid is driven positive. The coils are radio-frequency "chokes" to reduce the loss of high-frequency energy. Capacitor C is varied in position until the resonant action largely due to interelectrode capacitance C_{gp} and the inductance provided by the "shorted" line section establish the frequency desired. Feedback from the plate-to-grid circuit is provided by the interelectrode capacitive voltage-divider arrangement caused by C_{pk} and C_{gk}. A similarity with a Colpitts oscillator exists. This arrangement in effect divides the voltage between the resonant wires in Fig. 9-16 and impresses a portion between grid and cathode.

A balanced "push-pull" transmission-line oscillator with tuned-grid and tuned-plate circuits also is shown in Fig. 9-16. The basic principle of operation is as explained for the single-tuned type. The balanced type is particularly suited for feeding a balanced transmission line leading, perhaps, to a transmitting antenna. The circuits of Fig. 9-16 are hazardous because the direct plate voltage for the tubes is connected directly to the transmission-line wires. Modifications are readily made, however, for connecting the power source in a safe manner. For instance, the $+E_{bb}$ leads may be carried inside the metal tubes forming the transmission line, and the metal tubes connected through small fixed mica capacitors to the proper tube plate.

Coaxial Transmission-Line Oscillators. Statements made regarding the input impedance of parallel two-wire transmission-line sections apply in general to coaxial transmission lines, or cables, except that short coaxial sections of relatively large diameter tend to behave more like resonant cavities, having several possible modes of oscillation. Tubes for very high and ultra-high frequencies commonly are made for coaxial mounting (sections 8-23 and 8-24), and hence coaxial-type resonators are particularly useful in oscillators for several hundred to several thousand megacycles.

A coaxial-line oscillator, being an amplifier with positive feedback, results if the plate cavity of Fig. 8-36 is connected properly to the grid circuit. Also, a grid resistor must be added to obtain the correct direct voltage bias.[4,9]

9-12. *ELECTRON-COUPLED OSCILLATORS*

As was mentioned in section 9-4, a buffer, or isolating, tube must be used between an oscillator tube and the load to prevent variations in the load impedance from affecting the frequency of oscillation. Oscillators may be electron-coupled to the load in a manner such that the usual isolating stage may be omitted.[31]

Electron-coupled oscillators employ either tetrodes or pentodes. A circuit using a screen-grid tetrode is shown in Fig. 9-17. In this the cathode, control grid, and screen grid act as a conventional triode oscillator. For *alternating voltages,* the screen grid is effectively at ground potential through the capacitor C_1, and hence acts as a shield between the oscillator and the output circuit; that is, the plate.

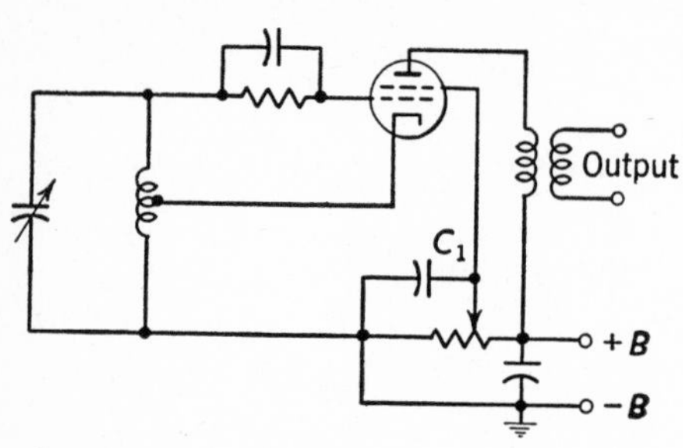

FIG. 9-17. An electron-coupled oscillator using a screen-grid tube.

When the plate is thus shielded from the oscillator circuit by the screen grid, the only coupling is due to the electron current flow which largely passes through the screen grid to the plate; hence the wording "electron coupled." When the plate voltage is sufficiently high, so that the plate current does *not* vary with plate voltage changes, then any changes occurring in the output or load circuit cannot be reflected back into the oscillator circuit and affect its frequency stability.

The variable voltage divider shown in Fig. 9-17 is for adjusting the ratio of the screen-grid to plate voltages. For some particular ratio of these voltages, the frequency will be found to be substantially independent of the supply voltage variations.

9-13. *POSITIVE-GRID TRIODE OSCILLATORS*

In the oscillators considered in the preceding pages the tubes were operated in the normal manner with grids negative and plates positive. Oscillators for producing ultrahigh frequencies have been developed in which the plate is negative and the grid is positive. Such oscillators were operated in communication systems at an early date.[27,28] For many years they were the usual source of ultrahigh frequencies. When certain tubes are operated with a *negative* plate and *positive* grid, *some* of the electrons attracted by the positive grid pass between the grid wires and their kinetic energy carries them on toward the plate. They never reach it, however, because it is negative, so they flow back toward the grid. They may strike the grid or may pass through the grid wires and travel on toward the cathode. The attraction of the positive grid will soon bring them to rest, and they will return toward the grid, which they may strike, or again pass through. These oscillations by *some* of the electrons between the plate and grid cause a very high-frequency voltage between these electrodes. When an electron approaches the plate it makes the plate less negative, and when it recedes it makes the plate more negative. The frequency of oscillation is determined, in a tube of given geometrical construction, by the electrode voltages. The high-frequency output and power efficiency are both low because many of the electrons are collected by the positive grid wires and do not oscillate, and because of other losses. Such oscillators are called **Barkhausen oscillators,** or **Barkhausen-Kurz oscillators,** after early experimenters. Similar devices called **Gill-Morell oscillators** have been developed in which a Lecher-wire system, or transmission line, regulates the frequency of oscillation.

9-14. *NEGATIVE RESISTANCE OSCILLATORS*

As mentioned in the opening paragraphs of this chapter, sine-wave oscillators are classified [1] as either feedback-amplifier oscillators, or as negative-resistance oscillators. Also as mentioned, negative resistance [29] supplies power to a circuit, and positive resistance (the common type) absorbs power from a circuit. The distinction between the two types of oscillators may not always be clear,[4] and some prefer to consider all oscillators from the negative-resistance viewpoint.

A **negative-resistance oscillator** is defined [1] as "an oscillator produced by connecting a parallel-tuned resonant circuit to a two-terminal negative-resistance device. (One in which an increase in voltage results in a decrease in current.)" The dynatron and transitron are cited [1] as examples of negative-resistance oscillators.

The Dynatron Oscillator. In the **dynatron oscillator** the negative resistance is [1] "derived between plate and cathode of a screen-grid tube operating such that secondary electrons produced by the plate are attracted to the higher-potential screen grid." This negative-resistance characteristic is shown in Fig. 5-3, in which the screen grid is held at +45 volts. As the plate voltage is increased positively, the plate current I_b increases positively until +10 volts is reached; a further *positive increase* in plate voltage causes a *decrease* in plate current. This is a negative-resistance effect between plate and cathode. Similar action would result for certain higher voltage combinations. Secondary emission usually is more pronounced in the older tetrodes.

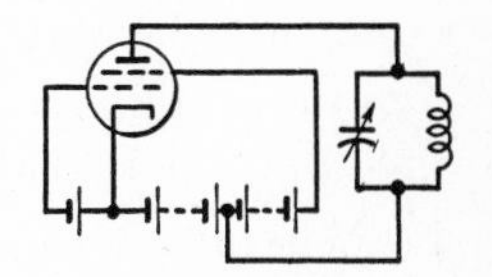

FIG. 9-18. Circuit of a simple dynatron oscillator.

The basic circuit of a dynatron oscillator, developed by Hull,[30] is shown in Fig. 9-18. If the magnitude of the negative resistance is less than the equivalent resistance of the resonant circuit, oscillations will start, and will increase in amplitude until the negative resistance offered at the new condition of oscillation is equal to the equivalent positive resistance of the resonant circuit.

The Transitron Oscillator. The **transitron oscillator,** although classified [1] as a negative-resistance oscillator, is, more exactly, a **negative-transconductance oscillator** [1] "employing a screen-grid tube with negative transconductance produced by a retarding field between the negative screen grid and the control grid which serves as the anode."

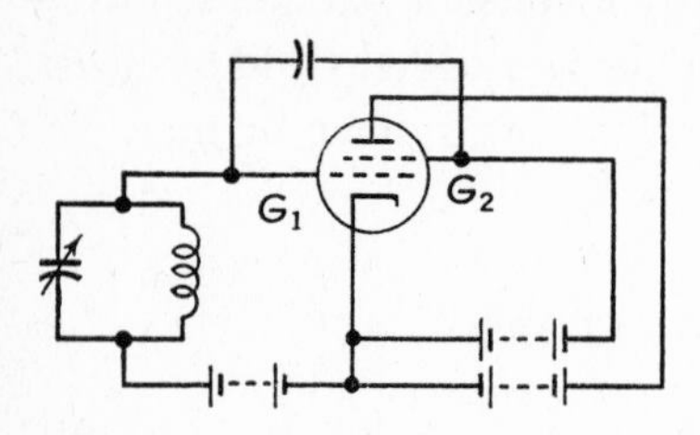

FIG. 9-19. Circuit using a tetrode as a transitron oscillator.

A transitron oscillator using a screen-grid tube would be connected as in Fig. 9-19. Negative transconductance between screen grid G_2 and control grid G_1 is obtained as follows: Some of the electrons from the cathode are captured by positive grid G_1, and some electrons pass through. Some of the latter are repelled back toward positive grid G_1 by the negative grid G_2. Many of these are then captured by positive grid G_1. If now the potential of grid G_2 is *increased* in a *positive* direction by making it less negative, fewer electrons will be repelled back to positive grid G_1, and the current taken by grid G_1 may *decrease*. Voltage change on one electrode affecting current in another in this way is a negative transresistance, the reciprocal of which is a negative transconductance. The preceding explanation could have been made on the basis of the effect produced on grid G_2 by changes on grid G_1. An advantage of the transitron over the dynatron is that the negative transconductance, or transresistance, in a transitron does not depend on secondary emission, which varies with different tubes and with tube life. In practice transitron oscillators usually employ pentodes.[4,9,31,32]

9-15. *CAVITY OSCILLATORS*

As explained in section 9-11, the input impedance of short-circuited transmission-line sections varies like the input impedance of resonant L–C circuits. Because of this, it was shown that short-circuited line sections could be used as resonators to control the frequency of oscillators (Fig. 9-16). A simple cavity is "generated" if the short-circuited transmission-line section of Fig. 9-20 is rotated as indicated by the arrow. Such a

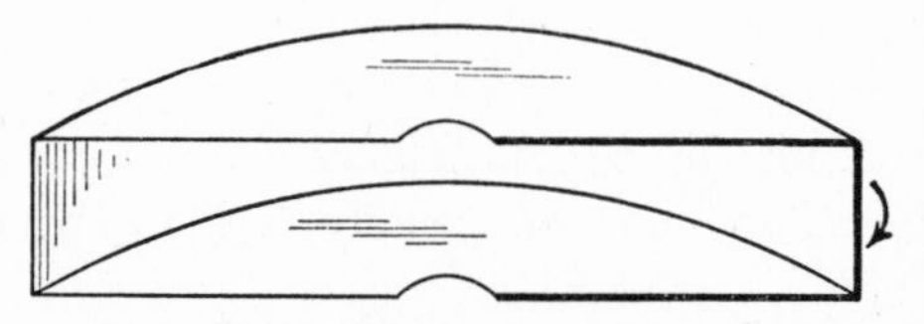

FIG. 9-20. The short-circuited transmission line at the right when rotated will "generate" a resonant cavity.

metal cavity will, at certain frequencies, behave much like a short-circuited line, and can be used as a **cavity resonator** to control the frequency of an oscillator. A cavity resonator will oscillate in different modes at various frequencies.[33] Detailed information on cavities is contained in the many articles and books on ultrahigh-frequency applications.

9-16. *VELOCITY-MODULATED TUBES*

The tubes considered in the preceding pages were of conventional type consisting of a cathode, one or more grids, and a plate (or anode). The desire for operation at ultrahigh frequencies led to the development of a tube of different type. This is the **electron-beam tube,**[34] whose performance "depends upon the formation and control of one or more electron beams." The electron beam originates in an **electron gun,**[34] which is an electrode structure "which produces, and may control, focus, and deflect an electron beam." In the tubes under consideration the electron gun produces, controls, and focuses the beam in such a manner that the electrons in the beam proceed with considerable velocities down the axis of the tube. In electron-beam tubes called **velocity-modulated tubes,**[35] the beam is influenced by electrodes as it passes down the tube. The electrons emerge from the gun with various velocities, and the electrodes accentuate these velocity variations, cause the beam to be **velocity modulated,** and cause the electrons to be **bunched.** It should be noted that electron transit time limits the operation of ordinary electron tubes, whereas velocity-modulated tubes depend on transit time for operation. Such tubes are used as amplifiers, modulators, and oscillators. This last use has been most extensive and now will be considered.

In a **velocity-modulated oscillator**[1] the velocity of an electron beam, or stream, is "varied (velocity-modulated) in passing through a resonant cavity called a buncher. Energy is extracted from the bunched electron stream at a higher energy level in passing through a second cavity resonator called the catcher. Oscillations are sustained by coupling energy from the catcher cavity back to the buncher cavity."

9-17. *THE KLYSTRON OSCILLATOR*

The **Klystron** is extensively used in the velocity-modulated oscillator of the preceding paragraph. This electron device was perfected about 1938 by the Varians in this country. (Regarding European developments, see page 527, reference 41.) An early form of the Klystron is shown in Fig. 9-21. Electrons are emitted from a massive thermionic cathode. The number leaving the cathode region is controlled by what is essentially control-grid action. The grid is shaped so that it also assists in forming an electron beam. It is positive with respect to the cathode and acts as an accelerator. The electron beam travels toward the two gridlike portions of a resonating cavity

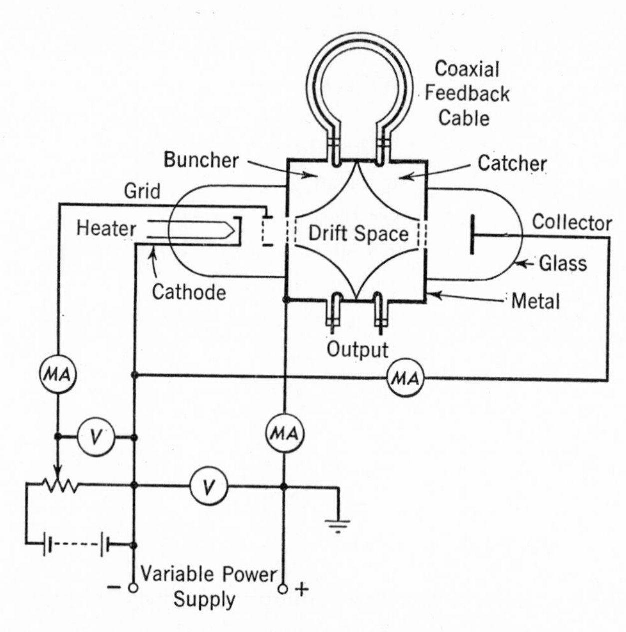

FIG. 9-21. Simplified diagram of a basic type of Klystron, and connections for oscillation. Note that the positive side of the power supply and the resonant metal cavities marked "buncher" and "catcher" are grounded. Also note that the grid is positive with respect to the cathode.

known as the **buncher.** After passing through the buncher the electrons flow through the electric field-free drift-space, and toward the gridlike portions of resonant cavity known as the **catcher.** After passing through the catcher they flow to the collector and back to the cathode. Of course some electrons are picked up by the grids of the buncher and catcher, but the action of the tube does not depend on these. The main beam flows from the control grid to the collector. The operation of the tube depends on the interaction between the gridlike portions of the buncher and catcher and the electrons in the beam.

It will be noted that coupling loops extend into the resonant cavities. These act as loop antennas connected to the ends of the coaxial cable. By analogy with short-circuited transmission-line theory, large currents are known to flow in the ends of the cavities, and intense magnetic fields will exist in these regions. The loop antennas make it possible to introduce signal energy into the cavities, or to abstract signal energy from the cavities. Connections as an oscillator are shown in Fig. 9-21. The metallic resonating cavities are at ground potential, and the positive side of the direct-current power source is grounded.

To explain Klystron operation, assume that a single high-speed electron has just passed through the first grid of the buncher. As this happens, it drives electrons out of the first grid of the buncher cavity and toward the second grid of the buncher. This will cause current flow in the walls of the cavity, and a difference of potential between the two grids of the buncher. When the electron reaches the center of the opening of the buncher cavity (when it is midway between the buncher grids), it will be producing an equal effect on each grid, and no resultant action exists. When the electron reaches the vicinity of the second grid of the buncher, it repels electrons from this grid and makes this grid positive and the first grid negative, reversing the action of the preceding paragraph. A similar action occurs when the electron considered passes through the catcher cavity.

If a *perfectly* uniform stream of electrons were passing through the tube, no resultant difference of potential between the grids of the buncher or the catcher would result. But, a stream of electrons is not uniform and differences in potential will result. If a signal is fed back from the catcher into the buncher, then the difference of potential caused by this signal between the buncher grids will speed up certain electrons and slow up others. In this way the beam becomes velocity-modulated. Also, when these bunched electrons flow through the catcher, they will produce large differences of potential between the grids of the catcher, thus providing high-frequency energy for feedback into the buncher, and for external purposes. A logical question is: How does the action start? The first transient electron-current flow through the tube would provide the starting impulse much as in any oscillator. Furthermore, an electron beam is not *perfectly* uniform; there are random fluctuations in an electron beam that would provide a starting transient.

The pickup loop of the catcher is coupled back through a coaxial cable into a similar loop in the buncher. The length of the loop and the accelerating voltage are such that the phase relations of the signal fed back into the catcher is correct. The buncher and the catcher are tuned to the same frequency. Only a part of the energy available at the catcher is fed back into the buncher, and thus high-frequency energy may be drawn off through a second coupling circuit for external purposes. A given Klystron will oscillate at certain voltages, and will cease to operate at other voltages. This is because the voltage affects the electron velocity, and for a given tube and adjustment, optimum velocities exist for correct phase relations. The phase of the signal feedback must be proper for sustained oscillations.

The Reflex Klystron.[37,38] The **reflex Klystron** has only one cavity and a **repeller** electrode that directs the electrons back into the cavity through which they have just passed. In this way, the use of two cavities is avoided, and no feedback loop is needed. The reflex Klystron oscillator is extensively used as a test source, and for other purposes, in the range from about 1000 to 30,000 megacycles.

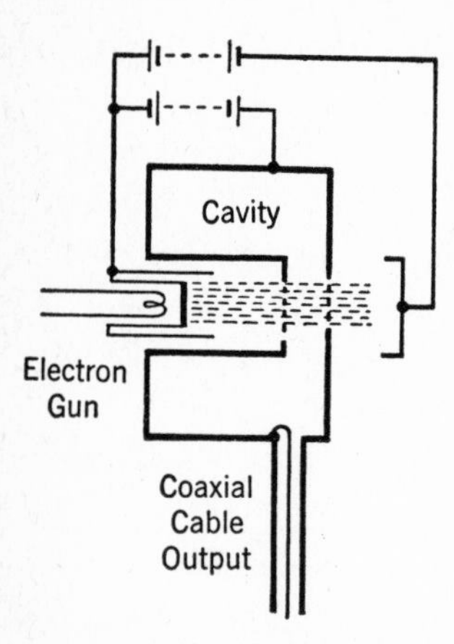

FIG. 9-22. Elements of a Klystron oscillator using a single cavity.

The **reflex Klystron oscillator** is shown in Fig. 9-22. Assuming that oscillation has been established, the radio-frequency voltage existing between the cavity grids acts on the stream of electrons flowing through these cavity grids and toward the repeller. Hence, when the electrons emerge from the cavity into the repeller space they are starting to bunch, and to be velocity-modulated, in accordance with the radio-frequency voltage between the grids. This bunching action continues in the repeller space as the negative repeller turns back the electrons and forces them back into the resonant cavity. These bunched, or velocity-modulated, electrons produce voltages between the cavity grids as explained for the two-cavity Klystron. Bunched electrons must arrive back at the cavity when the phase of the cavity voltage is such that the electrons are decelerated if oscillations are to be sustained. The resonant cavity design greatly affects the frequency of oscillation. Frequency also is affected by the repeller voltage. Frequency adjustment by this means has been referred to as **electron tuning.**

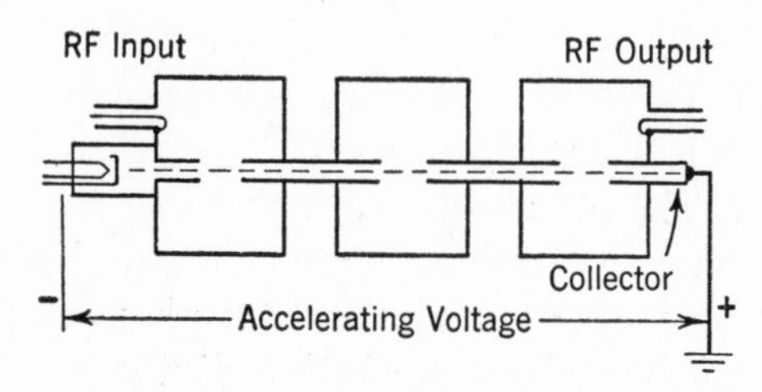

FIG. 9-23. Showing the basic principles of a multicavity Klystron that will handle many kilowatts of ultrahigh-frequency power.

The Multicavity Klystron. Since their invention in 1938, Klystrons have been used for many purposes other than oscillators, and have been treated in many articles and several books.[39,40,41] They may be modulated,[40] and have been used as amplifiers in point-to-point radio-relay telephone systems.[42] In general it may be said that triodes will perform these functions better, particularly at low power levels up to about 4000 megacycles. However, multicavity Klystrons appear at present (1955) to offer excellent possibilities as power amplifiers [43] to handle many kilowatts at frequencies from about 500 megacycles up to, perhaps, 2500 megacycles. The basic principle of a multicavity Klystron is illustrated by Fig. 9-23. Tubes of this type are being developed for power amplifiers in ultrahigh-frequency television.[43,44,45]

9-18. *THE MAGNETRON OSCILLATOR*

The magnetron is extensively used for producing large amounts of power for short time intervals at frequencies from about 1000 to 30,000 megacycles. A **cylindrical-anode magnetron** consisting of a straight filament thermionic cathode at the axis of a metal cylinder, and with an electromagnet arranged to produce a magnetic field between, and parallel to, the filament and cylinder was reported first by Hull [46] in 1921. Hull also discussed a feedback-type audio-frequency magnetron oscillator.[47]

The cylindrical anode of this early magnetron is made positive with respect to the cathode, and this establishes a radial electric field between these two electrodes. The static characteristics are depicted in Fig. 9-24. The radial electric field, and the axial magnetic field (assumed directed out), are omitted; also, space-charge effects are neglected. Note that magnetic field intensity is plotted along the X axis, and the magnitude of the electron current from cathode to anode is plotted on the Y axis. The probable path of an electron for several magnetic field intensities is shown. These paths result from the combined action of the electric and magnetic fields (sections 1-3 and 1-4). With the electric field at all times constant, but with *no* magnetic field, an electron will follow the radial path as in (a). With a *weak* magnetic field applied, the path will be slightly curved as shown at (b), and the electron current has the same value. With a stronger magnetic field,

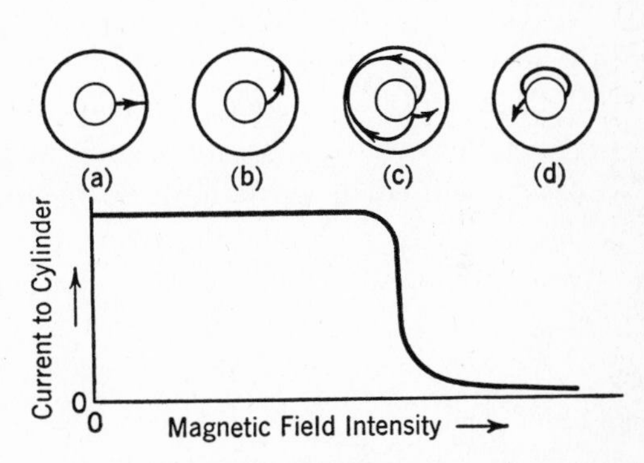

FIG. 9-24. Electron paths in a cylindrical anode for various magnetic field intensities. The cathode is at the center. The outer cylindrical anode is held at a constant positive potential with respect to the cathode. The magnitude of the current to the anode for each path is shown by the part of the curve immediately beneath. (Reference 47.)

the path is such that (c) an electron may miss the cylinder, and hence the current to the anode falls, reaching a theoretical zero for a critical cutoff magnetic field intensity. If the magnetic field is increased further (d), the electron path is further bent in toward the cathode, and the electron current remains at substantially zero. The mathematical theory applying is given in many places.[47,48,49,50,51]

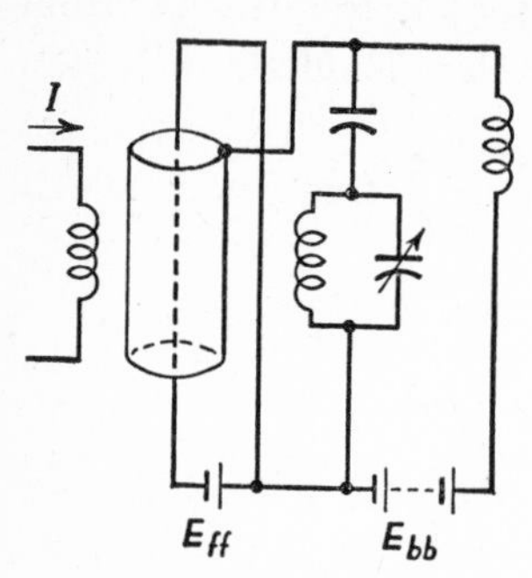

FIG. 9-25. An early magnetron oscillator.

Magnetron Feedback Oscillator. Using the simple magnetron just described in a feedback circuit as in Fig. 9-25, Elder, in 1925, obtained a power output of 8 kilowatts at 30 kilocycles.[49] One magnetic field is constant, and a second magnetic field is produced by the current in the inductor of the parallel resonant circuit. The circuit will oscillate when connections and resultant magnetic field strengths are such that maximum current flows during minimum cathode to anode voltage.[50] No extensive use resulted, perhaps because the triode oscillator was superior.

Magnetron Negative-Resistance Oscillator. A negative-resistance magnetron oscillator, developed[47,50,52] by Habann about 1924, is shown in Fig. 9-26. This uses a **split-anode magnetron.** If the two anode segments of the split-anode magnetron are held at the same positive potential, it will perform like the magnetron previously considered. However, the split-anode magnetron oscillator is not used in the same type of circuit, operates with a magnetic field intensity *much above cutoff,* and is a negative-resistance device for reasons now to be given.

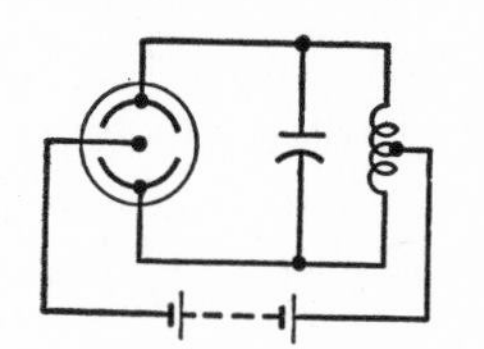

FIG. 9-26. Circuit of a split-anode magnetron oscillator. The heated cathode is at the center. The coil for producing the constant field is not shown. Because of the tuned circuit, the voltage from each segment to cathode will be pulsating, and will contain an alternating component.

If the two anode segments are held at different voltages as Fig. 9-27a indicates, the voltage distribution shown in Fig. 9-27b results, the numbers on the lines being volts positive with respect to the cathode. Now when an electron leaves the cathode, and because of the combined electric and magnetic field starts to spiral, the radius of curvature will be greater on the upper side than on the lower because of this peculiar voltage distribution. This will bend the path of an electron so that it is caused to strike the *segment of lower potential,*[47,50,] whereas without the magnetic field it would strike the segment of higher potential. The electron path of Fig. 9-27b was photographed by Kilgore.[52] If the difference of potential between the two segments is plotted on the X axis, and currents to the two segments are plotted on the Y axis, the curves of Fig. 9-27c result.

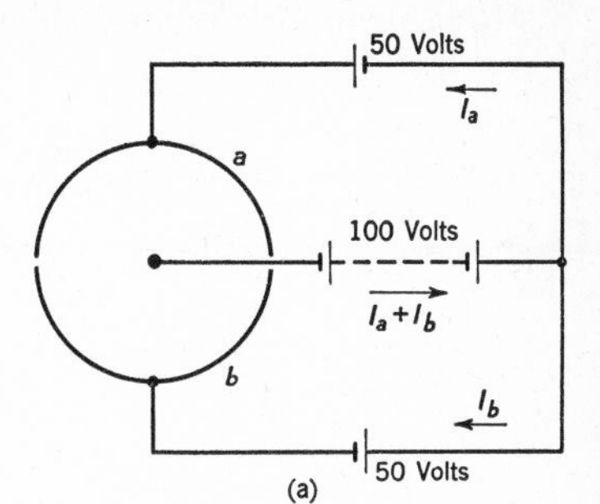

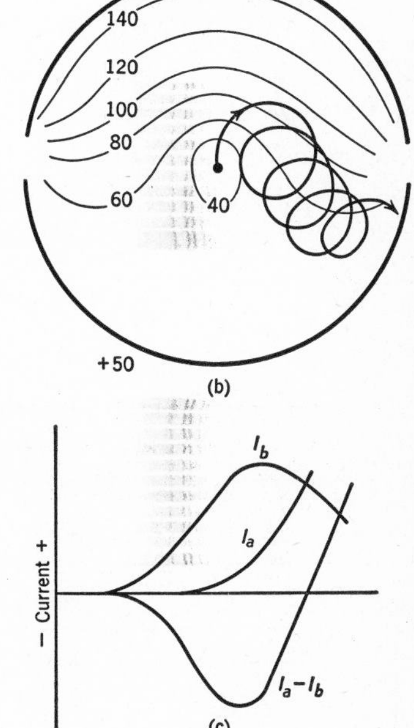

FIG. 9-27. The magnetron circuit is shown at (a). Segment *a* is made 100 volts more positive than segment *b*. This gives the voltage distribution (b). Under the combined influence of the electric and magnetic fields, an electron will take a path somewhat as shown, striking the *lower* segment. This gives the magnetron a negative-resistance characteristic, curve $I_a - I_b$ plotted against voltage in part (c).

The negative-resistance characteristic of the split-anode magnetron that makes oscillations possible is shown by the first negative part of curve I_a–I_b when plotted against segment voltage E_a–E_b. Referring to the oscillator circuit Fig. 9-26, the difference in the two currents flowing to the ends of the coil in the resonant circuit corresponds to I_a–I_b of Fig. 9-27c, and this decreases (increases negatively) as the oscillating voltage E_a–E_b across the coil and hence between magnetron segments increases positively, giving a negative-resistance characteristic that will sustain oscillations. It will be seen that an electron following the path of Fig. 9-27b first approaches the more positive anode segment and *absorbs* energy from the power supply but will strike the most negative segment and will *deliver* this energy to the less positive segment. If, at a given instant, the upper segment is more positive because of the alternating voltage across the tuned circuit, excess current flows to the lower segment. This sudden increase in current through the inductor of the resonant circuit will cause the potential of the lower segment (collecting the most electrons) to rise, and will cause the potential of the upper segment to fall; then the upper segment will collect the most current, and the cycle will continue. Transmission-line sections instead of inductors and capacitors commonly were used as resonators. The magnetron negative-resistance oscillator just discussed also is sometimes called a **dynatron oscillator,** or a **Habann oscillator.** Such oscillators were able to give continuous outputs of about 50 watts at frequencies as high as 600 megacycles.[52]

Magnetron Cyclotron-Frequency Oscillator. Early investigators, including [47,52,53] Yagi [54] and Okabe,[55] studied oscillations that occurred with split-anode magnetrons at magnetic field intensities *near the cutoff value,* instead of much greater than cutoff as with the oscillator just discussed.[47] Such operation is possible with either a continuous or a split anode, and is explainable in essentially the same manner for each case.[47] Oscillations of this type are discussed in most articles and books on magnetrons published since 1940, and the theory will be summarized briefly from reference 47.

Although the electrodes are cylindrical, assume that they have parallel faces as in Fig. 9-28. If a magnetic field exists as shown by vector B, and if *the electric field is zero,* then if an electron is thrown from the cathode with velocity v, its *path will be a circle* of radius given by equation 1-10, which also (section 1-4) can be written $r = mv/(eB)$, where m and e are the mass and charge of the electron. If t represents time, and d represents distance, then since $v = d/t$, $t = d/v$. In traversing one revolution of the circle, the time required is $t = 2\pi r/v$, and substituting the expression for radius gives

$$t = \frac{2\pi mv}{veB} \quad \text{and} \quad \omega = \frac{eB}{m} \tag{9-15}$$

where $t = 1/f$, and $\omega = 2\pi f$. This f is the so-called **cyclotron frequency.** (A fast-moving particle travels in a uniform magnetic field with a path of larger radius than that of a slow-moving particle; however, the time required for each particle to complete a revolution is the same. The Cyclotron, used in atomic physics, is based on this principle.)

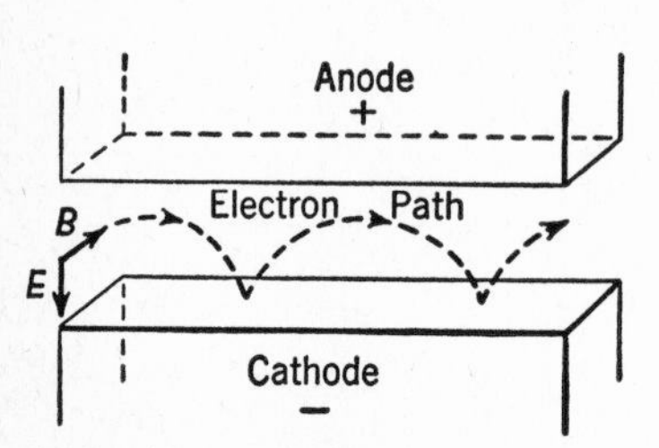

FIG. 9-28. Electron path in combined electric and magnetic fields between parallel-plane electrodes.

In a magnetron an electric field, indicated by vector E, exists between the anode and cathode as depicted in Fig. 9-28. The circular motion described in the preceding paragraph was for a magnetic field only; the path followed by an electron that is released from the cathode with zero initial velocity into a combined magnetic and electric field as shown by vectors B and E will be the cycloidal path of Fig. 9-28. This path results from the combination of the forces acting on the electron and can be mathematically determined.[56]

In an actual cyclotron-frequency magnetron oscillator there is a third force to be considered; this is due to the alternating "signal" voltage existing between segments of the split anode, these segments being, of course, connected to a resonator such as a transmission-line section. Electrons leaving the cathode will travel a path somewhat as explained in the preceding paragraph, it being remembered that in a split-anode magnetron the electric field is more tangential than radial [47] because of the differences of potential between anode segments. Because electrons leave the cathode at different times, they have, in their paths, different time-phase relations with the alternating-voltage components. Certain electrons will pass the vicinity of the gap between anode segments when the alternating voltage has raised the segment beyond to a *higher* potential than the applied direct-voltage value. These electrons may strike this segment and be withdrawn from the tube.

Other electrons will pass the gap and approach the segment beyond just when the alternating voltage has driven this segment *lower* in potential than the applied direct-voltage value. Because of their kinetic energies, such electrons may continue to approach, for an instant, and by approaching this segment (which is less positive than formerly) these electrons deliver energy to the segment and the connected resonating system such as a transmission-line section. Of course the energy delivered by an electron was absorbed from the direct-voltage source. The electrons may continue to absorb and deliver energy with decreasing amplitudes of oscillation until they are "spent." These "dead" electrons are withdrawn from the region between cathode and anode by slightly tilting the magnetic field, or by other means.[47] Frequencies of thousands of megacycles were produced by these oscillators, but the efficiency and power output were low.

Magnetron Traveling-Wave Oscillator. Oscillations in magnetrons were found to occur which did not depend on negative resistance, and which were not at cyclotron frequency. It early was recognized [47] that these oscillations were related to traveling waves within the magnetron. Also, early investigations were made of magnetrons with more than two anode segments. Samuel patented in 1936 an electron device [47,57] having cavities such as were used after the split-anode types became obsolete about 1940. Work was done by Alekseev and Malairov during 1936 and 1937 on magnetrons with cavities.[58] Randall and Boot are credited [59] with developing, about 1940, the cavity magnetron in the form (Fig. 9-29) widely used in the following years.[47,51]

Capacitance C_1 exists between the cathode and each portion of the anode between cavities. Capacitance C_2 exists between the parallel sides of the slot leading to the circular cavities, and inductance L is presented by each circular cavity.[57] As indicated in Fig. 9-30, if this equivalent circuit of a magnetron is "stretched out," a transmission line having the form (section 6-3) of a low-pass filter, and having a cutoff frequency, results. A more exact analysis [41] includes the effect of mutual inductance between cavities. Inexact though it may be, Fig. 9-30 indicates that the multicavity magnetron has the ability to transmit traveling electromagnetic waves of any frequency below the cutoff frequency. Such an electromagnetic wave (or waves) "travels" around the anode containing the cavities.

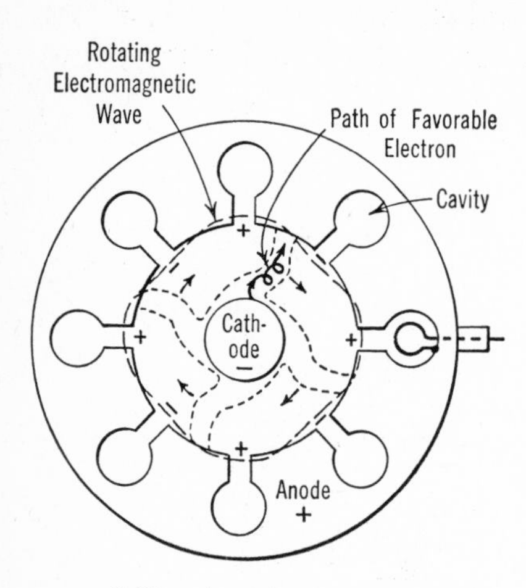

FIG. 9-29. A cavity-type magnetron. A constant radial electric-field component exists between the anode and cathode. A magnetic field parallel to the cathode (at right angles to the page) is produced by a permanent magnet (not shown). The "favorable" electrons follow the "spoke-like" regions which rotate as indicated. An electromagnetic wave also rotates, the instantaneous values being shown by the + and − signs in the cavities. (Adapted from reference 47.)

One explanation of the traveling-wave magnetron oscillator is based on Fig. 9-29. Electrons leaving the cathode in the correct phase to extract energy from the direct-voltage source, and to add this energy to the traveling wave, spiral to the anode somewhat as indicated by the path shown. This has been shown by calculations.[47] Furthermore, the calculations indicated that these "favorable" electrons (only) become arranged in "space-charge spokes," the edges of the spokes being shown by the dotted lines of Fig. 9-29. These electron spokes rotate as indicated by the arrow. Electrons leaving with "unfavorable" phases are driven back into the cathode. Other arrangements, or modes of operation, are possible, the one depicted in Fig. 9-29 being the commonly used π **mode.** For this mode the phase difference of the oscillations in adjacent cavities is π radians, or 180°. This is illustrated by positive and negative signs *in the cavity openings,* indicating relative phase relations of cavity oscillations (not electric charge). This traveling wave may be pictured as alternating fields with large tangential components across the cavity openings, the walls of which have a difference of potential between them. This rotating ultrahigh-frequency field, and the space-charge spoke electron-flow pattern, rotate together. The action has been likened to that in a polyphase electric generator.[60] The favorable electrons, after extracting energy from the direct-voltage source and delivering it to the ultrahigh-frequency wave, flow to the anode and are withdrawn. Ultrahigh-frequency electric energy is withdrawn from a magnetron oscillator by a loop of wire in a cavity (Fig. 9-29) or by coupling a wave guide to a cavity. The theory of a short-circuited transmission line predicts large current flow in the walls of the cavity most remote from the cavity opening.

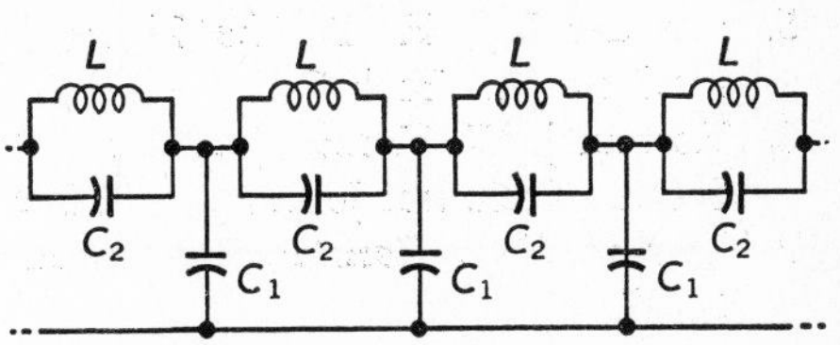

FIG. 9-30. Equivalent circuit of a cavity-type magnetron when "stretched out."

As previously mentioned, the traveling-wave magnetron oscillator commonly is operated in the π mode, and provisions sometimes are made to ensure that operation does not change to another mode. One method is to bond electrically, or strap, together alternate portions of the anode between cavities, which for the π mode should be at the same potential. Another scheme is to use the so-called "rising-sun" type of anode.[60]

Cavity magnetrons are extensively used in radar transmitters. Here they are pulse-operated by impressing anode voltage for perhaps a microsecond at the low-frequency rate of perhaps 1000 times per second. Frequencies of about 3000 to 10,000 megacycles commonly are generated, and instantaneous power outputs of hundreds and even thousands of kilowatts are produced efficiently. Such high peak powers place great emission requirements on the cathodes, and several theories have been advanced to explain the ability to meet these requirements.[47,51] Control grids have been used to modulate magnetrons.[61]

9-19. *OTHER OSCILLATORS*

As mentioned early in this chapter, the Standards[1] list 32 types of oscillators. Many of these have been discussed in the preceding pages. Others will be considered in the chapters that follow; for instance, the beat-frequency oscillator in Chapter 12, and the multivibrator in Chapter 14.

There are other oscillators of importance. For instance, the traveling-wave tube of the preceding chapter is used as an oscillator.[62] Another oscillator that has not been discussed is the Resnatron, an electron device that produces large amounts of continuous-wave ultrahigh-frequency power; for instance, 85 kilowatts at 600 megacycles.[63,64]

REFERENCES

1. Institute of Radio Engineers. *Standards on Antennas, Modulation Systems, and Transmitters. Definitions of Terms.* 1948.
2. Keithley, J. F. *Low frequency oscillator.* Electronics, Sept. 1948, Vol. 21, No. 9.

3. *Microwave bands standardized by National Bureau.* Electrical Engineering, May, 1946, Vol. 65, No. 5.
4. Edson, W. A. *Vacuum-Tube Oscillators.* John Wiley & Sons.
5. van der Bijl, H. T. *The Thermionic Vacuum Tube and Its Applications.* McGraw-Hill Book Co.
6. Horton, J. W. *Vacuum tube oscillators—A graphical method of analysis.* Bell System Technical Journal, April 1924, Vol. 3, No. 3.
7. Terman, F. E. *Resistance stabilized oscillators.* Electronics, July 1933, Vol 6, No. 7.
8. Terman, F. E., Buss, R. R., Hewlett, W. R., and Cahill, F. C. *Some applications of negative feedback with particular reference to laboratory equipment.* Proc. I.R.E., Oct. 1939, Vol. 27, No. 10.
9. Terman, F. E., and Pettit, J. M. *Electronic Measurements.* McGraw-Hill Book Co.
10. Ginston, E. L., and Hollingsworth, L. M. *Phase shift oscillators.* Proc. I.R.E., Feb. 1941, Vol. 29, No. 2.
11. Sorg, H. E., and Becker, G. A. *Grid emission in vacuum tubes.* Electronics, July 1945, Vol. 18, No. 7.
12. Llewellyn, F. B. *Constant frequency oscillators.* Proc. I.R.E., Dec. 1931, Vol. 19, No. 12.
13. Stevenson, G. H. *Stabilized feedback oscillators.* Bell System Technical Journal, July 1938, Vol. 17, No. 3.
14. Tournier, M. *History and application of piezoelectricity.* Electrical Communication, Apr. 1937, Vol. 15, No. 4.
15. Elbl, L. A. *Quartz crystal cuts.* Electronics, July 1943, Vol. 16, No. 7.
16. Heising, R. A. *Quartz Crystals for Electrical Circuits—Their Design and Manufacture.* D. Van Nostrand Co.
17. Mason, W. P. *Low temperature coefficient quartz crystals.* Bell System Technical Journal, Jan. 1940, Vol. 19, No. 1.
18. Mason, W. P. *A new quartz-crystal plate.* Proc. I.R.E., May 1940, Vol. 28, No. 5.
19. Terman, F. E. *Radio Engineering.* McGraw-Hill Book Co.
20. Mason, W. P. *New low-coefficient synthetic piezoelectric crystals for use in filters and oscillators.* Proc. I.R.E., Oct. 1947, Vol. 35, No. 10.
21. Meacham, L. A. *The bridge stabilized oscillator.* Proc. I.R.E., Oct. 1938, Vol. 26, No. 10.
22. Karplus, E. *Wide-range tuned circuits and oscillators for high frequencies.* Proc. I.R.E., July 1945, Vol. 33, No. 7.
23. Conklin, J. W., Finch, J. L., and Hansel, C. W. *New methods of frequency control employing long lines.* Proc. I.R.E., Nov. 1931, Vol. 19, No. 11.
24. Terman, F. E. *Resonant lines in radio circuits.* Electrical Engineering, July 1934, Vol. 53, No. 7.
25. Everitt, W. L. *Communication Engineering.* McGraw-Hill Book Co.
26. Dow, J. B. *A recent development in vacuum tube oscillator circuits.* Proc. I.R.E., Dec. 1931, Vol. 19, No. 12.
27. Karplus, E. *Communication on quasi-optical frequencies.* Electronics, June 1931, Vol. 2, No. 6.

28. McPherson, W. L., and Ullrich, E. H. *Micro-ray communication.* Electrical Communication, Apr. 1936, Vol. 14, No. 4.
29. Herold, E. W. *Negative resistance and devices for obtaining it.* Proc. I.R.E., Oct. 1935, Vol. 23, No. 10.
30. Hull, A. W. *The dynatron, a vacuum tube possessing negative resistance.* Proc. I.R.E., pp. 5–36, Vol. 6, 1918.
31. Brunetti, C. *The clarification of average negative resistance with extensions of its use.* Proc. I.R.E., Dec. 1937, Vol. 25, No. 12.
32. Brunetti, C. *The transitron oscillator.* Proc. I.R.E., Feb. 1939, Vol. 27, No. 2.
33. Bracewell, R. N. *Charts for resonant frequencies of cavities.* Proc. I.R.E., Aug. 1947, Vol. 35, No. 8.
34. Institute of Radio Engineers. *Standards on Electron Tubes: Definitions of Terms, 1950.* Proc. I.R.E., Apr. 1950, Vol. 38, No. 4.
35. Hahn, W. C. *Velocity-modulated tubes.* Proc. I.R.E., Feb. 1939, Vol. 27, No. 2.
36. Varian, R. H., and Varian, S. F. *A high-frequency oscillator and amplifier.* Jl. of Applied Physics, May 1939, Vol. 10.
37. Ginston, E. E., and Harrison, A. E. *Reflex-Klystron Oscillators.* Proc. I.R.E., Mar. 1946, Vol. 34, No. 3.
38. Pierce, J. R., and Shepard, W. G. *Reflex oscillators.* Bell System Technical Journal, July 1947, Vol. 26, No. 3.
39. Hamilton, D. R., Knipp, J. K., and Kupper, J. B. H. *Klystrons and Microwave Triodes.* McGraw-Hill Book Co.
40. Harrison, A. E. *Klystron Tubes.* McGraw-Hill Book Co.
41. Spangenberg, K. R. *Vacuum Tubes.* McGraw-Hill Book Co.
42. Friis, H. T. *Microwave repeater research.* Bell System Technical Journal, Apr. 1948, Vol. 27, No. 2.
43. Crosby, H. M. *Five-kw. Klystron UHF television transmitter.* Electronics, June 1951, Vol. 24, No. 6.
44. Learned, V., and Veronda, C. *Recent developments in high-power Klystron amplifiers.* Proc. I.R.E., Apr. 1952, Vol. 40, No. 4.
45. Preist, D. H., and Murdock, C. E. *High-power Klystrons at UHF.* Proc. I.R.E., Jan. 1953, Vol. 41, No. 1.
46. Hull, A. W. *The effect of a uniform magnetic field on the motion of electrons between coaxial cylinders.* Physical Review, Vol. 18, 1921.
47. Fisk, J. B., Hagstrum, H. D., and Hartman, P. L. *The magnetron as a generator of centimeter waves.* Bell Laboratories Record, Apr. 1946, Vol. 25, No. 2.
48. Hagstrum, H. D. *The generation of centimeter waves.* Proc. I.R.E., June 1947, Vol. 35, No. 6.
49. Elder, F. R. *The magnetron amplifier and power oscillator.* Proc. I.R.E., Vol. 13, 1925.
50. Sarbacher, R. I., and Edson, W. A. *Hyper and Ultrahigh Frequency Engineering,* John Wiley & Sons.
51. Collins, G. B. *Microwave Magnetrons.* McGraw-Hill Book Co.
52. Kilgore, G. R. *Magnetron oscillators for the generation of frequencies between 300 and 600 megacycles.* Proc. I.R.E., Aug. 1936, Vol. 24, No. 8.

53. Mouromtseff, I. E. *Development of electronic tubes.* Proc. I.R.E., Apr. 1945, Vol. 33, No. 4.
54. Yagi, H. *Beam transmission of ultra short waves.* Proc. I.R.E., Vol. 16, 1928, p. 715.
55. Okabe, K. *On the short-wave limit of magnetron oscillations.* Proc. I.R.E., Vol. 17, 1929, p. 652.
56. M.I.T. Staff. *Applied Electronics.* John Wiley & Sons.
57. U.S. Patent No. 2,063,342, Dec. 8, 1936.
58. Alekseev, N. F., and Malairov, D. D. *Generation of high-power oscillations with a magnetron in the centimeter band.* Proc. I.R.E., Mar. 1944, Vol. 32, No. 3.
59. Latham, R., King, A. H., and Rushforth, L. *The Magnetron.* Chapman & Hall.
60. Bronwell, A. B., and Beam, R. E. *Theory and Application of Microwaves.* McGraw-Hill Book Co.
61. Spencer, P. L. *Grid magnetron delivers modulated UHF output.* Electronics, May 1953, Vol. 26, No. 5.
62. Johnson, H. R., and Whinnery, J. R. *Traveling-wave oscillator tunes electronically.* Electronics, Aug. 1953, Vol. 26, No. 8.
63. Salisbury, W. W. *The Resnatron.* Electronics, Feb. 1946, Vol. 19, No. 2.
64. Garbuny, M. *Theory of the reflex Resnatron.* Proc. I.R.E., Jan. 1953, Vol. 41, No. 1.

QUESTIONS

1. How are oscillators classified?
2. What is negative resistance? How is it used in oscillators?
3. Does a dry cell have the characteristics of a negative resistance? Explain.
4. How can an amplifier be converted to an oscillator?
5. What is a tank circuit? A resonator? How are they used in oscillators?
6. How are positive and negative feedback used in oscillators?
7. During operation does the grid of an oscillator usually go positive? Why?
8. Are the tubes of an audio oscillator usually operated in class-A, -B, or -C? Why?
9. Why does the R–C oscillator operate at a single frequency?
10. Why does the phase-shift oscillator operate at a single frequency?
11. Why are grid resistors and capacitors commonly used in oscillators?
12. What causes intermittent operation of oscillators?
13. What causes an oscillator to block? Is blocking serious? Why?
14. How are crystals used in oscillators?
15. Are crystals used in pentode oscillators? If so, how?
16. What is a buffer? How can its use be avoided?
17. Fundamentally, what are the distinguishing features of a Hartley and a Colpitts oscillator?
18. In section 9-8 a crystal is likened to a series resonant circuit. If this is a correct comparison, just how does it control oscillations?
19. How does a butterfly resonator give variable inductance and capacitance?
20. How are the frequencies controlled in VHF and UHF oscillators?

21. It is stated that the interelectrode capacitances of Fig. 9-16 act as a voltage divider. Exactly how is this accomplished? Are the phases correct?
22. Why is it common practice to ground the cavities of Klystrons?
23. What are the basic types of Klystrons? What is electron tuning?
24. What are the chief applications of Klystrons?
25. Why does the magnetron possess negative resistance characteristics?
26. What are the basic types of magnetron oscillators?
27. Does electron bunching occur in magnetrons? Explain.
28. Under what conditions can an electron accept energy from a magnetic field? From an electric field?
29. Under what conditions can an electron deliver energy to a magnetic field? To an electric field?
30. What device will deliver ultrahigh-frequency energy at high energy level for short periods? Continuously?

PROBLEMS

1. A small oscillator is being connected as in Fig. 9-1. It is to oscillate from 500 kilocycles to 2 megacycles. A variable air capacitor having a capacitance of from 50 to 500 micromicrofarads is available. Can this capacitor be used with a constant inductance to cover the desired range? What will be the value of inductance? If more than one value of inductance must be used, calculate the values.
2. Derive equation 9-8.
3. The oscillator of Fig. 9-6 is being constructed. A bridge measurement at the frequency of resonance shows R_L, the equivalent resistance of the tuned circuit, to be 2450 ohms. The tube being used is a type 6J5 with 100 volts on the plate and a grid bias of -6 volts. About what should be the value of R, if it is to be a variable resistor?
4. About what will be the effective value of the voltage impressed on the grid of the second tube of problem 3? If the second tube also is a type 6J5 what will be the approximate output voltage of the oscillator?
5. Assume that the oscillator of problem 1 is to operate at 1,000,000 cycles, and determine the size of capacitor and coil to be used. Calculate the equivalent impedance of the tuned resonant circuit at the frequency of resonance, and at frequencies 5 per cent above and 5 per cent below this resonance value, when the resistance of the coil and the reflected load resistance give to the coil a Q of 12, and when combined resistance is such that the Q is 8. What conclusions are indicated?
6. Prove mathematically and verify with numerical solutions the statements made following equation 9-13 regarding the shape of the curve in Fig. 9-7.
7. Prove equation 9-14 to be correct.
8. Derive equation 9-6 on the basis of controlled feedback (section 8-19).

CHAPTER 10

SEMICONDUCTOR ELECTRON DEVICES[*]

Solid-state semiconductor electron devices have been studied and used for many years. The photoconductive properties of selenium were observed by Smith in 1873. The rectifying properties of selenium were discovered by Fritts in 1883. The "cat whisker" crystal detector was used in early radio receivers, but was largely supplanted, about 1920, by the vacuum tube. The present extensive use of semiconductor electron devices began with the discovery by Grondahl, in 1926, of the cuprous-oxide rectifier.

This device, commonly referred to as a **copper-oxide rectifier,** was soon adapted to many uses such as a rectifier of alternating electric power, as a rectifier for instruments, and in about 1937 as a modulator and demodulator in carrier-telephone circuits. Selenium rectifiers and copper-sulphide rectifiers also were developed during this period. About 1940 the so-called crystal detector came back into wide use in ultrahigh-frequency wave-guide systems. One important reason for this was its low capacitance.

During these years, the vacuum-tube almost exclusively was used for amplification. The truly revolutionary change in semiconductor applications came with the invention of the amplifying point-contact transistor by Bardeen and Brattain, about 1948, and of the junction transistor by Shockley in 1951.

In this chapter the electronic principles, and direct-current, or static, characteristics of semiconductor devices will be discussed. Circuit uses of these devices will be considered in subsequent chapters. In studying the devices it must be realized that the theories are new and in some instances have been

[*] Definitions will be found in IRE Standards on Electron Devices: Definitions of Semiconductor Terms, 1954, Proc. I.R.E., Oct. 1954, Vol. 42, No. 10.

proposed on a tentative basis. Also, it must be understood that semiconductor devices are manufactured in various ways, and that many different types exist. In studying semiconductor electron devices, the *exact* nature of the device under study must be known. Furthermore, care must be taken in applying the theory of one device to another. They may look the same physically, but may be different electronically.

Two-terminal semiconductor electron devices, such as crystal rectifiers and copper-oxide rectifiers, whose resistance (or impedance) varies with the magnitude or direction of voltage impressed or current passed, are called **varistors.**[1] There are two basic types: **Nonsymmetrical varistors,** or rectifiers, which will be discussed now, and **symmetrical varistors,** which will be considered in section 10-13.

10-1. *POINT-CONTACT CRYSTAL RECTIFIERS*

Early **crystal detectors,** as they were called, were made by pressing a fine wire of a material, such as tungsten, on one of several natural mineral crystals such as galena.[2] The fine wire was commonly known as a "cat whisker," and the wire was moved from point to point on the crystal until a "sensitive spot" giving good rectification was located. Around 1940 silicon crystal rectifiers were used widely, and several years later germanium rectifiers were introduced.[2,3,4] Both silicon and germanium crystal rectifiers, which are sometimes called **crystal diodes,** are now employed. The germanium point-contact rectifier will be discussed in the paragraphs immediately following.

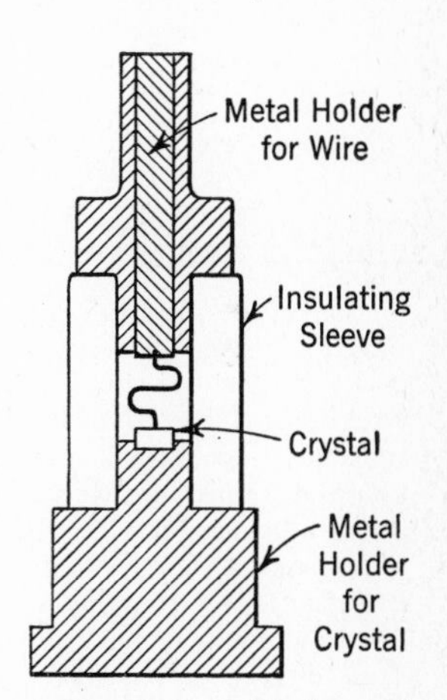

FIG. 10-1. One form of crystal rectifier, or crystal diode.

Physically, point-contact crystal rectifiers are made in many forms, one of which is shown in Fig. 10-1. The wire is a few thousandths of an inch in diameter, and often is of tungsten. The *n*-type germanium wafer on which the wire makes contact is about 0.05 by 0.05 inch, and about 0.025 inch thick, and may be considered to be a single crystal.[6] The fine wire is carefully pointed in a prescribed manner, and is pressed on the germanium and mechanically held in position. The point contact of most crystal rectifiers is then formed, or welded, to the germanium surface by passing, in the high-resistance direction through the contact, a current of about one ampere for about one second. The wire near the germanium surface reaches a temperature of about 900°C. This forming converts the *n*-type germanium in the immediate vicinity of the contact

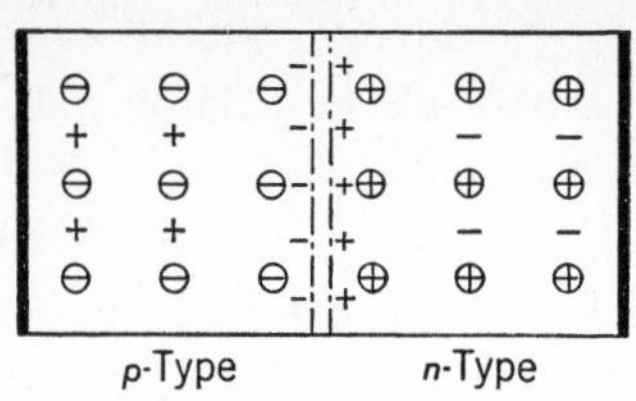

(a) No Applied Voltage. At Room Temperature

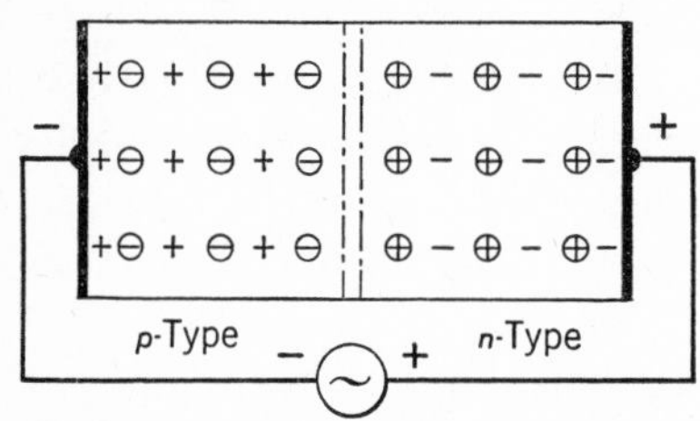

(b) Voltage Applied in Reverse Direction

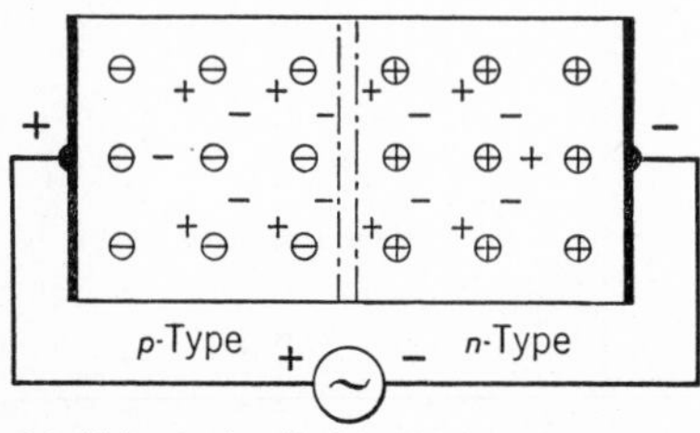

(c) Voltage Applied in Forward Direction

⊖ Acceptor atom + Positive hole
⊕ Donor atom − Negative electron

FIG. 10-2. In (a) are shown acceptor atoms in *p*-type germanium and donor atoms in *n*-type germanium. The acceptor atoms are circles with negative signs, indicating that they have accepted an electron from a germanium atom and have left a hole. The donor atoms are circles with positive signs, indicating that they have given up an electron. Holes and electrons have diffused across the junction (which has negligible thickness), producing a potential barrier. In (b) the "reverse" voltage has withdrawn charges from the junction, leaving a depletion layer of high resistance, essentially intrinsic germanium. In (c) the "forward" voltage is sweeping holes toward the negative terminal and electrons toward the positive terminal, resulting in high forward current flow. Positions of individual holes or electrons in the vicinity of atoms has no significance. Donor and acceptor atoms are at fixed positions in the crystal structure.

into *p*-type germanium.[6,7,8] The action is the result of the heating and rapid cooling.[9] There is evidence [8] that a *very* small island of *n*-type material may exist immediately under the contact even after forming.

10-2. *RECTIFICATION AT A p-n JUNCTION*

In a point-contact crystal rectifier of the type just discussed, the rectifying characteristics are considered by some to be caused largely by the *p-n* junction *within* the germanium and *not at* the point-surface contact.[6,7] A detailed study of the rectifying action at a *p-n* junction has been presented by Shockley.[7,10,11] The action may be summarized as follows.

For explaining the theory involved, in Fig. 10-2 is shown a continuous piece of germanium cut from a single crystal which was grown (section 10-14) in a manner such that half the crystal is *p*-type, and the other half is *n*-type. The actual *p-n* junction is very thin, but has been shown enlarged for purposes of illustration. The end electrodes are electrical contacts to the germanium, and serve no other purpose. In Fig. 10-2a the germanium is considered to be in thermal equilibrium at room temperature and to have no external voltages applied.

Under these conditions, and because of thermal agitation, there is an excess of positive holes in the *p*-type germanium and an excess of negative electrons in the *n*-type germanium, and there exists an electric field in the space between germanium atoms at the *p-n* junction. This field is produced because positive holes drift, or **diffuse**, across the junction from left to right, and negative electrons diffuse across the junction from right to left. By this action, a difference of potential and an electric field are established across the *p-n* junction, and the establishment of this field almost completely prevents further diffusion of both holes and electrons. Because of the diffusion that occurs, the right side of the junction is raised to a positive potential and the left side is lowered to a negative potential with respect to the "center" section of the junction. This difference of potential is given various names, such as a **potential barrier,** or **barrier layer.**

Reverse Direction of Applied Voltage. Now assume that the left end of the crystal is made negative and the right end positive as in Fig. 10-2b. Voltage applied in this manner is in the high-resistance **reverse direction,** or **back direction,** in which the crystal is essentially a nonconductor. This is because the excess positive holes in the *p*-type germanium are drawn to the left, and the excess negative electrons in the *n*-type are drawn to the right. As a result, the vicinity of the *p-n* junction becomes depleted of current carriers, and is sometimes called a **depletion layer,** or an **exhaustion layer.** Theoretically, the *p-n* junction layer would contain neither an excess of donors nor an excess of acceptors, and accordingly would be intrinsic in nature. Thus, when this junction layer is largely depleted of current carriers, the germanium crystal at room temperature becomes an intrinsic semiconductor of high resistance.

Forward Direction of Applied Voltage. Now assume that the left end of the crystal is made positive and the right end negative as in Fig. 10-2c. Voltage applied in this way is in the low-resistance **forward direction.** Excess negative electrons from the *n*-type germanium are now attracted to the left, and excess positive holes from the *p*-type germanium are attracted to the right. The *details* of the action [7] are not so simple as here depicted, and the *exact* nature of the phenomena involved is obscure. Nevertheless, from a practical standpoint, negative electrons and positive holes in the region of the *p-n* junction neutralize the electric-charge effects of each other, by recombination and perhaps by other means.[7] As a result, the potential barrier largely disappears, and the crystal readily passes current and hence becomes a low resistance.

In summarizing, in the preceding paragraphs the rectifying action of a *p-n* germanium junction was discussed. Because of the nature of the formed contact of a point-contact crystal rectifier, a *p-n* junction exists in this device.

Therefore, the explanations given for a *p-n* junction apply to the point contact, and explain, at least in part, why rectification occurs when a tungsten wire is welded to a wafer of *n*-type germanium. Regarding point-contact *silicon crystal rectifiers,* the following may be stated: Early theories were based on a **quantum-tunneling** effect in which it was assumed that electrons could readily penetrate, under favorable conditions, very thin potential barriers. This theory predicted rectification in the wrong direction, and was superseded by other theories.[2] In general, the theories that first were advanced explained rectification on the basis of a simple metal-semiconductor contact.[2] It appears probable that early theories neglected many important facts, such as the effect of surface treatments on the nature of the semiconducting material. Such treatments now are suspected of converting *n*-type material to *p*-type in the layers near the surface. Furthermore, it has been suggested [9] that the pressure of the pointed fine wire may result in surface changes from *n*-type to *p*-type material. The force *is* small, but so is the area, and it also should be recalled that a surface is not smooth. Hence, large pressures may result. Also, portions of the metal wire may diffuse into the crystal.[9] **Surface states,** or electric-charge distribution at the surface of semiconductors, are thought by some to influence rectification.[12]

In further summarizing point-contact rectification theory, attention is called to the fact that both *n*-type and *p*-type silicon rectifiers were early developed, and that these were not welded,[1,2] but that mechanical processes such as tapping the rectifier were employed after assembly.[5] It is possible that tapping may have assisted in causing changes in the crystal structure, at least near the surface and at the point. Most, if not all, modern point-contact silicon rectifiers are *p*-type, and the forward direction of current flow is from the silicon to the wire point. Because of the many factors involved, it does not appear advisable to attempt the presentation of an *all-inclusive* theory of point-contact rectification. Instead, it is felt that each basic type of rectification must be explained on an individual type basis, and that *all* the facts must be known regarding the exact nature of the contact before a sound explanation can be given.[1]

10-3. *STATIC CHARACTERISTICS OF POINT-CONTACT RECTIFIERS*

The static, or direct-current–direct-voltage characteristics of a point-contact rectifier give the relation between applied direct voltage and resulting current flow. The characteristics of a point-contact silicon rectifier of about 1940 are shown in Fig. 10-3. The dotted extension for the "silicon negative" condition indicates an unstable and often erratic negative resistance characteristic, in which a *decrease* in voltage results in an *increase* in current. As will be noted, the current magnitudes, and in particular the back voltage that may be impressed, are small, thus limiting the use to small signals.

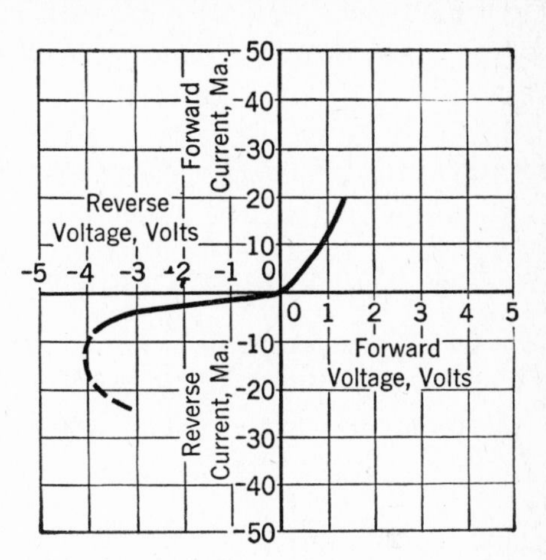

FIG. 10-3. Characteristics of an early point-contact silicon crystal rectifier with low back-voltage characteristic.

About 1944 it was discovered that suitably prepared *n*-type germanium when used in point-contact rectifiers gave greatly improved characteristics, as shown in Fig. 10-4. About a year later it was found that suitably prepared [2] silicon also gave improved characteristics (Fig. 10-4). In particular, these rectifiers could withstand a high back-voltage without passing appreciable current, and could be used to rectify much larger voltages than the early type of Fig. 10-3.

Current-Voltage Characteristic Curve. A brief summary of early theories of point-contact rectification was given in the preceding section. Based on these theories the following equation has been derived [2] for the current-voltage characteristic curve for an unwelded silicon rectifier.

$$I = A(\epsilon^{\alpha V} - 1) \qquad \text{10-1}$$

where I is the current, V the voltage *across the barrier,* and A and α are constants. The values of A and α are found by methods outlined in reference 2. Wide deviations exist between experimental and theoretical results using

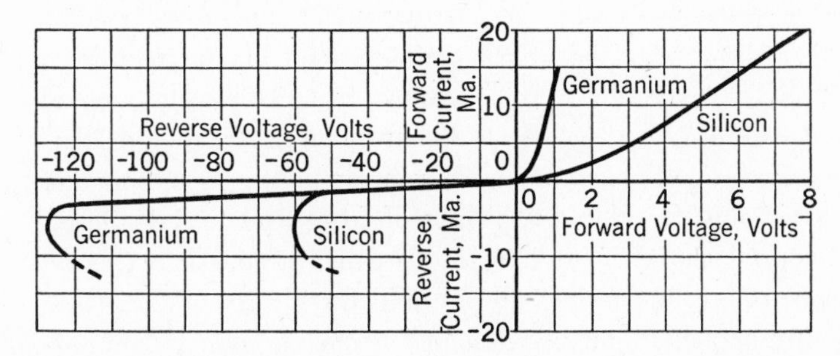

FIG. 10-4. Characteristics of point-contact crystal rectifiers designed for high reverse voltages. Note changes in voltage scales. It should not be assumed that all crystal rectifiers have these particular characteristics.

this equation, which is not surprising because much needs to be known regarding the exact nature of point-contact rectification. Attempts to explain the deviations lead to considerations of image-force effects, "leakage" of electrons by quantum tunneling, barrier space-charge distribution, and other effects,[2] such as surface conditions. As before indicated, it seems that the theory of the crystal rectifier cannot be explained in a few words on the basis of the differences in work functions between a metal and a semiconductor.

10-4. *EQUIVALENT CIRCUITS OF CRYSTAL RECTIFIERS*

An idealized curve for a crystal rectifier is given in Fig. 10-5, which also includes the equivalent circuit.[2] Resistor R is the component that results in the rectifying characteristics, and varies between zero and infinity for a theoretical rectifier. Resistor r is a "constant," or linear, **bulk resistance,** so called because it results from the "bulk" of the semiconductor material and not from the rectifying barrier layer. The term **spreading resistance** also is used, since resistance r is associated with spreading lines of current flow from the point contact, through the semiconductor, and to the metal contact fastened to the bottom of the semiconductor. Numerical values of $R + r$ are the reciprocals of the slopes of curves such as Figs. 10-3 and 10-4.

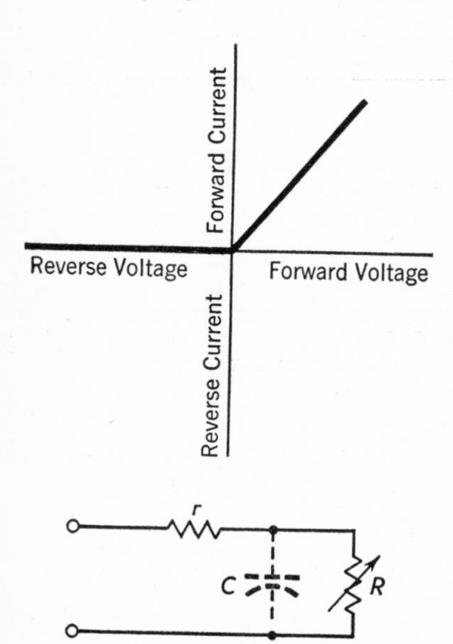

FIG. 10-5. Characteristic curve for an ideal point-contact crystal rectifier, and equivalent circuit. Capacitor C need be considered only at high frequencies.

At high frequencies, and in particular at frequencies of thousands of megacycles such as are used with wave-guide equipment, the equivalent circuit for the point-contact crystal rectifier must have capacitor C connected across the nonlinear barrier resistance R. This capacitor C represents the capacitance of the rectifying barrier, and is very low, values of from 0.02 to 1.0 micromicrofarad having been measured.[2] Thus, in the forward direction, R theoretically approaches zero, and C can be ignored; and for the reverse direction, R theoretically approaches infinity, and hence R can be ignored. The capacitance C is, accordingly, very important at wave-guide frequencies. The size and nature of the point contact, and other factors,[2] determine the value of C.

If a sinusoidal voltage of low-frequency is impressed on a point-contact germanium rectifier as in Fig. 10-6a, the rectified current flowing through the resistor, and the voltage across the resistor as observed on an oscillograph, are as in Fig. 10-6b. If, however, a sufficiently high frequency is used, the current and voltage will be as in Fig. 10-6c. Furthermore, if a pulse voltage instead of a sinusoidal voltage is impressed on the rectifier, the resulting current flow may be adversely affected by the rectifier, limiting its use. The effect is as if the rectifier were slow to become a nonconductor after the reverse voltage is applied.[6] A hole storage effect was used to explain this phenomenon, and experimental evidence [6] indicates that it is at least in part caused by **hole trapping,** the nature of which will be discussed later in this chapter.

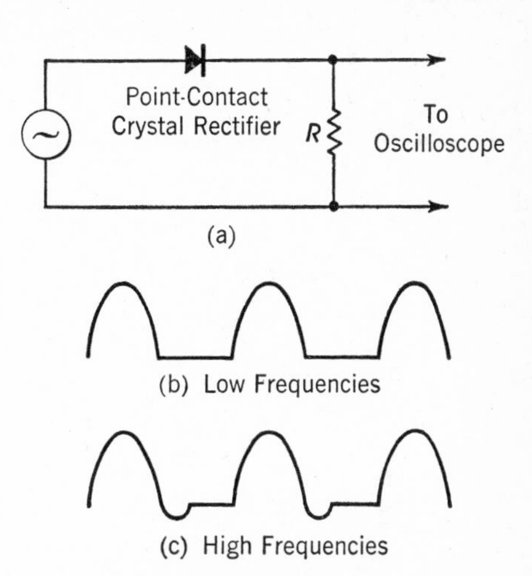

FIG. 10-6. At low frequencies, the current through resistor R and the voltage across R will be as shown by the upper curve, but at high frequencies it will be as the lower curve. This is interpreted as indicating trapping of positive holes. (Reference 6.)

Unfortunately, noise is developed by crystal rectifiers, the amount varying sometimes with individual units, with types, and with temperature and other factors.[2] Crystals also are subject to burnout, and to damage in handling.

A promising silicon *p-n junction* crystal rectifier [13] was announced in 1952. The junction area is estimated at 0.001 square centimeter, and the junction is formed by an alloying process. This device has excellent characteristics for certain purposes, and may supplant point-contact types for some applications.

10-5. *POINT-CONTACT TRANSISTORS*

The point-contact transistor is a semiconductor electron device that has amplifying properties. A typical point-contact transistor has two pointed phosphor bronze wires several thousandths of an inch in diameter pressed against a suitably prepared n-type germanium wafer called the **base.** The wafer is about 0.02 inch thick and 0.05 inch square, is cut from a single n-type crystal,[14] and is mounted with a low-resistance contact to a metal holder. The surface of the germanium crystal is processed by grinding (lap-

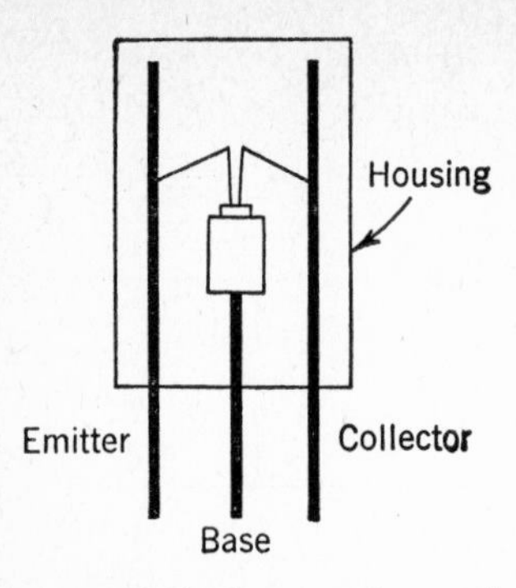

FIG. 10-7. Construction used in one point-contact transistor. The germanium wafer is the small portion on top of the holder attached to the base lead.

ping) and by chemical etching. The points of the wires are spaced several thousandths of an inch apart (Fig. 10-7), and one electrode, called the **collector,** is formed, or welded, to the *n*-type germanium; the other electrode, called the **emitter,** is not formed, but merely is pressed against the germanium. The assembly is suitably mounted in a metal cylinder as for the early **type A transistor,** or is enclosed in plastic, or is sealed in an air-tight metal envelope as in later types. Because three elements are used, this sometimes is called a **transistor triode.**

A small volume of the germanium in the immediate vicinity of *both* point contacts is considered to be *p*-type germanium.[15] Reasons for this assumption were discussed in section 10-1. It is well known that forming, or welding, converts *n*-type to *p*-type germanium. It is thought by some investigators that the lapping, chemical etching, and other processes convert surface layers from *n*-type to *p*-type. Thus, in Fig. 10-8, *p*-type portions are shown under both the emitter and collector. This diagram omits a very small *n*-type region considered by some [8] to be *immediately* under and in contact with the formed collector. The operation of the point-contact transistor and its ability to amplify are *not limited to action that occurs at the contacts,* but are considered largely to be the result of phenomena occurring at the *p-n* junctions indicated by the dotted lines in the body of the germanium.

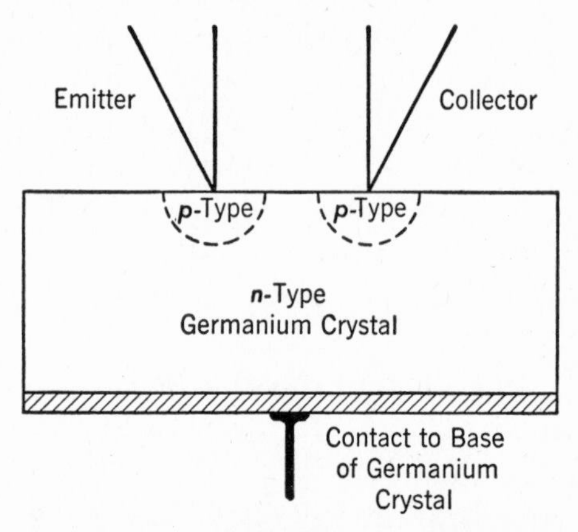

FIG. 10-8. Although the collector is formed, and the emitter is *not* formed, it is thought that the germanium immediately under *both* point contacts is *p*-type. The crystal as a whole is *n*-type. The illustration is not to scale, and the shapes of the points are not exact.

10-6. *CURRENT MULTIPLICATION IN POINT-CONTACT TRANSISTORS*

For typical operation the emitter is biased a few tenths of a volt positive with respect to the *n*-type germanium base. This is in the *forward* direction, and a (conventional) current of about one milliampere readily flows *into* the emitter and then into the germanium. The collector is biased negatively (with respect to the base) with a voltage of perhaps 20 volts connected between the collector and the germanium base. This is in the *reverse* direction, and a (conventional) current of about 3 milliamperes is all that can flow *out of* the collector.

Assume, for the moment, that the electrodes are free, that the potentials have *not* been applied, and that the transistor is in an equilibrium state at room temperature. In the vicinity of the emitter, positive holes from the *p*-type germanium have diffused across the dotted junction line of Fig. 10-8 to the *n*-type germanium. Also, electrons have diffused across the junction to the *p*-type region. This results in a potential barrier layer (section 10-2) with *p*-type negative and *n*-type positive, and further diffusion is greatly reduced.[7] In the vicinity of the collector similar action occurs, and a barrier of the same type is produced at the dotted junction.

Now assume that the emitter is made a few tenths of a volt positive with respect to the base. This will be in the direction to counteract the potential barrier at the dotted line under the emitter, and electrons readily will flow from the *n*-type germanium to the emitter, and holes readily will flow from the *p*-type region under the emitter to the *n*-type germanium. This accounts for the one milliampere that flows when a few tenths of a volt is applied as mentioned in the first paragraph of this section.

Now suppose that the collector is made about 20 volts negative with respect to the base. This will be in the direction to increase the potential barrier at the dotted line (Fig. 10-8) under the collector. As a result, negligible current will flow in the collector circuit. However, this relatively high negative potential will set up a strong field in the space between germanium atoms, and this electric field will attract positive holes toward the dotted junction line under the collector. Holes must migrate from atom to atom, and their mobilities are low. Of course, some holes will be neutralized by an ion absorbing an electron, but the loss of holes in this manner is small. Thus, most of the positive holes arrive in the vicinity of the collector, and the action is as if they accumulated there, forming, in effect, a positive space charge. This results in a neutralization of the barrier layer potential, and electrons are permitted to flow across the junction and from collector to base. If one arriving hole permits three electrons to flow, this would account for the collector current of 3 milliamperes mentioned in the first paragraph of this section.

It will be noted that the flow of collector current is controlled by holes released by a potential applied between emitter and base, and that one hole appears to attract more than one electron across the collector barrier layer. Hence, current multiplication, or amplification, controlled by signals impressed between emitter and base, is possible. In a point-contact transistor, a small change in the emitter current can produce a larger change in the collector current.

10-7. *THEORIES OF CURRENT MULTIPLICATION*

The phenomenon of current multiplication in point-contact transistors appears to occur in the immediate vicinity of the collector. The exact nature of the process is obscure.* The following theories have been presented.[7,16]

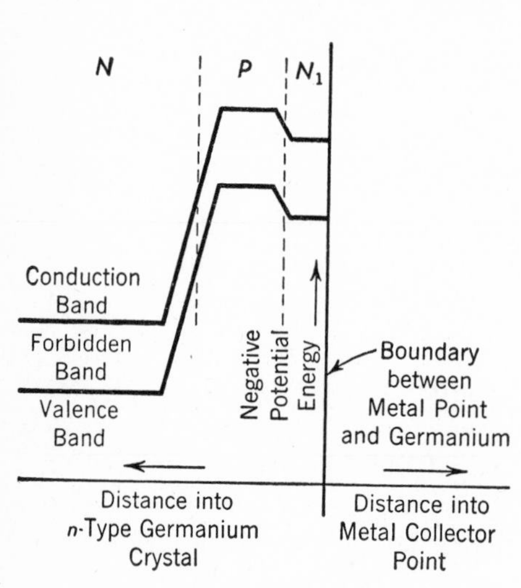

FIG. 10-9. The hook theory of current multiplication assumes a *very thin* layer N_1 of n-type material immediately under the collector point, then a thin layer P of p-type material, and then the bulk N of the n-type germanium crystal. The external voltage applied to the transistor makes the collector *negative* with respect to the germanium crystal. This accounts for the large rise in negative potential at the boundary N-P. The reason for the potential change at the P-N_1 boundary is explained in the text.

The p-n "Hook" Theory. It is believed that a *very* small n-type region N_1 exists *immediately* under the formed collector point and that this n-type region is surrounded by the thin layer of p-type material created by the collector forming process. The p-type region P is, in turn, surrounded by the n-type base germanium N. For normal point-contact transistor operation, the metal collector point contact and region N_1 are made negative, and the semiconductor germanium base and region N are made positive. Holes from the emitter region flow through N and arrive at region P. Diffusion of electrons from region N_1 has made region N_1 positive with respect to P. Holes tend to flow uphill in Fig. 10-9 because of the way potentials are plotted. The holes, accordingly, slow down and accumulate in region P because region N_1 has a positive charge, and attract electrons from N_1 into P. If the P layer is thin, so that little recombination occurs, the electrons readily pass through layer P, and then "downhill" to the positive base. The abrupt change

* For instance, consult the first paragraph of an article entitled "The effective surface recombination of a germanium surface with a floating barrier" by A. R. Moore and W. M. Webster, published in I.R.E. Proc. April 1955, Vol. 43, No. 4, and compare with the theories given here.

in potential at the N_1-P junction, and the resulting hook-shaped curve, give the theory its name.

The Trapping Theory. In this theory of current multiplication, it is considered that the crystal structure contains imperfections called **traps.**[7,16] A trap is assumed to capture a hole and hold it for an appreciable period before releasing it or permitting it to absorb an electron. Thus, if positive holes are trapped and held near the collector, this accumulation of positive electricity will reduce the existing potential barrier, and increase the flow of electrons from the collector into the base.

10-8. *STATIC CHARACTERISTICS OF POINT-CONTACT TRANSISTORS*

A circuit for determining the static, or direct-current, characteristics of a point-contact n-type germanium transistor is shown in Fig. 10-10. An arrowhead is placed on the emitter as shown. The assumed positive directions of currents and voltages are as indicated. The subscript ϵ sometimes is used to designate emitter currents and voltages, but probably e is used more often. The voltmeters shown must draw negligible current. Point-contact transistors usually are reasonably rugged, but transients, such as may occur in switching, should be avoided, unless it is known that they will not exceed rated values. It usually is advisable to adjust the circuit so that the current will be small when a transistor is inserted; also, the current should be small when a transistor is removed. Note that in Fig. 10-10 the emitter is made positive, and the collector negative.

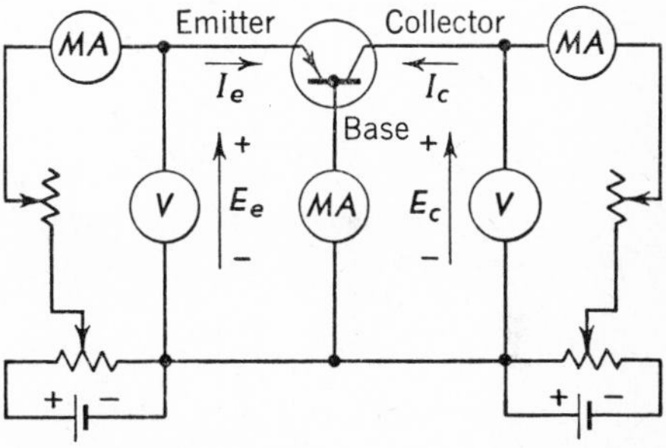

FIG. 10-10. Circuit for determining the static characteristics of a point-contact transistor. Accepted positive directions for currents and voltages are as indicated by the arrows. Note arrowhead designating the emitter of the point-contact transistor depicted.

The static curves for a typical point-contact n-type germanium transistor are shown in Fig. 10-11. These curves are taken much as for vacuum tubes. More curves are required to present transistor characteristics than with vacuum tubes, because for vacuum tubes the grid current may be assumed zero in many instances. Data for curves Fig. 10-11a and b are obtained by holding collector current I_c constant, varying emitter current I_e as desired, and observing emitter voltage E_e and collector voltage E_c. For curves (c) and (d), the emitter current I_e is held constant, the collector current I_c is varied, and values of E_e and E_c are observed. Note that the transistor is a current-controlled device, which is in contrast with a vacuum tube. The negative values of current I_c indicate that the current is flowing in the direc-

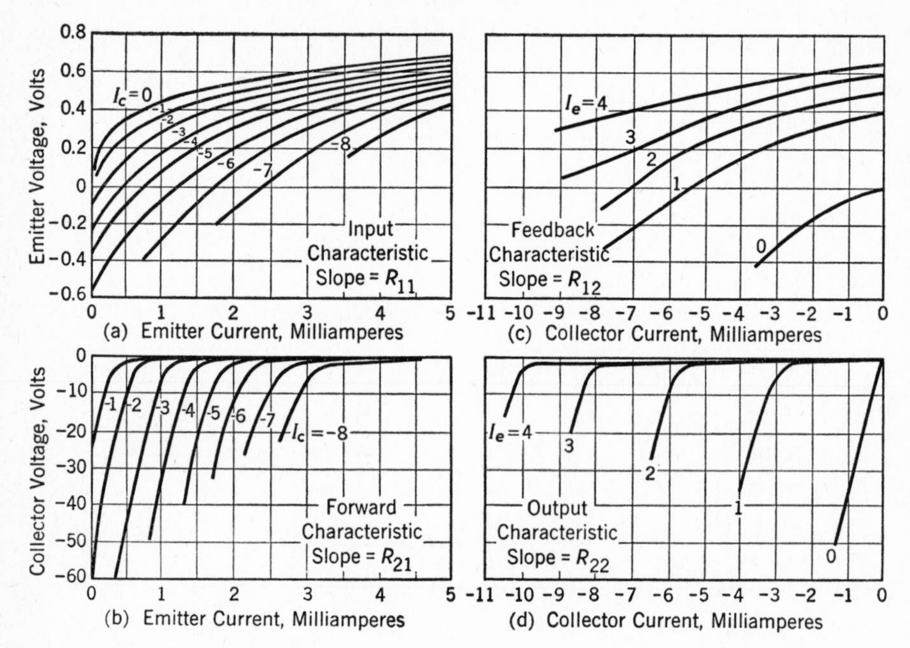

FIG. 10-11. Characteristics of a point-contact transistor. Numbers on curves are currents in milliamperes.

tion opposite to the assumed positive direction of Fig. 10-10. Also, a negative value of voltage indicates that the observed voltage was opposite to the assumed positive direction.

Input Characteristic. Figure 10-11a gives the input characteristics of the emitter because it gives the emitter-voltage variation with emitter current at different collector currents. The slope of a curve at any value of emitter and collector current gives the **emitter input resistance,** or merely **input resistance,** usually designated R_{11}. At $I_e = 1.0$ milliampere, and $I_c = 4.0$ milliamperes, the slope of the curve is 300, and R_{11} is 300 ohms. This value was determined graphically by drawing a tangent to the -4.0-milliampere curve at $I_e = 1.0$ milliampere, and finding the slope of the curve as for vacuum tubes. Expressed mathematically,

$$R_{11} = \left[\frac{\partial E_e}{\partial I_e}\right] I_c \text{ constant} \qquad 10\text{-}2$$

Forward Characteristic. Figure 10-11b gives transfer characteristics of the point-contact transistor in the forward direction (from emitter circuit to collector circuit). This curve gives the collector voltage variation with emitter current at different values of collector current. The slope of a curve at any value of emitter and collector current gives the **forward transfer resistance,** or merely **forward resistance,** usually designated R_{21}. For Fig. 10-11b at an emitter current of 1.0 milliampere and collector current of 4.0 milliamperes, $R_{21} = 100{,}000$ ohms, which was determined as previously explained. The equation for forward resistance is

$$R_{21} = \left[\frac{\partial E_c}{\partial I_e}\right] I_c \text{ constant} \qquad 10\text{-}3$$

Feedback Characteristic. Figure 10-11c gives the transistor feedback transfer characteristics, because the curves show how the emitter voltage varies with collector current. The slope of a curve at any point gives the **feedback transfer resistance,** or merely **feedback resistance,** usually designated R_{12}. For Fig. 10-11c at emitter and collector currents of 1.0 and 4.0 milliamperes, respectively, the slope, and hence the resistance, is $R_{12} = 100$ ohms. The relations for feedback resistance are

$$R_{12} = \left[\frac{\partial E_e}{\partial I_c}\right] I_e \text{ constant} \qquad 10\text{-}4$$

Output Characteristic. Figure 10-11d gives the **collector output resistance,** or **output resistance,** R_{22}. At emitter and collector currents of 1.0 and 4.0 milliamperes, the value is $R_{22} = 34{,}000$ ohms. The equation is

$$R_{22} = \left[\frac{\partial E_c}{\partial I_c}\right] I_e \text{ constant} \qquad 10\text{-}5$$

Current Multiplication. An important, and distinguishing, characteristic of the point-contact transistor is its current-multiplying, or current-amplifying properties. This **current multiplication factor,** α, can be found from the curves in several ways. For instance, in Fig. 10-11b, if the collector voltage is held constant at -30 volts and the collector current is changed from -3.0 to -5.0 milliamperes, the corresponding emitter current change will be from 0.65 to 1.35 milliamperes. Hence, the current multiplication, or amplification, is $\alpha = 2/0.7 = 2.9$ (approximately). The current multiplication factor *for low frequencies,* where reactive effects are negligible, is

$$\alpha = \left[\frac{\partial I_c}{\partial I_e}\right] E_c \text{ constant} \qquad 10\text{-}6$$

Using the values previously calculated, $\alpha = R_{21}/R_{22} = 100{,}000/34{,}000 = 2.9$

(approximately). Two theories for current multiplication were considered in section 10-7. From equations 10-3 and 10-5,

$$\alpha = \frac{\text{equation 10-3}}{\text{equation 10-5}} = \frac{R_{21}}{R_{22}} \qquad 10\text{-}7$$

It may appear surprising that α, the ratio of output current change to input current change, can be found from the ratio of two resistances. In considering this, it is of interest to reconsider the amplication factor μ for a vacuum tube as given in section 4-4. Here, μ is the ratio of output voltage change to input voltage change, and is shown to be equal to the product of plate resistance and transconductance. The reciprocal of transconductance is transresistance. Hence, for a vacuum tube,

$$\mu = r_p g_{pg} = \frac{r_p}{r_{pg}} \qquad 10\text{-}8$$

Thus, r_{21} is the forward transresistance of a transistor, in the same sense that r_{pg} is the forward transresistance of a vacuum tube. In fact, the word *transistor* is based on the transresistance concept. The curves of Fig. 10-11 indicate that values of the resistances considered in this section and of the current multiplication factor should vary widely with conditions of operation.

In their present form (1955), transistors are more sensitive to temperature variations than are vacuum tubes.[15,17] Also, germanium transistors are limited to use at temperatures below about 70°C. However, silicon transistors have been reported to operate at temperatures up to about 200°C. The characteristics of point-contact transistors are influenced by electrode spacing, type of germanium used, forming treatment, and other factors.[8,18]

10-9. *LARGE-AREA CONTACT SEMICONDUCTOR RECTIFIERS*

The first pages of this chapter treated semiconductor rectifiers with *point* contacts. Semiconductor rectifiers with contact areas that are large (compared to points) are widely used and will be discussed in the pages immediately following. These also are nonsymmetrical varistors. Three general types of such rectifiers are in use: copper-oxide rectifiers, selenium rectifiers, and copper-sulphide rectifiers. Other similar rectifiers have been developed experimentally. The copper-oxide rectifier and the selenium rectifier are used most widely, and only these two types, and a promising diffused-germanium type, will be considered. The copper-sulphide rectifier is sometimes used for low-voltage, high-current service.

Theories of Rectification for Large-Area Contacts. Early attempts were made to explain rectification for large-area contacts (as contrasted with points) on the basis of phenomena occurring at the contact between the surface of a metal and the surface of a semiconductor. The early theories included the quantum-mechanical "tunnel" effect, which assumed that electrons could pass "through" a thin potential barrier even though they had insufficient energies to pass "over" the barrier. This theory was discarded because it predicted rectification in the incorrect direction, and because evidence indicated that the potential-barrier layer was thicker than assumed.[2]

A theory advanced by Mott, and modified by Schottky [2,19] assumed that the barrier layer was thick compared to that considered in the electron-tunneling theory. It was considered that electron-tunneling was negligible, and that to pass across a junction, electrons must have sufficient thermal energies to pass over the junction potential barrier. It was assumed that electrons passed through the barrier space without excessive collisions. This was called the **diode theory,** because when once "free," electrons crossed the junction much as in a two-electrode vacuum tube.[2,19] Schottky's modification was that much ionization did occur in the junction, and that the space charges resulting were important; he predicted the existence of "reserve" layers, and "exhaustion" layers.[20] Such reasoning has led to a **diffusion theory.**[20] Experimental evidence did not always agree with the theoretical predictions of the preceding theories. In special cases agreement was found possible if it was assumed that the boundary potential was not uniform, but varied from point to point in **patches.**[19]

10-10. *COPPER-OXIDE RECTIFIERS* [21,22]

These consist essentially of a disk, or washer, of copper on which a layer of cuprous oxide is formed by a controlled heating and cooling process.[23] The copper forms one electrode of the rectifier. The rectification occurs at the copper–cuprous-oxide junction. A metallic electrode, for current-collecting purposes *only,* must be in contact with the outer side of the oxide layer. Vaporized metal contacts sometimes are employed.

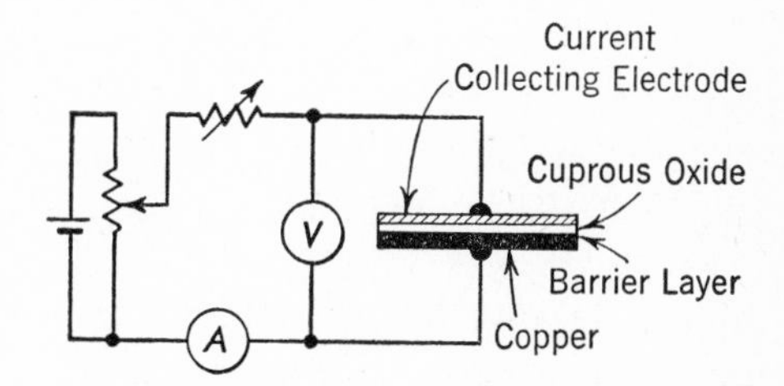

FIG. 10-12. Circuit for determining characteristics of a copper-oxide rectifier, or varistor. Battery connection shown is for forward, or low-resistance direction.

A circuit for studying the direct-current characteristics of a cop-

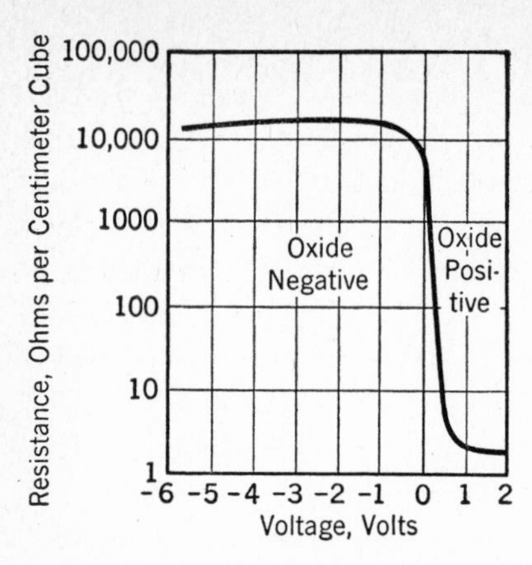

FIG. 10-13. Resistance variations of a typical copper-oxide rectifier. The lowest value is about 2 ohms.

per-oxide rectifier is shown in Fig. 10-12. From the current-voltage characteristic, the resistance curves can be calculated. Care should be taken not to exceed the voltage rating of the rectifier. These characteristics can be explained in a simplified way as follows: When the polarity is such that the copper (containing many free electrons) is negative and the oxide is positive, electrons flow readily from copper to cuprous oxide. Thus, the low-resistance, or forward, direction of *current* flow through the rectifier is from oxide to copper (Fig. 10-13). However, if the oxide is made negative and the copper positive, but few electrons are attracted from the oxide to the copper; hence, this is the high-resistance, or reverse, direction.

Cuprous oxide is an acceptor-impurity p-type semiconductor, having holes in the valence band. There is evidence that some donors also exist. It is stated [24] that the acceptors are probably vacant copper ion lattice sites, and that the nature of the donors is unknown. The mean free path is small compared with the rectifying potential barrier, and the diffusion theory, modified by the patch effect, is used to explain rectification.[24]

On the basis that the cuprous oxide is p-type, at room temperature some electrons will be excited to the acceptor atoms, leaving holes. With no applied voltage, some holes will diffuse into the copper, leaving the oxide negative. Almost all acceptor atoms adjacent to the metal become ionized. This leads to an **exhaustion layer** in the oxide adjacent to the metal as

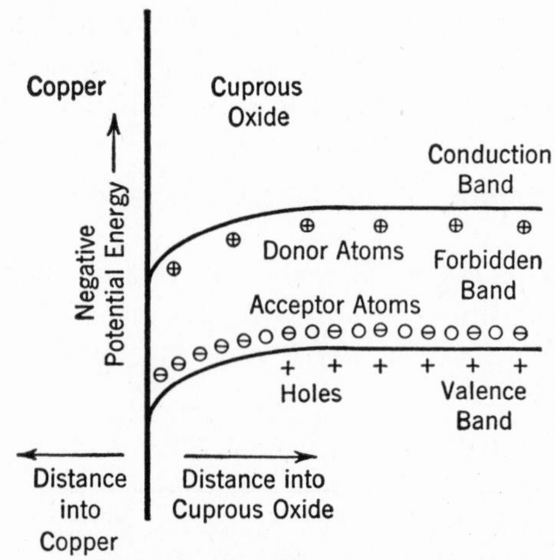

FIG. 10-14. Near the junction of copper and cuprous oxide the acceptor atoms take electrons from the valence bands of germanium atoms; this results in holes in the valence band. These holes diffuse to the copper, leaving a drop in potential energy in the oxide near the copper. This forms a potential-energy barrier for free electrons in the copper.

shown in Fig. 10-14. As there indicated, the absence of holes, and the presence of ionized negative acceptor atoms, produces a potential-energy barrier layer, with the exhaustion layer negative and the copper positive. Such a potential layer will keep electrons from the copper from flowing into the oxide.

If an external voltage is applied, making the copper more positive and the oxide more negative, the potential barrier will be enhanced and negligible current will flow because this is the high-resistance or reverse direction. However, if the external voltage is applied in the opposite direction so that it *tends* to make the copper negative and the oxide positive, this applied voltage will oppose the potential-barrier voltage, and a "large" current will flow. This is the low-resistance, or forward direction.

An equivalent circuit for a copper-oxide rectifier is as shown in Fig. 10-5. Capacitor C represents the capacitance of the junction, a typical value being 0.02 microfarad per square centimeter of disk area. A typical value of r_1 is 2.0 ohms per square centimeter. Resistor r_1 represents the spreading resistance (section 10-4). Resistor R represents the rectifying (forward and reverse) resistance, which varies as indicated in Fig. 10-13.

In general it can be stated that copper-oxide rectifier units may be used in any type of rectifier circuit, including those discussed in Chapter 6. Two

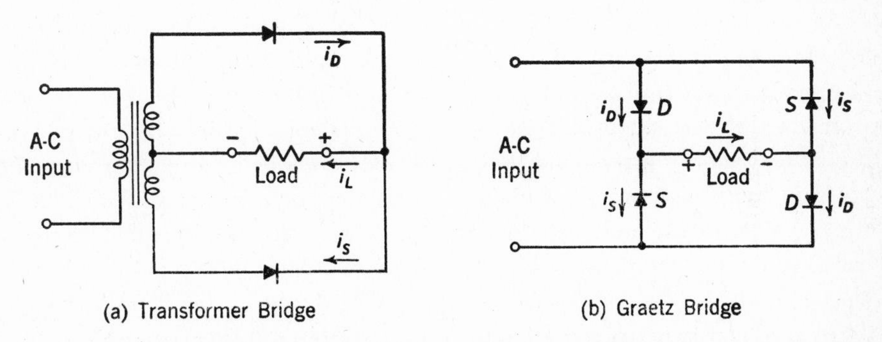

FIG. 10-15. Copper-oxide varistor rectifier circuits. Currents i_s flow on the negative half-cycles because the varistor units are imperfect, as Fig. 10-13 indicates. (From reference 25.)

common circuits are shown in Fig. 10-15. Because, as Fig. 10-13 shows, the resistance in the "nonconducting" direction is *not* infinite, currents i_S accordingly flow when the reverse voltage is applied. For a complete analysis, both the forward currents i_D and the reverse currents i_S must be considered.[25] At frequencies above a few hundred cycles, circuit conditions may require a consideration of capacitance C of Fig. 10-5, and of contact impedance,[26] instead of only contact resistance. Copper-oxide rectifiers are used extensively as nonlinear **varistors** for modulation and demodulation, discussed in Chapters 12 and 13.

10-11. *SELENIUM RECTIFIERS*

The semiconductor selenium is used in rectifiers as follows. A **back electrode**, of a metal such as steel or aluminum, is coated with a thin layer of selenium. Over this is placed a soft-metal alloy with low melting point. This metal is the **front electrode**, or **counter electrode.** By a controlled heat treatment and an electrical forming process a barrier layer is formed at the junction *between the selenium and the front electrode.*[27,28] The back electrode is so treated that *no* barrier layer is formed at its surface.

The selenium layer is considered to be *p*-type semiconductor material containing acceptors. The rectification occurs at the junction of selenium and the *front* soft-metal electrode. Thus the diffusion theory and the exhaustion-layer theory of Schottky [20] apply to the selenium rectifier as explained for the copper-oxide rectifier of the preceding section. However, the lack of agreement between theory and experiment has led to modifications, including the effect of field emission [20] (section 2-17). Hence, the theory of the preceding section applies only for the simplified case.[29,30]

Selenium rectifiers have been widely adopted, particularly for purposes of obtaining direct-current power from alternating-current sources. They are made in disks of various sizes, and the disks are stacked together and connected in series, parallel, and series-parallel combinations to provide rectifiers of various current and voltage capacities. Selenium rectifiers may be used in the various rectifier circuits that have been previously considered. Both copper-oxide and selenium rectifiers are sometimes called **dry-disk rectifiers.**

10-12. *DIFFUSED-JUNCTION GERMANIUM RECTIFIERS*

If an impurity metal, either donor or acceptor as desired, is placed on the surface of a germanium wafer cut from a germanium single crystal of the correct type, and if the wafer is properly heat-treated, the metal will diffuse into the wafer. By this means, excellent *p-n* junctions have been produced,[9] and devices using diffused junctions appear (1955) to have excellent possibilities. These junctions are mounted in a suitable holder, and the units are arranged in various combinations; for instance, in series, in parallel, or in series-parallel as desired.

These diffused junctions are characterized by low forward resistances, high back resistances, and high current-rectifying capacities. For example, a diffused-junction germanium rectifier using a wafer about one centimeter square was used with peak inverse voltages of 1000 volts and forward peak currents *as high as 800 amperes*. The temperature characteristics are good, and the rectifying efficiency is high.[31]

The theory of operation of these devices is the same as for any *p-n* junction; briefly, at room temperature, and with no external voltages applied, positive holes diffuse from the *p*-type to the *n*-type material, and negative electrons diffuse from the *n*-type to the *p*-type germanium. This action sets up a difference of potential across the junction, with the *n*-type positive and the *p*-type negative.

10-13. *SYMMETRICAL VARISTORS* [1]

Symmetrical varistors cannot be used for rectification because they pass currents equally well in both directions. Silicon carbide is an example of a symmetrical varistor, and for the following reasons: Silicon-carbide crystals have, individually, rectifying properties, and were used with a fine-wire contact to make early crystal detectors. However, in present silicon-carbide varistors, such as used as lightning arrestors, the silicon-carbide granules are mixed with a binder and fired to make a substance resembling a ceramic material. The granules then are in random arrangement, and their individual rectifying properties are lost and do not apply to the final substance.[1]

Devices made of disks or rods of this material are essentially high-voltage devices. They offer very high resistance at low voltages and currents, but the resistance drops to low values as the voltages and currents increase. The **Thyrite** lightning arrester is a silicon-carbide varistor.

10-14. *JUNCTION-TYPE TRANSISTORS*

Point-contact type transistors were considered in section 10-5. The junction-type transistor, now to be considered is, in a sense, in competition with the point-contact type. A junction-type "triode" transistor is made from two properly diffused *p-n* junctions (section 10-12), or is made from a single germanium crystal that has been "doped" with impurities as follows. During the crystal-growing process, as the growing single crystal of germanium is being withdrawn from the "melt," a pellet of an element that will make the crystal *n*-type, or that will make it *p*-type, as desired, is added to the melt.[9,14] Hence, the final crystal can be made to consist of *n*-type and *p*-type layers. Then, a piece of material can be cut from the germanium crystal so that it is n-type on each end and p-type in the center, or so that it is p-type on each end and n-type in the center. The center "slab" is very thin for reasons to be explained later.

10-15. *n-p-n JUNCTION TRANSISTORS*

An n-p-n junction transistor is shown in Fig. 10-16a. Because of thermal agitation, at room temperature there will be an initial diffusion of negative electrons from n-type to p-type germanium, and positive holes from p-type to n-type germanium. This initial diffusion will produce differences of potential across the junctions so that further diffusion is prevented, except as necessary to offset recombination that may occur. The n-type portions will be made positive with respect to the p-type center. These differences of potential establish potential barriers at the n-p junctions as shown by the vertical lines of Fig. 10-16b. No electrons in the n-type germanium have kinetic energies above the potential "hump" shown. If they had kinetic energies sufficiently high, they could cross the potential barriers into the p-type material. Hence, the n-type germanium is a region of low potential energy for electrons, and the p-type material is a region of high potential energy, as indicated in Fig. 10-16b.

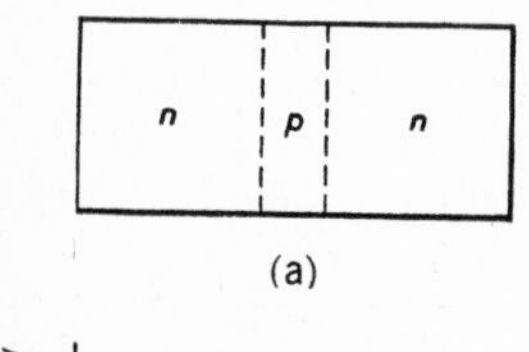

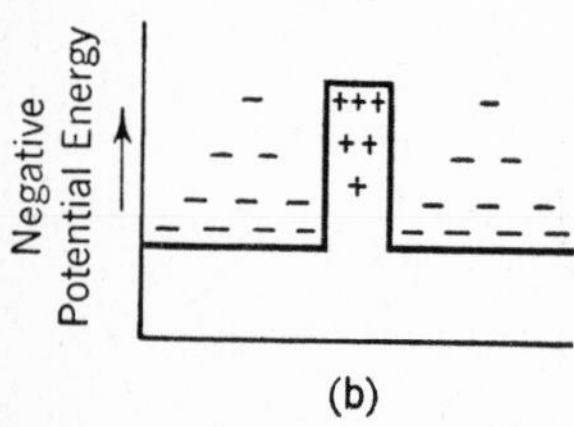

FIG. 10-16. A side view of a piece of single-crystal n-p-n germanium is shown in (a). The thickness of the p layer is exaggerated. In (b) is shown the results of diffusion at room temperature. Differences of potential are produced at the junctions so that further diffusion is prevented, except to offset the effects of recombination.

Now suppose that connections are made to the n-p-n transistor as in Fig. 10-17a. These connections serve *only* to join the germanium to the wire circuit. When the generator voltage is *zero*, the battery voltages alter the potential-energy diagram of Fig. 10-16b so that it appears as in Fig. 10-17b after a steady-state condition is reached. The negative terminal of battery E_{ee} raises the n-type material at the left end to a higher *negative* potential with respect to the p-type part. In other words, battery E_{ee} ***reduces*** the potential barrier. Hence, electrons can flow from the left n-type germanium to the center p-type material at room temperature, even though their energies are insufficient to carry them across the potential barrier with no battery E_{ee} connected as in Fig. 10-17a. Also some holes may flow from right to left across the barrier. Battery E_{cc} of Fig. 10-17a causes the n-type part at the right to be more *positive* (or less negative with respect to the *p*-type center) than before the battery was connected. Hence, electrons with sufficient velocities that may have entered the thin *p*-type material are quickly pulled out and flow through resistor *R*. The resulting current flow is indicated by *I*, and this causes an *IR* drop as shown by the positive and negative signs at the resistor. This drop subtracts from the battery voltage E_{cc}, so that only a part of this battery voltage is effective in increasing the potential barrier at the right junction.

The preceding discussion applies to an *n-p-n* junction transistor at room temperature, and in a static, or quiescent, condition. Now suppose that the alternating signal generator of Fig. 10-17a is connected, and that it is generating the *positive* half-cycle of voltage. This will oppose battery E_{ee}, and the *n*-type material at the left will not be forced to as high negative potential with respect to the *p*-type center. Hence, the electrons in the *n*-type material at the left must have greater thermal energies to cross the boundary, the number crossing will be reduced, current *I* will *fall*, and the potential of the *n*-type part at the right will be less negative.

When the signal generator produces the negative half-cycle, the negative potential of the *n*-type portion at the left will be increased, the electrons will need less thermal energy to cross the potential barrier, the current *I* will increase, the drop across the resistor will increase, and the potential of the *n*-type material will be more negative.

From the foregoing it is seen that an alternating voltage in series with the *n*-type portion at the left, called the **emitter** (as in the point-contact transistor), will cause current and voltage signal changes in the *n*-type portion at the right called the **collector.** As will be explained in the following chapter, the

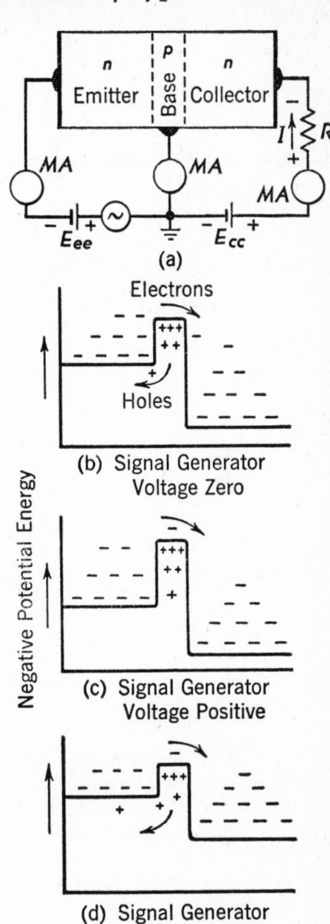

FIG. 10-17. When an *n-p-n* junction-type transistor is connected as in (a), some emitter and collector current will flow. The positive half-cycle of signal voltage makes the emitter less negative with respect to ground, and the collector current decreases. The negative half-cycle of signal voltage makes the emitter more negative, and the collector current increases. The magnitudes of the potential rises and falls are not, necessarily, to scale.

n-p-n transistor has amplifying properties, but not current-amplifying properties as in the point-contact type. For an *n-p-n* transistor, the ratio of collector to emitter current is less than unity. There is no current multiplication caused by "hooks" or "trapping." The *p*-type layer, called the **base,** is thin, and little recombination occurs in it. Most of the electrons that enter the base from the emitter flow into the collector. For a typical case, 97 per cent will flow to the collector, and 3 per cent will flow out the connection at the base.

10-16. *STATIC CHARACTERISTICS OF n-p-n JUNCTION TRANSISTORS*

A circuit for determining the static characteristics of an *n-p-n* junction-type germanium transistor is shown in Fig. 10-18. Note that the batteries have been reversed from the connections of Fig. 10-10; also, that the arrowhead on the emitter has been reversed. This is done to denote an *n-p-n* junction transistor, and to distinguish its graphical symbol from that for the *p-n-p* type.

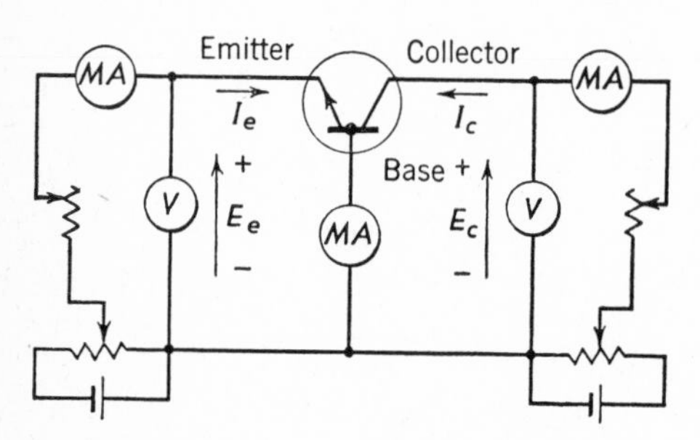

FIG. 10-18. A convenient circuit for determining the static characteristics of an *n-p-n* junction-type transistor. Conventional positive directions for currents and voltages are as indicated by the arrows. Note that the arrowhead on the emitter has been reversed from the direction used with the point-contact type, which is *p-n-p*.

For usual conditions of operation, the emitter current I_e is of approximately the same value as the collector current I_c, as explained in the preceding section. After the electrons from the emitter have crossed the emitter-to-base *n-p* junction, there is little opposition in their path to the collector. For this reason, the collector current is independent of collector voltage over wide limits (Fig. 10-19).

With a junction transistor, phenomena occur "deep" within the germanium crystal, but with the point-contact type, phenomena occur near the surface. Because of this (and because the emitter is biased in the forward direction), the input resistance of the *n-p-n* junction transistor is low compared to the point-contact type. Also, because the collector is biased in the reverse direction, the output resistance will be high in comparison. The curves shown in Fig. 10-19 are for the usual operating range as an amplifier.

By referring to Fig. 10-11 and the accompanying discussion it will be seen that the collector output resistance R_{22} can be found from the slope of a curve of Fig. 10-19; also, that the feedback resistance R_{12} can be found from the curves of Fig. 10-19. However, the slopes are such that graphical determinations are not reliable. For junction-type transistors, the characteristics accordingly are determined by bridge methods (section 4-8), much as for vacuum tubes.* Typical values for an *n-p-n* germanium transistor are $R_{11} = 266$ ohms; $R_{12} = 240$ ohms; $R_{21} = 13.1$ megohms, and $R_{22} = 13.4$ megohms. The way in which these characteristics vary, and other detailed information for an *n-p-n* junction-type transistor, are discussed in reference 32.

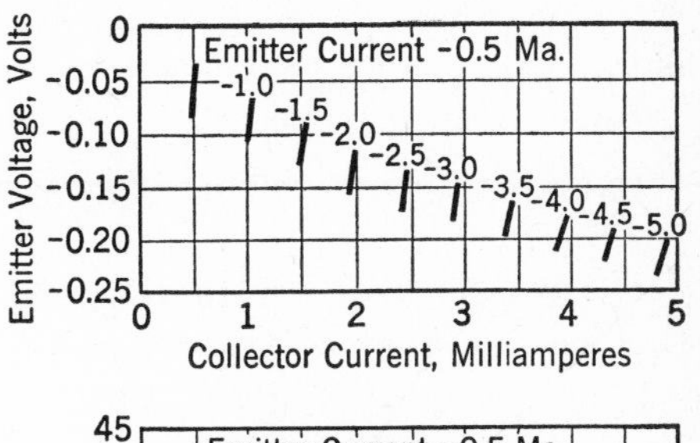

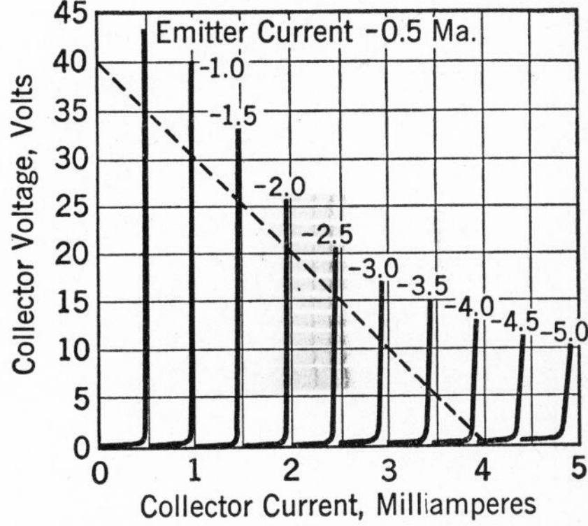

FIG. 10-19. Static characteristics of an *n-p-n* junction-type transistor. The broken line is a 10,000-ohm load line to which reference will be made in the following chapter.

10-17. *p-n-p JUNCTION TRANSISTORS*

If the battery connections of Fig. 10-18 are reversed, this circuit can be used to find the characteristics of a *p-n-p* transistor. Such a transistor can be made by diffusing a suitable metal into opposite faces of *n*-type germanium.[33] Of course, such a transistor could be made by a crystal growing process. For operation as an amplifier, the emitter should be biased positively, and the collector negatively, with respect to the base. The theory of operation can be explained much as for the *n-p-n* type, except that in Figs. 10-16 and 10-17 diagrams for holes, instead of electrons, would be plotted.[34]

An explanation for a *p-n-p* junction transistor predicts that the maximum ratio of collector to emitter current would be approximately unity, just as in the *n-p-n* type. This may be surprising when it is recalled that operation of the point-contact transistor, which has a current ratio greater than unity, also was explained on the *p-n-p* basis. It will be recalled that in section 10-7

* A circuit is described in Tele-Teck, page 75, February 1954, which determines the "R" parameters and also the "h" parameters sometimes used.

two theories for the current multiplication of the point-contact type were given, and that the matter is debatable. However, it is possible that the diffusion process is considerably different than the forming process, and it should be remembered that the emitter of the point-contact transistor is not formed.

10-18. *n-p-n* JUNCTION TRANSISTOR TETRODES

The two junction transistors previously considered are sometimes called junction transistor *triodes* because they have an emitter, a base, and a collector. Usually, *one* connection is made to each of these. In the so-called **transistor tetrode,**[35] *two* connections are made to the base as shown in Fig. 10-20.

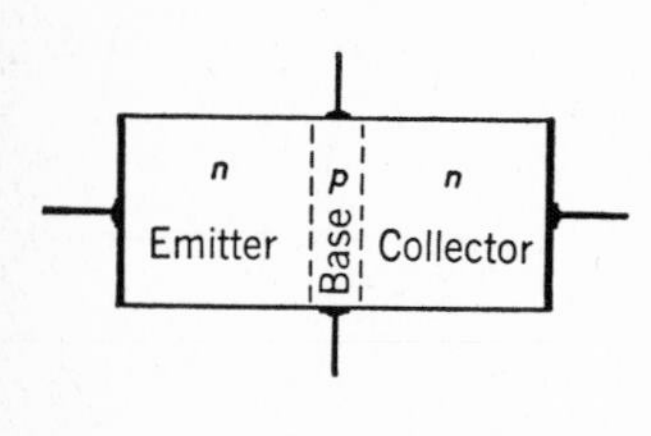

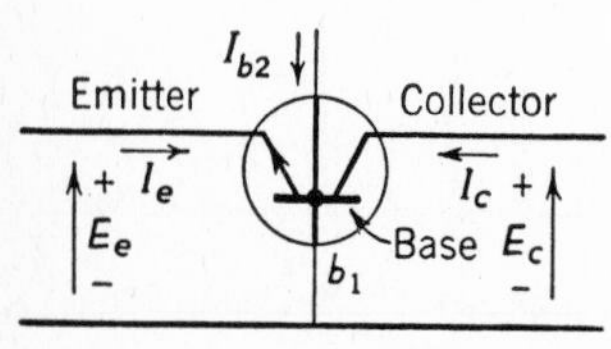

FIG. 10-20. The transistor tetrode has one more connection to the base than does the transistor triode. The *n-p-n* type is shown above. Conventional directions for currents and voltages are as indicated by arrows.

For typical operation, the connection to the *top* of the base is biased about 6 volts negative with respect to the *bottom* of the base. The emitter is biased about 0.1 volt negative with respect to the bottom of the base, and the collector is biased to a suitable positive voltage (Fig. 10-19). The −6 volts connected to the top of the *p*-type base forces adjacent portions of the emitter to a potential of *nearly* 6 volts negative with respect to the bottom of the base. The upper part of the base is at a negative potential with respect to adjacent portions of the emitter. This is the reverse direction, and electrons are repelled back into the emitter. In fact, all the normal transistor action occurs near the lower base contact,[35] which is biased about 0.1 volt positive with respect to the emitter.

As a result of the action just discussed, the base resistance r_b of the transistor equivalent diagram (Fig. 11-3) is reduced, which permits this *n-p-n* junction transistor tetrode to be used at frequencies ten times as great as otherwise would be possible.[35]

10-19. *p-n-p-n* JUNCTION TRANSISTORS

The *p-n-p-n* four-terminal transistor[36] may be constructed as in Fig. 10-21a, or may be produced, in effect, by connecting two transistors as shown in Fig. 10-21b. It will be recalled that the junction-type transistors that have been considered have current multiplication factors less than unity, and that the point-contact transistor has a factor greater than unity. One explanation for the fact that the collector current exceeded the emitter current in the point-contact transistor was the *p-n* "hook" theory considered in section 10-7. The *p-n-p-n* transistor also has a current multiplication greater than unity.

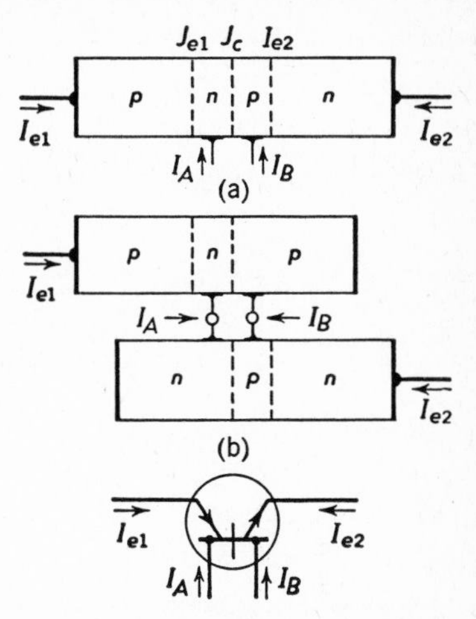

FIG. 10-21. The *p-n-p-n* junction transistor may be made from a single crystal of germanium, or from separate *p-n-p* and *n-p-n* transistors.

Briefly, the theory [36] is that the *p-n* junction J_{e1} on the left acts as an emitting junction for positive holes, the *n-p* junction J_{e2} on the right acts as an emitting junction for negative electrons, and junction J_c is a collector junction. Junction J_{e2} has current-multiplying properties, explained on the hook basis. The theory predicts that the current multiplication should be large, and values of 50 are readily obtained. The name **conjugate-emitter transistor** has been proposed.[36]

10-20. *p-n-i-p and n-p-i-n JUNCTION TRANSISTOR TRIODES*

Transistor theory indicates [37] that the presence of a thick layer of intrinsic germanium (an *i*-layer), positioned as specified in the designations *p-n-i-p* and *n-p-i-n*, would make a junction transistor triode operable to *3,000 megacycles.* Such transistors have been made, and have been found to have low base resistance, low collector capacitance, and high breakdown voltage. Although the first models did not operate at ultrahigh frequencies as predicted by theory, improvements are expected, and it is thought [37] that these transistors eventually will offer the vacuum tube serious competition over a greater range of frequencies, and at higher power levels, than before.

10-21. *THE SURFACE-BARRIER TRANSISTOR* [38]

A transistor was announced in 1953 that functions because of surface phenomena instead of action within the body of the semiconductor. A cross section of this **surface-barrier transistor** is shown in Fig. 10-22. A blank of single-crystal *n*-type germanium is etched by directing on opposite surfaces a metal salt solution carrying electric current in the direction such that germanium is removed. When the germanium blank has been etched to the correct thickness, the current direction is reversed, etching stops, and metal from the jets is deposited as electrodes on the germanium, providing emitter and collector electrodes. The body of the germanium is the base to which a current-conducting electrode is attached.

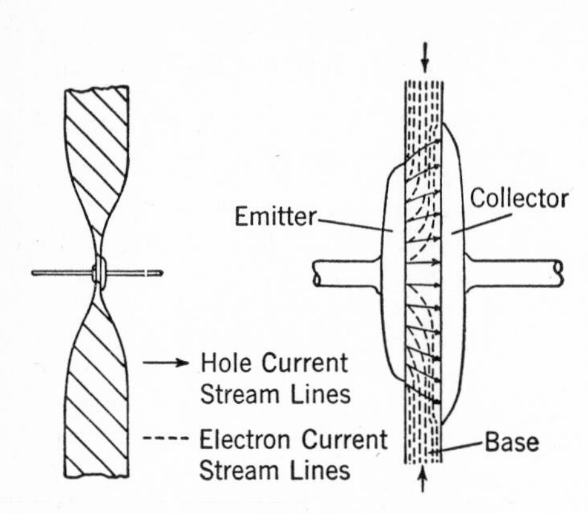

FIG. 10-22. At the left is shown the cross section of a surface-barrier transistor, and at the right the details of the center section. Single arrows at top and bottom of the figure at the right indicate directions of electron flow. (From reference 38.)

The operation is attributed to a potential barrier just beneath the crystal surface.[38] The barrier causes an electric field in a direction such that an electron would be forced into the germanium. The potential barrier is considered to be produced by phenomena such as incomplete bonds, because at the surface the orderly array of germanium atoms ends. If a metal electrode in contact with such a surface is made positive with respect to the germanium body beneath, the surface barrier will be reduced in thickness, and electrons may pass through. If the electrode is made negative, the potential barrier is thickened, and electrons largely are prevented from passing through. This explains rectification. It appears that the electrolytically deposited metal on the freshly etched surface tends to emit holes into the surface when the emitter electrode is made positive with respect to the n-type germanium. Because this polarity also would attract electrons, two currents flow as in Fig. 10-22. The surface-barrier transistor has been operated at 60 megacycles and above with excellent results. Its development is another indication of the remarkable progress to be expected in solid-state devices.

10-22. *THERMISTORS*

These are thermally sensitive varistors.[39] They are made of semiconducting materials, and are used extensively for control purposes.[39,40] Typical thermistors are composed of manganese and nickel oxides, of manganese, nickel, and cobalt oxides, and of iron and zinc oxides, and are made in several physical forms.[39] Since they are composed of semiconducting materials, it would be suspected that increasing temperature, and resulting thermal agitation, would cause a decrease in resistance. A comparison with a metal is shown in Fig. 10-23. Energy-level diagrams are used in explaining their characteristics.[39] The voltage-current relation for a thermistor of manganese and nickel oxides is shown in Fig. 10-24.

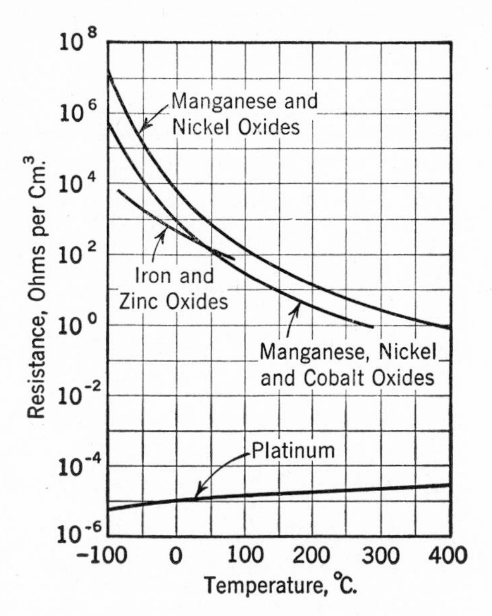

FIG. 10-23. Resistance variations of thermistors composed of different substances, and comparison with the metal platinum. (Data from reference 39.)

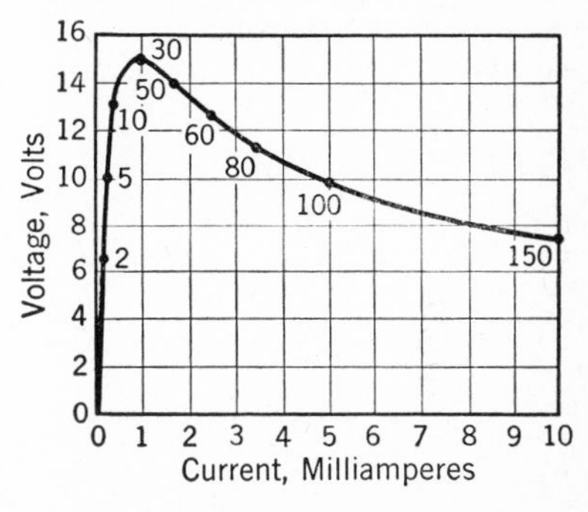

FIG. 10-24. Static characteristic curve for a thermistor. The numbers on the curve are for degrees centigrade temperature rise above surrounding air. (Reference 39.)

REFERENCES

1. Stansel, F. R. *Characteristics and applications of varistors.* Proc. I.R.E. April 1951, Vol. 39, No. 4.
2. Torrey, H. C., and Whitmer, C. A. *Crystal Rectifiers.* McGraw-Hill Book Co.
3. Cornelius, E. C. *Germanium crystal diodes.* Electronics, Feb. 1946, Vol. 19, No. 2.
4. Stephens, W. E. *Crystal rectifiers.* Electronics, July 1946, Vol. 19, No. 7.
5. Scaff, J. H., and Ohl, R. S. *Development of silicon crystal rectifiers for microwave radar receivers.* Bell System Technical Journal, Jan. 1947, Vol. 26, No. 1.
6. Waltz, M. C. *On some transients in the pulse response of point-contact germanium diodes.* Proc. I.R.E., Nov. 1952, Vol. 40, No. 11.
7. Shockley, W. *Transistor electronics: Imperfections, unipolar and analog transistors.* Proc. I.R.E., Nov. 1952, Vol. 40, No. 11.
8. Valdes, L. B. *Transistor forming effects in n-type germanium.* Proc. I.R.E., April 1952, Vol. 40, No. 4.
9. Jordan, J. P. *The ABC's of germanium.* Electrical Engineering, July 1952, Vol. 71, No. 7.

10. Shockley, W. *Electrons and Holes in Semiconductors.* D. Van Nostrand Co.
11. Shockley, W. *The theory of the p-n junctions in semiconductors and p-n junction transistors.* Bell System Technical Journal, July 1949, Vol. 28, No. 3.
12. Brattain, W. H. *Semi-conductor surface phenomena.* Bell System Monograph No. 1875, reprinted from Semi-Conducting Materials. Butterworth's Scientific Publications (London).
13. Pearson, G. L., and Sawyer, B. *Silicon p-n junction alloy diodes.* Proc. I.R.E., Nov. 1952, Vol. 40, No. 11.
14. Roth, L., and Taylor, W. E. *Preparation of germanium single crystals.* Proc. I.R.E., Nov. 1952, Vol. 40, No. 11.
15. Morton, J. A. *Present status of transistor development.* Proc. I.R.E., Nov. 1952, Vol. 40, No. 11.
16. Sittner, W. R. *Current multiplication in the type A transistor.* Proc. I.R.E., April 1952, Vol. 40, No. 4.
17. Coblenz, A., and Owens, H. L. *Variations of transistor parameters with temperature.* Proc. I.R.E., Nov. 1952, Vol. 40, No. 11.
18. Slade, B. N. *The control of frequency response and stability of point-contact transistors.* Proc. I.R.E., Nov. 1952, Vol. 40, No. 11.
19. *Semiconductor rectifiers.* Electrical Engineering, Oct. 1949, Vol. 68, No. 10.
20. Joffe, J. *Schottky's theories of dry solid rectifiers.* Electrical Communication, 1945, Vol. 22, No. 3.
21. Grondahl, L. O., and Geiger, P. H. *A new electronic rectifier.* Journal of A.I.E.E., March 1927, Vol. 46, No. 3.
22. Grondahl, L. O., *Twenty-five years of copper oxide rectifiers.* Trans. A.I.E.E., 1948, Vol. 67.
23. Brattain, W. H. *The copper oxide varistor.* Bell Laboratories Record, Jan. 1941, Vol. 19, No. 5.
24. Brattain, W. H. *The copper oxide rectifier.* Reviews of Modern Physics, July 1951, Vol. 23.
25. Huss, P. O. *An analysis of copper-oxide rectifier circuits.* Electrical Engineering, Sept. 1935, Vol. 56, No. 3.
26. Bardeen, J. *On the theory of the a-c impedance of a contact rectifier.* Bell System Technical Journal, July 1949, Vol. 28, No. 3.
27. Clarke, C. A. *Selenium rectifier characteristics, application and design factors.* Electrical Communication, 1941, Vol. 20, No. 1.
28. Yarmack, J. E. *Selenium rectifiers and principles of their design.* Electrical Communication, 1942, Vol. 20, No. 4.
29. Angello, S. J. *Electrical characteristics of the junction in a simplified selenium rectifier cell.* Trans. A.I.E.E., 1947, Vol. 66.
30. Henkels, H. W. *Experimental examination of rectifier theory as applied to the selenium rectifier.* Proc. National Electronics Conference, 1945, Vol. 5.
31. Ferguson, T. J. *The GIO germanium rectifier.* General Electric Review, July 1952, Vol. 55, No. 4.
32. Wallace, R. L., and Pietenpol, W. J. *Some circuit properties and applications of n-p-n transistors.* Bell System Technical Journal, July 1951, Vol. 30, No. 3.
33. Saby, J. S. *Fused impurity p-n-p junction transistors.* Proc. I.R.E., Nov. 1952, Vol. 40, No. 11.

34. Steele, E. R. *Theory of alpha for p-n-p diffused junction transistors.* Proc. I.R.E., Nov. 1952, Vol. 40, No. 11.
35. Wallace, R. L., Schimpf, L. G., and Dickten, E. *A junction transistor tetrode for high-frequency use.* Proc. I.R.E., Nov. 1952, Vol. 40, No. 11.
36. Ebers, J. J. *Four terminal p-n-p-n transistors.* Proc. I.R.E., Nov. 1952, Vol. 40, No. 11.
37. Early, J. M. *p-n-i-p and n-p-i-n junction transistor triodes.* Bell System Technical Journal, May 1954, Vol. 33, No. 3.
38. Bradley, W. E. *Principles of the surface-barrier transistor.* Tiley, J. W., and Williams, R. A. *Electrochemical techniques for fabrication of surface-barrier transistors.* Angell, J. B., and Keiper, E. P. *Circuit applications of surface-barrier transistors.* Kansas, R. *On the high-frequency performance of transistors.* Schwarz, R. F., and Walsh, J. F. *The properties of metal to semiconductor contacts.* Proc. I.R.E., Dec. 1953, Vol. 14, No. 12.
39. Becker, J. A., Green, C. B., and Pearson, G. L. *Properties and uses of thermistors—thermally sensitive resistors.* Trans. A.I.E.E., 1946, Vol. 65.
40. Bollman, J. H., and Kreer, J. G. *The application of thermistors to control networks.* Proc. I.R.E., Jan. 1950, Vol. 38, No. 1.

QUESTIONS

1. Why is it not advisable to explain point-contact crystal rectification on the basis of a simple metal-semiconductor contact?
2. How does welding, or forming, enter into the theory of operation of a point-contact germanium rectifier?
3. What is meant by the terms donor and acceptor?
4. Do donors and acceptors migrate within a crystal? Why?
5. Why are the acceptors and donors of Fig. 10-2 marked negative and positive? Do they produce electric fields that cause local space charges?
6. What effect does recombination in the *p-n* junction layer have on point-contact crystal rectification?
7. In discussing the depletion layer in section 10-2 it is stated that the junction layer becomes intrinsic in nature. What is the meaning, and importance, of this statement?
8. An equivalent circuit of a point-contact type of crystal rectifier is shown in Fig. 10-5. Is the capacitance shown important on both half-cycles of applied voltage? Why? What causes the capacitance?
9. If a rectifier is slow to become a nonconductor after the reverse voltage is applied, would the current curve at high frequencies be as shown in Fig. 10-6c?
10. Why is the region under the collector of Fig. 10-8 considered to be *p*-type germanium? Under the emitter?
11. Explain how it is possible for the collector current to exceed the emitter current in a point-contact transistor.
12. In section 10-7 it is stated that the crystal lattice structure contains imperfections. What are these imperfections?
13. Do similarities exist between α for a transistor and μ for a vacuum tube? What are they?

14. What is the meaning of the negative signs in Fig. 10-11?
15. Why is R_{12} regarded as a feedback resistance?
16. In Fig. 10-14 many acceptor atoms with negative signs marked on them are shown near the copper. Do these atoms produce the potential barrier?
17. In section 10-10 it is stated that some holes will diffuse into the copper, leaving the oxide negative. Is it possible for holes to cross over into the copper? What is your explanation of the action?
18. In the selenium rectifier, does the rectification occur at the junction of the metal back plate and the selenium?
19. What is a diffused junction? What are its important characteristics?
20. How is rectification explained in the diffused junction rectifier?
21. How are diffusion and recombination related in the junction transistor?
22. Why should the characteristics of a semiconductor device be expected to vary with temperature?
23. What are the phase relations between emitter and collector signal voltages? What simple proof can be given?
24. Four sets of curves are given to explain the operation of point-contact transistors, but only two sets are given for the junction type. Why?
25. From the theoretical standpoint, why should a thermistor have a characteristic curve such as Fig. 10-24?

PROBLEMS

1. Prepare a paper of about 300 words explaining why the forward direction of current flow in a *p*-type silicon crystal rectifier is from silicon to metal point.
2. Calculate the maximum values of resistance in the backward directions for the curves of Figs. 10-3 and 10-4. Calculate resistances in the forward direction.
3. Calculate the resistance values, and plot a curve of resistance on the X axis, and voltage on the Y axis for the germanium crystal rectifier of Fig. 10-4.
4. Prepare a paper of about 300 words explaining how to evaluate r, R, and C of Fig. 10-5.
5. Use the curves of Fig. 10-11, and calculate the various resistances for emitter currents of 1.0 milliampere, and collector currents of 3.0 milliamperes.
6. Prove mathematically that the relations given by equation 10-6 are correct.
7. For the transistor of Fig. 10-11, and at emitter and collector currents of 1.0 and 3.0 milliamperes respectively, calculate the power input in milliwatts to the emitter, and to the collector. What becomes of this power? If operated as a nondistorting amplifier (class-A) about what would be the maximum power output?
8. For the transistor of Fig. 10-19, and at emitter and collector currents of 1.5 milliamperes, calculate the power input in milliwatts to the emitter and to the collector.
9. Prepare a paper of about 300 words discussing current multiplication in *n-p-n* junction transistors.
10. Plot a curve of resistance versus current for the thermistor of Fig. 10-24.

CHAPTER 11

TRANSISTOR AMPLIFIERS AND OSCILLATORS

Both point-contact and junction transistors are used for amplification. For this purpose they have certain advantages over thermionic vacuum tubes; of importance is the fact that the transistor does not need a hot cathode. Also, transistors need no warm-up time, are rugged, and occupy little space. Among the present (1955) limitations of transistors is the fact that they are in the developmental state; however, they offer outstanding possibilities.

Amplifiers using vacuum tubes were treated in Chapter 7. The equations for making most calculations at low frequencies were found to be simple. One reason is that at low frequencies, such as within the audio range, the effects of inductances and capacitances in tube leads and electrodes may be neglected. Also, at these frequencies feedback within the tube is negligible. Another important reason is that for small-signal operation in class-A1, negligible grid current flows, thus making possible the simple equivalent circuit of Fig. 4-12.

With transistors, when used for small-signal operation as linear amplifiers, both feedback and emitter current must be considered even at low frequencies. Hence, the equivalent circuits and the design equations are more involved than for vacuum tubes. Transistor equivalent circuits based on the two terminal-pair generalized active network are available,[1] and are used in transistor amplifier calculations.[2,3,4] (Since this manuscript was completed, a book, reference 18, has been published on active networks.)

11-1. *EQUIVALENT CIRCUITS FOR SMALL-SIGNAL AMPLIFICATION* *

The discussion in the pages immediately following assumes that the frequency is such that reactances may be neglected, and that the signal magnitudes are so small that the electrode resistances calculated in the preceding chapter may be used. The transistor may be regarded as a current-operated device, with currents taken as the independent variables (Fig. 10-11). This is in contrast with vacuum-tube practice, where voltages are independent variables.

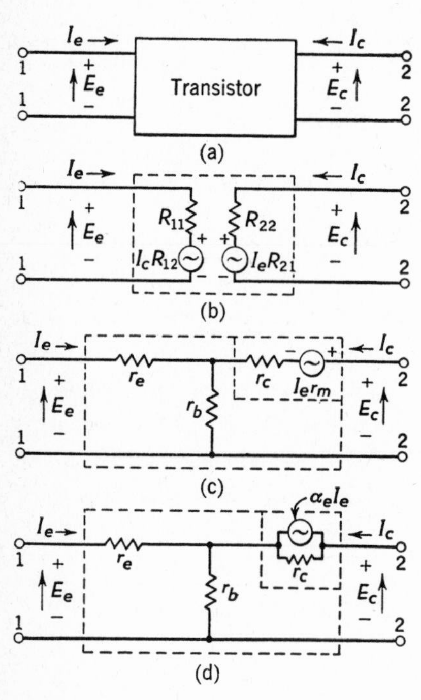

FIG. 11-1. Equivalent circuits for a transistor.

The transistor parameters R_{11}, R_{21}, R_{12}, and R_{22} may be determined graphically as explained in Chapter 10; or, they may be measured with a constant-current generator as a source and a high-impedance voltmeter calibrated in ohms; [3] or, by bridge methods similar to those used for vacuum tubes.[4]

In Fig. 11-1a is shown a "black box" representing a transistor, and in Fig. 11-1b an arrangement that is equivalent to the transistor.[1,2,3,4] The forward transfer resistance R_{21} multiplied by the emitter current I_e represents the amplified output voltage of the transistor. The feedback transfer resistance R_{12} multiplied by the collector current I_c gives the feedback voltage. The resistors R_{11} and R_{22} are the input and output resistances (section 10-8). Measurements made on Figs. 11-1a or b give identical results. The equations applying are

$$E_e - I_eR_{11} - I_cR_{12} = 0 \quad \text{and} \quad E_e = I_eR_{11} + I_cR_{12} \qquad 11\text{-}1$$

$$E_c - I_cR_{22} - I_eR_{21} = 0 \quad \text{and} \quad E_c = I_cR_{22} + I_eR_{21} \qquad 11\text{-}2$$

Another transistor equivalent circuit is shown in Fig. 11-1c. It is necessary to find relations among the parameters R_{11}, R_{21}, R_{12}, R_{22}, and r_e, r_b, and r_c.

* Useful information on this subject will be found in the last two articles and the two books listed in the references at the close of this chapter.

Assume R_{11} is being evaluated by the voltmeter method previously mentioned. The collector current I_c will be held *constant* at the desired value (which can be done by a constant-current generator connected between the output terminals). Then, $\partial E_e/\partial I_e$ and R_{11} will be determined (section 10-8). The same measurement made on the circuit of Fig. 11-1c will be the input resistance with the output terminals open.

The input resistance is, from the preceding paragraph, $\partial E_e/\partial I_e = R_{11}$, or

$$R_{11} = r_e + r_b \qquad 11\text{-}3$$

The output resistance is $\partial E_c/\partial I_c = R_{22}$ with I_e *constant* (section 10-8). This measurement made on the circuit of Fig. 11-1c yields

$$R_{22} = r_b + r_c \qquad 11\text{-}4$$

The forward transfer resistance is $\partial E_c/\partial I_e = R_{21}$ with I_c held constant (section 10-8). If a current I_e is impressed on the input of Fig. 11-1c, and if the *open-circuit* output voltage E_c is measured, then, since $I_c = 0$,

$$I_e r_b + I_e r_m = E_c, \qquad \frac{\partial E_c}{\partial I_e} = r_b + r_m, \qquad \text{and} \qquad R_{21} = r_b + r_m \qquad 11\text{-}5$$

The feedback transfer resistance is $\partial E_e/\partial I_c = R_{12}$ with I_e held constant (section 10-8). If a current I_c is impressed on the output terminals, and if the *open-circuit* voltage at the input terminals is measured, then, since $I_e = 0$,

$$I_c r_b = E_e, \qquad \frac{\partial E_e}{\partial I_c} = r_b, \qquad \text{and} \qquad R_{12} = r_b \qquad 11\text{-}6$$

Another transistor equivalent circuit is shown in Fig. 11-1d. This is obtained by substituting, in accordance with Norton's theorem (section 7-11), the current generator of Fig. 11-1d for the voltage generator of Fig. 11-1c. Thus, considering only the generator portion of Fig. 11-1c, the current when this portion is short-circuited is $I_{sc} = I_e r_m/r_c = \alpha_e I_e$, where

$$\alpha_e = \frac{r_m}{r_c} \qquad 11\text{-}7$$

Here α_e is not the same, theoretically, as α defined by equation 10-7.

For convenience, the relations just considered are summarized as follows:

$$\begin{aligned} R_{11} &= r_e + r_b & \qquad r_e &= R_{11} - R_{12} \\ R_{22} &= r_b + r_c & r_c &= R_{22} - R_{12} \\ R_{21} &= r_b + r_m & r_m &= R_{21} - R_{12} \\ R_{12} &= r_b & r_b &= R_{12} \end{aligned}$$

Typical values for a *point-contact transistor* determined graphically from curves of Fig. 10-11 are $R_{11} = 300$ ohms, $R_{22} = 34{,}000$ ohms, $R_{21} = 100{,}000$ ohms, $R_{12} = 100$ ohms, and $\alpha = 2.9$. The *true* value of current multiplication is given by equation 10-7, $\alpha = R_{21}/R_{22}$. However, since R_{12} is so small, the value for α is essentially the same as for α_e given by equation 11-7. From the preceding relations, $r_e = 200$ ohms, $r_c = 33{,}900$ ohms, $r_m = 99{,}900$ ohms, and $r_b = 100$ ohms. Since these values were determined graphically and the accuracy is not great, and because of the relative magnitudes involved, it is satisfactory in this instance to assume that $r_c = R_{22}$ and $r_m = R_{21}$.

11-2. *TRANSISTOR SMALL-SIGNAL AMPLIFIERS*

In amplifiers, transistors are operated with grounded bases, with grounded emitters, and with grounded collectors. When two or more transistors are operated with grounded bases, the term **common-base operation** sometimes is used. When two or more transistors are operated with grounded emitters, the term **common-emitter operation** is used, and when operated with grounded collectors the term **common-collector operation** is used. The term "grounded" as here employed means that *the circuit is arranged for grounding* a certain electrode, rather than that the electrode actually is grounded, a usage common in electronics.

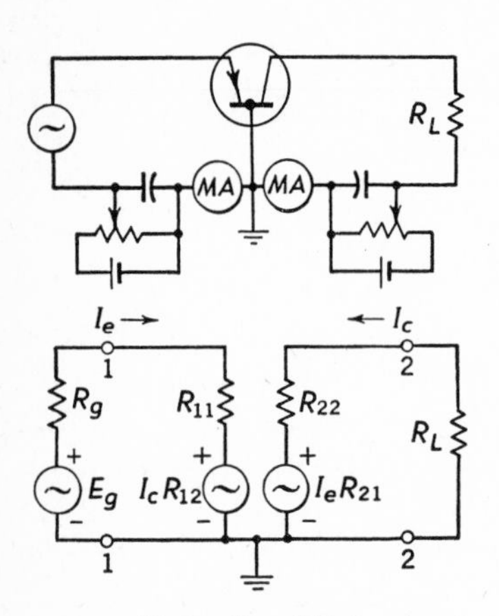

FIG. 11-2. Simplified circuit and one equivalent circuit for a point-contact transistor amplifier.

Before proceeding, it is of interest to review certain aspects of vacuum-tube amplifiers. When a tube is operated at low frequencies in class-A1, it is assumed that the grid draws no current from the signal source. For this reason, the internal impedance of the source does not enter into amplifier calculations. The equivalent circuit (Fig. 4-13) and design equations are simple. However, because the input resistance R_{11} of a common point-contact transistor is only a few hundred ohms, the transistor amplifier draws considerable current from the signal source, and hence the internal impedance of the source enters into calculations. For this reason, the design of a transistor amplifier is more involved than the design of an amplifier using vacuum tubes.

A simplified circuit of a single-stage point-contact transistor amplifier is shown in Fig. 11-2. The signal generator is assumed to contain an output transformer that will pass freely the emitter direct current; otherwise, a path must be provided. In the next section small-signal operation and constant transistor parameters are assumed.

11-3. *GROUNDED-BASE TRANSISTOR AMPLIFIERS*

Any of the three equivalent circuits of Fig. 11-1 may be used for considering the grounded-base transistor amplifier; the first part of this section will employ Fig. 11-1b and the second part Fig. 11-1c. Voltage equations for the two loops of Fig. 11-2 are

$$I_eR_g + I_eR_{11} + I_cR_{12} = E_g \qquad \text{11-8}$$

$$I_eR_{21} + I_cR_{22} + I_cR_L = 0 \qquad \text{11-9}$$

Solving for I_c in the second equation and substituting for I_c in the first equation gives the *emitter input signal current*:

$$I_e = \frac{E_g(R_{22} + R_L)}{(R_{11} + R_g)(R_{22} + R_L) - R_{12}R_{21}} = \frac{E_g(R_{22} + R_L)}{\Delta} \qquad \text{11-10}$$

Similarly, substituting the value I_e to obtain the *collector output signal current* I_c gives

$$I_c = -\frac{E_gR_{21}}{(R_{11} + R_g)(R_{22} + R_L) - R_{12}R_{21}} = -\frac{E_gR_{21}}{\Delta} \qquad \text{11-11}$$

As is evident, Δ, called [3] the **circuit determinant,** is

$$\Delta = (R_{11} + R_g)(R_{22} + R_L) - R_{12}R_{21} \qquad \text{11-12}$$

The magnitude of the *signal-current amplification* is

$$\frac{I_{out}}{I_{in}} = \frac{I_c}{I_e} = \frac{\text{equation 11-11}}{\text{equation 11-10}} = \frac{R_{21}}{R_{22} + R_L} \qquad \text{11-13}$$

The input signal voltage impressed between the emitter and base is the generator open-circuit voltage E_g minus the voltage drop inside the generator, or

$$E_{in} = E_g - I_eR_g = E_g - \frac{E_g(R_{22} + R_L)R_g}{\Delta} = E_g\left(1 - \frac{R_{22}R_g + R_LR_g}{\Delta}\right) \qquad \text{11-14}$$

The output signal voltage equals I_cR_L, thus

$$E_{out} = -\frac{E_gR_{21}R_L}{\Delta} \qquad \text{11-15}$$

The magnitude of the *signal-voltage amplification* is

$$\frac{E_{out}}{E_{in}} = \frac{\text{equation 11-15}}{\text{equation 11-14}} = \frac{R_{21}R_L}{\Delta - R_g(R_{22} + R_L)} \qquad 11\text{-}16$$

The *insertion power gain* (section 8-3) is determined from the ratio of the power delivered to the load after the transistor is inserted, to the power delivered before it is inserted. The power delivered *after* it is inserted is

$$P_L = E_L I_L = E_{out} I_c = \text{equation 11-15} \times \text{equation 11-11} = \frac{E_g^2 R_{21}^2 R_L}{\Delta^2} \qquad 11\text{-}17$$

The power delivered by the signal source to the load resistor R_L *before* the transistor is inserted is

$$P_L' = I^2 R_L = \frac{E_g^2 R_L}{(R_g + R_L)^2} \qquad 11\text{-}18$$

$$\text{Insertion power gain ratio} = \frac{\text{equation 11-17}}{\text{equation 11-18}} = \frac{R_{21}^2 (R_g + R_L)^2}{\Delta^2} \qquad 11\text{-}19$$

This ratio often is expressed in decibels.

The *operating power gain* is the ratio of power output from the transistor to the available power from the signal generator. The available power is calculated on the basis of the internal and load impedances being conjugates.[5] Since this discussion assumes that resistances only are involved, the available power from the signal generator under the matched conditions of $R_g = R_L$ is

$$P_g = I_L^2 R_L = \left(\frac{E_g}{2R_g}\right)^2 R_g = \frac{E_g^2}{4R_g} \qquad 11\text{-}20$$

$$\text{Operating power gain ratio} = \frac{\text{equation 11-17}}{\text{equation 11-20}} = \frac{4R_g R_L R_{21}^2}{\Delta^2} \qquad 11\text{-}21$$

Numerical calculations now will be made, using the *point-contact transistor* having the values in section 10-8, a signal generator of $E_g = 0.1$ volt and $R_g = 500$ ohms resistance, and a load resistance $R_L = 34{,}000$ ohms. From equation 11-12,

$$\begin{aligned}\Delta &= (300 + 500)(34{,}000 + 34{,}000) - 100 \times 100{,}000 \\ &= 44.4 \times 10^6 \text{ ohms}^2\end{aligned}$$

From equation 11-10,

$$\begin{aligned}I_e &= [0.1\,(34{,}000 + 34{,}000)]/44.4 \times 10^6 \\ &= 0.153 \times 10^{-3} \text{ ampere}\end{aligned}$$

From equation 11-11,

$$I_c = -(0.1 \times 100{,}000)/44.4 \times 10^6$$
$$= -0.225 \times 10^{-3} \text{ ampere}$$

From equation 11-13,

$$\frac{I_c}{I_e} = \frac{100{,}000}{34{,}000 + 34{,}000} = 1.47, \text{ or } \frac{I_c}{I_e} = \frac{0.225}{0.153} = 1.47$$

From equation 11-14,

$$E_{in} = 0.1\left(1 - \frac{34{,}000 \times 500 + 34{,}000 \times 500}{44.4 \times 10^6}\right)$$
$$= 0.0234 \text{ volt}$$

From equation 11-15,

$$E_{out} = -(0.1 \times 100{,}000 \times 34{,}000)/44.4 \times 10^6$$
$$= -7.65 \text{ volts}$$

From equation 11-16

$$\frac{E_{out}}{E_{in}} = \frac{100{,}000 \times 34{,}000}{44.4 \times 10^6 - 500 \times 34{,}000 - 500 \times 34{,}000} = 327$$

or

$$\frac{E_{out}}{E_{in}} = \frac{7.65}{0.0234} = 327$$

and

$$20 \log_{10} 327 = 50.2 \text{ decibels}$$

$$\text{Signal power input to transistor} = E_{in} I_e = 0.0234 \times 0.153 \times 10^{-3}$$
$$= 0.358 \times 10^{-5} \text{ watt}$$

$$\text{Signal power output from transistor} = E_{out} I_c = 7.65 \times 0.225 \times 10^{-3}$$
$$= 1.72 \times 10^{-3} \text{ watt}$$

$$\text{Actual power gain ratio} = \frac{\text{power output}}{\text{power input}} = \frac{1.72 \times 10^{-3}}{0.358 \times 10^{-5}} = 480$$

and

$$10 \log_{10} 480 = 26.8 \text{ decibels}$$

From equation 11-19,

$$\text{Insertion gain ratio} = \frac{(100{,}000)^2(500 + 34{,}000)^2}{(44.4 \times 10^6)^2} = 6040$$

and

$$10 \log_{10} 6040 = 37.8 \text{ decibels}$$

From equation 11-21,

$$\text{Operating power gain ratio} = \frac{4 \times 500 \times 34{,}000 \times (100{,}000)^2}{(44.4 \times 10^6)^2} = 345$$

and

$$10 \log_{10} 345 = 25.4 \text{ decibels}$$

The preceding derivations and calculations for a point-contact transistor were based on the equivalent circuit of Figs. 11-1b and 11-2. The derivations and calculations now to be made will employ the equivalent circuit of Fig. 11-1c, reproduced for convenience in Fig. 11-3. Voltage equations for the two loops are

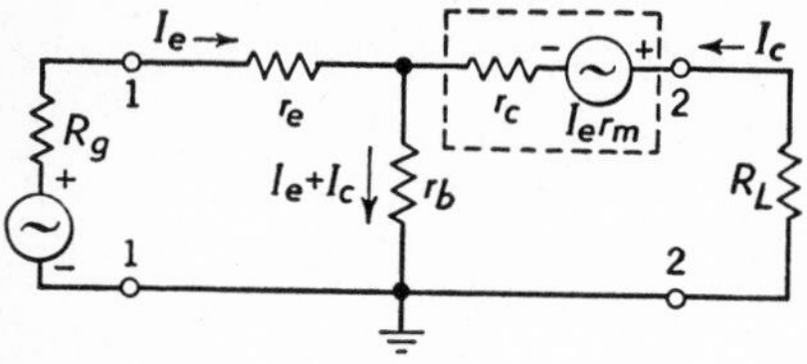

FIG. 11-3. One equivalent circuit of a grounded-base transistor amplifier.

$$I_eR_g + I_er_e + I_er_b + I_cr_b = E_g$$
$$I_er_b + I_cr_b + I_cr_c + r_mI_e + I_cR_L = 0$$

Solving for I_c in the second equation and substituting this expression for I_c in the first equation gives the emitter input signal current:

$$I_e = \frac{E_g(R_L + r_b + r_c)}{(R_g + r_b + r_e)(R_L + r_b + r_c) - r_b(r_b + r_m)} = \frac{E_g(R_L + r_b + r_c)}{\Delta} \qquad \text{11-22}$$

Similarly, substituting the value of I_e to obtain the collector output signal current I_c gives

$$I_c = \frac{-E_g(r_b + r_m)}{(R_g + r_b + r_e)(R_L + r_b + r_c) - r_b(r_b + r_m)} = -\frac{E_g(r_b + r_m)}{\Delta} \qquad \text{11-23}$$

The circuit determinant for Fig. 11-3 is

$$\Delta = (R_g + r_b + r_e)(R_L + r_b + r_c) - r_b(r_b + r_m) \qquad \text{11-24}$$

The magnitude of the *signal-current amplification* is

$$\frac{I_{out}}{I_{in}} = \frac{I_c}{I_e} = \frac{\text{equation 11-23}}{\text{equation 11-22}} = \frac{r_b + r_m}{R_L + r_b + r_c} \qquad \text{11-25}$$

The input signal voltage impressed between the emitter and the base is the generator open-circuit voltage E_g minus the voltage drop inside the generator, or

$$E_{in} = E_g - I_eR_g = E_g\left[1 - \frac{(R_L + r_b + r_c)R_g}{\Delta}\right] \qquad \text{11-26}$$

The output signal voltage equals the $-I_cR_L$ drop, thus

$$E_{out} = \frac{E_gR_L(r_b + r_m)}{\Delta} \qquad \text{11-27}$$

The magnitude of the *signal-voltage amplification* is

$$\frac{E_{out}}{E_{in}} = \frac{\text{equation 11-27}}{\text{equation 11-26}} = \frac{R_L(r_b + r_m)}{\Delta - R_g(R_L + r_b + r_c)} \qquad \text{11-28}$$

The *insertion power gain* is the ratio of the power P_L delivered to the load after the transistor is inserted, to the power P_L' delivered to the load before it is inserted:

$$P_L = E_L I_L = E_{out} I_c = \text{equation 11-27} \times \text{equation 11-23}$$

$$= \frac{R_L E_g^2 (r_b + r_m)^2}{\Delta^2} \qquad \text{11-29}$$

$$P_L' = I^2 R_L = \frac{E_g^2 R_L}{(R_g + R_L)^2} \qquad \text{11-30}$$

$$\text{Insertion power gain ratio} = \frac{\text{equation 11-29}}{\text{equation 11-30}} = \frac{(r_b + r_m)^2 (R_g + R_L)^2}{\Delta^2} \qquad \text{11-31}$$

The *operating power gain* is, as previously explained, the ratio of the output power from the transistor to the available power from the signal generator under conjugate impedance-match conditions. The power available from the signal generator is

$$P_g = I_L^2 R_L = \left(\frac{E_g}{2R_g}\right)^2 R_g = \frac{E_g^2}{4R_g} \qquad \text{11-32}$$

$$\text{Operating power gain ratio} = \frac{\text{equation 11-29}}{\text{equation 11-32}} = \frac{4R_g R_L (r_b + r_m)^2}{\Delta^2} \qquad \text{11-33}$$

Numerical calculations need not be made using the equations just derived and the data of section 10-8 for the same point-contact transistor previously considered. It can be seen from the preceding derivations that the relations summarized following equation 11-7 are valid, and that calculations using the equations just derived would give the same answers as the calculations previously made. In fact, had it been desired, all equations from 11-22 to 11-33 could have been obtained by substituting the equivalent values obtained from equations 11-3 to 11-6 in equations 11-10 to 11-21.

11-4. *GROUNDED-EMITTER TRANSISTOR AMPLIFIERS*

A simplified circuit for a point-contact transistor amplifier with the emitter grounded, and an equivalent circuit, are shown in Fig. 11-4. Using I_i for I_{in} and I_o for I_{out}, voltage equations are

$$I_i R_g + I_i r_b + (I_i + I_o) r_e = E_g \qquad \text{11-34}$$

$$(I_i + I_o) r_e + I_o r_c + I_e r_m + I_o R_L = 0 \qquad \text{11-35}$$

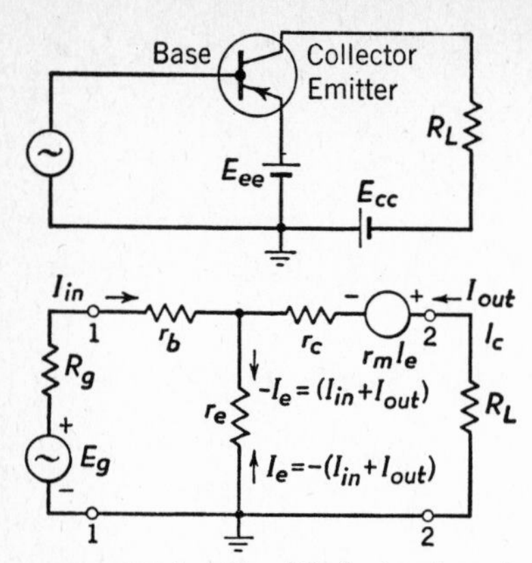

FIG. 11-4. Simplified circuit and equivalent circuit for a grounded-emitter transistor amplifier.

As will be noted, on the equivalent circuit, $I_e = -(I_{in} + I_{out})$. This is because I_e was taken *into* the emitter when voltage $I_e r_m$ was considered in the first equivalent circuits. Hence, in equation 11-35 this substitution for I_e must be made, and equation 11-35 becomes

$$(I_i + I_o)r_e + I_o r_c - (I_i + I_o)r_m + I_o R_L = 0 \qquad 11\text{-}36$$

If this equation is solved for I_o, and if this expression for I_o is substituted in equation 11-34, then

$$I_i = \frac{E_g(r_e + r_c - r_m + R_L)}{\Delta} \qquad 11\text{-}37$$

If equation 11-36 is solved for I_i and substitution made in equation 11-34,

$$I_o = \frac{E_g(r_m - r_e)}{\Delta} \qquad 11\text{-}38$$

where

$$\Delta = (R_g + r_b + r_e)(r_e + r_c - r_m + R_L) - r_e(r_e - r_m) \qquad 11\text{-}39$$

The magnitude of the *signal-current amplification* is

$$\frac{I_{out}}{I_{in}} = \frac{I_o}{I_i} = \frac{\text{equation 11-38}}{\text{equation 11-37}} = \frac{r_m - r_e}{r_e + r_c - r_m + R_L} \qquad 11\text{-}40$$

The signal voltage impressed between the base and the emitter equals $E_g - I_i R_g$, or

$$E_{in} = E_g - \frac{E_g(r_e + r_c - r_m + R_L)R_g}{\Delta}$$

$$= E_g\left[1 - \frac{R_g(r_e + r_c - r_m + R_L)}{\Delta}\right] \qquad 11\text{-}41$$

The signal-voltage output existing across R_L is $-I_o R_L$, or

$$E_{out} = -\frac{E_g(r_m - r_e)R_L}{\Delta} \qquad 11\text{-}42$$

The magnitude of the *signal-voltage amplification* is

$$\frac{E_{out}}{E_{in}} = \frac{\text{equation 11-42}}{\text{equation 11-41}} = \frac{R_L(r_m - r_e)}{\Delta - R_g(r_e + r_c - r_m + R_L)} \qquad 11\text{-}43$$

The signal power delivered to the load after the transistor is inserted between generator and load is

$$P_L = E_L I_L = E_{out} I_{out} = \text{equation 11-42} \times \text{equation 11-38}$$
$$= \frac{R_L E_g^2 (r_m - r_e)^2}{\Delta^2} \qquad \text{11-44}$$

The signal power delivered by the generator to the load before the transistor is inserted is given by equation 11-30.

$$\text{Insertion power gain ratio} = \frac{\text{equation 11-44}}{\text{equation 11-30}} = \frac{(r_m - r_e)^2 (R_g + R_L)^2}{\Delta^2} \qquad \text{11-45}$$

$$\text{Operating power gain ratio} = \frac{\text{equation 11-44}}{\text{equation 11-32}} = \frac{4R_g R_L (r_m - r_e)^2}{\Delta^2} \qquad \text{11-46}$$

11-5. *GROUNDED-COLLECTOR TRANSISTOR AMPLIFIERS*

A simplified circuit of a point-contact transistor with grounded collector and an equivalent circuit are shown in Fig. 11-5. Using I_i for I_{in} and I_o for I_{out}, the voltage equations are

$$I_i R_g + I_i r_b + (I_i + I_o) r_c - I_o r_m = E_g \qquad \text{11-47}$$

$$-I_o r_m + (I_i + I_o) r_c + I_o r_e + I_o R_L = 0 \qquad \text{11-48}$$

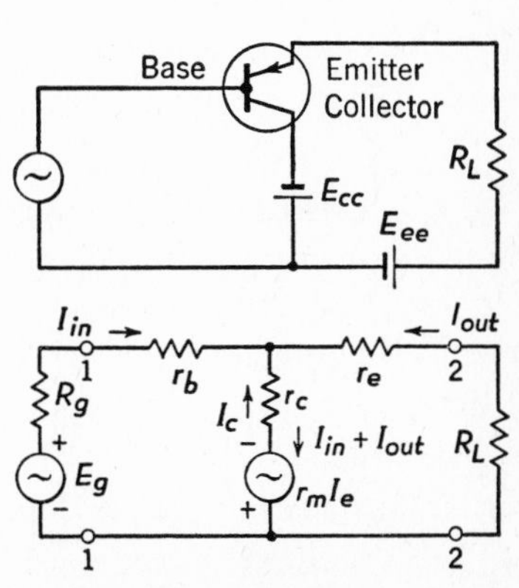

FIG. 11-5. Simplified circuit and equivalent circuit for a grounded-collector transistor amplifier.

Solving for I_o in the second equation and substituting in the first gives

$$I_i = \frac{E_g (R_L + r_e + r_c - r_m)}{\Delta} \qquad \text{11-49}$$

Solving for I_i in the second equation and substituting in the first gives

$$I_o = -\frac{E_g r_c}{\Delta} \qquad \text{11-50}$$

and

$$\Delta = (R_L + r_e + r_c - r_m)(R_g + r_b + r_c) - r_c (r_c - r_m) \qquad \text{11-51}$$

The magnitude of the *signal-current amplification* is

$$\frac{I_{out}}{I_{in}} = \frac{I_o}{I_i} = \frac{\text{equation 11-50}}{\text{equation 11-49}} = \frac{r_c}{R_L + r_e + r_c - r_m} \qquad \text{11-52}$$

The signal input voltage impressed between base and collector is

$$E_{in} = E_g - I_i R_g = E_g - \frac{R_g E_g (R_L + r_e + r_c - r_m)}{\Delta}$$

$$= E_g \left[1 - \frac{R_g (R_L + r_e + r_c - r_m)}{\Delta} \right] \qquad 11\text{-}53$$

The signal output voltage impressed across R_L is

$$E_{out} = -I_o R_L = \frac{E_g r_c R_L}{\Delta} \qquad 11\text{-}54$$

The magnitude of the *signal-voltage amplification* is

$$\frac{E_{out}}{E_{in}} = \frac{\text{equation 11-54}}{\text{equation 11-53}} = \frac{r_c R_L}{\Delta - R_g (R_L + r_e + r_c - r_m)} \qquad 11\text{-}55$$

Equations for power input, power output, insertion power gain ratio, and operating power gain ratio may be derived as in previous instances.

11-6. *AMPLIFIER STABILITY*

Because of feedback from output to input circuits, transistor amplifiers must be investigated to ascertain the conditions under which they are stable. If the output current has a finite value when the impressed generator voltage is zero, then a transistor amplifier will drive itself and produce sustained oscillations at a frequency determined by circuit conditions.

The first circuit to be considered is the grounded-base transistor of Fig. 11-3, for which from equation 11-23 the amplified signal output current is $I_c = -E_g(r_b + r_m)/\Delta$, from which $\Delta I_c = -E_g(r_b + r_m)$. For operation as an oscillator no external signal is impressed, and E_g becomes zero. The circuit will *not* oscillate if Δ is greater than zero. The circuit determinant Δ for the grounded-base transistor is, from equation 11-24,

$$\Delta = (R_L + r_c + r_b)(R_g + r_e + r_b) - r_b(r_b + r_m)$$

This may be written

$$\Delta = \frac{R_g + r_e}{r_b} + \frac{R_g + r_e}{R_L + r_c} + 1 - \frac{r_m}{R_L + r_c} \qquad 11\text{-}56$$

For stability, Δ must be greater than zero; hence in the above equation the sum of the positive terms must exceed the negative term. Thus, for stability,

$$\frac{R_g + r_e}{r_b} + \frac{R_g + r_e}{R_L + r_c} + 1 > \frac{r_m}{R_L + r_c} \qquad 11\text{-}57$$

Sometimes this is written [3]

$$\frac{R_E}{r_b} + \frac{R_E}{R_c} + 1 > \frac{r_m}{R_c} \tag{11-58}$$

where $R_E = R_g + r_e$, and $R_c = R_L + r_c$. In the grounded-base transistor, r_b acts to cause feedback, because the output current flows through it. (See also Fig. 11-2, and remember that $r_b = R_{12}$.) Thus a low value of r_b, and of its equivalent R_{12}, is desirable. From the stability standpoint, a low value of r_m also is advantageous.

The second circuit to be considered is the grounded-emitter circuit of Fig. 11-4. Here again, if an output signal current I_o exists when the applied signal voltage E_g equals zero, the circuit may oscillate. The current as given by equation 11-38 is $I_o = E_g(r_m - r_e)/\Delta$. The conditions for no oscillations, or stability, is $\Delta > 0$. From equation 11-39,

$$(R_g + r_b + r_e)(R_L + r_c + r_e - r_m) - r_e(r_e - r_m)$$

This may be written in the form

$$\frac{r_e}{R_L + r_c} + \frac{r_e}{R_g + r_b} + 1 > \frac{r_m}{R_L + r_c} \tag{11-59}$$

The third circuit to be examined for stability is the grounded-collector transistor of Fig. 11-5. The condition for stability is the same as for the preceding circuits, namely that the circuit determinant, equation 11-51, exceeds zero. Following previous methods, the condition for stability is

$$\frac{R_L + r_e}{r_c} + \frac{R_L + r_e}{R_g + r_b} + 1 > \frac{r_m}{r_c} \tag{11-60}$$

11-7. *COMPARISONS OF TRANSISTOR AMPLIFIERS*

Transistor amplifiers may be compared in several ways. In the following paragraphs they will be considered on the basis of input impedance, output impedance, and phase shift.

Grounded-Base Transistor Amplifier. The equivalent circuit for the grounded base transistor amplifier is shown in Fig. 11-3. The input impedance at terminals 1–1 and at low frequencies is

$$R_i = \frac{\text{equation 11-26}}{\text{equation 11-22}} = r_e + r_b - \frac{r_b(r_b + r_m)}{R_L + r_b + r_c} \tag{11-61}$$

The output impedance, which is the internal impedance of the circuit of Fig. 11-3 looking back into the circuit from terminals 2–2, is found as follows: The open-circuit voltage $E_{oc} = I_e'r_m + I_e'r_b$, where I_e' is the current flowing through r_b when terminals 2–2 are open. This current is $I_e' = E_g/(R_g + r_e + r_b)$. Hence, the open-circuit voltage is

$$E_{oc} = \frac{E_g(r_m + r_b)}{R_g + r_e + r_b} \qquad \text{11-62}$$

This is the equivalent internal voltage when Fig. 11-3 is regarded as a generator of open-circuit voltage E_{oc} at terminals 2–2 and internal impedance (or output impedance) R_o. When the circuit of Fig. 11-3 is connected to load R_L, it can be written that $I_c = E_{oc}/(R_o + R_L)$ and that $E_{oc}/I_c = R_o + R_L$. Thus, equation 11-62 divided by equation 11-23 gives $R_o + R_L$, and

$$R_o = \frac{E_{oc}}{I_c} - R_L = \frac{\text{equation 11-62}}{\text{equation 11-23}} - R_L$$

$$= r_c + r_b - \frac{r_b(r_b + r_m)}{R_g + r_b + r_e} \qquad \text{11-63}$$

The phase relation between the signal input and the amplified signal output of a grounded-base point-contact transistor may be determined from Fig. 11-2. The positive half-cycle of the signal voltage applied between the emitter and base will cause an increase in the current *into* the emitter. The holes flowing to the region of the collector will increase in number, resulting in an increase in current *out* of the collector. This results in a rise in voltage across load resistor R_L. Thus, the input signal voltage and current are *in phase* at low frequencies where reactances and transit times are negligible.

An examination of equations 11-62 and 11-63, or a computation using numerical values, will show that the grounded-base point-contact transistor amplifier has low input impedance, high output impedance, and at low frequencies no phase shift. For these reasons, this amplifier is said [3] to be analogous to the grounded-grid vacuum-tube amplifier of section 8-21.

Grounded-Emitter Transistor Amplifier. The equivalent circuit is shown in Fig. 11-4. The input impedance is

$$R_i = \frac{\text{equation 11-41}}{\text{equation 11-37}} = r_b + r_e - \frac{r_e(r_e - r_m)}{R_L + r_e + r_c - r_m} \qquad \text{11-64}$$

The output impedance is found as for the grounded-base amplifier. The open-circuit voltage at terminals 2–2 will be $E_{oc} = -I_{in}'r_m + I_{in}'r_e$, where I_{in}' is the current flowing through r_e when terminals 2–2 of Fig. 11-4 are open. The voltage drop $-I_{in}'r_m$ is negative because the emitter current is flowing in the wrong direction. This current is $I_{in}' = E_g/(R_g + r_b + r_e)$. Hence, the open-circuit voltage is

$$E_{oc} = \frac{E_g(r_m + r_e)}{R_g + r_b + r_e} \qquad \text{11-65}$$

When the circuit of Fig. 11-4 is connected to load resistor R_L, $I_{out} = E_{oc}/(R_o + R_L)$ and

$$R_o = \frac{E_{oc}}{I_{out}} = \frac{\text{equation 11-65}}{\text{equation 11-38}} - R_L = r_c + r_e - r_m - \frac{r_e(r_e - r_m)}{R_g + r_b + r_e} \qquad 11\text{-}66$$

The phase relations between input and output signals may be determined from Fig. 11-4 as follows: If the negative half-cycle of signal voltage makes the *upper* end of the signal generator more negative, the current *into* the emitter will increase, causing an increase in current out of the collector, and an increase in voltage with respect to ground of the *upper* end of R_L. This action causes a phase shift of 180 degrees. Thus, the grounded-emitter point-contact transistor has relatively high input impedance, low output impedance, a phase shift of 180°, and is analogous to the familiar grounded-cathode vacuum-tube amplifier.

Grounded-Collector Transistor Amplifier. This is shown in Fig. 11-5. The input impedance is

$$R_i = \frac{\text{equation 11-53}}{\text{equation 11-49}} = r_b + r_c - \frac{r_c(r_c - r_m)}{R_L + r_e + r_c - r_m} \qquad 11\text{-}67$$

Following in general the preceding methods of determining the output impedance, the open-circuit voltage at terminals 2–2 will be $E_{oc} = I_{in}'r_c$, there being no current I_c for the open-circuited condition. The current input for this condition is $I_{in}' = E_g/(R_g + r_b + r_c)$. Hence,

$$E_{oc} = \frac{E_g r_c}{R_g + r_b + r_c} \qquad 11\text{-}68$$

$$R_o = \frac{E_{oc}}{I_{out}} = \frac{\text{equation 11-68}}{\text{equation 11-50}} = r_e + r_c - r_m - \frac{r_c(r_c - r_m)}{R_g + r_b + r_c} \qquad 11\text{-}69$$

The phase relations are as follows: In Fig. 11-5, if the *upper* signal generator terminal is made more *negative* with respect to ground, there will be an increase of emitter current *into* the base, and more current will flow *out* of the collector and up into R_L. Hence, the *upper* end of R_L will be made more negative by the drop in the resistor, and there is no phase shift at low frequencies.

The grounded-collector point-contact transistor amplifier has high input impedance, low output impedance, and no shift in phase. Accordingly, it is analogous to the "cathode-follower," or grounded-plate, vacuum-tube circuit (section 8-21).

A consideration of Fig. 11-5 will indicate that this circuit is bilateral; that is, a signal will flow through it, and it will amplify, in either direction. Hence, a *single* point-contact transistor will accomplish amplification in both directions, a feat not possible with a single-element vacuum tube in a circuit of corresponding simplicity.

For reasons explained in this section, a comparison between a point-contact transistor and a vacuum tube is as follows: The emitter is analogous to the cathode; the base is analogous to the grid; and the collector is analogous to the plate. [3]

11-8. *COMPARISON OF TRANSISTORS AND VACUUM TUBES*

In Fig. 11-6 are shown characteristic curves of a triode vacuum tube and of a point-contact transistor. Suppose that the grid bias on the tube is −10 volts, that the direct voltage from plate to cathode is +100 volts, and that the

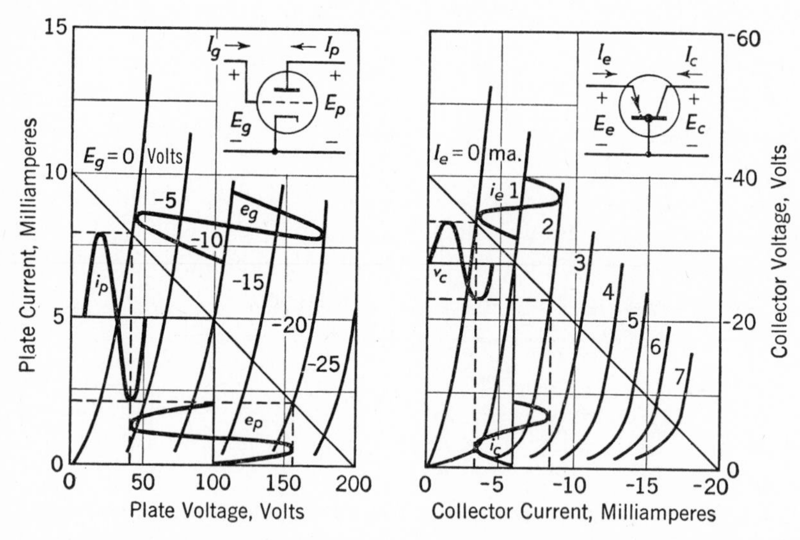

FIG. 11-6. At the left are the characteristics of a vacuum tube being operated as a class-A amplifier with a direct grid bias of −10 volts, with a plate voltage of +100 volts, with plate load resistance of 20,000 ohms, and with a grid-voltage signal swing e_g of 10 volts maximum; the resulting plate-current variation is i_p, and plate-voltage variation is e_p. At the right are the characteristics of a point-contact transistor being operated with a direct emitter bias of 1.0 milliampere, a direct collector current of −6 milliamperes, a collector load resistance of 2000 ohms, and an emitter-current signal swing i_e of 1.0 milliampere maximum; the resulting collector-current variation is i_c, and collector-voltage variation is v_c.

load resistance is 20,000 ohms, giving the load line (section 8-4) as drawn. If the signal voltage from grid to cathode causes the grid to vary from zero to −20 volts, the plate current will vary between approximately 7.7 and 2.3 milliamperes as indicated.

Now consider the transistor to be biased with an emitter current of +1.0 milliampere, to have a direct collector current of −6.0 milliamperes, and to have a load resistance of 2000 ohms. To find the load line, −6 milliamperes × 2000 ohms = −12 volts. A line (not shown in Fig. 11-6) is then drawn from −6 milliamperes on the *X* axis to −12 volts on the *Y* axis. The load line shown in Fig. 11-6 is then drawn through the point of operation, and parallel to the line (not shown) previously drawn from −6 milliamperes to −12 volts. If the signal current from the signal generator causes the emitter current to vary between zero and 2.0 milliamperes, the voltage between collector and base will vary between about −34 and −24 volts. It is apparent that signal-voltage input for vacuum tubes corresponds to signal-current input for transistors, and that signal-current output for tubes corresponds to signal-voltage output for transistors.

As a further comparison, the grid of a grounded-cathode vacuum-tube amplifier normally is biased in the *reverse* direction with a *negative voltage*, and the grid-current flow is negligible. For the grounded-base transistor amplifier, however, the emitter is biased in the *forward* direction with a *positive voltage*, and the emitter-current flow is considerable. The grid input impedance of a vacuum tube is *high* when operated as just explained, and the emitter input impedance of a transistor is low. It is evident in Fig. 11-6 that plate-current cutoff for a tube corresponds to collector-voltage cutoff for a transistor.

The amplification factor μ of a triode vacuum tube is the ratio of a small change in plate voltage to a small change in grid voltage under the condition that the plate-to-cathode current remains constant (section 4-4). The current multiplication factor α of a transistor is the ratio of a small change in collector current to a change in emitter current under the condition that the collector-to-base voltage remains constant (section 10-8).

A comparison of the power-supply sources is of interest. A vacuum tube requires constant sources of direct grid voltage and direct plate voltage. A constant voltage source must have zero internal impedance, or be provided with a voltage regulator. A transistor requires constant sources of direct current for the emitter and collector. A constant current source must have infinite internal impedance, or be provided with a current regulator.

Because of the comparisons just enumerated, a transistor often is considered to be the **dual** of a vacuum tube, a term used in network theory.[6] For instance, the dual of resistance is conductance (and vice versa); inductance is the dual of capacitance, etc. One element is the dual of another when the role of current in one is performed by voltage in the other, and vice versa.[7,8] The principle of duality is useful in transistor circuit design.

11-9. *TRANSISTOR LARGE-SIGNAL AMPLIFIERS*

In the transistor amplifiers previously considered it was assumed that the input and output signals were so small that the transistor parameters remained constant, and that no distortion occurred. In the transistor amplifiers now to be considered, it is assumed that the input and output signals are large, that the transistor parameters do not remain constant, and that distortion does occur. Only grounded-base amplifiers will be treated, and as in the preceding discussions, the explanations will be for the point-contact type having the characteristics given in Fig. 10-11.

The circuit under consideration is shown in Fig. 11-7a. Direct current for the emitter is supplied by battery E_{ee} and resistor R_e. The resistance of R_e is assumed so large that the emitter direct current is essentially $I_e = E_{ee}/R_e$. Similarly, the collector direct current is essentially $I_c = E_{cc}/R_c$. Emitter and collector direct currents remain fixed. The two blocking capacitors are assumed to have negligible capacitive reactances. From an alternating-current standpoint, the circuit is as in Fig. 11-7b.

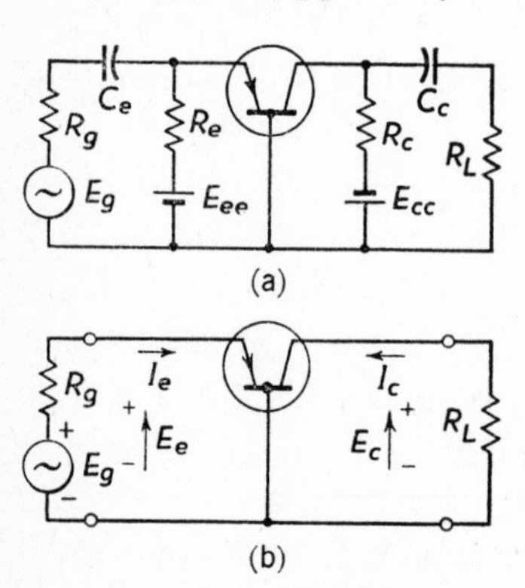

FIG. 11-7. Simplified circuit and equivalent circuit for a large-signal point-contact transistor amplifier.

When large signals are amplified by this transistor, distortion may occur either in the emitter circuit or in the collector circuit. Furthermore, because of feedback, the load in the collector circuit affects the operation of the emitter circuit. A transistor large-signal amplifier may be likened to a vacuum-tube amplifier with positive grid bias, and driven by an alternating signal voltage having a peak signal voltage less than the positive bias voltage. The grid will draw current, and the signal voltage appearing from grid to cathode will be the generator voltage minus the drop inside the generator. Since this current will, probably, be distorted, the voltage between grid and cathode will *not* be a replica of the signal voltage. Hence, distortion may occur in both the grid and plate circuits of the tube amplifier just as it may occur in both the emitter and collector circuits of a transistor amplifier.

A method of predicting the performance of a large-signal transistor amplifier is as follows.* A point-contact transistor having the collector charac-

* This method was communicated (in a memorandum prepared in September 1952) to the writer by his friend and former colleague, W. S. Burdic. See also L. P. Hunter, "Graphical Analysis of Transistor Characteristics." Proc. I.R.E., Dec. 1950, Vol. 38, No. 12.

teristics of Fig. 11-8a is to be used, with the direct emitter current held constant at 1.0 milliampere and the direct collector current held at 4.0 milliamperes. This gives the operating point P in Fig. 11-8a. The load resistance to be used is 10,000 ohms, giving the load line shown. This is a line passing through operating point P, and *parallel to* the dotted line having intercepts of -4 milliamperes on the X axis, and $-4 \times 10{,}000 = -40$ volts on the Y axis.

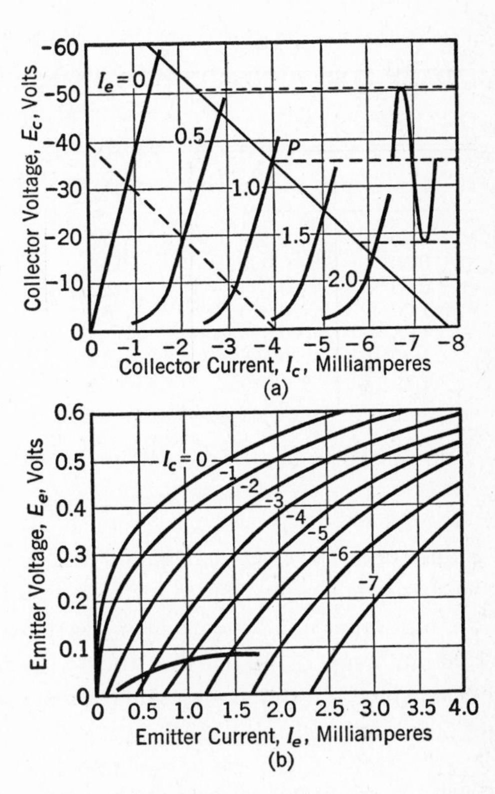

FIG. 11-8. Curves for a point-contact transistor used as a large-signal amplifier. Numbers on curves are milliamperes. (Numerical values given in text were obtained from the original, and larger, transistor characteristic curves, and are more accurate than values obtained from these curves.)

The next step is to transfer the effect of the collector 10,000-ohm load to the emitter-circuit characteristic curves. This is done as follows: In Fig. 11-8a, when the emitter current is 0.25 milliampere, the collector current (with a load resistance of 10,000 ohms) is -2.2 milliamperes as read on the X axis. On Fig. 11-8b a point is plotted at 0.25 milliampere emitter current on the (estimated, not shown) -2.2-milliampere collector-current curve. Again from the curves of Fig. 11-8a, when the emitter current is 0.5 milliampere, the corresponding value on the X axis is -2.9 milliamperes, and this value is plotted on Fig. 11-8b, after estimating the location of the -2.9-milliampere collector-current curve. At an emitter current of 0.75 milliampere (curve estimated), the collector current is -3.5 milliamperes, and this point is plotted on Fig. 11-8b. At an emitter current of 1.25 milliamperes, the corresponding collector current is -4.5 milliamperes, which is plotted on Fig. 11-8b. At 1.5- and 1.75-milliampere emitter currents, the corresponding collector currents are -5.1 and -5.6 milliamperes, and these also are plotted, giving the curve in the lower left of Fig. 11-8b.

The curve just obtained is used to find the relation between emitter current and emitter voltage when there is a resistor of 10,000 ohms in the collector circuit, and when the direct emitter and collector currents are 1.0 and 4.0 milliamperes, respectively. When $I_e = 0.25$ milliampere, $E_e = 0.01$ volt; when $I_e = 0.5$, $E_e = 0.03$; when $I_e = 0.75$, $E_e = 0.055$; etc. These and succeeding values are listed in Table 9.

Table 9

DATA FOR POINT-CONTACT TRANSISTOR LARGE-SIGNAL AMPLIFIER

(*Current values in milliamperes, voltages in volts*)

I_e	E_e	ΔI_e	$\Delta I_e R_g$	ΔE_e	e_g
0.25	0.010	−0.75	−0.375	−0.060	−0.435
0.50	0.030	−0.50	−0.250	−0.040	−0.290
0.75	0.055	−0.25	−0.125	−0.015	−0.140
1.00	0.070	0	0	0	0
1.25	0.075	+0.25	+0.125	+0.005	+0.130
1.50	0.080	+0.50	+0.250	+0.010	+0.260
1.75	0.085	+0.75	+0.375	+0.015	+0.390

From Fig. 11-8b, the instantaneous signal-generator open-circuit voltage is

$$e_g = \Delta I_e R_g + \Delta E_e \qquad 11\text{-}70$$

Values of ΔI_e next are calculated and recorded in Table 9. This is done on the basis that when the transistor is in operation as a large-signal amplifier, the maximum emitter-current swing that is permitted is a peak current of 0.75 milliampere about the direct-current value of $I_e = 1.0$ milliampere. Then, since the signal generator is assumed to have an internal impedance of 500 ohms resistance, $\Delta I_e R_g$ is calculated and recorded in the table.

Values of ΔE_e for use in equation 11-70 then are calculated and recorded in Table 9. In Fig. 11-8b, when the point of operation P is $I_e = 1.0$ milliampere, $I_c = -4.0$ milliamperes, and the load in the collector circuit is 10,000 ohms, the value of the emitter-to-base voltage is $E_e = 0.07$ volt. When ΔI_e is -0.75 milliampere, from Table 9 $E_e = 0.01$ volt, and $\Delta E_e = 0.07 - 0.01 = 0.06$, or -0.06 volt with respect to 0.07 volt as the no-signal reference value. Also from Table 9, when ΔI_e is -0.5 milliampere, $E_e = 0.03$ volt and $\Delta E_e = -0.04$ volt, and when $\Delta I_e = 0.25$ milliampere, $\Delta E_e = -0.015$ volt, etc. Values of ΔE_e added to $\Delta I_e R_g$ give e_g (of Table 9), the instantaneous values of signal generator closed-circuit output voltage.

The computations and curves thus made indicate that if the emitter current *is* sinusoidal, the voltage between emitter and base, and also the open-circuit generator voltage, will *not* be sinusoidal. It is common, however, to use for test purposes signal generators that *do* produce sinusoidal open-circuit voltages. Thus the curve of Fig. 11-9 is plotted, and from this the shape of the emitter-current wave can be determined for the condition of sinusoidal open-circuit generator voltage under study.

If the effective value of the signal generator open-circuit voltage is 0.30 volt, the peak value will be 0.424 volt. If this sinusoidal signal voltage is applied to Fig. 11-9, the resulting instantaneous emitter-current variations ΔI_e are obtained. Thus in Fig. 11-9, ΔI_e varies from $+0.82$ milliampere to -0.73 milliampere with respect to the direct current of $I_e = 1.0$ milliampere. If these variations are applied to Fig. 11-8a, the corresponding voltage changes will be from -18 to -51 volts as shown producing an output voltage wave of 16 volts peak value for one half cycle and 15 volts peak for the other half. This is the output voltage across the 10-000-ohm load resistor for the point-contact transistor operated as has been prescribed. Because the two half-cycles of the output voltage wave are not identical, the transistor causes some distortion. Reference to the discussion in section 8-4 will show that this is largely second-harmonic distortion. The power output and percentage distortion can be found by equations similar to those in section 8-4.

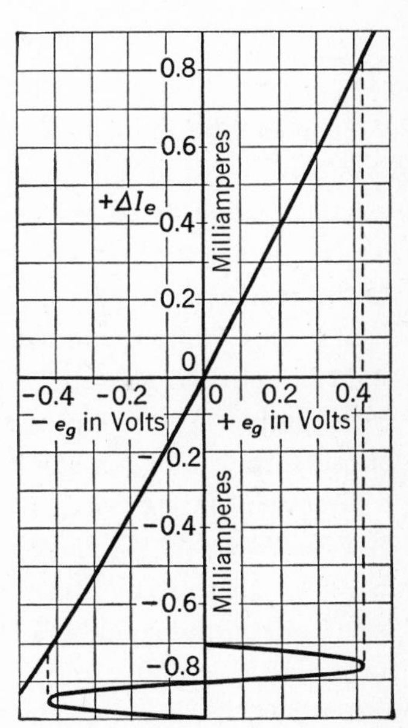

FIG. 11-9. For determining the emitter current that flows when a sinusoidal open-circuit voltage is impressed on the point-contact transistor having the characteristics of Fig. 11-8.

11-10. *JUNCTION TRANSISTOR AMPLIFIERS* [9,10,11]

The preceding pages have been devoted to the basic theory of transistor amplifiers, and in general the discussions apply to both the point-contact and the junction types. Numerical calculations were made for only the point-contact type. In this section the emphasis will be on the junction type.

The electronic principles of junction-type transistors were considered in the preceding chapter. Both *n-p-n* and *p-n-p* types, as well as others of special nature, were discussed. In this section only the *n-p-n* junction transistor will be treated. A close analogy exists between junction-type transistors and vacuum tubes.[12]

The characteristics of a typical *n-p-n* junction transistor are shown in Fig. 10-19. Typical values for this type of transistor are given in section 10-16. From these,[10] and the relations in section 11-1, $r_e = 26$ ohms, $r_b = 240$ ohms.

$r_m = 13.1$ megohms, $r_c = 13.4$ megohms, and $\alpha = 0.9785$. Variations in these parameters are shown in Fig. 11-10.

The equations previously derived in this chapter for grounded base, grounded emitter, and grounded collector apply to the junction transistor. Because of the great differences among the numerical values, certain values may sometimes be neglected, leading to some simplification. The characteristics of the several types of junction transistor amplifiers will now be briefly summarized for *small-signal operation.*[10]

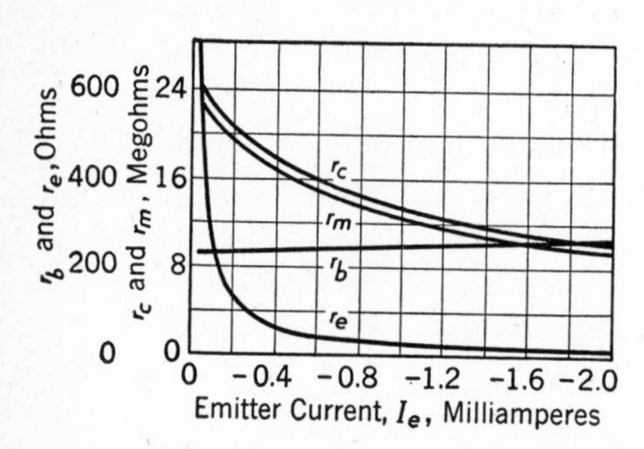

FIG. 11-10. Variations in the parameters of an *n-p-n* junction transistor. (Data from reference 10.)

Grounded-Base n-p-n Junction Transistor Amplifier.[4,10] A practical circuit is shown in Fig. 11-11. Note that the arrowhead on the emitter is reversed in direction as compared with point-contact transistors. This is the convention for the *n-p-n* junction transistor, but for a *p-n-p* junction transistor the arrowhead is in the same direction as for the point-contact type. Note also that the emitter battery and collector battery are reversed as compared to the point-contact type. If the emitter-battery voltage $E_{ee} = -1.5$ volts, and $R_e = 1500$ ohms, then $I_e = 1.0$ milliampere, and will be constant, the combination acting as a constant direct-current source because from Fig. 10-19 the emitter voltage drop is negligible. The value of the collector-battery voltage E_{cc} will depend on output circuit. For transformer coupling, a voltage of $E_{cc} = 4.5$ volts will give a direct collector current of about 1.0 milliampere. The equivalent circuits of Figs. 11-1 to 11-4 and the equations accompanying these circuits apply. There is no shift in phase between input and output signals. If the transformer ratio is unity, and if a resistance of several hundred thousand ohms is connected to the transformer, the voltage amplification is very high. The transformer may be employed in the usual way for voltage changing and impedance matching.

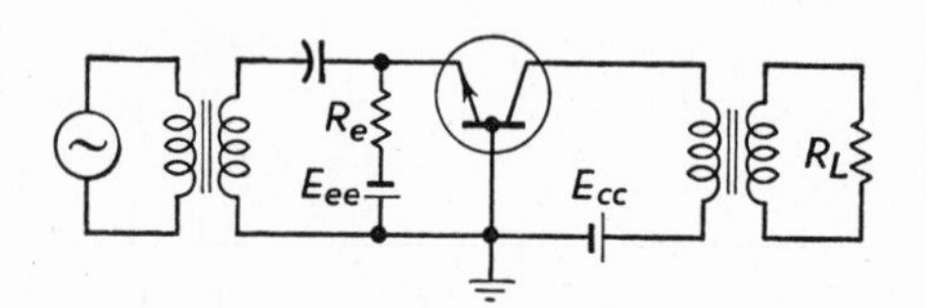

FIG. 11-11. Circuit for a grounded-base *n-p-n* junction transistor amplifier.

Grounded-Emitter n-p-n Junction Transistor Amplifier.[4,10] The input impedance is higher and the output impedance is lower if a *n-p-n* junction transistor is operated with the emitter instead of the base grounded. Also, there is a 180° shift in phase. Thus the grounded-emitter amplifier is similar to a grounded-cathode vacuum-tube amplifier. The equations previously derived (section 11-4) apply. Many circuits [4,10] may be used with the grounded-emitter *n-p-n* junction transistor, one being shown in Fig. 11-12. Grounded-emitter operation is the most promising method for many purposes.

Grounded-Collector n-p-n Junction Transistor Amplifier.[4,10] The *n-p-n* junction transistor operated with grounded collector has relatively high input impedance, low output impedance, and low gain. There is no phase shift, and the circuit is similar to the vacuum-tube cathode-follower. The equations for the point-contact transistor with grounded emitter apply. One circuit arrangement is shown in Fig. 11-13. If the resistor across the capacitor in the base circuit is removed, and the base allowed to float, the input impedance will be increased. Sometimes, the resistor shown is connected from base to the negative battery terminals,[10] giving a lower collector current than in Fig. 11-13.

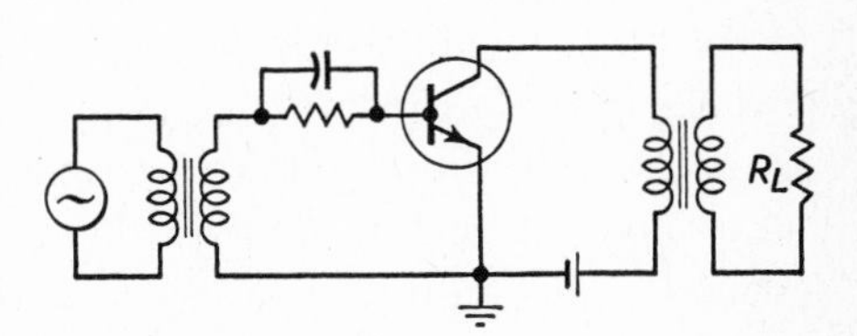

FIG. 11-12. Circuit for a grounded-emitter *n-p-n* junction transistor amplifier.

Large-signal Amplification with n-p-n Junction Transistors. The method applied to the point-contact transistor operated with large signals also may be applied to the junction transistor. However, r_e of the equivalent circuit is low (26 ohms, as previously considered), and if it is assumed that the junction transistor is driven by a *constant-current* signal source, power-output calculations may be simplified as follows: [10]

The curves for the collector circuit of an *n-p-n* junction transistor are shown in Fig. 10-19. Suppose that the emitter direct current is $I_e = -2$ milliamperes, that direct voltage on the collector is $E_c = 20$ volts, that the collector direct current is $I_c = 1.92$ milliamperes, and that the load resistance in the collector circuit is 10,000 ohms. The direct-current power input to the collector circuit is 39.4 milliwatts. If the signal to be amplified varies the emitter current between -0.1 and -3.9 milliamperes, from Fig. 10-19 the collector current will vary between approximately 0.15 and 3.65 milliamperes, and the collector voltage will vary between approximately 38 and 2 volts. From equation 8-5, the amplified signal power output is

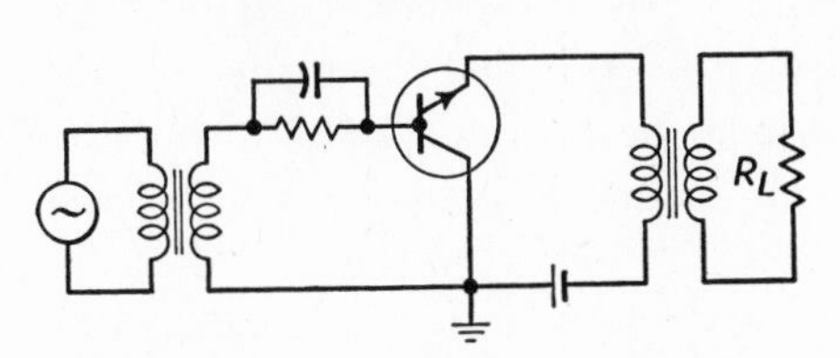

FIG. 11-13. Circuit for a grounded-collector *n-p-n* junction transistor amplifier.

$$\text{Power output} = \frac{(I_{\text{max.}} - I_{\text{min.}})(E_{\text{max.}} - E_{\text{min.}})}{8} = \frac{(3.65 - 0.15)(38 - 2)}{8}$$

$$= 15.75 \text{ milliwatts}$$

The efficiency is 15.75/39.4 = 40.0 per cent, and higher efficiencies are possible.[10]

Characteristics of Junction-Type Transistor Amplifiers. The power output of the *n-p-n* junction-type transistor just considered is low. However, as explained in the preceding chapter, junction-type triode transistors with relatively large power-handling capacities are available and others will be developed. At present (1955) the following statements can be made: In general, junction-type transistor *triodes* are not as satisfactory at high-frequencies as are the point-contact type; one reason is the effect of transit time and dispersion in the junction; another is that electrode capacitive effects become of importance.[4,10] In general, however, junction-type transistors produce less noise, operate at lower direct voltages, and have greater power-handling capacity (50 watts has been reported unofficially) and efficiency than contact types. Although this discussion has been limited to the *n-p-n* transistor, the *p-n-p* type also is used. Operation of transistor amplifiers in class-A, class-B, and class-C is possible, and tuned amplifiers are used.[4] The transistor amplifiers that have been described are for audio-frequencies and the lower radio-frequencies. Junction transistor tetrodes, described in section 10-18 are used in amplifiers and oscillators up to about 100 megacycles.[13]

11-11. *TRANSISTOR OSCILLATORS* [4,14]

As explained in Chapter 9 the two basic types of oscillators are negative-resistance oscillators and feedback oscillators. As will be explained at the close of Chapter 14, certain transistors may be operated so as to be two-terminal negative-resistance devices. Only feedback oscillators will be considered in this section.

A *vacuum-tube feedback* oscillator is essentially a tuned amplifier with positive voltage feedback, and with a nonlinear characteristic in the circuit to limit the magnitude of oscillations. A *transistor feedback oscillator* differs in this important detail: positive *current* feedback must be provided. Whereas a vacuum tube amplifies signal voltage that is impressed between control grid and cathode, a transistor amplifies when a signal current is injected into the emitter. Many circuit arrangements may be used for transistor oscillators.[4,14,15]

A grounded-base transistor oscillator circuit is shown in Fig. 11-14. The electrode potentials are for operation as a nondistorting amplifier. Oscillations increase until the emitter is driven negative, or until the collector is driven positive. Distortion may be reduced by connecting a resistor in series with the collector, and another from emitter to base.[4] Transistor oscillators may be crystal-controlled.[4]

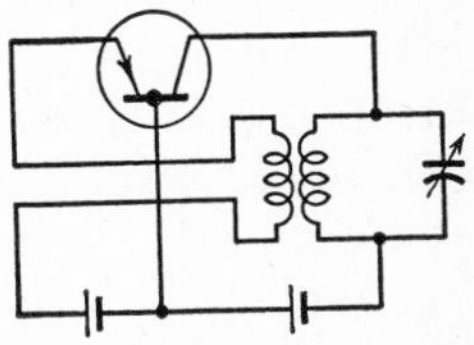

FIG. 11-14. A simple transistor oscillator. Battery polarities are for a point-contact transistor, or for a *p-n-p* junction type.

11-12. *HYBRID PARAMETERS* [19]

In the preceding pages transistor calculations have been made using *resistance* parameters such as R_{11}, r_e, etc. It is possible to determine *conductance* parameters such as g_{11}, g_{22}, etc., and to make calculations using equations derived to use these parameters. Another system has been developed; this system employs hybrid parameters, so called because the units of measure are not the same. Referring to the generalized "black box" of Fig. 11-1 with input terminals 1-1 and output terminals 2-2, the parameters are as follows: If a generator is connected at terminals 1-1, if terminals 2-2 are short-circuited, if E_{11} is a small-signal voltage impressed across the transistor, and if I_{11} is the current into the transistor, then $h_{11} = E_{11}/I_{11}$ (ohms). If I_{22}, the current in the short-circuited output, is measured, then $h_{21} = I_{22}/I_{11}$ (no unit of measure). If a small-signal voltage is applied at output terminals 2-2, if terminals 1-1 are open, then $h_{22} = I_{22}/E_{22}$ (mho). If the voltage at the input terminals 1-1 is measured under these same conditions, then $h_{12} = E_{11}/E_{22}$ (no unit of measure). Because these parameters are measured in ohms, mhos, or are dimensionless, the term hybrid parameters is used. Small-signal values must be used to determine the parameters at a *point* of operation.

REFERENCES

1. Peterson, L. C. *Equivalent circuits of linear active four-terminal networks.* Bell System Technical Journal, Oct. 1948, Vol. 27, No. 4.
2. Shockley, W. *Electrons and Holes in Semiconductors.* D. Van Nostrand Co.
3. Ryder, R. M., and Kircher, R. J. *Some circuit aspects of the transistor.* Bell System Technical Journal, July 1949, Vol. 28, No. 3.
4. Shea, R. F. *Transistor Circuits.* John Wiley & Sons.
5. Institute of Radio Engineers. *Standards on Electron Tubes: Definitions of Terms,* 1950. Proc. I.R.E., April 1950, Vol. 38, No. 4.
6. Guillemin, E. A. *Introductory Circuit Theory.* John Wiley & Sons.
7. Wallace, R. L., and Raisbeck, G. *Duality as a guide in transistor circuit design.* Bell System Technical Journal, April 1951, Vol. 30, No. 2.

8. Raisbeck, G. *Transistor circuit design.* Electronics, Dec. 1951, Vol. 24, No. 12.
9. Shockley, W., Sparks, M., and Teal, G. K. *p-n junction transistors.* Physical Review, July 1951, Vol. 83.
10. Wallace, R. L., and Pietenpol, W. J. *Some circuit properties and applications of n-p-n transistors.* Bell System Technical Journal, July 1951, Vol. 30, No. 3; also, Proc. I.R.E., July 1951, Vol. 39, No. 7.
11. Smith, K. D. *Properties of junction transistors.* Tele-tech, Jan. 1953, Vol. 12, No. 1.
12. Giacoletto, L. J. *Junction transistor equivalent circuits and vacuum-tube analogy.* Proc. I.R.E., Nov. 1952, Vol. 40, No. 11.
13. Wallace, R. L., and Schimpf, L. G. *A junction transistor tetrode for high-frequency use.* Proc. I.R.E., Nov. 1952, Vol. 40, No. 11.
14. Oser, E. A., Enders, R. O., and Moore, R. P. *Transistor oscillators.* R.C.A. Review, Sept. 1952, Vol. 13, No. 3.
15. Thomas, D. E. *Low drain transistor audio oscillators.* Proc. I.R.E., Nov. 1952 Vol. 40, No. 11.
16. Giacoletto, L. J. *Terminology and equations for linear active four-terminal networks including transistors.* R.C.A. Review, March 1953, Vol. 14, No. 1.
17. Stansel, F. R. *Transistor equations.* Electronics, March 1953, Vol. 26, No. 3.
18. Rideout, V. C. *Active Networks.* Prentice-Hall, Inc.
19. Coblenz, A., and Owens, H. L. *Transistors: Theory and Applications.* McGraw-Hill Book Co.

QUESTIONS

1. Why, and in what way, is the equivalent circuit for a transistor as a linear amplifier different from the equivalent circuit of a vacuum-tube amplifier?
2. What is Norton's theorem referred to just preceding equation 11-7?
3. Following equation 11-7 it is stated that α and α_e are not the same. Why? When are they approximately equal numerically?
4. What important assumption has been made regarding the signal generator of Fig. 11-2?
5. What is meant by the term circuit determinant?
6. Why consider the available power from the signal generator when determining the operating power gain?
7. Section 11-3 refers to the "grounded base." What is the meaning of this? Must the base actually be connected to ground for the term to apply?
8. In the first paragraph of section 11-6 a statement is made regarding the conditions for oscillation. Can there be exceptions to this statement?
9. In section 11-6 is the statement "The circuit will *not* oscillate if Δ is greater than zero." Do you agree with this statement?
10. How do the circuits of Figs. 11-3, 11-4, and 11-5 compare in stability?
11. If these circuits tend to oscillate, how can they be stabilized?
12. Why is it said that the transistor emitter, base, and collector are equivalent respectively to the cathode, grid, and plate of a vacuum tube?
13. What is the meaning of the statement that the transistor is the dual of the vacuum tube?
14. What important facts must be considered in designing point-contact tran-

sistor class-A amplifiers that need not be considered in vacuum-tube class-A amplifier design?

15. Will the magnitude of the load resistor of a point-contact transistor amplifier affect distortion caused in the emitter circuit? Why?
16. Just prior to section 11-10 the statement is made that the distortion largely is caused by second harmonics. What circuit could be used to reduce this distortion?
17. Do the explanations made for point-contact transistor amplifiers apply to *n-p-n* junction transistor amplifiers? If modifications are required, what are they? What simplifications are possible?
18. Are the parameters for *n-p-n* junction transistors similar in magnitude to those for the point-contact type?
19. Are there advantages of grounded-emitter operation for *n-p-n* junction transistor amplifiers over other types of operation? If so, name them.
20. Generally speaking, which operates better at high frequencies, point-contact or junction transistors?
21. What is the maximum power efficiency that theoretically can be obtained with junction *n-p-n* transistors?
22. What equations for the vacuum tube are similar to equations 11-1 and 11-2 for the transistor?
23. Generally speaking, is a point-contact transistor or a junction transistor the noisier?
24. In what important details does a transistor oscillator differ from a similar vacuum-tube oscillator?
25. Must consideration be given to phase relations in transistor oscillator design? Explain.

PROBLEMS

1. Prove that $\alpha_e = \alpha + (\alpha - 1)r_b/r_c$.
2. Write definitions for class-A, class-B, and class-C transistor operation.
3. Repeat the calculations made in section 11-3 for a signal generator having an internal impedance of 750 ohms resistance.
4. In Fig. 11-5 insert the generator E_g, R_g, in series with the emitter and the load R_L in series with the base, and prove that the circuit will, or will not, amplify.
5. Derive equations 11-57 and 11-58.
6. Repeat the calculations of section 11-9 when the load is 15,000 ohms instead of 10,000 ohms.
7. Repeat the calculations of section 11-10 when the load is 20,000 ohms instead of 10,000 ohms.
8. Use the same transistor as in section 11-3 and calculate the input and output resistances as discussed in section 11-7.
9. Use the same data as in problem 8, and determine as explained in section 11-6 if the amplifiers are stable.
10. Compare the results of problems 8 and 9. If the circuits are unstable, explain how to prevent oscillations.

CHAPTER 12

MODULATORS

The purpose of a communication system is to transmit, from one point to another, speech, music, and other signals of various types. By the process called modulation these signals are translated to various locations in the frequency spectrum for transmission. At the receiving point the signal desired is selected on the basis of the frequency band occupied during transmission.

Modulation is defined [1] as the "process or result of the process whereby some characteristic of one wave is varied in accordance with another wave." This definition does not appear to be related to the statements of the preceding paragraph. When modulation is investigated further, however, it is found that when some characteristic (such as amplitude or frequency) of one wave is varied in accordance with a second wave, *a third wave is produced.* This new wave is a complex wave containing components of many frequencies, some of which were created by the process of modulation. Among these created components exist the modulating signal, but at a new position in the frequency spectrum. This signal may be selected on the basis of its frequency (for instance by a parallel resonant circuit) and then may be transmitted to the distant station.

For instance, the process of modulation makes possible the *simultaneous* transmission of many conversations over one pair of wires of a telephone transmission line or telephone cable in the following way: [2] Although each separate conversation originally exists within substantially the *same* voice-frequency band, *each* of these bands may be used to modulate individually a wave of different frequency. As a result of modulation, each message is raised, or translated, to a new and *different* band of frequencies for transmission. At the receiving station each message is selected on the basis of the frequency band to which it was translated. Then, by the process of demodulation (Chapter 13), each message is returned to the voice-frequency band originally occupied. Hundreds of conversations may be transmitted simultaneously over a coaxial cable.

Regarding modulation in radio, a voice-frequency signal covering a band from several hundred to several thousand cycles would not radiate well, and must be raised to a higher band of radio-frequencies for transmission. Also, transmission and reception at various radio-frequency bands is possible by the process considered in the preceding paragraph. Thus, many radio systems may operate simultaneously. This is defined [1] as **frequency-division multiplex,** a process "for the transmission of two or more signals over a common path by using a different frequency band for each signal."

12-1. *THE NATURE OF SPEECH AND MUSIC*

In modulation the signal causing the modulation often is speech or music. Thus, before studying modulation further, it is advisable to investigate the nature of speech and music. It has been found [3,4] that the band of frequencies covered in speaking and singing by the voices of men and women extends from about 60 to 10,000 cycles. Average speech power is about 10 microwatts, with peak powers of over 100 microwatts. For musical instruments it was found that a band of from 40 to 15,000 cycles would contain most of the tones. An orchestra may produce peak powers of 60 to 70 watts. It was found [3,4] that if a band of frequencies of about 50 to 8000 cycles and a volume range of 40 decibels was provided for in the transmission of musical programs, musicians could *detect no important change* in quality. A later study [5] showed that radio listeners prefer either a narrow tonal range (120 to 4000 cycles) or a medium tonal range (60 to 6000 cycles) instead of a wide tonal range (30 to 10,000 cycles). These findings aroused much discussion,[6] and led to a second paper [7] on the subject.

An investigation [8] in which observers listened to musicians directly (rather than over a sound-reproducing system as in the previous tests here reported) indicated that listeners prefer wide-range music. Although such matters are controversial, it has been found that a band of from about 200 to 3500 cycles gives excellent intelligibility and naturalness for telephone communication purposes; furthermore, although wider channels are available, it is probable that most radio-broadcast network programs (including programs by symphony orchestras) are transmitted over program channels covering a band of frequencies of about 100 to 5000 cycles. Of course, radio programs direct from studio performances *may* be of wider range, depending on the equipment used.

In conformity with the preceding statements, the modulating wave for telephone purposes only will be considered to cover a band from about 200 to 3500 cycles, and the modulating wave for radio-broadcast program purposes will be considered to cover a band of from 100 to 5000 cycles, it being recognized that there are exceptions, and that the figures merely are typical. Furthermore, the modulating signal will, for convenience, be considered sinusoidal. Actually, a speech- or music-modulating wave is complex and contains many frequency components. However, if a circut will transmit satisfactorily sine waves distributed throughout the band occupied by a complex wave, then for telephone and most radio purposes, the circuit will transmit satisfactorily a complex wave.

12-2. *TYPES OF MODULATION*

As stated in the second paragraph of this chapter, modulation may be considered as a *process* whereby some characteristic of one wave (carrier) is varied in accordance with another wave (modulating wave). The **carrier** is defined [1] as "a wave suitable for being modulated," and in radio commonly is a sinusoidal wave of, for example, 1.0 megacycle. The early choice of the term carrier is unfortunate because it implies that this wave *carries the signal* to the distant station, a viewpoint that is to be avoided, as the following pages will indicate. The **modulating wave** is defined [1] as "a wave which causes a variation of some characteristic of a carrier." The human voice, as it operates a microphone, produces a familiar modulating wave.

The following three basic types of modulation are defined: [1]

Amplitude modulation (AM): "Modulation in which the amplitude of a carrier is the characteristic varied." This is the system used almost exclusively until about 1940.

Angle Modulation: "Modulation in which the angle of a sine-wave carrier is the characteristic varied from its normal value." As will be explained later in this chapter, phase and frequency modulation are types of angle modulation.

Pulse Modulation: "Modulation of one or more characteristics of a pulse carrier." Several systems of pulse modulation will be considered at the close of this chapter.

12-3. *TYPES OF AMPLITUDE MODULATION*

The study of modulation (and demodulation, Chapter 13) is handicapped by the lack of uniformity in meaning of the technical terms used. Also, the terms used, and certain of the definitions, do not always describe the fundamental processes involved. For example, **cathode modulation** is defined [9] as "amplitude modulation accomplished by application of the modulating voltage to the cathode circuit"; **grid modulation** is [9] modulation "produced by the introduction of the modulating signal into the control-grid circuit of any tube in which the carrier is present"; and **plate (anode) modulation** is [9] "modulation produced by introducing the modulating signal into the plate circuit of any tube in which the carrier is present." These statements merely specify *where the signal is injected* and do not specify how the process of modulation is accomplished. Misunderstandings often arise, because for certain types of modulation the signals may be *injected* into the grid circuit, but the actual process of modulation occurs *in the plate circuit* (section 12-11).

Systems of modulation may be classified on the basis of the power level involved. Thus, **high-level modulation** is [9] modulation "produced at a point in a system where the power level approximates that at the output of the system," and **low-level modulation** is [9] modulation "produced at a point in a system where the power level is low compared with the power level at the output of the system." These two systems were depicted in Fig. 8-20. Two viewpoints commonly are used in considering amplitude modulation.

The *first viewpoint* is that the amplitude of a single-frequency carrier merely is caused to vary in accordance with the instantaneous amplitude of the modulating signal. This *is* in accordance with the definition of amplitude modulation; however, it must be remembered that as a result of the process of modulation a new wave is created, and although this new wave *appears* (on an oscillograph) merely to be the original single-frequency carrier wave varying in amplitude, actually this modulated wave contains several new frequency components. These components were *not* present in the input to the modulator.

The *second viewpoint* is analytical; it stresses the fact that these new frequency components are created in the process of modulation, and that certain of these components are the original modulating signal wave translated to a higher position in the frequency spectrum. From the analytical viewpoint, amplitude modulation may be regarded as a form of nonlinear distortion. This is because distortion sometimes is defined [9,10] as "a change in wave form" (although the definition [11] an "*undesired* change in wave form" also is used); and further because nonlinear distortion is defined [9] as distortion "which occurs in a system when the ratio of voltage to current therein is a function of the magnitude of either." As a result of nonlinear distortion, frequency components exist in the output of a distorting device that were not present in the input. If a carrier wave and a modulating wave are impressed simultaneously on a device that will cause nonlinear distortion, among the frequency components created by the distortion will exist sum frequencies and difference frequencies (section 8-13). In fact, the principle of modulation is used in **intermodulation methods**[12] of measuring the nonlinear distortion of electronic equipment.

Regardless of viewpoints and definitions, the fact remains that the purpose of amplitude modulation is to translate the intelligence to be transmitted from signals occupying a low-frequency band, to signals occupying a band at higher frequencies. Furthermore, this is done by simultaneously impressing a single-frequency wave and the modulating signal wave on a device the output wave of which *is not* a replica of the two original modulating signals, but is a new wave containing various components, among which is the modulating signal wave, in frequency bands close to the frequency of the carrier. Everitt has stated[13] that "modulation is obtained whenever a component of the output wave is made proportional to the product of two input waves."

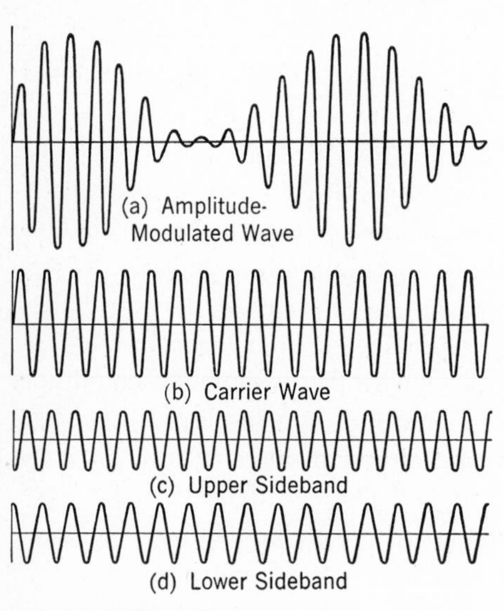

FIG. 12-1. In (a) is shown a 100 per cent amplitude-modulated wave, the modulating signal being sinusoidal. In (b) is shown the carrier component, in (c) is the upper-sideband component, and in (d) the lower-sideband component of the amplitude-modulated wave. These curves are approximately to scale, and the three lower curves when added point-by-point give the amplitude-modulated wave at the top. The three lower curves are a plot of equation 12-4.

12-4. *ANALYSIS OF AMPLITUDE-MODULATED WAVE*

A portion of an amplitude-modulated wave, created by modulating a sinusoidal carrier wave with a sinusoidal modulating wave, is shown in Fig. 12-1a. If such a wave is analyzed, it will be found to contain *three* separate frequency components. These waves are: (b) the original carrier wave; (c) a wave having a frequency equal to the sum of the carrier frequency and the modulating frequency, sometimes called the **upper side frequency**; and (d) the carrier minus the modulating frequency, or the **lower side frequency.** Thus, if the carrier is a 20,000-cycle sinusoidal wave as it might be in a carrier-telephone system,[2] and if the modulating frequency is a 1000-cycle sinusoidal wave, then the upper side frequency will be a 21,000-cycle sine wave, the carrier will exist as before, and the lower side frequency will be a 19,000-cycle sine wave. A single frequency, such as 1000 cycles, often is taken to represent the modulating voice signals in studying modulation.

If the modulating signal is *not* a single frequency, but is a signal covering a band width of from 100 to 5000 cycles, then sidebands instead of side frequencies will be created. **Sidebands** are defined [1] as "frequency bands on both sides of the carrier frequency within which fall the frequencies of the waves produced by the process of modulation."

Amplitude-modulation broadcast equipment is so designed that the maximum variation of the modulated wave during operation should just reach zero. The **modulation factor** is [1] the "ratio of the peak variation actually used to the maximum design variation." The **percentage modulation** is [1] the "modulation factor expressed as a percentage." For an amplitude-modulated wave the percentage modulation is [9] the "ratio of half the difference between the maximum and minimum amplitudes to the average amplitude."

An amplitude-modulated wave having 100 per cent modulation is shown in Fig. 12-1a, and the components that make up the wave are shown below. These waves are approximately to scale and a point-by-point addition will prove that the modulated wave is, in fact, composed of the wave beneath it. It is important to note that the wave of carrier frequency (that is, the carrier) *does not vary in amplitude during modulation.* Graphs of the three components of a wave modulated 100 per cent by a sinusoidal signal are shown in Fig. 12-2.

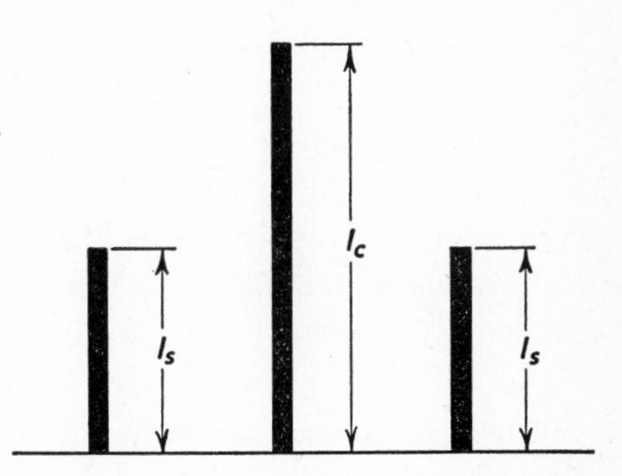

FIG. 12-2. Showing the relative heights of the carrier component I_c and the two sideband components I_s for 100 per cent modulation.

An analysis of the amplitude-modulated wave of Fig. 12-1a is as follows: The instantaneous value of any sinusoidal current is

$$i = I_{\text{max.}} \sin \omega t \qquad \text{12-1}$$

Now suppose the amplitude $I_{\text{max.}}$ is made to vary sinusoidally (by a signal of lower frequency) by any one of the several systems of modulation, the percentage modulation being m. Then, at any instant

$$I_{\text{max.}} = I_c(1 + m \sin \omega_v t) \qquad \text{12-2}$$

If this value is substituted in equation 12-1, then

$$i = I_c(1 + m \sin \omega_v t) \sin \omega_c t \tag{12-3}$$

in which ω equals 2π times the frequency f, and the subscripts v and c represent the voice-frequency modulating signal and carrier components, respectively. Expanding equation 12-3 gives

$$i_{am} = I_c \sin \omega_c t + \frac{mI_c}{2} \cos (\omega_c - \omega_v)t - \frac{mI_c}{2} \cos (\omega_c + \omega_v)t \tag{12-4}$$

This equation shows that the wave of Fig. 12-1a contains three frequency components, the carrier represented by $I_c \sin \omega_c t$, the lower sideband represented by $[mI_c/2][\cos (\omega_c - \omega_v)t]$, and the upper sideband represented by $[mI_c/2][\cos (\omega_c + \omega_v)t]$. For 100 per cent modulation the maximum value of each sideband is $I_c/2$. For any percentage modulation m, the maximum value is $mI_c/2$.

From the preceding discussion it should not be inferred that the carrier and two sidebands are the only frequency components existing in the output of a modulator. The analysis just given applies only to an amplitude-modulated wave as impressed on an antenna *after* the undesired components have been removed.

As was previously explained, for 100 per cent modulation the amplitude of each sideband is half that of the carrier. The power contained in each component is proportional to the amplitude squared, since power varies as the current squared. If the amplitude of the carrier equals 1.0, and *each* sideband 0.5, then the corresponding power relations will be 1.0 and 0.25. That is, if P represents the total power in the modulated wave, then 66⅔ per cent will be in the carrier, and 33⅓ per cent in the *two* sidebands (16⅔ per cent in each one).

The *intelligence* or program signal power is *entirely in the sidebands. No intelligence whatever is conveyed by the carrier wave.* Although the carrier wave and both sidebands are at present broadcast for entertainment purposes, in other systems only the carrier and one sideband, or only one sideband is transmitted. The final power-amplifying stages handle only useful power, if a single sideband is transmitted.

From an examination of Figs. 12-1 and 12-2, it can be proved that the continuous (or average) power output with 100 per cent modulation is 1.5 times the output with zero modulation. Also, for modulation peaks the instantaneous power output is four times the unmodulated carrier. This is because power is proportional to current or voltage squared.

12-5. *TYPES OF VACUUM-TUBE AMPLITUDE MODULATORS*

A **modulator** is defined [1] as "a device to effect the process of modulation." It is important to understand just what part of the circuit is the modulator. Sometimes the last audio-frequency stage in a transmitter is called the modulator, when in reality this stage is an audio-frequency **modulating amplifier** and it drives the modulator; that is, the modulating amplifier *drives* the modulator that effects the process of modulation and causes the sidebands to be created. There are two distinct types of vacuum-tube modulators, classified on the basis of the part of the tube in which the modulation *initially is produced.* These are [14] **plate-circuit modulators** and **grid-circuit modulators.**

12-6. *THE VAN DER BIJL SYSTEM OF AMPLITUDE MODULATION*

This system, named after its inventor, was developed early in the history of vacuum tubes.[15] It was used in carrier-telephone systems.[16,17] In this modulator distortion and modulation occur *in the plate circuit* because of the nonlinear relation between plate current and grid voltage. In the van der Bijl system, the tube is not biased to cutoff. This is, accordingly, a **class-A modulator.**[9] Because this system well illustrates the theory of modulation, it will be treated in detail.

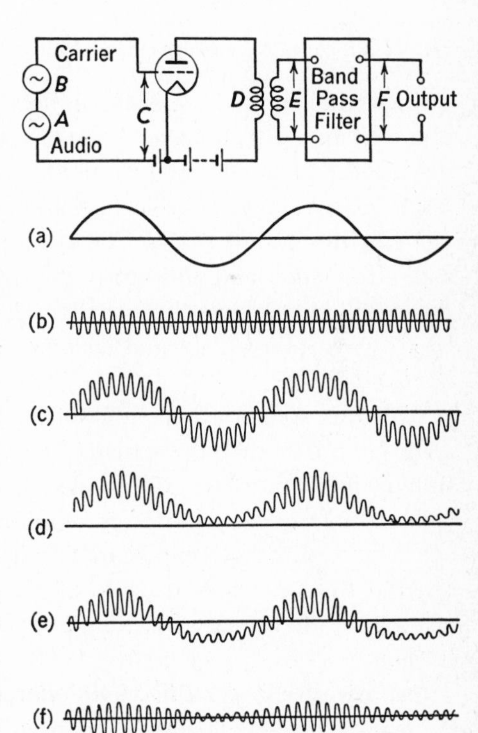

FIG. 12-3. Circuit for the van der Bijl system of modulation, and an analysis of the currents and voltages in various parts of the circuit.

The circuit is arranged as in Fig. 12-3, and the operation is as follows: Assume that the audio-frequency sinusoidal signal (a) is to modulate the carrier-frequency signal (b). These two frequencies are impressed simultaneously between grid and cathode of the tube, which is biased so that operation is on the nonlinear part of the grid-voltage–

plate-current curve as indicated in Fig. 12-4. The combined action of the carrier- and audio-frequencies will cause the potential of the grid to vary as indicated by (c) of Fig. 12-3, and as in Fig. 12-4.

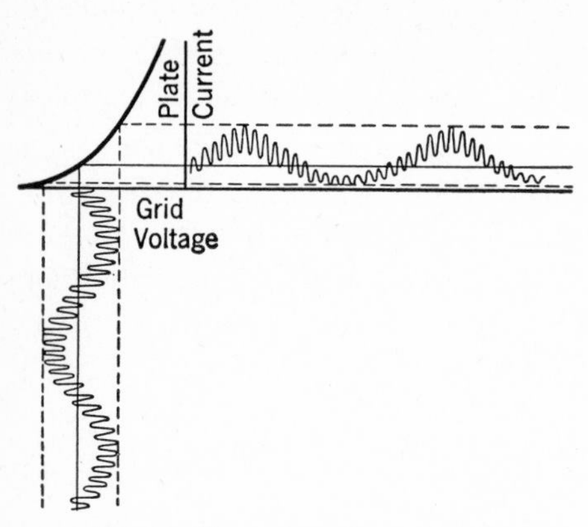

FIG. 12-4. Illustrating the method of distortion used in the van der Bijl modulator of Fig. 12-3 for creating the desired sidebands.

If the bias were such that the grid-voltage variations were on the linear portion of the grid-voltage–plate-current curve, *amplification alone would occur,* and the plate current would be a replica of the grid-voltage variations. Since the high negative grid bias shifts the operation to the nonlinear portion, the plate current is distorted as indicated by Fig. 12-4, and also by curve (d) of Fig. 12-3. This distorted curve will contain a large number of frequency components, among which are the audio-frequency, the carrier-frequency, and the upper and lower sidebands. It will also contain a direct-current component.

After the plate current (d) of Fig. 12-3 passes through the output transformer, the voltage induced in the secondary will vary as shown by curve (e). On the output side of the band-pass filter [2] (inserted to remove the audio-frequency and unwanted higher frequency components) the wave shape will be as shown by curve (f). This is the familiar "amplitude-modulated" wave (Fig. 12-1) and *contains the carrier and the upper and lower sidebands only.*

With this system a low value of load resistance should be placed in the plate circuit so that the dynamic curve of Fig. 12-4 will not approach a straight line. Since the grid always is negative, negligible carrier-frequency and audio-frequency power is required from the driving amplifiers. The efficiency and the percentage modulation are low. This system is sometimes referred to as a modulated class-A amplifier. Although amplification does take place, the primary function is *to distort the waves and produce modulation.*

Analysis of van der Bijl Modulation. Over the *lower curved portion* of the grid-voltage–plate-current curve of Fig. 12-4, the plate current may be represented approximately by the Taylor series:

$$I_p = ae_g + be_g^2 + \cdots \qquad 12\text{-}5$$

All terms of this series higher than those given by equation 12-5 may be neglected. Because of equation 12-5, this method sometimes is called **square-law modulation.**

If the instantaneous value of the carrier-frequency voltage is $E_c \sin \omega_c t$, and that of the audio-frequency or voice-frequency voltage is $E_v \sin \omega_v t$, then the instantaneous grid potential will be

$$e_g = E_c \sin \omega_c t + E_v \sin \omega_v t \qquad 12\text{-}6$$

Substituting this value of instantaneous grid potential in equation 12-5, the instantaneous plate current is equal to

$$i_p = aE_c \sin \omega_c t + aE_v \sin \omega_v t + bE_c^2 \sin^2 \omega_c t + bE_v^2 \sin^2 \omega_v t + 2bE_cE_v \sin \omega_c t \sin \omega_v t \qquad 12\text{-}7$$

This equation can be expanded by the relation

$$\sin \omega_c t \sin \omega_v t = \left[\frac{1}{2} \cos (\omega_c - \omega_v)t - \frac{1}{2} \cos (\omega_c + \omega_v)t\right] \qquad 12\text{-}8$$

After these substitutions are made, equation 12-7 becomes

$$i_p = aE_c \sin \omega_c t + aE_v \sin \omega_v t + bE_c^2 \left(\frac{1}{2} - \frac{1}{2} \cos 2\omega_c t\right) + bE_v^2 \left(\frac{1}{2} - \frac{1}{2} \cos 2\omega_v t\right) + 2bE_cE_v \times \left[\frac{1}{2} \cos (\omega_c - \omega_v)t - \frac{1}{2} \cos (\omega_c + \omega_v)t\right] \qquad 12\text{-}9$$

Performing the multiplications and dropping constant terms (alternating components are of present interest) gives

$$i_p = aE_c \sin \omega_c t + aE_v \sin \omega_v t - \frac{bE_c^2}{2} \cos 2\omega_c t - \frac{bE_v^2}{2} \cos 2\omega_v t + bE_cE_v \cos (\omega_c - \omega_v)t - bE_cE_v \cos (\omega_c + \omega_v)t \qquad 12\text{-}10$$

The carrier frequency is $f_c = \omega_c/(2\pi)$ and the audio-frequency is $f_v = \omega_v/(2\pi)$. Thus, the following alternating components exist in the input to the filter (curve *e* of Fig. 12-3): (1) the *carrier frequency* f_c; (2) the audio-frequency f_v; (3) twice the carrier frequency $2f_c$; (4) twice the audio-frequency $2f_v$; (5) the *lower sideband* $(f_c - f_v)$; and (6) the *upper sideband* $(f_c + f_v)$.

The band-pass filter will pass the carrier and the two sidebands and will reject all other components, resulting in the amplitude-modulated wave (*f*) of Fig. 12-3. If the carrier frequency f_c is 20,000 cycles and the voice signals f_v occupy a band from 200 to 3500 cycles, then the frequency components at the output of the filter will be the carrier $f_c = 20{,}000$ cycles, the lower sideband $f_c - f_v$ extending from 16,500 to 19,800 cycles, and the upper sideband $f_c + f_v$ from 20,200 to 23,500 cycles. These figures are typical of a carrier-telephone circuit for operation on a telephone line. In such systems the entire modulated wave of Figs. 12-1 and 12-3 is not transmitted. Since *each* sideband contains all the information to be transmitted, most carrier-telephone systems *suppress* the carrier and one sideband, and *transmit only one sideband to the distant station.*

12-7. *AMPLITUDE MODULATION WITH BALANCED MODULATORS*

As explained in the preceding section, in carrier telephony only one sideband is transmitted. The other components must, therefore, be removed by some means. Band-pass filters will accomplish this, theoretically. A difficulty is that a margin of only 200 cycles lies between the carrier to be suppressed and the sideband to be transmitted and that the amplitude of the carrier is much greater than that of the sideband. Special filters of the quartz-crystal type [18] would probably be necessary for such sharp discrimination.

It is possible to suppress *the carrier* in the modulator circuit if the **balanced modulator** of Fig. 12-5 is employed. Then, the filter need only discriminate between the two sidebands. For each individual tube, modulation will occur as in the van der Bijl system of the preceding section. The two sidebands will exist (with other products of modulation) in the output of the transformer.

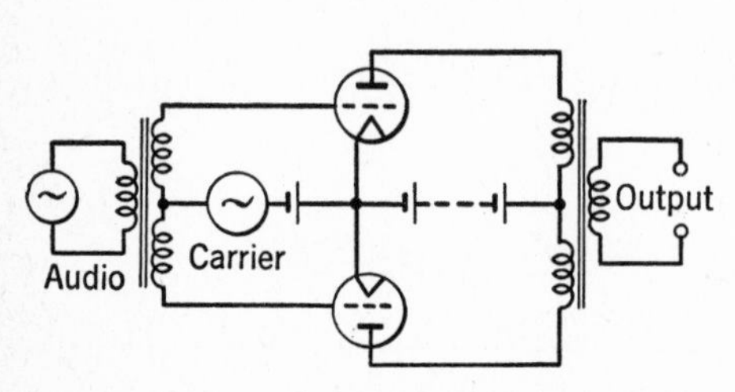

FIG. 12-5. A balanced modulator which suppresses the carrier component.

The carrier frequency is suppressed in the following manner: As Fig. 12-5 shows, the carrier-frequency voltage is impressed on the two grids in phase. That is, for the positive half-cycle of the carrier, *both* grids are driven more positive; for the negative half-cycle, *both* grids are driven more negative. Thus, the plate-current changes that occur for *carrier-frequency* variations will be the same for each tube, and will be *balanced out in the primary of the output transformer.*

12-8. *GRID-CIRCUIT AMPLITUDE MODULATION*

As previously mentioned, vacuum-tube modulators may be classified as *plate-circuit* modulators or as *grid-circuit* modulators. The types that previously have been considered are plate-circuit modulators; grid-circuit modulators will be considered in this section.[14]

Referring to Fig. 4-9, it is seen that the grid-voltage–grid-current curve is nonlinear. Over the lower part of the grid-current curve, equation (12-5) and the accompanying derivations, showing how the sidebands are generated, apply. It is apparent that this nonlinear grid-current characteristic can be used to produce modulation *in the grid circuit,* just as the nonlinear part of the plate-current curve can be used to produce modulation *in the plate circuit.*

A simplified circuit of a grid-circuit or grid-current modulator is shown in Fig. 12-6. The tube may be operated with no bias, or with the bias either *slightly* positive or *slightly* negative, depending on conditions. The carrier and audio voltages are impressed on the grid with a resistor R_g of about 1,000,000 ohms in series. The load resistor R_L has about the same magnitude as the plate resistance of the tube. The grid bias and plate voltages are so selected that distortion occurs *in the grid circuit,* and amplification occurs *in the plate circuit.*

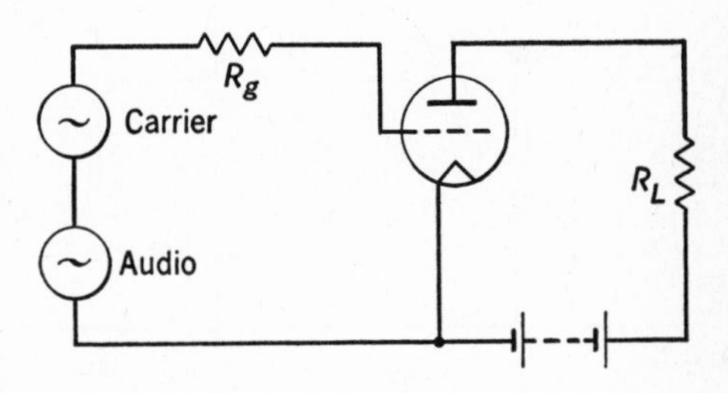

FIG. 12-6. Simplified circuit for grid-circuit modulation.

With these adjustments, the operation is described[14] as follows: The resultant potential, caused by the combined action of the carrier-frequency and the audio-frequency voltages, will be somewhat as shown by curve (c) of Fig. 12-3. With the bias at about zero value, so long as the grid-potential variations maintain the grid negative, no grid current flows; but, when the grid is driven positive by the combined action of the carrier and the audio signals, grid current flows and a *voltage drop occurs* across the high resistance R_g. Thus, when grid current flows and a voltage drop occurs across R_g, the voltage actually impressed between grid and cathode is distorted, and contains the desired (and also undesired) products of modulation; that is, the sidebands. With proper adjustments, modulation occurs in the *grid circuit,* and amplification in the *plate circuit.* The efficiency of the system is high, but it has had little application. Filters in the output are used to suppress the undesired frequency components.

12-9. *AMPLITUDE MODULATION WITH PLATE-MODULATED CLASS-C OSCILLATOR*

This method [15] is often called **Heising modulation,** so named after its inventor. In the past, this method was widely used in radio, but it has been largely superseded because of frequency instability and other limitations. Other names such as **constant-current modulation** are applied.[9]

A circuit of a plate-modulated class-C oscillator is shown in Fig. 12-7. The audio-frequency input from a microphone is fed into the grid circuit of the

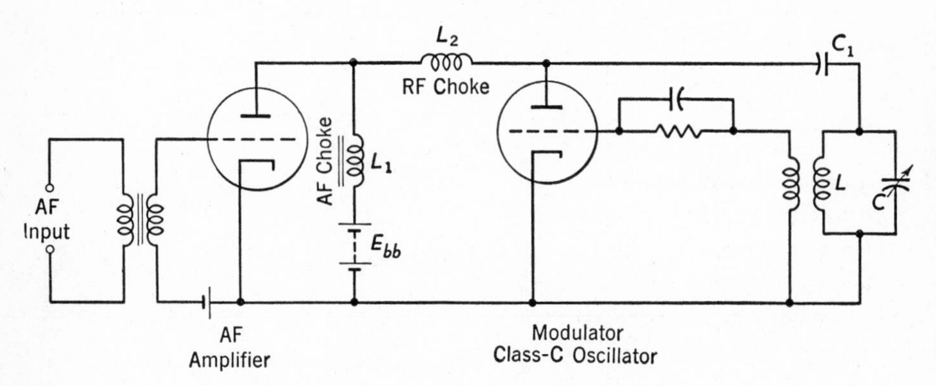

FIG. 12-7. Circuit for a plate-modulated class-C oscillator.

audio-frequency amplifier. This tube sometimes incorrectly is called the modulator tube. It is merely an audio-frequency, class-A, impedance-coupled, power-amplifier tube. No (intentional) distortion occurs in its plate circuit.

The modulated class-C oscillator tube receives its plate-supply voltage from the same source E_{bb}, and through the same coil L_1, as does the amplifier tube. Otherwise, the circuit is that of a conventional class-C oscillator. Connections from the antenna would be made to the resonant parallel circuit L–C. The radio-frequency choke coil L_2 prevents the flow of the high-frequency currents to the left. The capacitor C_1 prevents the flow of current from the plate-supply battery E_{bb} around through the circuit L–C. Note in particular that *only audio-frequency components exist to the left of the radio-frequency choke* L_2.

The operation of the plate-modulated oscillator of Fig. 12-7 is essentially as follows: Negligible alternating current flows through the audio-frequency choke coil L_1; hence, the plate-supply direct current to each tube is essentially constant. The oscillator tube is the alternating-current load for the amplifier tube. When an audio-frequency signal is impressed on this amplifying tube, the potential of the plate of the amplifying tube will vary with the audio signal just as in any amplifying tube. Since the two plates are tied together, the plate potential of the oscillator tube also must vary in accordance with the audio signals.

The alternating radio-frequency carrier voltage generated by the oscillator varies almost directly with the magnitude of the plate-supply voltage. Since the plate voltage is varying in accordance with the audio signal, the radio-frequency carrier voltage must also follow the audio variations, as Fig. 12-8 indicates.

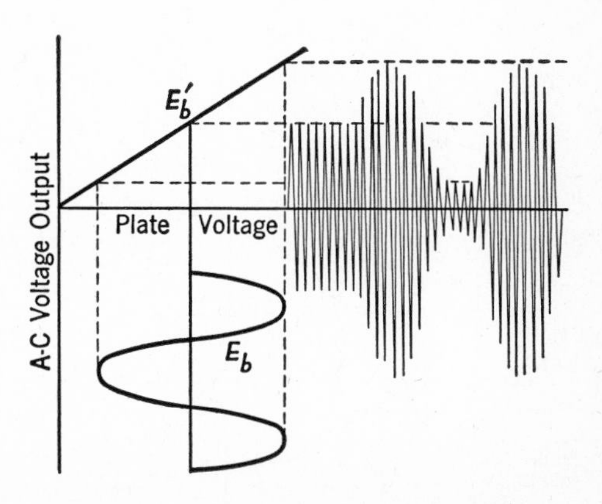

FIG. 12-8. The magnitude of the alternating voltage output of the oscillator depends on the instantaneous magnitude of the direct voltage E_b'. When this is caused to vary in accordance with the audio voltage variations E_b, the alternating radio-frequency voltage across the resonant tuned circuit L–C will be as shown. The envelope of the radio-frequency output voltage has the shape of the audio signal E_b. This does not mean that any appreciable audio-frequency component exists across the circuit L–C. This circuit is tuned to the carrier frequency and thus the only appreciable voltages existing across it are due to the carrier and the two sidebands. These components are impressed on the antenna.

The output circuit for the oscillator is through the small capacitor C_1 to the parallel circuit L–C which is resonant and offers a high impedance to the carrier frequency. Thus, a large voltage drop will exist across this parallel circuit for the carrier and for the upper and lower sidebands, generated when the oscillator output is caused to vary at the audio rate. Other frequency components are produced, but these do not appear to any great extent as voltages across L–C, because this tuned circuit offers low impedance to them. The fact that C_1 is small prevents the audio components from passing around through the circuit L–C instead of through the load offered to the amplifying tube by the plate-circuit resistance of the oscillator tube. The capacitor C_1 also prevents the flow of direct current.

This modulator now is seldom employed for at least two reasons. *First,* if 100 per cent modulation is attempted, then the oscillator plate voltage must approach zero and hence oscillations may stop. *Second,* the magnitude of the plate-supply voltage is changing at an audio rate, and this will cause the *frequency* of the oscillator to vary. Frequency modulation is undesired.

12-10. *AMPLITUDE MODULATION WITH PLATE-MODULATED CLASS-C AMPLIFIER*

If a class-C *amplifier* instead of a class-C *oscillator* is used, the difficulties just mentioned do not present themselves. In the plate-modulated class-C amplifier, the carrier-frequency grid exciting voltage usually is produced by a crystal oscillator and associated buffer amplifier. Thus, the carrier-frequency is held constant and the output of the modulator is not (theoretically) frequency modulated. Also, this device may be 100 per cent modulated.

This type modulator is widely used. Since the class-C oscillator and the class-C amplifier differ only in the fact that one is self-excited and the other externally excited, the discussions of the preceding section can be readily extended to cover the plate-modulated class-C amplifier.

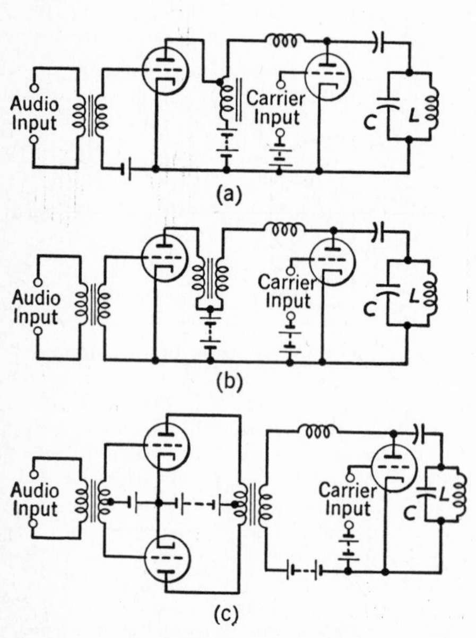

FIG. 12-9. Circuits for plate-modulated class-C amplifiers. The modulated signal composed of the carrier and the two sidebands exists as a voltage across the tuned circuit L–C. The antenna would be coupled to this tuned circuit.

The circuit of Fig. 12-7 *could* be used for a plate-modulated class-C amplifier by separately exciting the oscillator and using it as an amplifier. This circuit is not desirable because 100 per cent modulation is impossible. The two plates are in parallel, and for complete modulation the plate-potential variations of the class-A amplifier must be from zero to twice the direct plate-supply voltage; this cannot be achieved practically. The circuit can be modified by placing a resistor in series with the choke coil L_2. The resistor is of such value that the direct voltage applied to the plate of the modulated class-C amplifier tube is reduced to a value such that the modulating class-A amplifier tube can swing the plate from zero to twice the direct plate-supply voltage. This resistor must be shunted by a large capacitor so that the audio-frequency currents from the class-A amplifier tube have a low-impedance path to the plate circuit of the class-C amplifier tube.

The three circuits shown in Fig. 12-9a, b, and c often are used for plate-modulated class-C amplifiers. In each of these, the tuned circuit *L–C* would be coupled to additional amplification for low-level modulation, or to the antenna for high-level modulation systems.

In the first circuit, 100 per cent modulation is possible because the autotransformer used for coupling (instead of a choke) increases the voltage impressed on the modulated class-C amplifier tube. Also, this autotransformer changes the magnitude of the load resistance offered by the plate circuit of the second tube. By selecting the proper ratio of transformation, the correct load for the class-A tube can be obtained. This is also constant-current (Heising) modulation.

Of all the possible schemes of coupling the driving audio amplifier to the plate-modulated class-C amplifier, the last two circuits of Fig. 12-9, employing transformers, are the most satisfactory. In the first of these two, a common plate supply can be used as shown, or the plates of the two tubes may be separately supplied, thus increasing the flexibility of conditions of operation. In the last of these systems, the audio-frequency driving tubes are operated in push-pull class-A, AB, or B as desired. In fact, the circuits are often adjusted so that class-A operation obtains for small inputs, but for large audio-frequency signals the tubes operate as class-AB or as class-B. These drivers usually are audio-frequency *power* amplifiers (since they usually drive power-consuming class-C stages).

Equivalent Circuit for Modulated Amplifier. The circuits of Fig. 12-9 may be represented by the equivalent circuit of Fig. 12-10. Although the audio-frequency signal voltage is not in series with the tuned resonant circuit *L–C* in Fig. 12-9, the action is equivalent to the simplified circuit shown. The grid is biased to about twice cutoff, so that with a sinusoidal carrier-frequency voltage applied to the grid, less than half-cycles of carrier current will flow (Fig. 12-10a) in the plate circuit *with no audio-frequency voltage applied.* This wave is badly distorted, but will contain a large carrier-frequency component (equation 6-1) to which the *L–C* circuit is tuned. The impedance of this *L–C* circuit will be a large value of pure resistance *to the carrier* frequency (and sidebands) but will offer low impedance to all other components. Thus, the voltage appearing across the circuit *L–C* will be largely a carrier-frequency voltage, and this will be impressed on an antenna coupled with *L–C*.

When only an audio-frequency signal voltage is impressed, the resulting plate potential will be that of the plate supply plus the plate signal voltage. If the magnitude of the peak values of this signal voltage is equal to the direct plate-supply voltage, the potential of the plate will vary with the audio-frequency signals from zero to twice the direct plate-supply voltage. Assuming a sinusoidal audio-frequency signal voltage, the plate potential will vary as in Fig. 12-10b.

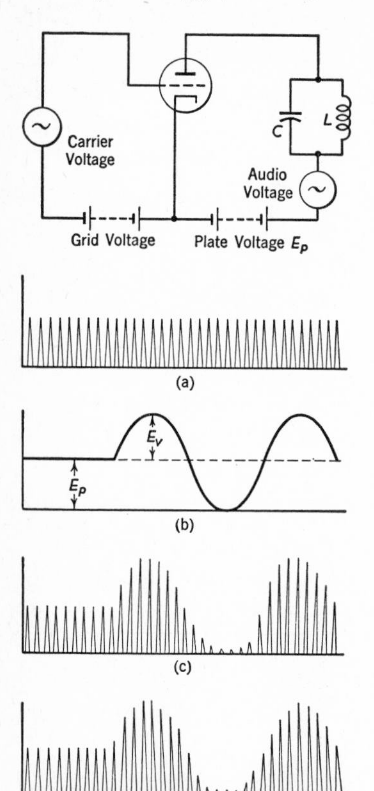

FIG. 12-10. Equivalent circuit and curves for a modulated class-C amplifier.
(a) Plate current with only carrier impressed.
(b) Direct plate voltage E_p and audio voltage E_v.
(c) Plate current when carrier and audio voltage are impressed.
(d) The voltage across the tuned circuit is an amplitude-modulated wave containing the carrier and the two sidebands.

When the carrier-frequency voltage and the audio-frequency signal voltages are impressed simultaneously, the plate current will vary as in Fig. 12-10c. This plate current is distorted. Instead of containing only harmonics of the carrier voltage as in Fig. 12-10a, it now also contains (among other frequencies) the carrier frequency, the carrier frequency *plus* the audio-frequency, and the carrier frequency *minus* the audio-frequency. These two last components are the desired *upper and lower sidebands which convey the intelligence or musical program.* The carrier-frequency component is constant and conveys no information.

This modulated plate current of Fig. 12-10c must flow through the parallel circuit L–C which is tuned to the carrier frequency and *also to the sidebands.* Thus, the voltage across the parallel L–C circuit will be due to the combined effect of the carrier frequency and the sidebands, and this is the signal *impressed on the antenna.* This wave shape, both before and after modulation, is shown in Fig. 12-10d. The tuned circuit offers low impedance to all the other components of the complex wave of Fig. 12-10c, and hence they are not radiated in the same magnitude as are the desired frequencies. If it were not for the selective action of this tuned circuit, the transmitter would radiate at various frequencies, instead of radiating at the frequencies of the sidebands and the carrier.

It is of interest to investigate the power relations in a plate-modulated class-C amplifier. As mentioned in section 12-4, the power relations in a completely modulated wave are as follows: If the power output of the carrier is assumed 1.0, then that of each sideband is 0.25, and the total *average* power output of the modulated wave would be 1.5. The carrier wave *is unaffected* by the process of modulation, the power output being 1.0 before and after modulation. The power supplied the sidebands must, therefore, be furnished by the audio-frequency modulating amplifier.

12-11. *AMPLITUDE MODULATION WITH GRID-MODULATED CLASS-C AMPLIFIER* [19,20]

In the modulators just described, the carrier was impressed between grid and cathode, and the audio-frequency variations between plate and cathode. Modulation was effected in the plate circuit. In the system now to be described, *both frequencies* are impressed *in the grid circuit* of a class-C amplifier, and modulation is effected *in the plate circuit.* This often is called grid modulation, a term not in agreement with section 12-8. It is plate modulation by *grid injection.* A simplified equivalent circuit is shown in Fig. 12-11.

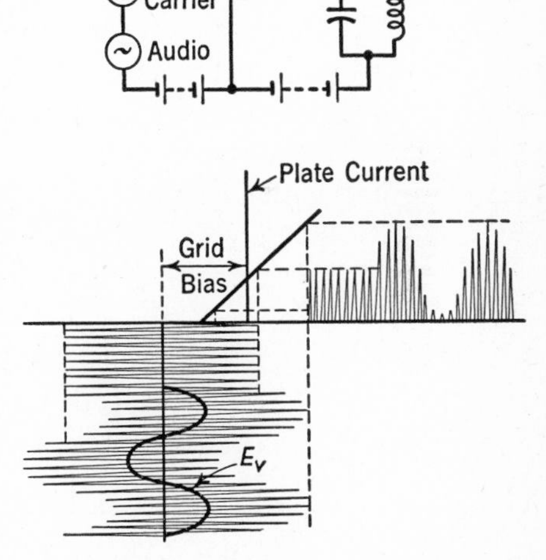

FIG. 12-11. Circuit for the grid-modulated class-C amplifier, and analysis of operation. The signal E_v is the audio signal voltage used to modulate the carrier.

In this figure also is shown the method of distorting the carrier and audio-frequency impulses to produce the sidebands and thus accomplish modulation. As indicated, the tube is biased to twice cutoff. The magnitude of the carrier voltage depends on the type of operation desired. The magnitude of the carrier voltage may be so adjusted that the grid is never driven positive, even on modulation peaks. Or, the magnitude of the carrier voltage may be such that the grid may be driven slightly positive even with no audio-frequency excitation. This later scheme is shown in Fig. 12-11.

The variations in grid potential when the carrier and audio-frequency voltages are impressed simultaneously are shown in Fig. 12-11. The plate current will be distorted as shown. As in previous instances, this current will contain, among other values, the carrier and the two sidebands. These will cause a large voltage drop across the tuned parallel circuit as explained in the preceding section, and this is the modulated signal.

When operated so that the grid is always negative and draws no current, negligible power is taken from the carrier-frequency and the audio-frequency amplifiers. All the power in the modulated wave, including that due to the presence of the sidebands, comes from the plate supply of the class-C tube. When the grid-modulated class-C amplifier is operated so that the grid draws current, then both the carrier-frequency and the audio-frequency amplifiers must supply power. This power is supplied only to the grid circuit, however, and the modulating audio-frequency *does not* supply the power to generate the sidebands as in the plate-modulated scheme. The efficiency of a grid-modulated amplifier is lower than the plate-modulated amplifier, and hence the former is best suited only for low-level modulation.

12-12. *SUPPRESSOR-GRID AMPLITUDE MODULATION*

This system is used in low-power radio-telephone transmitters such as are employed by many amateurs.[21] A suppressor-grid small-signal pentode, as distinguished from the power pentode, is used. In the former tube the shielding effect of the screen grid is very good, and self-oscillations of the tube are prevented without the use of neutralizing circuits.

The circuit is arranged as in Fig. 12-12. Both the suppressor grid and the control grid are made negative, and the carrier wave *C* and the modulating voice wave *V* are introduced as indicated. The electron current flowing to the plate is affected by both these voltages, and the desired sidebands are generated. The radio-frequency choke coils, *L′* offer high impedance to the radio-frequency components, which accordingly flow through the tuned resonant circuit *L–C*. A voltage will exist across this circuit for the carrier and the sidebands, and this signal is impressed on the antenna. The suppressor grid may be driven positive on modulation peaks and hence take power from the audio-frequency amplifier driving it.

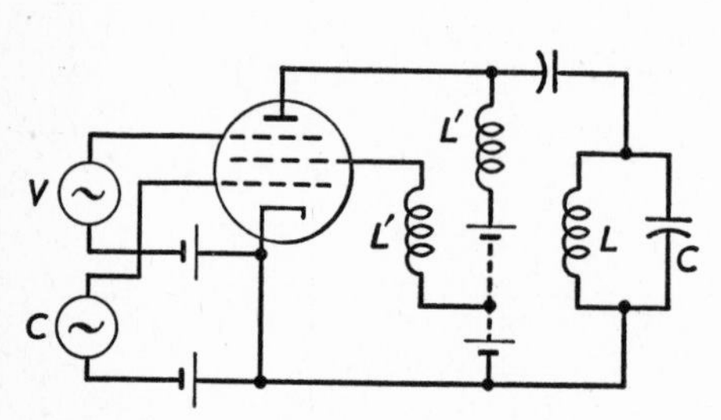

FIG. 12-12. Circuit for suppressor-grid amplitude modulation.

12-13. *MISCELLANEOUS AMPLITUDE-MODULATION SYSTEMS*

The most important systems of amplitude modulation have been considered in the preceding pages. Several other systems will be mentioned; additional information will be found in references 22 and 23.

Cathode modulation sometimes is used, in which the modulating signal is impressed between cathode and ground of a class-C radio amplifier tube.

Screen-grid modulation sometimes is used, in which the modulating signal is impressed in series with the screen direct-voltage source.

Absorption modulation [9] is a method of producing amplitude modulation by "means of a variable-impedance device inserted in or coupled to the output circuit." The modulating signal is used to control the variable impedance. One successful method uses transmission-line elements.[24]

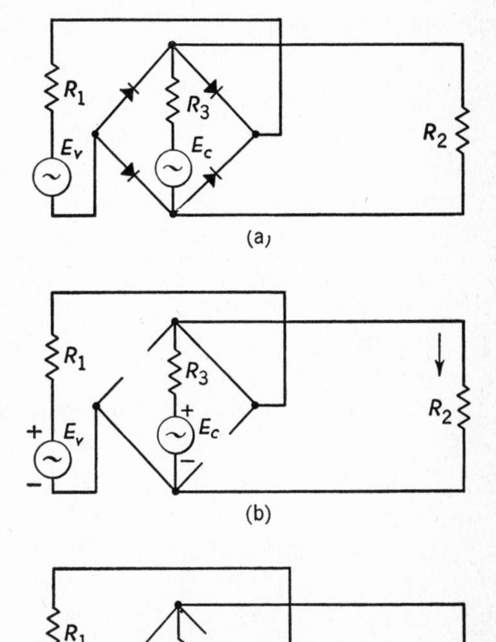

FIG. 12-13. In (a) is shown a simplified circuit of a copper-oxide bridge modulator. In (b) and (c) are shown the equivalent circuits with the polarities of the carrier voltage E_c as indicated. These equivalent circuits are based on the rectifier characteristics of Fig. 10-13. When the copper is positive and the oxide negative, the resistance is high, and when the polarities are reversed, the resistance is low.

12-14. *AMPLITUDE MODULATION WITH COPPER-OXIDE VARISTORS*

Copper-oxide varistors were discussed in section 10-10, and typical characteristic curves were given in Fig. 10-13. As an approximation, if the oxide is made positive and the copper negative, the varistor has zero impedance, and if the oxide is made negative and the copper positive, the varistor has infinite impedance. These characteristics are utilized in modulators (and demodulators, Chapter 13) for carrier-telephone circuits; in fact, varistors are used almost exclusively.[25,26] In the explanations that follow it is assumed that the carrier-frequency voltage is the control voltage, that it is larger than the modulating voice voltage, and that the carrier fixes the point of operation on the varistor characteristic curve of Fig. 10-13, causing the varistor to offer essentially zero or infinite impedance to the modulating signal. Note that the carrier voltage E_c of Fig. 12-13 always can force a carrier current component through R_2.

Many varistor modulator circuits are possible.[27] The one shown in Fig. 12-13 suppresses the modulating signal (the voice) and transmits the carrier and the two sidebands. As previously mentioned, the carrier-frequency voltage of Fig. 12-13 is used to control the impedance of the path offered to the modulating voice signal shown in Fig. 12-13. Note that many cycles of carrier voltage occur during one cycle of modulating voice voltage (assumed to be a single frequency of, say, 1000 cycles). When the carrier voltage has the polarity shown in Fig. 12-13b, the varistor bridge is caused to offer the circuit indicated, and the voice voltage will force a current *down* through resistor R_2, representing the output of the modulator. When the carrier voltage is as in Fig. 12-13c, the bridge is as indicated, and the voice modulating voltage will force a current *up* through the output resistor R_2. Because of this switching action, one cycle of modulating voice voltage will cause the modulated signal output shown in Fig. 12-14.

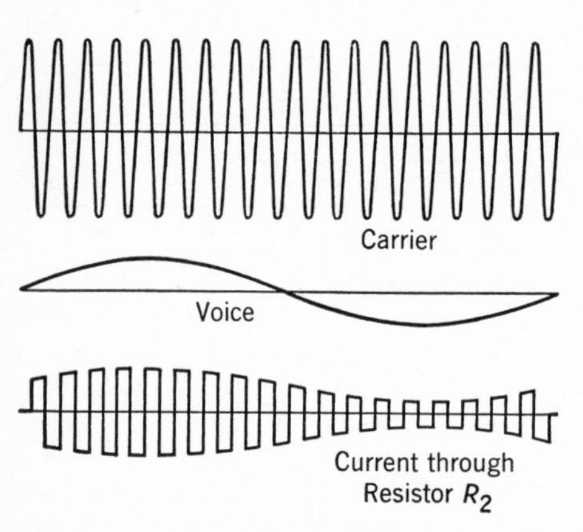

FIG. 12-14. Because of the "switching" action of the carrier wave on the copper-oxide varistor, and because the source of carrier voltage always is connected across resistor R_2 representing the load in Fig. 12-13, the current through R_2 will appear approximately as shown if examined with an oscillograph.

The current I_2 that E_v causes to flow through resistor R_2 is determined by the **transfer impedance** between the source of voice-frequency voltage E_v and the load resistor R_2. It is in reality this transfer impedance which is controlled by the carrier frequency impressed on the bridge circuit. If it is assumed that the bridge circuit varies as indicated by (b) and (c) of Fig. 12-13, then the transfer impedance between the source of audio voltage E_v and the load resistor R_2 varies as a rectangular wave between equal maximum positive and negative values.

The *maximum* value of the transfer impedance may be found as follows: The transfer impedance $Z_t = E_v/I_2$, where I_2 is the current through the resistor R_2 which represents the line (or load) to which modulated energy is to be fed. Thus, using either Fig. 12-13b or c, and assuming that the bridge elements are of either zero or infinite resistance, an expression for the current I_2 through R_2 is written, and when this is divided into E_v, the transfer impedance becomes

$$Z_t = \frac{(R_1R_2 + R_2R_3 + R_1R_3)}{R_3} \qquad \text{12-11}$$

For solving mathematically, it is more convenient to write the expression for the transfer admittance and use this instead of the transfer impedance. The transfer admittance merely is the reciprocal of the transfer impedance, and also varies between equal positive and negative maximum values in accordance with the impressed carrier frequency.

The equation for a rectangular recurring wave is

$$y = \frac{4Y}{\pi}\left(\sin \omega t + \frac{1}{3}\sin 3\omega t + \frac{1}{5}\sin 5\omega t + \cdots\right) \qquad \text{12-12}$$

where y is the instantaneous value and Y is the maximum value. For representing the variations in the admittance of the circuit, Y is the maximum value as explained in the preceding paragraph, and ωt becomes $\omega_c t$, and $\omega_c = 2\pi f_c$, where f_c is the carrier frequency. It is, therefore, possible to write an expression for the admittance of the circuit at any instant, as this circuit is varied at the carrier-frequency rate.

The instantaneous expression for the voice-frequency signal is

$$e_v = E_v \sin \omega_v t \qquad \text{12-13}$$

where $\omega_v = 2\pi f_v$, and f_v is the voice frequency. When the expressions for the varying transfer admittance and the voice-frequency voltage are multiplied, an equation is obtained for the instantaneous current through the load circuit R_2. In this equation will be found the upper and lower sideband terms. These sidebands have been created in the circuit by a process of distortion caused by the varying admittance. Of course the preceding action is ideal, and neglects the existence of undesired distortion, some of which is removed by filters.[27] Although only copper-oxide varistors have been considered, it is important to note that germanium varistors are used in similar circuits for modulation; also, modulation (and demodulation) is possible with transistors but at present (1955) rarely is used. (See reference 7 at the end of Chapter 11.)

12-15. *BEAT-FREQUENCY OSCILLATORS*

These were not considered previously because they involve the principle of modulation. In this device two vacuum-tube circuits oscillate at slightly different frequencies; for example 100,000 and 101,000 cycles. These two voltages simultaneously are impressed on a *vacuum-tube modulator,* and sum and difference frequencies are created by the modulating or distorting process. The output of the modulator is connected to a low-pass filter which passes the lower sideband. In the instance here considered this would be a 1000-cycle wave. Usually the selected sideband would be amplified.

The words *vacuum-tube modulator* are stressed because it often is stated that the two frequencies are "beat together and the beat frequency is extracted by the detector tube." Such statements are meaningless, because merely "beating" two waves or "mixing them together" *does not create a new frequency* unless the process of modulation is involved. This principle is stressed because it is the cause of confusion. Because the frequencies generated by the two oscillators are high, small coils and capacitors can be used. The frequency of one oscillator only need be changed. In most instances it is only necessary to vary the setting of an air capacitor. The frequency range can be made continuously variable, an important factor.

12-16. *ANGLE MODULATION*

As defined,[1] modulation is "the process or result of the process whereby some characteristic of one wave is varied in accordance with another wave." As explained in the preceding pages, in amplitude modulation the instantaneous magnitude of a sinusoidal carrier wave is caused to vary in accordance with the voice.

In **angle modulation,**[1] the "angle of a sine-wave carrier is the characteristic varied from its normal value." For convenience assume that the modulating wave is a sinusoidal "voice" wave, the instantaneous value being

$$e_v = E_{mv} \sin \omega_v t \qquad 12\text{-}14$$

where E_{mv} is the maximum value, and $\omega_v = 2\pi f_v$, f_v being the voice frequency. The equation for the instantaneous value of the sinusoidal carrier wave to be modulated is

$$e_c = E_{mc} \sin 2\pi f_c t \qquad 12\text{-}15$$

where E_{mc} is the maximum value of the carrier and $2\pi f_c t$ is the angle of the carrier varying with time, the frequency being f_c. Carrier frequency f_c is the frequency assigned to a radio transmitter; *this frequency must be held constant.* For explaining angle modulation the statement cannot be made that angle $2\pi f_c t$ is varied, and in explaining frequency modulation the statement cannot be made that f_c is varied; instead, equation 12-15 must be written

$$e_c = E_{mc} \sin (\omega_c t + \phi) \qquad 12\text{-}16$$

This equation is the basis for explaining both phase and frequency modulation, which are forms of angle modulation.

12-17. *PHASE MODULATION*

In phase modulation (PM)[1] the "angle of a sine-wave carrier is caused to depart from the carrier angle by an amount proportional to the instantaneous value of the modulating wave." Thus in equation 12-16, $\omega_c t$ is the carrier angle and ϕ is varied with respect to it. If m_p is the phase deviation,[1] or "the peak difference between the instantaneous angle of the modulated wave and the angle of the sine-wave carrier," then m_p is the greatest phase-shift angle that can be caused by the sinusoidal modulating signal of equation 12-14. At any time t, the magnitude of the instantaneous phase shift will be $\phi = m_p \sin \omega_v t$. Then equation 12-16 becomes

$$e_{pm} = E_{mc} \sin (\omega_c t + m_p \sin \omega_v t) \qquad 12\text{-}17$$

and this is the equation for a phase-modulated wave in accordance with the definition given.

12-18. *FREQUENCY MODULATION*

In **frequency modulation (FM)**,[1] for a sine-wave carrier "the instantaneous frequency of the modulated wave differs from the carrier frequency by an amount proportional to the instantaneous value of the modulating wave." The **instantaneous frequency** [1] is the "time rate of change of the angle of an angle-modulated wave." The instantaneous frequency of the angle-modulated wave (equation 12-16) is

$$f_{inst} = \left(\frac{1}{2\pi}\right)\left(\frac{d}{dt}(\omega_c t + \phi)\right) = \frac{\omega_c}{2\pi} + \left(\frac{1}{2\pi}\right)\left(\frac{d\phi}{dt}\right) \qquad 12\text{-}18$$

This applies because $\omega_c = 2\pi_c$ is a constant, f_c being the carrier frequency assigned to the station, and *is fixed.* In frequency modulation only $d\phi/dt$ can be caused to vary by the modulating voice wave.

By definition,[1] the **frequency deviation** (here designated by f_d) is "the peak difference between the instantaneous frequency of the modulated wave and the carrier frequency." Thus, f_d is the maximum change in frequency between the modulated wave and the assigned carrier frequency. The instantaneous value of the frequency change caused by the modulating voice wave is $f_d \sin \omega_v t$, and from equation 12-18

$$\frac{1}{2\pi}\frac{d\phi}{dt} = f_d \sin \omega_v t \qquad \text{12-19}$$

and

$$\phi = 2\pi \frac{f_d}{\omega_v} \int \sin \omega_v t \omega_v \, dt = -\frac{f_d}{f_v} \cos \omega_v t \qquad \text{12-20}$$

When this is substituted in equation 12-16, the expression for a frequency-modulated wave becomes

$$e_{fm} = E_{mc} \sin (\omega_c t - m_f \cos \omega_v t) \qquad \text{12-21}$$

where $m_f = f_d/f_v$, and is called the **modulation index,** defined [1] as "the ratio of the frequency deviation to the frequency of the modulating wave." If desired, the equation for the modulating wave may be specified as $e_v = E_{mv} \cos \omega_v t$ instead of as equation 12-14; then equation 12-21 becomes

$$e_{fm} = E_{mc} \sin (\omega_c t + m_f \sin \omega_v t) \qquad \text{12-22}$$

A frequency-modulated wave appears as in Fig. 12-15 if the modulating wave is sinusoidal; an amplitude-modulated wave is shown for comparison. The extent, or *magnitude,* of the instantaneous frequency swing is determined (for a certain FM radio transmitter) by the *amplitude* of the modulating signal at that instant. The *number of times* per second that the frequency changes is determined by the frequency of the modulating signal. The *amplitude* of a frequency-modulated wave *is constant.*

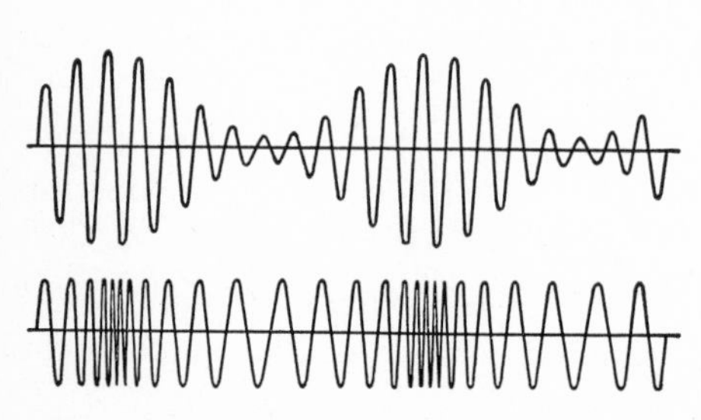

FIG. 12-15. Comparison of an amplitude-modulated wave (above) and a frequency-modulated wave (below).

An amplitude-modulated wave was shown to contain a carrier wave and two sidebands. A frequency-modulated wave contains many sidebands; theoretically an infinite number. This can be illustrated by expanding equation 12-22 and evaluating by Bessel functions.[28] Thus, equation 12-22 becomes

$$e_{fm} = A\{J_0(m_f)\sin\omega_c t + J_1(m_f)[\sin(\omega_c+\omega_v)t - \sin(\omega_c-\omega_v)t]$$
$$+ J_2(m_f)[\sin(\omega_c+2\omega_v)t + \sin(\omega_c-2\omega_v)t]$$
$$+ J_3(m_f)[\sin(\omega_c+3\omega_v)t - \sin(\omega_c-3\omega_v)t]$$
$$+ J_4(m_f)[\sin(\omega_c+4\omega_v)t + \sin(\omega_c-4\omega_v)t] + \cdots\} \qquad 12\text{-}23$$

The values of J are plotted in Fig. 12-16; from this, or from tables, the magnitudes of the sidebands can be found. Thus, if $m_f = f_d/f_v = 2.0$, the maximum magnitude of the carrier component will be approximately $0.3E_{mc}$, where E_{mc} is the maximum value of the unmodulated carrier output. Also, at $m_f = 2.0$, the first-order sidebands have magnitudes of about $0.57E_{mc}$, and this pair of sidebands will have frequencies $(f_c + f_v)$ and $(f_c - f_v)$, respectively, the second-order sidebands will have approximate magnitudes of $0.34E_{mc}$ and frequencies $(f_c + 2f_v)$ and $(f_c - 2f_v)$, etc. In this manner the frequency spectrum for any modulation index may be found. It is of interest to recall that in an amplitude-modulated wave the component of carrier frequency *does not vary* in amplitude, and to note that in a frequency-modulated wave the component of carrier frequency *does vary* in amplitude.

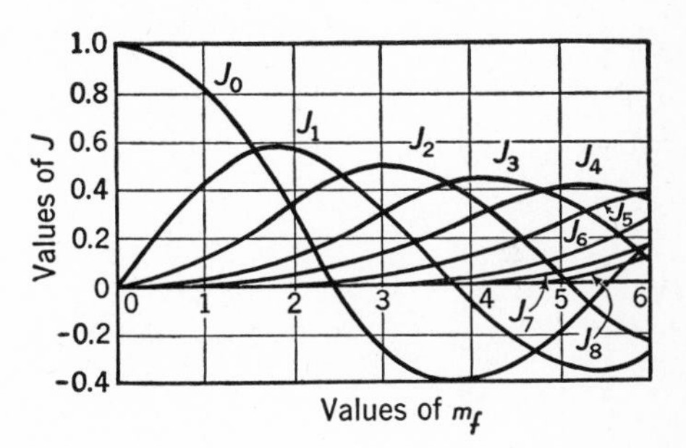

FIG. 12-16. Curves for evaluating J of equation 12-23.

12-19. *COMPARISON OF AMPLITUDE AND FREQUENCY MODULATION*

The expression for an amplitude-modulated voltage wave can be obtained from equation 12-4. If the wave is 100 per cent modulated, and if the maximum value of the unmodulated carrier voltage output is assumed 1.0, then

$$e_{am} = 1.0\sin\omega_c t + 0.5\cos(\omega_c - \omega_v)t - 0.5\cos(\omega_c + \omega_v)t \qquad 12\text{-}24$$

If f_c is the carrier frequency, $\omega_c = 2\pi f_c$; likewise, if f_v is the modulating voice frequency, $\omega_v = 2\pi f_v$.

The expression for a frequency-modulated wave is given by equation 12-23. If $m_f = f_d/f_v = 1.0$, then from Fig. 12-16, $J_0 = 0.8$, $J_1 = 0.4$, and

$J_2 = 0.1$ approximately. If J_2, J_3, etc., are assumed zero, and if A is assumed to be 1.25, then from equation 12-23 the frequency-modulated wave will be

$$e_{fm} = 1.0 \sin \omega_c t + 0.5 \sin (\omega_c + \omega_v)t - 0.5 \sin (\omega_c - \omega_v)t \qquad 12\text{-}25$$

These expressions for amplitude-modulated and frequency-modulated waves now will be studied. In Fig. 12-17 are shown diagrams [28] for amplitude modulation, diagrams for frequency modulation, and the sine-wave modulating signal voltage. *For the conditions specified,* equation 12-24 for

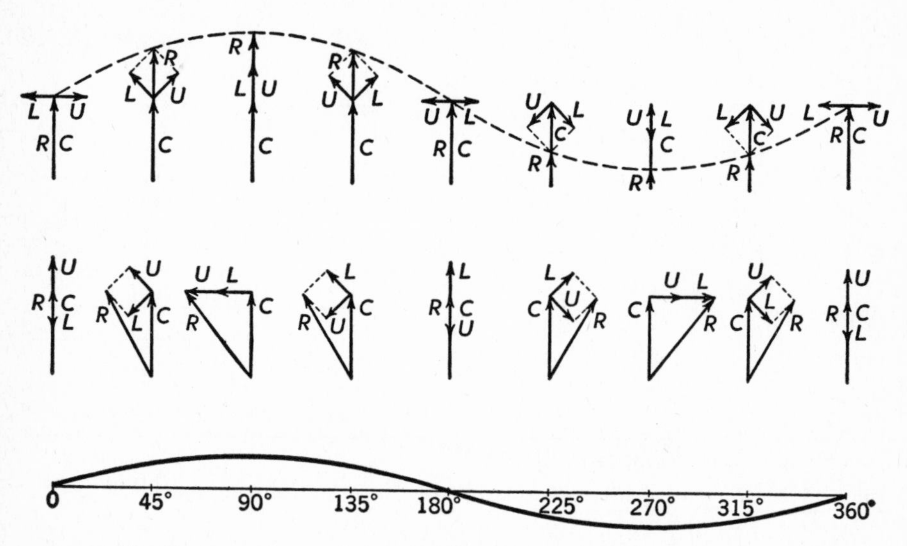

FIG. 12-17. Rotating phasors (vectors) consisting of carrier component C, upper sideband U, lower sideband L, and R, the resultant of the three components. The upper diagram is for amplitude modulation; the center diagram represents frequency modulation for a small value of m_f; and the lower curve represents the sine-wave modulating signal. A comparison of the lengths of the resultants of the center figure will show a small amplitude variation indicating some (unwanted) amplitude modulation as well as frequency modulation. This is because only the first sidebands were considered.

the amplitude-modulated wave consists of a carrier component of length 1.0 and angular velocity ω_c, an upper sideband component of length 0.5 and angular velocity $\omega_c + \omega_v$, and a lower sideband of length 0.5 and angular velocity $\omega_c - \omega_v$. Also, *for the conditions specified,* the frequency-modulated wave will contain the same components, with this difference: All components of the frequency-modulated wave are sine terms, whereas the amplitude-modulated wave contains both sine and cosine terms.

Thus in Fig. 12-17, the diagram at the left of the upper series representing amplitude modulation is composed of carrier component C, lower sideband L, and upper sideband U. Because these are positive cosine and negative cosine terms, they are at right angles to the carrier and point in opposite directions. Also, because lower sideband L has lower angular velocity than carrier C (which also rotates at its angular velocity but here is maintained stationary for reference), and because upper sideband U has greater angular velocity than carrier C, at successive intervals of the modulating wave, the three components will combine as indicated by resultant R. The lengths of this resultant give the instantaneous magnitudes of the amplitude-modulated wave. In the center series of diagrams for frequency modulation, because the upper sideband U is a positive sine term, and the lower sideband L is a negative sine term, the initial positions (for zero modulating voltage) are as shown at the left. As these components rotate at their different angular velocities about the end of carrier C, the resultant R gives the instantaneous magnitudes of the frequency-modulated wave. It will be noted that resultant R is not always exactly the same length, and for this reason the frequency-modulated wave will contain a small amount of amplitude modulation.

Of interest is a comparison of the band-width requirements of systems of amplitude modulation and frequency modulation. Radio-broadcast applications will be considered. The modulating signal is assumed to cover the band 50 to 10,000 cycles. If the carrier to be amplitude-modulated is 1.0 megacycle, the modulated signal will consist of a lower sideband of 990,000 to 999,950 cycles, the carrier frequency of 1.0 megacycle, and the upper sideband from 1,000,050 to 1,010,000 cycles, the total band width required being 20,000 cycles. For frequency-modulation broadcast a typical *radiated* carrier frequency is 100 megacycles, and for peak modulation, the maximum frequency swing is $\pm 75{,}000$ cycles. Assuming maximum frequency modulation, the frequency deviation is $f_d = 75{,}000$ cycles, and for a modulating frequency of 50 cycles, $m_f = f_d/f_v = 75{,}000/50 = 1500$. For a modulating frequency of 10,000 cycles, $m_f = 75{,}000/10{,}000 = 7.5$. A study of the discussion following equation 12-23 will indicate that for $m_f = 1500$ an extremely large number of sidebands of appreciable magnitude will exist; however, the sidebands will be close together. For $m_f = 7.5$ the sidebands of appreciable magnitude are relatively few, but they are comparatively far apart. For any modulating frequency it will be found that all important sidebands will lie with a band width of 200,000 cycles, 100,000 below and 100,000 cycles above the carrier. Thus high-quality AM broadcast requires a frequency band of 20,000 cycles, and FM broadcast a 200,000-cycle band. It should be mentioned that FM systems, for communication purposes only, do not use frequency swings as great as $\pm 75{,}000$ cycles, and do not require transmission bands 200,000 cycles wide.

12-20. *SYSTEMS OF FREQUENCY MODULATION*

Many systems for producing frequency-modulated waves have been developed. Three of these will be discussed in this section; others are presented in references 29, 30, and 31.

Frequency Modulation by Wave Synthesis. This method early was discussed by Roder,[32] and first used by Armstrong.[33] It is based on several important facts: First, Roder showed[32] that for a *small* modulation index $m_f = f_d/f_v$ (such as 0.5), a frequency-modulated wave and an amplitude-modulated wave were similar, but that the sidebands were 90° out-of-phase with the carrier for one type, and in phase for the other type. This was discussed in the preceding section. Roder also pointed out that a frequency-modulated wave *would result* if the audio signal that was being used to drive a phase modulator was first distorted so that the magnitudes of the various components of the modulating signal were made inversely proportional to frequency of that component. This follows from an examination of equations 12-17 and 12-21, which differ in that m_p is a constant, but

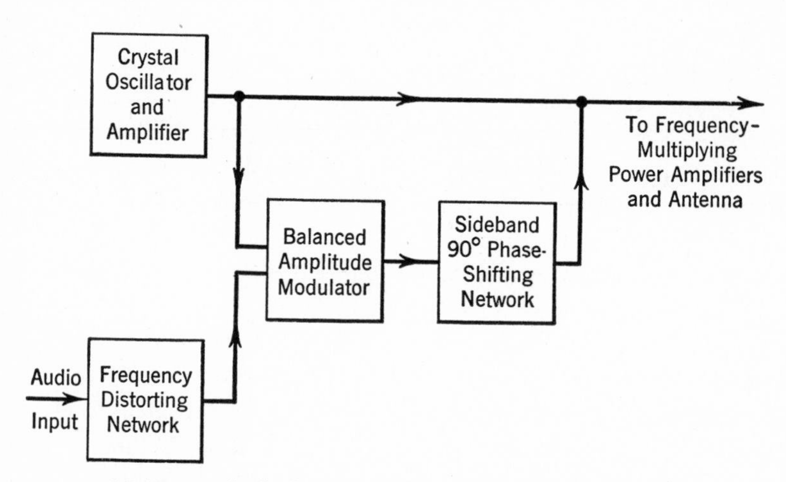

FIG. 12-18. Method of synthesizing a frequency-modulated wave.

$m_f = f_d/f_v$. Such an arrangement for producing frequency-modulated waves is shown in Fig. 12-18. The balanced amplitude modulator (Fig. 12-5) suppresses the carrier. Then, if the modulator is driven by an audio signal that passes through a distorting network (as just explained) and if the sidebands are shifted in phase 90° and combined with carrier signal direct from the oscillator, the output is frequency-modulated in accordance with statements in the preceding section. It is important to note that the modulation index must be small—so small, in fact, that an amplitude-modulated wave and the frequency-modulated wave are much the same. If such frequency-modulated signals were transmitted in a system of radio, the main advantage of frequency-modulation (improved reception during severe electrical-circuit noise conditions) would be reduced considerably (Chapter 13). However, since the processes outlined in Fig. 12-18 best can be accomplished at frequencies of about a megacycle, and since frequency-modulation systems often transmit at much higher frequencies (sometimes hundreds of megacycles), considerable frequency multiplication (section 8-17) is required prior to radiation. The effect of this is to multiply also the frequency swing and produce a frequency-modulated wave having the same characteristics as those resulting if a large modulation index had been used. Fortunately, the wave synthesis and frequency multiplication can be achieved at low power levels using small-signal tubes.

Frequency Modulation by Reactance Tubes. If the current in a vacuum-tube circuit can be made to lag the voltage across the circuit by 90°, and if by varying some tube electrode potential the input current can be made to vary, then the vacuum-tube circuit will have the property of variable inductive reactance. Or, if the current in a vacuum-tube circuit can be made to lead the voltage by 90°, and if by varying some electrode potential the current magnitude can be varied, then the vacuum-tube circuit will have the property of variable capacitive reactance. In reactance-tube modulation a vacuum-tube circuit is made to control the frequency of an oscillator by acting as a variable inductor, or as a variable capacitor.

The basic circuit of a reactance-tube frequency modulator is shown in Fig. 12-19. If connections are *not* made at points 1–2, the parallel circuit C–L determines the frequency of the oscillator. The series circuit composed of C_1 and R_1 is connected across tuned circuit C–L. This series circuit has high impedance because C_1 has low capacitance; resistor R_1 is of low resistance. Tuned circuit C–L and the oscillator tube act as a load of essentially pure resistance across the reactance tube.

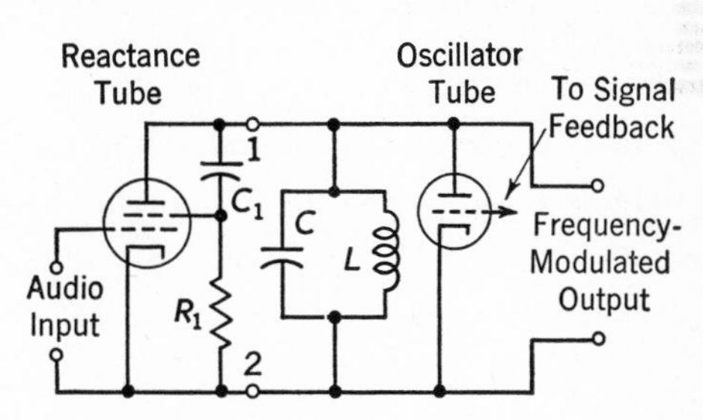

FIG. 12-19. Simplified circuit of a reactance-tube modulator.

Assume that the tube at the right is oscillating; then, a voltage at the frequency of oscillation will exist across tuned circuit C–L and points 1–2. A very small current will flow through high-impedance circuit C_1 and R_1, and because of the low capacitance of C_1, this current will *lead* the voltage across points 1–2 by approximately 90°. The voltage drop across R_1 will be in phase with this current, and this voltage is impressed between a control grid and cathode of a multielectrode tube having two "control" grids of about equal sensitivity. This voltage will be amplified and will cause an alternating plate current to flow. Because the plate current of a vacuum tube is in phase with the grid voltage (section 7-8), the alternating plate current of the reactance tube at the left of Fig. 12-19 will lead the voltage across the tuned circuit. Thus, the reactance-tube circuit appears to terminals 1–2 to be essentially pure capacitance, and this will alter the oscillator frequency. Furthermore, the varying audio-frequency modulating voltage impressed between the other control grid and cathode can be made to alter the magnitude of the reactive current in accordance with the voice. In this way, the apparent capacitance that the reactance tube in effect connects across the oscillator tuned circuit will vary with the voice, and the oscillator output will be frequency-modulated.

The reactance-tube circuit of Fig. 12-19 can be made to present variable inductance between points 1–2 in the following manner: A blocking capacitor of negligible reactance is placed in series at point 1; a resistor is substituted for C_1, and a capacitor (shunted with a resistor of several megohms so that the grid will not be isolated) is substituted for R_1. This new R–C series combination must have high impedance, and the resistance must greatly exceed the capacitive reactance. Then, the small current through the R–C branch will be almost in phase with the voltage across points 1–2, the voltage across the grid capacitor will lag the current by almost 90°, and the amplified plate current delivered to terminals 1–2 will lag the voltage across this circuit, in effect placing a variable inductance across it.

Reactance tubes are used extensively for modulation in FM transmitters, balanced circuits being commonly employed.[29,30,34,35,36] It was originally used for frequency control.[37] In fact, the reactance-tube circuit is an impedance element of basic importance.[38] If the oscillator of Fig. 12-19 is disconnected and an impedance bridge is connected across points 1–2, measurements will show that the reactance tube does offer capacitive or inductive reactance.

Frequency Modulation Using Special Tubes. Special electron devices have been developed for frequency modulation, a particularly successful device being the **Phasitron.**[39] Its operation will be summarized briefly. Electrons are emitted from a cathode that is between special focusing electrodes and two anodes, coaxial construction being used. The electrons flow to the anodes in the form of a *thin electron disk.* Below this "disk" are deflector wires having a three-phase radio-frequency voltage applied so that a rotating electric field exists beneath the electron disk. This three-phase rotating alternating electric field causes the electron disk to have a "scalloped" edge that rotates with the field. The two anodes are constructed and arranged so that the "scalloped electron disk" delivers an alternating current to the anodes. The tube is placed in a coil that carries the modulating current, and the magnetic field of the coil causes the scalloped edge to advance and retire from the normal position in accordance with the variations of the modulating signal. This causes phase modulation. However, the modulating signal input is distorted (as has been explained), and thus the Phasitron produces a frequency-modulated wave.

12-21. *PULSE MODULATION*

As defined [1] in section 12-2, **pulse modulation** is "modulation of one or more characteristics of a pulse carrier." A pulse carrier may be a series of pulses having several characteristics that may be altered by different modulating methods. Pulse modulation has had limited application since about 1940, and still is in the developmental state. The subject will be summarized in the following paragraphs.

In a sense, pulse modulation is not new; the early Morse telegraph used a pulse code, and more recently, printing-telegraph systems use a pulse code. However, in the systems to be considered, speech, and even television programs, have been transmitted by pulse methods. In discussing pulse modulation the Standards will be followed closely to avoid confusion.

Pulse-Amplitude Modulation. In **pulse-amplitude modulation (PAM)**[39] a pulse carrier is amplitude-modulated as in Fig. 12-20.

Pulse-Code Modulation. In **pulse-code modulation (PCM)**[1,39] a pulse code consisting of a modulated train of pulses is used. A system of pulse-code modulation will be considered in the following section.

Pulse-Time Modulation. In ***pulse-time modulation*** [1,39] "the time of occurrence of some characteristic of a pulse carrier is varied from the unmodulated wave." Several types of pulse-time modulation are possible.

In **pulse-duration modulation** [1,39] the duration of a pulse is varied by the modulating signal. (Fig. 12-20.) This also is called **pulse-length modulation** and **pulse-width modulation.**[1,39]

In **pulse-frequency modulation (PFM)**[1,39] the pulse repetition rate is the characteristic varied by the modulating wave (Fig. 12-20). A more precise term is **pulse repetition-rate modulation.**[1,39]

In **pulse-interval modulation** [1,39] the pulse spacing is varied.

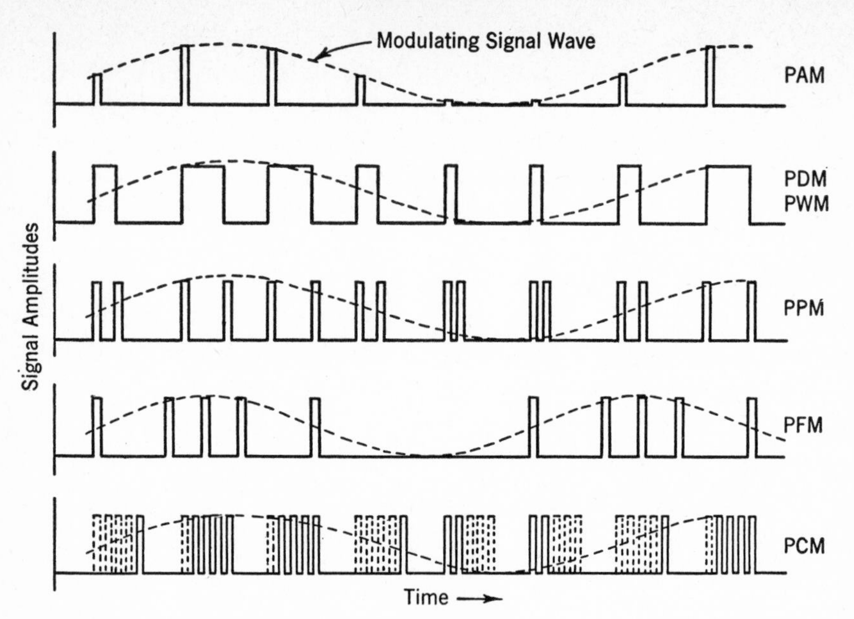

FIG. 12-20. Approximate signals in pulse-modulated waves when modulated with a sinusoidal signal. These diagrams are to convey a general picture; relative heights are of no significance. (Reference 40.)

In **pulse-position modulation (PPM)**[1,39] the "position in time of a pulse is varied" by the modulating signal (Fig. 12-20). This has been called **pulse-phase modulation.**[1,39]

12-22. *PULSE-CODE MODULATION (PCM)*

In pulse-code modulation a train, or sequence, of pulses is coded in the modulation process so that information can be transmitted. This has been called[40] **pulse-count modulation.** An elementary system is shown in Fig. 12-20 where the modulating sine wave is sampled at several points, and information specifying the signal magnitude at each point is transmitted by a pulse code; that is, a certain magnitude is transmitted by sending one pulse out of five, etc.

A pulse-code system having promising characteristics has been developed. This system has been used to transmit speech, music, and television programs.[41,42,43] The system can be described briefly as follows:[44] The signal to be transmitted is *sampled* at regular intervals, and the amplitude of the signal at each interval is determined. The amplitudes are *quantized* by placing them in discrete ranges of amplitudes. Information regarding the quantized range into which a given sampled amplitude falls is transmitted by a code similar to PCM of Fig. 12-20. Of much interest is the electron-beam tube that quantizes the signals.[43] At the receiving end the pulse signals are decoded and the original modulating signal is reconstructed.

If a sine wave is sampled and quantized, and coded pulses are transmitted merely to reconstruct a sine wave at the receiving station, it is evident that a comparatively wide frequency band is required to transmit a single frequency. To transmit a signal having a given band width requires a channel of much greater band width. Nevertheless, the system has important advantages. One is that communication can be maintained under excessive interference conditions. For example, in a microwave radio-relay system, if the signal received at a repeater point is badly distorted, but if the mere presence and absence of the pulses can be determined, the signal can be reconstructed, amplified, and transmitted in its original shape on to the next repeater point. Reference 45 is suggested as a source of additional information on pulse-code modulation systems.

REFERENCES

1. Institute of Radio Engineers. *Standards on Modulation Systems: Definitions of Terms,* 1953. Proc. I.R.E., May 1953, Vol. 41, No. 5.
2. Albert, A. L. *Electrical Communication.* John Wiley & Sons.
3. Clark, A. B., and Green, C. W. *Long distance cable circuit for program transmission.* Bell System Technical Journal, July 1930, Vol. 9, No. 3.
4. Snow, W. B. *Audible frequency range for music, speech, and noise.* Journal Acoustical Society of America, July 1931, Vol. 3.
5. Chinn, H. A., and Eisenberg, P. *Tonal-range and sound-intensity preference of broadcast listeners.* Proc. I.R.E., Sept. 1945, Vol. 33, No. 9.
6. Proc. I.R.E., Oct. 1946, Vol. 34, No. 10, pp. 757–761.
7. Chinn, H. A., and Eisenberg, P. *Influence of reproducing system on tonal-range preference.* Proc. I.R.E., May 1948, Vol. 36, No. 5.
8. Olson, H. F. *Frequency-range preference for speech and music.* Electronics, Aug. 1947, Vol. 20, No. 8.
9. Institute of Radio Engineers. *Standards on Antennas, Modulation Systems, and Transmitters: Definitions of Terms.* 1948.
10. American Institute of Electrical Engineers. *American Standard Definitions of Electrical Terms.* 1941.
11. Institute of Radio Engineers. *Standards on Receivers: Definitions of Terms.* 1952. Proc. I.R.E., Dec. 1952, Vol. 40, No. 12.
12. Terman, F. E., and Pettit, J. M. *Electronic Measurements.* McGraw-Hill Book Co.
13. Everitt, W. L. *Communication Engineering.* McGraw-Hill Book Co.
14. Peterson, E., and Keith, C. R. *Grid current modulation.* Bell System Technical Journal, Jan. 1928, Vol. 7, No. 1.
15. van der Bijl, H. J. *Thermionic Vacuum Tubes.* McGraw-Hill Book Co.

16. Colpitts, E. H., and Blackwell, O. B. *Carrier current telephony and telegraphy.* Trans. A.I.E.E., 1921, Vol. 40.
17. Affel, H. A., Demarest, C. S., and Green, C. W. *Carrier systems on long distance telephone lines.* Bell System Technical Journal, July 1928, Vol. 17, No. 3.
18. Mason, W. P. *Electrical wave filters employing quartz crystals as elements.* Bell System Technical Journal, July 1934, Vol. 13, No. 4.
19. Kishpaugh, A. W. *A low-power broadcast transmitter.* Bell Laboratories Record, Oct. 1932, Vol. 11, No. 2.
20. Oswald, A. A. *A short-wave single-side-band radiotelephone system.* Proc. I.R.E., Dec. 1938, Vol. 26, No. 12.
21. American Radio Relay League. *The Radio Amateur's Handbook.*
22. Terman, F. E. *Radio Engineering.* McGraw-Hill Book Co.
23. Terman, F. E. *Radio Engineers' Handbook.* McGraw-Hill Book Co.
24. Parker, W. N. *A unique method of modulation for high-fidelity television transmitter.* Proc. I.R.E., Aug. 1938, Vol. 26, No. 8.
25. Green, C. W., and Green, E. I. *A carrier-telephone system for toll cables.* Electrical Engineering, May 1938, Vol. 57, No. 5.
26. Kendall, B. W., and Affel, H. A. *A twelve-channel carrier telephone system for open-wire lines.* Electrical Engineering, July 1939, Vol. 58, No. 7.
27. Caruthers, R. S. *Copper oxide modulators in carrier telephone systems.* Bell System Technical Journal, Apr. 1939, Vol. 18, No. 2.
28. Everitt, W. L. *Frequency modulation.* Electrical Engineering, Nov. 1940, Vol. 59, No. 11.
29. Hund, A. *Frequency Modulation.* McGraw-Hill Book Co.
30. Black, H. S. *Modulation Theory.* D. Van Nostrand Co.
31. Wrathhall, L. R. *Frequency modulation by non-linear coils.* Bell Laboratories Record, March 1946, Vol. 25, No. 3.
32. Roder, H. *Amplitude, phase, and frequency modulation.* Proc. I.R.E., Dec. 1931, Vol. 19, No. 4.
33. Armstrong, E. H. *A method of reducing disturbances in radio signaling by a method of frequency modulation.* Proc. I.R.E., May 1936, Vol. 24, No. 5.
34. Crosby, M. G. *Reactance tube frequency modulators.* RCA Review, July 1940, Vol. 5, No. 1.
35. Morrison, J. F. *A new broadcast transmitter circuit designed for frequency modulation.* Proc. I.R.E., Oct. 1940, Vol. 28, No. 10.
36. Hund, A. *Reactance tubes in F-M applications.* Electronics, Oct. 1942, Vol. 15, No. 10.
37. Travis, C. *Automatic frequency control.* Proc. I.R.E., Oct. 1935, Vol. 23, No. 10.
38. Reich, H. J. *The use of vacuum tubes as variable impedance elements.* Proc. I.R.E., June 1942, Vol. 30, No. 6.
39. Institute of Radio Engineers. *Standards on Pulses: Definitions of Terms.* Part 1, Proc. I.R.E., June 1951, Vol. 39, No. 6; Part 2, Proc. I.R.E., May 1952, Vol. 40, No. 5.
40. Grieg, D. D. *Pulse-count modulation.* Electrical Communication, Sept. 1947, Vol. 24, No. 3.

41. Goodall, W. M. *Telephony by pulse-code modulation.* Bell System Technical Journal, July 1947, Vol. 26, No. 3.
42. Feldman, C. B. *A 96-channel pulse-code modulation system.* Bell Laboratories Record, Sept. 1948, Vol. 26, No. 9.
43. Meacham, L. A., Peterson, E., and Sears, R. W. *Pulse-code modulation using electron-beam tubes.* Bell System Technical Journal, Jan. 1948, Vol. 27, No. 1.
44. Oliver, B. M., Pierce, J. R., and Shannon, C. E. *The philosophy of PCM.* Proc. I.R.E., Nov. 1948, Vol. 36, No. 1.
45. Oxford, A. J. *Pulse-code modulation systems.* Proc. I.R.E., July 1953, Vol. 41, No. 7.

QUESTIONS

1. Why is modulation necessary in radio and in telephony?
2. What is a carrier wave? Does it carry the signal?
3. What band width is required for speech? For radio programs?
4. By what two different methods is amplitude modulation accomplished with vacuum tubes?
5. What type of amplitude modulation is used most in radio broadcasting?
6. A wave analyzer is tuned to the carrier frequency of an amplitude-modulated radio transmitter. The magnitude of the carrier component is measured. Will the indication of the wave analyzer vary when the modulation is varied? Why?
7. Is the grid-modulated class-C amplifier actually a grid-circuit modulated device?
8. Do the analyses made in sections 12-4 and 12-6 verify the statement at the close of section 12-3?
9. Why is the Heising system called constant-current modulation?
10. Why is a resistor sometimes placed in series with L_2 of Fig. 12-7?
11. Why are frequency and phase modulation classed as angle modulation?
12. What fundamental difference exists between phase and frequency modulation?
13. Does the magnitude of the carrier vary in frequency modulation? Does the frequency of the carrier vary in frequency modulation?
14. How does the reactance-tube system of frequency modulation operate?
15. How can you prove that the circuit to the left of points 1–2 of Fig. 12-19 acts like inductance or capacitance?
16. If the top row of diagrams of Fig. 12-17 represents amplitude modulation, why does it not define both halves of an amplitude-modulated wave?
17. Why does a frequency-modulated signal as used in broadcast occupy a radio-spectrum band width of about 200,000 cycles?
18. How could you determine when m_f was about 2.5 in frequency modulation?
19. Does the amount of power radiated by a frequency-modulation transmitter vary during modulation?
20. From the standpoint of the radio-spectrum band width required, how do amplitude-modulation, frequency-modulation, phase-modulation, and pulse-code modulation systems of similar quality compare?

PROBLEMS

1. Verify the statements made in section 12-4 regarding the power division for a 100 per cent amplitude-modulated wave, and for the average power output and the power output on modulation peaks.
2. Referring to section 12-4, determine the relative power relations in the carrier and sidebands for 50 per cent modulation. What is the average power output and the output on modulation peaks as compared with no modulation?
3. A plate-modulated class-C amplifier puts out a one-kilowatt carrier when it is unmodulated. Calculate the power output from the modulating amplifier for 25, 50, 75, and 100 per cent modulation. Write a brief discussion of this method as compared with the grid-modulated class-C amplifier, giving the merits of each system.
4. Prove mathematically the statements made immediately following equation 12-13.
5. A frequency-modulation communication system uses a maximum frequency change, or frequency deviation, of 30,000. During the modulation process at an instant the modulation index is 5.0. Draw a graph showing the frequency spectrum. Does this mean that the modulating wave had a frequency of 6,000 cycles? Explain.

CHAPTER 13

DEMODULATORS

The preceding chapter was devoted to a study of modulation, a process by which low-frequency audio or other signals are translated, or moved, to higher-frequency bands for transmission over wires or through space to the distant receiving station. This chapter will consider the methods used at the receiving station for returning these signals to their original bands of low frequencies.

Demodulation is defined [1] as "the process of recovering the modulating wave from the modulated carrier." In considering this definition, however, it should be remembered that, as explained in the preceding chapter, an amplitude-modulated wave is composed of a carrier component and two sidebands. A **demodulator** is a device to effect the process of demodulation. Sometimes demodulation is called **detection,** and a demodulator is called a **detector.** The demodulation of both amplitude-modulated waves and frequency-modulated waves will be considered in this chapter. The demodulation of pulse-modulated signals is discussed in the references given in the preceding chapter.

13-1. *THE DEMODULATION OF AMPLITUDE-MODULATED WAVES*

Demodulation has been defined [2] as "the process of *modulation* when carried out in such a manner as to recover the original signal." For amplitude modulation, modulation and demodulation are *the same fundamental process,* as this definition indicates. Thus, if an audio-frequency signal covering the band 200 to 3500 cycles, and a 20,000-cycle carrier simultaneously are impressed on an amplitude modulator, the information in the modulating audio signal will be translated to two sidebands, the lower sideband of 16,500 to 19,800 cycles, and the upper sideband of 20,200 to 23,500 cycles. Now, if either, or both, these sidebands and the carrier simultaneously are impressed on a modulator exactly the same as the one considered, sum and difference frequencies again will be created. Of interest are the difference frequencies 20,000 − (16,500 to 19,800), giving a band from 200 to 3500 cycles, and (20,200 to 23,500) − 20,000 giving a second band from 200 to 3500 cycles. These two bands add in intensity to give the final demodulated signal. Hence, *one basic circuit,* which may be regarded as a *modulator,* can be used to translate the information in a low-frequency audio signal to a higher-frequency region, or, the same circuit can be used to translate information back to the original low-frequency position. Fundamentally, modulation and demodulation are the same, a point often overlooked.

As has been stressed, after modulation the sidebands, *not the carrier,* contain the information to be transmitted. Furthermore, *each* sideband is complete within itself; each contains all the characteristics of the original modulating signal. It is, therefore, possible to transmit the desired information to the distant station by sending only *one* sideband. The carrier component may, or may not, be transmitted. Demodulation when the *carrier and both sidebands* are transmitted was considered in the preceding paragraph, and as explained, two components result from the process and these add to give a final signal twice the magnitude of each. If the *carrier and one sideband* are transmitted, then only one demodulated component results. If *one sideband* only is transmitted, then a locally generated signal of the same frequency as the original carrier and the received sideband are impressed on the demodulator, and one demodulated component of the original modulating frequencies will be created.

In amplitude-modulation radio-broadcast systems the carrier and both sidebands are transmitted. However, in many systems used for both wire [3,4] and radio [5] communication (not "broadcast"), only one sideband is transmitted. In some systems a greatly reduced carrier is transmitted with one sideband, the carrier signal being used for control of the local oscillator. Savings are made in equipment and power costs, and in band width when only one sideband is transmitted.

Demodulation is accomplished by the use of crystal rectifiers, by vacuum tubes, and by copper-oxide varistors. Two basic methods commonly are used, so-called "linear" detection, and square-law detection. The pages immediately following will be devoted to the demodulation of amplitude-modulated waves; the demodulation of frequency-modulated signals will be treated in the last pages of the chapter.

13-2. *DEMODULATION WITH "LINEAR" RECTIFIERS (AM)*

The term "linear" as used in radio will be considered to prevent confusion. A **linear power amplifier** is defined [6] as an amplifier for which the "signal output voltage is directly proportional to the signal input voltage." A **linear rectifier** is defined [6] as "a rectifier, the output current or voltage of which contains a wave having a form identical with the envelope of an impressed signal wave." If a sine-wave signal voltage is impressed on a linear *amplifier,* the output wave will be a perfect replica of the input wave. However, if a sine-wave signal voltage is impressed on a linear rectifier, the output wave will be a distorted version of the input; the output will be half-cycles (Fig. 6-1).

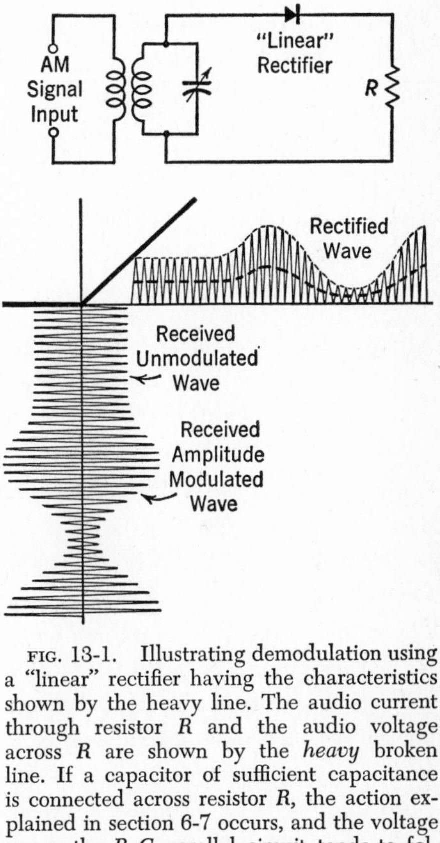

The characteristics of an ideal linear rectifier are shown in Fig. 13-1. If the incoming signal is an amplitude-modulated wave, the rectified current flowing through resistor R, and the voltage across resistor R, will be as shown. Assuming that the original modulating signal was a sine wave of frequency f_v (perhaps 1000 cycles), the rectified wave will have an "envelope" that is exactly like the "envelope" of the *positive side* of the impressed signal. This envelope is *not* the original modulating signal; in fact, the envelope is not a component of the signal: it is merely an outline of the wave. The received radio signal consists of a carrier component that *does not* vary, and two sidebands that *do* vary in accordance with the modulating signal. None of the original modulating voice-frequency wave is transmitted or received; the

FIG. 13-1. Illustrating demodulation using a "linear" rectifier having the characteristics shown by the heavy line. The audio current through resistor R and the audio voltage across R are shown by the *heavy* broken line. If a capacitor of sufficient capacitance is connected across resistor R, the action explained in section 6-7 occurs, and the voltage across the R–C parallel circuit tends to follow the "envelope" of the rectified wave. This envelope, shown by the *light* broken line, does not exist as a current or voltage, but is the instantaneous summation of the many components actually present in the rectified wave.

voice-frequency wave is recreated in the process of rectification, which process correctly may be regarded as distortion, when distortion is defined [6] as "a change in wave form." The recreated signal, a replica of the original modulating signal, is represented by the heavy broken line in Fig. 13-1.

An analysis of linear demodulation, or detection, is as follows: [7] The signal-voltage wave that modulated the carrier is assumed to have been a sine wave of voice frequency f_v, having an instantaneous value

$$e_v = E_v \sin \omega_v t \tag{13-1}$$

where E_v is the *maximum* value of the signal-voltage wave, and $\omega_v = 2\pi f_v$. The carrier-voltage wave that was modulated is assumed to have been a sine wave of frequency f_c, having an instantaneous value

$$e_c = E_c \sin \omega_c t \tag{13-2}$$

where E_c is the maximum value of the unmodulated carrier, and $\omega_c = 2\pi f_c$. The equation for the received amplitude-modulated wave containing only the carrier and two sidebands is (section 12-4)

$$e_{am} = E_c(1 + m \sin \omega_v t) \sin \omega_c t \tag{13-3}$$

where m is the percentage modulation and is less than 1.0 in Fig. 13-1.

Assume that a "linear" rectifier is being used in the process of **linear detection** now being considered. As shown in Fig. 13-1, a linear rectifier passes only the positive half-cycles of the received signal wave. When the signal is not modulated, the rectified voltage wave across resistor R is as given by equation 6-1,

$$e = \frac{E_c}{\pi} + \frac{E_c}{2} \sin \omega_c t - \frac{2E_c}{3\pi} \cos 2\omega_c t \cdots \tag{13-4}$$

When the received wave *is* modulated, the maximum value of the rectified wave varies in accordance with the term $1 + m \sin \omega_v t$ as in equation 13-3. Using only the first two terms of equation 13-4, it is seen that the equation for the *rectified* modulated wave is

$$e_{rec\ am} = \frac{E_c}{\pi}(1 + m \sin \omega_v t) + \frac{E_c}{2}(\sin \omega_c t)(1 + m \sin \omega_v t) \tag{13-5}$$

$$e_{rec\ am} = \frac{E_c}{\pi} + \frac{mE_c}{\pi} \sin \omega_v t + \frac{E_c}{2}(\sin \omega_c t) + \frac{mE_c}{2}(\sin \omega_c t \sin \omega_v t) \tag{13-6}$$

The *first right-hand term* of equation 13-6, E_c/π, is the direct voltage component; the *second term*, $(mE_c/\pi) \sin \omega_v t$, is the desired original modulating signal of voice frequency $f_v = \omega_v/2$ (this component was suppressed at the transmitter and only the carrier and sidebands were transmitted; it has been recreated by the distortion); the *third term*, $(E_c/2) \sin \omega_c t$, is the carrier-frequency component; and the *fourth term* when expanded (section 12-6) will give audio-frequency components. The carrier and other unwanted radio-frequency components are separated easily from the desired voice-frequency signal (section 13-12).

13-3. *DEMODULATION WITH "SQUARE-LAW" DETECTORS (AM)*

The van der Bijl system of modulation explained in section 12-6 utilizes the curved "square-law" portion of the plate-current curve of a vacuum tube

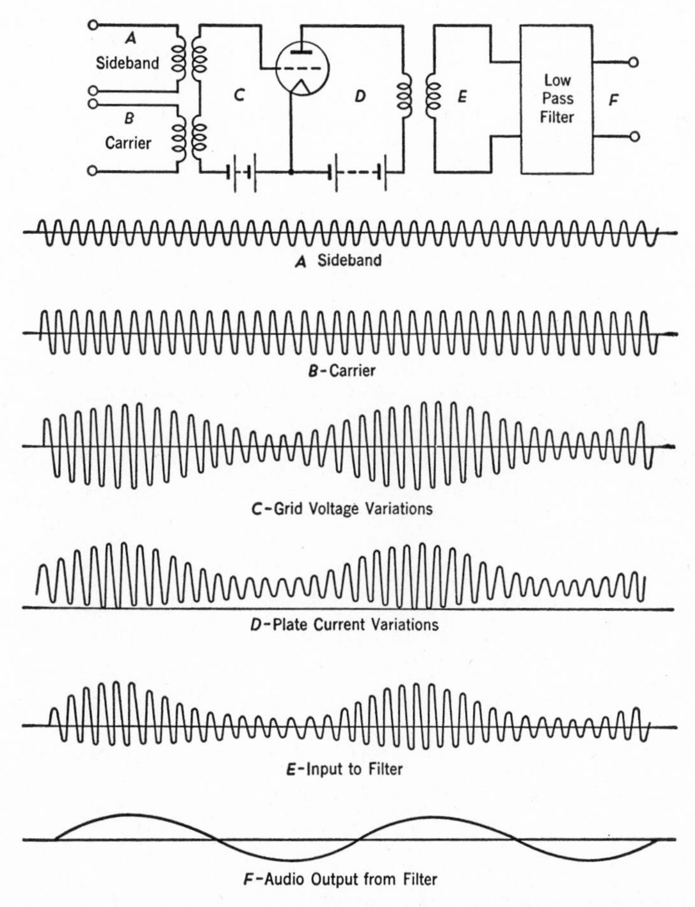

FIG. 13-2. Illustrating plate-circuit square-law detection. Curve *C* represents the resultant wave when a carrier and *one* sideband are combined. The shapes of the waves will be slightly different than when a carrier and *two* sidebands are combined. This distinction is not evident in the illustrations, but is easily demonstrated experimentally.

to cause distortion and create the desired sidebands; because of the curvature, sum and difference components are created. This same principle is used in demodulation, illustrated by Fig. 13-2. It is assumed that at the transmitting end a carrier wave of 20,000 cycles was amplitude-modulated with a voice-frequency wave of 1000 cycles; that the carrier and the lower sideband were suppressed; and that *only* the upper sideband of 21,000 cycles was transmitted to the receiving end for demodulation. This upper sideband is wave *A*. For demodulation, a wave (13-2*B*) having the frequency of the missing carrier must be produced by a local oscillator and introduced as shown. The combined action of waves *A* and *B* causes the grid to vary with respect to the cathode as shown by *C*. Because the tube is biased as indicated in Fig. 13-3, and because the signal swing is small, the output wave is distorted as indicated by Fig. 13-2*D*. This distorted wave will contain the recreated 1000-cycle component doing the modulation at the transmitting end, and this is the "transmitted" signal. Undesired components also are created, some of which are removed by filters.

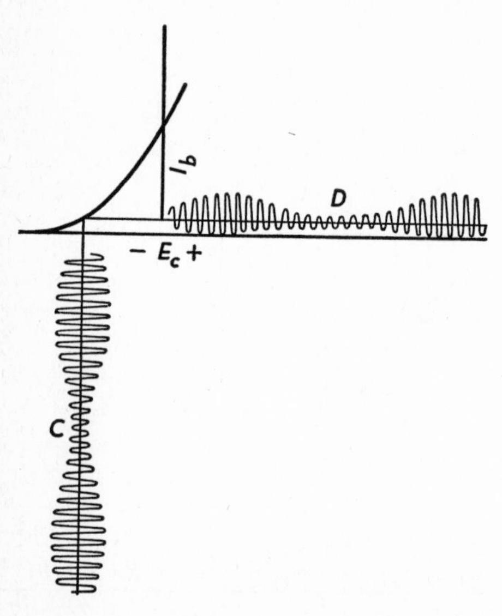

FIG. 13-3. When the tube of Fig. 13-2 is biased so that the relation between grid voltage and plate current is as shown, the impressed voltage *C* will cause distorted plate current *D* to flow. This current will contain the desired audio-frequency component.

If f_c represents the carrier, and f_v represents the voice-frequency modulating signal, the frequency of the received upper sideband will be $f_c + f_v$. If the two frequencies f_c and $f_c + f_v$ are demodulated, an analysis such as section 12-6 will show that the difference frequency $(f_c + f_v) - f_c = f_v$ exists in the output. If the carrier f_c and *two* sidebands $f_c + f_v$ and $f_c - f_v$ are transmitted as in radio broadcast, then a second audio term $f_c - (f_c - f_v) = f_v$ will be created. The two voice components f_v combine to give the desired audio output.

13-4. *CRYSTAL DEMODULATORS*

The point-contact crystal rectifiers described in section 10-1 are used for demodulation, or detection. Because of their low internal capacitance, small size, and suitability for mounting in wave-guide equipment, these crystals are especially useful at frequencies of about 1000 megacycles and above. Actual characteristics are shown in Fig. 10-4, and an ideal curve in Fig. 10-5. Because these curves are similar to those of Fig. 13-1, it follows that a point-contact crystal (often called a crystal diode) can be used for linear demodulation, or detection.

Crystal Frequency Converters. Amplification is difficult to achieve at ultrahigh frequencies (300 to 3000 megacycles) and above. Hence, in such systems the received signal usually is translated, or converted, to an **intermediate-frequency (IF)** [1] band at about 50 megacycles before amplifying. Frequency translation, or conversion, may be accomplished in the **crystal frequency converter** of Fig. 13-4. Although tuned *L–C* input circuits are

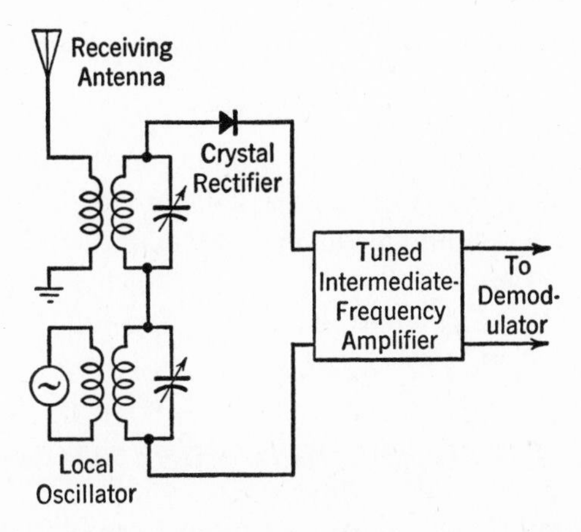

FIG. 13-4. Illustrating the principle of frequency conversion (translation) using a crystal diode rectifier.

indicated, wave-guide inputs probably would be used. The crystal frequency converter also is called a first detector, or a crystal mixer, terms that obscure the true nature of the process.

If an amplitude-modulated wave, consisting of a carrier and two sidebands and designated f_{am}, is impressed with a single-frequency f_{osc} from a **local oscillator** [1] on the crystal converter, the output will contain the terms f_{am}, $f_{osc} + f_{am}$, and $f_{osc} - f_{am}$. If the *carrier* of the original received amplitude-modulated signal is 3000 megacycles, and the local-oscillator frequency is 3050 megacycles, the difference term $f_{osc} - f_{am}$ will be the complete received amplitude-modulated wave translated, or converted, to the intermediate-frequency band centered at 50 megacycles. This translated wave will be an exact replica of the received ultrahigh-frequency signal. This signal, now centered at 50 megacycles, will be accepted and amplified by the sharply tuned intermediate-frequency amplifier, and all other components created in the process will be rejected.

Crystal Detectors. If an amplitude-modulated signal *only* is impressed on a point-contact crystal rectifier, demodulation will occur essentially as in Fig. 13-1, because the crystal may be considered as a linear rectifier. Among the created components will be the original modulating voice-frequency wave f_v in accordance with the explanation in section 13-2. The amplitude-modulated signal may be at the frequencies originally received, or at an intermediate frequency. Note that *no* signal is introduced simultaneously from a local oscillator. This method of demodulation is not widely used.

13-5. *VACUUM-TUBE DIODE DEMODULATORS (AM)* [8]

A two-electrode vacuum tube, or diode (section 3-2) has the rectifying characteristic shown in Fig. 3-3. This is similar to the linear rectifying characteristic indicated in Fig. 13-1; hence, a diode may be used as a so-called linear detector of amplitude-modulated waves in accordance with section 13-3. In radio receivers the diode element often is part of a multipurpose tube also containing a triode portion. The details of this method will be considered in section 13-12.

13-6. *VACUUM-TUBE TRIODE DEMODULATORS (AM)* [9,10]

With vacuum-tube triodes, distortion and the resulting demodulation may occur either in the grid circuit, or in the plate circuit. The following discussion is for triodes, but similar operation is possible with tetrodes and pentodes.

Triode Grid-Circuit (or Grid-Current) Square-Law Demodulation. The grid-current curve of a triode was discussed in section 4-11, and is shown in Fig. 13-5. If the grid bias is about zero, if the plate voltage is at a favorable value, and if the amplitude-modulated signal wave is small, then operation will be over the lower nonlinear part of the curve. The theory of square-law demodulation of section 13-3 applies. The difference terms, which are the desired demodulated voice-frequency signal, will exist as a component of the distorted grid current that flows.

A vacuum tube is a voltage-operated device, and hence a resistor R_g (Fig. 13-6), of perhaps a megohm, must be placed between the grid and cathode so that the voice-frequency signal-current component in passing through this

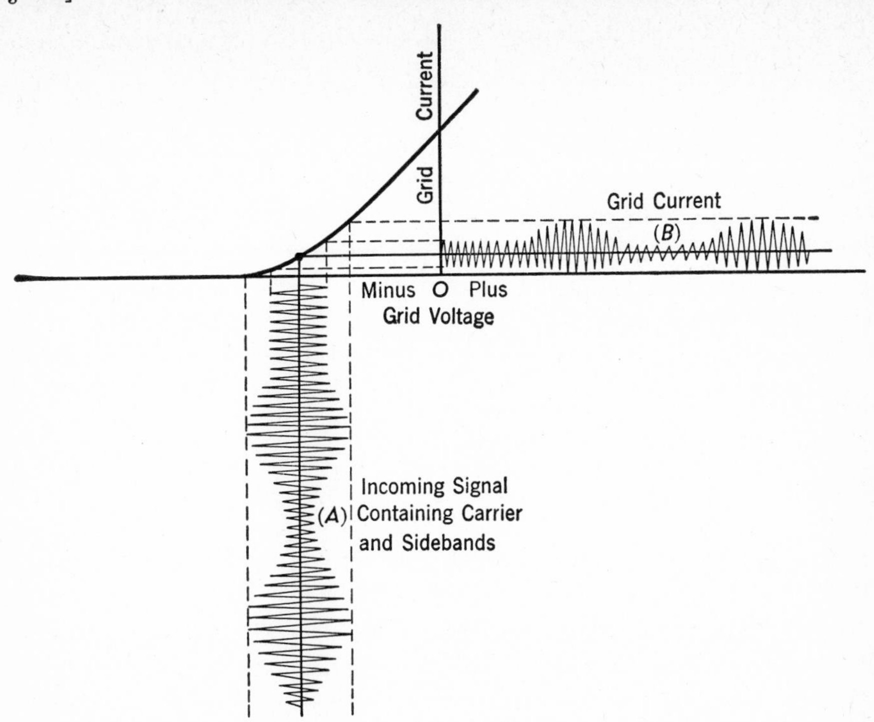

FIG. 13-5. Diagram for the grid-circuit square-law detector. The incoming signal A is distorted in the *grid circuit*, producing grid current B which contains the desired audio component. This current component flows through resistor R_g of Fig. 13-6, causing an audio-frequency voltage drop between grid and cathode. This voltage is amplified by the tube, and the plate current will consist largely of the desired audio signal.

resistor will cause a corresponding signal voltage between grid and cathode for *amplification in the plate circuit.* A direct-current component also exists in the distorted grid current, and flows through this resistor, biasing the grid as indicated. The magnitude of the direct plate voltage, the magnitude of the applied amplitude-modulated wave, and the resistance of R_g must be such that operation is on the curved part of the characteristic. Capacitor C_g provides a path of low impedance for the current components of radio-frequency so that essentially the entire received signal voltage is impressed between grid and cathode. The size of C_g (perhaps 250 micromicrofarads) must be such that the reactance offered to the desired voice-frequency component is high. Capacitor C_p (perhaps 0.001 microfarad) in the plate circuit is to offer a low-impedance by-pass path for radio-frequency components that may exist in the plate circuit. Prior to about 1930 this method of demodulation was widely used, one reason being that it is "sensitive," because with correct operation demodulation is obtained in the grid circuit, and *amplifica-*

tion in the plate circuit. However, with modern radio signals having large percentage modulation the audio distortion is excessive.

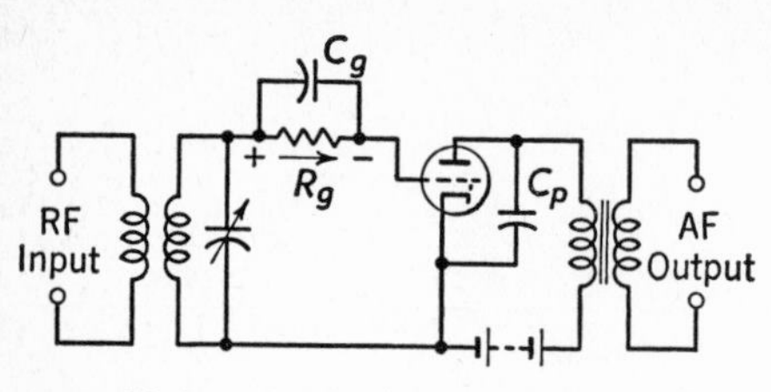

FIG. 13-6. Circuit for grid-current square-law detection.

Triode Grid-Circuit* (*or Grid-Current*) *Linear Demodulation. As just explained, for grid-circuit square-law demodulation a small signal is impressed between grid and cathode so that operation is over the *curved* portion in Fig. 13-5. For so-called linear demodulation (section 13-2), the circuit of Fig. 13-6 also is used, but a relatively large received amplitude-modulated signal of several volts is used. With the correct direct plate potential, with R_g equal to about 0.25 megohm, and with C_g about 100 micromicrofarads, the rectified direct grid-current component flowing through R_g will bias the tube about as in Fig. 13-7, and linear demodulation will occur. The desired demodulated voice-frequency component also will flow through R_g, because C_g offers high reactance to audio-frequencies. This will cause the grid to vary in accordance with the audio signal, and amplification in the plate circuit will result. Capacitor C_g offers low reactance to the radio-frequency grid-current components, and hence almost the entire received amplitude-modulated signal appears between grid and cathode for rectification.

Triode Plate-Circuit* (*or Plate-Current*) *Square-Law Demodulation. This was analyzed in section 13-3, and illustrated in Figs. 13-2 and 13-3. Although these illustrations and much of the discussion is for a received signal consisting of a single sideband,[11] demodulation of a received signal composed of a carrier and two sidebands is considered at the close of section 13-3. Square-law demodulation results in excessive distortion of the voice-frequency wave, and this method is not widely used.

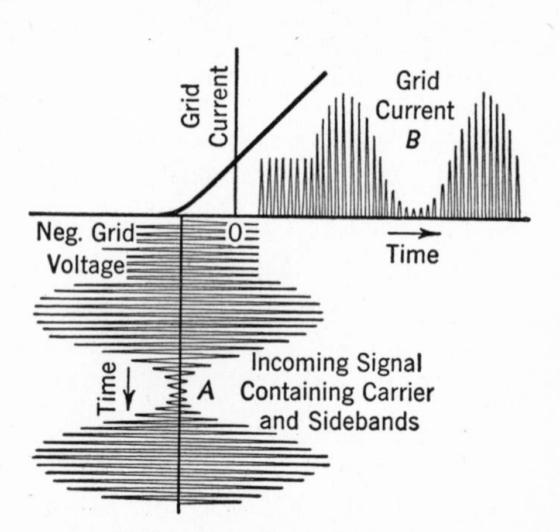

FIG. 13-7. Illustrating "linear" grid-circuit, or grid-current, demodulation, or detection.

Triode Plate-Circuit* (*or Plate-Current*) *Linear Demodulation.[12] A typical circuit is shown in Fig. 13-8, and a graphical analysis in Fig. 13-9. The grid is biased to cutoff by the direct plate-current flow through resistor R_2. Capacitor C_2 offers a path of low reactance to all alternating components. Inductor L_1 is a radio-frequency choke, and capacitor C_1 is a radio-frequency by-pass, the combination filtering all radio-frequency components from the output. Because the grid is biased to cutoff, and because a large amplitude-modulated signal is impressed, so-called linear rectification and demodulation result (section 13-2).

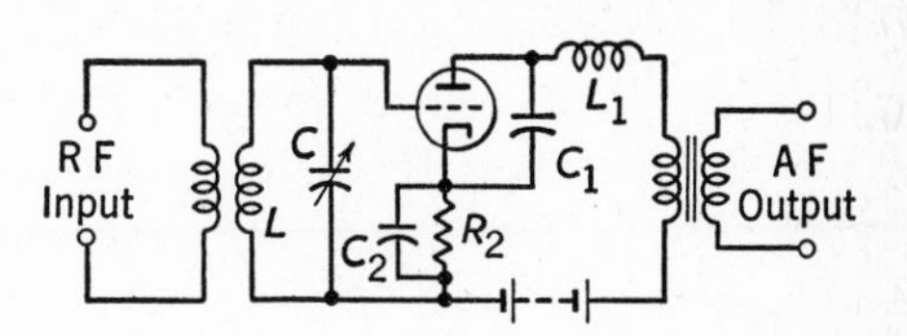

FIG. 13-8. Circuit for triode plate-circuit "linear" detection.

13-7. *A CLASSIFICATION OF DEMODULATORS*

The following classification, made [10] by Everitt, is of fundamental importance.

Driving-Point Impedance Detectors. In such detectors demodulation occurs because of a nonlinear relation (*over an entire cycle*) between the amplitude-modulated signal voltage impressed between a pair of terminals of a device, and the current that flows through the *same pair* of terminals. Devices having such characteristics include the point-contact crystal rectifier, the diode vacuum tube, and the triode vacuum tube when operated as a grid-circuit, or grid-current, demodulator. Such devices usually must have an external impedance, such as a resistor-capacitor combination (Fig. 13-6) connected in series with the terminals so that the desired voice-frequency

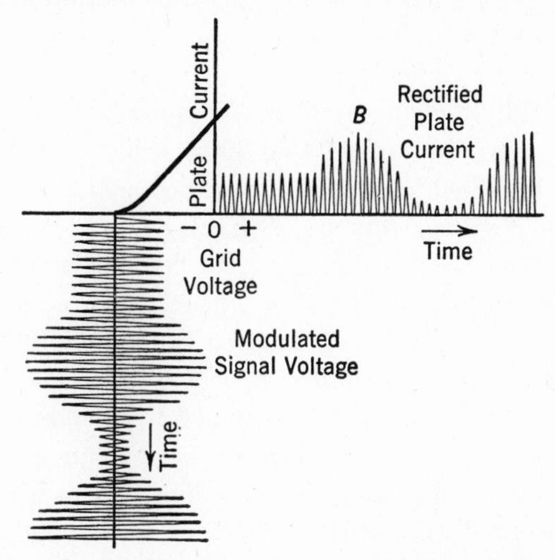

FIG. 13-9. Illustrating triode plate-circuit, or plate-current "linear" detection.

component of the distorted current will produce a voice-frequency voltage for amplification.

Transfer-Impedance Detectors. In these detectors demodulation occurs because of the nonlinear relation (over an entire cycle) between the amplitude-modulated signal voltage applied to one pair of terminals and the current that flows through a *different pair* of terminals. The plate-circuit detectors of the preceding section are examples of transfer-impedance detectors.

13-8. *SPEECH INVERSION* [10,13,14]

One of the important requirements of any system of communication is *secrecy*, and this must be possible with radio as well as with wire systems. If broadcasting methods were used in transmitting transoceanic radiotelephone conversations, the lack of secrecy would render the system almost valueless. Secrecy methods of several types have been developed; these include a **frequency inverting** or **speech scrambling process**, which will now be considered. Although the details may differ with various systems, the process is somewhat as follows: Suppose that a 50,000-cycle wave is modulated with a 3000-cycle wave, and that the *lower sideband only* is selected with a filter. The output of the filter will be a 47,000-cycle wave. Now suppose this 47,000-cycle wave and a 42,000-cycle wave are impressed on a modulator, and the *lower* sideband *only* is selected by a filter. The output of this second filter will be a 5000-cycle wave. By the same process, if the original 50,000-cycle wave is modulated by 200 cycles, the output of the *second* filter will be a 7800-cycle wave.

To apply this to an audio-frequency wave, assume that the 50,000-cycle wave is modulated by the voice, which is considered in this instance to contain frequency components from 200 to 3000 cycles. By the double-modulation system just described, the low-frequency components of 200 cycles will be high-frequency components of 7800 cycles, and the high-frequency components of 3000 cycles will become the low-frequency components of 5000 cycles in the *output of the second filter*. In this wave the speech is *inverted* or *scrambled*, and the conversation is unintelligible except to those equipped with special apparatus for re-inverting the frequencies. The system could be adjusted at intervals so that the frequency bands would be changed and reception rendered even more difficult. For making reception still more difficult, the inverted frequencies could be divided into several groups, and each of these could be transmitted at different wave lengths.

13-9. *COPPER-OXIDE VARISTOR DEMODULATORS*

Modulation with copper-oxide varistors was discussed in section 12-14. As has been stressed, modulation and demodulation are fundamentally the same process. Thus, the explanations given for modulation by varistors apply also to demodulation. This can be proved by assuming that a single sideband, varying in accordance with the modulating voice-frequency wave, is impressed at E_v of Fig. 12-13. A locally generated carrier component of constant amplitude is injected at E_c to control the "switching" of the bridge circuit. In some carrier-telephone systems the *same* varistor bridge operates both as a modulator and demodulator. Generally speaking, copper-oxide varistors are used in carrier-telephone systems, although they do find application in low-frequency modulators of radio transmitters.[15] Crystal rectifiers can be used in bridge circuits such as Fig. 12-13, and sometimes are employed in carrier-telephone equipment, and also for radio-frequencies at which the copper-oxide varistor is unsuited.[15]

13-10. *RADIO RECEIVERS FOR AMPLITUDE-MODULATED SIGNALS*

The **radio receiver** is[1] a "device for converting radio waves into perceptible signals." As has been explained, information is transmitted through space by wireless means at frequencies from about 15 kilocycles[16] to tens of thousands of megacycles. The radio receiver must select, amplify, and demodulate these radio signals so that the received information is made available at audio-frequencies. Although other types[9] of radio receivers are possible, the discussion will be limited to two basic types.

Tuned Radio-Frequency Radio Receivers. The block diagram of Fig. 13-10 will be used to describe the operation of a tuned radio-frequency

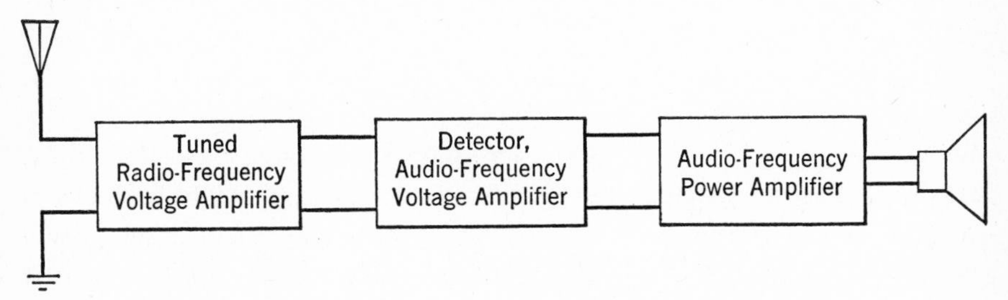

FIG. 13-10. Block diagram of a tuned radio-frequency radio receiver.

receiver. Appreciable signal voltages at various frequencies are induced in the receiving antenna by many distant radio transmitters. Some tuning may exist in the antenna circuit, but in general, the received amplitude-modulated radio signals are impressed as voltages at the input to the tuned radio-frequency amplifier. This commonly consists of several tetrode or pentode small-signal vacuum tubes operated as tuned radio-frequency voltage amplifiers (section 7-19), hence the name tuned radio-frequency receiver. These tuned amplifier stages must select and amplify the signal from the desired station and reject all other signals. Therefore, these amplifier stages must be tunable over the entire frequency range to be covered by the receiver. Grid-circuit square-law demodulation was extensively used in these receivers.

Superheterodyne Radio Receivers. The invention of this set was by Armstrong, about 1920. It sometimes is called a "double-detection" set, but neither this term nor "superheterodyne" describes the operation. The superheterodyne block diagram of Fig. 13-11 shows a tuned radio-frequency

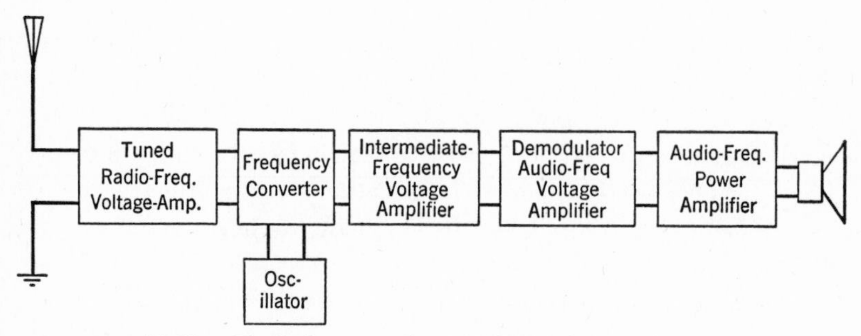

FIG. 13-11. Block diagram of a superheterodyne radio receiver.

voltage amplifier at the input; this is not included in small sets. The principle of the superheterodyne was described in section 13-4 when discussing crystal frequency converters.

When using the superheterodyne, the tuned radio-frequency amplifier is tuned to the station desired. This amplifier is not sharply tuned, but does amplify the signal from the desired station more than it amplifies undesired signals. The purpose of the frequency converter is to translate, or move, the received carrier and sidebands to an intermediate frequency (section 13-4). Suppose that at the distant station a program in the band 100 to 5000 cycles is amplitude-modulating a 1,000,000-cycle carrier. The received signal will have three components: the 1,000,000-cycle carrier, an upper sideband from 1,000,100 to 1,005,000 cycles, and a lower sideband from 995,000 to 999,900 cycles. The over-all band width is 995,000 to 1,005,000 cycles. The tuning of the radio-frequency amplifier is coupled mechanically to the local oscillator. In many sets, when the amplifier is tuned to 1,000,000 cycles the local oscillator is adjusted to generate a sine wave of 1,465,000 cycles. This wave is amplitude-modulated in the frequency converter by the received and amplified signal, and sum and difference components are created. The difference component is the *entire* desired signal translated to an intermediate-frequency band, with the carrier at 465,000 cycles, the upper sideband at 465,100 to 470,000 cycles, and the lower sideband at 460,000 to 464,900 cycles.

The intermediate-frequency amplifier is a tuned transformer-coupled amplifier that will select and amplify only those frequencies within the band 460,000 to 470,000 cycles. Thus, of all the signals induced in the antenna, *only* the signal from the station with a 1,000,000-cycle carrier will be accepted, amplified, and passed on by the intermediate-frequency amplifier. This amplifier is designed and adjusted in the factory to operate *only* over the band 460,000 to 470,000 cycles; it is not adjusted by the person using the set. The characteristics may, therefore, be far superior to those of a tuned radio-frequency amplifier that must be tunable over a wide range, as in the tuned radio-frequency receiver. It is possible for the signal from a station outside the standard amplitude-modulation broadcast band (550 to 1600 kilocycles) to be modulated by the signal from the local oscillator. Thus a station sending a 1930-kilocycle signal into the frequency converter and the 1465-kilocycle signal from the local oscillator will give a signal at 465 kilocycles, and this will pass through the intermediate-frequency amplifier to the demodulator, and will be heard in the loud-speaker. The tuned radio-frequency amplifier and tuned antenna circuits reduce such **image-frequency** [1] reception.

13-11. *VACUUM-TUBE FREQUENCY CONVERTERS*

As explained, a frequency converter translates, or moves, the received radio signals to the frequency of the intermediate-frequency amplifier, the principle employed being modulation by nonlinear elements

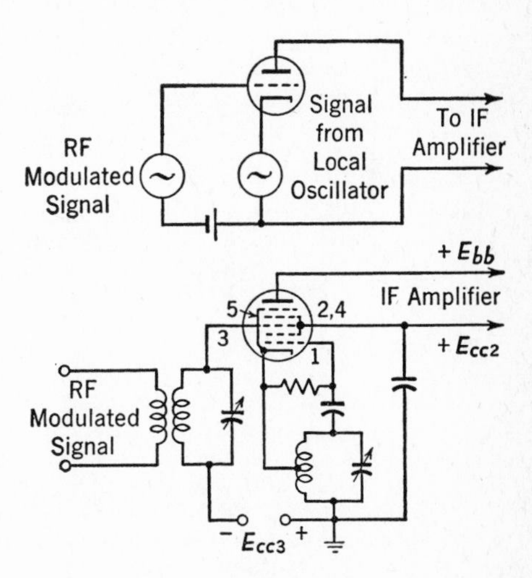

FIG. 13-12. Equivalent circuit (above) and simplified circuit (below) of a frequency converter (also called first detector and mixer). Grid 5 is the suppressor grid internally connected to the cathode.

such as a vacuum tube operated as a distorting amplifier. A frequency converter is shown in Fig. 13-12. The amplitude-modulated signal from the antenna, and a single-frequency wave from the local oscillator simultaneously are impressed in series between a control grid and cathode. The grid is highly biased so that the plate current is distorted, and the desired intermediate-frequency band created. A single tube often is used both as an oscillator and a modulator. Thus, in Fig. 13-12 grid 1 is the control grid of the oscillator, and is biased by grid current flowing through the resistor between grid and cathode. Grid 2 acts as a plate for the oscillator, and also as a shield between grids 1 and 3. Grid 4 acts as a screen grid between grids 3 and 5. The voltage across the part of the oscillator coil between cathode and ground constitutes the signal from the oscillator, and is in series with the signal from the antenna impressed between control grid 3 and ground. The variable capacitor in the antenna and the variable capacitor in the oscillator are linked mechanically.

13-12. *AUTOMATIC VOLUME CONTROL AND DIODE DETECTION* (*AM*)

In Fig. 13-13 is shown a tube containing a diode section and a high-mu triode section. This tube and the associated circuit provide an automatic volume-control voltage, act as a diode "linear" demodulator (often called a second detector), and provide audio-frequency amplification.

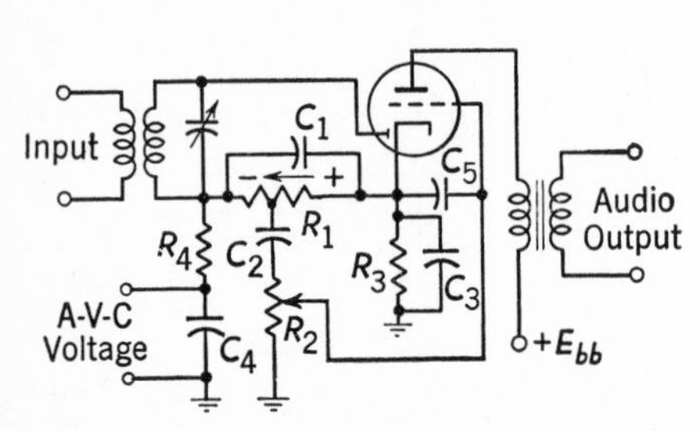

FIG. 13-13. Method of obtaining an automatic-volume-control voltage from the diode-detector part of the tube. Typical values are: $C_1 = 150$ micromicrofarads; $R_1 = 0.5$ megohm, $C_2 = 0.0025$ microfarad, $R_2 = 1.0$ megohm, $C_4 = 0.05$ microfarad, and $R_4 = 2.2$ megohms.

Automatic Volume Control. The carrier component of the amplitude-modulated signal radiated by a transmitter *does not vary* in accordance with the magnitude of the transmitted signal. However, because of changes in the transmission path between transmitter and receiver,[9,17] the strength of the received carrier component may fluctuate. This is called **fading.** During fading the output of a radio receiver is maintained almost constant by an automatic-volume-control voltage obtained as follows: Assume that an *unmodulated* carrier is being received. This will be rectified by the diode section of the tube, and a direct-current component will flow through R_1, causing the voltage drop indicated by the arrow. This voltage will cause capacitor C_4 to charge slowly through high resistor R_4.

Now assume that the characteristics of the path between transmitter and receiver improve; a stronger unmodulated carrier will be received, a stronger rectified direct current will result, and the charge on capacitor C_4, and the voltage across it, will increase. If the transmission path becomes poorer, less rectified current will result, and the voltage across C_4 will decrease. The voltage across capacitor C_4, which is called an automatic volume-control voltage, is used to bias the control grid of one or more variable-mu amplifier tubes (section 5-16) preceding the diode. When the carrier signal is strong, the bias is large, and the amplification of the tube is reduced. When the carrier is weak, the bias is small, and the amplification is increased. By proper design, the signal impressed on the diode can be made almost constant for large fluctuations in received signal strength. The preceding explanation was based on the reception of an unmodulated signal. This was because the time constant of the R_4–C_4 combination is so great that the automatic volume-control voltage across C_4 will vary only for relatively long-period changes caused by fading, and will not follow modulation changes caused by speech and music.

Diode Detection. If the received signal impressed on the diode section of Fig. 13-13 is an amplitude-modulated wave, demodulation will be in accordance with sections 13-2 and 13-5. A direct-current component and a voice-frequency component both will flow through resistor R_1. The values of C_2 and R_2 are such that the voice-frequency components readily flow to ground and back to cathode. Thus, a voice-frequency, or audio, signal voltage will exist across R_2. This is a manual volume control, and its adjustment determines the magnitude of the audio signal voltage impressed between the grid and cathode for voltage amplification in the triode section of the tube.

13-13. *DEMODULATION OF FREQUENCY-MODULATED WAVES*

The amplitude of a frequency-modulated wave is constant, the instantaneous frequency is determined by the strength of the audio signal, and the rate at which the frequency varies depends on the frequency of the modulating wave. In early receivers [18] two steps were involved in demodulation; *first,* the frequency variations were changed to amplitude variations, and *second,* the resulting amplitude-modulated wave was demodulated by means previously studied. One of the following methods usually is employed for demodulating frequency-modulated waves.

The Discriminator. This device must be preceded by a limiter,[1] which may be an amplifier "whose output is constant for all inputs above a critical value." One method is to use a small-signal sharp-cutoff pentode in a voltage-amplifying circuit, with direct electrode potentials such that if the frequency-modulated signal driving voltage exceeds several volts, grid current flows through a grid resistor and limits the amplification. An ideal limiter removes from a frequency-modulated signal all amplitude variations such as might result from atmospheric noise, or "bursts" of static. For this reason, an important characteristic of a frequency-modulation system is essentially noise-free reception (section 13-14). An amplitude-modulation system oper-

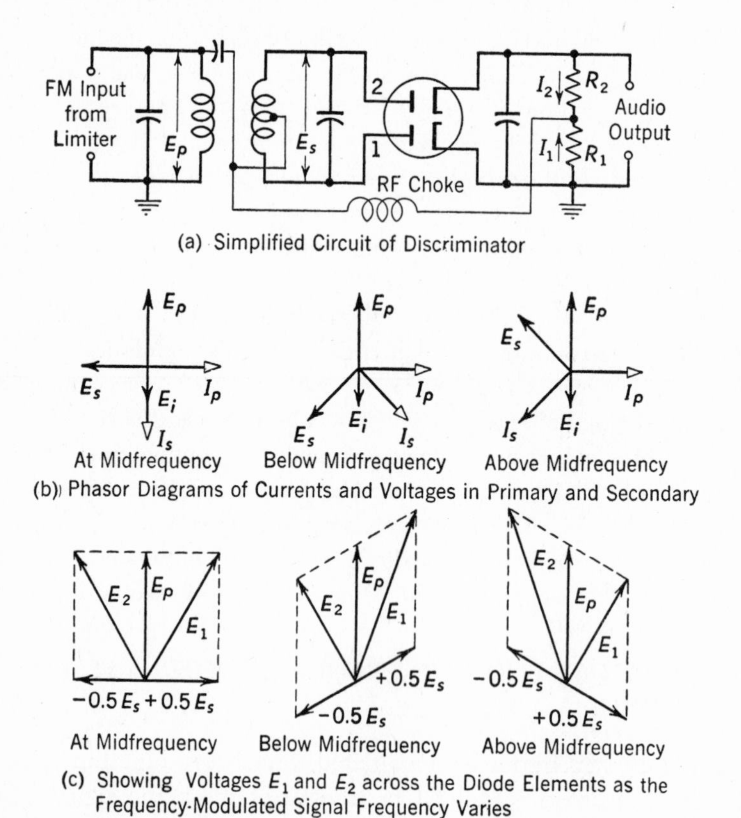

FIG. 13-14. Circuit and phasor diagrams for a discriminator used to demodulate frequency-modulated signals.

ates because of changes in signal amplitude, and is sensitive to magnitude changes caused by disturbances such as static. The **discriminator** demodulates the frequency-modulated wave and recreates the original voice-frequency signal wave. The discriminator of Fig. 13-14 is based on frequency-control circuits.[20]

Discriminator Action; Received Signal at Midfrequency. The primary is tuned to midfrequency, and, if at the instant under consideration the frequency is at midfrequency, the diagram at the left in (b) applies. The impressed voltage E_p causes I_p to flow in the primary. This current induces E_i *in series* in the secondary that also is tuned to midfrequency. The secondary current I_s will be in phase with induced voltage E_i at midfrequency. A voltage drop results across the secondary tuning capacitor; this lags the secondary current by 90°. The connections are such that the voltage between each anode and cathode is the sum of the primary voltage E_p and one-half the secondary voltage E_s. The capacitor between the primary and the center of the secondary has negligible reactance. When the signal is at midfrequency, E_1 and E_2 are equal, the rectified currents I_1 and I_2 are equal, and since R_1 equals R_2, no resultant voltage exists at audio output.

Discriminator Action; Received Signal Below Midfrequency. For this frequency the secondary will be out of resonance, capacitive reactance will predominate, I_s will lead E_i, and voltage E_s will be as indicated. This causes diode voltage E_1 to exceed diode voltage E_2, rectified current I_1 will exceed I_2, and the lower audio-output terminal will be positive and the upper terminal will be negative.

Discriminator Action; Received Signal Above Midfrequency. The secondary will be out of resonance, inductive reactance will predominate, and current I_s will lag voltage E_i, placing E_s as indicated. Diode voltage E_2 will exceed E_1; also, rectified current I_2 will exceed I_1, and the upper audio-output terminal will be positive and the lower negative.

At the frequency-modulation transmitter, one cycle of modulating audio signal voltage causes a frequency shift below midfrequency and a shift above midfrequency. This will cause one cycle of audio-output discriminator voltage. At the transmitter a strong modulating voltage will cause a large frequency shift, and, at the receiver, a large frequency shift will result in a greater lead, or lag, and a stronger audio output.

In some circuits the capacitor connecting the primary and the secondary of Fig. 13-14 is omitted, and a coil is placed in the wire leading to the center of the secondary. This coil is coupled to the primary and serves to pick up a reference voltage that corresponds to E_p of Fig. 13-14 (*b*) and (*c*).

The discriminator must be preceded by a limiter. Because the limiter reduces the signal to the point at which all normal amplitude variations are eliminated, the intermediate-frequency amplifier preceding the limiter must have much gain. Two limiters sometimes are used for more complete action.

The Ratio Detector.[21] The **ratio detector** of Fig. 13-15 requires no limiter. Note that one diode is reversed and that other changes have been made from Fig. 13-14. Also, the capacitor between primary and secondary often is omitted, and, as for some discriminators, another secondary coil is placed in series with the wire to the center of the secondary. The phasor diagrams of Fig. 13-14 apply to the ratio detector.

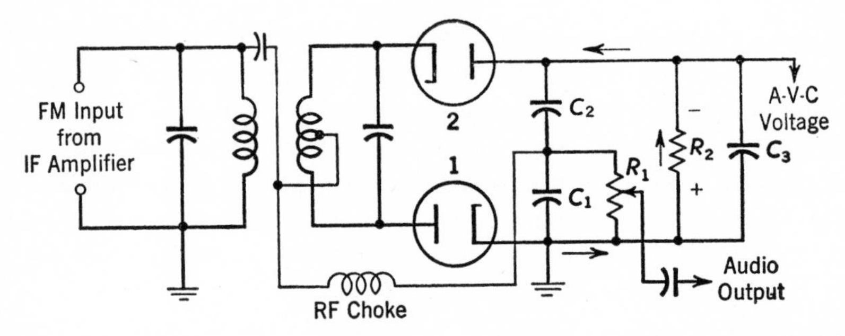

FIG. 13-15. Circuit of the ratio detector, a demodulator for frequency-modulated signals.

Capacitor C_3 of Fig. 13-15 is about 10 microfarads, and the circuit is such that the voltage across C_3 can vary only slowly. Assume that the received station is unmodulated and is, accordingly, on midfrequency. If the station produces a strong signal (the limiter is omitted when a ratio detector is used), a large rectified current will flow as shown by the arrows, and capacitor C_3 will charge slowly to a relatively large voltage value. If the station produces a weak signal, a small rectified current will flow, and capacitor C_3 will discharge slowly to a low value. This makes available an automatic-volume-control voltage between the terminal A-V-C and ground. Note that at the midfrequency no direct current flows through the radio-frequency choke or resistor R_1.

If the transmitter is frequency-modulated and the signal falls below the midfrequency, the radio-frequency voltage across diode 1 will increase and that across diode 2 will decrease (Fig. 13-15). This will cause the current in diode 2 to fall and that in diode 1 to rise; this causes current flow through the RF choke and through resistor R_1. The voltage across capacitor C_2 will fall, and that across C_1 will rise, but the sum of these voltages will equal that across C_3. When the frequency rises above the midfrequency, the action is reversed. Thus, an audio voltage, which corresponds to the frequency shifts in the received-modulated signal, exists across resistor R_1. If a *sudden* change occurs in the *magnitude* of the received frequency-modulated signal, the magnitude of the demodulated output voltage remains fixed because capacitor C_3 holds the voltage constant across capacitors C_1 and C_2. Only the ratio of these voltages can change. Thus a limiter need not be used.

Frequency-Modulation Receivers. The basic circuit for receiving frequency-modulated signals is similar to the superheterodyne of Fig. 13-11 for amplitude-modulated signals. Input circuit details differ because the radio signals received are at about 100 megacycles. An intermediate frequency centered at 10.7 megacycles is common. The frequency converter moves, or translates, the *entire* received frequency-modulated to the intermediate-frequency band. A ratio detector often replaces the limiter and discriminator. A deemphasis circuit is included in broadcast frequency-modulation receivers to compensate for the effect of a preemphasis circuit used in the transmitter to emphasize the weak high-frequency program components so that the signal-to-noise ratio is greater.

13-14. *BAND WIDTH AND RADIO NOISE*

In 1928 Hartley stated (in discussing transmission over a system of given band width) that [22] "the total amount of information which may be transmitted over such a system is proportional to the product of the frequency range which it transmits by the time during which it is available for transmission." About twenty years later the effect of noise was considered, and the original **Hartley law** was modified [23] and expressed mathematically

$$C = w \log_2 \left(1 + \frac{S}{N}\right) \qquad 13\text{-}7$$

where C is the communication-channel capacity to carry information per unit time, w is the band width of the channel, and S/N is the signal-to-noise power ratio. It will be of interest to compare amplitude-modulation systems with frequency-modulation systems with respect to the modified Hartley law.[24]

The speech-input circuits of a frequency-modulation radio-broadcast system are required [25] to pass up to 15,000 cycles, and as explained in section 12-19, such a station requires a radio-frequency band width of 200 kilocycles. If an amplitude-modulation transmitter were designed to pass the same upper frequency, the band width required would be 30 kilocycles (section 12-19). If these two systems were tested under comparable conditions, it would be found that the audio signal from the frequency-modulation system had a higher signal-to-noise ratio than the audio signal from the amplitude-modulation system. Low noise reception may be achieved at the "expense" of band width. When a narrow-band frequency-modulation system such as used for communication only is compared [26] with an amplitude-modulation system on an impartial basis, it is found that each system has certain advantages. It is of interest to note that pulse-code modulation systems (section 12-21) that have band widths greater than frequency-modulation systems are even less susceptible to noise than the latter.

13-15. *MAGNETIC AND DIELECTRIC AMPLIFIERS*

The theory of modulation and demodulation may be applied to at least certain magnetic and dielectric amplifiers. In each of these amplifiers the nonlinear characteristic of an impedance is utilized. As is well known, the impedance of a coil with a closed ferromagnetic core may vary considerably with the magnitude of the current through the coil. For capacitors, it commonly is stated that the magnitude of the impedance is independent of the magnitude of the impressed voltage over the usual working range. This *is* true for dielectrics such as mica, but *is not* true for capacitors with dielectrics such as titanium dioxide, barium titanate, strontium titanate, and other substances.[27] The dielectric constant, and hence the capacitance of a capacitor using such material, vary much with the magnitude of the impressed voltage; in fact, a decrease in capacitance of about 50 per cent was measured when the voltage across a nonlinear capacitor was increased from zero to 180 volts.

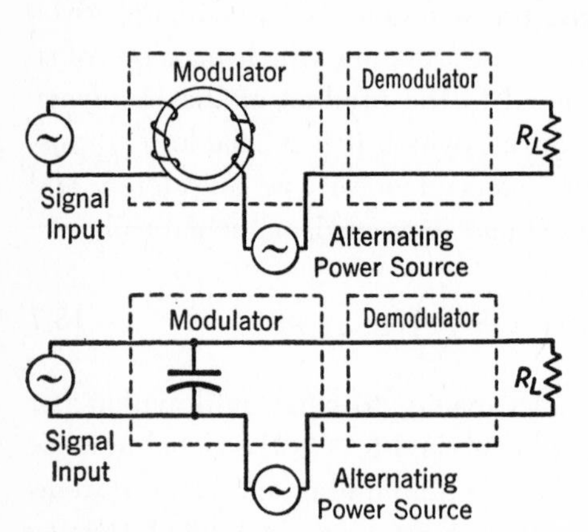

FIG. 13-16. Circuits for studying, from the modulation viewpoint, the principles of magnetic amplifiers (above), and dielectric amplifiers (below).

From the viewpoint of modulation and demodulation,[28,29] elementary magnetic and dielectric amplifiers are shown in Fig. 13-16. These do *not* represent practical circuits but are for explaining the modulation-demodulation viewpoint only. The upper circuit contains a nonlinear inductor, and the lower one a nonlinear capacitor. In the upper diagram a source of direct current might be added in series with the signal input to "bias" the nonlinear coil to the optimum operating magnetic-flux value. In the lower diagram, a direct voltage might be added in series with the signal input to "bias" the nonlinear capacitor to the optimum operating electric-flux value. In either instance, the alternating-power input (of single frequency) will cause alternating-current flow through the load resistor. When the signal to be amplified is impressed, the reactance of the inductor, or of the capacitor, will vary with the signal, and the instantaneous magnitude of the current flowing through the coil will vary in accordance with the signal to be amplified.

Thus in effect, amplitude modulation occurs in both circuits of Fig. 13-16 when the signal to be amplified is impressed as shown. Power will flow into the modulator from the alternating-power input circuit, and sidebands will be created. To obtain an *amplified* version of the *input signal* it is necessary to demodulate this amplitude-modulated wave, usually with selenium rectifiers. With proper design the output signal is a good replica of the input signal, and at a higher power level. Again, the circuits of Fig. 13-16 are simplified, are for illustrating the principle, and probably would not be used in practice. They do illustrate the modulation-demodulation viewpoint, however. The more generally used viewpoint regarding magnetic amplifiers is given in Chapter 15. Dielectric amplifiers and magnetic amplifiers may be considered as duals. Dielectric amplifiers are not now (1955) used in practice to the extent that further consideration is justified. There are, nevertheless, reasons to believe that dielectric amplifiers will become useful devices in the future.

REFERENCES

1. Institute of Radio Engineers. *Standards on Receivers: Definitions of Terms.* 1952. Proc. I.R.E., Dec. 1952, Vol. 40, No. 12.
2. Institute of Radio Engineers. *Standards on Radio Receivers.* 1943.
3. Green, C. W., and Green, E. I. *A carrier-telephone system for toll cables.* Electrical Engineering, May 1938, Vol. 57, No. 5.
4. Kendall, B. W., and Affel, H. A. *A twelve-channel carrier telephone system for open-wire lines.* Electrical Engineering, July 1939, Vol. 58, No. 7.
5. Klenk, L. M., Munn, A. J., and Nedeleka, J. *A multichannel single-sideband radio transmitter.* Proc. I.R.E., July 1952, Vol. 40, No. 7.
6. Institute of Radio Engineers. *Standards on Antennas, Modulation Systems, and Transmitters.* 1948.
7. Hellerman, H., and Cahn, C. R. *Linear detection of amplitude-modulated signals.* Proc. I.R.E., Oct. 1952, Vol. 40, No. 10, p. 1251.
8. Kilgour, C. E., and Glessner, J. M. *Diode detection analysis.* Proc. I.R.E., July 1933, Vol. 21, No. 7.
9. Terman, F. E. *Radio Engineering.* McGraw-Hill Book Co.
10. Everitt, W. L. *Communication Engineering.* McGraw-Hill Book Co.
11. Polkinghorn, F. A., and Schlaak, N. F. *A single-sideband short wave system for transatlantic telephony.* Proc. I.R.E., July 1935, Vol. 23, No. 7.
12. Ballantine, S. *Detection at high signal voltages.* Proc. I.R.E., July 1929, Vol. 17, No. 7.
13. Roberts, W. W. *Speech scrambling methods.* Electronics, Oct. 1943, Vol. 16, No. 10.

14. Koros, L. L. *An improved speech inverter system.* Electronics, May 1950, Vol. 23, No. 5.
15. Kerwien, A. E. *Design of modulation equipment for modern single-sideband transmitters.* Proc. I.R.E., July 1952, Vol. 40, No. 7.
16. Walter, J. C. *A million-watt naval communication transmitter.* Electrical Engineering, Nov. 1953, Vol. 72, No. 11.
17. Albert, A. L. *Electrical Communication.* John Wiley & Sons.
18. Armstrong, E. H. *A method of reducing disturbances in radio signaling by a method of frequency modulation.* Proc. I.R.E., May 1936, Vol. 24, No. 5.
19. Everitt, W. L. *Frequency modulation.* Electrical Engineering, Nov. 1940, Vol. 59, No. 11.
20. Foster, D. E., and Seeley, S. W. *Automatic tuning, simplified circuits, and design practice.* Proc. I.R.E., Mar. 1937, Vol. 25, No. 3.
21. Seeley, S. W., and Avins, J. *The ratio detector.* R.C.A. Review, June 1947, Vol. 8, No. 2.
22. Hartley, R. V. L. *Transmission of Information.* Bell System Technical Journal, July 1928, Vol. 7, No. 3.
23. Shannon, C. E. *Communication in the presence of noise.* Proc. I.R.E., Jan. 1949, Vol. 37, No. 1.
24. Fink, D. G. *Bandwidth vs. noise in communication systems.* Electronics, Jan. 1948, Vol. 21, No. 1.
25. Federal Communications Commission. *Standards of Good Engineering Practice Concerning FM Broadcast Stations.* Sept. 20, 1945.
26. Toth, E. *A-M and narrow-band F-M in UHF communication.* Electronics, Feb. and Mar. 1949, Vol. 22, Nos. 2 and 3.
27. Shirane, G., Jona, F., and Pepinsky, R. *Some aspects of ferroelectricity.* Proc. I.R.E., Dec. 1955, Vol. 43, No. 12.
28. Manley, J. M. *Some general properties of magnetic amplifiers.* Proc. I.R.E., Mar. 1951, Vol. 39, No. 3.
29. Shaw, G. S., and Jenkins, J. L. *Nonlinear capacitors for dielectric amplifiers.* Electronics, Oct. 1953, Vol. 26, No. 10.

QUESTIONS

1. Why may modulation and demodulation be considered to be the same basic phenomenon?
2. To transmit information, must the carrier component be included in the signal?
3. Why may the term "linear rectification" cause confusion?
4. Do communication systems transmit only one sideband? Only two sidebands? What important principles are involved?
5. Why may demodulation be considered as distortion? Does this apply to frequency-modulation reception?
6. What is meant by the envelope of a modulated signal? If such a signal is demodulated in a linear rectifier, is the envelope of the rectified wave actually the audio component?
7. Why is "linear detection" superior to "square-law detection"?
8. Just prior to section 13-4 a statement is made regarding the way the two voice components combine. Can you prove this to be true?

9. How does a frequency converter operate? Why may the term "mixer" be misleading?
10. Why are crystals sometimes better than vacuum tubes for frequency conversion?
11. Why are resistor R_g and capacitor C_g used in Fig. 13-6?
12. What are typical values of C_1, L_1, C_2, and R_2 of Fig. 13-8? Why is inductor L_1 used?
13. How are the classifications of section 13-7 illustrated in Fig. 13-6?
14. What is meant by image frequencies? How do they affect receiver design?
15. How are copper-oxide varistors used in demodulation?
16. Why are point-contact crystal rectifiers, but not copper-oxide varistors, used at radio-frequencies?
17. In section 13-10 is the statement that tetrodes or pentode tubes are used as tuned radio-frequency voltage amplifiers. Why are triodes not used?
18. Just prior to section 13-12 is the statement that certain capacitors are linked mechanically. Why is this so?
19. How are demodulation, automatic volume control, and amplification accomplished by a single tube?
20. What is a limiter, and how does it operate?
21. In discussing discriminators the statement is made that the primary is tuned. Does this cause variations in the magnitude of E_p of Fig. 13-15 as the instantaneous frequency varies?
22. How is automatic volume control achieved in a receiver using a discriminator?
23. How could a frequency-modulation radio receiver be used to receive a phase-modulated signal?
24. Why are preemphasis and deemphasis used in frequency-modulation systems?
25. In a radio-broadcast receiver for both amplitude-modulated and frequency-modulated waves, what major circuit portions may be used in common?

PROBLEMS

1. Expand equation 13-6 and explain the effect of terms other than the component f_v. Prove that a perfect "linear" rectifier can, or cannot, cause audio distortion if the load is resistance. Can you name a type of filter to select the desired audio component that would present a resistance input?
2. An amplitude-modulated wave has a peak value of 10 volts. It was modulated by a 1000-cycle sinusoidal signal, the percentage modulation being 0.75 per cent. If this is impressed on the circuit of Fig. 13-13, approximately what will be the effective value of the audio voltage across resistor R_1?
3. Prove mathematically the statements in section 13-9 that copper-oxide varistors can be used to demodulate amplitude-modulated waves.
4. Prove graphically that a copper-oxide varistor will demodulate amplitude-modulated waves.
5. Certain typical values are given for Fig. 13-13. Make calculations for the R_4–C_4 combination and compare with the statements made in section 13-12. Repeat for the R_1–C_1 combination. What is the purpose of the R_2–C_2 branch? What are typical values? Repeat for R_3–C_3.

CHAPTER 14

WAVE-SHAPING AND CONTROL CIRCUITS

Early electron tubes were used almost entirely as rectifiers, amplifiers, and oscillators, and for similar purposes. In general, theory, design, and operation were explained from the sine-wave viewpoint, although often the tubes were used with nonsinusoidal waves such as speech. Electron-tube devices for generating nonsinusoidal signals such as square waves and pulses, and for utilizing such signals for control and other purposes, actually were developed at an early date. Relatively speaking, however, they were used to a limited extent until about 1940 when the extensive applications in television, radar, and the general field of control began.

This chapter will consider certain electron devices used primarily for wave-shaping and control. Wherever possible, definitions and descriptive terms from the Standards will be employed. This is to avoid the confusion existing because the same electron devices are used in different and somewhat isolated fields. Many basic circuits have been developed, and extensive variations exist. Only the most common types will be considered. In the pages immediately following, sinusoidal input voltages will be used, but square waves and pulses could be used in many instances. The effects of tube reactances, transit time, stray and residual circuit reactances, and the reactances of coupling capacitors will be neglected. Also, in many instances the internal impedance of sources will be neglected; the explanations can be extended easily to consider these impedances if desired.

14-1. *DIODE LIMITERS*

A **limiter** [1] is "a transducer whose output is constant for all inputs above a critical value." Such devices may be thought of as *magnitude limiters,* and are useful in wave-shaping circuits.

Series Diode Limiters. The diode of the upper circuit of Fig. 14-1 in series with resistor R acts as a rectifier and permits current flow through R, and voltage drop across R, only for the positive half-cycle of applied voltage. Ideal rectifier characteristics are shown by the broken line and actual characteristics by the solid line. Voltage drop e_d exists within the diode, and the output voltage $e_o = iR$ is $e_o = e_i - e_d$. If the diode is reversed, negative voltage will exist across R. The circuit of Fig. 14-1 is a *magnitude limiter* in the sense that the output magnitude is limited to zero when the impressed voltage is negative. This circuit is a *polarity limiter* in the sense that the output is limited to a positive voltage across R with respect to ground for Fig. 14-1, or is limited to a negative voltage if the diode is reversed. The impressed voltage may be **pulses,** a term [2] that "covers almost any transient phenomenon."

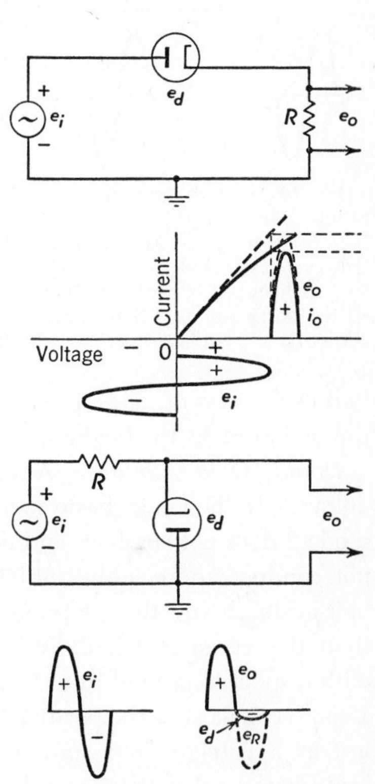

FIG. 14-1. A series limiter (above), and a shunt limiter (below). The internal impedance of the source is assumed zero.

Shunt Diode Limiters. If connections are made as in the lower circuit of Fig. 14-1, the diode does *not* conduct for the *positive* half-cycle, and the output voltage e_o equals the input voltage e_i. For the negative half-cycle the diode conducts and most of the impressed voltage will be lost across resistor R, assumed to be a large value of resistance compared with the internal resistance of the diode. The output voltage e_o will equal e_d, the drop across the diode. If the diode is reversed, no current will flow for the *negative* half-cycle, and the negative applied voltage will exist as an output voltage e_o. The positive output voltage will be the drop across the diode e_d. The circuit of Fig. 14-1 also may be regarded as a *polarity limiter,* because the output largely is limited to positive voltage, or to negative voltage if the diode is reversed.

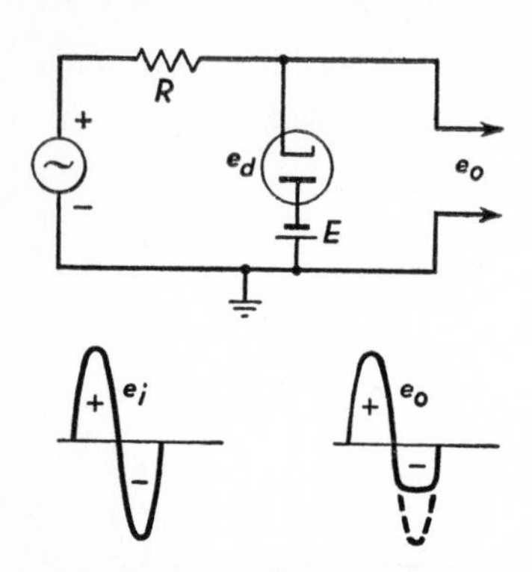

FIG. 14-2. A shunt diode magnitude limiter. If the diode and battery are reversed, the negative half-cycle is passed and the positive half-cycle is limited. The dotted portion of e_o is the voltage across R.

If a diode is connected in shunt as in Fig. 14-2, the entire positive impressed voltage exists between the output terminals. For the *negative* half-cycle, diode current cannot flow until the plate and cathode are at about the same potential with respect to ground, which will occur approximately when the generator drives the cathode as far negative as direct voltage E holds the plate negative. When current flows, if R is large compared with the diode internal resistance, then most of the applied voltage is lost across R. Hence, for *positive* impressed voltage $e_o = e_i$, and for *negative* impressed voltage $e_o = E$ approximately. If the diode and the direct voltage E of Fig. 14-2 are reversed, then for the *positive* half-cycle $e_o = E$, and for the *negative* half-cycle $e_o = e_i$. This is *magnitude limiting* action in accordance with the definition at the beginning of the section.

Double-Diode Limiter. A method of limiting both positive and negative voltages is shown in basic form in Fig. 14-3. For the positive half-cycle, diode 2 does not conduct, and diode 1 does not conduct appreciably until the positive half-cycle drives the plate more positive than the cathode, which is held positive with respect to ground by $+E$. When diode 1 conducts, most of the positive half-cycle is lost as a voltage drop across R, which is large compared to the internal resistance of the tube, and the output e_o is limited as shown. For the negative half-cycle diode 2 conducts, and the voltage drop in R limits the negative half-cycle. If e_i is large compared to e_o, the output is essentially a square wave. If the limiter of Fig. 14-3 is followed by an amplifier, and if this is followed by a second limiter driving a power amplifier, the combination is an excellent source of square waves. If desired, $+E$ and $-E$ may have different magni-

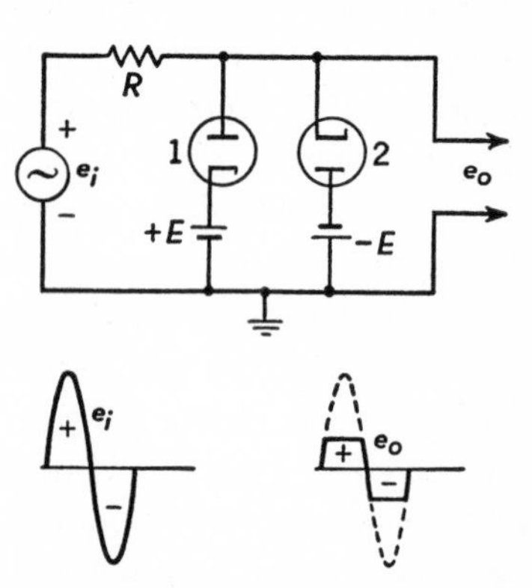

FIG. 14-3. A double-diode limiter. The dotted portions of the output wave exist across the resistor.

tudes, and hence $+e_o$ and $-e_o$ will be different in size. *Series-shunt diode limiters* are a possibility.

14-2. *DIODE CLIPPERS*

A **clipper** [1] is "a transducer which gives output only when the input exceeds a critical value." The shunt diode limiters of Figs. 14-2 and 14-3 are clippers *if* they are arranged so that the output is taken across resistor R as in the upper circuit of Fig. 14-4, which is a *single-diode clipper*. For the positive half-cycle of applied voltage, appreciable current flows only during the interval t_1 to t_2 when e_i exceeds E; no current flows for the negative half-cycle. A double-diode clipper is shown in the lower part of Fig. 14-4. As indicated, the output voltage e_o consists of the portions "clipped" from both the positive and negative half-cycles. Resistor R may be removed for some types of loads.

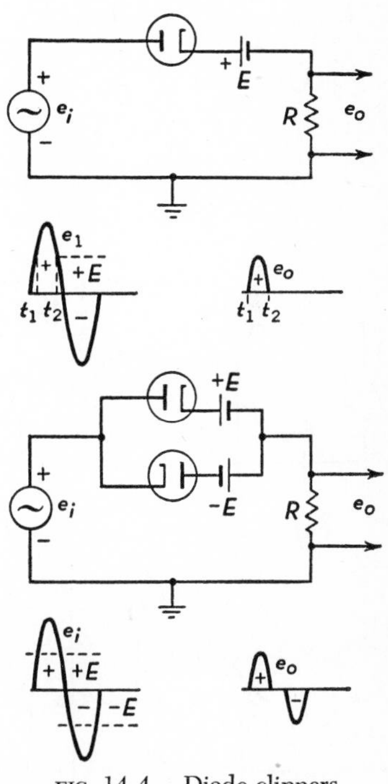

FIG. 14-4. Diode clippers.

14-3. *DIODE CLIPPER LIMITERS*

A **clipper limiter** [1] is a "transducer which gives output only when the input lies above a critical value and a constant output for all inputs above a second higher critical value." The names **amplitude gate** and **slicer** have been used. In the circuit of Fig. 14-5, direct voltage E_2 exceeds E_1. As the impressed voltage e_i rises positively, no current flows through load resistor R_L until the impressed voltage e_i exceeds E_1, when clipping occurs. Then, current will flow until the impressed voltage e_i exceeds E_2. At this point the current through diode 2 causes a voltage drop across R, no current flows through diode 1, and the output voltage e_o is limited. No output occurs for the negative half-cycle.

14-4. *TRIODE LIMITERS AND CLIPPERS*

Triodes, tetrodes, and pentodes are used as limiters and clippers. The basic principles are similar, and triodes will be used as illustrations.

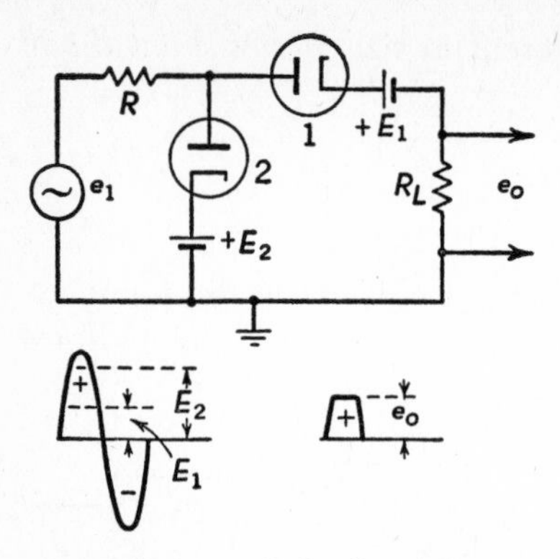

FIG. 14-5. A diode clipper limiter. Only that part of the positive half-cycle that is between broken lines E_1 and E_2 reaches output terminal e_o.

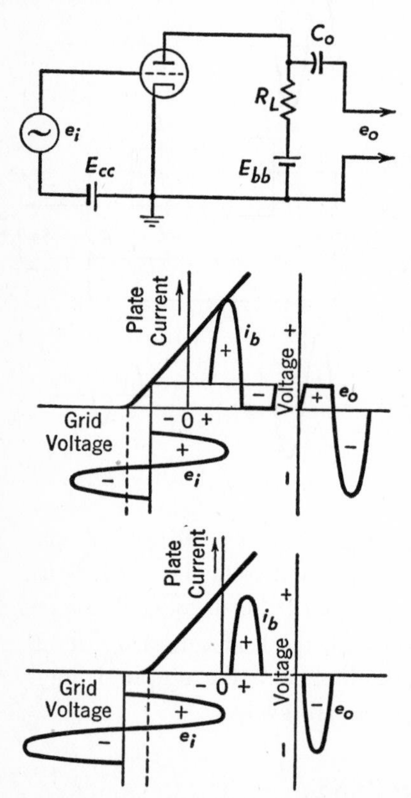

FIG. 14-6. Illustrating triode plate-current cutoff limiting (above) and clipping (below). Plate-current rise causes plate-voltage fall.

Triode Plate-Current Cutoff Limiters and Clippers. As shown in Fig. 14-6, the signal impressed between grid and cathode exceeds the negative grid-bias voltage. The plate current is shown by i_b. The output voltage is limited as shown by e_o. For a given signal, the magnitude of the limiting can be controlled by varying the grid-bias voltage. This is illustrated by the first set of curves of Fig. 14-6. If the negative bias is increased until the tube is biased beyond cutoff, plate current flows for part of the positive half-cycle of grid-signal voltage. When no current flows, e_o is zero, and when current flows, output voltage e_o goes negative as shown by the lower curves of Fig. 14-6. This may be regarded as clipping.

Triode Grid-Current Limiters. The triode of Fig. 14-7 has a resistor of perhaps one megohm in the grid circuit. If *E_c is zero,* then the grid will be driven positive by the impressed positive portions of the grid to cathode signal voltage, grid current will flow through R_g, and much of the impressed signal voltage will be lost as a voltage drop across R_g. Thus, but little *increase* in plate current results for the positive half-cycle of grid-signal voltage, little instantaneous increase in plate current occurs, the drop in output voltage e_o is small, and hence the negative half-cycle of output voltage is limited to a low value. For the negative half-cycle of applied signal voltage, negligible grid current flows, the plate current falls, and output voltage e_o rises, producing a full positive half-cycle. If the grid bias E_c of Fig. 14-7 *is not zero,* and if the grid *is not* driven beyond cutoff, the negative half-cycle will be limited to a value other than almost zero.

Triode Plate-Current Saturation Limiters. If the grid of the triode of Fig. 14-8 is driven increasingly positive, above a certain grid-driving voltage "region" the tube saturates (section 3-3), and the plate-current curve "flattens," and reaches a value that is reasonably constant. With zero bias and a sinusoidal applied grid-signal voltage of the magnitude indicated, the voltage output e_o will be limited as shown.

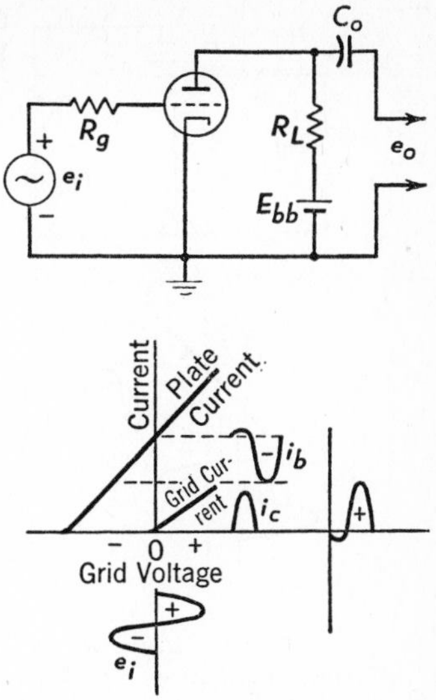

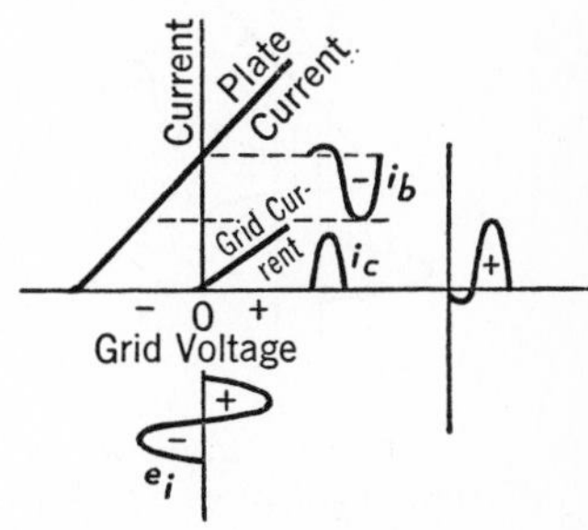

FIG. 14-7. Illustrating the triode grid-current limiter. Output voltage variations are indicated at the right.

Triode Grid-Current and Plate-Current Cutoff Limiters. The limiting action to be discussed occurs in class-A2 amplifiers if the signal voltage between grid and cathode is large compared with the grid-bias voltage, and if resistance exists in the grid input circuit, including the signal source. The limiting is caused by the action shown in Figs. 14-6 and 14-7, when the grid of the latter circuit is biased negatively. As indicated in Fig. 14-9, grid-current flow and the drop of signal voltage in R_g causes the *negative* half-cycle of e_o to be "rounded off," and plate-current cutoff "squares off" the positive half-cycle of e_o. This difference, although slight, is a convenient means of identifying the two half-cycles when the signal is displayed on an oscillograph tube [3] (or oscilloscope as popularly called). This method is useful in generating square waves. Information on signal shaping by limiting and clipping is given in references 4 to 8, inclusive.

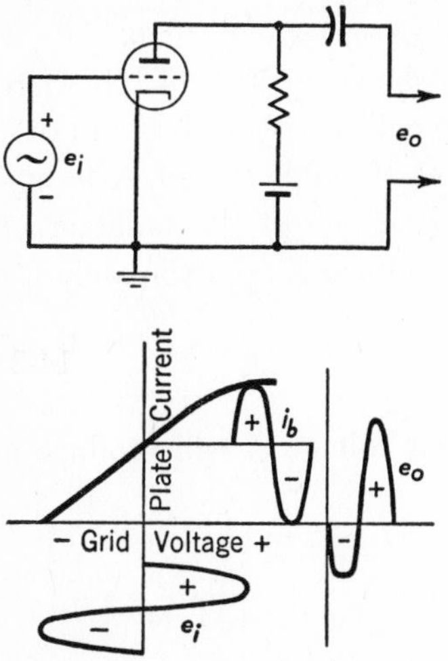

FIG. 14-8. Triode plate-current saturation limiter.

14-5. *DIFFERENTIATING NETWORKS*

A differentiating network [1] is a "network whose output is the time derivative of its input wave form." A *capacitor-resistor differen-*

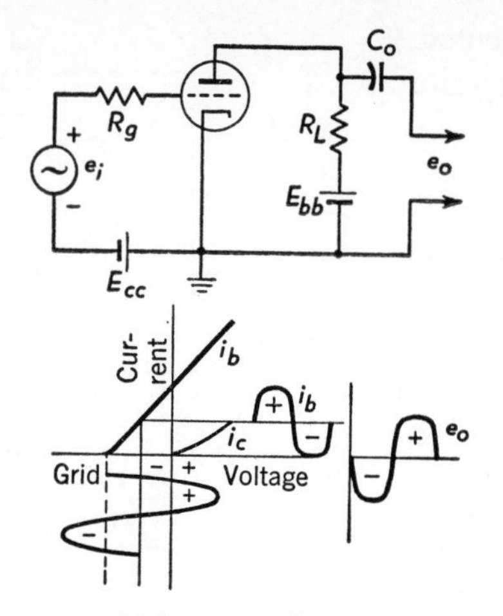

FIG. 14-9. A triode circuit utilizing both grid-current limiting and plate-current limiting.

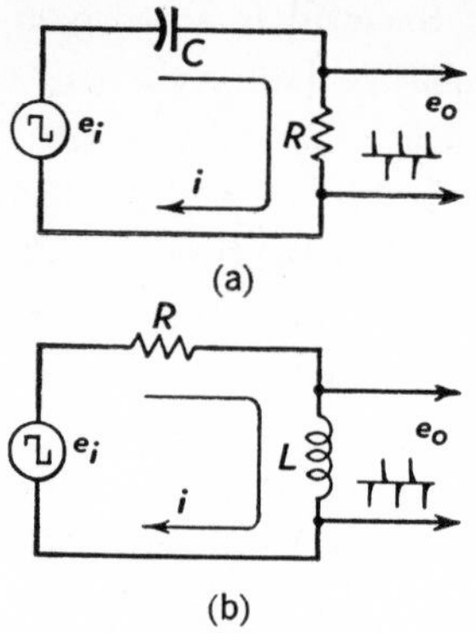

FIG. 14-10. Differentiating networks.

tiating network is shown in Fig. 14-10a, in which X_c greatly exceeds R in magnitude. The output voltage e_o is observed on a cathode-ray tube, the input impedance to which is assumed infinite. The instantaneous current through C and R is approximately $i = C(de/dt)$, and the output voltage is approximately

$$e_o = iR = RC\frac{de}{dt} \tag{14-1}$$

If a square-wave voltage is impressed as shown in Fig. 14-10a, the output voltage e_o will be "spikes" as indicated, because an output voltage will exist only when the magnitude of the impressed voltage varies with time.

A *resistor-inductor differentiating network* is shown in Fig. 14-10b. It is assumed that R greatly exceeds X_L, and that the coil is without loss and distributed capacitance. The instantaneous current through R and L is approximately $i = e_i/R$. The instantaneous voltage e_o across the inductor is $e_o = L(di/dt)$; hence, the instantaneous output voltage is approximately

$$e_o = \left(\frac{L}{R}\right)\left(\frac{de_i}{dt}\right) \tag{14-2}$$

This equation indicates that a square-wave input voltage e_i will produce a "spiked" output voltage e_o.

14-6. *INTEGRATING NETWORKS*

An **integrating network** [1] is a "network whose output wave form is the time integral of the input wave form." Such networks sometimes are used in angle-modulation systems. A *resistor-capacitor integrating network* is shown in Fig. 14-11a, in which R greatly exceeds X_c. The current is approximately $i = e_i/R$. But, the equation for the current through a capacitor is $i = C(de_o/dt)$. Hence, $de_o = (i/C)\,dt$, $de_o = (e_i/RC)dt$, and the output voltage is approximately

$$e_o = \frac{1}{RC}\int e_i\,dt \qquad 14\text{-}3$$

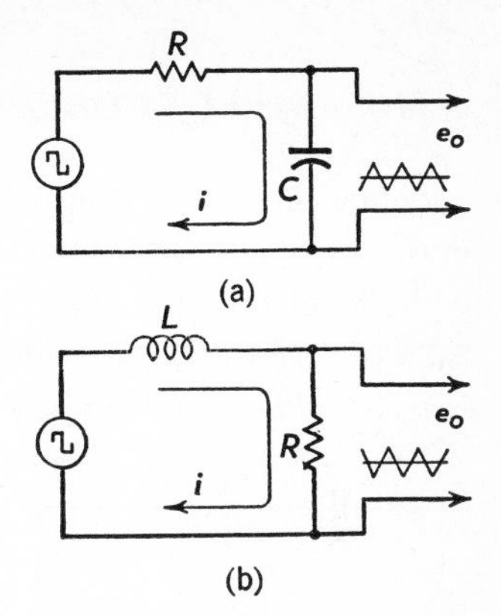

FIG. 14-11. Integrating networks.

If the input voltage e_i is a square wave, the output voltage e_o will be as shown in Fig. 14-11a. Impressing a square-wave voltage on a circuit with R much greater than X_c will cause constant-current flow for each half-cycle, and hence a rise in capacitor voltage constant with time.

An *inductor-resistor integrating network* is shown in Fig. 14-11b. The inductor is assumed perfect, and X_L greatly exceeds R. Thus, $e_i = L(di/dt)$, and $di = (e_i/L)\,dt$. The output voltage is $e_o = iR$, or approximately

$$e_o = \frac{R}{L}\int e_i\,dt \qquad 14\text{-}4$$

If the input voltage is a square wave, the output voltage is as shown in Fig. 14-11b.

14-7. *VACUUM-TUBE DIFFERENTIATORS AND INTEGRATORS*

The circuit of Fig. 14-12 provides amplification in addition to giving a differentiated output. In this circuit controlling resistance R of Figs. 14-10b and 14-11a is provided by the high plate resistance of the pentodes shown. As an approximation, the instantaneous plate current is $i_p = e_i g_{pg}$. For Fig. 14-12, $e_o = L(di/dt)$, and the output voltage is

$$e_o = g_{pg}L\frac{de}{dt} \qquad 14\text{-}5$$

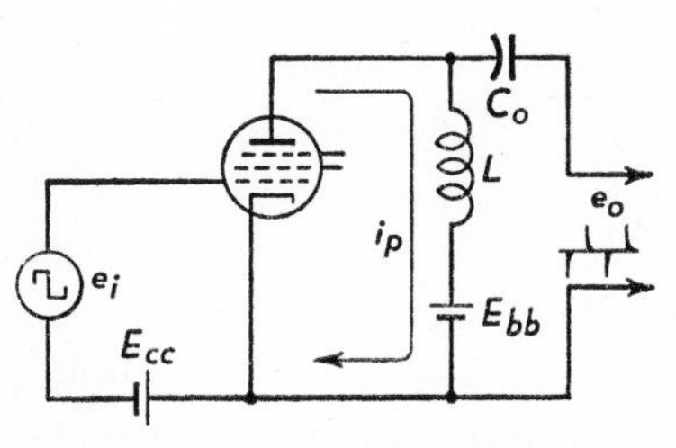

FIG. 14-12. A differentiating circuit using a pentode.

The circuit of Fig. 14-12 can be modified to integrate the applied signal voltage.

14-8. *DIRECT-CURRENT RESTORERS*

Sometimes it is necessary to transmit and amplify a pulsating direct-current signal. Often the simplest means is to remove the direct-current component, amplify the information-containing alternating component, and then restore the desired amount of direct current or voltage. A **d-c restorer** [9] is a "device, used in an ac transmission system, to establish dc transmission." The d-c restorer is important in television.[10] Sometimes it is called a **clamper,** because it "fixes" or "clamps" the signal at a desired direct-current or voltage level.

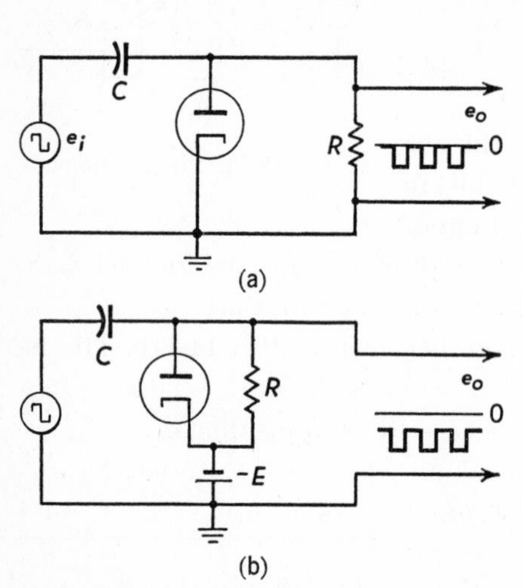

FIG. 14-13. In (a) is shown a direct-current restorer that "clamps" the signal at zero, and in (b) a restorer that clamps at $-E$ volts.

The *negative* d-c restorer of Fig. 14-13a indicates that a diode is used; however, in practice this may be a grid-cathode circuit.[10,11] Suppose C is about 1.0 microfarad, R about 1.0 megohm, and the square-wave input voltage e_i is about 100 volts peak-to-peak at 1000 cycles. If C initially is uncharged, then the first *positive* half-cycle will charge C through the diode which, when conducting, has low anode-to-cathode resistance of about 1000 ohms. During the *negative* half-cycle the diode will not conduct, and capacitor C cannot discharge appreciably through the 1.0 megohm resistor R. After several cycles of the applied square-wave voltage, the capacitor will be fully charged and the voltage across it will remain essentially constant. For the *positive* half-cycle the output voltage e_o is near zero because the conducting diode almost short-circuits resistor R. For the *negative* half-cycle (during which the diode does not conduct) the voltage across R is the sum of the impressed voltage and the capacitor voltage, or approximately 100 volts, and the upper end of R is *negative.* Hence, the d-c restorer output e_o is a pulsating negative voltage, "clamped" at essentially zero volts. If the diode is reversed, output voltage e_o will be a positive pulsating voltage clamped at zero.

If a constant source of voltage is added as in Fig. 14-13b, then the upper output terminal always will be at least about $-E$ volts negative with respect to ground. Thus, the output signal e_o is "clamped" at $-E$ volts, and appears as in Fig. 14-13b. Reversing the diode and the direct-voltage source causes the output to be a positive pulsating voltage clamped at $+E$ volts.

14-9. *TRIGGERS, SWITCHES, AND KEYERS*

An important feature of electron devices is the ability of one device, or circuit, to control another. Such control requires the use of triggers, switches, keyers, and similar devices that often utilize electron tubes. Triggers, switches, and keyers (*as electron devices*) now will be considered.

A **trigger** is an electron device, or circuit, that initiates an action, such as switching, in another circuit, the action continuing after the triggering impulse has disappeared.

A **switch** is an electron device, or circuit, that when "closed" initiates a change in a circuit, the change remaining in effect until the switch is "opened."

A **keyer** [1] is "a device which changes the output of a transmitter from one value of amplitude or frequency to another in accordance with the intelligence to be transmitted." This definition also is used in the broad sense of changing the output of any electron device. The terms **triggering, switching,** and **keying** sometimes are used loosely to cover the same, or similar, functions. A related term is **gating,** used in the sense of "opening" a circuit so that a signal may pass through, or out.

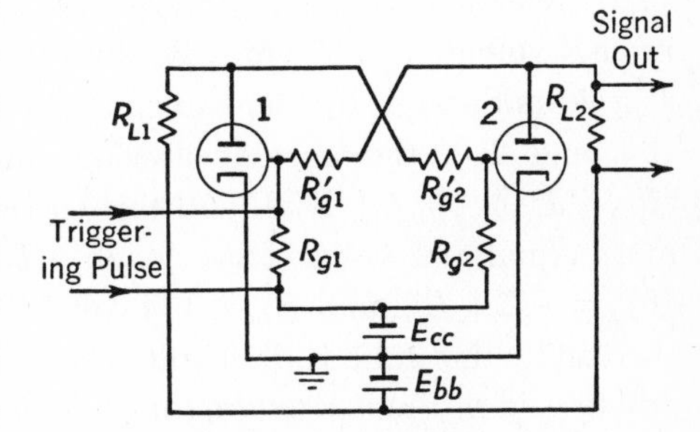

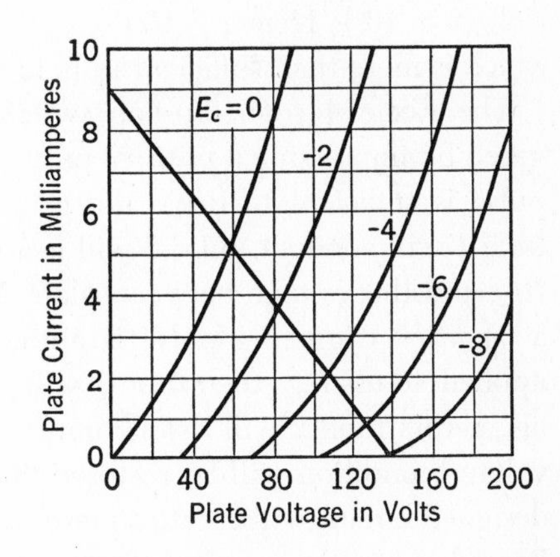

FIG. 14-14. An Eccles-Jordan switching circuit and tube characteristics showing the 15,000-ohm load line.

14-10. *FLIP-FLOP CIRCUITS* [12,13]

Certain electronic circuits have more than one condition of stability. A **flip-flop** is defined [14] as "an electronic circuit having two stable states and ordinarily two input terminals (or types of input signals) each of which corresponds with one of the two states. The circuit remains in either state until caused to change to the other state by application of the corresponding signal." The **Eccles-Jordan circuit**, developed in 1919, is an important circuit of the flip-flop type (Fig. 14-14). The principle of operation will be described, using data for a typical triode.

In the circuit of Fig. 14-14, R_{L1} and R_{L2} each equals 15,000 ohms, and R_{g1}, R'_{g1}, R_{g2}, and R'_{g2} each equals 1 megohm. Voltage E_{bb} is 135 volts, and the switched voltage is to have a magnitude of 55 volts. For the 15,000-ohm load line of Fig. 14-14, when a tube *is not conducting* grid-to-cathode voltage must be -8 volts (or more) to cause cutoff. Then, the plate-to-cathode voltage is $+135$ volts. If the tube *is conducting,* and the bias is -2 volts, the plate voltage falls to $+80$ volts. This gives a voltage change of 55 volts, which will appear across the plate load resistors of 15,000 ohms as the desired "switched" voltage. Suppose *tube 2 is conducting,* and that the plate-to-ground voltage is $+80$ volts. Because *tube 1 is not* conducting, no *plate* current flows through R_{L1}, and since this resistance is negligible compared with R'_{g2} and R_{g2}, the current flowing through these two resistors will be $(E_{bb}+E_{cc})/(R_{g2}'+R_{g2})$, and this current multiplied by R_{g2} and minus E_{cc} must equal -2 volts. Thus $[(135+E_{cc})\times 1.0\times 10^6]/(1.0\times 10^6+1.0\times 10^6)=-2+E_{cc}$, and $E_{cc}=139$ volts. If $E_{cc}=139$ volts, and if the plate of the conducting tube is $+80$ volts above ground, then the voltage from cathode to grid of the *nonconducting tube* is $-139+I_gR_{g1}$, where $I_g=(+80+139)/(2\times 10^6)$. Hence $-139+[(80+139)\times 10^6]/(2\times 10^6)=-29.5$ volts, which is more than sufficient to hold tube 1 below cutoff.

The Eccles-Jordan flip-flop switching circuit of Fig. 14-14 may be triggered by impressing a *positive* pulse across R_{g1} as indicated. If the triggering pulse is sufficient to drive the grid to cathode voltage above cutoff, then tube 1 will conduct, tube 2 will cease to conduct, no drop will occur across R_{L2}, and the voltage between plate 2 and ground will rise to $+135$ volts. If a *negative* triggering pulse is applied across R_{g1}, or if a *positive* pulse is applied across R_{g2}, then tube 1 will be cut off, tube 2 will conduct, the voltage across tube 2 will drop from $+135$ volts to $+80$ volts, and the original voltage condition will be restored. Switching occurs rapidly, and a properly designed circuit will create square waves or pulses when triggered correctly. Pentodes may be used instead of triodes. Flip-flop devices in conjunction with differentiating circuits are used to "scale down" the number of pulses when pulses are to be counted. Thus, the term **scale-of-2 circuit** sometimes is applied. Many variations of the basic Eccles-Jordan circuit are used.[7,8] In practice a power supply and a voltage divider would be used instead of the batteries of Fig. 14-14. The flip-flop circuit fundamentally is a **bistable multivibrator** (section 14-16).

14-11. *START-STOP MULTIVIBRATOR* [4,6,7,15]

When a triggering pulse is applied to the Eccles-Jordan circuit of the preceding paragraph, the circuit is transferred from one stable state to another, in which state it remains until a second triggering pulse is applied, which returns it to the original state. In the **start-stop multivibrator** [2] (also called **one-shot multivibrator**) now to be considered, a triggering pulse starts a sequence of operation, and after the sequence is completed the circuit will have returned to its original condition.

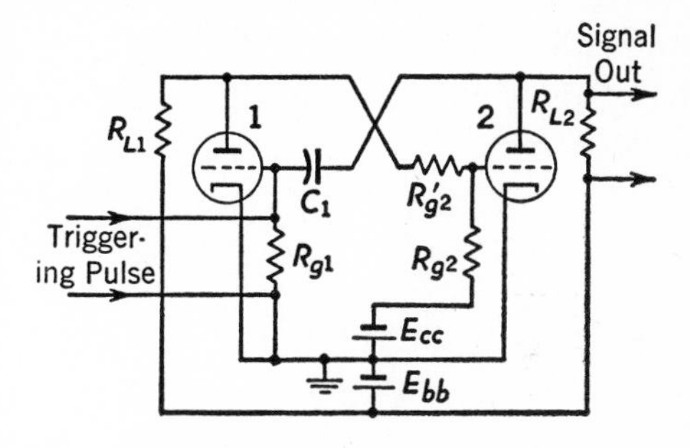

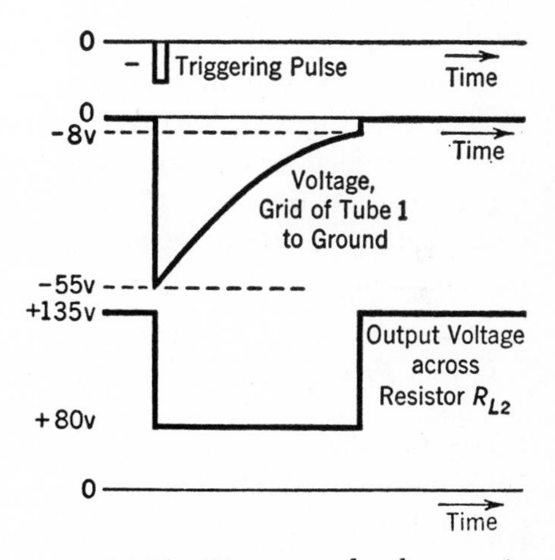

FIG. 14-15. Diagram and voltage variations for a start-stop, or one-shot, multivibrator. Voltage values *not* to scale.

A start-stop, or one-shot, multivibrator results if Fig. 14-14 is modified as shown in Fig. 14-15. Tube 1 will conduct because the grid is connected to the cathode and will be at approximately the cathode potential. If E_{bb} is 135 volts, and R_{L1} and R_{L2} each is 15,000 ohms, the plate current will be about 5.0 milliamperes (Fig. 14-14). The plate-to-ground voltage of tube 1 will be $135 - (5 \times 10^{-3} \times 15{,}000) = +60$ volts. Tube 2 does not conduct because of bias voltage $-E_{cc}$. The voltage from plate to ground of tube 2 is +135 volts because this tube is not conducting, and capacitor C_1 will be charged to about 135 volts.

Now if a negative triggering pulse of sufficient magnitude and of short duration is applied across R_{g1}, the grid of tube 1 will be driven negative with respect to cathode, and tube 1 will stop conducting. When this occurs, the voltage drop across R_{L1} suddenly is removed, and the plate-to-cathode voltage of tube 1 suddenly rises to +135 volts. The voltage from grid to ground of tube 2 rises toward zero, and tube 2 conducts. The voltage between plate and cathode of tube 2 then falls to about +80 volts as in section 14-10. But, capacitor C_1 previously was charged to +135 volts, and it must discharge from +135 to +80 volts through R_{g1}. The voltage drop caused by the transient discharge current flowing in R_{g1} places initially a *negative* voltage of 135 − 80 = 55 volts between grid and cathode of tube 1, and keeps it from conducting after the negative triggering pulse has disappeared. Tube 1 cannot again conduct until the drop across R_{g1}, which is its grid-to-cathode voltage, has increased to −8 volts. If C_1 of Fig. 14-15 is 0.001 microfarad, if R_{g1} is 1.0 megohm as in the preceding section, and if the effect of R_{L2} is neglected, then

$$i = \frac{E}{R_{g1}} \epsilon^{-t/(R_{g1}C_1)} \quad \text{and} \quad iR_{g1} = -E\epsilon^{-t/(R_{g1}C_1)} \qquad \text{14-6}$$

The time required for the bias voltage on tube 1 to rise from −55 to −8 volts so that tube 1 can again conduct is

$$-8 = -55\epsilon^{-t/(1.0\times10^6\times0.001\times10^{-6})} \quad \text{and} \quad t = 0.00193 \text{ second}$$

a value determined by transferring −55 to the other side of the equation and then taking the natural logarithm of each side. During this discharging period, which may be varied as desired, the output voltage from plate of tube 2 to ground remains at +80 volts. At the end of the discharging period it suddenly increases to +135 volts. Neglecting minor transient effects, the voltages in the circuit are as shown in Fig. 14-15. The start-stop multivibrator fundamentally is a **monostable multivibrator** (section 14-16).

14-12. *FREE-RUNNING MULTIVIBRATORS* [4,6,7,15]

The start-stop multivibrator of the preceding section required a triggering pulse for each cycle of operation. The circuit to be considered in the next section does not require triggering. It is a **relaxation oscillator,** defined [1] as an oscillator "whose fundamental frequency is determined by charging or discharging a capacitor or inductor through a resistor, producing wave forms which may be rectangular or sawtooth." The **multivibrator** [1] is a "relaxation oscillator employing two tubes to obtain the in-phase feedback voltage by coupling the output of each to the input of the other through, typically, resistance-capacitance elements. The fundamental frequency is determined by the time constants of the coupling elements and may be further controlled by an external voltage." The multivibrator first was described [4] in 1918 by

Abraham and Bloch. The free-running multivibrator is an **astable multivibrator.**

If both resistors R'_{g1} and R'_{g2} of Fig. 14-14 are replaced by capacitors of, for instance, 0.001 microfarad, and if the voltage E_{cc} is removed, the **free-running multivibrator** of Fig. 14-16 results. The multivibrator sometimes is considered to be a two-stage resistance-capacitance coupled amplifier with positive feedback.[6] After being energized, oscillations start, and build up to a "steady-state" condition. The arrangement of Fig. 14-16 is used to explain operation, and two important points should be kept in mind. *First,* either tube 1 or tube 2 is conducting, and after an initial switching period, the direct voltage from grid to cathode of the conducting tube is essentially zero. From the curves of Fig. 14-14, the plate current then is about 5.0 milliamperes, the drop across resistor R_{L1} or R_{L2} is $0.005 \times 15{,}000 = 75$ volts, and the voltage from plate to cathode of the conducting tube is $+135 - 75 = +60$ volts. *Second,* when a tube is not conducting, no voltage drop occurs in its plate circuit, and the plate-to-cathode voltage is $+135$ volts. When a tube is not conducting, it is because its grid-to-cathode potential is more negative than -8 volts; if it is less than -8 volts, the tube will conduct.

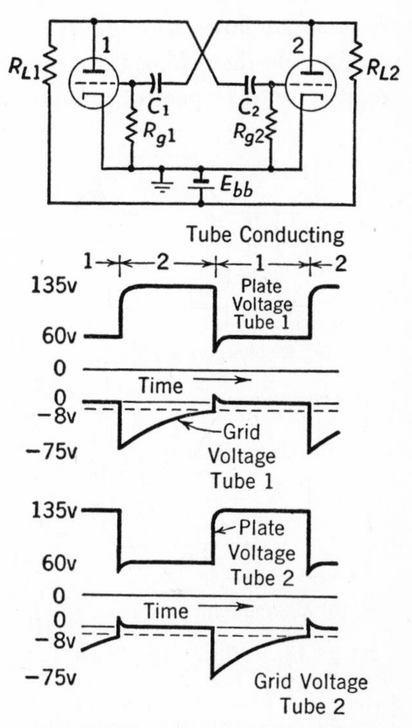

FIG. 14-16. Voltages from plate to ground and from grid to ground for a free-running multivibrator with grids connected to cathodes through grid resistors. Broken lines at -8v indicate cutoff values. Curves approximately to scale.

Multivibrators with Grids Connected to Cathodes. The circuit of Fig. 14-16 has the grids connected through resistors to the cathodes. Assume that tube 1 is conducting, that the grid-to-cathode voltage is zero, and that its plate-to-cathode voltage is $+60$ volts. Since tube 2 is not conducting, its plate-to-cathode voltage is $+135$ volts. The voltage between grid and cathode of tube 2 is negative, is below cutoff, but is rising toward cutoff as Fig. 14-16 shows. The reason for this rise will be explained later.

When the rising voltage from grid to cathode reaches -8 volts, tube 2 starts to conduct and its plate-to-cathode voltage starts to fall because of the voltage drop across resistor R_{L2}. At this same instant, capacitor C_1, which is connected to the plate of tube 2, starts to discharge through R_{g1} because this capacitor previously was charged to +135 volts, and this voltage now is falling because tube 2 is starting to conduct and there is a drop across R_{L2} as previously explained. The circuit through which C_1 discharges from +135 to +60 volts is shown in Fig. 14-17a, the discharge current being i_d. Because this current flows through R_{g1}, a large voltage drop will exist across this resistor, and the grid of tube 1, which just stopped conducting, will be driven *negative* with respect to ground and cathode.

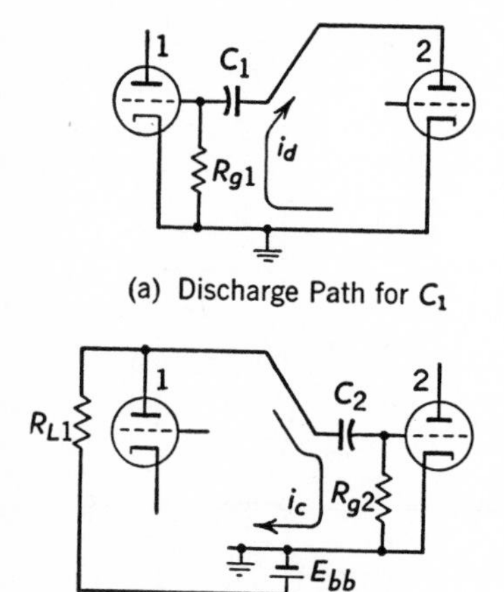

FIG. 14-17. When tube 1 stops conducting and tube 2 starts, capacitor C_1 must discharge as shown by current i_d. This current flowing through R_{g1} biases tube 1 far beyond cutoff. Also, capacitor C_2 must charge as shown, and current i_c flowing through R_{g2} biases the grid of tube 2 positively. Capacitor C_1 discharges slowly, and C_2 charges rapidly as explained in text.

At the instant under consideration when tube 1 stops conducting and tube 2 starts conducting, the voltage drop across R_{L1} disappears and C_2 starts to charge from +60 to +135 volts. The path for this is shown in Fig. 14-17b, the charging current being designated i_c. This current will (initially) flow through resistor R_{g2}, and the resulting i_cR_{g2} voltage drop will drive the grid of tube 2 *positive* with respect to ground and cathode. This causes the grid-to-cathode circuit of tube 2 to pass current, causes the grid-to-cathode resistance to drop to, perhaps, 1000 ohms, and this, in effect, short-circuits resistor R_{g2}. For this reason, capacitor C_2 quickly charges from +60 to +135 volts through the relatively low resistance of R_{L1}; thereafter the grid-to-cathode voltage is essentially zero. However, during the time that the grid is positive, a larger plate current flows in tube 2 than when the grid is zero. This larger current causes an increased voltage drop in resistor R_{L2}, and causes the plate-to-cathode voltage of tube 2 to fall below the value it will have when its grid is at zero with respect to cathode. Also, this same phenomenon causes the plate voltage of tube 2 to be "rounded off" and not to rise immediately to the final value it has when C_2 charges to +135 volts. Also, the grid of tube 1 is driven slightly more negative.

The discussion now must return to Fig. 14-17a, and to a consideration of the phenomena when C_1 discharges from +135 to +60 volts. The rate of discharge largely is controlled by R_{g1}, which is one megohm. Hence, the time required is large compared with the charging time considered in the preceding paragraph. However, after sufficient time has elapsed, the voltage from grid to cathode of tube 1 rises to cutoff, tube 1 conducts, and tube 2 is blocked by the current through R_{g2} caused by capacitor C_2 discharging from +135 to +60 volts.

The time required for the multivibrator to pass through a cycle of operation is chiefly determined by C_1, R_{g1}, C_2, and R_{g2}, because the conducting interval for tube 2 largely is fixed by the discharge of C_1 through R_{g1}, and the conducting interval for tube 1 by the discharge of C_2 through R_{g2}. If the assumption is made that when switching occurs the grid of the newly conducting tube is *not* driven positive, and the grid of the newly nonconducting tube falls *only* to the potential of $-(135-60)$ or -75 volts, the time required for each half-cycle of operation is given by equation 14-6 and the accompanying calculations. Or, the time for one cycle can be found as follows.

It is assumed that the grid-voltage rise of Fig. 14-16 is exponential as in applying equation 14-6. The cutoff voltage is taken as E_b/μ (section 4-4). Voltage change E_o is the output voltage variation across resistors R_{L1} and R_{L2} (that is, $135 - 60 = 75$ volts in preceding calculations). Then, from equation 14-6,

$$\frac{E_b}{\mu} = E_o \epsilon^{-t/(R_{g1}C_1)} \quad \text{and} \quad \frac{E_b}{\mu E_o} = \epsilon^{-t/(R_{g1}C_1)} \qquad 14\text{-}7$$

$$\epsilon^{-t/(R_{g1}C_1)} = \frac{E_b}{\mu E_o} \quad \text{and} \quad \epsilon^{t/(R_{g1}C_1)} = \frac{\mu E_o}{E_b} \qquad 14\text{-}8$$

Taking the natural logarithm of both sides gives

$$\frac{t}{R_{g1}C_1} = \log_\epsilon \frac{\mu E_o}{E_b} \quad \text{and} \quad t = R_{g1}C_1 \log_\epsilon \frac{\mu E_o}{E_b} \qquad 14\text{-}9$$

which is the time for one *half*-cycle. For an entire cycle

$$t = (R_{g1}C_1 + R_{g2}C_2) \log_\epsilon \frac{\mu E_o}{E_b} \qquad 14\text{-}10$$

where t will be in seconds when the units are ohms, farads, and volts. The basic frequency of a multivibrator is the reciprocal of the time in seconds given by equation 14-10. If R_{g1} does not equal R_{g2}, or C_1 does not equal C_2, an **unsymmetrical multivibrator** results, and the time interval for the two parts of the cycle will not be the same. These intervals can be computed separately by equation 14-9.

Multivibrators with Grids Connected to Positive Bias. The grid resistors of the multivibrators just considered were connected to the cathodes. After the grid has been driven negative, the voltage rises along the curves of Fig. 14-16, reproduced as curve *A* of Fig. 14-18. For reasons explained later, sometimes it is advisable to connect the grid resistors to a variable source of voltage (for frequency control [16]) *positive* with respect to the cathodes, or to a constant source of positive voltage, as results if the grid resistors are connected to $+E_{bb}$ of Fig. 14-16. Assuming that the grid resistors are returned to $+E_{bb}$, then when changeover occurs from one tube to the other, the grid of the tube just ceasing to conduct will be driven to a negative voltage essentially as before. However, the voltage *will not rise* along curve *A* of Fig. 14-18 toward zero as a final voltage, but *will rise* along curve *B* toward $+E_b$ as a final value. Thus, the time required for the grid to rise from the maximum negative value to the cutoff value (that is, the time for one half-cycle) will be different, as indicated in Fig. 14-18.

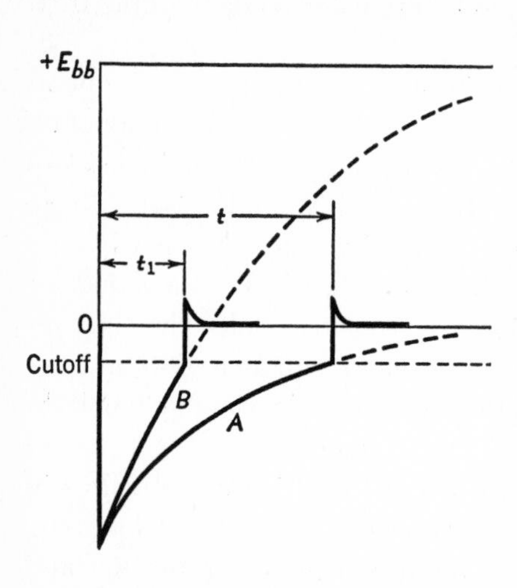

FIG. 14-18. Curve *A* shows how the grid potential of a tube rises when connected through resistors directly to the cathode, and curve *B* shows how the grid potential of a tube rises when the grid resistors are connected to a positive voltage such as E_{bb}. Note that the two time intervals t and t_1 are different.

Synchronizing Multivibrators. Although the frequency of a multivibrator largely is determined by circuit parameters as explained, to an extent the frequency may be controlled by an injected signal. Suppose that a sinusoidal signal is injected so that the instantaneous grid voltage varies as in Fig. 14-19. Then, it will be the signal that drives the grid above cutoff and that causes switching from one tube to the other. If the signal frequency is varied within limits, the frequency of the multivibrator will "follow" or be "drawn." If the difference between the signal frequency and the inherent multivibrator frequency becomes too great, the signal will lose control. However, if the signal frequency is further varied, a new frequency is reached at which the multivibrator "locks in" and becomes synchronized. A study of Figs. 14-18 and 14-19 shows the following: At a given distance to the left of cutoff a greater injected voltage will be required to drive the grids of a biased multivibrator beyond cutoff than is required for the unbiased type.

This is advantageous, because a positively biased multivibrator is less likely to be triggered by stray pickup and is accordingly more stable than the unbiased type. From this discussion it follows that if a multivibrator is biased with a variable positive voltage, some frequency control is possible by varying the voltage.

The gas-triode relaxation oscillator, and other circuits used to generate a saw-tooth wave for oscillographs, will be considered in Chapter 17.

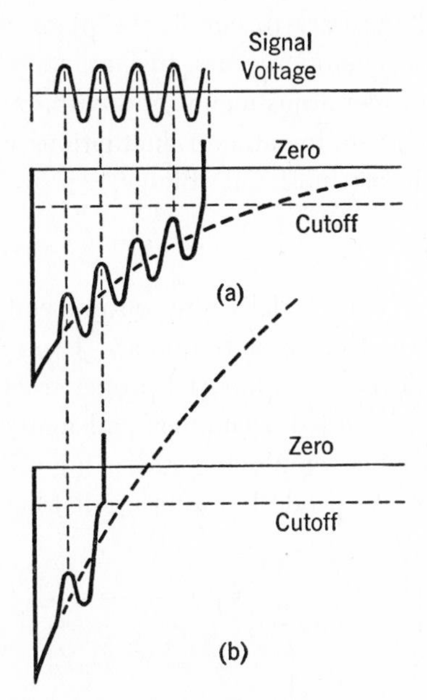

FIG. 14-19. Showing how a sinusoidal signal voltage may be used to "trip" a free-running multivibrator with (a) grid returned to cathode, and (b) grid returned to $+E_{bb}$.

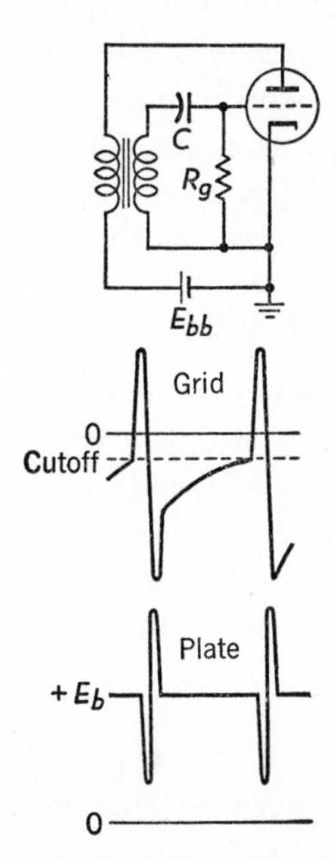

FIG. 14-20. Circuit, and variations in voltages from grid to ground and plate to ground, for a blocking oscillator. Curves not to scale, and shapes are approximate.

14-13. *BLOCKING OSCILLATORS*

The **blocking oscillator** (or **squegging oscillator**) is defined [1] as "an electron-tube oscillator operating intermittently with grid bias increasing during oscillation to a point where oscillations stop, then decreasing until oscillation is resumed." A blocking oscillator is a transformer-coupled feedback oscillator that operates intermittently for the following reasons. The coupling between grid and plate circuits of Fig. 14-20 is "tight," providing much positive feedback. Transformer connections are such that an increase in plate current induces in the grid circuit a voltage that drives the grid positive. This action causes the plate current to increase until current saturation occurs. When the grid is positive, the grid to cathode resistance is low, and capacitor C quickly charges.

When the plate current no longer increases, the grid voltage starts to drop, the plate current is decreased, resulting in a further grid-voltage drop, etc. The decrease in plate current results in a plate-voltage rise. Capacitor C discharges through R_g, and the resulting voltage drop across R_g holds the grid below cutoff. When the grid voltage rises to cutoff, the plate current starts to increase, and another cycle of oscillation occurs. Instantaneous plate and grid voltages are indicated. With proper adjustments, and with, perhaps, some negative feedback included, the plate-to-cathode fluctuations can be made to approximate closely one cycle of sinusoidal voltage.

14-14. *FREQUENCY DIVIDERS*

A **frequency divider** is defined [1] as a "device delivering output voltage at a frequency that is a proper fraction of the input frequency. Usually, the output frequency is an integral submultiple or an integral proper fraction of the input frequency." This device also is called a **counter,** and many types have been developed.[7,8] A frequency divider employing a blocking oscillator is shown in Fig. 14-21. The cathode of the triode blocking oscillator is held

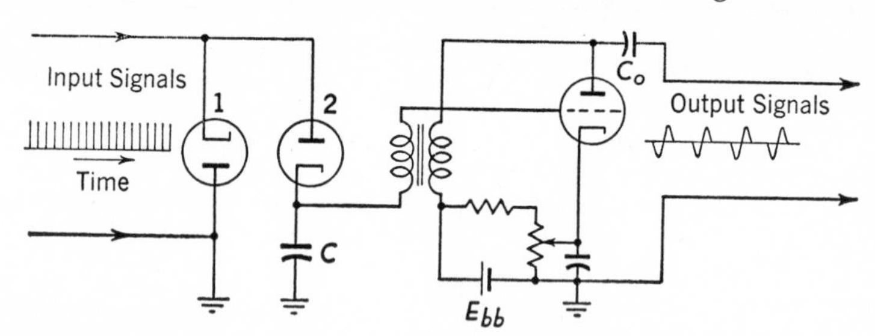

FIG. 14-21. A frequency-dividing circuit that divides by 5. This sometimes is called a counting circuit.

positive with respect to ground by the voltage divider across E_{bb}. The capacitor across the divider grounds the cathode for pulses. *Negative* pulses impressed on the input cause diode 1 to conduct, which in effect places a low resistance to ground across the input, and "grounds" negative pulses. *Positive* pulses are passed by diode 2 and charge capacitor C, each pulse raising the grid voltage positively above ground. When the grid-to-cathode voltage reaches cutoff (the cathode is biased positively), the blocking oscillator starts a cycle of operation. When the grid is driven positive with respect to cathode, capacitor C discharges through the tube. When the grid is driven negative with respect to the cathode, diode 2 and diode 1 both conduct; therefore, the grid always returns to essentially ground potential, thus stabilizing the frequency divider. Gas triodes sometimes are used in frequency dividers.

14-15. GATES

In certain electronic applications a portion of a signal is to be passed and all other portions are to be rejected. **Gating** [1] is "the process of selecting those portions of a wave which exist during one or more selected time intervals, or which have magnitudes between selected limits." The electronic circuit that selects the signal is a **gate,**[14] a "circuit having an output and a multiplicity of inputs so designed that the output is energized when and only when a certain definite set of input conditions are met." In computer work, gates often are called **"and" circuits.**[14] The term gate sometimes is applied to the signal that operates a gate circuit. The terms "enabling" gate and "disabling" gate also are used.[5] An example of gating is shown in Fig. 14-22.

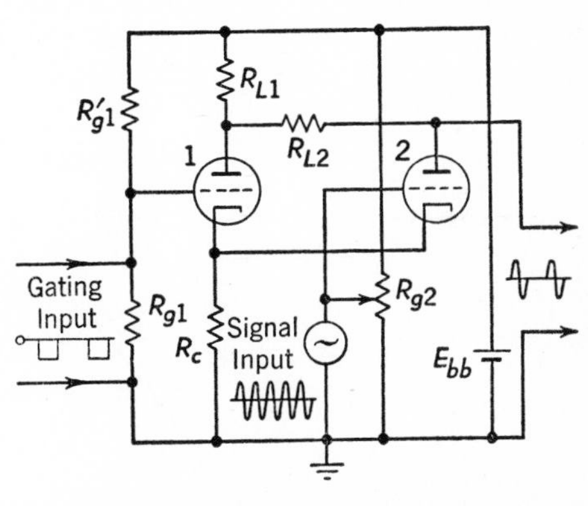

FIG. 14-22. In this gate, a sinusoidal signal is impressed on the grid of tube 2, which is blocked until the circuit is gated by negative pulses that remove the blocking potentials and permit tube 2 to pass single cycles of the impressed signal. Less than one cycle, or more than one cycle may be passed.

The voltage divider formed of R_{g1} and R_{g1}' is proportioned so that the grid of tube 1 is positive to ground. The load in the plate circuit of tube 1 is R_c and R_{L1}, and direct plate current I_b flows. The I_bR_c voltage drop makes the cathode positive to ground, but *not* sufficiently positive to block tube 1. The voltage divider R_{g2} is adjusted so that when tube 1 is conducting, tube 2 is blocked. The signal that is to be gated is connected as indicated; since tube 2 is blocked, no output results. However, if a negative gating pulse is applied, the grid of tube 1 will be blocked for the duration of the pulse, and the following occur: When tube 1 blocks, plate current in tube 1 ceases, the voltage drops caused by this current will disappear, the plate potential on tube 2 will rise, the cathode potential will fall, and with proper adjustment of R_{g2}, tube 2 will operate as a class-A amplifier and will pass the signal for the duration of the gating pulse applied to tube 1.

14-16. *POINT-CONTACT TRANSISTOR SWITCHING CIRCUITS* [17,18,19,20]

An important characteristic of a transistor is its ability to function in switching, and similar, circuits. In particular this is true of the point-contact type. This device possesses the properties of current multiplication and negative resistance (Chapter 10). Certain point-contact transistor circuits (to which this discussion is confined) have two different stable conditions, or states, separated by an unstable negative-resistance region. By the application of a triggering signal, such as a pulse, the circuit may be changed, or switched, from one stable state to another. When the **monostable transistor circuit** is triggered, it changes from one stable state to another, and immediately returns to the original state. When the **bistable transistor circuit** is triggered, it changes from one state to another and remains in the second state until triggered again. An **astable transistor circuit** oscillates between states.

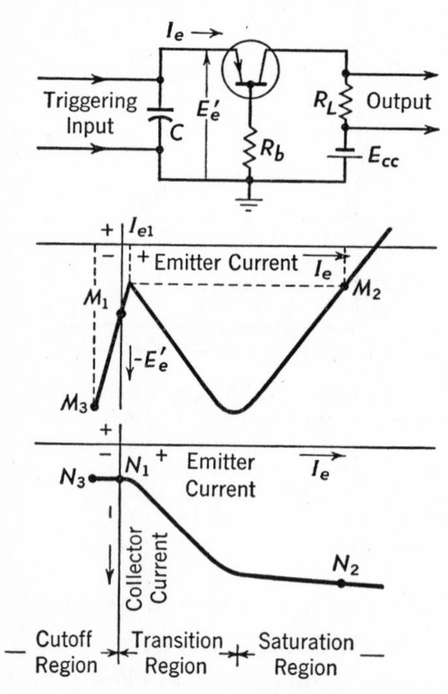

FIG. 14-23. A monostable point-contact transistor switching circuit. Characteristics at the left, in the vicinity of zero emitter current, are exaggerated for clarity.

Monostable Point-Contact Transistor Circuits. A basic monostable circuit [17] is shown in Fig. 14-23. This may be modified to supply emitter direct current.[18] The upper curve of Fig. 14-23 determines the mode of operation of the circuit; the lower curve determines the shape of the output signal. These curves are obtained using circuits similar to Fig. 10-10, but with resistance connected in series with the collector and the base. The voltage E_e' is measured as indicated. Emitter current i_e is plotted as the independent variable. The curves show three distinct parts called [17] a low-conduction *cutoff region,* a negative-resistance *transition region,* and a high-conduction *saturation region.* The curves have been modified in the region of zero emitter current to emphasize the phenomena. Because the characteristic in each region is essentially a straight line, the transistor and its circuit may be represented by three equivalent circuits, one for each region.[17] One point of stability is M_1 of Fig. 14-23, a point at which the current through the capacitor and emitter is zero.

If a triggering source sends a positive current pulse into the emitter that drives the emitter beyond current I_{e1}, the current amplification of the transistor and the positive feedback characteristic of the transistor and base resistor R_b, then cause (by their own action and independently of the triggering pulse) the emitter-to-ground voltage E_e' first to fall to a more negative value, and then to rise to the value at point M_2. The transient *increase* in emitter current charges capacitor C so that the upper terminal is *negative* with respect to ground, and when point M_2 is reached, there is nothing to maintain emitter current I_{e2}, so the emitter current falls, and voltage E_e' varies as shown in Fig. 14-23. Because the capacitor discharges a "negative" current through the emitter, circuit conditions pass to point M_3, and then back to the *original point* M_1 when the capacitor is discharged completely. During this interval the collector current has varied according to Fig. 14-23 from N_1 to N_2, to N_3, and back to N_1, causing a voltage change to be "switched" and to appear as a pulse across resistor R_L.

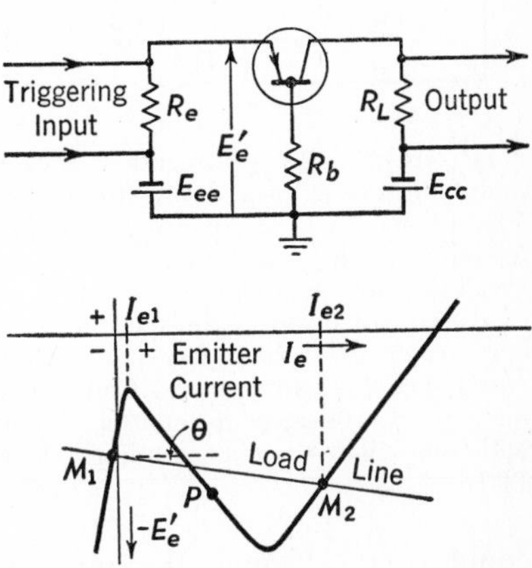

FIG. 14-24. A bistable point-contact transistor switching circuit. The circuit is stable at points M_1 and M_2. For collector currents corresponding to points M_1 and M_2, consult Fig. 14-23. The curve has been modified slightly in the zero emitter-current region so that the changes may be shown.

Bistable Point-Contact Transistor Circuits. When the bistable transistor switching circuit of Fig. 14-24 is triggered, it changes from one stable state to another, and remains in the new state until triggered again, when it returns to the original state. The emitter characteristic curve and *emitter* load line determine the triggering requirements. This bistable circuit is OFF when the emitter current and voltage are at point M_1, and ON when at point M_2. Note that emitter voltage is negative; that is, in the high-resistance direction for a point-contact transistor. The load line slope equals $-R_e$; that is, the magnitude of arc tan θ is $-E_e/I_e$. The *lower* curve of Fig. 14-23 is the transfer characteristic, showing how collector current varies with emitter current when load resistor R_L is in the collector circuit.

Assume that the circuit is in the OFF state, at point M_1. A very small emitter current, and a small collector current flow. Now suppose a triggering pulse *positive* with respect to ground drives the emitter above point I_{e1}. The current then will follow the characteristic curve and reach the ON position M_2, giving emitter current I_{e2}. The current through R_L now is much larger than formerly, and a larger "switched" output voltage exists across R_L than formerly. The circuit remains in the ON position until triggered by a negative triggering pulse that drives the emitter current to a value below that corresponding to the "valley" of the curve; then, the circuit returns to the OFF position and remains in this state until triggered by a suitable positive pulse.

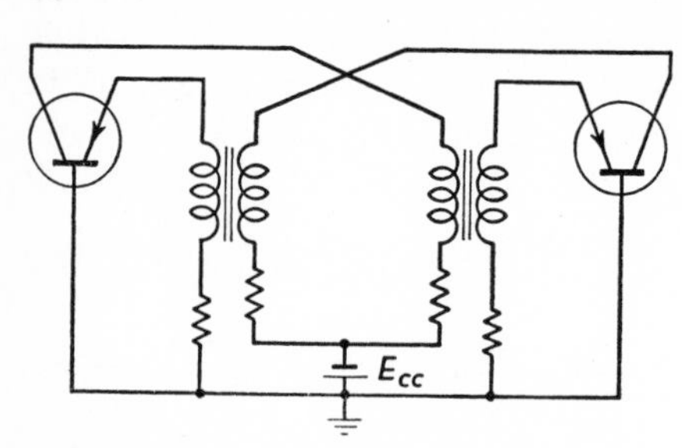

FIG. 14-25. A point-contact transistor multivibrator. Arrows designate the emitters. In one such multivibrator unity-ratio 500-ohm audio transformers, 10,000-ohm variable resistors, and $E_{cc} = 45$ volts were used. Resistors were adjusted for emitter currents of 1.0 milliamperes and collector currents of 2.5 milliamperes. Compare this method of collector to emitter coupling with the plate-to-grid coupling of Fig. 14-16.

Astable Point-Contact Transistor Circuits. If the emitter bias of Fig. 14-24 is changed in polarity and R_e is chosen so that operation is "at" point P on the curve shown in Fig. 14-24, then the circuit will operate as a negative-resistance device and will oscillate (Chapter 9). If a capacitor C is connected across the input terminals, then nonsinusoidal R–C oscillations result; if an L–C *series* resonant circuit is connected, then a sinusoidal output may be obtained at resistor R_L.[17] Transistor feedback oscillators were considered in section 11-11. Almost all, if not all, vacuum-tube circuits have their transistor counterpart.[21] Thus, the one-shot multivibrator of Fig. 14-15 performs a switching function much like the transistor monostable circuit of Fig. 14-23; also, the free-running vacuum-tube multivibrator circuit of Fig. 14-16 is similar to the free-running point-contact transistor multivibrator[21] of Fig. 14-25.

REFERENCES

1. Institute of Radio Engineers. *Standards on Antennas, Modulation Systems, Transmitters,* 1948.
2. Institute of Radio Engineers. *Standards on Pulses: Definitions of Terms—Part 1,* 1951. Proc. I.R.E., June 1951, Vol. 39, No. 6.
3. Institute of Radio Engineers. *Standards on Electron Tubes: Definitions of Terms,* 1950. Proc. I.R.E., April 1950, Vol. 38, No. 4.
4. Seeley, S. *Electron-Tube Circuits.* McGraw-Hill Book Co.
5. Terman, F. E., *Radio Engineering.* McGraw-Hill Book Co.

6. Terman, F. E., and Pettit, J. M. *Electronic Measurements.* McGraw-Hill Book Co.
7. Chance, B., Hughes, V., MacNichol, E., Sayer, D., and Williams, F. *Waveforms.* McGraw-Hill Book Co.
8. Puckle, O. S. *Time Bases.* John Wiley & Sons.
9. Institute of Radio Engineers. *Standards on Television, Definitions of Terms,* 1948.
10. Fink, D. G. *Television Engineering.* McGraw-Hill Book Co.
11. Wendt, K. R. *Television dc component.* RCA Review, March 1948, Vol. 9, No. 1.
12. Johnston, R. F., and Ratz, A. G. *A graphical method of flip-flop design.* Communication and Electronics Papers of A.I.E.E., March 1953.
13. Pressman, R. *How to design bistable multivibrators.* Electronics, April 1953, Vol. 26, No. 4.
14. Institute of Radio Engineers. *Standards on Electronic Computers: Definitions of Terms,* 1950. Proc. I.R.E., March 1951, Vol. 39, No. 3.
15. Abbot, A. E. *Multivibrator design by graphic methods.* Electronics, June 1948, Vol. 21, No. 6.
16. Sturtevant, J. M. *A voltage controlled multivibrator.* Electronics, Oct. 1949, Vol. 22, No. 10.
17. Lo, A. W. *Transistor trigger circuits.* Proc. I.R.E., Nov. 1952, Vol. 40, No. 11.
18. Anderson, A. E. *Transistors in switching circuits.* Proc. I.R.E., Nov. 1952, Vol. 40, No. 11.
19. Hunter, L. P., and Fleisher, H. *Graphical analysis of some transistor switching circuits.* Proc. I.R.E., Nov. 1952, Vol. 40, No. 11.
20. Schultheiss, P. M., and Reich, H. J. *Some transistor trigger circuits.* Proc. I.R.E., June 1951, Vol. 39, No. 6.
21. Wallace, R. L., and Raisbeck, G. *Duality as a guide in transistor circuit design.* Bell System Technical Journal, April 1951, Vol. 30, No. 2.

QUESTIONS

1. Do limiters and clippers accomplish the same thing? Give a definition of each type.
2. Is the upper circuit of Fig. 14-1 a "linear" rectifier? If not, how can this condition be approached?
3. If the biasing voltages $+E$ and $-E$ of Fig. 14-3 are interchanged, how does the circuit function? What would be the output voltage if $+E$ and $-E$ were made somewhat greater than the maximum value of e_i?
4. How does the double-diode clipper of Fig. 14-4 operate?
5. In discussing Fig. 14-7, the statement is made that if the bias is not zero, the negative half-cycle is not limited to almost zero. Why?
6. In discussing circuits such as Fig. 14-6 and 14-7, output capacitor C_o is assumed to have negligible effect on the wave shape of output voltage e_o. What will be the result if the capacitance is so low that the effect of C_o cannot be neglected?
7. In discussing Fig. 14-8, the statement is made that saturation occurs. Has

a name been applied to this type of saturation? How should triodes be operated so that saturation occurs readily?

8. If a triangular wave of positive and negative pulses of short duration and about three times the magnitude of e_i is impressed between grid and cathode of Fig. 14-8, what will be the appearance of the output wave?
9. What advantages and disadvantages are offered by the circuits of Figs. 14-3 and 14-9 for generating essentially square waves?
10. If a sine-wave voltage is impressed on a differentiating network, what will be the characteristics of the output voltage wave? Why is this network sometimes used in angle-modulation systems?
11. What is the significance of the minus sign in equations 14-2 and 14-4? Does this sign influence the drawing of Figs. 14-10b and 14-11b? Are these figures drawn correctly?
12. Referring to Fig. 14-10, and assuming perfectly square-wave inputs, what will be the shapes of the output voltages if R of the upper circuit has inductance, and if L of the lower circuit has resistance?
13. Are the output waves of Fig. 14-11 indicated correctly with respect to the driving-voltage changes?
14. In section 14-10 triggering by applying pulses to the grid circuit is discussed. Are there other methods of triggering? What are they?
15. How are flip-flop circuits used to scale down the number of received pulses for counting as mentioned at the end of section 14-10?
16. Why are the two expressions listed as equation 14-8 equal?
17. What is meant by the term unsymmetrical multivibrator? Does equation 14-10 apply to such multivibrators?
18. How can the instantaneous grid voltages be made to vary as in Figs. 14-19?
19. Why did not the capacitor C of Fig. 14-20 hold the grid positive just as it holds the grid negative?
20. Why does the blocking oscillator plate-to-cathode voltage rise above the direct voltage as shown in Fig. 14-20?
21. Why are diodes 1 and 2 used in Fig. 14-21? May one, or both, these diodes be eliminated? What circuit, or circuits, do these diodes form?
22. How can emitter direct current be provided in Fig. 14-23?
23. What will be the shape of the voltage wave from collector to ground before, during, and after the monostable circuit of Fig. 14-23 has been triggered?
24. The circuit of Fig. 14-24 must be triggered each time it changes from one stable state to another. A "stray" pulse *could* trigger the circuit. What, if any, relationship exists between the magnitude of R_e and switching reliability?
25. Is there any relationship between transistor parameters R_{12}, R_b, and the switching circuit of Fig. 14-24?

PROBLEMS

1. Germanium point-contact crystal diodes having the characteristics of Fig. 10-4 are used in the limiting circuit of Fig. 14-3. Assume a sinusoidal input voltage and select a value of R such that the output voltage is one-half the peak input voltage.
2. The generator of Figs. 14-1, 14-2, and 14-3 was assumed to have negligible

internal impedance. Assume that $R_g = 0.25R$ and that e_i of the generator is a sinusoidal open-circuit voltage, and redraw the figures if this new condition causes important changes.

3. Explain how to modify Fig. 14-12 so that the output voltage is the integral of the input voltage. Mathematically prove this to be true.
4. For the d-c restorer of Fig. 14-13, and using the numerical values given, calculate the following: The voltage across the capacitor at the end of the first positive half-cycle; the voltage across this capacitor at the end of the first negative half-cycle; the length of time it takes for capacitor C to become fully charged from a practical viewpoint; and the fluctuations in the voltage across it after this time.
5. What must be the sign and magnitude of the voltage pulse applied across R_{g1} that will trigger the flip-flop circuit of Fig. 14-14 and cause tube 1 to conduct and tube 2 to cut off? What must be the sign and magnitude of the voltage pulse applied across R_{g2} that will cause tube 2 to conduct and tube 1 to cutoff?
6. Calculate the time required for one cycle, and the basic frequency of a multivibrator with tubes having the characteristics of Fig. 14-14 used in the circuit of Fig. 14-17 modified so that R_{L1} and R_{L2} each equals 10,000 ohms.
7. Explain fully what determines the frequency of the output produced by the blocking oscillator of Fig. 14-20.
8. Make a complete graphic analysis of Fig. 14-21 when the circuit is adjusted so that five incoming pulses are required for each cycle of oscillator operation. Show all voltages with respect to ground.
9. Draw a diagram showing how a gas triode may be used instead of the vacuum triode of Fig. 14-21. Make a complete graphic analysis of the voltages with respect to ground.
10. Determine the equivalent circuits that apply for each region of Fig. 14-23, and develop an explanation for transistor switching circuits based on these equivalent circuits.

CHAPTER 15

*MAGNETIC AMPLIFIERS**

by

J. J. WITTKOPF

Magnetic amplifiers, and magnetic control devices, have received much attention since about 1940. Such equipment is not new, the saturable-core reactor having been used as early as 1885.[1] Although these magnetic devices are not "electronic" in the usual sense, they have both supplanted and supplemented electron tubes for many control applications. It is important that those persons in the field of electronics understand magnetic amplifiers and devices. Magnetic and dielectric amplifiers were considered briefly in section 13-15, the paragraphs expressing a point of view rather than a comprehensive discussion, as in this chapter.

15-1. *MAGNETIC MATERIALS* [2]

The tremendous increase in the use of magnetic amplifiers during the last decade can be attributed to the great improvement in magnetic materials. The theory of ferromagnetism has been extensively studied in the past, but even today not all ferromagnetic effects fully are understood. Briefly, **ferromagnetism** may be explained as follows. In an atom electrons are considered to move in orbits around the nucleus. Some of the electrons have an addi-

* This chapter (starting at section 15-1) was written by my friend and former colleague J. J. Wittkopf, who for the past several years has been actively engaged in the development and application of magnetic amplifiers and magnetic devices in the important field of control. Mr. Wittkopf was at Oregon State College for several years before entering industry. He was in charge of courses in basic electronics and industrial electronics, and is therefore well qualified by both teaching and industrial experience to prepare this chapter. I consider it a privilege to be permitted to include his chapter on magnetic amplifiers in this book.

Arthur L. Albert

tional motion which is a **spin** about their own axes. A ferromagnetic material is characterized by having electrons that spin in a preferred direction. The axis of this preferred spin is the direction in which magnetization is most easily established.

The most common crystal structures of magnetic materials are the body-centered cubic lattice (as in iron) and the face-centered cubic lattice (as in nickel). The direction of easy magnetization in the iron crystal is along one of the six edges of the cube, and in the nickel crystal it is parallel to one of the cube diagonals. These crystalline structures are shown in Fig. 15-1.

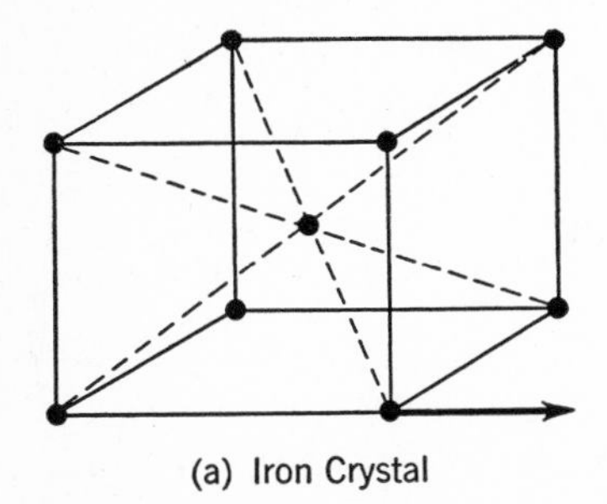

(a) Iron Crystal

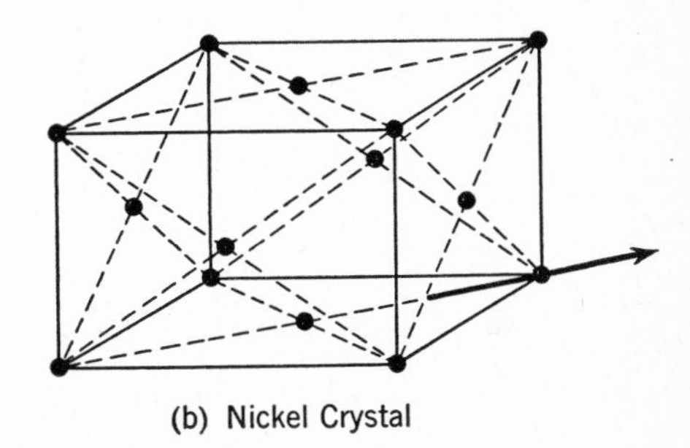

(b) Nickel Crystal

FIG. 15-1. Crystal lattice for ferromagnetic materials; arrows indicate direction of easy magnetization. Heavy dots represent atoms.

Since the magnetic behavior of a ferromagnetic material is largely determined by the crystal lattice structure, it is **magnetically anisotropic.** In an unmagnetized ferromagnetic crystal, the random electron spins result in no noticeable external magnetic effects. However, if a small amount of magnetizing force is applied externally along the direction of easy magnetization, some of the electron spins will become aligned and increase the magnetization in this direction. Within the crystalline structure, certain groups of neighboring electrons have parallel spin axes and axes of a group change alignment together when an external field is applied. These groups, or regions, are called **domains.** As the external field strength is increased, the domains that require greater magnetizing force become aligned until finally all the domains allowable are aligned. The crystal thus is saturated in the direction of applied magnetizing force. The sizes of domains vary considerably, depending upon the alloy, heat treatment, and mechanical stresses.

If the usually observed magnetizing curve of a ferromagnetic material is examined minutely, it will be found that the flux density increases in small incremental steps. This is caused by the fact that a small finite change in magnetizing force must be accomplished before domain rotation occurs. When domain rotation occurs, the flux density suddenly increases to a new

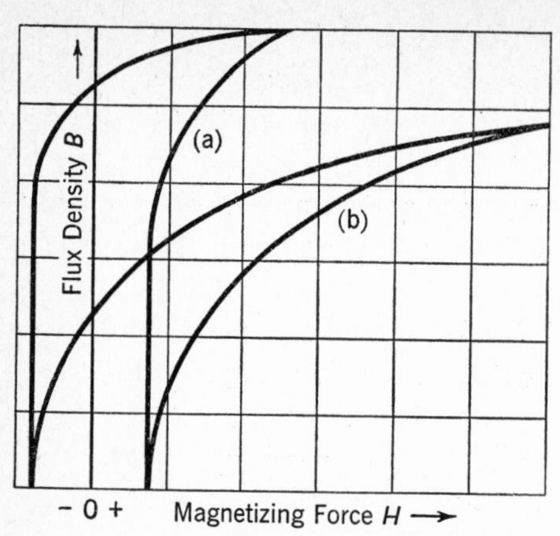

FIG. 15-2. Comparison between (a) grain-oriented, and (b) non-grain-oriented silicon steel.

level. This phenomenon was noted by Barkhausen and circuit noise produced by these sudden flux changes is called **Barkhausen noise.**[3] This noise usually is very small and need be considered only in circuits of very low energy-level.

Materials having excellent anisotropic magnetic characteristics are produced by orienting the crystals in the same direction so that a small magnetizing force is sufficient to align all domains. These materials have hysteresis loops that are almost rectangular, and are extensively used for magnetic amplifiers. The grain orientation is mainly achieved by cold-rolling of the magnetic material. The finished product is a ribbon of ferromagnetic material that is usually wound into toroidal cores and annealed in hydrogen. The thickness of the ribbon is determined from a consideration of the frequency. High-frequency applications may require a thickness of 0.00025 inch; more commonly, the thickness varies between 0.001 and 0.002 inch for 400-cycle use, and 0.004 and 0.005 inch for 60-cycle applications. To utilize the extreme anisotropic characteristics of these materials, it is essential that the magnetizing force be applied in the direction of cold-rolling. The hysteresis loops of the same material are shown in Fig. 15-2a for the grain-oriented condition, and in Fig. 15-2b for the non-grain-oriented condition. The rectangularity of the hysteresis loop may be improved further if the annealing is done in a magnetic field.

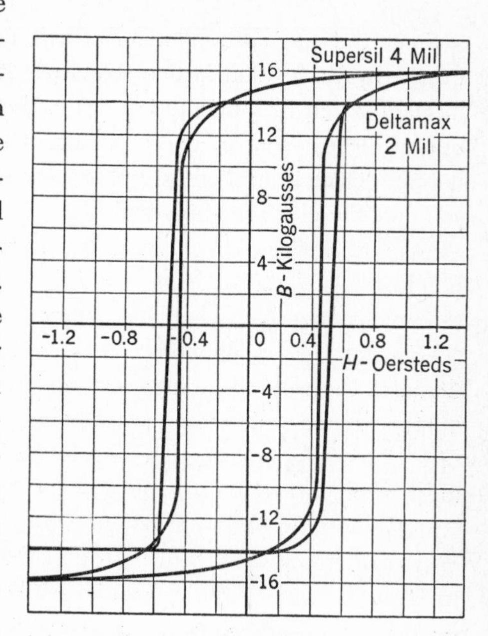

FIG. 15-3. Hysteresis loops for Supersil and Deltamax at 400 cycles.

For magnetic amplifiers and related devices, two types of magnetic alloys are most frequently used. These are the iron-silicon alloys and the nickel-iron alloys. The first type, which has been used many years in transformers, has been improved by grain orientation, and this material (containing 3 to 4 per cent silicon) is suitable for use in magnetic amplifiers. The outstanding features of an iron-silicon alloy are the high saturation flux density B and low coercive force H. The most common trade names for this material are **Hypersil, Silectron,** and **Supersil.** In the second group, probably the most widely used material for magnetic amplifiers is a grain-oriented alloy containing 50 per cent nickel and 50 per cent iron. This material has a very marked anisotropy and has the most rectangular hysteresis loop ever achieved in a polycrystal material. The saturation flux density is only slightly less than for the forementioned iron-silicon alloy and the coercive force is about the same; however, the very sharp saturation characteristic makes this material almost ideal for magnetic amplifier applications. The most common trade names for this material are **Deltamax, Hypernik, Orthanol,** and **Permenorm 5000Z.** The 400-cycle dynamic hysteresis loops of Supersil and Deltamax are shown in Fig. 15-3.

15-2. *SATURABLE REACTORS*

The saturable reactor is the basic element of the magnetic amplifier. It consists of a magnetic core that has one or more coils wound thereon. The core material usually selected has a distinct nonlinearity near the saturation level like the Supersil and Deltamax materials of Fig. 15-3. Consider the saturable reactor shown in Fig. 15-4, which has N turns, no resistance, and an instantaneous voltage e_r appearing across it. The relationship between flux ϕ and instantaneous voltage e_r is given by

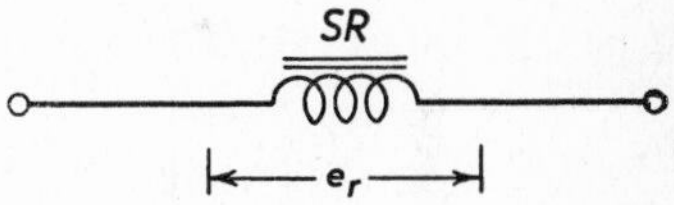

SR

e_r

FIG. 15-4. A saturable reactor will be indicated by the letters *SR.*

$$e_r = N\frac{d\phi}{dt} \qquad 15\text{-}1$$

If it is assumed that the flux changes from some level ϕ_1 to ϕ_2 in the time interval t_1 to t_2, this equation may be solved to yield the following expression.

$$N\int_{\phi_1}^{\phi_2} d\phi = \int_{t_1}^{t_2} e_r\,dt \qquad 15\text{-}2$$

Performing the integration gives

$$N(\phi_2 - \phi_1) = \int_{t_1}^{t_2} e_r\, dt \qquad 15\text{-}3$$

The change in flux $(\phi_2 - \phi_1)$ may be written as $\Delta\phi$; then

$$N\,\Delta\phi = \int_{t_1}^{t_2} e_r\, dt \qquad 15\text{-}4$$

Equation 15-4 indicates that the change in flux linkages $(N\Delta\phi)$ in the reactor is equal to the area under the voltage-time curve in the interval t_1 to t_2. The units associated with the integral in equation 15-4 are voltage and time. Thus the flux-linkage change in a reactor is equal to the volt-seconds applied to the reactor terminals. In this regard a reactor is similar (actually the dual) to a capacitor where the change in charge is equal to the applied ampere-seconds. This is expressed in equation 15-5 where ΔQ is the change in charge for a capacitor during the time interval t_1 to t_2:

$$\Delta Q = \int_{t_1}^{t_2} i_c\, dt \qquad 15\text{-}5$$

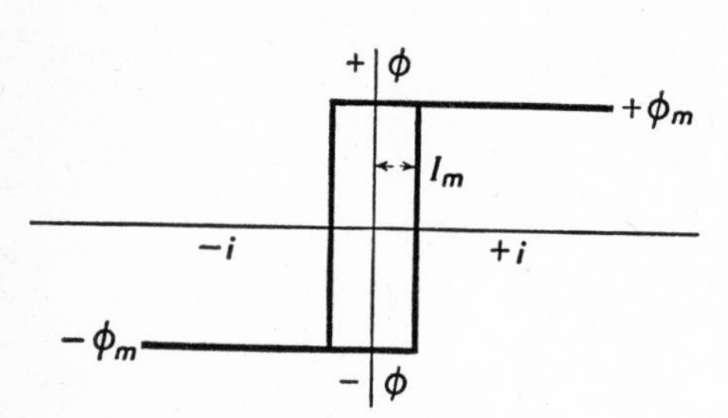

FIG. 15-5. Rectangular hysteresis loop for a saturable reactor.

Throughout this chapter, core materials having a rectangular hysteresis loop will be considered for simplicity. Actually the nickel-iron alloys like Deltamax and others very closely approximate a truly rectangular characteristic. The hysteresis loop for a particular reactor may be plotted with core flux as a function of coil current as shown in Fig. 15-5. If the flux in this reactor were to change from $-\phi_m$ to $+\phi_m$, the volt-seconds that would be absorbed by the reactor before positive saturation occurred would be equal to the total flux-linkage change; that is,

$$\int e_r\, dt = N[\phi_m - (-\phi_m)] = 2N\phi_m \qquad 15\text{-}6$$

It is important to understand that the flux in the core is dependent *only* upon the voltage-time integral and is *not* dependent upon the magnetomotive force. In any saturable reactor, the instantaneous current that flows in the coil is determined by the shape of the hysteresis loop and the manner in which the voltage-time integral varies with time. If material with a rectangular hysteresis loop is used in the reactor (Fig. 15-5), the coil current during the time when the flux is changing from $-\phi_m$ to $+\phi_m$ is constant and independent of voltage wave form. If saturation occurs, then the current is determined by the other elements in the circuit.

15-3. *SATURABLE REACTORS WITH RESISTANCE LOADS*

In the series circuit of Fig. 15-6 the source voltage essentially all appears across the saturable reactor during the time it is not saturated. A small voltage drop occurs across R_L as a result of the small magnetizing current I_m that flows in the circuit. This current ordinarily is very small compared to the load current that flows *following saturation* of the reactor when the alternator voltage essentially all appears across R_L. Then, there will be a small voltage appearing across the reactor owing to the coil resistance and the "air-core inductance" of the reactor.

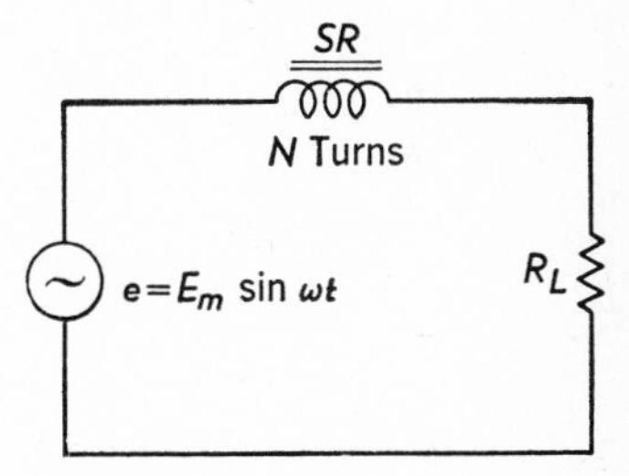

FIG. 15-6. A saturable reactor of N turns in series with a load resistor R_L.

Assume that at time $t = 0$ the flux in the core is $-\phi_m$. Then the following analysis may be made.

$$N \int_{-\phi_m}^{+\phi_m} d\phi = \int_0^{t_s} E_m \sin \omega t \, dt \qquad 15\text{-}7$$

The time t_s is the time during the first half-cycle when the core reaches positive saturation; hence

$$2N\phi_m = \frac{E_m}{\omega}(1 - \cos \omega t_s) \qquad 15\text{-}8$$

$$t_s = \frac{1}{\omega}\left[\cos^{-1}\left(1 - \frac{2\omega N \phi_m}{E_m}\right)\right] \qquad 15\text{-}9$$

Equation 15-9 indicates the time after the beginning of a half-cycle at which the reactor becomes saturated. The voltage and current wave forms for the circuit of Fig. 15-6 are shown in Fig. 15-7. It is assumed that the alternator voltage is large enough to cause saturation in Fig. 15-7a. The current that flows prior to saturation is the reactor magnetizing current I_m. After saturation the current is limited by the resistance R_L. If the alternator voltage is not large enough to cause saturation, the wave forms will be those shown in Fig. 15-7b.

15-4. *SELF-SATURATING CIRCUITS*

The **self-saturating circuit** of Fig. 15-8 differs from the circuit of Fig. 15-6 just considered by the addition of a rectifier. It will be assumed throughout

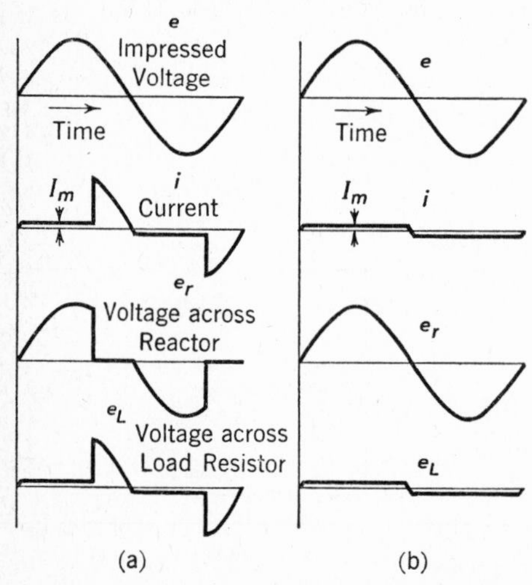

FIG. 15-7. Voltage and current wave forms for Fig. 15-6. In (a) the alternator voltage is sufficiently large to cause saturation; in (b) the alternator voltage does not cause saturation.

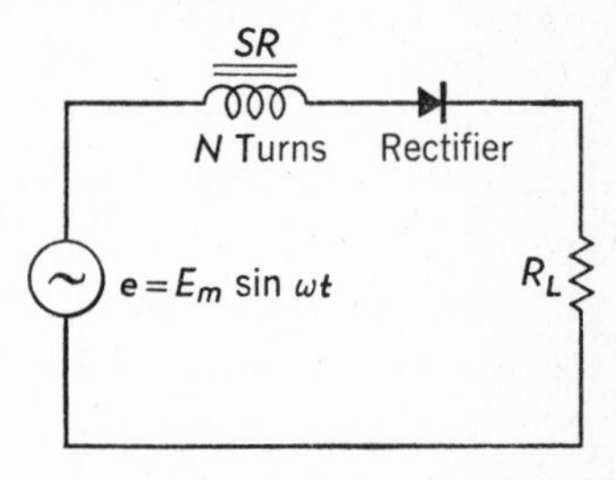

FIG. 15-8. A self-saturating circuit.

these discussions that all rectifiers are ideal. An **ideal rectifier** is one that has zero voltage drop when it is conducting in the forward direction and zero current through it in the inverse direction. During the time when the alternator voltage is negative, the entire voltage appears as inverse voltage across the rectifier and the current in the circuit is zero. During the time when the alternator voltage is positive, the circuit conditions are dependent upon the magnetic history, or previous conditions, of the flux in the reactor core. Assume that the following conditions exist:

$$\phi = -\phi_m \qquad (\text{at } t = 0) \qquad 15\text{-}10$$

$$\int_0^{\pi/\omega} E_m \sin \omega t \, dt > 2N\phi_m \qquad 15\text{-}11$$

The second condition means that the reactor will saturate at $t = t_s$ in the first *half-cycle* as shown in Fig. 15-9. During the remainder of the first half-cycle the voltage all appears across the load R_L and the current is determined by Ohm's law. During the negative half-cycle the voltage all appears across the rectifier and the flux condition in the core remains just as it was at the end of the first half-cycle. Then at the beginning of the next half-cycle, when the alternator voltage again is positive, the reactor core already is saturated at $+\phi_m$. Therefore the reactor cannot support voltage and consequently all the voltage appears across the load.

Now consider the case where equation 15-11 is altered so that

$$\int_0^{\pi/\omega} E_m \sin \omega t \, dt = 0.4(2N\phi_m) \qquad \text{15-12}$$

This means that during the first *half-cycle* (when the alternator voltage is positive) the flux will change only 0.4 of the amount it can change before saturation occurs. Then, because the negative voltage during the second *half-cycle* appears across the rectifier, the flux in the reactor core will remain at the level existing at the end of the first half-cycle. At the start of the third *half-cycle* (when the alternator voltage is again positive) the flux starts changing from this level and will change to 0.8 of the total allowable change, $2\phi_m$. During the time the reactor is absorbing volt-seconds, the current in the circuit is the magnetizing current I_m. When the fifth *half-cycle* begins, the flux starts to change again from the level existing at the end of the third half-cycle. Since the flux will reach the positive saturation value at the mid-point of this fifth half-cycle, the alternator voltage will then appear across the load. The current from this time on will be determined by Ohm's law during the positive half-cycles and will be zero during negative half-cycles. The flux in the reactor core will remain at $+\phi_m$. The wave forms for this condition are shown in Fig. 15-10 and the flux condition in Fig. 15-11.

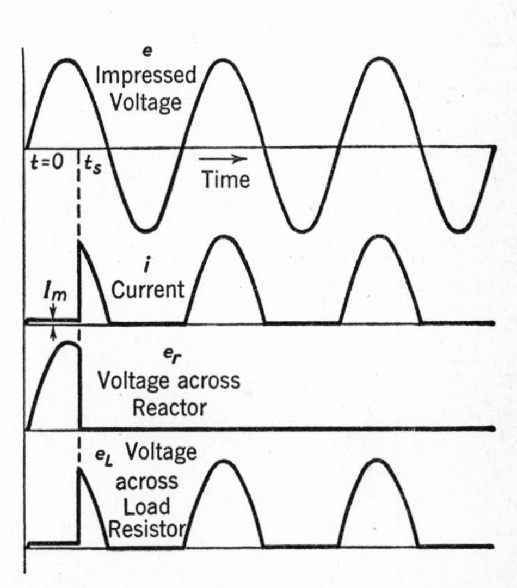

FIG. 15-9. Voltage and current wave forms for Fig. 15-8.

Further thought will lead to the conclusion that for any value of applied voltage (so that $E_m > R_L I_m$), this type of circuit will in time arrive at the condition where the reactor remains saturated and has no further effect upon circuit operation. Thus the circuit is aptly described by the words, *self-saturating* circuit.

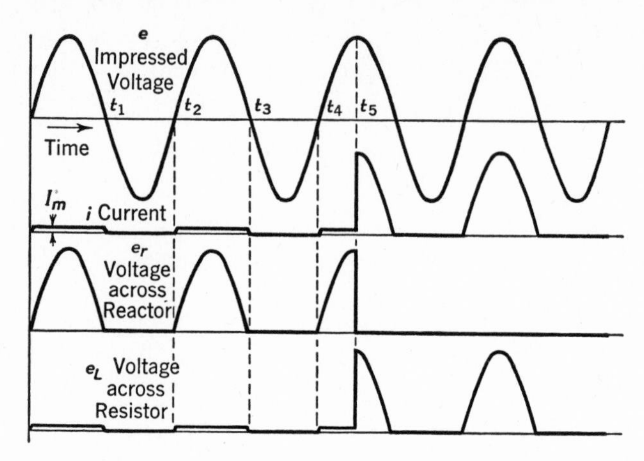

FIG. 15-10. Wave forms for the self-saturating circuit of Fig. 15-8 when conditions are as specified by equation 15-12.

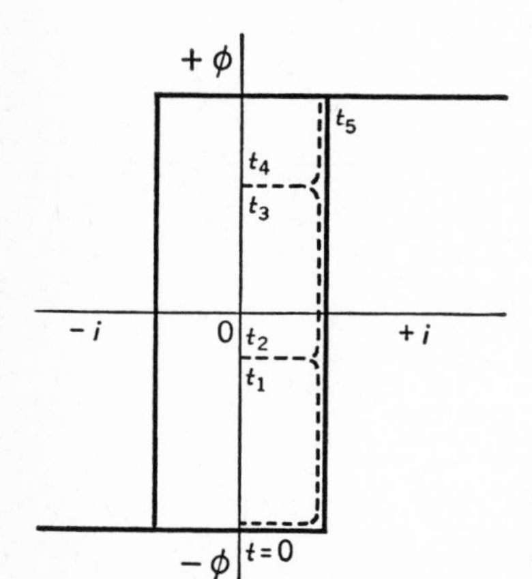

FIG. 15-11. Indicating how the flux rises in the core of the self-saturating reactor of Fig. 15-8 in accordance with equation 15-12, and as shown in Fig. 15-10.

15-5. *CONTROL OF SELF-SATURATING CIRCUITS* [4,5,6,7,8]

In the preceding section it was shown that the reactor would saturate and remain saturated in a self-saturating circuit. For this circuit to be useful, it is necessary to provide some sort of control to "reset" the flux in the reactor core to some predetermined level during the negative half-cycle. Then, the reactor would not be saturated at the start of the succeeding positive half-cycle. The time at which saturation occurs will be dependent upon the

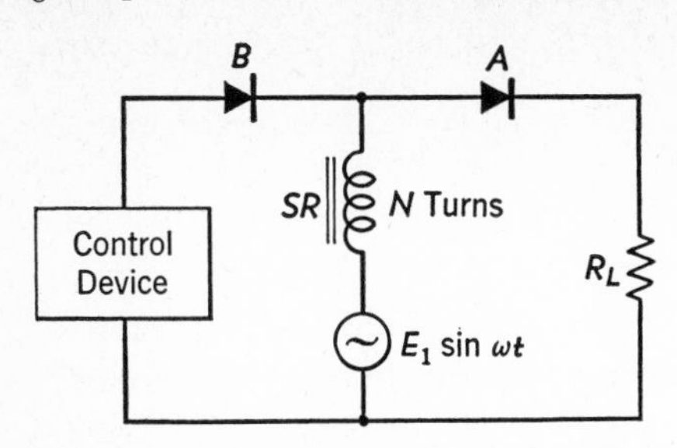

FIG. 15-12. Half-wave Ramey-type magnetic amplifier.

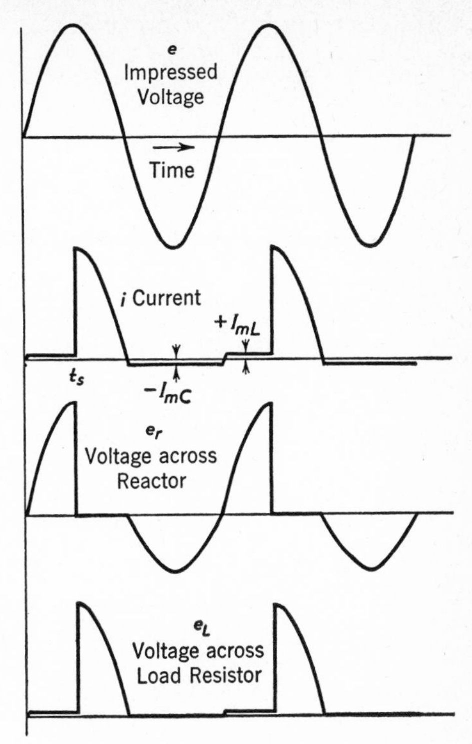

FIG. 15-13. Voltage and current wave forms of the circuit in Fig. 15-12 for conditions $\alpha = 0.5$.

amount by which the flux was reset during the negative half-cycle by the control circuit.

Assume that the self-saturating circuit of Fig. 15-8 is designed in accordance with equation 15-11 so that during the first positive half-cycle the alternator voltage will cause the reactor to just reach, but not exceed, saturation; it becomes saturated and remains that way. Now suppose a control circuit is added as in Fig. 15-12. This is a basic magnetic amplifier, often called a **Ramey circuit.**[7,8] When the alternator voltage is positive, current flows through rectifier *A*, but rectifier *B* appears to be an open circuit and the control circuit is not affected. During the negative half-cycle, rectifier *A* appears to be an open circuit, but rectifier *B* can conduct, so the control circuit is active during this time. Assume for this example that the control device is of such a nature that it causes only a portion of the alternator voltage to appear across the reactor. Since this voltage is negative, it will cause the flux in the reactor core to be reset by the amount indicated in equation 15-13.

$$\Delta\phi = \frac{1}{N}\int_{\pi/\omega}^{2\pi/\omega} \alpha E_1 \sin \omega t \, dt \qquad 15\text{-}13$$

The quantity α is the per-unit amount of the alternator voltage that appears across the reactor during the negative half-cycle. Some thought will lead to the conclusion that (since the shapes of the current and voltage wave forms are different) a control device that accomplishes such a voltage division is not easily realizable physically. However, this approach will be used for this example and then extended to other control devices.

If it is assumed that $\alpha = 0.5$, the reactor will be reset each negative half-cycle by an amount equal to one-half the maximum permissible excursion, $2\phi_m$. This is so because the reactor design in this example is such that the flux change with the full alternator voltage applied to the reactor terminals is $2\phi_m$. The voltage and current wave forms for the condition that $\alpha = 0.5$ are shown in Fig. 15-13.

If the useful output is the average value of the voltage appearing across the load, the output may be calculated as a function of the control characteristic, α. The volt-seconds absorbed during the reset-time is given by the equation

$$N\Delta\phi = \int_{\pi/\omega}^{2\pi/\omega} \alpha E_1 \sin \omega t \, dt \qquad \text{15-14}$$

The volt-seconds absorbed by the reactor during the positive half-cycle before saturation occurs is equal in magnitude to the reset volt-seconds.

$$\int_0^{t_s} E_1 \sin \omega t \, dt = -\int_{\pi/\omega}^{2\pi/\omega} \alpha E_1 \sin \omega t \, dt \qquad \text{15-15}$$

The volt-seconds that appear across the load in any one cycle, then, is that indicated in the equation

$$\text{Volt-seconds across load per cycle} = \int_{t_s}^{\pi/\omega} E_1 \sin \omega t \, dt \qquad \text{15-16}$$

This quantity is equal to the total volt-seconds in a positive half-cycle of alternator voltage minus the reset volt-seconds absorbed by the reactor during the negative half-cycle, or

$$\int_{t_s}^{\pi/\omega} E_1 \sin \omega t \, dt = \int_0^{\pi/\omega} E_1 \sin \omega t \, dt - \int_0^{t_s} E_1 \sin \omega t \, dt \qquad \text{15-17}$$

If the last integral in equation 15-17 is replaced by the equal quantity given in equation 15-15, the following is obtained:

$$\int_{t_s}^{\pi/\omega} E_1 \sin \omega t \, dt = \int_0^{\pi/\omega} E_1 \sin \omega t \, dt + \int_{\pi/\omega}^{2\pi/\omega} \alpha E_1 \sin \omega t \, dt \qquad \text{15-18}$$

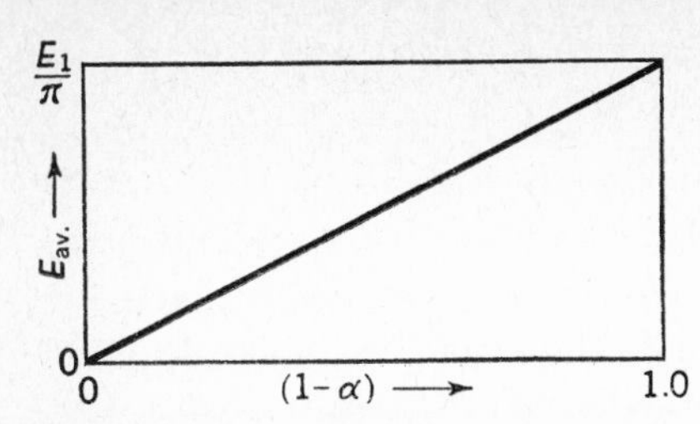

FIG. 15-14. Control characteristics for Ramey-type circuit shown in Fig. 15-12.

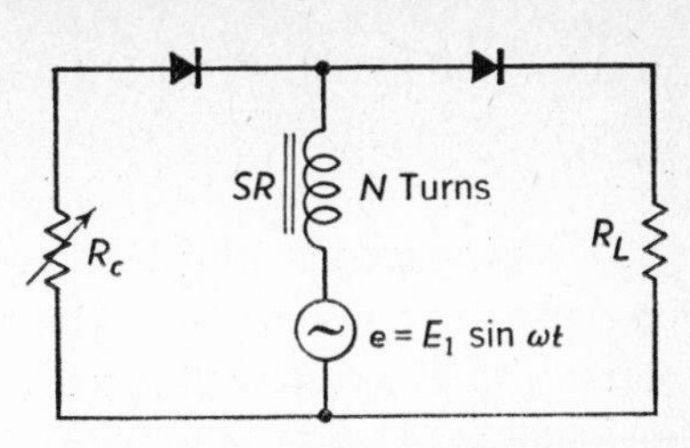

FIG. 15-15. Half-wave Ramey-type magnetic amplifier with variable resistance control.

The average voltage across the load is

$$E_{av.} = \frac{1}{\frac{2\pi}{\omega}} \int_{t_s}^{\pi/\omega} E_1 \sin \omega t \, dt$$

Then, the average load voltage is

$$E_{av.} = \frac{1}{\frac{2\pi}{\omega}} \left(\int_0^{\pi/\omega} E_1 \sin \omega t \, dt + \int_{\pi/\omega}^{2\pi/\omega} \alpha E_1 \sin \omega t \, dt \right) \quad 15\text{-}19$$

After the indicated integrations in equation 15-19 have been performed, the average value of output voltage is determined as a function of α as follows:

$$E_{av.} = \frac{(1-\alpha)E_1}{\pi} \quad 15\text{-}20$$

A plot of this function is shown in Fig. 15-14. It should be remembered that α may have values only between 0 and $+1.0$.

Now consider the circuit shown in Fig. 15-15, where the control element is a variable resistor R_c. During the negative half-cycle, alternator voltage e will at first all appear across R_c because the reactor current will be less than the reactor magnetizing current I_m. When the reactor current is constrained to a value less than that required for flux change, no voltage will appear across the reactor. When the applied alternator voltage e has increased to the value where the control circuit current is equal to the reactor magnetizing current I_m, the current remains at this value until the voltage falls again

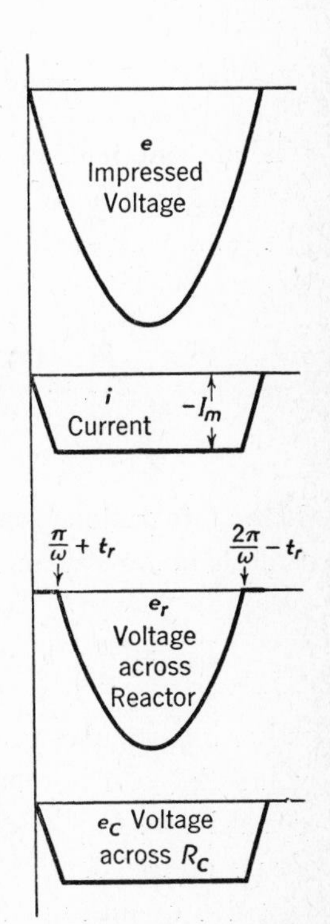

FIG. 15-16. Negative half-cycle wave forms for the magnetic amplifier of Fig. 15-15.

to this value near the end of the negative half-cycle. During the intervening time the reactor absorbs reset volt-seconds. The voltage and current wave forms during the negative half-cycle are shown in Fig. 15-16. In this, t_r is the time in seconds after the beginning of the negative half-cycle when the control-circuit current has increased until it is equal to the magnetizing current I_m. The reset volt-seconds absorbed by the reactor may be calculated as follows.

$$N\Delta\phi = \int_{\frac{\pi}{\omega}+t_r}^{\frac{2\pi}{\omega}-t_r} (E_1 \sin \omega t - R_c I_m)\,dt \qquad \text{15-21}$$

The time t_r may be determined because it is the value of time when the instantaneous control-circuit current is equal to the reactor magnetizing current; hence

$$\frac{E_1 \sin \omega t_r}{R_c} = I_m \qquad \text{and} \qquad t_r = \frac{1}{\omega} \sin^{-1} \frac{R_c I_m}{E_1} \qquad \text{15-22}$$

The arc sine term in equation 15-22 is always (by definition) the first quadrant angle. The average output voltage across the load may be written in the same way as was equation 15-19 in the foregoing example. Thus

$$E_{\text{av.}} = \frac{1}{\frac{2\pi}{\omega}} \left[\int_0^{\pi/\omega} E_1 \sin \omega t\,dt + \int_{\frac{\pi}{\omega}+\frac{1}{\omega}\sin^{-1}\frac{R_c I_m}{E_1}}^{\frac{2\pi}{\omega}-\frac{1}{\omega}\sin^{-1}\frac{R_c I_m}{E_1}} (E_1 \sin \omega t - R_c I_m)\,dt \right] \qquad \text{15-23}$$

If the integrations are performed, the resulting expression indicates the control characteristic, and

$$E_{\text{av.}} = \frac{E_1}{\pi}\left[1 - \sqrt{1 - \left(\frac{R_c I_m}{E_1}\right)^2} - \left(\frac{R_c I_m}{E_1}\right)\left(\frac{\pi}{2} - \sin^{-1}\frac{R_c I_m}{E_1}\right)\right] \qquad \text{15-24}$$

For a particular case, I_m and E_1 are both constant. If these are known, the value $E_{\text{av.}}$ can be computed for any value of R_c. It is convenient for the general case to plot equation 15-24 as shown in Fig. 15-17 to illustrate the shape of the control characteristic. It must be remembered that I_m is a negative quantity in the third term of equation 15-24 because the value of this quantity affects the control circuit only during negative half-cycles.

Another method of controlling the amount of reset makes use of a direct control voltage as shown in Fig. 15-18. During the negative half-cycle, the rectifier in the control circuit remains open until the instantaneous value of the alternator voltage exceeds the control voltage E_c. During the time the control-circuit rectifier is conducting, the reset voltage across the reactor is

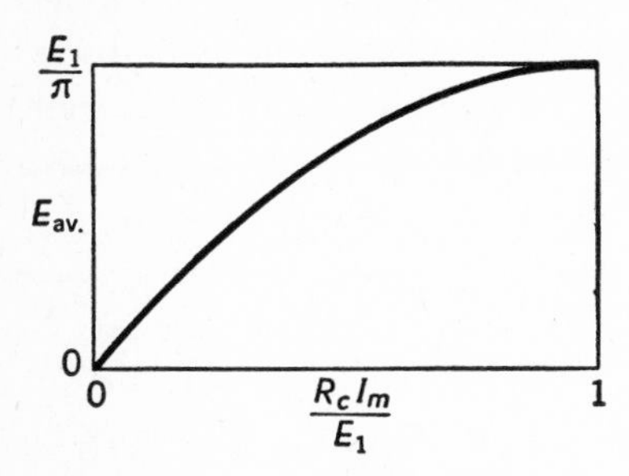

FIG. 15-17. Control characteristics for Ramey circuit shown in Fig. 15-15.

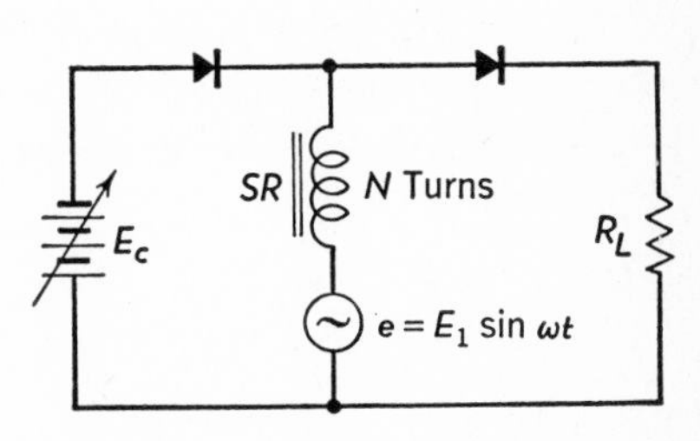

FIG. 15-18. Half-wave Ramey-type magnetic amplifier with direct control voltage E_c.

$e_r = E_1 \sin \omega t - E_c$. It will be noted that this example is exactly like the foregoing one if the quantity $R_c I_m$ is replaced by E_c. The operation is different but the equation of performance is identical in form (see equation 15-24).

$$E_{av.} = \frac{E_1}{\pi}\left[1 - \sqrt{1 - \left(\frac{E_c}{E_1}\right)^2} - \left(\frac{E_c}{E_1}\right)\left(\frac{\pi}{2} - \sin^{-1}\frac{E_c}{E_1}\right)\right] \qquad 15\text{-}25$$

The third term (in brackets) of equation 15-25 will be negative since E_c is negative and the arc sine always is in the first quadrant by definition.

It will be noted that this type of control results in a nonlinear gain characteristic as shown in Fig. 15-17. Actually the gain for small changes in input voltage falls off toward zero as the output approaches maximum.

15-6. *FULL-WAVE RAMEY-TYPE MAGNETIC AMPLIFIER*

The circuits discussed in the preceding section may be connected for full-wave output for normal single-phase operation. When full-wave connections are used, the right-hand side of equations 15-20, 15-24, and 15-25 should be multiplied by 2. Usually some device such as a vacuum tube or a transistor is used with the Ramey type of circuit as a control element. The circuit shown in Fig. 15-19 utilizes a triode vacuum tube as a control element. When the grid voltage on the triode is zero, nearly all of the transformer voltage (half of the total secondary) appears as reset voltage on the reactor that has negative voltage applied. Thus the output voltage across the load is small, nearly zero. When the grid voltage is such that the triode is cut off all the time, the transformer voltage all appears across the triode and the reactors are not reset at all. Then the output voltage is maximum. For intermediate values of grid voltage, the output voltage varies between zero and maximum. The shape of the control characteristic is dependent upon the characteristics of the reactors, the rectifiers, and of course the triode.

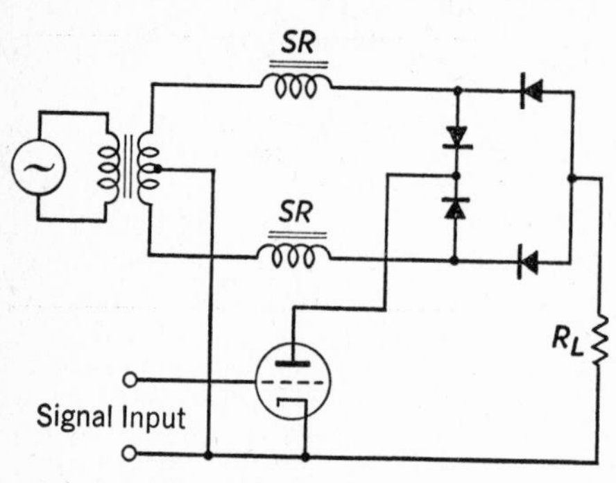

FIG. 15-19. Full-wave Ramey-type magnetic amplifier with vacuum-tube control.

If the rectifiers used in the load circuit are not ideal, the control circuit will not be isolated from the load circuit during the reset period as was assumed in the preceding section. Suppose that the triode is operating at cutoff so that the output voltage across the load is maximum. Now if the load circuit rectifiers are not ideal, they will conduct in the inverse direction and the passive reactor will be partially reset owing to voltage appearing across it in a negative direction. This will cause a reduction in the maximum output voltage obtainable from a particular circuit. A high-quality rectifier should therefore be used in this circuit.

The power gain of such a circuit is extremely large because of the high grid-circuit impedance of the vacuum tube. Hence, the input power is very small. The voltage gain largely is dependent on the tube characteristics because the magnetic-amplifier part contributes a gain of unity if the control is effected in the load winding as shown in Fig. 15-19. However, separate control windings can be wound on each reactor to increase (or decrease) the voltage gain.

The response time of this type of magnetic amplifier is very fast because any change in control signal immediately affects the reset conditions in the reactors. Study will indicate that after a step change in input, the output will have reached the new steady-state condition within one cycle of the supply frequency. This type of circuit provides the fastest response time that can be obtained with magnetic amplifiers that actually have amplification.

15-7. *MAGNETIC AMPLIFIERS WITH CONTROL WINDINGS*

A common type of magnetic amplifier is shown in Fig. 15-20. The output is an alternating voltage, which may be used as such or it may be rectified in a bridge rectifier circuit and filtered to obtain direct voltage. The winding terminals, which are of the same polarity, are each indicated by a dot. This means that if the battery E_c were suddenly connected as shown, terminals 4, 8, 2, and 6 would all become instantaneously positive (assuming, of course, that neither reactor was saturated). The resistance R_c is the total control-circuit resistance, which includes the control winding resistance and any additional resistance that may be connected in the circuit.

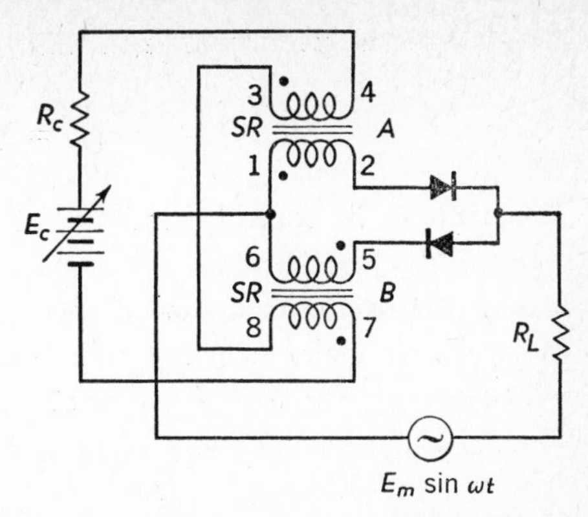

FIG. 15-20. Magnetic amplifier with control winding and with alternating voltage output.

The operation of this circuit is explained by considering the voltage and current wave forms for one condition of steady-state operation as shown in Fig. 15-21. Assume that the reactors are designed so that

$$2N\phi_m = \int_0^{\pi/\omega} E_m \sin \omega t \, dt \qquad 15\text{-}26$$

Reactor A absorbs volt-seconds during the first half-cycle until it saturates and then the current is limited by the load resistance R_L. During this interval the flow of current through the load winding of reactor B is constrained to zero by the presence of the series rectifier. During the next half-cycle reactor B absorbs volt-seconds until it saturates and during this interval the flow of current through the load winding of reactor A is constrained to zero by the series rectifier. Since the circuit is the self-saturating type, the reactors must be reset during the inverse half-cycles. Reset is accomplished as follows. Before saturation when reactor A of Fig. 15-20 is absorbing volt-seconds, terminal 1 is positive. By transformer action, terminal 3 of the control winding also becomes posi-

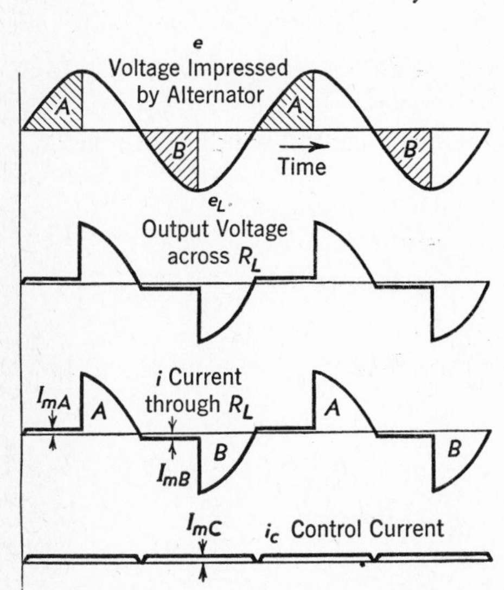

FIG. 15-21. Wave forms for the magnetic amplifier of Fig. 15-20 with alternating voltage and current output to load resistor R_L.

tive and the induced voltage in this winding is

$$e_{cA} = N_c \frac{d\phi_A}{dt} \qquad 15\text{-}27$$

This voltage in control winding A is applied to the control winding of reactor B so that terminal 8 is positive. This causes a flux change in reactor B that is opposite to that which occurs when it absorbs volt-seconds in the forward half-cycle. Consequently, the volt-seconds absorbed from transformer action in reactor B are reset volt-seconds. The current in the control winding is the magnetizing current I_{mc}. Some thought will indicate that if there were no leakage flux in either reactor, and if there were no resistance in the control circuit, reactor B would be reset by the same number of volt-seconds as was absorbed by reactor A before saturation occurred. In the next half-cycle reactor A will be reset by transformer action in reactor B.

In an actual case, the leakage flux will be very small because of the construction generally used. The main difference between e_{cA} and e_{cB} is the voltage drop caused by the magnetizing current I_{mc} flowing through the control-circuit resistance. In this case the reactor is not completely reset by transformer action in the forward conducting reactor. Rather, some of the reset volt-seconds must be supplied by an external source. Consider the reset of reactor B. Equation 15-27 gives the induced voltage in the control winding of reactor A. The flux reset in reactor B is

$$\phi_B = \frac{1}{N_c}\int e_{cB}\, dt \qquad 15\text{-}28$$

Where e_{cB} is the component of voltage in the control winding of reactor B that is caused by the flux change. Thus

$$e_{cB} \cong e_{cA} - R_c I_m \qquad 15\text{-}29$$

and

$$e_{cB} \cong N_c \frac{d\phi_A}{dt} - R_c I_m \qquad 15\text{-}30$$

Hence

$$\phi_B \cong \frac{1}{N_c}\int \left(N_c \frac{d\phi_A}{dt} - R_c I_m\right) dt \qquad 15\text{-}31$$

and

$$\phi_B \cong \phi_A - \frac{1}{N_c}\int R_c I_m dt \qquad 15\text{-}32$$

During steady-state conditions, it is necessary that $\phi_B = \phi_A$. This means that each reactor will be reset the same amount each time so that every cycle is like every other cycle. This requires that an additional voltage, e_c, be added in the control circuit so that during the inverse half-cycle it will contribute a number of volt-seconds equal to the second term in the right-hand side of equation 15-32.

$$\int e_c\,dt \cong \frac{1}{N_c}\int R_c I_m\,dt \tag{15-33}$$

The wave form of the voltage e_c is not important so long as equation 15-33 is satisfied. Since the magnetizing current is constant in a reactor having core material with a rectangular hysteresis loop, the quantity $R_c I_m$ is a constant. Therefore, if a d-c voltage source were used in the control circuit as shown in Fig. 15-20, steady-state conditions would result when

$$E_c \cong R_c I_m \tag{15-34}$$

An approximate equality is indicated because leakage flux was assumed to be negligible. This will usually be the case for ordinary values of R_c, but if a very low-resistance control circuit were used, the leakage-flux effect might be appreciable. There is no simple way to compute this effect.

Now suppose that with a finite value of E_c the operating conditions in the circuit of Fig. 15-20 were as shown in Fig. 15-21. What will occur in the circuit, if the voltage E_c is suddenly reduced to zero at the beginning of the forward half-cycle for reactor A? During this first half-cycle, reactor A will absorb the normal steady-state amount of volt-seconds before saturation occurs. However, reactor B will not be completely reset, as equation 15-35 indicates.

$$\phi_B' = \phi_A - \frac{1}{N_c}\int R_c I_m\,dt \tag{15-35}$$

It will be reset less than normal by the amount due to the $R_c I_m$ drop. Therefore, during the next half-cycle it will absorb fewer than normal volt-seconds before saturation, and the output will increase. During this period reactor A will be reset even a lesser amount.

$$\phi_A' = \phi_B' - \frac{1}{N_c}\int R_c I_m dt \qquad \text{and} \qquad \phi_A' = \phi_A - 2\left(\frac{1}{N_c}\int R_c I_m\,dt\right) \tag{15-36}$$

Again, the output will increase because reactor A will absorb a still lesser number of volt-seconds before saturation. This will continue until the reset flux becomes zero, that is

$$0 = \phi_A - n\left(\frac{1}{N_c}\int R_c I_m\,dt\right) \tag{15-37}$$

The number of half-cycles that must elapse before the output has increased to its final value is denoted by n. The smaller the value of R_c, the greater will be the quantity n. The response time of this type of amplifier is dependent upon the control-circuit resistance. The lower this value of resistance, the longer the response time will be. However, it is also noted that the smaller the value of R_c, the smaller the control voltage required. Hence, for high gain (both voltage and power) the control-circuit resistance should be small. This, however, causes the amplifier to have a long response time.

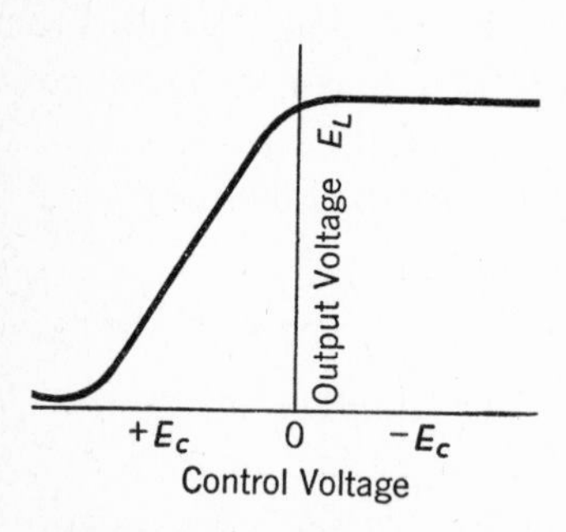

FIG. 15-22. Control characteristic of a magnetic amplifier using control windings, and with control voltage polarity as shown in Fig. 15-20.

By using control windings of many turns to reduce the magnetizing current required, and by keeping the control-circuit resistance low, extremely large voltage and power gains can be obtained. The operation then is such that when the control voltage is zero the output is maximum, and when the control voltage is maximum the output is reduced to the magnetizing current, which is usually very small compared to the normal load current. A study of the circuit operation will indicate that the inverse voltage on the rectifiers is very small for reasonably low values of control-circuit resistance. This is an advantage because the inverse characteristics of the rectifiers then have but little effect upon circuit performance. The control characteristic of a typical circuit is shown in Fig. 15-22.

15-8. *THE DESIGN OF SELF-SATURATING AMPLIFIERS*

The design of a magnetic amplifier of the self-saturating type primarily involves the design of the reactor. Other considerations include selection of the rectifiers, determination of the input voltage to provide the desired output, and the selection of a control method. The load winding of the reactor may be designed as follows. Assume that the supply voltage is sinusoidal. Then the relationship between reactor flux and supply voltage may be written

$$E_m \sin \omega t = N_L \frac{d\phi}{dt} \qquad 15\text{-}38$$

If the reactor is designed so that it will not saturate with the indicated supply voltage, the flux will change from $-\phi_m$ to $+\phi_m$ during one half-cycle of the supply voltage. Hence

$$\int_0^{\pi/\omega} E_m \sin \omega t \, dt = \int_{-\phi_m}^{+\phi_m} N_L \, d\phi \qquad 15\text{-}39$$

$$E_m = \omega N_L \, \phi_m \quad \text{and} \quad \phi_m = B_m \, AS \qquad 15\text{-}40$$

where B_m = saturation flux density of core material, A = cross-sectional area of core, and S = stacking factor for core. Also,

$$E_m = \sqrt{2} E_{\text{rms}}, \quad \omega = 2\pi f, \quad \text{and} \quad E_{\text{rms}} = 4.44 f N_L B_m AS \qquad 15\text{-}41$$

Any self-consistent system of units may be used in equation 15-41. Often the equation is written

$$E_{\text{rms}} = 4.44 \times 10^{-8} f N_L B_m AS \qquad 15\text{-}42$$

where E_{rms} = volts (rms), f = cycles per second, N_L = turns on load winding, B_m = saturation flux density in gausses, A = cross-sectional area of core in square centimeters, and S = stacking factor, a dimensionless quantity.

The hysteresis loop for a core material having a rectangular hysteresis loop (Deltamax) is shown in Fig. 15-3. Usually these cores are wound in toroidal form and placed in a nylon case to avoid subjecting the core to mechanical stresses when it is wound and impregnated. The data listed in Table 10 gives the standard Deltamax core sizes.

Table 10 *

STANDARD TAPE WOUND CORES ENCASED IN NYLON CONTAINERS

AECo. CASE NO.	CORE SIZE (INCHES)			CASE SIZE (INCHES)		
	O.D.	I.D.	H.	O.D.	I.D.	H.
5340	0.750	0.500	0.125	0.810	0.440	0.195
5515	0.900	0.650	0.125	0.960	0.590	0.195
4168	1.250	1.000	0.125	1.330	0.920	0.195
5504	1.125	0.750	0.188	1.205	0.670	0.262
4635	1.375	1.000	0.250	1.445	0.930	0.320
4179	2.000	1.625	0.250	2.110	1.525	0.330
5387	1.750	1.250	0.250	1.820	1.180	0.320
5233	1.500	1.000	0.375	1.570	0.930	0.445
4178	2.500	2.000	0.500	2.652	1.860	0.610
4180	3.000	2.500	0.500	3.152	2.360	0.610
5320	2.500	1.500	0.500	2.600	1.400	0.600
5468	3.500	2.500	1.000	3.630	2.370	1.120
5581	4.500	3.000	1.500	4.680	2.820	1.645
5582	5.850	3.100	1.375	6.030	2.920	1.605

* Data courtesy of Arnold Engineering Co.

If a control winding is used, the design is accomplished as follows. The magnetizing force in the control winding H_c is obtained from the hysteresis loop in Fig. 15-3. Since the loop is nearly rectangular, the average value of H_c is that value of magnetizing force where the curve crosses the zero flux-density axis. The magnetizing force is related to the ampere turns in the control winding by

$$H_c = \frac{0.495 N_c I_c}{l} \qquad 15\text{-}43$$

where H_c = magnetizing force in oersteds, N_c = turns in control winding, I_c = control current in amperes, and l = mean length of magnetic path in inches. If the core has been selected, both H_c and l are known. Then

$$N_c I_c = \frac{H_c l}{0.495} \qquad 15\text{-}44$$

The control voltage will be very nearly equal to the product of control-circuit resistance R_c and I_c, or $E_c = R_c I_c$. Therefore,

$$\frac{N_c E_c}{R_c} = \frac{H_c l}{0.495} \qquad 15\text{-}45$$

In many cases the magnitude of the control voltage is known. Hence the ratio of control turns to control-circuit resistance can be calculated.

$$\frac{N_c}{R_c} = \frac{H_c l}{0.495 E_c} \qquad 15\text{-}46$$

Therefore the number of turns, the winding resistance, and any other control-circuit resistance may be compromised to yield an optimum design from the viewpoint of weight, size, appearance, and ease of manufacture. If control current instead of control voltage is specified, equation 15-44 is solved for the number of turns N_c.

15-9. *EXAMPLE OF DESIGN OF SELF-SATURATING MAGNETIC AMPLIFIER*

An example of the design of a self-saturating magnetic amplifier having control windings is as follows. The specifications to be fulfilled are: maximum output voltage, 50 volts direct; output current, 1 ampere direct; signal voltage, 0 to 0.5 volt direct; signal-source resistance, 5 ohms; and supply frequency, 400 cycles. The circuit of Fig. 15-23 will be used. First, the rectifiers are selected. They must have adequate current-carrying capacity and must withstand the inverse voltage. The assumption is made that the rectifier and resistance drops in the load circuit cause a 5-volt reduction in load voltage at maximum current. When the output is maximum, the voltage wave shape will be a full-wave rectified wave. The transformer secondary voltage must be $E_s = 1.11(50 + 5) = 61$ volts (rms). The load windings on the reactor then will be designed for this voltage. Equation 15-42 is used to compute the load-winding turns. In this example $E_{\text{rms}} = 61$ volts, $f = 400$ cycles per second, $B_m = 14{,}000$ gausses (Deltamax, Fig. 15-3), and $S = 0.8$ for 0.002-inch thick ribbon in toroidal cores. Solving equation 15-42 for $N_L A$ yields the following: $N_L A = 61 \times 10^{-8}/(4.44 \times 400 \times 14{,}000 \times 0.8) = 307$. A Deltamax 5233 core is selected (Table 10), for which $A = 0.605$ square centimeter and $N_L = 307/0.605 = 507$ turns. The wire size is determined in accordance with current and other requirements. A current density of 500 circular mils per ampere (rms) commonly is used. Wire circular mils $= 500 \times (\pi/4) \times 1 = 393$, and no. 24 AWG wire will be used for the load windings.

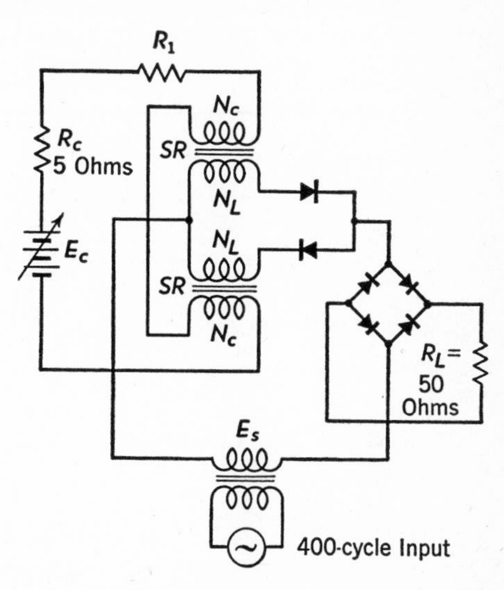

FIG. 15-23. Circuit for design example of section 15-9. The designation SR indicates that the devices are used to obtain saturable-reactor characteristics.

The control winding can now be determined by equation 15-46. From Fig. 15-3 for Deltamax, $H_c = 0.5$ oersted; for core 5233 of Table 10, the mean length determined from the outside diameter and the inside diameter is $l = 3.93$ inches, and from design specifications, $E_c = 0.5$ volt. Hence, from equation 15-46, $N_c/R_c = (0.5 \times 3.93)/(0.495 \times 0.5) = 7.95$. If 250 turns are selected for N_c, the control-circuit resistance is $R_c = 250/7.95 = 31.4$ ohms. Since the control winding resistance will be small (10 to 20 ohms, depending upon wire size used), additional resistance R_1 will be required in the control circuit; that is, $R_1 = 31.4 - (5 + R_c)$. The maximum control current will be $I_c = (0.5/31.4) \times 1000 = 15.9$ milliamperes. The wire size can be determined as before, but generally a larger size is used for winding ease, probably no. 32 AWG in this example.

The actual construction of a magnetic amplifier using toroidal cores is usually accomplished in the following way. The load coils are each wound on their respective cores, using a distributed winding around the entire circumference of the toroid. Then the two toroids (with the load windings thereon) are stacked together and a single control winding is wound on both cores as if they were one. When the control voltage is zero, this amplifier will deliver to the load a direct voltage of 50 volts. When the control voltage is 0.5 volt, the output current to the load will be the sum of the magnetizing current and the reflected control-winding current, or $I_{mL} = (0.5 \times 3.93)/(0.495 \times 507) + (0.0159)(250/507) = 0.0157$ ampere. The minimum output voltage is $E_L = 50 \times 0.0157 = 0.785$ volt.

15-10. *NON-SELF-SATURATING MAGNETIC AMPLIFIERS*

Non-self-saturating magnetic amplifiers, as in Fig. 15-24, have been used since the turn of the century in numerous applications. Although the circuit seems simple compared to the self-saturating amplifier of Fig. 15-20, the

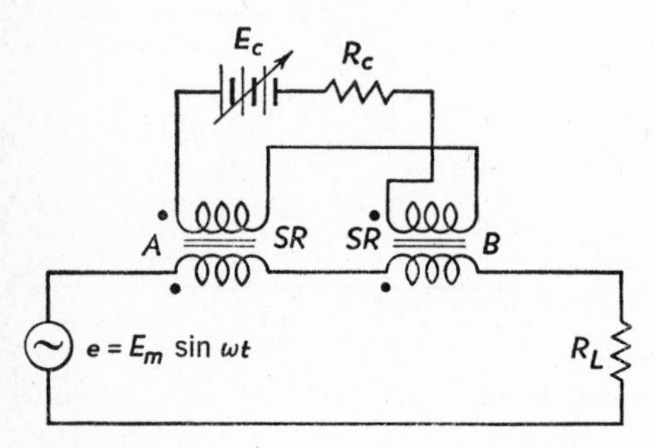

FIG. 15-24. Non-self-saturating magnetic amplifier.

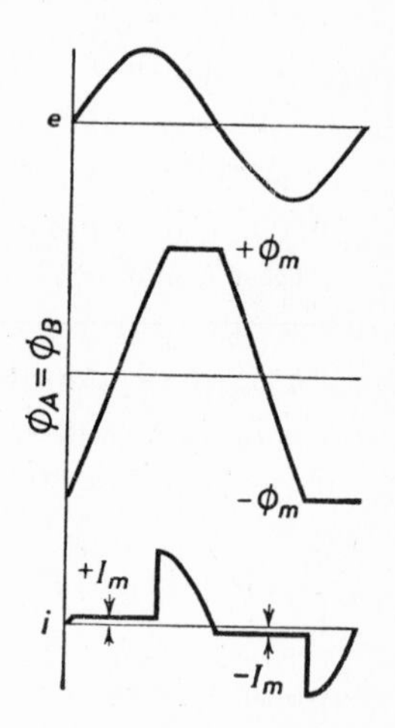

FIG. 15-25. Approximate wave forms for Fig. 15-24, with control circuit open.

operation is more complex. In discussing Fig. 15-24 it will be assumed that the core material has a rectangular hysteresis loop, that the control circuit has load windings connected in series, and that the control windings are connected in series opposing. When the control circuit is open, this circuit behaves just like the circuit of Fig. 15-6 except that there are two reactors in series. Hence before saturation occurs, the alternator voltage divides between the two load windings, and only the magnetizing current I_m flows in the load. Since the two reactors are identical, they will both saturate simultaneously, and the alternator voltage will then appear across the load. For this condition of operation, the voltage, current, and flux wave forms are shown in Fig. 15-25. Usually, however, the reactors are designed so that the flux just reaches, but does not exceed, the saturation level.

If the control winding is closed, the operation will not be different than described if E_c is zero. Since the control windings are connected series opposing, the induced voltages oppose and no current flows. However, if the control voltage is not zero, the conditions are very different. Since the control voltage causes the fluxes to change in opposite directions, the two core fluxes will no longer be equal as in Fig. 15-25. One core will have a flux increase and the other a flux decrease as the result of applying a control voltage.

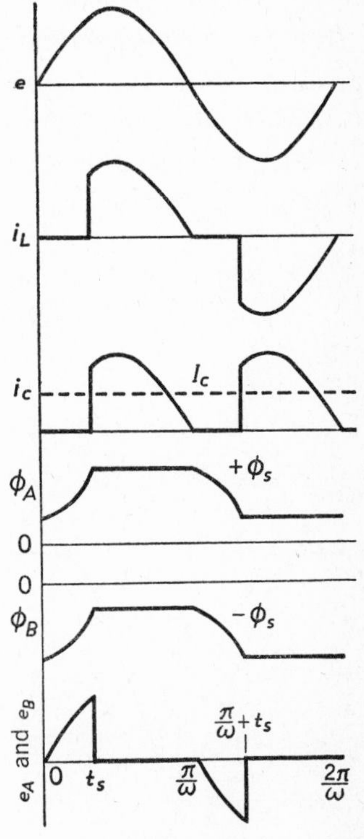

FIG. 15-26. Wave forms for Fig. 15-24 when control circuit is closed and a control voltage E_c is applied. $I_c = E_c/R_c$ and is the average value of the control current.

The operation of this circuit is best explained by examining one complete cycle of steady-state operation. The wave forms for voltage, current, and flux are shown in Fig. 15-26. During the time from zero to t_s neither reactor is saturated. Since the two reactors are coupled together through a low-impedance control circuit (due to transformer action), the flux changes and therefore the reactor voltages are constrained to equality. The alternator voltage divides equally between the two reactors, and the load voltage is essentially zero. Only the magnetizing current i_m flows in the load during this time, and it is assumed to be negligible in this example.

At time t_s the flux in reactor A reaches saturation level and hence it can no longer support voltage. The control winding on reactor A then looks like a short-circuit to the control winding on reactor B. Reactor B acts like a current transformer with a short-circuited secondary. Since no voltage will appear across the load winding of reactor B (because of its short-circuited control winding), the reactor does not further absorb volt-seconds, so the flux does not change. The alternator voltage all appears across the load until the end of the first half-cycle, when the voltage changes polarity and the flux in reactor A again can change.

During the time from π/ω to $(\pi/\omega) + t_s$ the alternator voltage again divides equally between the two reactors. Since they are now absorbing negative volt-seconds, the fluxes are changing in the negative direction. At time $(\pi/\omega) + t_s$ the flux in reactor B reaches the negative saturation level and the alternator voltage again appears across the load. A study of the instantaneous polarities in the circuit (Fig. 15-24) will indicate that the direction of the control current i_c is the same during the two periods of saturation, one where reactor A is saturated and one where reactor B is saturated. The wave form of the control current is shown in Fig. 15-26. The average value of this current is $I_c = E_c/R_c$.

The instantaneous current in the control windings is always related to the current in the load windings by

$$N_c i_c = N_L i_L \qquad 15\text{-}47$$

Since the current i_c is unidirectional, it follows that the average value of i_c and the rectified average value of i_L must also obey the same expression. Hence,

$$N_c I_c = N_L I_{L\,(\mathrm{av.})} \qquad \text{and} \qquad I_{L\,(\mathrm{av.})} = \frac{N_c}{N_L} I_c \qquad 15\text{-}48$$

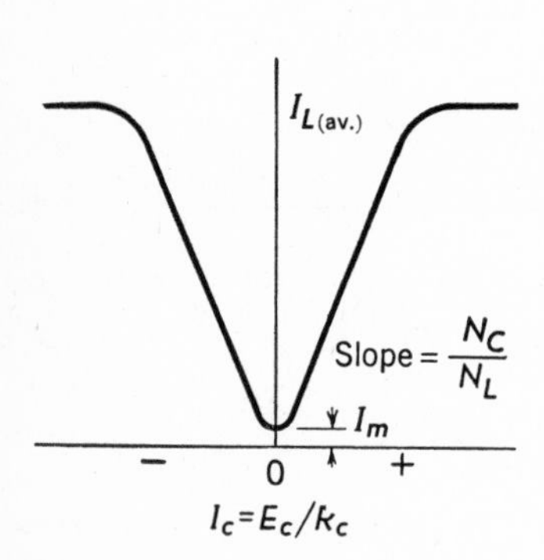

FIG. 15-27. Control characteristics of a non-self-saturating magnetic amplifier.

Thus, equation 15-48 indicates the control characteristic of this type of magnetic amplifier. When I_c (and E_c) are zero, only the magnetizing current I_m flows in the load. When I_c is so large that one or the other of the two reactors is always saturated, the output is maximum. A typical control characteristic is shown in Fig. 15-27. The characteristics of magnetic amplifiers can be altered greatly by feedback.[4] This may be accomplished by rectifying the output current and causing it to flow through feedback windings similar to control windings. References not cited in the preceding pages are listed as 9 to 13 inclusive.

REFERENCES

1. *Magnetic Amplifiers.* Bureau of Ships, Department of the Navy.
2. Bozorth, R. M. *Ferromagnetism.* D. Van Nostrand Co.
3. Kraumhansel, J. A., and Beyer, R. T. *Barkhausen noise in magnetic amplifiers.* Journal of Applied Physics, Vol. 20, 1949, p. 528.
4. Johnson, W. C. *The magnetic amplifier.* Electrical Engineering, July 1953, Vol. 72, No. 7.
5. Finzi, L. A., and Pittman, G. F. *Methods of magnetic amplifier analysis.* Electrical Engineering, Aug. 1953, Vol. 72, No. 8.

6. Ramey, R. A. *Magnetic amplifier circuits and applications.* Electrical Engineering, Sept. 1953, Vol. 72, No. 9.
7. Ramey, R. A. *On the Control of Magnetic Amplifiers.* Naval Research Laboratory Report 3869, 1951.
8. Ramey, R. A. *On the mechanics of magnetic amplifier operation.* Trans. A.I.E.E., Vol. 70, 1951.
9. Miles, J. G. *Types of magnetic amplifiers-survey.* Communication and Electronics, A.I.E.E., July 1952, p. 229.
10. Storm, H. F. *Saturable reactors with inductive dc load.* Communication and Electronics, A.I.E.E., Nov. 1952, p. 335.
11. Ramey, R. A. *The single-core magnetic amplifier as a computer element.* Communication and Electronics, A.I.E.E., Jan. 1953, p. 442.
12. Finzi, L. A., and Mathias, R. A. *An application of magnetic circuits to perform multiplication and other analytical operations.* Communication and Electronics, A.I.E.E., Sept. 1953, p. 454.
13. House, C. B. *Flux preset high-speed magnetic amplifiers.* Communication and Electronics, A.I.E.E., Jan. 1954, p. 728.

QUESTIONS

1. What is the meaning of ferromagnetism? Of magnetically anisotropic?
2. What are the important characteristics of a body-centered cubic lattice? Of a face-centered cubic lattice?
3. From the standpoint of domain alignment, what is the importance of a rectangular hysteresis loop?
4. In section 15-1 is the statement "it is essential that the magnetizing force be applied in the direction of cold-rolling." Why?
5. For Fig. 15-3, by what factor should values in oersteds be multiplied to convert to ampere-turns per meter? From kilogausses to webers per square meter?
6. What is the meaning of duality? Why is a reactor the dual of a capacitor, as stated just preceding equation 15-5?
7. What is the meaning of "air core inductance" (just preceding equation 15-7)?
8. What is the meaning of "volt-second"? Does this represent stored energy? Why?
9. What are the characteristics of an ideal rectifier for magnetic amplifier applications?
10. How is equation 15-11 interpreted if the first term equals the second?
11. What is a Ramey circuit? What are its important characteristics?
12. What is the meaning of the quantity α?
13. Why are transistors or vacuum tubes often used in conjunction with magnetic amplifiers?
14. What is meant by reset? Reset volt-seconds? Reset time?
15. How does a control winding function?
16. What factors must be known for magnetic amplifier design?
17. Why are magnetic cores enclosed in nylon containers? In what form is the nylon?
18. How may very high voltage and power gains be obtained from magnetic amplifiers?

19. How do the basic characteristics of self-saturating and non-self-saturating magnetic amplifiers differ?
20. Why is feedback used in magnetic amplifiers? How is it applied?

PROBLEMS

1. For Fig. 15-28a calculate the instantaneous current as a function of time after the switch is closed, assuming that the flux in the reactor core initially is $-\phi_m$, that the core is Deltamax 5233 (2 mil), that $N = 600$ turns, and that the coil resistance is negligible.
2. What value of applied 400-cycle voltage will cause the flux in the reactor core to just reach (but not exceed) the saturation value in the circuit of Fig. 15-28b having a Deltamax 5320 (2 mil) core, $N =$ 500 turns, and negligible coil resistance.

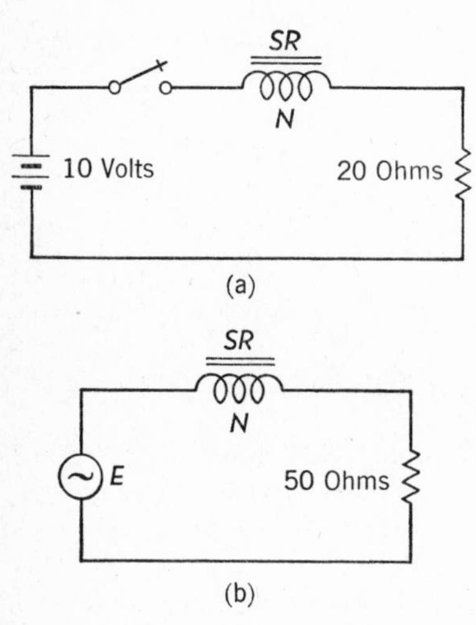

FIG. 15-28. Circuits for problems 1 and 2.

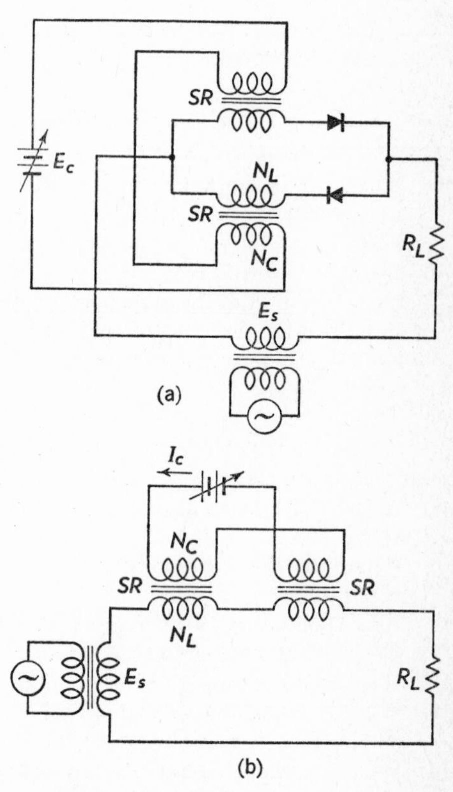

FIG. 15-29. Circuits for problems 4 and 5.

3. For the circuit of problem 2, calculate and plot the following quantities when the alternator voltage E is 250 volts (rms) at 400 cycles: current, reactor voltage, and load voltage.
4. Design the self-saturating magnetic amplifier shown in Fig. 15-29a to meet the following specifications, using Deltamax core material: supply frequency = 400 cycles, $R_L = 50$ ohms, minimum load voltage not greater than 5 volts (rms), maximum load voltage = 115 volts (rms), and control voltage E_c is 0 to 1 volt direct voltage.

5. Design the non-self-saturating magnetic amplifier shown in Fig. 15-29b to meet the following specifications, using Deltamax core material: supply frequency = 400 cycles, $R_L = 50$ ohms, minimum load voltage not greater than 5 volts (rms), maximum load voltage = 115 volts (rms), and control direct current I_c is 0 to 0.1 ampere.

CHAPTER 16

PHOTOELECTRIC DEVICES

When electromagnetic radiations, such as visible light, strike certain materials, phenomena termed **photoelectric effects** occur. This general term includes the following types of photoelectric phenomena: First, by a **photoemissive effect,** electromagnetic radiations may cause electrons to be liberated from a photosensitive surface. Second, by a **photovoltaic effect,** radiations may cause a difference of potential to exist between the terminals of a suitably constructed cell. Third, by a **photoconductive effect,** light shining on certain substances may cause their conductance (and resistance) to vary. This chapter will consider photoelectric devices operating on these principles.

16-1. *PHOTOTUBES* *

A **phototube** is defined [1] as a "vacuum tube in which one of the electrodes is irradiated for the purpose of causing electron emission." The radiation liberates the electrons by a photoemissive effect. The radiant energy of the light (or other radiation) imparts energy to a few of the electrons of the metal so that they are able to escape through the surface forces of the metal to the region outside (section 2-4). The essential elements of a phototube are an evacuated glass bulb containing an electron-emitting *cathode* and a positive electron-collecting *anode.* The electron-emitting light-sensitive material may be on a small electrode at the center of the bulb (central-cathode type), or the electron-collecting anode may be at the center (central-anode type).[2] The bulb may be highly evacuated or may contain gas. The **vacuum phototube** is one which is evacuated to such a degree that its electrical char-

* See also *Definitions of Phototube Terms.* Proc. I.R.E., August 1954, Vol. 42, No. 8.

acteristics are essentially unaffected by gaseous ionization. A **gas phototube** is one in which the pressure of the contained gas or vapor is such as to affect substantially the electrical characteristics of the tube.

Vacuum Phototubes.[3] In many phototubes the light-sensitive material is a cesium-oxygen-silver layer formed on a cylindrically shaped metal electrode. The anode is a wire at the center. The arrangement, showing a top view of the phototube, and a circuit for studying its characteristics, are shown in Fig. 16-1. The intensity of the light striking the phototube is variable. The anode voltage can be changed, and the current can be measured. Curves such as are shown in Fig. 16-2 are obtained. The energy in the

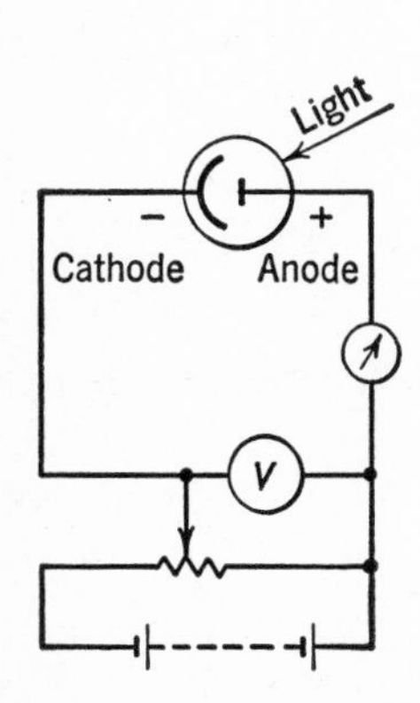

FIG. 16-1. Circuit for studying a vacuum phototube. A protective current-limiting resistance commonly is used when testing a gas tube.

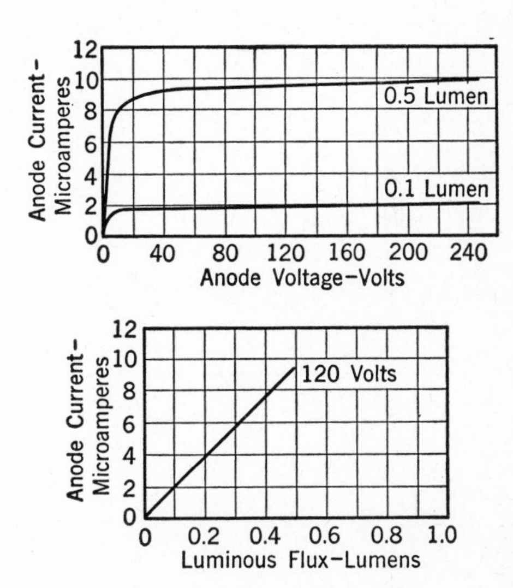

FIG. 16-2. Characteristic curves for a vacuum phototube. (Data from reference 3.)

radiation striking the light-sensitive layer causes electrons to be emitted. Some of these are just able to escape from the surface of the composite metallic layer, but others have considerable initial velocities. These liberated electrons are influenced by the electric field existing between the cathode and the anode, this field being *largely* the result of the positive potential between anode and cathode. Since any phototube may be (gas phototubes should be) operated with high resistances in the anode circuit, a correction for the IR drop across this resistance must be applied.

The output of a phototube is greater if the light intensity striking the photosensitive material is increased. The curves of Fig. 16-2 show that the tube soon becomes saturated, after which the current to the anode is limited (for given conditions) by the photoemission. The lower curve of Fig. 16-2 indicates that the current output of a phototube varies *directly* with the luminous flux from the light source. It is this important characteristic which makes possible the use of the phototube in sound motion pictures, facsimile, and other applications.

Gas Phototubes.[3] The physical construction of these phototubes is essentially the same as that of the vacuum type. In making the gas phototube, after the evacuating process a small amount of inert gas (such as argon at a pressure of about 150 microns) is admitted. The function of the gas is as follows: The electrons emitted from the light-sensitive cathode are drawn to the positive anode. If the voltage is sufficient, the electrons may attain a velocity sufficient to cause ionization by collision when they strike neutral gas atoms. As a result of this ionization, additional electrons are produced, and these in turn flow on to the positive anode. Some of these liberated electrons will also cause ionization by collision, thus further increasing the current through the tube for a given light intensity. Massive, slow-moving, positive ions also are produced by the ionization process. These move toward the negative cathode. They increase the current by knocking electrons out of the cathode surface. It is conceivable that a space-charge reducing influence is exerted by the positive ions in gas phototubes. Since ionization may become cumulative and large currents may flow, a gas phototube should always *be operated with a protective current-limiting resistor in the anode circuit,* unless other control provisions are made. Curves illustrating the characteristics of a gas phototube are shown in Fig. 16-3. After ionization occurs the current increases rapidly and would damage the tube if a protective resistor were not used. The lower curves of Fig. 16-3 indicate that for the gas

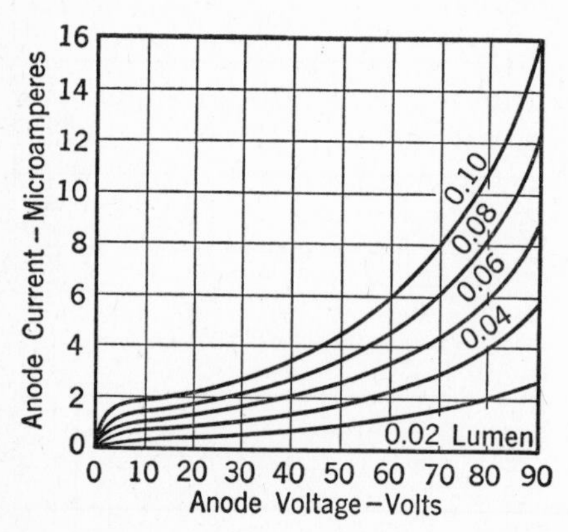

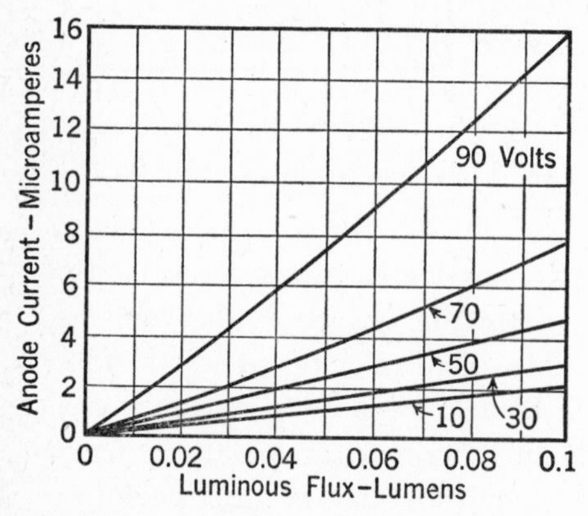

FIG. 16-3. Characteristic curves for a gas phototube. (Data from reference 3.)

phototube (as well as the vacuum type) the output current substantially is *directly* proportional to the luminous flux from the light source. This characteristic is of great importance.

16-2. *PHOTOMETRIC DEFINITIONS*

Before proceeding with the discussion of phototubes, it is advisable to define certain photometric terms from reference 4.

Light. This is defined as "radiant energy evaluated according to its capacity to produce visual sensation."

Luminous Flux. This is defined as "the time rate of flow of light." The unit is the **lumen.**

Lumen. This is equal to the "flux through unit solid angle (steradian) from a uniform point source of one candle, or to the flux on a unit surface all points of which are at unit distance from a uniform point source of one candle." Assuming that the lamp is a point source, the luminous flux F in lumens falling on the *projected* cathode area (sometimes called the window area) is

$$F = \frac{AC}{d^2} \qquad \text{16-1}$$

where A is the projected area (not the *entire* area, because the cathode is cylindrical), C is the luminous intensity or candle power of the source,[4] and d is the distance (in the same system of measurement as A) from the source to the cathode. These relations are shown in Fig. 16-4.

Phototube Definitions. The following definitions[5] are useful in phototube studies.

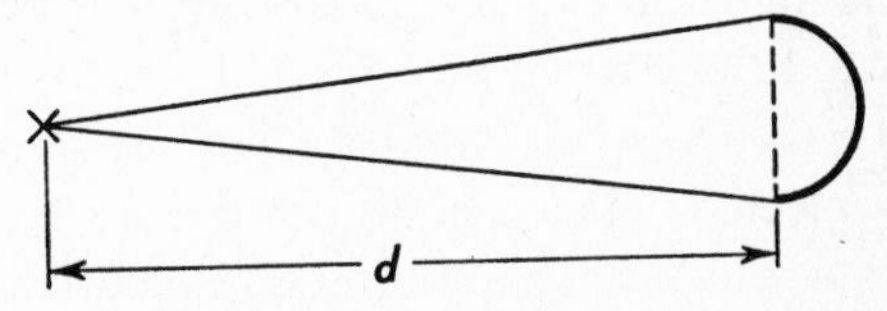

FIG. 16-4. Illustrating the terms of equation 16-1. The projected area of the phototube cathode is indicated by the dotted line.

Static Sensitivity. This is the "quotient of the direct anode current by the incident radiant flux of constant value."

Dynamic Sensitivity. This is the "quotient of the alternating component of anode current by the alternating component of incident radiant flux."

Luminous Sensitivity. This is the "quotient of the anode current by the incident luminous flux."

Gas Amplification Factor. This is the "factor of increase in the sensitivity of a gas phototube due solely to the ionization of the contained gas."

16-3. *COMPARISON OF VACUUM AND GAS PHOTOTUBES*

The characteristics of a vacuum and of a gas phototube are plotted in Fig. 16-5. These data are for identical phototubes with the exception that one contains argon gas at a pressure of 150 microns and the other is a vacuum type. These curves show that the characteristics are (almost) identical until ionization by collision occurs in the gas tube. They also indicate that the operating voltage needed on gas tubes is considerably less than for the vacuum type. The gas amplification factor of a typical phototube is less than 10. Vacuum phototubes are especially suited for precise measurements. The vacuum phototube, although not so sensitive as the gas type, is more stable. It is not likely to be damaged by drawing high instantaneous currents.

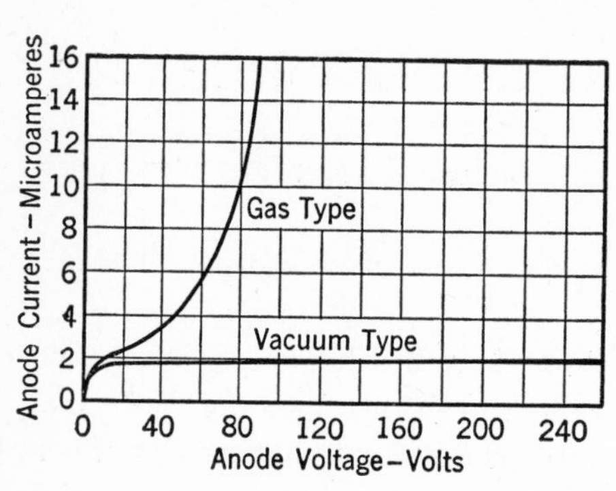

FIG. 16-5. Comparison of vacuum-type and gas-type phototubes at 0.1 lumen. (From Figs. 16-2 and 16-3.)

16-4. *TESTING PHOTOTUBES* [5]

For determining the current-voltage characteristic, the illumination is held constant, the voltage is varied, and the current observed. For the current-illumination tests, the voltage is held constant, the luminous flux is varied, and the current measured. To make these measurements, a calibrated light source and a photometer box are necessary. The type of light source to be selected depends on the nature of the light-sensitive cathode and the purpose for which the phototube is to be used. For example, if the phototube is to be used where the light source is a lamp with a tungsten filament, it should be tested with such a lamp and with the filament at the same operating temperature. In using a tungsten-filament lamp to test a phototube, the current should be measured accurately because a 1 per cent variation in current may cause a *6 per cent* variation in the light flux.

The photometer box should exclude all extraneous light, and all interior parts should be painted a dull black so that light will not be reflected. One or more masks should be arranged to ensure that the only light striking the cathode comes directly (and not by reflection) from the calibrated lamp. This lamp may be mounted on a carriage so that d of equation 16-1 may be varied. The test circuit is shown in Fig. 16-6. The protective resistor R may be omitted so that corrections for the voltage drop across it are not necessary. If this is done, care must be taken to prevent damaging gas phototubes. With the arrangement of Fig. 16-6 the curves of Figs. 16-2, 16-3, and 16-5 may be obtained. For any location of the calibrated lamp, the luminous flux in lumens is given by equation 16-1.

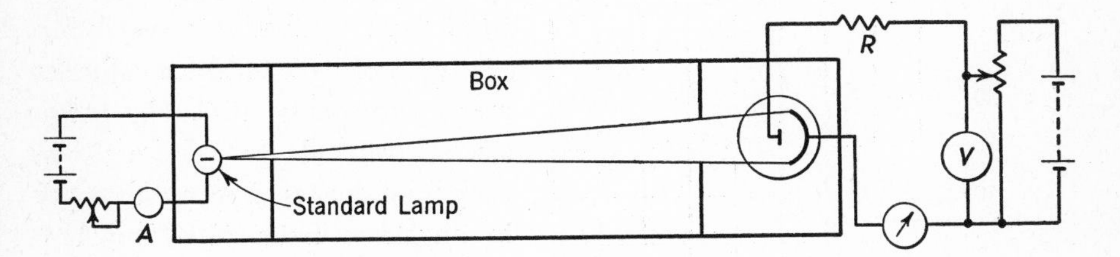

FIG. 16-6. Arrangement for studying a phototube. Either the lamp or the phototube may be moved in the light-tight box.

16-5. *THE CESIUM-OXYGEN-SILVER CATHODE*

These cathodes are extensively used in phototubes of both the vacuum and the gas types for purposes where either ordinary incandescent lamps or daylight furnish the illumination. The cathode is formed in one process [6] from a cylindrical section of silver. After a thorough pumping, with all metallic parts of the phototube at a high temperature to free occluded gases, pure oxygen is admitted and the silver cathode is oxidized by a glow discharge produced between the cathode and the anode. An exact amount of oxygen is combined with the silver. The correct amount of cesium is then formed on the cathode. A pellet containing cesium chromate, chromic oxide, and powdered aluminum, which was placed in the tube, is heated by a high-frequency induction furnace, and the cesium is evaporated and is condensed on the glass walls of the tube. It is then necessary to transfer the cesium to the cathode. This is accomplished by placing the phototube in a stream of hot air. This heats the glass without a corresponding rise in temperature of the cathode. The cesium is evaporated from the glass walls and condenses on the cooler cathode, where it combines with the silver oxide. The tube is then sealed from the exhaust pump if a vacuum phototube is desired, or the correct amount of gas is introduced for the gas type. The process is so controlled, and the quantities of cesium and oxygen so adjusted, that after the reactions occur there is a surface layer of cesium approximately one atom thick. This often is termed a **cesium–cesium-oxide–silver,** or **Cs-CsO-Ag, cathode.**

The direct-current output of a phototube at various wave lengths of luminous flux (the **spectral sensitivity**) is an important characteristic. This characteristic depends on at least two factors; first, the nature of the light-sensitive cathode surface, and second, the light-transmitting nature of the enclosing bulb. The characteristics of phototubes with two different types of light-sensitive cathodes are shown in Fig. 16-7. The potassium-hydride layer was used in early phototubes. The wave length in angstrom units (10^{-10} meter) and the visible spectrum are both shown. Phototubes are available with cathodes other than the cesium-oxygen-silver type, and with bulbs of special characteristics.[7] Such phototubes are most sensitive to other than visible light. The current-wave-length characteristics are determined almost entirely by the cathode surface and the light-transmitting qualities of the envelope.

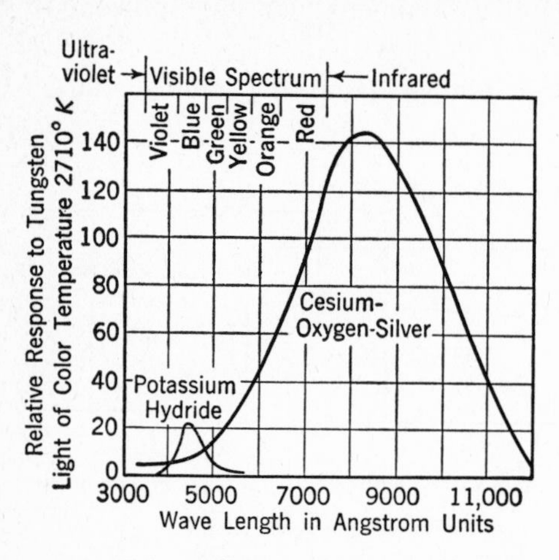

FIG. 16-7. Relative response of an early phototube as compared with the cesium-oxygen-silver type. (Reference 6.)

16-6. *PHOTOTUBE APPLICATIONS* [8]

Broadly considering the uses of phototubes, three classifications can be made. *First,* phototubes are used to open or close relays and control mechanical operations when light beams falling upon their cathodes are interrupted or restored. *Second,* phototubes are used as transducers [9] to cause electrical currents to vary in accordance with light-beam variations in sound motion picture reproduction and similar applications. *Third,* phototubes are used to measure light intensities and for related purposes.

Relay Operations. Suppose that it is desired that a phototube close a relay and operate a counter when a box on a conveyer belt passes between a light source and a phototube. Examination of Figs. 16-2 and 16-3 will indicate the impracticability of attempting to build for commercial purposes a relay that could be operated *directly* by these phototubes, which could change the relay current only a few microamperes. Thermionic-vacuum tubes therefore are used with such phototubes to increase the current change available for relay operation (Fig. 16-8). With the phototube *out* of the circuit, the grid bias on the amplifier tube is determined by the voltage drop across R_1 and R_2. With the phototube *in* the circuit, but *with the light beam intercepted,* a current will flow through the phototube because of residual room illumination, and this current will cause an opposing voltage drop across R_1 and R_2, *reducing* the grid bias on the amplifier tube and allowing a *larger* plate current to flow. When the light beam *is not intercepted,* the phototube current will greatly increase, thus reducing the grid-bias voltage considerably, and causing the plate current of the amplifier tube to increase greatly. This will operate the counter.

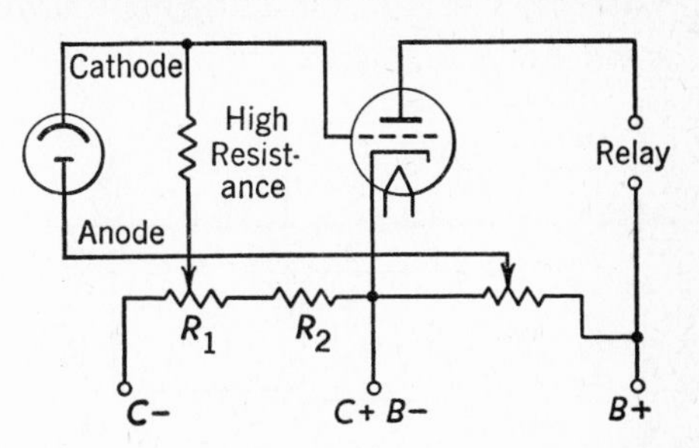

FIG. 16-8. Circuit for operating a relay with a phototube.

The design of Fig. 16-8 is simple. The phototube currents with residual illumination, and with the light beam shining on the cathode, must be known. From these values the grid-bias voltage variations can be calculated. Also, the current on which the counter relay will *close* and the current on which it will *open* must be known. An amplifier tube is then selected that will furnish these currents at available bias voltages. In case much power is needed to operate the counter or other device, a sensitive relay controlling a local source of power, such as a battery, for operating the larger device can be used. Gas phototubes often are used in the circuit just considered.

Sound Reproduction. In the reproduction of sound on films, a constant light beam is directed onto the sound track on the edge of the moving film. The light that passes through and falls on the cathode of the phototube will, accordingly, vary in intensity, as does the recorded speech and music variations on the edge of the film. The corresponding phototube current is, therefore, a faithful reproduction of the original sounds recorded in the studios. For sound reproduction in the theater the weak phototube output must be amplified to a high level for driving the loud-speakers.

Matching Measurements.[9] An interesting circuit for matching brightness, color, turbidity, and other qualities is shown in Fig. 16-9. Vacuum phototubes are used in this circuit because they are better suited to precise measurements than are gas types. One phototube receives light from the lamp or from (by reflection from, or transmission through) the material to be studied. The other phototube receives light directly from a standard lamp, or material. The two phototubes are in series and so *must* pass the same current. If the two tubes are identical, and if the circuit is properly adjusted, the voltage across each tube will be the same *for the same illumination.* If the illumination is *different* in any respect, then the voltage across each tube will *not* be the same. This will cause the bias on the thermionic vacuum tube to change, and the microammeter reading to vary (Fig. 16-9). The characteristic curve of one phototube has been plotted reversed. A small change in the illumination of one tube causes a large voltage change to be applied to the control grid of the amplifying tube. This will cause a large change in the plate current.

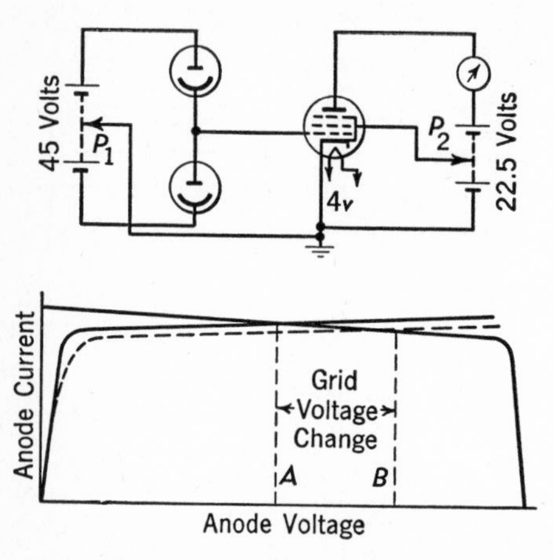

FIG. 16-9. A sensitive circuit using vacuum phototubes for color matching. With equal illumination the grid-bias voltage will be at *A*. For unequal illumination on the two phototubes the bias shifts to *B*. The broken line indicates the characteristics of the phototube with slightly less illumination. (Adapted from reference 3.)

Facsimile Systems. These are used to transmit fixed graphic material, including photographs, by wire lines or by radio. This first wire facsimile facilities [10] were opened in 1925, and later were replaced by an improved system.[11] A typical system operates somewhat as follows: A beam of light passes over each portion of the film, and the light passing through the film at a given instant depends on the density of the film at that point. This light beam falls on a phototube, and the electrical output will correspond to the characteristic of the film at that point. These electrical impulses are transmitted to the distant station, where they are to operate a **light valve** [11] that exposes a photographic film in such a manner that, when processed, it is a reproduction of the film at the sending station.

16-7. *ELECTRON-MULTIPLIER PHOTOTUBES* [12]

As the preceding discussions indicated, the output of a phototube is low. Amplification is required in most applications before the output of a phototube is sufficient for utilization. This amplification can be accomplished with vacuum tubes. If the output of a phototube is very low, however, it may be of the same order of magnitude as the noise developed by the vacuum tubes and associated amplifier circuit. Thus, after amplification the signal would be unsatisfactory for many purposes. It is possible to amplify a phototube signal (and other signals as well) with a device known as an **electron multi-**

plier. The electron multiplier makes use of secondary emission (section 2-16) to produce amplification.

In an electron multiplier tube a number of electrodes are arranged so that the electrons from one are attracted to the next, and so on. Each electrode is held at a positive potential higher than the one preceding it. If electrons are liberated by some means near the first electrode, they will be attracted to it, and under conditions favorable to secondary emission they will liberate electrons from it. A ratio of 5 to 1 is a reasonable expectation. The electrons will be attracted by the second electrode held at a higher positive potential and will liberate electrons from it by secondary emission, and so on. It is usual to have eight or ten such steps in a single tube. Such an arrangement is

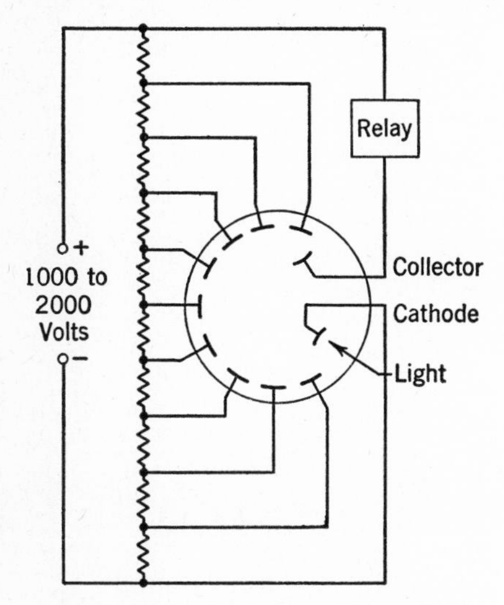

FIG. 16-10. Connections of an electron-multiplier phototube. Light striking the cathode emits electrons. These are attracted from one anode to the next, emitting electrons at each step. The final current is gathered by the collector, and this current operates the relay or other device to be controlled. (Adapted from reference 12.)

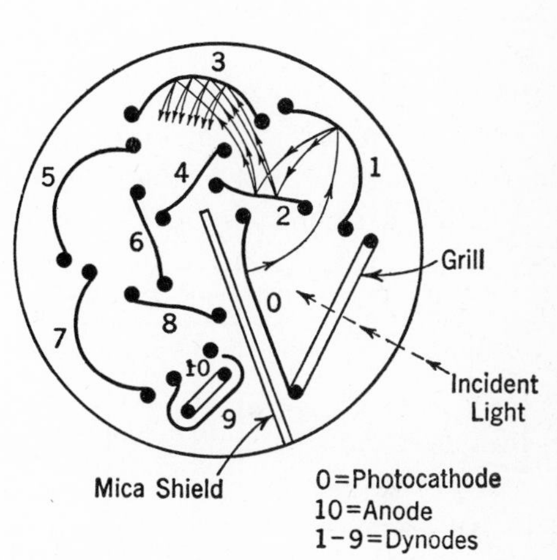

FIG. 16.11. Showing how the electrodes of Fig. 16-10 are arranged, and the electron paths. (Adapted from reference 12.)

shown schematically in Fig. 16-10. This is *not* the actual arrangement of the electrodes. Actually, they are large, and are so "interleaved" that the assembly fits into a small glass bulb (Fig. 16-11).

In Fig. 16-10, the electron multiplier is used to amplify the output from a light-sensitive cathode, the entire device being called an **electron-multiplier phototube.**[12] The output is large compared with the ordinary phototube, and amplification by vacuum tubes is not necessary for many applications. The signal-to-noise ratio is high because no thermionic cathodes are used. Generally speaking, amplification produced with thermionic vacuum tubes is noiser than that produced with secondary-electron multipliers.

16-8. *PHOTOELECTRIC DEVICES IN TELEVISION*

A television system requires that light reflected by an object or scene be changed to electric signals by a photoelectric device. Early systems include the **Nipkow disk,**[13] the **Farnsworth image dissector,**[13,14] and the **Zworykin Iconoscope.**[15] The first system used phototubes in simple form. Although a mechanical scanning disk was employed, television in both monochrome and color was demonstrated.[16,17]

The Image Dissector. In this photoelectric device light reflected by the object or scene to be televised is optically focused on a photoelectric screen, and an "image" in liberated photoelectrons is produced. As these electrons proceed to a positive anode, they are influenced by a magnetic field in a way such that the various portions of the electron image are moved in successive "lines" across a small hole in the anode. Electrons constituting that portion of the image in front of the hole at a given instant pass through the hole and into a secondary-electron multiplier, where the signal containing information regarding the appearance of that portion of the image or scene is amplified.[13,14] The image dissector is not used in modern television.

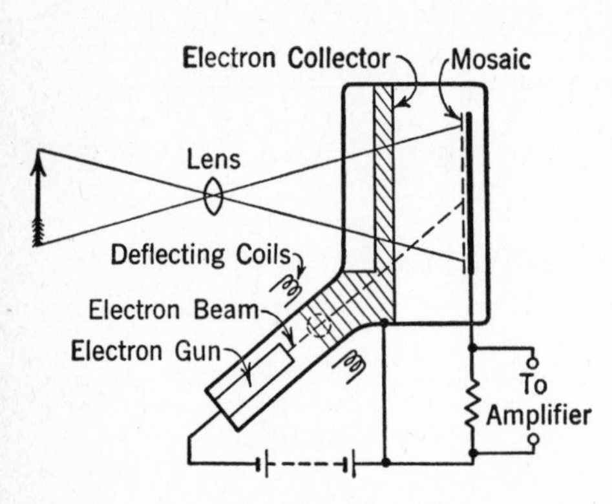

FIG. 16-12. Simplified diagram of an Iconoscope. There are two sets of deflecting coils situated at right angles with each other.

The Iconoscope. This tube sometimes is used (1955) in television for transmitting programs from motion-picture films. The **Iconoscope** [1] is a "television camera tube in which a high-velocity electron beam scans a photoactive mosaic that has electrical storage capability." The mosaic contains a large number of isolated, and insulated, globules of photosensitive material deposited on a mica sheet, the opposite side of which is coated with a conducting layer called a **signal plate.** Each globule forms with the signal plate a miniature capacitor. If secondary emission and related phenomena *are neglected,*[18] the operation of the Iconoscope is as follows: The optical system of Fig. 16-12 focuses on the photoactive mosaic the light reflected by the image or scene to be televised. Each of the many globules emits electrons in proportion to the intensity of the light striking that particular globule. By this emitting action each minute capacitor formed by a globule and the back plate becomes electrically charged, the magnitude of the charge being determined by the intensity of the light from a particular portion of the image. Thus, an "image" in electric charges on globules is produced on the mosaic.

The electron gun of Fig. 16-12 focuses an electron beam on the mosaic, and this beam is caused by deflecting coils to sweep across the mosaic in lines much as the eyes of a person scan a printed page in reading. As the beam of negative electrons passes over a globule, the charge thereon is altered in accordance with the degree of illumination to which it has just been exposed. This change in charge causes an instantaneous current flow through the resistor of Fig. 16-12, and this instantaneous current magnitude contains the required information regarding that specific portion of the image. The picture at the distant receiving station is reconstructed from such bits of information received in sequence. The Iconoscope provides excellent picture resolution, but requires intense illumination, an important factor limiting its use for general studio camera purposes.

The Image Orthicon.[19,20,21,22] The **Image Orthicon**[1] is a "camera tube in which an electron image is produced by a photoemitting surface and focused on a separate storage target, which is scanned on its opposite side by a low-velocity electron beam." As indicated in Fig. 16-13, the light reflected by the

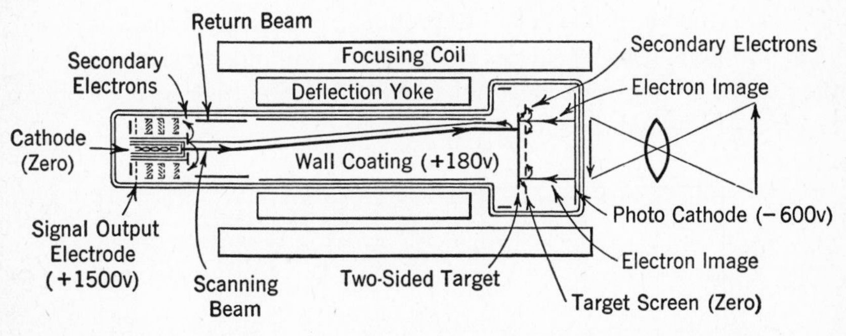

FIG. 16-13. Illustrating the principles involved in the Image Orthicon.

object or scene to be televised is focused on a photocathode of thin glass with a continuous electrically conducting photosensitive surface deposited on the side opposite to that on which the image is focused. The photosensitive surface emits photoelectrons in accordance with the light from the image, forming an "electron image" of the object or scene. The two-sided target and the target screen are at +600 volts with respect to the photocathode, and the electron image is attracted to this target. A focusing coil maintains the electrons of the "image" in parallel rays. The two-sided target is of very thin semiconducting glass, and the electron image striking it releases electrons by secondary emission; this forms an image charge-pattern on the side of the target toward the photocathode. The secondary electrons are collected by the target screen.

Because the glass of the two-sided target is semiconducting and thin, the charge pattern produced as explained also appears on the side of the target toward the electron gun. Negligible spreading and "blurring" of the charge pattern occurs because of the thinness of the semiconducting glass. A beam of low-velocity electrons is caused to move back and forth and scan the side of the target toward the gun. If there is no positive charge (caused by the impinging electron image) on an area being scanned at a given instant, the electrons in the scanning beam will be repelled at that instant. If there is a positive charge in a given area, however, then some electrons will be absorbed from the scanning beam and the remainder will be repelled. In this way the returned electron beam is modulated in accordance with the image or scene to be televised. This repelled modulated electron beam strikes a comparatively large area around the gun aperture from which it was shot. Secondary electrons are released, and these enter a secondary-electron multiplier, where much amplification occurs. The output is an electric signal conveying, by its instantaneous magnitude, information regarding each small area of the scanned image. The Image Orthicon is used almost exclusively (1955) as a **camera pickup tube** [1] to produce a video signal that may extend from about 30 to 4,500,000 cycles in television. An important reason for this is the fact that for a wide range of intensities the output is reasonably independent of the ambient illumination intensity.

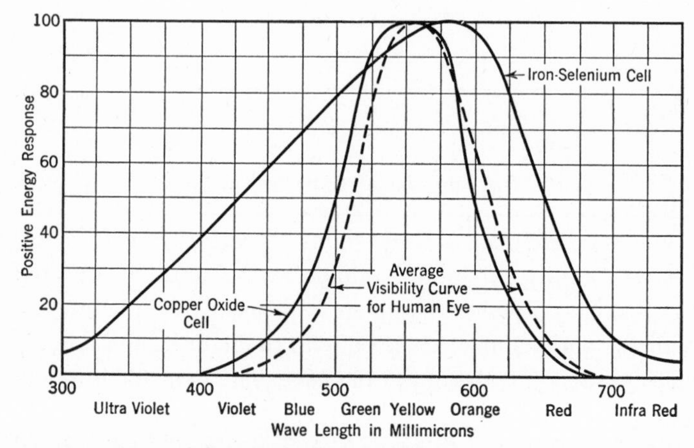

FIG. 16-14. Spectral sensitivity curves for photocells and for the human eye.

16-9. *PHOTOVOLTAIC CELLS*

The phototubes just considered are photoemissive devices and a voltage must be *applied* between cathode and anode; in a sense the device acts like a valve the internal resistance of which varies with the illumination. For the photovoltaic cell now to be considered, a *voltage is generated* by the absorbed radiant energy. This voltage forces electrons around the wire circuit; no external voltage is required. Becquerel discovered this effect in 1839. The term **barrier photocell** sometimes is applied. Barrier rectifiers were considered in section 10-9. In photocells it appears that absorbed radiant energy gives certain electrons the energy required to cross the potential barrier and flow around the connecting circuit in the high-resistance direction.

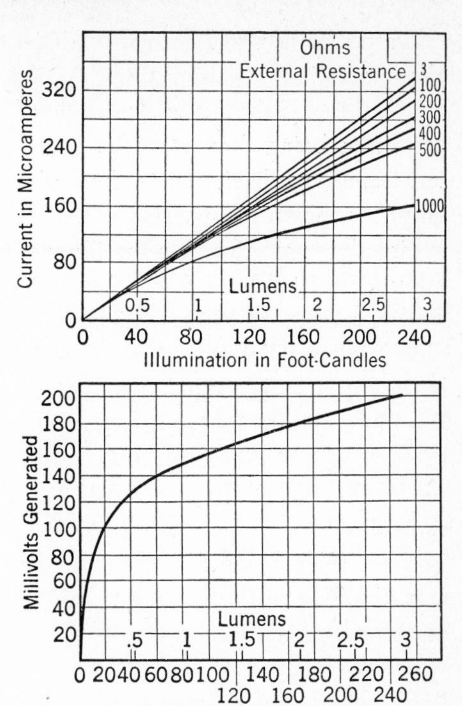

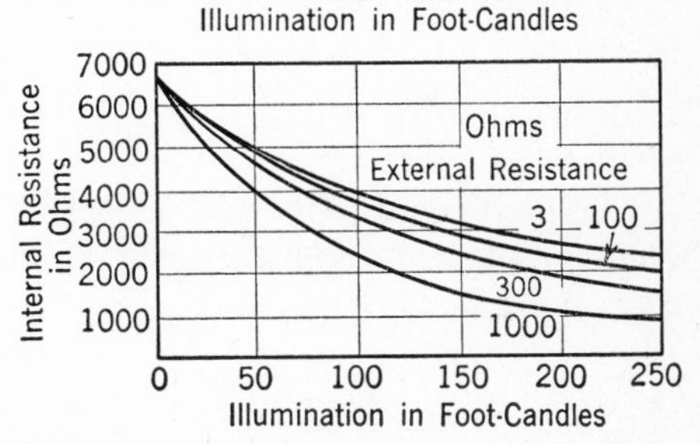

FIG. 16-15. Characteristics of an iron-selenium photocell. The open-circuit voltage is shown at the center. Light source is tungsten lamp at 3000°K.

Photovoltaic cells may be of the electrolytic type, or the dry or "electronic" type; the former is obsolete. Photovoltaic cells of the dry type may be constructed with a thin layer of suitable material on a metal disk, with the light passing through the thin layer to the barrier layer (back-effect cell), or may be constructed with a layer of suitable material on a metal disk and with a very thin light-transparent layer of a conducting material like silver on the material (front-effect cell). For this cell the photovoltaic action seems to be most pronounced at the front electrode. The material of the layers is either cuprous oxide formed on copper, or iron selenide deposited on iron.

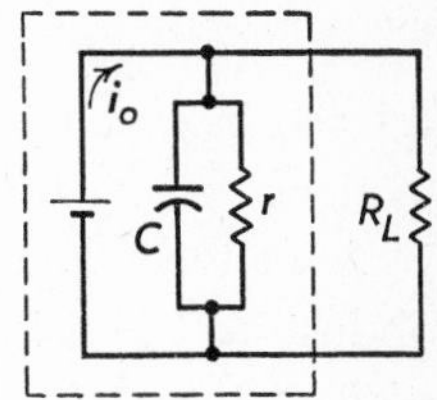

FIG. 16-16. Equivalent circuit for a photovoltaic cell connected to a load resistance R_L.

Characteristics of Photovoltaic Cells. The two types of cells have similar characteristics. The iron-selinide (or iron-selenium) cell, extensively manufactured under the name **Photronic cell,** will be discussed, the spectral characteristics being shown in Fig. 16-14. By the addition of a filter the characteristics are made similar to those of the eye. The effect of load resistance on output is shown in Fig. 16-15. For low resistance (only), the output current is directly proportional to the illumination. The open-circuit voltage and the internal resistance also are shown in this figure. The equivalent circuit is shown in Fig. 16-16, in which the capacitor may be omitted for constant illumination. The generated current is represented by i_o flowing from the battery at the left. This current is exactly proportional to the intensity of the illumination. Part of the current flows within the cell and not in the external circuit; thus, internal resistance r is in parallel with the current source. The value of r varies with the illumination and the external resistance as in Fig. 16-15. The internal parallel resistance of the Photronic cell is high. To have a large percentage of the generated current flow to an external circuit, the load resistance must be low. The open-circuit voltage between the output terminals is *not* directly proportional to the illumination except at very low intensities (Fig. 16-15). The photovoltaic cell is used extensively in light meters, and to operate relays and alarms.

16-10. *PHOTOCONDUCTIVE DEVICES*

The **photoconductive cell** is a resistance element, the resistance of which changes when the cell is illuminated. The effect was discovered by Smith in 1873; he noticed that the resistance changed when light fell on a selenium resistance. The phenomenon has been called a **photoconductive effect.** Semiconductors other than selenium exhibit these effects.[23,24,25,26] Photons of light are thought to release electrons *within* the semiconductor and thus cause its resistance to fall. Early photoconductive cells had a detrimental time lag, which was much reduced in later types. These cells commonly are made of a thin layer of selenium between two parallel conductors, sometimes placed in evacuated bulbs. Photoconductive cells are satisfactory for applications such as operating relays.

A photoconductive television camera tube, the **Vidicon,** was developed for industrial television and certain remote pickup purposes.[27,28,29] A cross section of this device is shown in Fig. 16-17, the operation being as follows: Electrons are emitted by the thermionic cathode and are accelerated by grid 2, which is at about +300 volts with respect to the cathode. Grid 1 operates at voltages from 0 to −100 volts and controls the magnitude of the beam

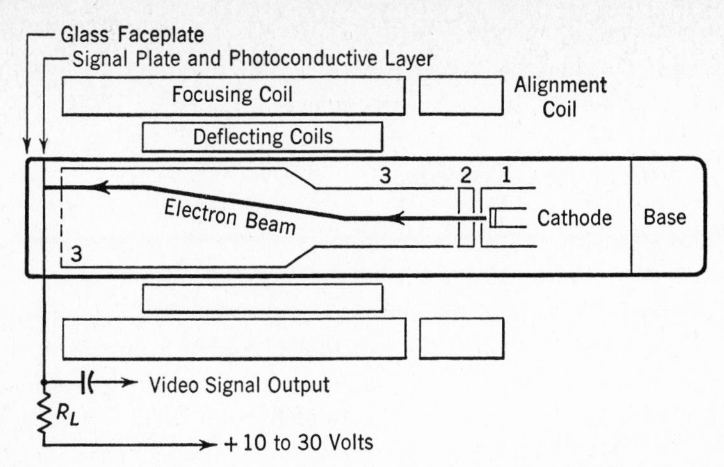

FIG. 16-17. Cross section of the Vidicon tube used in television cameras. The alignment, focusing, and deflection coils encircle the tube which is about an inch in diameter, and about 6 inches long.

current. Grid 3 is a long cylindrical electrode at slightly less than +300 volts and has a mesh across the end opposite the cathode. Since this mesh is at a lower potential than accelerating grid 2, it decelerates the electrons in the beam. The scanning beam passes through holes in this decelerating mesh and impinges with low velocity on the photoconductive layer. The layer (amorphus selenium and antimony trisulfide have been used [29]) is placed on a light-transparent, electrical-conducting layer that acts as a signal plate. The photoconductive layer and the conducting signal plate are on the inside of the glass faceplate closing the end of the Vidicon. The electron beam is aligned by the alignment coil, focused by the focusing coil (section 17-5), and caused to scan the photoconductive layer by the horizontal and vertical deflecting coils.

Assume that the photoconductive layer is *not* illuminated. Electrons passing through the mesh of grid 3 impinge on the surface of the layer and charge it negatively with respect to the signal plate. As a limit, the potential of the surface approaches that of the cathode.[27] Negligible current flows to the positive signal plate because of the low conductivity of the nonilluminated layer. An object or scene to be televised and a focusing lens would be at the left in Fig. 16-17. Reflected light from the object passes through the lens and is focused on the transparent signal plate and photoconductive layer. Illuminated *portions* of the layer become conducting and current flows through these portions, and during the 1/30 second between successive scans the scanned surface of illuminated portions becomes positively charged a volt or two with respect to the cathode.[27] When the scanning beam passes over such an area it will deposit electrons to neutralize the accumulated positive charge. When this occurs, current will flow through R_L of Fig. 16-17, and this constitutes the instantaneous television video current.

16-11. *PHOTOTRANSISTORS*

The **phototransistor** is a photoconductive device [30] the basic arrangement of which is given in Fig. 16-18. A small depression is ground in a germanium wafer, the final thickness of the germanium after etching being about 0.003 inch. A phosphor bronze wire electrode at the center of the depression serves as a collector. Light directed on the germanium wafer as indicated imparts energy to certain valence electrons, causing them to enter the conduction band. The action is the same fundamentally as the action of thermal energy considered at several points in Chapter 10; also, the same basic types of potential-energy diagrams apply.[30] The light must fall in the *immediate* vicinity of a spot on the germanium opposite the collector wire. In practice a lens, sometimes built into the end of the metal tube containing the phototransistor element, focuses the light on the sensitive spot. In the circuit of Fig. 16-18 some "dark" current will flow, and when the device is correctly illuminated the current increases in proportion (Fig. 16-18) to the amount of light falling on the cell. This current increment is the photocurrent.[30] Phototransistors using *p-n* junctions instead of point contacts also have been developed.[31,32] Phototransistors promise to be very important light-sensitive devices.

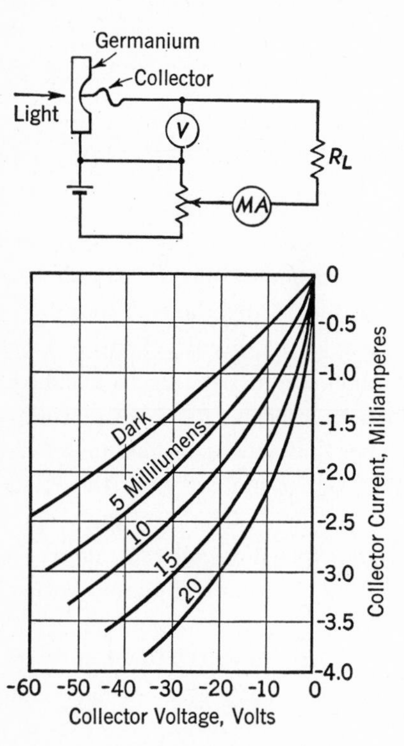

FIG. 16-18. Circuit for studying a phototransistor, and static characteristics. (Adapted from reference 30.)

REFERENCES

1. Institute of Radio Engineers. *Standards on Electron Tubes: Definitions of Terms,* 1950.

2. Ives, H. E. *The alkali metal photoelectric cell.* Bell System Technical Journal, April 1926, Vol. 5, No. 2.
3. R.C.A. Manufacturing Co., *Pamphlet describing 917 phototube;* also, *Pamphlet describing 918 phototube.*
4. American Institute of Electrical Engineers. *American Standard Definitions of Electrical Terms.* 1941.
5. Institute of Radio Engineers. *Standards on Electronics,* 1943.
6. Kelley, M. J. *The caesium-oxygen-silver photoelectric cell.* Bell System Technical Journal, July 1932, Vol. 11, No. 3.
7. Glover, A. M. *A review of the development of sensitive phototubes.* Proc. I.R.E., Aug. 1941, Vol. 29, No. 8.
8. Westinghouse Electric Corp. *Industrial Electronics Reference Book.* John Wiley & Sons.
9. Institute of Radio Engineers. *Standards on Transducers: Definitions of Terms,* 1951. Proc. I.R.E., Vol. 39, No. 8.
10. Ives, H. E., Horton, J. W., Parker, R. D., and Clark, A. B. *The transmission of pictures over telephone lines.* Bell System Technical Journal, April 1925, Vol. 4, No. 2.
11. Reynolds, F. W. *A new telephotograph system.* Electrical Engineering, Sept. 1936, Vol. 55, No. 9.
12. Rajchman, J. A., and Snyder, R. L. *An electrically focused multiplier phototube.* Electronics, Dec. 1940.
13. McGee, J. D. *Distant electric vision.* Proc. I.R.E., June 1950, Vol. 38, No. 6.
14. Brolly, A. H. *Television by electric methods.* Electrical Engineering, Aug. 1934, Vol. 53, No. 8.
15. Zworykin, V. K. *The Iconoscope–A modern version of the electric eye.* Proc. I.R.E., Jan. 1934, Vol. 22, No. 1.
16. *Two-way television: A group of three papers on television.* Bell System Technical Journal, July 1930, Vol. 9, No. 3.
17. Ives, H. E., and Johnsrud, A. L. *Television in colors by a beam scanning method.* Journal of the Optical Society of America, Jan. 1930, Vol. 20.
18. Zworykin, V. K., Morton, G. A., and Flory, L. E. *Theory and performance of the Iconoscope.* Proc. I.R.E., Aug. 1937, Vol. 25, No. 8.
19. Rose, A., Weimer, R. K., and Law, H. B. *The Image Orthicon–A sensitive television pickup tube.* Proc. I.R.E., July 1946, Vol. 34, No. 7.
20. Janes, R. B., Johnson, R. E., and Handel, R. R. *A new Image Orthicon.* R.C.A. Review, Dec. 1949, Vol. 10, No. 4.
21. Janes, R. B., Johnson, R. E., and Handel, R. R. *Producing the 5820 Image Orthicon.* Electronics, June 1950.
22. Fink, D. G. *Television Engineering.* McGraw-Hill Book Co.
23. Zworykin, V. K., and Wilson, E. D. *Photocells and Their Applications.* John Wiley & Sons.
24. Hughes, A. L., and DuBridge, L. A. *Photoelectric Phenomena.* McGraw-Hill Book Co.
25. Hughes, A. L. *Fundamental laws of photoelectricity.* Electrical Engineering, Aug. 1934, Vol. 53, No. 8.
26. Pakswer, S. *Lead sulfide photoconductive cells.* Electronics, May 1949.

27. Weimer, R. K., Forgue, S. V., and Goodrich, R. R. *The Vidicon photoconductive camera tube.* Electronics, May 1950.
28. Webb, R. C., and Morgan, J. M. *Simplified television for industry.* Electronics, June 1950.
29. Vine, B. H., Janes, R. B., and Veith, F. S. *Performance of the Vidicon, a small developmental television camera tube.* R.C.A. Review, March 1952, Vol. 13, No. 1.
30. Shive, J. N. *The phototransistor.* Bell Laboratories Record, Aug. 1950, Vol. 28, No. 8.
31. Shive, J. N. *Properties of the M-1740 p-n junction photocell.* Proc. I.R.E., Nov. 1952, Vol. 40, No. 11.
32. Lehovec, K. *New photoelectric devices utilizing carrier injection.* Proc. I.R.E., Nov. 1952, Vol. 40, No. 11.

QUESTIONS

1. What are the principal parts of a phototube? How are these arranged?
2. Why is gas used in some phototubes?
3. What precautions should be taken in operating gas phototubes?
4. What is the magnitude of the gas amplification factor?
5. In testing phototubes why should the illumination be varied by changing distance instead of by changing the lamp current?
6. Into what classes can phototube uses be grouped?
7. What is the meaning of facsimile?
8. What is the principle of operation of an electron-multiplier phototube?
9. What are important advantages of amplification with electron multipliers as compared with vacuum-tube amplifiers?
10. In secondary emission what is the ratio of secondary to primary electrons?
11. What are several important phototube applications?
12. What is the principle of operation of the image dissector?
13. What is the principle of operation of the Iconoscope?
14. What is the principle of operation of the Image Orthicon?
15. Discuss the role of secondary emission in the above camera tubes.
16. What is meant by back-effect and front-effect cells?
17. What are the basic differences among phototubes, photovoltaic cells, and photoconductive cells?
18. Why is the term *photovoltaic* used, when the current and not the voltage is directly proportional to the illumination?
19. For what applications should a low-resistance load be used with a photovoltaic cell?
20. How does a photoconductive cell operate? How is it used?
21. Is an electron multiplier used in the Vidicon? Why?
22. How does the Vidicon operate? Where is it used?
23. What is meant by the term industrial television?
24. What is a phototransistor, and how does it operate?
25. Will the phototransistor probably replace or supplement the phototube? Why?

PROBLEMS

1. The anode-to-cathode voltage drops for typical vacuum and gas phototubes are shown in Figs. 16-2 and 16-3. For the 0.1 lumen curves calculate the direct-current resistance at 10-volt increments from 10 to 90 volts. Plot these resistance values against anode-to-cathode voltage. What conclusions can be drawn?
2. The two phototubes whose characteristics are shown in Fig. 16-5 are substantially identical except that one contains gas. Calculate the gas amplification factor at several values of anode voltage.
3. The gas phototube of Fig. 16-3 is used in a circuit such as Fig. 16-8. The triode used is a type 31, with +135 volts on the plate with the relay connected. The circuit is adjusted so that the phototube anode voltage is +80 volts when the light intensity is 0.1 lumen. The "C" battery voltage is 45 volts, and R_1 and R_2 are each 5000 ohms. The high resistance in the phototube circuit is 2 megohms, and the slider of R_1 is set at the center. When light falls freely on the cathode the intensity is 0.1 lumen. When the beam is interrupted the intensity is 0.02 lumen. What will be the current through the relay for these two conditions?
4. A phototube gives 3 microamperes through 5 megohms with 90 volts on the anode when it is in room illumination with a beam of light interrupted, and gives 10 microamperes when the light beam directed toward it is not interrupted. The relay it is to operate closes on 12 milliamperes and opens on 4 milliamperes. Design a *complete* circuit so that the relay will be closed when the light beam is interrupted.
5. The multiplier phototube of Fig. 16-10 is operated so that the ratio of secondary to primary electrons is 3.0 for each electrode except, of course, the cathode and the collector. What will be the multiplication ratio for the tube? Repeat, assuming a ratio of 5.0 at each electrode. Why are the cathode and collector excluded in this computation?

CHAPTER 17

CATHODE-RAY TUBES

Electron devices that utilized streams of electrons in the form of electron beams, or electron rays, were considered at several points in the preceding chapters. These include the traveling-wave tube (section 8-25), and the television camera tubes of Chapter 16. These devices, and the tubes to be considered in this chapter, are **electron-beam tubes** defined [1] as "an electron tube the performance of which depends upon the formation and control of one or more electron beams." The **cathode-ray tube** to be considered in this chapter is of this type, and further is defined as "an electron-beam tube in which the beam can be focused to a small cross section on a surface and varied in position and intensity to produce a visible pattern."

In a cathode-ray tube an electron beam, or ray, commonly impinges on a fluorescent screen. The kinetic energy of the electrons causes the screen to glow at the spot where the electrons strike. The ratio of the electron charge to mass is great (section 1-2) and for this reason electrons are accelerated readily. Thus an electron beam may be deflected easily by an electric field produced by a signal voltage applied between deflecting electrodes, usually within the tube. Or, the electron beam is readily deflected by a magnetic field produced by a signal current flowing through suitable coils. Deflecting the beam causes the visible spot to move, making possible the observance of the wave form of an electric signal. Because of such properties, the cathode-ray tube has become a measuring device of much importance. Furthermore, the cathode-ray tube is used as the image-reproducing picture tube in television receivers. This alone is an application of supreme importance.

17-1. *EARLY CATHODE-RAY TUBES* [2,3,4,5,6]

The use of the cathode ray in measurements was suggested by Hess in 1894. The modern cathode-ray tube is based on a device developed by Braun in 1897, and shown in Fig. 17-1. It consisted essentially of a *cold* cathode *A*, a *positive* anode *B*, a beam-forming diaphragm *C*, a fluorescent screen *D*, and contained air at low pressure. The positive potential of *B* drew electrons (caused by natural ionization, section 1-12) to it, and these were accelerated

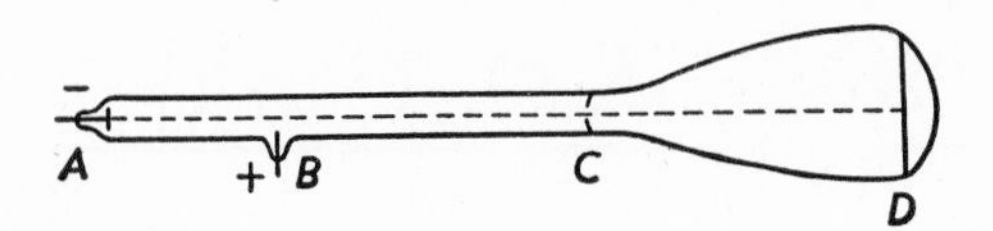

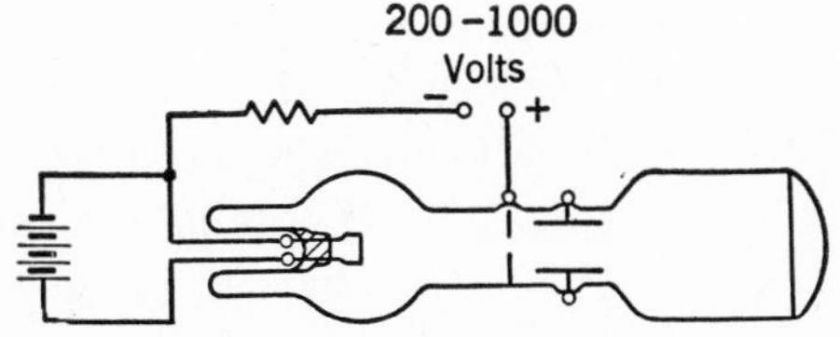

FIG. 17-1. The early cathode-ray tubes of Braun (above), and of Wehnelt (below). (From reference 2.)

sufficiently to cause ionization by collision with gas atoms of the residual air. The positive ions were attracted to the negative cathode and a cold-cathode discharge (section 2-17) was established. Because of the high velocities reached (the potential from the electrostatic generator being 50,000 volts or more) the electrons were carried beyond positive electrode *B*, and some passed through the hole in *C* and to the zinc-sulfide fluorescent screen at *D*. Since these electrons traveled at high velocities, their mutual repulsion did not cause the beam to spread appreciably. The cathode ray striking the fluorescent screen could be deflected magnetically by coils placed to the right of the diaphragm *C*, or electrically by placing the tube in an electric field between the plates of a capacitor. This tube almost immediately found application in the communication and power fields.[4]

A development of much importance was made by Wehnelt in 1905 when he produced a thermionic cathode-ray tube with a heated, lime-coated filament. This tube could be operated from low-voltage 220-volt power circuits. A diagram of this tube, showing two deflecting plates to which the voltage to be studied is connected, is given in Fig. 17-1. A contribution of importance was made by Dufour in 1914. In his tube no fluorescent screen was used; instead, a photographic plate was placed *within* the tube, and the electron beam impinged directly on this plate. In this way, records of rapidly changing phenomena readily could be made. Previously it had been necessary to place the photographic plate near the fluorescent screen, and expose the plate by the weak light from the screen. For the satisfactory photographing of a rapidly occurring electrical phenomenon many repetitions of the phenomenon sometimes were required.

17-2. *GAS THERMIONIC CATHODE-RAY TUBES*

The first cathode-ray tube generally available was the thermionic gas-focused tube described by Johnson.[2,6] Because a hot cathode supplied the electrons for the beam, a high operating voltage was not required as for the cold-cathode type. This was important because at this time operation from a few radio "B" batteries was accepted practice, rectifiers not being available generally. Low accelerating voltage gave low beam velocities, and the electrons tended to spread excessively. Gas focusing, suggested by van der Bijl,[6] was used to focus the beam the principle being as follows.

The tube is filled with argon at about 0.01 millimeter pressure. The beam electrons traveling along the axis of the tube causes ionization by collision, producing massive immobile positive ions, and highly mobile negative ions (electrons). After ionization the positive ions move but little. However, the electrons are deflected readily from the region of ionization. As a result, a column of positive ions forms near the center of the tube, and the negative ions accumulate around this column. This produces a radial electric field about the axial electron beam; this field focuses the beam by forcing the outer beam electrons that are trying to spread (because of mutual repulsion) back toward the center of the tube. Such focusing is satisfactory in principle, and a low-velocity beam is deflected readily by an electric signal for study; however, the presence of positive ions is bothersome. An important reason is that these massive ions tend to travel to the negative cathode, and bombard it, causing its destruction unless special construction is used.[2,6]

17-3. *VACUUM THERMIONIC CATHODE-RAY TUBES*

Because of its many limitations, and because of the development of rectifiers and amplifiers, the gas type was replaced by the vacuum cathode-ray tube about 1935.[3] For general measurement purposes thermionic vacuum cathode-ray tubes operate at accelerating voltages of about 1000 to 2000 volts. The trace on the fluorescent screen is more brilliant than for the low-voltage type, and the beam has less tendency to spread; also, the beam is less affected by stray fields, such as the magnetic field of the earth.

Thermionic cathode-ray tubes contain an **electron gun**,[1] which is "an electrode structure which produces, and may control, focus, and deflect an electron beam." Electron guns differ in detail, but each has a thermionic cathode to produce electrons, a positive electron-**accelerating electrode** [1] to "increase the velocity of the beam electrons," and commonly a control grid [1] to regulate the number of electrons in the beam. Included within the electron-gun structure *may be* a **focusing electrode,**[1] which is "an electrode to which a potential is applied to control the cross-sectional area of the electron beam." By this means is achieved **electrostatic focusing,**[1] a "method of focusing an electron beam by the action of an electric field." Also, in the electron-gun structure *may be included* **deflecting electrodes,**[1] meaning electrodes "the potential of which provides an electric field to produce deflection of the electron beam." The *italics* were used because many cathode-ray tubes use a **focusing coil** or **focusing magnet,**[1] an "assembly producing a magnetic field for focusing an electron beam." Also, some tubes use a **deflecting yoke,**[1] an "assembly of one or more coils the current through which provides a magnetic field to produce deflection of an electron beam."

17-4. *CATHODE-RAY FOCUSING WITH ELECTRIC FIELDS*

The electrons from the cathode must be formed into a beam directed toward the fluorescent screen, and must be focused so that a beam of small cross section strikes the screen, thereby producing a sharp brilliant spot. The

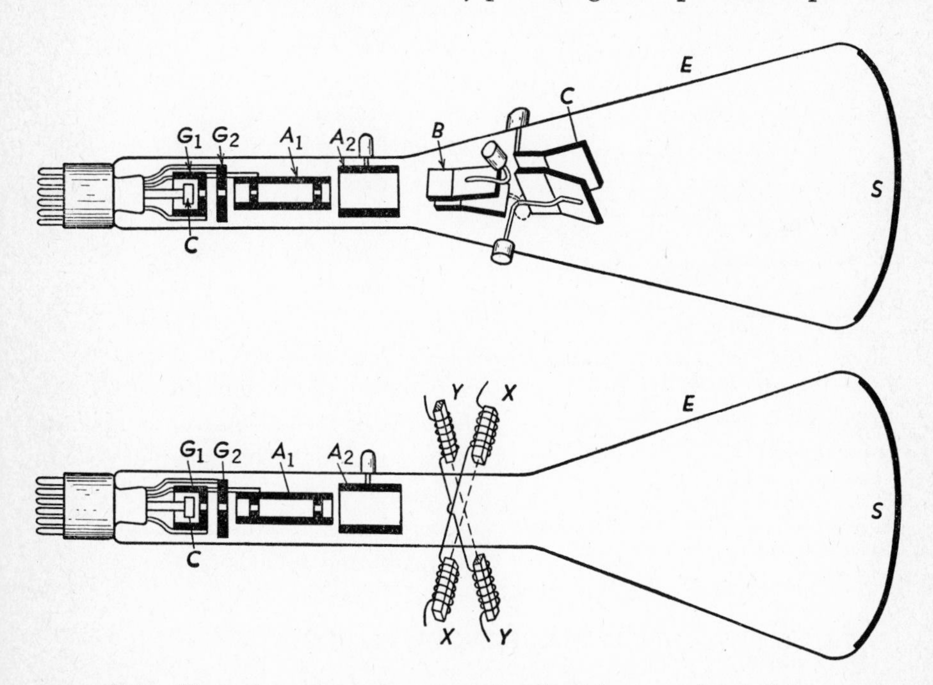

FIG. 17-2. Illustrating an arrangement of the electrodes in medium-voltage vacuum cathode-ray tubes. The upper tube uses an electric field, and the lower a magnetic field, for deflecting the beam. (From reference 7.)

design of a focusing system is a subject in **electron optics.** The details of a focusing system vary, a typical arrangement being shown in Fig. 17-2. Cathode C is indirectly heated. The hole in control grid G_1 permits the passage of only electrons with velocities directed along the axis; also, this grid is at an adjustable negative voltage with respect to the cathode, and serves to control the magnitude of the beam current. Grid G_2 is a positive accelerator grid, and also acts as an electron lens, further forming an electron beam. Anode A_1 and anode A_2 accelerate and focus the electrons, anode A_1 being more positive than A_2.

Beam Focusing with Metal Disk. A metal disk with a small hole in the center focuses the beam as follows: If G_2 of Fig. 17-2 is E volts positive with respect to G_1, if anode A_1 is E' volts positive with respect to G_2 and if E' exceeds considerably E, then there will be a *resultant* electric field in, and near, the hole in the disk somewhat as shown in Fig. 17-3. Some of the electrons will strike the metal disk and will be lost to the beam. The remaining electrons will tend to follow the lines of electric force. Electrons tending to diverge will be directed back into the beam. The focusing action can be controlled by varying either E or E'.

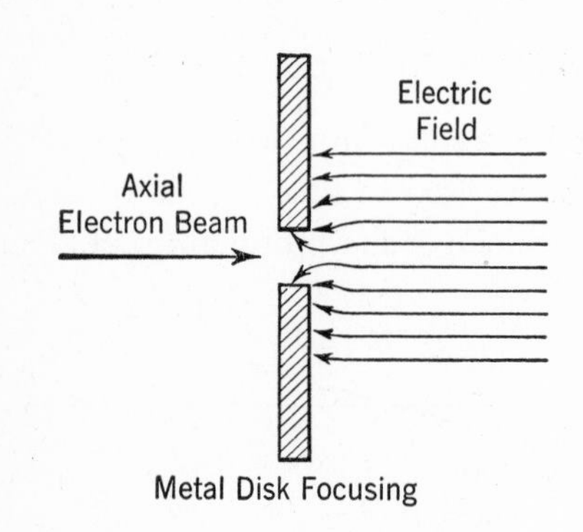

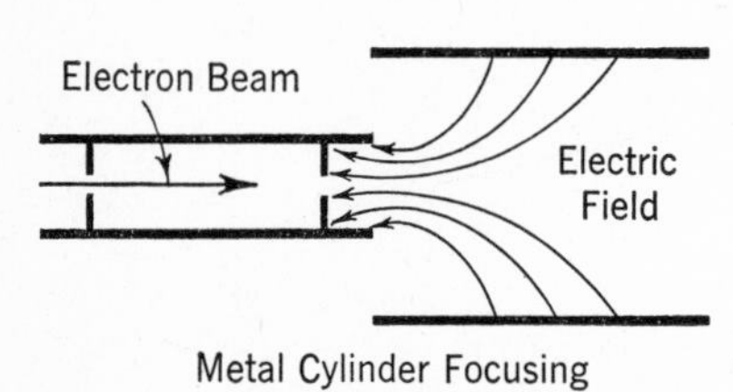

FIG. 17-3. Methods of focusing electron beams in cathode-ray tubes by electric fields. Sometimes constant field-strength contours are used to illustrate the focusing instead of lines of electric force as shown here.

Beam Focusing with Metal Cylinders. Assume that in Fig. 17-2 anode A_2 is more positive than A_1. An electric field will exist between the two anodes somewhat as shown in Fig. 17-3. Because of the shape of the field, electrons at the edges of the beam will be forced toward the axis of the tube. By varying the voltage between the electrodes, the focusing can be controlled. The design of electron lenses such as discussed is a specialized subject treated in reference 8, which also lists books and articles on the subject.

17-5. *CATHODE-RAY FOCUSING WITH MAGNETIC FIELDS*

The Image Orthicon and the Vidicon of the preceding chapter were shown within a relatively long magnetic focusing coil. Magnetic focusing also is used extensively with cathode-ray tubes, particularly in television receivers. However, with these a relatively short coil, or a short permanent-magnet focusing assembly, is employed.[10]

Assume that a magnetic field exists *parallel* to the axis of a cathode-ray tube. From section 1-4 a moving electron will experience an accelerating force *only* if the electron has a component of velocity at right angles to the magnetic field. Thus, if an electron is moving along the axis of a cathode-ray tube because of an accelerating electric field parallel to the axis, and if the electron tends to move away from the axis because of the repulsion of other beam electrons, then the electron will experience a force (section 1-4) tending to return it to the center of the beam. The combined effect of the electric and magnetic fields will cause the electron to follow a spiral path.[8,10] An analysis based on equations for electron motion in a magnetic field will show that the angular velocity is independent of the transitional velocity, and that all electrons complete a circle in the same time. Therefore, all spiraling electrons can be allowed to complete one or more revolutions as they travel down the tube, and they then will focus on the screen (Fig. 17-4). The explanation for the focusing action produced by a short coil, or by a magnet, will be similar except that the "twisting" action is confined to a small region.[8,10]

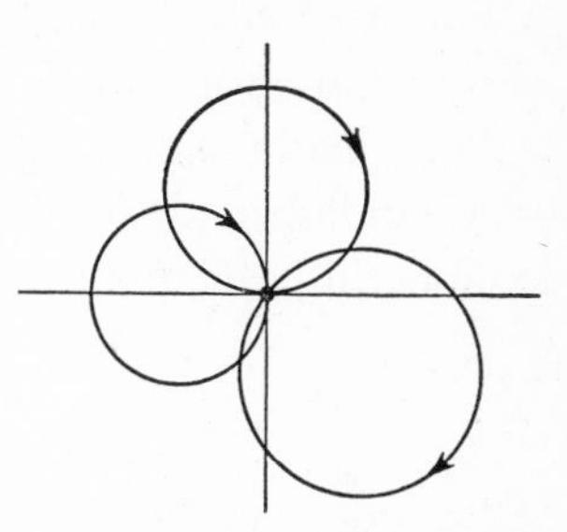

FIG. 17-4. End view, showing paths (in planes at right angles to axis of tube) of electrons with velocity components at right angles to the beam at the axis of the tube.

17-6. *FLUORESCENT SCREENS*

The term **fluorescent screen** is used popularly to designate the material at the viewing end of a cathode-ray tube. It commonly is stated that the fluorescent material emits light when the beam electrons impinge on it. Actually, the material is a **phosphor,** defined [1] as a "substance capable of luminescence." The term **luminescence** applies (with reservations [9]) to all forms of visible radiation. Luminescence includes [9] **fluorescence,** which is luminescence emitted by a phosphor during excitation as with an electron beam, and **phosphorescence,** which is luminescence emitted by a phosphor after excitation has ceased. Many substances are used as phosphors, depending on the spectral and time characteristics [9] desired. A phosphor commonly used is **zinc orthosilicate,** or **willemite,** which gives the familiar bright green spot.

17-7. *BEAM DEFLECTION BY ELECTRIC FIELDS*

The upper cathode-ray tube of Fig. 17-2 has two sets of plates used to deflect the electron beam by an electric field produced by voltages impressed between plates of a set. In the following discussion it will be assumed that the entire plate area is flat and that the lines of electric force between plates of a set are parallel. An analysis of the deflecting action is as follows.[2]

In operation a voltage of E volts is impressed between cathode C and positive anode A_2 of Fig. 17-2. Assuming an electron starts from rest and travels unimpeded along the axis of the tube, the final velocity in meters per second will be (equation 1-7)

$$v = \sqrt{\frac{2Ee}{m}} = 5.95 \times 10^5 \sqrt{E} \qquad 17\text{-}1$$

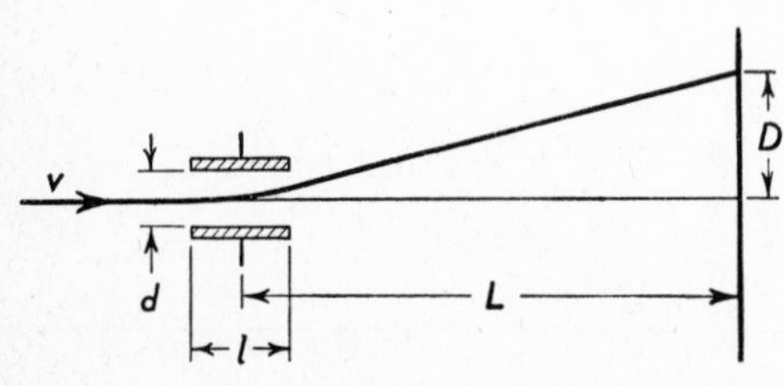

FIG. 17-5. Diagram for studying the deflection of the beam in a cathode-ray tube by an electric field. (Adapted from reference 2.)

where e and m are as given in section 1-2. From Fig. 17-5, if the *deflecting* voltage E' volts is impressed between the plates d meters apart, the voltage gradient will be E'/d volts per meter. The deflecting acceleration in meters per second per second experienced by an electron passing between the plates and along the axis is (equation 1-4) $a = E'e/(dm)$. Since velocity in general is $v = at$, the transverse velocity component v' of an electron (component at right angles to the axis) will be

$$v' = \frac{E'et}{md} = \frac{E'el}{mdv} \qquad 17\text{-}2$$

where l/v is substituted for time t, because an electron traveling with a velocity component v along the tube axis will remain between the plates for the interval $t = l/v$.

After emerging from between the deflecting plates, the electron beam travels to the screen, where a bright spot is formed on the phosphor a distance D from the center. The distance D is related to L of Fig. 17-5 by

$$\frac{D}{L} = \frac{v'}{v} = \frac{E'el}{mdv^2} = \frac{lE'}{2dE} \quad \text{and} \quad D = \frac{lLE'}{2dE} \qquad 17\text{-}3$$

In this derivation the first part of equation 17-1 is substituted for v, and equation 17-2 for v'. In equation 17-3 distance D will be in the same units as l, L, and d, and E' and E may be in any units provided they are the same.

17-8. *BEAM DEFLECTION BY MAGNETIC FIELDS*

The lower cathode-ray tube of Fig. 17-2 has two sets of coils arranged to produce magnetic fields at right angles to the electron beam. Also, the fields from these two sets of deflecting coils are mutually at right angles. A diagram for studying the theory is Fig. 17-6, in which B represents the magnetic field from one set of coils, the field being directed *out* from the page as indicated by the dots. From equation 1-10, the radius of curvature of the path of an electron in the field is

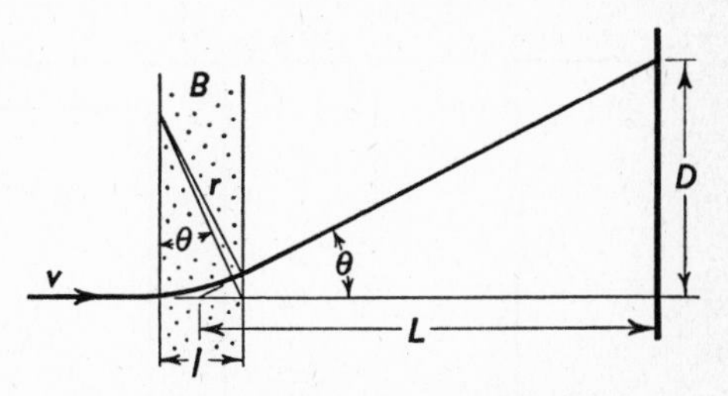

FIG. 17-6. Diagram for studying the deflection of the beam in a cathode-ray tube by a magnetic field. The magnetic field is shown by the dots, representing lines of magnetic force coming out from the page. (From reference 2.)

$$r = \frac{5.65 \times 10^{-12} v}{B} \qquad 17\text{-}4$$

where r is in meters, v is in meters per second, and B is in webers per square meter. When equation 17-1 is substituted,

$$r = \frac{3.36 \times 10^{-6}\sqrt{E}}{B} \qquad 17\text{-}5$$

From Fig. 17-6, $\tan \theta = D/L = l/r$ approximately. Thus

$$D = \frac{Ll}{r} = \frac{LlB}{3.36 \times 10^{-6}\sqrt{E}} = \frac{0.3LlB10^6}{\sqrt{E}} \quad \text{(approx.)} \qquad 17\text{-}6$$

In these expressions D is in meters when L and l are in meters, and the other units are as previously stated.

17-9. *CATHODE-RAY OSCILLOGRAPHS*

The **cathode-ray oscillograph** or **cathode-ray oscilloscope** is an electron device for studying the shapes of electric signals, or waves, of almost any type. The signal is caused to deflect the electron beam of an **oscillograph tube** (**oscilloscope tube**) defined [1] as "a cathode-ray tube used to produce a visible pattern, which is the graphical representation of electrical signals, by variations of the position of the focused spot or spots in accordance with these signals." Although the term oscilloscope has wide use, there is an increasing tendency to use the term oscillograph, as was done in early literature.[2,3,6] Also, the term oscillograph is used in I.R.E. Standards.

Oscillograph Connections. The basic connections for a simple cathode-ray oscillograph are given in Fig. 17-7. The electrodes are similar to Fig. 17-2, but are shown for convenience as disks, with holes. Grid G_2 is omitted in this tube. Certain characteristics of the tube are shown in Figs. 17-8 and 17-9.

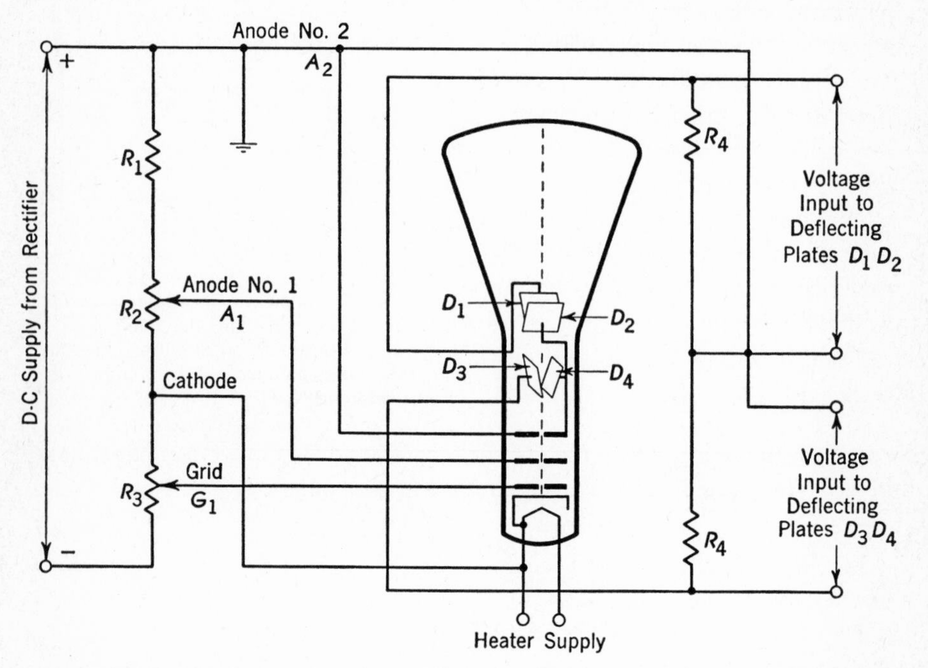

FIG. 17-7. Basic connections for a cathode-ray tube used to observe voltage phenomena. (From reference 7.)

As indicated in Fig. 17-7, anode A_2, one deflecting plate of each set, and the positive supply terminal, are grounded. The voltages between anode A_1 and grid G_1 are varied by adjusting R_2 and R_3. Varying R_3 controls the number of electrons in the beam (Fig. 17-8) and the intensity of the spot. The focusing is controlled by R_2. The signal to be studied is impressed across either of the identical resistors R_4. To investigate an electric current, the current is passed through a resistor of low resistance and the voltage drop across the resistor is impressed across the deflecting plates; or, the current may be passed through deflecting coils. Vacuum-tube amplifiers usually are between the input terminals and the deflecting plates so that weak signals may be displayed. The modern cathode-ray oscillograph is a highly developed instrument, incorporating a surprisingly large variety of features, and can be used in many ways.[11]

Spectral Characteristics and Persistence Time.[5,8,10,12,13,14] As previously mentioned, the phosphor, willemite, emits a bright green light where the electron beam strikes it. The eye is sensitive to this radiation,[8] and this phosphor is excellent for visual observation. It has medium persistence. Other phosphors with different spectral and persistence characteristics, adapting the tubes to various special uses, are available.[12,14] Methods of testing cathode-ray tubes are given in reference 15.

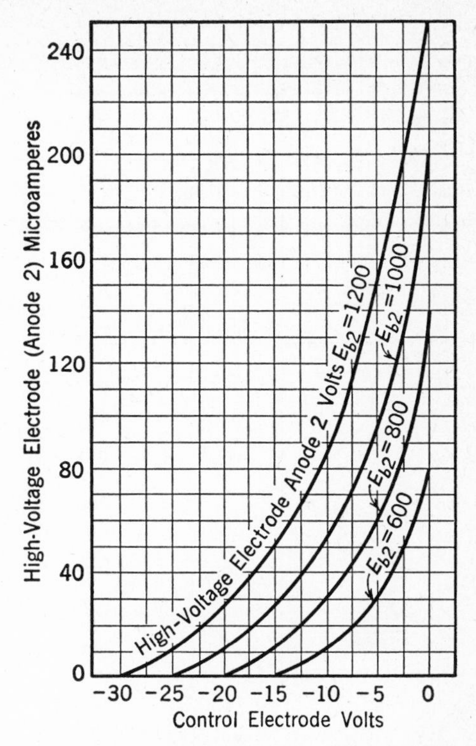

FIG. 17-8. Illustrating the effect of the control-grid voltage G_1 on the beam current in a cathode-ray tube, when potential A_1 is adjusted for focus. (From reference 7.)

17-10. *CATHODE-RAY OSCILLOGRAPH PATTERNS*

In the conventional tube are two sets of deflecting plates ordinarily referred to as the **vertical deflecting plates** and the **horizontal deflecting plates.** The deflection of the cathode ray or beam, and the resulting trace on the fluorescent screen, are controlled by the voltages impressed on these plates. Thus, if an alternating voltage is impressed on the *horizontal* deflecting plates (which is the pair in the *vertical* plane) of the tube of Fig. 17-7, the beam will trace a straight *horizontal* line. If an alternating voltage is

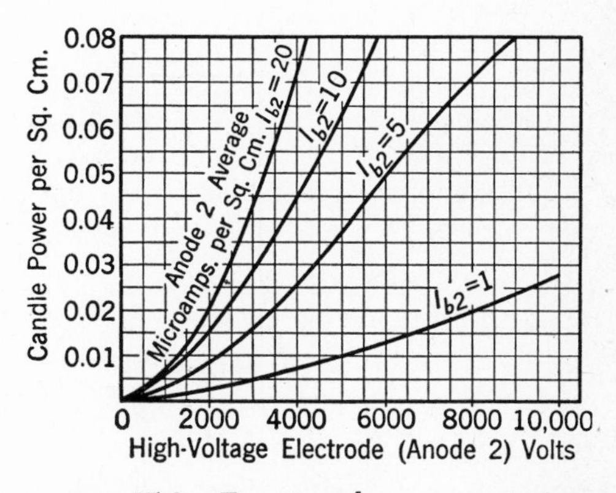

FIG. 17-9. For given beam currents in a cathode-ray tube, the light emitted by the fluorescent screen depends on the voltage of anode 2. (From reference 7.)

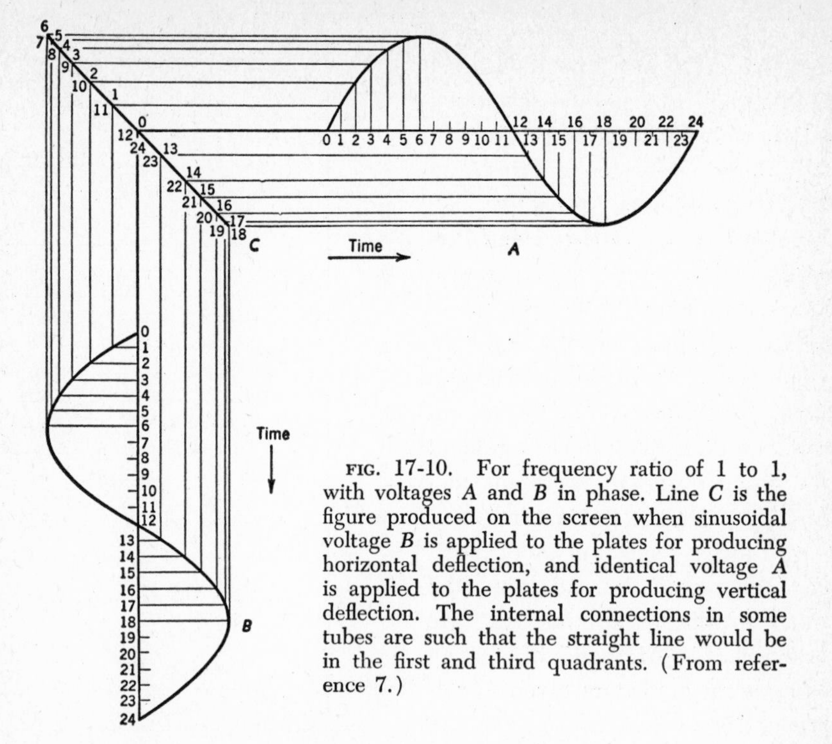

FIG. 17-10. For frequency ratio of 1 to 1, with voltages *A* and *B* in phase. Line *C* is the figure produced on the screen when sinusoidal voltage *B* is applied to the plates for producing horizontal deflection, and identical voltage *A* is applied to the plates for producing vertical deflection. The internal connections in some tubes are such that the straight line would be in the first and third quadrants. (From reference 7.)

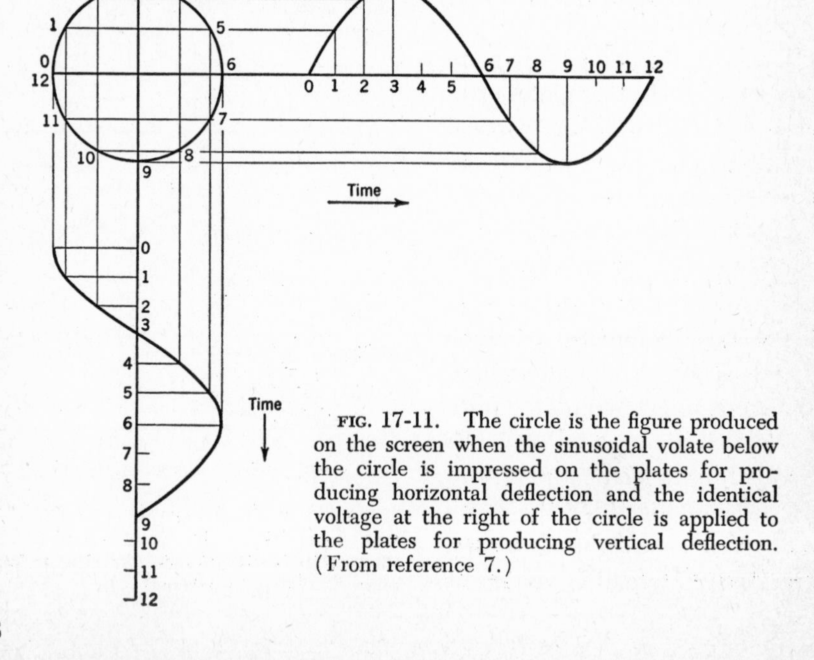

FIG. 17-11. The circle is the figure produced on the screen when the sinusoidal volate below the circle is impressed on the plates for producing horizontal deflection and the identical voltage at the right of the circle is applied to the plates for producing vertical deflection. (From reference 7.)

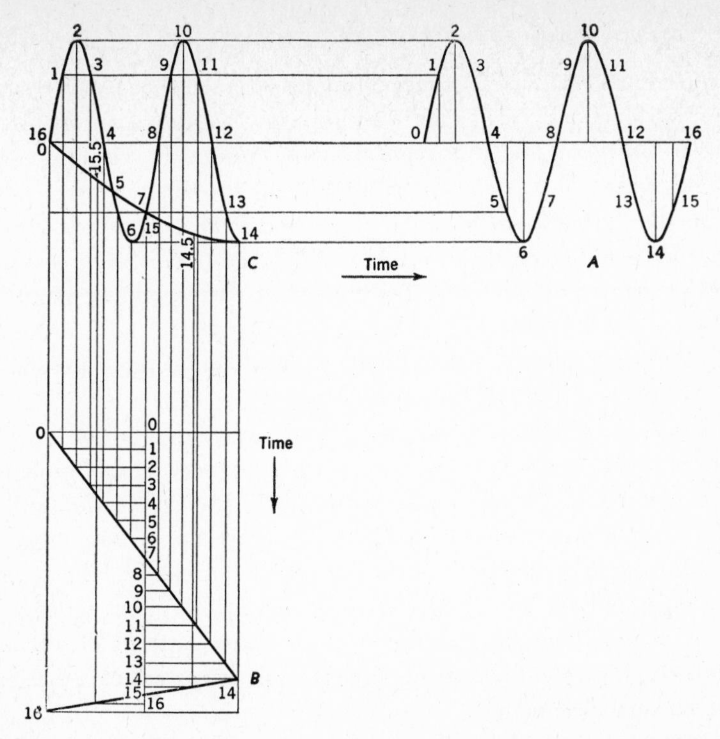

FIG. 17-12. Frequency ratio 1 to 2, with sinusoidal voltage *A* connected to produce vertical deflection, and saw-tooth voltage *B* used to produce horizontal deflection. (From reference 7.)

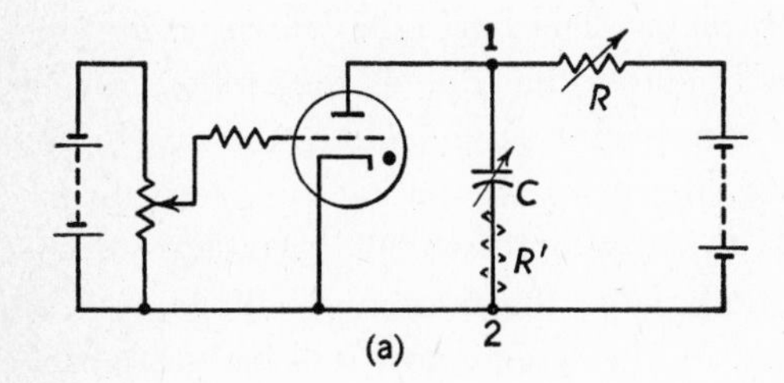

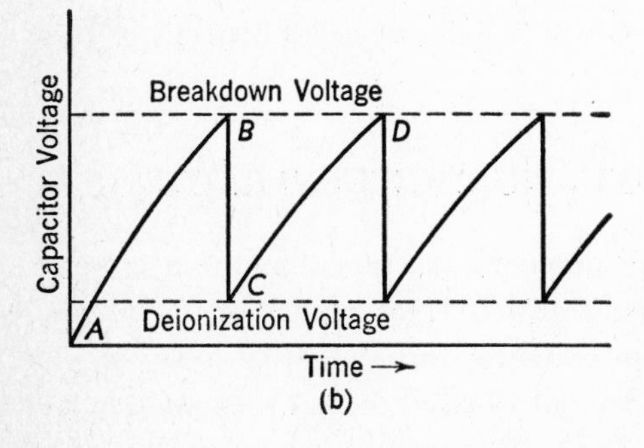

FIG. 17-13. Sweep-circuit oscillator using a gas triode to generate a saw-tooth wave for producing the horizontal time axis on a cathode-ray tube. The sweeping voltage exists across points 1-2.

impressed on the *vertical* deflecting plates (which lie in the horizontal plane), the beam will trace a *vertical* line. Now suppose that *equal* alternating voltages that are *in phase* and that have the *same* frequency are impressed on *each* set of plates. This is shown by Fig. 17-10 in which *A* represents the voltage on the *vertical* plates and *B* that on the *horizontal* set. There are now *two forces* acting on the beam, and the trace *C* on the screen will be produced by the *resultant* force. By projecting corresponding points from these two curves, the figure *C* on the screen is proved to be a *straight line* inclined at an angle of 45°. (Sometimes internal connections are such that the line is in the first and third quadrants.)

The conditions of the preceding paragraph were for equal in-phase voltages of the same frequency. If the two voltages are now equal in magnitude and frequency, but 90° out of phase as in Fig. 17-11, a projection will show that the resulting wave trace *C* will be a *circle*. The possibilities of the cathode-ray tube as a phase-measuring device are apparent. The sinusoidal voltages impressed on a cathode-ray tube may differ in magnitude, frequency, or phase. It is possible, therefore, to have patterns of almost any conceivable type on the fluorescent screen. Such traces are called **Lissajous figures,** after an experimenter.

Suppose that a sinusoidal voltage to be observed is impressed across the vertical deflecting plates. As previously mentioned, this will result in a straight vertical line being traced on the fluorescent screen. Now suppose that the vertical plates are disconnected and that a "saw-tooth" wave is impressed on the horizontal deflecting plates. The horizontal force acting on the beam will deflect the beam slowly across the tube as the voltage gradually increases, but will sweep the beam back *very quickly* as the saw-tooth wave rapidly dies to zero. If now the sinusoidal alternating voltage is again connected to the vertical plates and the saw-tooth sweeping voltage simultaneously is connected to the horizontal plates, the sinusoidal voltage, which formerly merely caused a vertical line on the screen, now will be "stretched out" and will appear as a sinusoidal trace (or otherwise if the first voltage is not sinusoidal) on the fluorescent screen. A point-by-point analysis of this action is shown in Fig. 17-12.

17-11. *SWEEPING, OR SCANNING, VOLTAGE GENERATORS*

The saw-tooth voltage wave in the preceding section sometimes is generated by a circuit employing a grid-controlled gas triode or Thyratron (section 4-23). The basic circuit of an oscillator for generating a saw-tooth wave is shown in Fig. 17-13. The gas triode is biased so that the anode must be quite positive before the tube will break down and conduct. At the instant of starting, assume that the capacitor is uncharged, and the tube is not conducting. The capacitor will start to charge through resistor *R*. The voltage across the capacitor will rise along curve *A–B*, following the usual law of a *R–C* circuit. At point *B*, the voltage across the capacitor, which is also the voltage across the gas tube, has reached the breakdown value and the tube suddenly conducts. This puts a virtual short circuit across the capacitor, and its voltage suddenly drops from *B* to *C*. When the voltage drops to *C*, the tube deionizes, and the action is repeated, the voltage on the next pulse being from *C* to *D*, and similarly for succeeding pulses. This device sometimes is called a **relaxation oscillator.**

The voltage across the capacitor at points 1–2 is the saw-tooth voltage wave desired for the sweep circuit of a cathode-ray tube. The frequency can be regulated by varying *C* and *R*. The line *C–D* is not straight, but by using a relatively small section of the curve, it is approximately straight. If a constant-current element is connected in place of *R*, the rate of charge will be constant and the portion *C–D* of Fig. 17-13 will be essentially straight. A pentode has constant-current characteristics, because the plate current is little affected by the plate voltage, and pentodes sometimes are used in place of resistor *R*.

Some arrangement must be made to synchronize the sweep-circuit oscillator with the signal to be studied or the trace will "wander" across the screen. A simple method is to select with a voltage-dividing arrangement a small part of the signal voltage that is being impressed for observation on the deflecting plates, and impress this voltage on the grid of the gas tube. Then, as the capacitor voltage builds up, and the tube is *about* to conduct, it will be triggered each time by the peak of the alternating signal voltage. In this way the two circuits are synchronized or "locked" together.

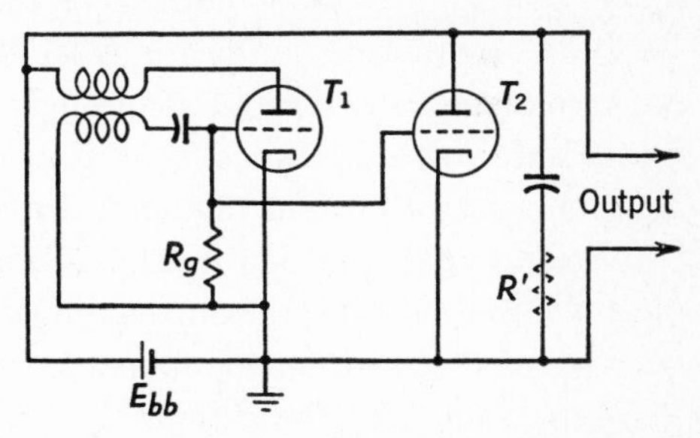

FIG. 17-14. Scanning, or sweep-circuit generator employing a blocking oscillator.

A vacuum tube may be used in a circuit like Fig. 17-13 to provide a saw-tooth voltage wave. The tube is biased considerably beyond cutoff, and capacitor *C* normally is charged. A pulse of short duration is inserted in the grid circuit so that the grid is driven highly positive. The tube then conducts and quickly discharges the capacitor. As soon as the pulse disappears, the tube is blocked and capacitor *C* slowly charges through *R*, giving a saw-tooth voltage wave. The multivibrator and blocking oscillator circuits of Chapter 14 form the basis for some sweep circuits (Fig. 17-14).

17-12. *SWEEPING, OR SCANNING, CURRENT GENERATORS*

If a cathode-ray tube is arranged for magnetic deflection, then since the deflecting magnetic field will follow closely the current, a saw-tooth sweep current must be provided for the coil. The coil possesses both resistance and inductance, and hence the current through it will not have the same wave form as the voltage across it. In Fig. 17-15a is shown the shape of the desired sweep current. This current flowing through the resistance only of the deflecting coil would cause a voltage drop as shown by (b), and flowing through the inductance only would cause a voltage drop as shown by (c). The total voltage drop, and the impressed voltage, must be the sum of these as indicated by Fig. 17-15d. The instantaneous expression for the voltage across the deflecting coil is

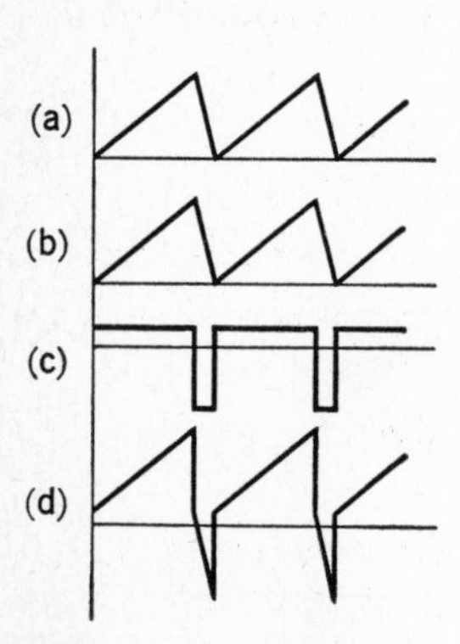

FIG. 17-15. Shapes of current and voltages for magnetic deflection.

$$e = ir + \frac{Ldi}{dt} \qquad 17\text{-}7$$

where r is the resistance and L is the inductance of the deflecting coil. Thus, if the current is a saw-tooth wave, the impressed voltage must be as shown by Fig. 17-15d.

A voltage of the desired shape may be obtained by connecting a resistor R' in series with the capacitor of Fig. 17-13. When the capacitor is charging with a constant current regulated by resistor R, there is a constant voltage drop across R' as needed in Fig. 17-15d. Furthermore, when capacitor C of Fig. 17-13 discharges, the current flows through R' and the required negative voltage of Fig. 17-15d is obtained.

As mentioned at the close of the preceding section, a blocking oscillator is used in sweep circuits. If the blocking oscillator of Fig. 14-20 is connected as in Fig. 17-14, the grid of T_2 will vary (Fig. 14-20) as that of tube T_1. Tube T_2 will conduct for brief intervals, thus discharging capacitor C, and will be blocked for relatively long intervals, permitting capacitor C to charge from E_{bb} through R. This will give a saw-tooth voltage output for *electric-field* deflection; however, if resistor R' is added, voltage d of Fig. 17-15 will be produced for *magnetic-field* deflection. The blocking oscillator of Fig.

17-14 may be synchronized by injecting a pulse in the grid circuit. Amplifiers may follow the sweep generators if needed.

17-13. *MEASUREMENTS WITH CATHODE-RAY TUBES*

This versatile device has been adapted to a wide variety of measurements. Usually the tube with deflecting plates is used and this type only will be considered.

Voltage Measurements. The unknown voltage to be measured is impressed on one pair of deflecting plates, and the resulting deflection compared with that caused by a known voltage. A sweep voltage may or may not be used.

Current Measurements. The current to be measured can be passed through a resistor and the voltage drop applied to one set of plates. The deflection (either with or without a sweep applied) is observed on the screen. A cross-section background for the screen is advantageous but not necessary. The deflection can be compared with that produced by a known current; or, if the tube has been calibrated in volts and the resistance is known, the current can be calculated.

Power Measurements.[16] Figure 17-10 shows that the pattern of the cathode ray is a straight line when two *in-phase* voltages of the same magnitude and frequency are impressed simultaneously on the two sets of deflecting plates; but as Fig. 17-11 shows, the figure is a circle when the two voltages are *90° out of phase.* Since power is equal to the product $EI \cos \theta$, from the area of the figure the power taken by any circuit can be determined.

Phase Measurements.[17] An arrangement to measure the phase shift in a vacuum-tube amplifier is shown in Fig. 17-16. If the R–C circuit is omitted, and if the input and output signal voltages are impressed on the two sets of deflecting plates, the resulting figure on the oscilloscope screen will be an ellipse, the area depending on the amount of phase shift. If the R–C combination is inserted and the voltage across the capacitor or resistor (if R and C are interchanged) is impressed as indicated, the two voltages on the two sets of plates can be brought in phase as indicated by a straight line on the screen. The phase shift angle then can be determined from the settings of R and C. Phase measurements also can be made from measurements made on the elliptical patterns as indicated by Fig. 17-17. For these figures the magnitude of the phase angle is

$$\sin \theta = \frac{b}{a} \qquad \text{17-8}$$

Several methods [17] have been developed to determine if the angle is "leading" or "lagging." Perhaps the simplest method is to insert an adjustable phase shifter in series with one set of plates and vary the phase shifter until the ellipse becomes a line.

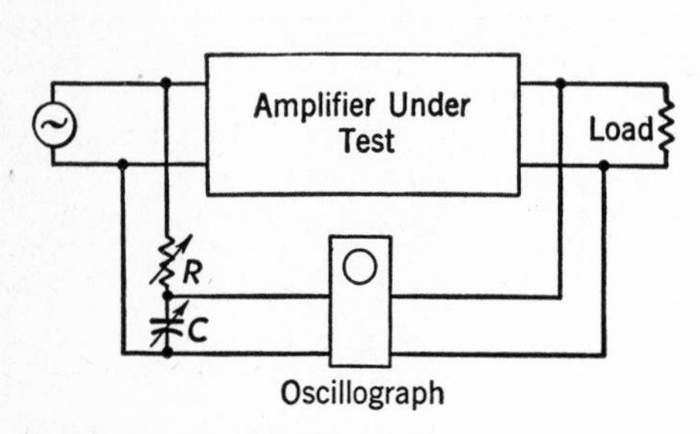

FIG. 17-16. Method of measuring phase shift of an amplifier.

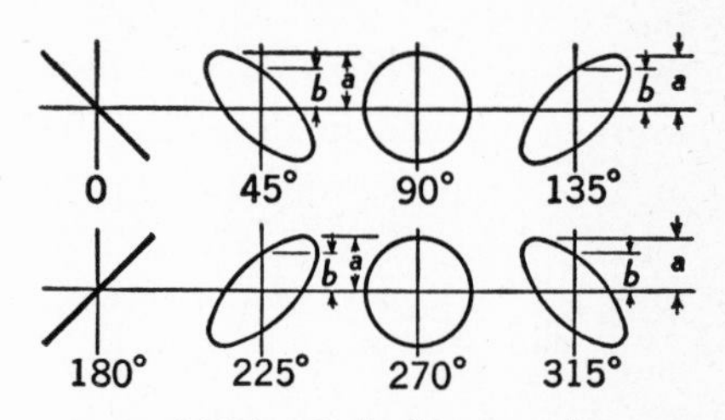

FIG. 17-17. Cathode-ray tube patterns for various phase angles. For these the angles between the impressed voltages are found from $\sin \theta = b/a$. (See statement regarding internal connections, Fig. 17-10.)

Frequency Measurements. An unknown frequency can be compared with a known frequency by impressing the two frequencies on the two sets of deflecting plates. From the patterns that result, the frequency of the unknown source can be determined. It is difficult to specify simple rules by which this can be accomplished, because the patterns become complicated. The simplest method is to use a variable source of known frequency, and vary it until a 1 to 1 ratio is obtained (see Figs. 17-10 and 17-11).

Hysteresis Measurements.[18] The hysteresis loss per cycle in a magnetic material is proportional to the area of the hysteresis loop for that material. The hysteresis loop can be reproduced by a cathode-ray tube and the power loss determined.

Vacuum-Tube Characteristics.[17] The cathode-ray tube can be used to study both the static and load characteristics of vacuum tubes and transistors and to study their uses in amplifiers.

Modulation Measurements.[17,19] There are at least two ways of studying the output of a modulator. The **trapezoidal method** employs an audio-frequency sweep on one set of plates and the modulated signal voltage on the other. The **modulated-envelope method** uses a saw-tooth sweep on one set of plates and the modulated signal on the other. With the first method a trapezoidal figure is observed, but with the second the modulated signal itself is reproduced on the screen.

17-14. *MISCELLANEOUS CATHODE-RAY DEVICES*

Dufour Tube. This consists of a cold cathode and an anode, with a voltage of 50,000 to 60,000 volts applied between electrodes. The electron beam passes through deflecting plates and strikes the photographic film *directly.* A viewing window for observing the film is provided on one side. The device must be opened for inserting films. This necessitates the continuous use of an evacuating system. In most instances a mechanical arrangement is provided so that films may be changed without opening the device until several films are exposed. When the anode is made positive with respect to the cathode, some of the free electrons in the residual gas in the tube are drawn rapidly toward the anode, and ionization by collision with residual gas atoms results. A beam is formed, and this beam exposes the photographic film by impinging on it. Some Dufour tubes use magnetic and some use electric fields to concentrate the beam. One factor limiting the frequency is the high speed with which the electron beam moves across the screen. If the speed is too great, the illumination becomes low. With the Dufour tube, since the electrons themselves strike the film, high beam speeds produce good exposures. Experimenters have obtained recording speeds of one-fifth the velocity of light.[3] The Dufour oscillograph has been used in high-voltage investigations to study transients of short time duration. By its use, a study of lightning discharges has been possible.[3] Such surges may last only a few microseconds.

Lenard Tube. It is apparent that the necessity for inserting the film *into the tube* complicates the operation. Evacuating troubles and other difficulties arise. Thus an attempt was made to shoot the electron beam *out of the tube* into the air and to have the electrons impinge on a photographic film *outside the tube.*[3] A beam of high-velocity electrons will pass through a thin metal foil, such as aluminum. In the Lenard tube a window of thin aluminum foil is provided. Many of the high-velocity electrons find interatomic paths through the foil. The foil must be backed by a strong grid to stand the pressure of the air. Voltages of about 75,000 volts were used on this tube to obtain high electron velocities.[20] It has never achieved wide application as an oscillograph tube.

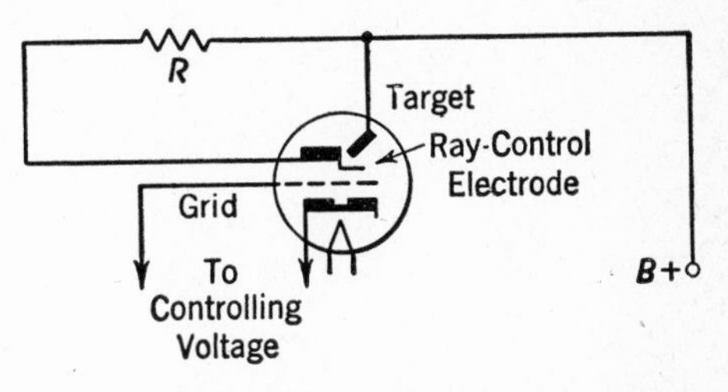

FIG. 17-18. The cathode-ray indicator tube and its associated circuit.

Indicator Tube. The indicator tube of Fig. 17-18 consists of a *long* cathode, a grid, a ray-control electrode (which includes the plate), and a target. The *inside* of the target is coated with a fluorescent material. Note that this material is *not* coated on the end of the glass bulb. A light shield for the upper portion of the cathode is provided. When the grid is at a low negative potential, a large current will flow to the plate (the ray-control electrode) and since this current must flow through the resistor *R*, the plate becomes *less positive* than the target. That is, a negative voltage will exist between the control electrode and the target. When the grid potential becomes *more* negative, *less* current flows to the control electrode, *less* drop occurs across the resistor *R*, and a *smaller* negative potential will exist between the ray-control electrode and the target. The shaded portions of the cathode represent the thermionic oxide coating. Around the grid is the cylindrical plate. An extension from the plate goes up through the target as indicated. This extension is the thin offset part of the ray-control electrode of Fig. 17-19. The potential variations with respect to the target control the area of the target struck by electrons given off by the *upper* end of the cathode. Thus, when a *large* negative potential is placed on the grid of the lower triode portion, a large area of the target glows and the **shadow angle** is small. When a *small* negative potential is placed on the electric grid, the ray-control electrode and the target are at a high difference of potential and the shadow angle is large (Fig. 17-20). The negative voltages for controlling the "electric eye" as a sharpness of tuning indicator are obtained from the voltage across capacitor C_4 of a diode detector (Fig. 13-13). When the radio set is tuned, the signal is the maximum value, and the direct voltage across C_4 will be maximum. This will place a *large* negative voltage on the grid, which will cause the shadow angle to be *small*, indicating that the set is tuned. An indicator tube has been developed that has two ray-control electrodes mounted on opposite sides of the cathode. With this tube two symmetrically opposite shadow angles may be obtained.

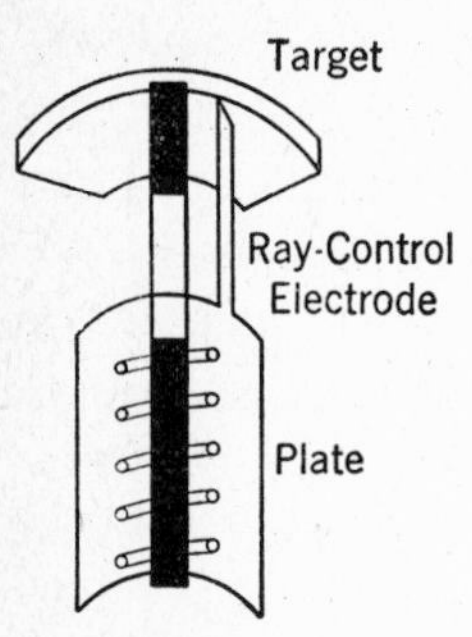

FIG. 17-19. Showing the essential features of a cathode-ray indicator tube. The light-shield is omitted.

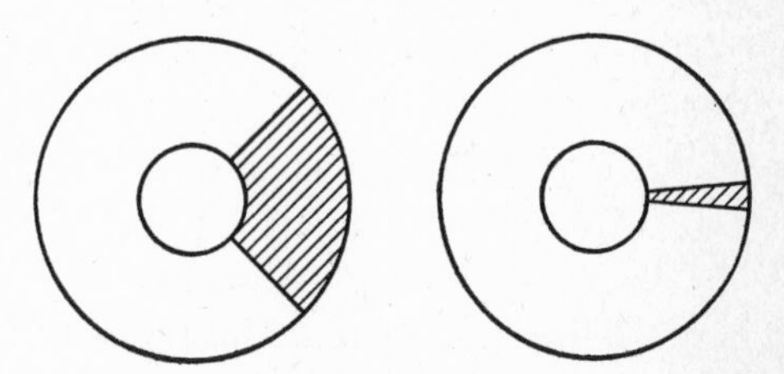

FIG. 17-20. Showing the effect of grid bias on the shadow angle of a cathode-ray indicator tube.

17-15. *TELEVISION CATHODE-RAY PICTURE TUBES*

A cathode-ray tube is used in television receivers to display the image or scene transmitted. This is called a **picture tube,** defined [21] as "a cathode-ray tube used to produce an image by variation of the beam intensity as the beam scans a raster." This tube sometimes is called a **Kinescope,** and once was called an **Oscillite.** In reproducing a televised image the electron beam progresses over the phosphor on the end of the tube in exact synchronism with the scanning beam in the distant camera pickup tube. The instantaneous value of the control-grid voltage of the picture tube varies in magnitude with the received signal. Thus the instantaneous intensity of the beam and the resulting visible spot on the picture-tube screen vary in accordance with the light reflected by the corresponding spot on the image, and the image is "painted" on the picture-tube screen.

Deflection Methods. As explained earlier in this chapter, there are two fundamental ways of deflecting an electron beam. One method is *electric-field deflection.* Two sets of deflecting plates are placed within the tube, and the electron beam passes between plates of a set. In television the vertical sweep voltage that determines which horizontal line is scanned is impressed between the set of plates that lie in the horizontal plane. The horizontal sweep voltage that determines the motion along the horizontal line being scanned is impressed between the set of plates that lie in the vertical plane. The other method is *magnetic field deflection.* No deflecting plates are placed within the tube; instead, a **deflecting yoke** [21] consisting of "an assembly of one or more coils, whose magnetic field deflects the electron beam" is used. The yoke is placed at the neck of the tube near the electron gun. The details of electric and magnetic deflection were discussed in sections 17-7 and 17-8. Electric deflection is common in television picture tubes using beam accelerating voltages of less than 7000 volts and deflection angles of less than 45 degrees, and magnetic deflection is used above these values.[10] When all factors are considered, the two systems cost about the same.[10]

Ion Traps. In addition to electrons, both positive and negative massive ions exist in a cathode-ray tube. These ions have charges that usually are the same magnitude as the charge on an electron, but sometimes are small multiples of this unit charge. However, the mass of these ions is from a few thousand to a few hundred thousand times the mass of an electron.[10] These ions are thought to originate as follows: Both positive and negative massive ions emanate from the hot cathode.[22] These may be regarded as charged metallic atoms. In a cathode-ray tube the positive ions return to the negative cathode, but the electrons and negative ions proceed to the positive plate. A cathode-ray tube cannot be *completely* evacuated; also, gas is liberated during use from the hot portions of the tube. Ionization by collision will occur, and massive positive ions thus formed will be drawn toward the cathode, bombarding it and releasing additional ions.[23]

In a picture tube with *electric* deflection these ions travel the same path and are deflected the same amount as the electrons constituting the beam. This is indicated by equation 17-3 in which neither charge, nor mass, enter. In a picture tube with *magnetic* deflection, the massive ions experience a deflection different from the electrons, as will be shown. From equation 1-9, the force acting on a particle is $F = evB$. Equating this to the centrifugal force as in section 1-4 gives $evB = mv^2/r$, and $r = mv/(eB)$. Substituting for v in the first expression gives

$$r = \frac{1}{B}\sqrt{\frac{2Em}{e}} \qquad 17\text{-}9$$

and when this is substituted in equation 17-6, $D = Ll/r$,

$$D = LlB\sqrt{\frac{e}{2Em}} \qquad 17\text{-}10$$

As this equation shows, particles having the same charge but different mass will be deflected differently *by a magnetic field.* Hence, when magnetic deflection is used in a tube, either an **ion trap** must be used, or the tube must have an aluminized screen. If these precautions are not taken, then the massive negative ions, which will be deflected but little, will impinge repeatedly in a small area near the center of the screen. This bombardment will cause the phosphor to become insensitive, and will cause a dark blemish to exist. Often an ion trap and an aluminized screen both are used.

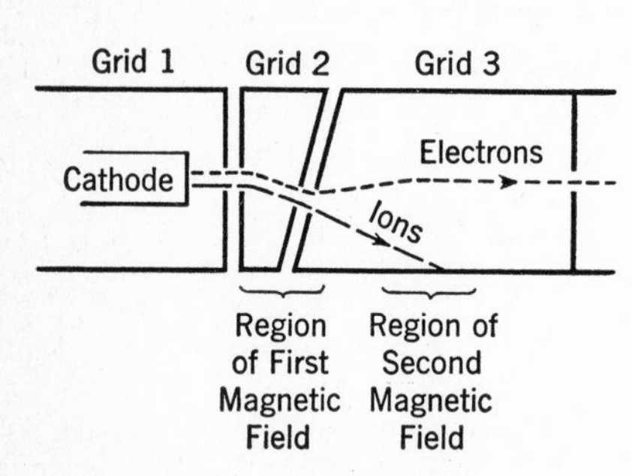

FIG. 17-21. A negative-ion trap employing two magnetic fields. The number of dashes do not represent relative numbers of particles.

Several methods have been developed for trapping, and removing, the massive negative ions from the electron beam, one [23] being shown in Fig. 17-21. The ends of grid 1 and grid 2 are at an angle with the axis. This distorts the *electric* field between the two grids (which are at a difference of potential), giving a tilted lens effect,[23] which bends downward the mixed beam of both electrons and massive negative ions. A *magnetic* field of the correct strength in the region shown readily deflects the electrons upward, but has little influence on the massive ions in accordance with equation 17-9. Hence this arrangement "traps" the ions and removes them from the beam. A second magnetic field positioned as indicated corrects the path of the deflected electrons and directs them down the axis.

Aluminized Screens. If a very thin layer of aluminum is deposited[26] on the side of the phosphor toward the electron gun, an ion trap theoretically need not be used. Electrons pass through this layer and excite the phosphor, but "large" massive ions do not pass through and cannot cause a blemish on the screen. Other advantages of the aluminum coating are as follows: If an aluminum backing is *not* used on the phosphor, about 50 per cent of the light generated by the impinging electrons is emitted toward the electron gun, about 20 per cent is lost by reflection in the glass tube face, and only about 30 per cent is directed toward the observer as useful light transmitting the televised image.[27] Also, some of the light that is radiated toward the electron gun is reflected toward the screen, limiting picture contrast.[27] The aluminum coating is deposited by evaporation on the screen. It reflects the radiant light energy outward, as Fig. 17-22 shows. Also, operation is improved in other ways.[27]

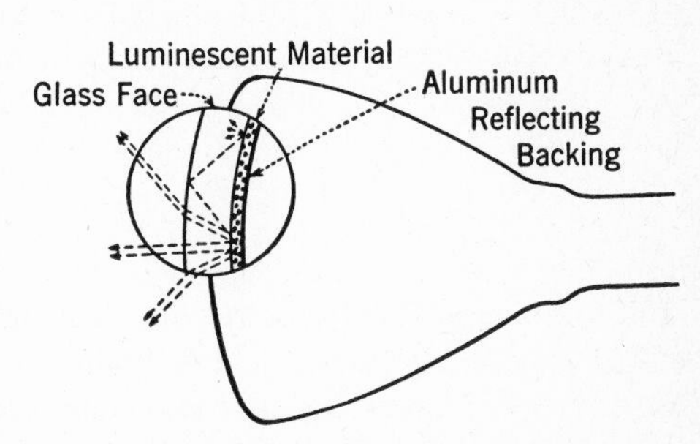

FIG. 17-22. Illustrating the action of an aluminized screen. (Courtesy American Institute of Electrical Engineers.)

17-16. *MULTIPLE-GUN CATHODE-RAY TUBES*

Cathode-ray tubes with two, and with three, electron guns focused on the same luminescent screen have been developed. With such tubes having separate deflecting systems for each gun, two or more phenomena may be observed simultaneously in their correct time relationships. Such tubes for measurement purposes are relatively uncommon. If two or more phenomena are to be observed "simultaneously," an electron switch[28] may be used to connect one, and then the other, to the deflecting system. In color television *either* a one-gun picture tube,[29] or a three-gun picture tube is used.[30,31]

REFERENCES

1. Institute of Radio Engineers. *Standards on Electron Tubes: Definitions of Terms*, 1950. Proc. I.R.E., April 1950, Vol. 38, No. 4.

2. Johnson, J. B. *The cathode-ray oscillograph.* Bell System Technical Journal, Jan. 1932, Vol. 11, No. 1.
3. Kuehni, H. P., and Ramo, S. *A new high-speed cathode-ray oscillograph.* Electrical Engineering, June 1937, Vol. 56, No. 6.
4. Lewis, G., and Mann, F. J. *Ferdinand Braun—Inventor of the cathode-ray tube.* Electrical Communication, Dec. 1948, Vol. 25, No. 4.
5. Mouromtseff, I. E. *Development of electron tubes.* Proc. I.R.E., April 1945, Vol. 33, No. 4.
6. Johnson, J. B. *A low-voltage cathode-ray oscillograph.* Bell System Technical Record, Nov. 1922, Vol. 1, No. 2.
7. R.C.A. Manufacturing Co. *Cathode Ray Tubes and Allied Types. Pamphlet TS-2.*
8. Spangenberg, K. R. *Vacuum Tubes.* McGraw-Hill Book Co.
9. Perkins, T. B. *Cathode-ray tube terminology.* Proc. I.R.E., Nov. 1935, Vol. 23, No. 11.
10. Fink, D. G. *Television Engineering.* McGraw-Hill Book Co.
11. Tektronix, Inc. Literature describing cathode-ray oscillographs.
12. R.C.A. Tube Department. *Tube Handbook HB-3, Cathode-Ray Tube Section.*
13. Reference Sheets. Electronics, April and Dec. 1938.
14. Zworykin, V. K., and Morton, G. A. *Television.* John Wiley & Sons.
15. Institute of Radio Engineers. *Standards on Electron Tubes: Methods of Testing,* 1950. Sept. 1950, Vol. 38, No. 9.
16. Ryan, H. J. *A power diagram indicator for high-tension circuits.* Trans. A.I.E.E., 1911, Vol. 30.
17. Terman, F. E., and Pettit, J. M. *Electronic Measurements.* McGraw-Hill Book Co.
18. Johnson, J. B. *A Braun tube hysteresigraph.* Bell System Technical Journal, April 1929, Vol. 8, No. 2.
19. Terman, F. E. *Radio Engineering.* McGraw-Hill Book Co.
20. Stinchfield, J. M. *Cathode-ray tubes and their application.* Electrical Engineering, Dec. 1934, Vol. 53, No. 12.
21. Institute of Radio Engineers. *Standards on Television: Definitions of Terms,* 1948.
22. Michaelson, H. B. *Metallic-ion emission in vacuum tubes.* The Sylvania Technologist, July 1948, Vol. 1, No. 3.
23. Steier, H. P., Kelar, J., Lattimer, C. T., and Faulkner, R. D. *Development of a large metal kinescope for television.* R.C.A. Review, March 1949, Vol. 10, No. 1.
24. Szegho, C. S., and Noskowicz, T. S. *Improved electron-gun ion traps.* Tele-Tech, June 1951, Vol. 10, No. 6.
25. Bowie, R. M. *The negative-ion blemish in a cathode-ray tube and its elimination.* Proc. I.R.E., Dec. 1948, Vol. 36, No. 12.
26. Ewald, E. R. *Manufacturing metallized picture tubes.* Electronics, Feb. 1950, Vol. 23, No. 2.
27. Epstein, D. W., and Pensak, L. *Improved cathode-ray tubes with metal-backed luminescent screens.* R.C.A. Review, March 1936, Vol. 7, No. 1.

28. George, R. H., Heim, H. J., Mayer, H. F., and Roys, C. S. *A cathode-ray oscillograph for observing two waves.* Electrical Engineering, Oct. 1935, Vol. 54, No. 10.
29. Law, R. R. *A one-gun shadow-mask color Kinescope.* Proc. I.R.E., Oct. 1951, Vol. 39, No. 10.
30. Law, H. B. *A three-gun shadow-mask color Kinescope.* Proc. I.R.E., Oct. 1951, Vol. 39, No. 1.
31. Moodey, H. C., and Van Ormer, D. D. *Three-beam guns for color Kinescopes.* Proc. I.R.E., Oct. 1951, Vol. 39, No. 10.

QUESTIONS

1. What are the principal parts of a cathode-ray tube?
2. How goes gas focusing function in a cathode-ray tube? Why is it not used in modern tubes?
3. What methods of focusing are used in modern cathode-ray tubes?
4. What may occur if a cathode-ray tube is in operation, but the beam is not continuously moving over the screen?
5. Why may the camera tubes of the preceding chapter be classed as cathode-ray tubes?
6. How does electric-field focusing operate?
7. How does magnetic-field focusing operate?
8. What is the meaning of each of the following terms: phosphor, luminescence, flourescence, and phosphorescence?
9. The derivation of equation 17-3 was based on the electron. Does this apply to massive ions?
10. Why is a blemish sometimes caused on the phosphor of a cathode-ray tube?
11. Why is a scanning, or sweep, voltage or current required?
12. Are the wave shapes of scanning voltages and currents the same?
13. How do the circuits of Figs. 17-13 and 17-14 operate when resistor R' is added?
14. What is an important advantage of electric-field focusing over magnetic-field focusing?
15. How is synchronism maintained in cathode-ray oscillographs?
16. What is an ion trap, why is it used, and how does it function?
17. Why are the spectral characteristics and time of persistence important?
18. What methods of synchronizing may be used with Figs. 17-13 and 17-14?
19. How can it be determined if the relations in Fig. 17-17 are for current leading or current lagging?
20. Why are aluminized screens used? What are their advantages?
21. If an aluminized screen is used, must the tube contain an ion trap?
22. Why are multiple-gun tubes used?
23. What application may make such tubes reasonably common?
24. Are aluminized screen tubes commonly used in oscillograph tubes?
25. How may more than one phenomenon be "simultaneously" displayed on a tube with a single gun?

PROBLEMS

1. Calculate the approximate velocity and energy with which an electron strikes the screen in a cathode-ray tube when the accelerating voltage is 1500 volts, and when it is 9000 volts.
2. In the cathode-ray tube of Fig. 17-5, $l = 1.25$ centimeters, $L = 18.5$ centimeters, $d = 0.5$ centimeter, $E = 1200$ volts, and $E' = 50$ volts. If the phosphor covers a circle 3 inches in diameter, will the beam be deflected off the phosphor? What voltage will deflect it to just the edge of the phosphor?
3. Prove that as stated in equation 17-3, $D/L = v'/v$.
4. Draw a diagram similar to Fig. 17-11 for two voltages equal in magnitude, 90° out of phase, and with one voltage twice the frequency of the other.
5. Draw phasor (or vector) diagrams for Fig. 17-16 as connected, and with R and C interchanged. Also, write equations for the phase angles for both conditions. Which circuit should be used for determining the phase angle of a vacuum-tube amplifier at low frequencies?

INDEX